KU-647-394

Contents

Preface

The fundamental aim underlying the writing of *Biological Science* was the desire to emphasise the unifying scientific nature of biological systems despite the amazing diversity in structure and function seen at all levels of biological organisation.

Books 1 and 2 comprise a complete text for the A-level student, following all syllabuses in Biological Sciences and incorporating all the topic areas recommended by the GCE Interboard Working Party on the A-level common core in Biology (published 1983). The text will also be relevant to all first-year University and Further Education College students studying the Biological Sciences.

Each chapter is designed to provide comprehensive, up-to-date information on all topics in Biological Sciences, and the accuracy and relevance of this information has been checked by leading authorities in the appropriate fields and by practising teachers and examiners. The text includes:
– clearly written factual material,
– a carefully selected series of thoroughly pretested practical investigations relevant to the A-level course,
– a variety of types of question designed to stimulate an enquiring approach and answers to them.

Whilst it is recognised that the study of Biological Science follows no set pattern, the content of books 1 and 2 has been arranged so that each book contains material approximating to each year of a two-year course.

The appendices, which provide information and techniques vital to the study of Biological Science at this level, recognise that many students do not study Chemistry and Physics to the same level. Mathematical, physical and chemical concepts related to Biological Sciences are emphasised throughout the text, as appropriate.

Preface to second edition

The second edition of Biological Science has incorporated information which is relevant to recent curriculum innovations. These include:
- the introduction of Advanced Supplementary (AS) Level syllabuses and examinations,
- revised Advanced Level syllabuses which emphasise the social, environmental and technological relevance of biological science,
- the introduction of the Sixth Form Entrance Papers for the purposes of university entrance,
- the requirements associated with the development of the Advanced Level modular curriculum.

Rather than dramatically increase the extent of the text, which is already comprehensive, some of the original text and illustrations have been removed as they are no longer considered necessary for students studying at these levels. New information has been introduced which both updates existing material and enables the books to justify their claim as the most comprehensive texts in the biological sciences at these levels.

In chapter 12, greater attention is paid to the effects of humans and other organisms on the environment. For example pollution is linked with disruption of biogeochemical cycles, and issues such as the 'greenhouse effect', eutrophication, the ozone layer and acid rain are discussed in detail. More applications of population ecology are included, for example fisheries management, and there are new sections on agricultural and horticultural practices, pesticides and conservation.

Recent advances in our knowledge of the appearance of HIV and its associated disease, AIDS, have resulted in the incorporation of additional information. Theories of phloem translocation have been substantially revised; similarities in the energetics of mitochondria and chloroplasts have been stressed; the usefulness of ecological pyramids is reassessed; the original meaning of the term 'symbiosis' has been adopted in line with current trends; and classification of relationships between organisms is discussed, including research on the importance of mycorrhizas. These are just a few examples of the many changes, some major some minor, which have been made.

Another important influence in preparing this second edition was the *Report of Recommendations on Biological Nomenclature* published by the Institute of Biology and the Association for Science Education in 1989. One of the Report's most significant recommendations concerns classification and this has resulted in a major revision of chapters 2, 3 and 4. Recommendations on the terms, units and symbols used in describing water relations have been adopted throughout.

As in the first edition, each chapter is designed to provide comprehensive, up-to-date information on the relevant topics in the Biological Sciences, and the accuracy and – a relevance of the information has been checked by leading authorities in the appropriate fields and by practising teachers and experienced examiners. The text includes:
- clearly written factual material,
- a carefully selected series of pretested practical investigations,
- a variety of questions designed to stimulate an enquiring approach and answers to them.

The content of books 1 and 2 are arranged so that each book contains course material approximating to each year of a two-year course. The appendices, which provide information and techniques vital to the study of biological sciences at this level, recognise that many students do not study Physics and Chemistry to the same level. Mathematical, physical and chemical concepts related to Biological Sciences are emphasised throughout the text, as appropriate.

Acknowledgements

The authors and publisher wish to acknowledge the many friends, colleagues, students and advisers who have helped in the production of *Biological Science*.

In particular, we wish to thank:
Dr R. Batt, Dr Claudia Berek, Professor R.J. Berry, Dr. A.C. Blake, Dr John C. Bowman, Mr R. Brown, Dr Fred Burke, Mr Richard Carter, Dr Norman R. Cohen, Dr K.J.R. Edwards, Mr Malcolm Emery, Nr Nick Fagents, Dr James T Fitzsimons, Dr John Gay, Dr Brij L. Gupta, Vivienne Hambleton, Dr David E. Hanke, the late Dr R.N. Hardy, the late Reverend J.R. Hargreaves, Dr S.A. Henderson, Mr Michael J. Hook, Mr Colin S. Hutchinson, Illustra Design Ltd, Dr Alick Jones, Mrs Susan Kearsey, Dr Simon P. Maddrell FRS, Professor Aubrey Manning, Dr Chris L. Mason, Mrs Ruth Miller, Dr David C. Moore, A.G. Morgan, Dr Rodney Mulvey, Dr David Secher, Dr John M. Squire, the late Professor James F. Sutcliffe, Dr R.M. Taylor, Stephen Tomkins, Dr Eric R. Turner, the late Dr Paul Wheater, Dr Brian E.J. Wheeler, Dr Michael Wheeler.

The authors are particularly indebted to Mrs Adrienne Oxley, who patiently and skilfully organised the pretesting of all the practical exercises. Her perseverance has produced exercises that teachers, pupils and laboratory technicians can depend upon.

However, the authors accept full responsibility for the final content of these books.

Finally, the authors wish to express their thanks to their wives and families for the constant support and encouragement shown throughout the preparation and publication of these books.

We also wish to thank the following for permission to use their illustrations, tables and questions.
Figures: 2.4, 2.5, 2.19*a*, 2.23*c*, 3.5*c*, 3.16*b*, 3.29*c*, 3.29*d*, 7.6, 7.10, 7.15, 7.19*b*, 7.21, 7.22, 8.2*e*, 8.2*f*, 8.3*a*, 8.3*b*, 8.4*d*, 8.5*e*, 8.6, 8.8*c*, 8.8*d*, 8.9*b*, 8.11*b*, 8.11*c*, 8.11*e*, 8.12*b*, 8.12*d*, 9.3, 9.6, 10.30*c*, Biophoto Associates; 2.9, Professor L. Caro/Science Photo Library; 2.13*b*, Dr H.G. Pereira (1965) *Journal of Molecular Biology*, **13**; 2.13*c*, R.W. Horne, I. Pasquali-Ronchetti & Judith M. Hobart (1975) *J. Ultrastruct. Res.*, **51**, 233; 2.14, from *The structure of viruses*, R.W. Horne, copyright © 1963 by Scientific American Inc., all rights reserved; 2.15*a*, R.W. Horne (1974) *Virus structure*, Academic Press, London; 2.15*b*, Dr Lee D. Simon/Science Photo Library; 2.17, Dr Thomas F. Anderson & Dr Lee D. Simon/Science Photo Library; 2.26*a*, 2.26*b*, National Institute for Research in Dairying, Reading; 3.5*a*, 8.1*b*, 8.4*c*, 8.5*d*, 8.11*f*, 9.13, 13.1, 13.2, 13.6, 13.10, Centre for Cell and Tissue Research, York; 3.28*b*, Roy Edwards; 3.28*c*, 3.29*e*, 3.39*a*, 3.39*b*, 10.3. Heather Angel; 3.19*a*, 3.19*b*, Dr Lawrence Bannister; 4.1, Jane Burton/Bruce Coleman Ltd; 4.26, R. Buchsbaum (1948) *Animals without backbones*, vol.2, University of Chicago Press; 4.29, 4.30, H.G.Q. Rowett (1962) *Dissection guides*, John Murray, London; 4.31*b*, 4.31*c*, 4.33*b*, 4.33*c*, C. James Webb; 4.31*d*, Barnabys Picture Library; 4.32*b*, 4.32*c*, 4.34*d*, Stephen Dalton/Natural History Photographic Agency; 4.33*d*, E.J. Hudson/Frank W. Lane; 4.34*b*, 4.34*c*, Shell International Petroleum Co; 4.41, Oxford Local Examinations, A62-P, special sheet 2, summer 1978; 5.10*c*, Nigel Luckworth; 5.33, D-G. Smyth, W.H. Stein & S. Moore (1963) *J. Biol. Chem.*, **238**, 227; 5.36, 5.39, R.E. Dickerson & I. Geis (1969) *The structure and action of proteins*, W.A. Benjamin, California; 5.38*b*, 5.38*d*, Sir John Kendrew; 5.38*c*, R.E. Dickerson (1964) *The proteins*, ed. H. Neurath, 2nd ed., vol.2, Academic Press Inc., New York; 5.45, Dr J.M. Squire, Biopolymer Group, Imperial College; 5.49, Professor M.H.F. Wilkins, Biophysics Department, King's College, London; 6.3, reprinted by permission from *Nature*, vol. 213, p. 864, copyright © 1967 Macmillan Journals Limited; 6.5, C.F. Stoneman & J.C. Marsden (1974) *Enzymes and equilibria*, Scholarship Series in Biology, Heinemann Educational Books, London; 7.5, Dr Glenn L. Decker, School of Medicine, John Hopkins University; 7.11*b*, Cancer Research Campaign and Paul Chantrey; 7.14, from *Biochemistry*, 2nd ed. by L. Stryer, W.H. Freeman and Company, copyright © 1981; 7.19*a*, M.A. Tribe, M.R. Erant & R.K. Snook (1975) *Electron microscopy and cell structure*, Cambridge University Press;; 7.20*a*, 8.13*b*, 8.14*b*, 8.15*c*, 8.16*b*, 8.17*c*, 8.18*b*, 8.19, 8.21, 8.22, 8.23, 8.24, 8.25, 8.26, 8.27, 8.28*a*, 8.29, 8.32*a*, 8.32*b*, 8.33, 8.34, 8.37, 8.39, 8.40, 8.41, 10.22*b*, 10.23, 10.25, 10.26 10.28*a*, 10.28*b*, 10.30, 11.15*d*, 11.31, Dr Paul Wheater; 7.25, Dr Klaus Weber; 7.26, Dr Elias Lazarides, California Institute of Technology; 7.27, E. Frei & R.D. Preston FRS; 8.2*d*, Rothamsted Experimental Station; 8.17*d*, 10.30*b*, Mr P. Crosby, Department of Biology, University of York; 8.20, W.H. Freeman & B. Bracegirdle (1967) *An atlas of histology*, 2nd ed., Heinemann Educational Books, London; 8.31, John Currey (1970) *Animal skeletons*, Studies in Biology no. 22, Edward Arnold, London; 9.4, Gene Cox; 9.8, 9.16, 9.29*b*, Dr A.D. Greenwood; 9.18, D.O. Hall & K.K. Rao (1972) *Photosynthesis*, 1st ed., Studies in Biology no. 37, Edward Arnold, London; 9.28, Dr Alex B. Novikoff, Albert Einstein School of Medicine, & Saunders College Publishing; 9.29*a*, C.C. Black (1971) *Plant Physiology*, **47**, 15–23, with permission of the publisher; 10.6, J.P. Harding; 10.7*a*, Kim Taylor/Bruce Coleman Ltd.; 10.7*b*, Dr Brad Amos/Science Photo Library; 10.8, Topham; 10.9*b*, Dr Tony Brain/Science Photo Library; 10.12, Griffin & George; 10.13, Nuffield Biology Text III, *The maintenance of life* (1970), Longman; 10.17*a*, Charles Day; 10.17*b*, King's College School of Medicine and Dentistry, London; 10.21*a*, 10.21*b*, 10.21*c*, 10.21*d*, Dr C.A. Saxton, Unilever Research; 10.22*a*, Dr L.M. Beidler/Science Photo Library; 10.32, Nuffield Text *Maintenance of the organism* (1970) Nuffield Foundation, Longman; 10.33, from *An introduction to human physiology*, 4th ed., by J.H. Green, published by Oxford University Press 1976; 11.14*a*, 11.14*b*, 11.15*e*, Dr Brij L. Gupta, Zoology Department, Cambridge University; 11.16, M.A. Tribe & P. Whittaker (1972) *Chloroplasts and mitochondria*, 1st ed., Studies in Biology no. 31, Edward Arnold, London; 11.17, Dr Ernst F.J. van Bruggen, State University of Groningen; 11.31, reproduced with permission from G.M. Hughes, *The vertebrate lung* (2nd edn) 1979, Carolina Biology Reader Series, copyright Carolina Biological Supply Company, Burlington, North Carolina, USA; 11.32, Philip Harris Biological Ltd; 11.33, B. Siegwart, P. Gehr, J. Gil & E.R. Weibel (1971) *Respir. Physiol.*, **13**, 141–59; 12.2, from *Ecology*, 2nd ed. by Eugene P. Odum, copyright © 1975 by Holt,

Rinehart & Winston, reprinted by permission of Holt, Rinehart & Winston, CBS Publishing; 12.6*b*, Dr E.J. Popham; 12.7*a*, 12.7*e*, S. Cousins (1985) *New Scientist*, 4/7/85, p.51; 12.8, 12.9, 12.11, 12.19, from *Fundamentals of ecology*, 3rd ed. by Eugene P. Odum, copyright © 1971 by W.B. Saunders Company, reprinted by permission of Holt, Rinehart & Winston, CBS Publishing; 12.10, 12.13, 12.33, M.A. Tribe, M.R. Erant & R.K. Snook (1974) *Ecological principles*, Basic Biology Course 4, Cambridge University Press; 12.15, from *The biosphere*, G. Evelyn Hutchinson, copyright © 1970 by Scientific American Inc.; 12.16, R.J. Chorley & P. Haggett (eds.) (1967) *Physical and information models in geography*, Methuen, London; 12.17, A. Crane & P. Liss (1985) *New Scientist*, 21/11/85; 12.18, M. McElroy (1988) The challenge of global change, *New Scientist*, **119**, 1623, 34–6; 12.20, C.F. Mason (1981) *Biology of freshwater pollution*, Longman; 12.29, W.D. Billings (1972) *Plants, man and the ecosystem*, 2nd ed., Macmillan, London; 12.31, A.G. Tansley (1968) *Britain's green mantle*, 2nd ed., George Allen & Unwin, London; 12.35, B.D. Collier, G.W. Cox, A.W. Johnson & P.C. Miller, *Dynamic ecology* © 1973, p. 321, reprinted by permission of Prentice-Hall Inc., New Jersey; 12.37, A.S. Boughey (1971) *Fundamental ecology*, International Textbook Co.; 12.38, T.R.E. Southwood (1974) *Am. Nature* **108**, 791–804; 12.39, W.G. Abrahamson & M. Gadgil (1973) *Am. Nat.* **107**, 651–61; 12.40, 12.42, 12.43, Open University Foundation Course (S100) unit 20, copyright © 1971 The Open University Press; 12.41, D. Lack (1966) *Population studies of birds*, Clarendon Press, Oxford; 12.44, C.B. Huffaker (1958) *Experimental studies on predation: dispersion factors and predator–prey oscillations*; 12.47, 12.48, M. Graham (1956) *Sea fisheries – their investigation*, Arnold; 12.49, R.V. Tait (1981) *Elements of marine ecology*, 3rd ed., Butterworths; 12.50, Richard North, *The Independent*, February 1988; 12.51, J.P. Dempster (1968) The control of *Pieris rapae* with DDT. II Survival of the young stages of *Pieris* after spraying, *J. Appl. Ecol.* **5**, 451–62; 12.52, 12.53, N.W. Moore (1987) *A synopsis of the pesticide problem* in J.B. Cragg (ed.) *Advances in ecological research*, Blackwell; 12.54, after M. Markkula, K. Tiittanen & M. Nieminen (1972) *Ann. Agr. Fenn.* **11**, 74–8; 12.55, C. Rose (1985) Acid rain falls on British woodlands, *New Scientist*, **108**, 1482, 52–7; 12.56, J. H. Ottaway (1980) *The biochemistry of pollution*, Studies in Biology 123, Arnold; 12.57, F. Pearce (1986) Unravelling a century of acid pollution, *New Scientist*, **111**, 1527, 23; 13.5, John Edward Leigh; 13.22, D.A.S. Smith (1970) *School Science Review*, Association for Science Education.

Tables: 2.1, A2.4, E.A. Martin(ed.) (1976) *a dictionary of life sciences*, Pan Books, London; 5.1, based on A.L. Lehninger (1970) *Biochemistry*, Worth, New York with permission of Plenum Publishing Corporation, copyright Plenum Publishing Corporation; 6.3, A. Wiseman & B.J. Gould (1971) *Enzymes, their nature and role*, Century Hutchinson Limited, London; 10.4, 10.5, *Manual of nutrition* (1976), reproduced by permission of the Controller of Her Majesty's Stationery Office; 11.3, John E. Smith (1988) *Biotechnology*, 2nd ed. New Studies in Biology, Edward Arnold; 12.1, 12.2, from *Fundamentals of ecology*, 3rd ed. by Eugene P. Odum, copyright © (1971) by W.B. Saunders Company, reprinted by permission of Holt, Rinehart & Winston, CBS Publishing; 12.3, A.N. Duckham & G.B. Masefield (1970) *Farming systems of the world*, Chatto & Windus, London; 12.4, 12.5, 12.7, C.F. Mason (1981) *Biology of freshwater pollution*, Longman; 12.6, B. Moss (1980) *Ecology of fresh waters*, Blackwell Scientific; 12.8, 12.9, by permission of Griffin & George; 12.10, by permission from The Open University Press; 12.15, 12.16, *1981 World population data sheet*, Population Reference Bureau Inc., Washington D.C.; 12.19, Open University Science Foundation Course (S100) unit 20, copyright © 1971 The Open University Press.

Questions: 2.3, University of Oxford Delegacy of Local Examinations (OLE); 12.11, modified from M.A. Tribe, M.R. Erant & R.K. Snook (1974) *Ecological principles*, Basic Biology Course 4, Cambridge University Press; 12.19, 12.21, Open University Science Foundation Course (S100) Unit 20, copyright © (1971) The Open University Press.

Cover: The cover photograph shows a green lacewing *Chrysopa carnea* performing a vertical take-off from a hawthorn leaf (Stephen Dalton/NHPA).

Chapter One

Introduction to the subject

Biology (*bios*, life; *logos*, knowledge) is a science devoted to the study of living organisms. Science has progressed by breaking down complex subjects of study into their component parts so that today there are numerous branches of biology devoted to specialised study of the structure and function of select organisms, as shown in fig. 1.1. This principle is often called the 'reductionist' principle and, carried to its logical conclusions, it has focussed attention on the most elementary forms of matter in living and non-living systems. This approach to study seeks fundamental understanding by looking at the parts rather than the whole. An opposing approach, based upon the 'vitalist' principle, considers that 'life' is something special and unique, and maintains that life cannot be explained solely in terms of the laws of physics and chemistry, having properties which are special to the system as a whole. The aim of biology must ultimately be to explain the living world in terms of scientific principles, although appreciating that organisms behave in ways which often seem beyond the capabilities of their component parts. Certainly the consciousness of living organisms cannot be described in terms of chemistry and physics even though the neuro-physiologist can describe the working of the single neurone in physico-chemical terms. Consciousness may be the collective working of millions of neurones and their electrochemical states, but as yet we have no real concept of the chemical nature of thought and ideas. Nor do we understand completely how living organisms originated and evolved. There have been many attempts to answer this question from theological to biological and chapters 22–25, in book 2, attempt to put the different viewpoints but with the emphasis on the possible biological explanations.

Thus we are reduced to the position that we cannot define precisely what life is or whence it came. All that we can do is to describe the observable phenomena that distinguish living matter from non-living matter. These are as follows.

Nutrition (chapters 9 and 10)

All living organisms need food, which is assimilated and used as a source of energy, and materials for living processes such as growth. The distinction between the majority of plants and animals is based upon their methods of obtaining food. Most plants photosynthesise, incorporating the energy of light, to make their food. This is a form of autotrophic nutrition. Animals and fungi, on the other hand, obtain their food from other organisms, breaking down their organic compounds by enzymes and absorbing the products. This is known as heterotrophic nutrition. Most bacteria are also heterotrophic but some are autotrophic.

Respiration (chapter 11)

All life processes require energy and much of the food obtained by autotrophic and heterotrophic nutrition is used as a source of this energy. The energy is released during the breakdown of certain energy-rich compounds by the process of respiration. The energy released is stored in molecules of adenosine triphosphate (ATP) and this compound has been found to occur in all living cells.

Irritability (chapters 15 and 16)

Living organisms have the ability to respond to changes in both the internal and external environments and thus ensure that they maximise their chances of survival. For example the dermal blood vessels in the skin of a mammal dilate in response to a rise in body temperature, and the consequent heat loss brings about a restoration of the optimum temperature of the body. A green plant on a window sill in a room grows towards one-sided light coming through the window thus ensuring maximum exposure to light for photosynthesis.

Movement (chapter 17)

Animals are distinguished from plants by their ability to move from place to place, that is they locomote. This is necessary in order for them to obtain their food, unlike plants which can manufacture their own food from raw materials obtained in one place. Nevertheless, movement occurs in plants, both within their cells and indeed within

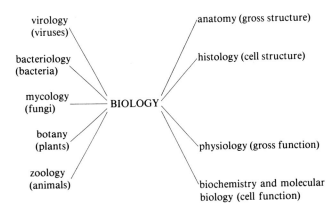

virology
(viruses)

bacteriology
(bacteria)

mycology
(fungi)

botany
(plants)

zoology
(animals)

BIOLOGY

anatomy (gross structure)

histology (cell structure)

physiology (gross function)

biochemistry and molecular
biology (cell function)

Fig 1.1

whole structures, although at a much slower rate than animals. Some bacteria and unicellular plants are also capable of locomotion.

Excretion (chapter 19)

Excretion is the removal from the body of waste products of metabolism. For example the process of respiration produces toxic waste products which must be eliminated. Animals take in an excess of protein during nutrition and, since this material cannot be stored, it must be broken down and excreted. Animal excretion is, therefore, largely nitrogenous excretion. In addition, the removal of certain unusable materials absorbed by the organism, such as lead, radioactive dust and alcohol can be regarded as excretion.

Reproduction (chapter 20)

The life span of organisms is limited, but they all have the ability to perpetuate 'life', thereby ensuring the survival of the species. The resulting offspring have the same general characteristics as the parents, whether such individuals are produced by asexual or sexual reproduction. The 'reductionist' search for the explanation of this inheritance has now revealed the existence of molecules known as nucleic acids (deoxyribosenucleic acid, DNA, and ribosenucleic acid, RNA) which appear to contain the coded information passed between organisms of succeeding generations.

Growth (chapter 21)

Non-living objects, such as a crystal or a stalagmite, grow by the addition of new material to their outside surface. Living organisms, however, grow from within, using food that they obtain by autotrophic or heterotrophic nutrition. The molecules are formed into new living material during the process of assimilation.

These seven characteristics can be observed to a greater or lesser extent in all living organisms and are our only means of indicating whether life exists or not. However, they are only the *observable* characteristics of the all-important properties of living material, that is, extracting, converting and using energy from the environment. In addition, living material is able to maintain and even increase its own energy content. In contrast to this, dead organic matter tends to disintegrate as a result of the chemical and physical forces of the environment. In order to maintain themselves and prevent this disintegration, organisms have an inbuilt self-regulating system to ensure that there is no net energy loss. This control is referred to as homeostasis and operates at all levels of biological organisation, from the molecular level to the community level.

The characteristics of life outlined above are dealt with in detail in the chapters indicated. Inevitably, many of these chapters extend the explanations in terms of physical and chemical concepts, for it is in these fields that the major research and additions to our knowledge have come in recent years. The study of protein synthesis, DNA, ATP, enzymes, hormones, viruses, antigen–antibody reactions and many other examples, all provide some explanation of what is happening in the cells and bodies of organisms.

In the appendix, in book 2, you will find some basic information required by a biologist, including biochemistry, scientific method, the experimental approach, a glossary of terms and so on. The appendix is designed to supply information to those students who may be lacking in one or more of these areas. With this knowledge the student must strive to develop powers of critical observation and description which are part of the thinking processes underlying scientific enquiry.

Chapter Two

Variety of life – prokaryotes, viruses and fungi

Chapter 2 is concerned with three groups of organisms: bacteria (prokaryotes), viruses and fungi. Although, as will become clear, they are placed in three very different groups taxonomically, they are often studied together in the branch of biology known as **microbiology**. This is because the techniques used in their study are similar in many respects, involving particularly microscopy and sterile (aseptic) procedures. They are of tremendous ecological and economic importance, and are intimately associated with the developing areas of biotechnology and genetic engineering.

2.1 Prokaryotes compared with eukaryotes

All cellular organisms so far studied fall naturally into one of two major groups, the prokaryotes and eukaryotes. **Prokaryotes** appeared about 3 500 million years ago and comprise a variety of organisms collectively known as bacteria. **Eukaryotes** include protoctists, fungi, green plants and animals. Eukaryotes appeared first in the late Pre-Cambrian period, about 2 000 million years ago, and probably evolved from prokaryotes.

The cells of prokaryotes (*pro*, before; *karyon*, nucleus) lack true nuclei. In other words, their genetic material (DNA) is not enclosed by nuclear membranes, and lies free in the cytoplasm. The cells of eukaryotes (*eu*, true) are

Table 2.1 Major differences between prokaryotes and eukaryotes.

Feature	Prokaryote	Eukaryote
Cell size	Average diameter 0.5–5μm.	Up to 40 μm diameter common; commonly 1 000–10 000 times volume of prokaryotic cells.
Form	Unicellular or filamentous.	Unicellular, filamentous or truly multicellular.
Genetic material	Circular DNA lying naked in the cytoplasm. No true nucleus or chromosomes. No nucleolus.	Linear DNA associated with proteins and RNA to form chromosomes within a nucleus. Nucleolus in nucleus.
Protein synthesis	70S ribosomes (smaller). No endoplasmic reticulum involved. (Many other details of protein synthesis differ, including susceptibility to antibiotics, e.g. prokaryotes inhibited by streptomycin.)	80S ribosomes (larger). Ribosomes may be attached to endoplasmic reticulum.
Organelles	Few organelles. None are surrounded by an envelope (2 membranes). Internal membranes scarce; if present usually associated with respiration or photosynthesis.	Many organelles. Envelope-bound organelles present, e.g. nucleus, mitochondria, chloroplasts. Great diversity of organelles bounded by single membranes, e.g. Golgi apparatus, lysosomes, vacuoles, microbodies, endoplasmic reticulum.
Cell walls	Rigid and contain polysaccharides with amino acids. Murein is main strengthening compound.	Cell walls of green plants and fungi rigid and contain polysaccharides. Cellulose is main strengthening compound of plant walls, chitin of fungal walls.
Flagella	Simple, lacking microtubules. Extracellular (not enclosed by cell surface membrane). 20 nm diameter.	Complex, with '9+2' arrangement of microtubules. Intracellular (surrounded by cell surface membrane). 200 nm diameter.
Respiration	Mesosomes in bacteria, except cytoplasmic membranes in blue-green bacteria.	Mitochondria for aerobic respiration.
Photosynthesis	No chloroplasts. Takes place on membranes which show no stacking.	Chloroplasts containing membranes which are usually stacked into lamellae or grana.
Nitrogen fixation	Some have the ability.	None have the ability.

Based on *A Dictionary of Life Sciences*, E. A. Martin (ed.), (1976), Pan Books.

much more complex and are characterised by a true nucleus, that is genetic material enclosed by membranes (the **nuclear envelope**) to form a definite, easily recognisable structure.

Many other fundamental differences exist between prokaryotes and eukaryotes, the more important of which are summarised in table 2.1. Some of the cell structures mentioned are discussed in more detail in chapter 7. Figs 2.3, 7.3 and 7.4 illustrate typical prokaryote and eukaryote cells.

Fig 2.1 summarises the classification of living organisms used in this book. Five kingdoms are used (see introduction to chapter 3).

Kingdom Prokaryotae

2.2   Bacteria

Bacteria are the smallest organisms having a cellular structure, their average diameter being about 1 µm, enough room for 200 average-sized globular protein molecules (of 5 nm diameter) to fit across the cell. Such a molecule in solution can diffuse about 60 µm per second; thus no special transport mechanisms are needed for these organisms. Bacteria range in length from about 0.1 to 10 µm. They are unicellular, and hence can only be seen individually with the aid of a microscope, that is they are **micro-organisms**. The study of bacteria, or **bacteriology**, is one branch of **microbiology**, the latter also including the study of viruses (**virology**), fungi (**mycology**) and other micro-organisms. Many of the techniques used to handle these organisms are similar.

Bacteria occupy many environments, such as soil, dust, water, air, in and on animals and plants and they can even be found in hot springs at temperatures of 60 °C or higher. Their numbers are enormous; one gram of fertile soil is estimated to contain 100 million and 1 cm³ of fresh milk

may contain more than 3 000 million. Together with fungi their activities are vital to all other organisms because they cause the decay of organic material and the subsequent recycling of nutrients. In addition, they are of increasing importance to humans, not only because some cause disease, but because they can be utilised in many economically important processes. Their importance is discussed further in sections 2.5 and 2.6.

2.2.1   Classification

There are a number of distinct groups of organisms at this level of organisation, the smallest and simplest of which are little more complex than viruses. The only group of concern here is the 'true bacteria' or Eubacteria. Fig 2.2 illustrates the range of organisms that are generally classified as bacteria.

2.2.2   Structure

Fig 2.3 shows the structure of a generalised bacterium. Fig 2.4 is an electron micrograph of a section through a rod-shaped bacterium and reveals how little structure is visible compared with a eukaryotic cell (for instance figs 7.5 and 7.6).

Capsules and slime layers

Capsules and slime layers are slimy or gummy secretions of certain bacteria which show up clearly after negative staining (when the background, rather than the specimen, is stained). A **capsule** is relatively thick and compact, whereas a **slime layer** is diffuse. In some cases these layers unite bacteria into colonies. Both offer useful additional protection to bacteria, for example capsulate strains of pneumococci grow in their human hosts causing pneumonia, whereas non-capsulate strains are easily attacked and destroyed by phagocytes, and are therefore harmless.

Cell wall

The cell wall confers rigidity and shape and can be clearly seen in a section (fig 2.4). As with plant cells, it prevents the cell from swelling and bursting as a result of osmosis when, as often occurs, it is in a medium of higher water potential (section 14.1.3). Water, various ions and small molecules can pass freely through tiny pores in the wall, but larger molecules like proteins and nucleic acids are excluded. The wall also has antigenic properties, caused by both proteins and polysaccharides.

Bacteria fall into two natural groups according to their cell wall structure. Some are stained with Gram's stain and are termed **Gram positive**, others do not retain the stain

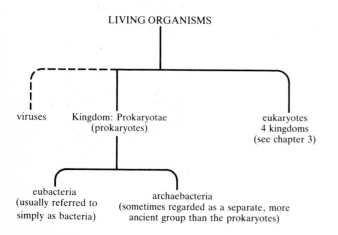

Fig 2.1   *Summary classification of living organisms*

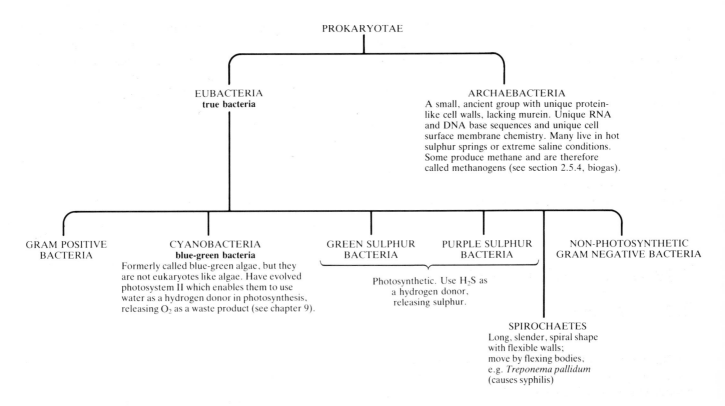

PROKARYOTAE

EUBACTERIA
true bacteria

ARCHAEBACTERIA
A small, ancient group with unique protein-like cell walls, lacking murein. Unique RNA and DNA base sequences and unique cell surface membrane chemistry. Many live in hot sulphur springs or extreme saline conditions. Some produce methane and are therefore called methanogens (see section 2.5.4, biogas).

GRAM POSITIVE BACTERIA

CYANOBACTERIA
blue-green bacteria
Formerly called blue-green algae, but they are not eukaryotes like algae. Have evolved photosystem II which enables them to use water as a hydrogen donor in photosynthesis, releasing O_2 as a waste product (see chapter 9).

GREEN SULPHUR BACTERIA

PURPLE SULPHUR BACTERIA

Photosynthetic. Use H_2S as a hydrogen donor, releasing sulphur.

NON-PHOTOSYNTHETIC GRAM NEGATIVE BACTERIA

SPIROCHAETES
Long, slender, spiral shape with flexible walls; move by flexing bodies, e.g. *Treponema pallidum* (causes syphilis)

Other organisms resembling bacteria, but of doubtful affinity, include **mycobacteria** (slime bacteria) which move by gliding; **mycoplasmas** or **pleuropneumonia-like organisms** (**PPLO**s), extremely small parasites lacking cell walls; **rickettsias**, organisms similar to large viruses which cause typhus.

Fig 2.2 *Classification of prokaryotes*

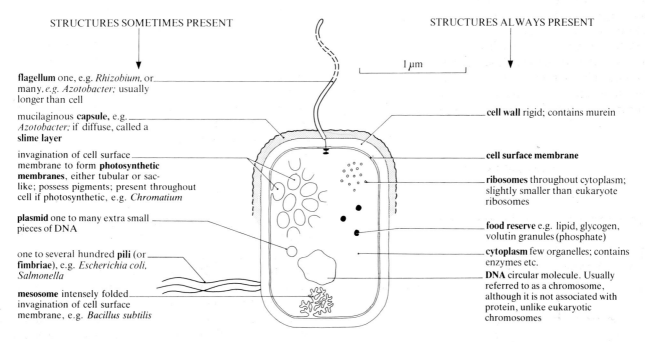

STRUCTURES SOMETIMES PRESENT

STRUCTURES ALWAYS PRESENT

1 μm

flagellum one, e.g. *Rhizobium*, or many, *e.g. Azotobacter;* usually longer than cell

mucilaginous **capsule**, e.g. *Azotobacter;* if diffuse, called a **slime layer**

invagination of cell surface membrane to form **photosynthetic membranes**, either tubular or sac-like; possess pigments; present throughout cell if photosynthetic, e.g. *Chromatium*

plasmid one to many extra small pieces of DNA

one to several hundred **pili** (or **fimbriae**), e.g. *Escherichia coli, Salmonella*

mesosome intensely folded invagination of cell surface membrane, e.g. *Bacillus subtilis*

cell wall rigid; contains murein

cell surface membrane

ribosomes throughout cytoplasm; slightly smaller than eukaryote ribosomes

food reserve e.g. lipid, glycogen, volutin granules (phosphate)

cytoplasm few organelles; contains enzymes etc.

DNA circular molecule. Usually referred to as a chromosome, although it is not associated with protein, unlike eukaryotic chromosomes

Fig 2.3 *A generalised rod-shaped bacterium*

during the decolourising procedure (section 2.7) and are termed **Gram negative**. Both types of wall have a rigid, unique framework of **murein**, a molecule consisting of parallel polysaccharide chains cross-linked in a regular fashion by short peptide chains. Each cell is thus effectively surrounded by a net-like sac which is really one molecule. (The polysaccharide portion is described in table 5.7.)

In Gram positive bacteria, such as *Lactobacillus*, the murein net is infilled with other components, mainly polysaccharides and proteins, to form a relatively thick, rigid box. The walls of Gram negative bacteria, such as *Escherichia coli* and *Azotobacter*, are thinner but more complex. Their murein layer is coated on the outside with a smooth, soft, lipid-rich layer. This protects them from **lysozyme**, an antibacterial enzyme found in tears, saliva and other body fluids and egg white. Lysozyme breaks the polysaccharide backbone of murein by catalysing hydrolysis of certain sugar linkages. The wall is thus punctured and lysis (osmotic swelling and bursting) of the cell occurs if the organism is in a solution of higher water potential. The same lipid-rich layer also confers resistance to penicillin, which attacks Gram positive bacteria by interfering with the cross-linking in growing cells, thus making the walls less rigid and more susceptible to osmotic shock.

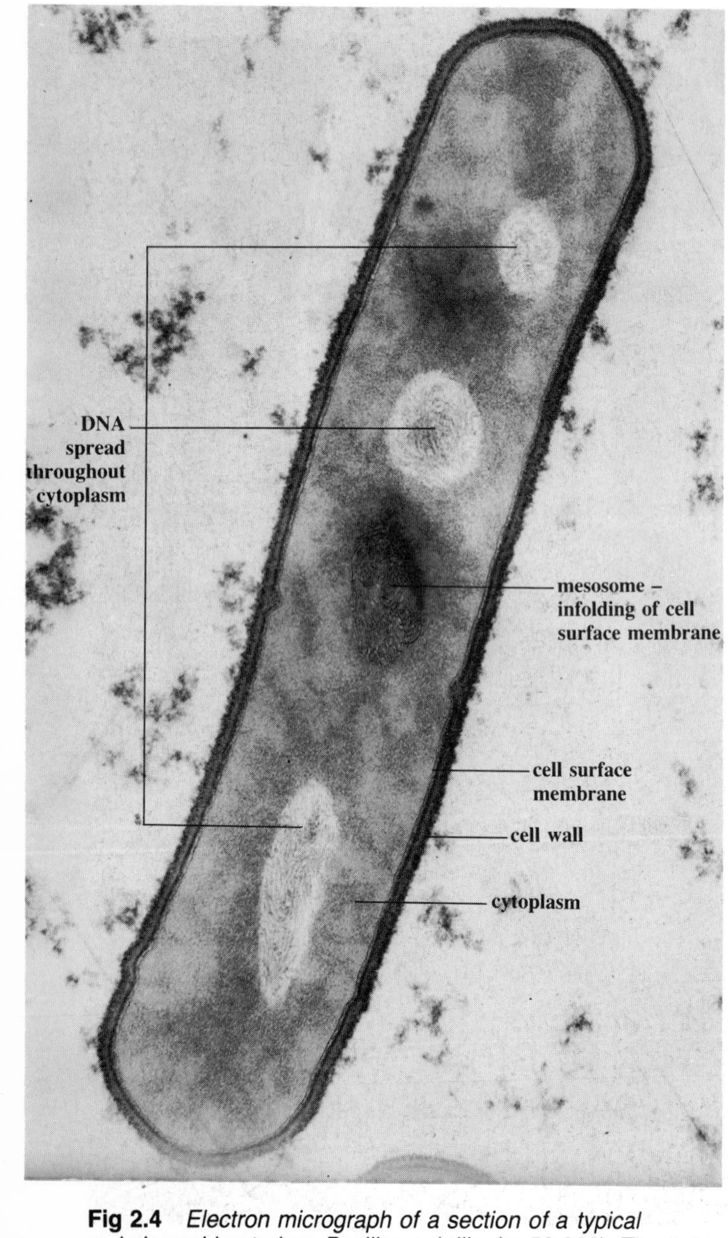

Flagella

Many bacteria are motile due to the presence of one or more flagella.

Bacterial flagella are much simpler in structure than those of eukaryotes (section 17.6.2, table 2.1) resembling just one of the microtubules of eukaryotic flagella. They are made of identical spherical subunits of a protein called **flagellin**, similar to the actin of muscle and arranged in eleven helical spirals to form a hollow cylinder, about 10–20 nm in diameter. The flagellum is rigid, though shaped into a wave, and has a unique mechanism of action. The base apparently rotates on ring-shaped bearings so that the flagellum does not beat but performs a corkscrew motion to propel the cell along. This is apparently the only structure in nature where the principle of the wheel is used. Another interesting feature is that a solution of flagellin subunits will associate spontaneously into helical threads. Spontaneous self-assembly is an important feature of many complex biological structures and in this case is due entirely to the particular sequence of amino acids (primary structure) of the protein.

Motile bacteria can move in response to certain stimuli, that is show tactic movements. For example, aerobic bacteria will swim towards oxygen (positive aerotaxis) and motile photosynthetic bacteria are positively phototactic (that is swim towards light).

Flagella are most easily seen with the electron microscope if the technique of metal shadowing, described in section A2.5, is used (see fig 2.5).

Pili (sing. pilus) or fimbriae (sing. fimbria)

Projecting from the walls of some Gram negative bacteria are protein rods called **pili** or **fimbriae** (fig 2.5). They are shorter and thinner than flagella and are concerned with cell to cell, or cell to surface, attachments, conferring a specific 'stickiness' on those strains that possess them. Various types occur, but of particular interest is the F pilus, coded for by a plasmid (section 2.2.4) and involved in sexual reproduction.

Fig 2.4 *Electron micrograph of a section of a typical rod-shaped bacterium,* Bacillus subtilis *(× 50 000). The light areas contain DNA*

DNA spread throughout cytoplasm

mesosome – infolding of cell surface membrane

cell surface membrane

cell wall

cytoplasm

Cell surface membrane, mesosomes and photosynthetic membranes

Like all cells, the living material of bacterial cells is surrounded by a partially permeable membrane. The structure and functions of the cell surface membrane are similar to those in eukaryotic cells (section 7.2.1). It is also the site of some respiratory enzymes. In addition, in some bacteria it forms mesosomes and/or photosynthetic membranes.

Mesosomes are infoldings of the cell surface membrane (figs 2.3 and 2.4) which may be artefacts created by the techniques used in specimen preparation for electron microscopy. They appear to be associated with DNA during cell division, facilitating the separation of the two daughter molecules of DNA after replication and aiding in the formation of new cross-walls between the daughter cells.

Among photosynthetic bacteria, sac-like, tubular or sheet-like infoldings of the cell surface membrane contain the photosynthetic pigments, always including bacterio-chlorophyll. Similar membranes are associated with nitrogen fixation.

Genetic material (bacterial 'chromosome')

Bacterial DNA is a single circular molecule of about 5×10^6 base pairs and of length 1 mm. The total DNA (the **genome**), and hence the amount of information encoded, is much less than that of a eukaryotic cell: typically it contains

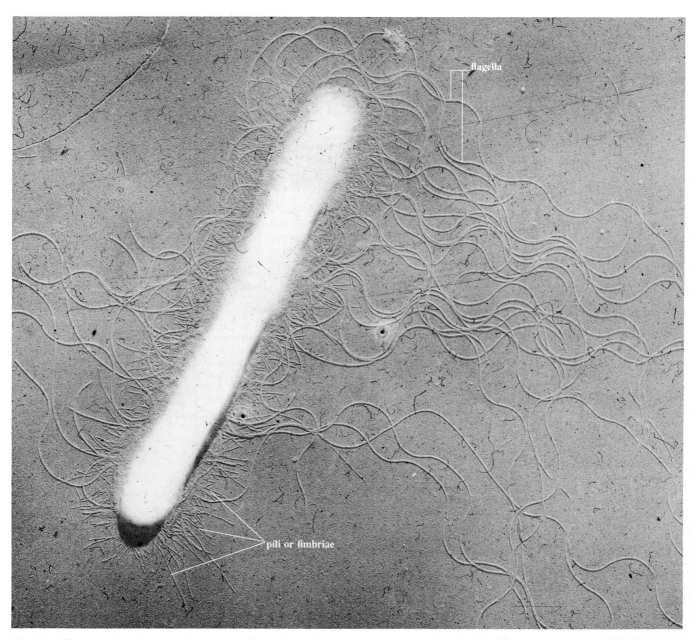

Fig 2.5 Transmission electron micrograph of rod-shaped bacterium to show shape, wall, fimbriae and long wavy flagella ($\times$ 28 000)

several thousand genes, about 500 times fewer than a human cell. (See also table 2.1 and fig 2.3.)

Ribosomes

See table 2.1 (protein synthesis) and fig 2.3.

Spores

Some bacteria, mainly of the genera *Clostridium* and *Bacillus*, form endospores (spores produced inside cells). They are thick-walled, long-lived and extremely resistant, particularly to heat and short-wave radiations. Their position in the cell is variable and is of importance in recognition and classification (see fig 2.6). If a whole cell forms a dormant, resistant structure it is called a cyst, as in some *Azotobacter* species.

2.2.3   Form

Bacterial shape is an important aid to classification. The four main shapes found are illustrated in fig 2.6. Examples of both useful and harmful bacteria are given.

2.2.4   Growth and reproduction

Growth of individuals and asexual reproduction

Bacteria have a large surface area to volume ratio and can therefore gain food sufficiently rapidly from their environment by diffusion and active transport mechanisms. Therefore, providing conditions are suitable, they can grow very rapidly. Important environmental factors affecting growth are temperature, nutrient availability, pH and ionic concentrations. Oxygen must also be present for obligate aerobes and absent for obligate anaerobes.

On reaching a certain size, dictated by the nucleus to cytoplasm ratio, bacteria reproduce asexually by binary fission, that is by division into two identical daughter cells. Cell division is preceded by replication of the DNA and while this is being copied it may be held in position by a mesosome (figs 2.3 and 2.4). The mesosome may also be attached to the new cross-walls that are laid down between the daughter cells, and plays some role in the synthesis of cell wall material. In the fastest growing bacteria such divisions may occur as often as every 20 min; this is known as the **generation time**.

---

(1) COCCI (sing. coccus) spherical

Cocci

Staphylococci (like a bunch of grapes)

e.g. *Staphylococcus aureus*, lives in nasal passages; different strains cause boils, pneumonia, food poisoning and other diseases

Streptococci (chains)

e.g. many *Streptococcus* spp.; some infect upper respiratory tract and cause disease, e.g. *S. pyogenes* causes scarlet fever and sore throats; *S. thermophilus* gives yoghurt its creamy flavour; *S. lactis*, see section 2.3.4

Diplococci (pairs)

the pneumococci (*Diplococcus pneumoniae*) are the only members; cause pneumonia

(2) BACILLI (sing. bacillus) rod-shaped

single rods

e.g. *Escherichia coli*, common gut-living symbiont; *Lactobacillus*, see section 2.3.4; *Salmonella typhi* causes typhoid fever

rods in chains

e.g. *Azotobacter*, a nitrogen-fixer; *Bacillus anthracis* causes anthrax

Bacilli with endospores showing various positions, shapes and sizes of spores

oval spore

central not swollen e.g. *Bacillus anthracis*, causes anthrax

spherical spore

terminal swollen e.g. *Clostridium tetani*, causes tetanus

subterminal swollen e.g. *Clostridium botulinum* (spores may also be central), causes botulism

(4) VIBRIOS comma-shaped

e.g. *Vibrio cholerae*, causes cholera single flagellum

(3) SPIRILLA (sing. spirillum) spiral-shaped

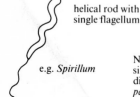

helical rod with single flagellum

e.g. *Spirillum*

NB body of spirochaetes is similar in form but locomotion differs, e.g. *Treponema pallidum* causes syphilis

Fig 2.6   *Forms of bacteria, illustrated by some common useful and harmful types*

Population growth

2.1 Consider the situation where a single bacterium is placed in a nutrient medium under optimal growth conditions. Assuming it, and its descendants, divide every 20 min, copy table 2.2 and complete it.

Using the data from the table, draw graphs of number of bacteria (graph A) and $\log_{10}$ number of bacteria (graph B) on the vertical axes against time (horizontal axis). What do you notice about the shapes of the graphs?

Table 2.2 Growth of a model population of bacteria.

Time (in units of 20 min)	0	1	2	3	4	5	6	7	8	9	10
A Number of bacteria											
B Log₁₀ number of bacteria (to one decimal place)											
C Number of bacteria expressed as power of 2											

The kind of growth shown in table 2.2 is known as logarithmic, exponential or geometric. The numbers form an exponential series. This can be explained by reference to line C in table 2.2 where the number of bacteria is expressed as a power of 2. The power can be called the logarithm or exponent of 2. The logarithms or exponents form a linearly increasing series 0, 1, 2, 3, etc., corresponding with the number of generations.

Returning to table 2.2, the numbers in line A could be converted to logarithms to the base 2 as follows:

A Number of bacteria	1	2	4	8	16	32	64	128	256	512	1024
D Log₂ number of bacteria	0	1	2	3	4	5	6	7	8	9	10

Compare line C with line D. However, it is conventional to use logarithms to the base 10, as in line B. Thus 1 is 10^0, 2 is $10^{0.3}$, 4 is $10^{0.6}$, etc.

The curve in graph A is known as a **logarithmic** or **exponential curve**. Such growth curves can be converted to straight lines by plotting the logarithms of growth against time. Under ideal conditions, then, bacterial growth is theoretically exponential. This mathematical model of bacterial growth can be compared with the growth of a real population. Fig 2.7 shows such growth. The growth curve shows four distinct phases. During the **lag phase** the bacteria are adapting to their new environment and growth has not yet achieved its maximum rate. The bacteria may,

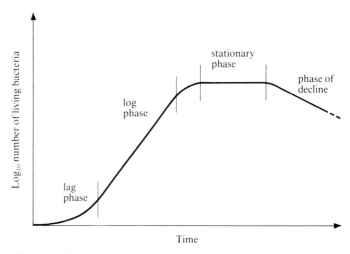

Fig 2.7 *Typical growth curve of a bacterial population*

for example, be synthesising new enzymes to digest the particular spectrum of nutrients available in the new medium.

The **log phase** is the phase when growth is proceeding at its maximum rate, closely approaching a logarithmic increase in numbers when the growth curve would be a straight line. Eventually growth of the colony begins to slow down and it starts to enter the **stationary phase** where growth rate is zero, and there is much greater competition for resources. Rate of production of new cells is slower and may cease altogether. Any increase in the number of cells is offset by the death of other cells, so that the number of living cells remains constant. This phase is a result of several factors, including exhaustion of essential nutrients, accumulation of toxic waste products of metabolism and possibly, if the bacteria are aerobic, depletion of oxygen.

During the final phase, the **phase of decline**, the death rate increases and cells stop multiplying. Methods of counting bacteria are described in the practical work at the end of this chapter.

Table 2.3 Culture of bacteria at 30°C.

Time/h	Number of cells in millions	
	living	living and dead
0	9	10
1	10	11
2	11	12
5	18	20
10	400	450
12	550	620
15	550	700
20	550	850
30	550	950
35	225	950
45	30	950

Sexual reproduction or genetic recombination

Bacteria exhibit a primitive form of sexual reproduction which differs from eukaryote sexual reproduction in that there are no gametes and cell fusion does not occur. However, the essential feature of sexual reproduction, namely exchange of genetic material, does take place and is called **genetic recombination**. Part (rarely all) of the DNA from the donor cell is transferred to the recipient cell whose DNA is genetically different. Parts of the donor DNA can replace parts of the recipient DNA, the process involving breakage and reunion of DNA strands by certain enzymes. The DNA formed contains genes from both parent cells and is called **recombinant DNA**. The offspring, or **recombinants**, will show variation as a result of the mixing of genes. This is the basic advantage of sexual reproduction because variation contributes to the process of evolution.

Three methods are known by which recombination can be achieved. In order of their discovery they are transformation, conjugation and transduction.

In **transformation** the donor and recipient do not come into contact. The process was discovered by Griffith in 1928. He was working with a pneumococcus, a bacterium causing pneumonia. The colonies were of two types, rough(R) and smooth(S) in appearance. The former were non-pathogenic and had no capsules; the latter were pathogenic and had large capsules (see section 2.2.2). Griffith discovered that if a mouse was injected with living R cells and dead (heat-killed) S cells it would die within a few days, and from its blood living S cells could be isolated. He concluded that the dead S cells had released a factor which enabled R cells to develop capsules and thus resist destruction by the host. This 'transformation' proved to be heritable, and since the molecule of inheritance was at that time unknown, though suspected to be protein, great efforts were made to identify the transforming factor.

In 1944, Avery, MacLeod & McCarty succeeded in isolating and identifying the factor and were surprised to find that it was DNA, not protein. This was the first direct evidence that the genetic material is DNA.

It is now known that during transformation a short piece of DNA is released by the donor and actively taken up by the recipient, in which it replaces a similar, though not necessarily identical, piece of DNA. It only occurs in a few genera, including some pneumococci, in so-called 'competent' strains where DNA can penetrate the recipient. A possible mechanism for transformation is shown in fig 2.8.

Conjugation involves DNA transfer between cells in direct contact. In contrast to transformation and transduction, large fractions of the donor DNA may be exchanged. It was discovered in 1946 in *Escherichia coli* in the following experiment. Normally *E. coli* can make all of its own amino acids, given a supply of glucose and mineral salts. Random

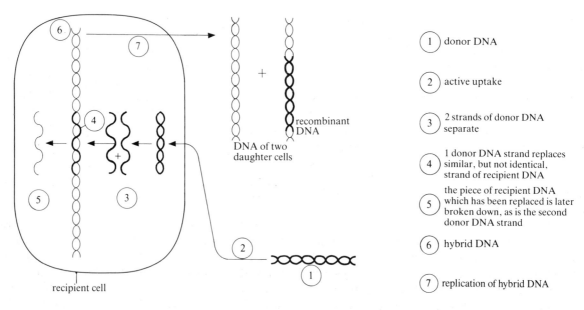

1. donor DNA
2. active uptake
3. 2 strands of donor DNA separate
4. 1 donor DNA strand replaces similar, but not identical, strand of recipient DNA
5. the piece of recipient DNA which has been replaced is later broken down, as is the second donor DNA strand
6. hybrid DNA
7. replication of hybrid DNA

Fig 2.8 *A possible method for transformation. The precise method for active uptake of donor DNA is not known*

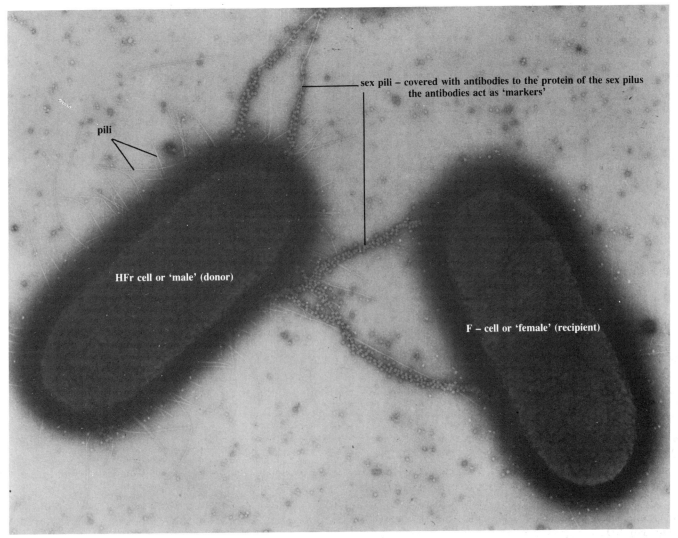

Fig 2.9 *Transmission electron micrograph of conjugating bacteria, one 'male' with two 'females' (× 19475). The second 'female' cell is beyond the top of the photograph*

mutations were induced by exposure to radiation and two particular mutants selected. One could not make biotin (a vitamin) or the amino acid methionine. Another could not make the amino acids threonine and leucine. About 10^8 cells of each mutant were mixed and cultured on media lacking all four growth factors. Theoretically, none of the cells should have grown, but a few hundred colonies developed, each from one original bacterium, and these were shown to possess genes for making all four growth factors. Exchange of genetic information had therefore occurred, but no chemical responsible could be isolated. Eventually it was shown with the electron microscope that direct cell to cell contact, that is conjugation, can occur in *E. coli* (fig 2.9).

The ability to serve as a donor is determined by genes in a small circular piece of DNA called the sex factor, or **F factor** (F for fertility), a type of plasmid (see below) which codes for the protein of a special type of pilus, the F pilus or sex pilus. The sex pilus enables cell to cell contact to be established. DNA is a double-stranded molecule and

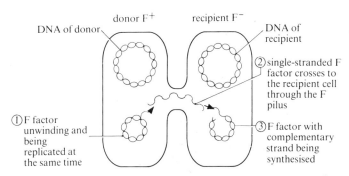

Fig 2.10 *Conjugation and transfer of F factor between cells. 1, 2 and 3 represent successive stages in transfer*

during conjugation one of the two strands of F factor DNA passes through the sex pilus from the donor (F^+) to the recipient (F^-). The process is summarised in fig 2.10 which shows that the donor retains the F factor while the recipient also receives a copy. 'F+ness' can therefore be spread throughout a population. Donor cells may become F^- cells by spontaneously losing the F factors.

The F factor is particularly important because in a few cases, about 1 in 100 000, it becomes integrated with the rest of the DNA in the host cell. In such cases, the process of conjugation involves transfer of not only the F factor, but also the rest of the DNA. This takes about 90 min and separation may occur before exchange is complete. Such strains consistently donate all or large portions of their DNA and are called Hfr strains (H = high, f = frequency, r = recombination), because the donor DNA can recombine with the recipient DNA.

During **transduction** a small, double-stranded piece of DNA is transferred from donor to recipient by a bacteriophage (a virus, see section 2.3). The probable mechanism is shown in fig 2.11.

Some viruses have the ability to integrate their DNA with bacterial DNA; it is replicated at the same time as their host's DNA and passed from one bacterial generation to the next. Occasionally it becomes active and codes for the production of new viruses. The host (bacterial) DNA breaks down and odd pieces may get included inside new virus coats, sometimes to the exclusion of the viral DNA. The new 'viruses', or **transducing particles**, then carry the DNA to other bacteria.

Plasmids and episomes

Plasmids and episomes are small pieces of DNA separate from the bulk of the DNA. They often, though not always, replicate in step with the host DNA and are not essential for the survival of the host.

Originally a distinction was drawn between plasmids, which could not integrate with the host DNA, and episomes, which could do so. Episomes include F factors and temperate phages (section 2.3.4). In recent years the general term plasmid has been adopted for both. Plasmids are now known to be extremely common and can be regarded as subcellular parasites or symbionts even simpler than viruses. The question of whether viruses are living is discussed in section 2.3.2; with plasmids, which are only

DNA molecules, the question is even more difficult to answer.

Plasmids confer various abilities on their hosts. Some are 'resistance factors' (R plasmids or R factors), that is factors that confer resistance to antibiotics. An example is the penicillinase plasmid of staphylococci which can be transduced by bacteriophages. This contains a gene for the enzyme penicillinase which breaks down penicillin, thus conferring resistance to penicillin. Spread of such factors by sexual reproduction among bacteria has important implications for medicine. Other plasmid genes confer resistance to disinfectants; cause disease, such as staphylococcal impetigo; are responsible for the fermentation of milk to cheese by lactic acid bacteria; and confer ability to utilise complex substances as food, such as hydrocarbons, with potential applications in cleaning oil spills and producing protein from petroleum.

To conclude, it should be said that all forms of sexual reproduction in bacteria are rare, but significant because of the vast numbers of cells occurring in bacterial colonies. Sexual reproduction is much less organised than in eukaryotes and the complete genome (total DNA) can only be exchanged during conjugation of bacteria and then it is a rare event. It has particular significance in spreading resistance to antibiotics and disinfectants.

2.2.5 Nutrition

At the beginning of chapter 9, and summarised in table 9.1, organisms are placed into four nutritional categories. There are examples of bacteria in all four categories, as shown in table 2.4. The most important group is the chemoheterotrophic bacteria. In their feeding strategies these bacteria resemble fungi. As with fungi there are three groups, namely saprotrophs, mutualists and parasites (see sections 2.4.1 and 10.1).

A **saprotroph** is an organism that obtains its food from dead and decaying matter. The saprotroph secretes

Fig 2.11 *Mechanism of transduction*

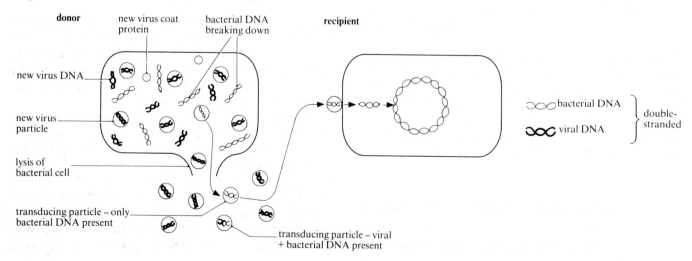

Table 2.4 The four nutritional categories of bacteria and some of their characteristics.

	Autotrophic (carbon source is carbon dioxide (inorganic))		Heterotrophic (carbon source is organic compounds made by other organisms)	
	Photoautotrophic*	Chemoautotrophic (chemosynthetic)	Photoheterotrophic*	Chemoheterotrophic
Energy source	Light	Chemical – from oxidation of inorganic substances during respiration	Light	Chemical – from oxidation of organic substances during respiration
Types	Only green and blue-green bacteria, purple sulphur bacteria and some purple non-sulphur bacteria	Nitrifying bacteria, sulphur bacteria and others	Only purple non-sulphur bacteria, extremely few in number	Most bacteria – important as saprotrophs, parasites and mutualists, utilising enormous range of chemical substances as food
Further information	section 9.9	section 9.10	—	section 2.2.5

* photoautotrophic + photoheterotrophic types comprise the *photosynthetic* bacteria.

enzymes onto the organic matter, so that digestion is outside the organism. Soluble products of digestion are absorbed and assimilated within the body of the saprotroph.

Saprotrophic bacteria and fungi constitute the **decomposers** and are essential in bringing about decay and recycling of nutrients. They produce humus from animal and plant remains, but also cause decay of materials useful to humans, especially food. Their importance in the biosphere is stressed in section 2.5, and also in chapter 12.

Mutualism is the name given to any form of close relationship between two living organisms in which both partners benefit. Examples of bacterial mutualists are *Rhizobium*, a nitrogen-fixer living in the root nodules of legumes, such as pea and clover, and *Escherichia coli*, which inhabits the gut of humans, and probably contributes vitamins of the B and K groups.

A **parasite** is an organism that lives in or on another organism, the **host**, from which it obtains its food and, usually, shelter. The host is usually of a different species and suffers harm from the parasite. Parasites which cause disease are called **pathogens**. Some examples are given in section 2.6. Some parasites can only survive and grow in living cells and are called **obligate parasites**. Others can infect a host, bring about its death and then live saprotrophically on the remains; these are called **facultative parasites**. It is a characteristic of parasites that they are very exacting in their nutritional requirements, needing 'accessory growth factors' that they cannot manufacture for themselves but can only find in other living cells.

2.2.6 Useful and harmful bacteria/Practical work

See sections 2.5 and 2.6, and 2.7 respectively.

2.3 Viruses

2.3.1 Discovery

In 1852, the Russian botanist D. J. Ivanovsky prepared an infectious extract from tobacco plants that were suffering from mosaic disease. When the extract was passed through a filter able to prevent the passage of bacteria, the filtered fluid was still infectious. In 1898 the Dutchman Beijerink coined the name 'virus' (Latin for poison) to describe the infectious nature of certain filtered plant fluids. Although progress was made in isolating highly purified samples of viruses and in identifying them chemically as nucleoproteins (nucleic acids combined with proteins), the particles still proved elusive and mysterious because they were too small to be seen with the light microscope. As a result, they were among the first biological structures to be studied when the electron microscope was developed in the 1930s.

2.3.2 Characteristics

Size

Viruses are the **smallest living organisms**, ranging in size from about 20 nm to 300 nm; on average they are about 50 times smaller than bacteria. As stated above they cannot be seen with the light microscope (since they are usually smaller than half a wavelength of light) and they pass through filters which retain bacteria.

The question is often posed, 'Are viruses living?'. If, to be defined as living, a structure must possess genetic material (DNA or RNA), and be capable of reproducing itself, then the answer must be that viruses are living. If to be living demands a cellular structure then the answer is that they are not. It should also be noted that viruses are not capable of reproducing outside the host cell. They are on the borderline between living and non-living and remind us that there is a continuous spectrum of increasing complexity from simple molecules to the elaborate enclosed systems of cells.

Habit

Since viruses can only reproduce themselves inside living cells, they are all obligate parasites. They usually cause obvious signs of disease. Once inside the host cell, they 'switch off' (inactivate) the host's DNA and, using their own DNA or RNA, instruct the cell to make new copies of the virus (section 2.3.3). Viruses are transmitted from cell to cell as inert particles.

Structure

Viruses have a very simple structure consisting of a length of genetic material, either DNA or RNA, forming a **core** surrounded and protected by a coat of protein called a **capsid**. The fully assembled, infective particle is called a **virion**. A few viruses, such as herpes and influenza viruses, have an additional lipoprotein **envelope** derived from the surface membrane of the host cell. Unlike all other organisms viruses are non-cellular.

Virus coats are often built up of identical repeating subunits called **capsomeres**. These form highly symmetrical structures that can be crystallised, enabling information about their structure to be obtained by X-ray crystallography as well as electron microscopy. Once the subunits of a virus have been made by the host, they show the property of self-assembly into a virus. Self-assembly is characteristic of many other biological structures and is of fundamental importance in biology. Fig 2.12 shows a simplified, generalised structure of a virus.

Icosahedron and dodecahedron (such as adenovirus, polyoma/papilloma virus, polio virus). An icosahedron has 20 triangular faces with 12 corners and 30 edges. Fig 2.13a shows a regular icosahedron. Using the technique of negative staining the detailed structure of viruses can be observed because the stain can penetrate between, and show up, all the surface features. For example figs 2.13b and c reveal that, in the case of the

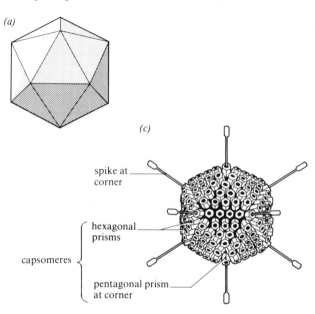

(a)

(c)

spike at corner

hexagonal prisms

capsomeres

pentagonal prism at corner

(b)

× 480 000

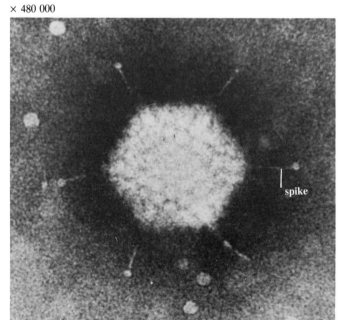

spike

Fig 2.13 *(a) Solid model of an icosahedron. (b) Adenovirus particle, an icosahedral virus showing spikes at the corners. Electron micrograph of a negatively stained preparation. (c) Drawing from a three-dimensional model of an adenovirus. The capsid is made of 252 capsomeres, 12 at corners, 240 on faces and edges. Adenoviruses are DNA viruses which have been isolated from a variety of mammals and birds. They infect lymphoid tissue in humans and cause respiratory disease*

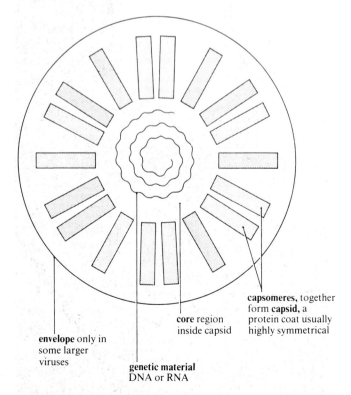

capsomeres, together form capsid, a protein coat usually highly symmetrical

core region inside capsid

envelope only in some larger viruses

genetic material DNA or RNA

Fig 2.12 *Generalised section of a capsomere-possessing virus*

14

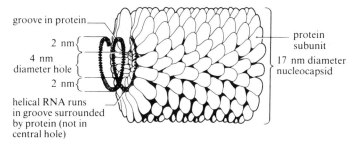

groove in protein

2 nm
4 nm diameter hole
2 nm

helical RNA runs in groove surrounded by protein (not in central hole)

protein subunit

17 nm diameter nucleocapsid

Fig 2.14 *Structure of tobacco mosaic virus. Drawing of part of the rod-shaped virus based on X-ray diffraction, biochemical and electron microscope data*

Helical symmetry. This is best illustrated by the tobacco mosaic virus (TMV), an RNA virus (fig 2.14). The 2 130 identical protein subunits, together with the RNA, form an integrated structure called a **nucleocapsid**. In some viruses the nucleocapsid is enclosed in an envelope, for instance the mumps and influenza viruses.

Bacteriophage type. Viruses that attack bacteria form a group called bacteriophages. Some of these have a distinct icosahedral head, with a tail showing helical symmetry (fig 2.15).

Complex types. Some viruses have a complex structure, such as rhabdoviruses and pox viruses.

adenovirus, each of the 20 faces is made up of a number of capsomeres. The overall number of capsomeres is 252 (240 hexagonal, 12 pentagonal at the corners). This number varies between viruses, for instance herpes has 162, polyoma 42, $\Phi \times 174$ bacteriophage 12. All have 12 pentagonal capsomeres, the latter having no hexagonal capsomeres and forming a shape called a **dodecahedron**.

2.3.3 Life cycle of a bacteriophage

The life cycle of a typical bacteriophage is shown in fig 2.16; fig 2.17 shows the scale of the phage in relation to the bacterium. *E. coli* is a typical host and it can be attacked by at least seven strains of phage, known as T_1 to T_7. A T-even phage (for instance T_2) is illustrated in figs 2.15 and 2.16.

(a)

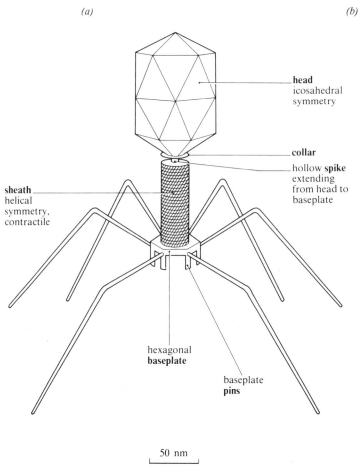

head
icosahedral symmetry

collar

hollow **spike** extending from head to baseplate

sheath
helical symmetry, contractile

hexagonal **baseplate**

baseplate **pins**

50 nm

(b)

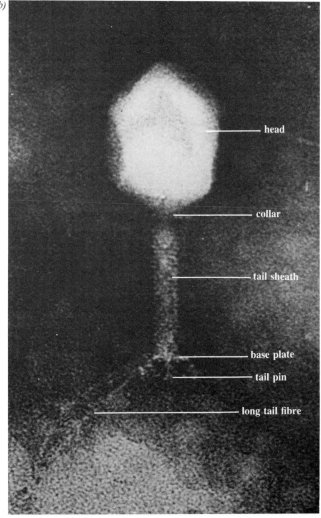

head

collar

tail sheath

base plate

tail pin

long tail fibre

Fig 2.15 *(a) Structure of a bacteriophage. (b) Electron micrograph of negatively stained bacteriophage*

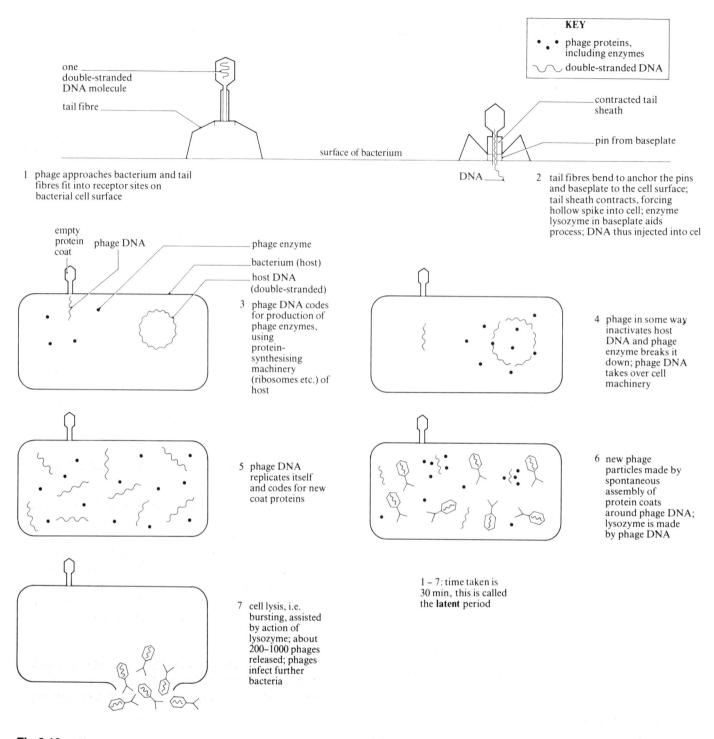

KEY

• • • phage proteins, including enzymes

⌇⌇⌇ double-stranded DNA

one double-stranded DNA molecule

tail fibre

surface of bacterium

1 phage approaches bacterium and tail fibres fit into receptor sites on bacterial cell surface

contracted tail sheath

pin from baseplate

DNA

2 tail fibres bend to anchor the pins and baseplate to the cell surface; tail sheath contracts, forcing hollow spike into cell; enzyme lysozyme in baseplate aids process; DNA thus injected into cel

empty protein coat

phage DNA

phage enzyme

bacterium (host)

host DNA (double-stranded)

3 phage DNA codes for production of phage enzymes, using protein-synthesising machinery (ribosomes etc.) of host

4 phage in some way inactivates host DNA and phage enzyme breaks it down; phage DNA takes over cell machinery

5 phage DNA replicates itself and codes for new coat proteins

6 new phage particles made by spontaneous assembly of protein coats around phage DNA; lysozyme is made by phage DNA

1 – 7: time taken is 30 min, this is called the **latent** period

7 cell lysis, i.e. bursting, assisted by action of lysozyme; about 200–1000 phages released; phages infect further bacteria

Fig 2.16 *Life cycle of a bacteriophage*

2.3.4 Life cycles of other viruses

A similar life cycle probably occurs in most viruses. The penetration process differs in bacterial, plant and animal viruses because bacterial and plant viruses have to penetrate cell walls. This does not always involve the injection process described in fig 2.16 and protein coats are not always left outside the cell.

Some phages do not replicate once inside the host cell but instead their nucleic acid becomes incorporated into the DNA of the host cell. They may then remain without influence through several generations, being replicated

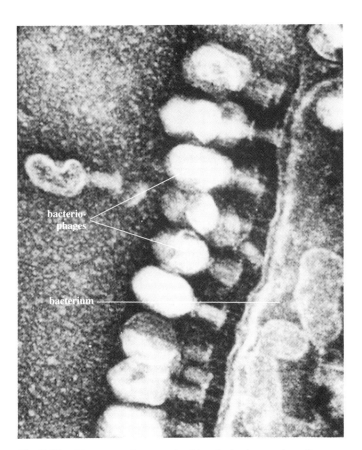

Fig 2.17 *Electron micrograph of bacteriophages heavily infesting a bacterium* Escherichia coli

Kingdom Fungi

2.4 Fungi

The fungi are a large and successful group of organisms of about 80 000 named species. They range in size from the unicellular yeasts to the large toadstools, puffballs and stinkhorns, and occupy a very wide range of habitats, both aquatic and terrestrial. They are also of major importance for the essential role that they play in the biosphere, and for the way in which they have been exploited by humans for economic and medical purposes.

Fungi include the numerous moulds growing on damp organic matter (such as bread, leather, decaying vegetation and dead fish), the unicellular yeasts which are abundant on the sugary surfaces of ripe fruits and many parasites of plants. The latter cause some economically important diseases of crops, such as mildews, smuts and rusts. A few fungi are parasites of animals, but are less significant in this respect than bacteria.

The study of fungi is called **mycology** (*mykes,* mushroom). It constitutes a branch of microbiology because many of the handling techniques used, such as sterilising and culturing procedures, are the same as those used with bacteria.

2.4.1 Characteristics and classification of Fungi

Fungi are eukaryotes that lack chlorophyll, and are therefore heterotrophic, like animals. However, they have rigid cell walls and are non-motile, like plants. Traditionally, they have been regarded as plants*, but more modern classifications, such as that shown in fig. 3.1, place them in a separate kingdom. Their characteristics and classification are summarised in fig 2.18 and table 2.5. The two largest and most advanced groups are the Ascomycota and the Basidiomycota.

> **2.5** Using those features of the kingdom Fungi given in table 2.5, prepare a table of differences between fungi and chlorophyll-containing plant cells.

Structure

The body structure of the fungi is unique. It consists of a mass of fine, tubular branching threads called **hyphae** (singular, hypha), the whole mass being called a **mycelium.** Each hypha has a thin, rigid wall whose chief component is chitin, a nitrogen-containing polysaccharide also found as a

only as the host replicates its own DNA. Such phages are known as **temperate phages** and the bacteria that harbour them **lysogenic**, meaning that they have the potential for lysis, but that this is not shown until the phage resumes activity. The inactive phage is called a **prophage** or **provirus**.

2.3.5 Evolutionary origin of viruses

The most plausible hypothesis for the origin of viruses is that they represent escaped nucleic acid, that is nucleic acid that has become capable of replicating itself independently of the cell from which it originated, even though it means using (parasitising) the machinery of that, or other, cells. Viruses would thus be derived from cellular organisms and should not be regarded as primitive forerunners of cellular organisms.

How common such 'escapes' have been cannot be easily judged, but it is likely that our increasing knowledge of genetics will reveal more variations on the theme of parasitic nucleic acids.

2.3.6 Viruses as agents of disease

See section 2.6.2.

---

* At one time fungi were given the status of a class, and together with the class Algae formed the division Thallophyta of the plant kingdom. The **Thallophyta** were those plants whose bodies could be described as a thallus. A **thallus** is a body, often flat, which is not differentiated into true roots, stems and leaves and lacks a true vascular system.

structural component in the exoskeletons of arthropods (section 5.2.4). The hyphae are not divided into true cells. Instead, the protoplasm is either continuous or interrupted at intervals by cross-walls called **septa** which divide the hyphae into compartments similar to cells. Unlike normal cell walls their formation is not a consequence of nuclear division, and a pore normally remains at their centre allowing protoplasm to flow between compartments. Each compartment may contain one, two or more nuclei, which are distributed at more or less regular intervals along the hyphae. Hyphae lacking cross-walls are called **non-septate** (**aseptate**) or **coenocytic**, the latter term applying to any mass of protoplasm containing many nuclei and not split into cells. Hyphae having cross-walls are called **septate**. Within the cytoplasm the usual eukaryote organelles are found, such as mitochondria, Golgi apparatus, endoplasmic reticulum, ribosomes and vacuoles. In the older parts, vacuoles are large and cytoplasm is confined to a thin peripheral layer. Sometimes hyphae aggregate to form more solid structures such as the fruiting bodies of the Basidomycota.

Nutrition

Fungi are heterotrophic, that is require an organic source of carbon. In addition, they require a source of nitrogen, usually organic such as amino acids; inorganic ions such as K^+ and Mg^{2+}; trace elements such as Fe, Zn and Cu; and organic growth factors such as vitamins. The exact range of nutrients required, and hence substrates on which they are found, is variable. Some fungi, particularly obligate parasites, require a wide range of ready-made components; others can synthesise their own requirements given only a source of carbohydrate and mineral salts. Some may synthesise most of their requirements but need one or more particular amino acids or vitamins. The nutrition of fungi can be described as **absorptive** because they absorb nutrients directly from outside their bodies. This is in contrast to animals, which normally **ingest** food, and then **digest** it *within their bodies* before **absorption** takes place.

With fungi, digestion, if necessary, is external using extracellular enzymes.

Fungi obtain their nutrients as saprotrophs, parasites or mutualists. In this respect they are like most bacteria and the three terms have already been defined in section 2.2.5. (See also section 10.1.)

Saprotrophs. Fungal saprotrophs produce a variety of digestive enzymes. If they secrete the three main classes of digestive enzymes, namely carbohydrases, lipases and proteases, they can utilise a wide range of substrates, for example the *Penicillium* species which form green and blue moulds on substrates such as soil, damp leather, bread and decaying fruit.

The hyphae of saprotrophic fungi are usually chemotropic, that is they grow towards certain substrates in response to chemicals diffusing from these substrates (section 15.1.1).

Fungal saprotrophs usually produce large numbers of light, resistant spores. This allows efficient dispersal to other food sources. Examples are *Mucor*, *Penicillium* and *Agaricus*.

Saprotrophic fungi and bacteria together form the **decomposers** which are essential in the recycling of nutrients. Especially important are the few that secrete the enzyme cellulase, which breaks down cellulose. Cellulose is an important structural component of plant cell walls, and the rotting of wood and other plant remains is achieved partly by decomposers secreting cellulase.

Some fungal saprotrophs are of economic importance, such as *Saccharomyces* (yeast) and *Penicillium* (section 2.4.3).

Parasites. Fungal parasites may be facultative or obligate (section 2.2.5), and more commonly attack plants than animals. Obligate parasites do not normally kill their hosts, whereas facultative parasites frequently do, and live saprotrophically off the dead remains. Obligate parasites include the powdery mildews, downy mildews,

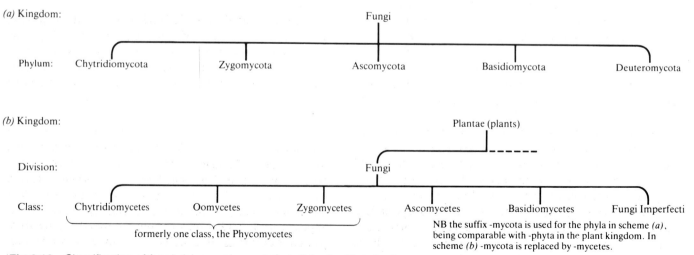

Fig 2.18 *Classification of fungi: (a) a modern scheme, (b) a traditional scheme*

Table 2.5 Classification and characteristics of fungi.

Kingdom Fungi

General characteristics
Heterotrophic nutrition because they lack chlorophyll and therefore non-photosynthetic. They can be parasites, saprotrophs or mutualists
Rigid cell walls containing chitin as the fibrillar material
Body is usually a mycelium, a network of fine tubular filaments called hyphae
If carbohydrate is stored, it is usually as glycogen, not starch
Reproduce by means of spores
Non-motile

Phylum Zygomycota	*Phylum Ascomycota*	*Phylum Basidiomycota*
Sexual reproduction by conjugation, involving fusion of two gametangia to produce a zygospore	Sexual reproduction involves production of spores (ascospores) inside a special structure called an ascus	Sexual reproduction involves production of basidia which bear spores (basidiospores) externally
Asexual reproduction by conidia or sporangia containing spores. No zoospores	Asexual reproduction by conidia. No sporangia	Asexual reproduction by formation of spores. Not common
Non-septate hyphae and large well-developed branching mycelium	Septate hyphae	Septate hyphae
e.g. *Rhizopus stolonifer*, common bread mould, a saprotroph. *Mucor*, common moulds, saprotroph	e.g. *Penicillium* and *Aspergillus*, saprotrophic moulds. *Saccharomyces* (yeast), unicellular saprotrophs. *Erysiphe*, obligate parasites causing powdery mildews, e.g. of barley. *Ceratocystis ulmi*, parasite causing Dutch elm disease	e.g. *Agaricus campestris*, field mushroom, saprotroph

Phylum Chytridiomycota
A small group of microscopic, often unicellular, fungi, e.g. *Synchytrium endobioticum*, a parasite causing wart disease of potatoes.
Phylum Deuteromycota (Fungi Imperfecti)
Fungi in which sexual reproduction has never been observed, and whose classification is uncertain, e.g. *Trichophyton* which causes athlete's foot and ring-worm.

NB: Phyla end in '-mycota'

rusts and smuts, and are usually restricted to a narrow range of hosts from which they require a specific range of nutrients. Facultative parasites are usually less specialised and may grow on a variety of hosts or substrates.

If the host is a plant, hyphae penetrate through stomata, directly through the cuticle and epidermis, or through wounds. Once inside the plant, hyphae normally ramify between cells, sometimes producing pectinases which digest a path through the middle lamellae. The fungus may be systemic, that is spread throughout the host, or it may be confined to a small part of the host.

Facultative parasites commonly produce sufficient pectinases to cause 'soft rot' of the tissue, reducing it to a mush. Subsequently cells may be invaded and killed with the aid of cellulase which digests the cell walls. Cell constituents may be absorbed directly or digested by secretion of further fungal enzymes. Obligate parasites possess specialised penetration and absorption devices called **haustoria**. Each haustorium is a modified hyphal outgrowth with a large surface area which pushes into living cells without breaking

their cell surface membranes, and without killing them. The success of the parasite depends on the continued life of the host. Haustoria are rarely produced by facultative parasites.

The life cycles of parasitic fungi are sometimes complex. This is particularly true of obligate parasites, such as rust fungi, whose life cycles involve several stages and more than one host. Obligate parasites usually produce resistant spores by sexual reproduction to coincide with the deaths of their hosts. In this way they may overwinter.

Mutualism. Two important types of mutualistic union are made by fungi, namely lichens and mycorrhizae. **Lichens** are mutualistic associations between fungi and algae. The fungus is an ascomycote or a basidiomycote, while the alga is a green alga (or blue-green bacterium). Lichens are commonly encrusted on exposed rocks and trunks of trees; they also hang from trees in wet forests. It is believed that the alga contributes organic food from photosynthesis, while the fungus is

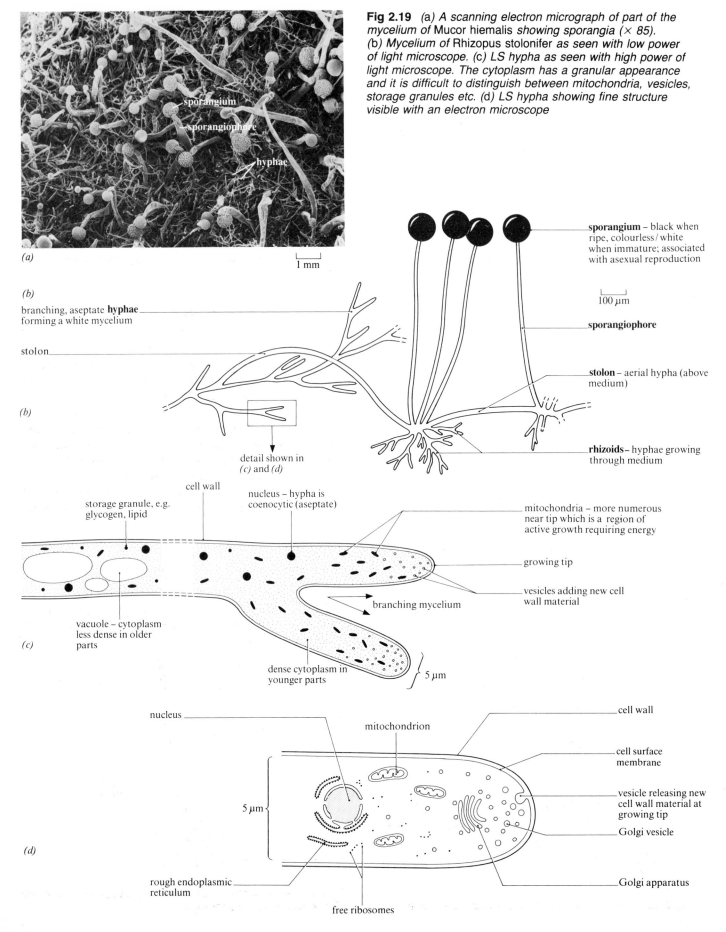

Fig 2.19 *(a) A scanning electron micrograph of part of the mycelium of* Mucor hiemalis *showing sporangia (× 85). (b) Mycelium of* Rhizopus stolonifer *as seen with low power of light microscope. (c) LS hypha as seen with high power of light microscope. The cytoplasm has a granular appearance and it is difficult to distinguish between mitochondria, vesicles, storage granules etc. (d) LS hypha showing fine structure visible with an electron microscope*

(a)

1 mm

(b)

branching, aseptate **hyphae** forming a white mycelium

stolon

(b)

detail shown in *(c)* and *(d)*

sporangium – black when ripe, colourless / white when immature; associated with asexual reproduction

100 μm

sporangiophore

stolon – aerial hypha (above medium)

rhizoids – hyphae growing through medium

storage granule, e.g. glycogen, lipid

cell wall

nucleus – hypha is coenocytic (aseptate)

mitochondria – more numerous near tip which is a region of active growth requiring energy

growing tip

vesicles adding new cell wall material

branching mycelium

vacuole – cytoplasm less dense in older parts

dense cytoplasm in younger parts

5 μm

(c)

nucleus

mitochondrion

cell wall

cell surface membrane

vesicle releasing new cell wall material at growing tip

Golgi vesicle

5 μm

Golgi apparatus

rough endoplasmic reticulum

free ribosomes

(d)

protected from high light intensity and able to absorb water and mineral salts. The fungus can also conserve water, enabling some lichens to grow in dry conditions where no other plants exist.

2.4.2   Phylum Zygomycota

Characteristics of the Zygomycota are given in table 2.5. They are a small group of fungi which are regarded as ancestral to the two main divisions, Ascomycota and Basidiomycota.

An example, *Rhizopus*, is a very common saprotroph, similar in structure and appearance to *Mucor*, but more widespread. Both *Rhizopus* and *Mucor* are called pin moulds for the reason given below (in asexual reproduction). *Rhizopus stolonifer* is a common species and is the common bread mould. It may also grow on apples and other fruit in storage, causing soft rot.

Structure

The structure of the mycelium and individual hyphae is shown in fig 2.19. The mycelium is profusely branching and aseptate. Unlike *Mucor*, the mycelium develops aerial stolons which arch above the medium and produce hyphae called **rhizoids** where they touch down. From these points sporangiophores develop.

Life cycle

The life cycle of *Rhizopus stolonifer* is summarised diagrammatically in fig 2.20.

Asexual reproduction

After two or three days in culture, *Rhizopus* begins to produce vertically growing hyphae called **sporangiophores** (*-phore*, stalk). These are negatively geotropic. Each sporangiophore tip swells into a **sporangium** which becomes separated from the sporangiophore by a domed cross-wall called the **columella** as shown in fig 2.21. Inside the sporangium the protoplasm splits up into portions, each

of which acquires a wall and becomes a spore containing several nuclei. The appearance of the sporangiophores and sporangia resembles a collection of pins; hence *Rhizopus* and the other closely related fungi such as *Mucor* are called **pin moulds**. As the sporangium matures it becomes black and dries, the wall eventually cracking unevenly to expose a dry, powdery mass of spores. The columella collapses, as shown in fig 2.21, providing a wide platform from which spores are easily blown away and dispersed. Under wet conditions, the sporangia would not dry and crack, thus preventing release of spores when conditions are unfavourable for dispersal. The haploid spores germinate if they land on a suitable substrate and produce a new mycelium.

2.6   What is the purpose of the sporangiophores?

Sexual reproduction

Many fungi exist in two different mating strains. Sexual reproduction can only occur between different strains, even if both produce male and female sex organs. Such self-sterile fungi are called **heterothallic** and the two strains are usually designated plus and minus (*not* male and female). They are structurally identical but physiologically slightly different. Fungi in which there is only one strain, and which are therefore self-fertile, are called **homothallic**. The advantage of heterothallism is that cross-fertilisation occurs and greater variation is ensured.

Rhizopus stolonifer is heterothallic. The events of sexual reproduction are summarised in fig 2.22. The initial events

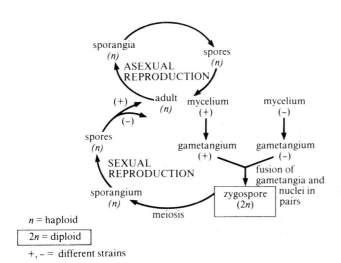

n = haploid

2*n* = diploid

+, − = different strains

Fig 2.20 *Diagrammatic life cycle of* Rhizopus stolonifer

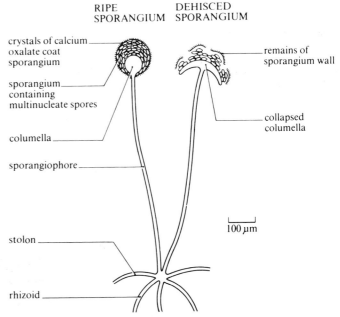

Fig 2.21 *Asexual reproduction in* Rhizopus stolonifer, *showing LS ripe and dehisced sporangia*

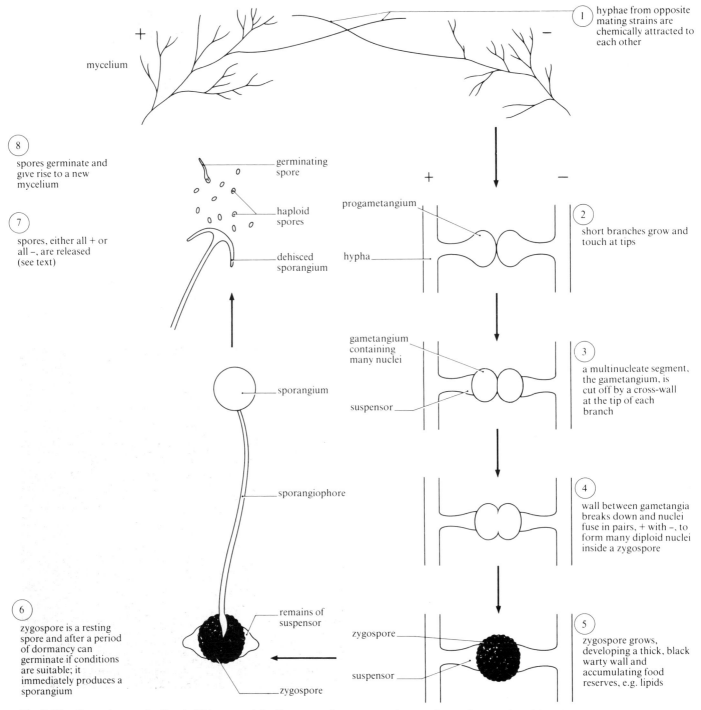

Fig 2.22 *Sexual reproduction in* Rhizopus stolonifer, *+ and − represent opposite mating strains, (1) to (8) sequence of events*

are caused by diffusion of hormones between the strains. The hormones stimulate growth of long hyphae between the colonies. These probably release volatile chemical signals which attract the opposite strain, a form of chemotropism.

There are no typical gametes and fertilisation is a process of nuclei fusing in pairs as described in fig 2.22. Since the gametangia are of equal size, the process of sexual reproduction is described as **isogamy**.

The product of nuclear fusion is a zygospore containing many diploid nuclei. It is thought that all of these degenerate except for one. This divides by meiosis to form four haploid nuclei only one of which survives. It is a matter of chance whether this is of a plus or minus strain.

The zygospore, unlike the asexually produced spores, is not specialised for dispersal but rather for a period of dormancy, since it has food reserves and a thick protective wall. Dispersal is achieved immediately after germination, when asexual reproduction occurs by production of a sporangium as shown in fig 2.22. Within the germinating

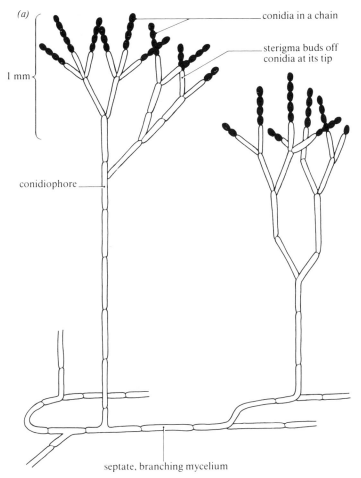

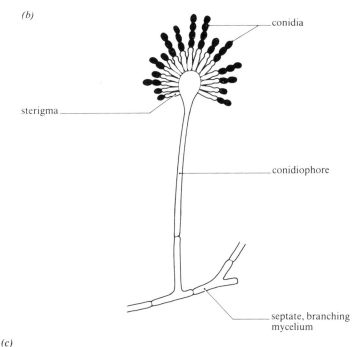

(a) conidia in a chain

sterigma buds off conidia at its tip

1 mm

conidiophore

septate, branching mycelium

(b) conidia

sterigma

conidiophore

septate, branching mycelium

(c) conidia

sterigma

conidiophore

Fig 2.23 *Two common members of the Ascomycota showing asexual reproduction. (a)* Penicillium, *with a brush-like arrangement of conidia. (b)* Aspergillus, *with a spherical mass of conidia. (c) Scanning electron micrograph (SEM) of conidiophore of* Aspergillus niger *(× 1 372)*

structure, the remaining haploid nucleus divides by mitosis and repeated divisions form many haploid nuclei, each of which can form the nucleus of a spore in the sporangium. Thus all the spores are of the same strain. Fig 2.20 gives a non-pictorial summary of sexual reproduction.

2.4.3   Phylum Ascomycota

Characteristics of the Ascomycota are given in table 2.5. They are the largest group of fungi and are relatively advanced, showing more complexity than the Zygomycota, especially in their sexual structures. They

include the yeasts, some common moulds, powdery mildews, cup fungi, morels and truffles.

Penicillium is a widespread saprotroph, forming blue, green and sometimes yellow moulds on a variety of substrates. It reproduces asexually by means of **conidia**. Conidia are spores formed at the tips of special hyphae called **conidiophores**. They are not enclosed in a sporangium, but are naked and free to be dispersed as soon as they mature. The structure of *Penicillium* is shown in fig 2.23a. Its mycelium forms a circular colony of small diameter and its spores give colour to the colony so that the young outer edge of the mycelium is usually white, whereas the more mature central portion, where spores have developed, appears coloured. The economic importance of *Penicillium* species is discussed in section 2.5.4.

Aspergillus generally grows on the same substrates as *Penicillium* and resembles it closely. It forms black, brown, yellow and green moulds. Fig 2.23*b* shows an asexually reproducing mycelium for comparison with *Penicillium*.

2.4.4 Phylum Basidiomycota

Characteristics of the Basidiomycota are given in table 2.5. They are almost as large a group as the Ascomycota, together with which they form the 'higher' fungi, that is the most advanced fungi. Their large 'fruiting bodies' make them some of the most conspicuous fungi, including those commonly known as mushrooms or toadstools,* puffballs, stinkhorns and bracket fungi. The group also contains important obligate parasites, namely the smuts and rusts.

Agaricus (*Psalliota*) belongs to the group of gill-bearing toadstools. The toadstool, or mushroom, is a short-lived 'fruiting body'. The mycelium grows saprotrophically on organic matter in the soil and may live for a number of years. It forms thick strands called **rhizomorphs** in which the hyphae are compacted to form tissues. Rhizomorphs can resist adverse conditions, becoming dormant until favourable conditions return. They grow from the tips and are responsible for vegetative spread of the fungus. The external features of *Agaricus* are shown in fig 2.24, together with the structure of the gills.

The 'fruiting body', or **sporophore**, appears above ground in the autumn in temperate regions and is made entirely from hyphae which compact to form tissues. The edges of the gills are made of **basidia**, which produce spores (**basidiospores**). The gills exhibit positive geotropism so hang down vertically. The spores, which are produced in large numbers (about half a million per minute from a large mushroom), are forcibly ejected from the basidia and drop down between the gills to be carried away by air currents.

2.4.5 Useful and harmful fungi

See sections 2.5 and 2.6.

2.5 Benefits and uses of micro-organisms

Micro-organisms are important for their natural roles in the biosphere, the fact that they can be deliberately exploited by humans in a number of ways, and because they are sometimes harmful, particularly as disease-causing agents (pathogens). Examples of these

three main areas of importance are discussed below in sections 2.5.1 to 2.5.4 and in section 2.6, with reference to bacteria, viruses and fungi. Protozoans and some algae are also micro-organisms; the latter are dealt with in section 3.2.

2.5.1 Micro-organisms and soil fertility

Micro-organisms play an important part in soil fertility. Below is a summary of information discussed in more detail elsewhere.

Decay and formation of humus. The formation of humus from the litter and fermentation layers above it is discussed in chapter 12. Humus is a layer of decayed organic matter which, besides containing nutrients, has important physical and chemical properties, such as water-retaining ability. The action of saprotrophic bacteria and fungi (together known as decomposers) in decaying organic matter is also discussed in 9.11.1. Inorganic products which can be recycled as a result include carbon dioxide, ammonia, mineral salts (for instance phosphates and sulphates) and water.

Nutrient or biogeochemical cycles. The nitrogen, sulphur and phosphorus cycles are discussed in section 9.11. The nitrogen cycle involves:

(*a*) nitrogen-fixing bacteria, including free-living saprotrophs, such as *Azotobacter*, mutualists, such as *Rhizobium*, and the free-living photosynthetic blue-green bacteria.
(*b*) nitrifying bacteria which can convert organically combined nitrogen (for instance protein) to nitrate, such as *Nitrosomonas* and *Nitrobacter*.
(*c*) denitrifying bacteria which can convert nitrate to nitrogen gas, such as *Thiobacillus*.

Further information concerning bacteria and fungi in the nitrogen cycle is given in section 9.11.1.

2.5.2 Sewage disposal

The activities of decomposers in sewage works closely resemble those of decomposers in the soil because they break down organic matter to harmless, soluble, inorganic materials. The sewage, separated into liquid and sludge in settling tanks, is often digested in several stages by a combination of aerobic and anaerobic bacteria. Methane gas produced by anaerobic bacteria is sometimes utilised as fuel to run the machinery at the works. The products of a sewage farm are a liquid, which is normally recycled by emptying into rivers, and a sludge that contains harmless organic and inorganic materials and micro-organisms (mainly bacteria and protozoans) and can be dried and used as fertiliser unless contaminated with heavy metals. Saprotrophic fungi and bacteria, together with protozoans, are part of the jelly-like film of living organisms covering the stones of 'filterbeds' at sewage works.

---

* Mushrooms and toadstools are really synonymous terms, although edible species are sometimes called mushrooms and poisonous species toadstools.

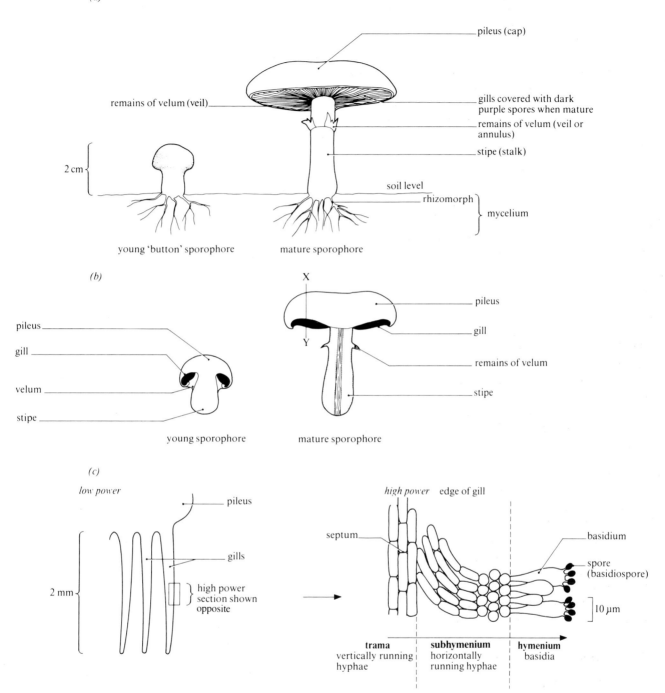

Fig 2.24 *Structure of* Agaricus campestris, *the field mushroom.* Agaricus bisporus, *the cultivated mushroom, is very similar but the basidia bear two, not four, spores.*
(a) Entire sporophores with mycelium. (b) VS sporophores.
(c) Part of VS pileus from X to Y

2.5.3   Mutualistic micro-organisms

Mammals and other animals cannot digest cellulose since they lack the enzyme cellulase. For herbivores, cellulose constitutes a large proportion of the diet, and they have cellulose-digesting bacteria and protozoa living mutualistically in the gut. In rabbits, the bacteria occupy the caecum and the appendix; in cows and sheep, they are present in the rumen. This is of indirect significance to humans who use these animals as a source of food.

Of more direct use is the 'flora' of human intestines. Many bacteria inhabit the intestines and some, such as *E. coli*, synthesise vitamins of the B group and vitamin K.

Some skin micro-organisms of humans offer protection against invasion by pathogenic organisms.

Some important further examples of mutualism involving micro-organisms are given in table 10.1.

2.5.4 Biotechnology and genetic engineering

Biotechnology may be defined as the application of organisms, biological systems or processes to the manufacturing and service industries (The Royal Society, 1981). Where whole organisms are involved, these are generally micro-organisms, such as bacteria, viruses, fungi and algae. The estimated market for biotechnology is about £40 billion per annum by the end of the century. Some of the products will be produced in bulk at relatively low cost, others will be 'fine' products, produced in small quantities for relatively high prices.

At the heart of biotechnology are the 'fermentation' processes. Originally the term 'fermentation' was reserved for anaerobic activity, but it is now applied more loosely to any process in which microbes are cultured in containers (**fermenters** or **bioreactors**). The use and design of fermenters is described in section 11.3.10. The traditional fermentation industries include brewing, baking, cheese and butter manufacture, but there is now an enormous range of products, as indicated in table 11.3.

Brewing

The oldest fermentation industry is that of brewing. Beer is brewed from barley which has been partially germinated to convert its starch store to the sugar maltose. Gibberellins (section 15.2.6) are used to speed up this process and to control it precisely. The subsequent fermentation is carried out in a large vat and is brought about by the unicellular fungus yeast *Saccharomyces* (for instance *S. cerevisiae*, *S. carlsbergensis*). During this process sugar is converted to carbon dioxide and alcohol which reaches a final concentration of 4–8%. Hops are added at an earlier stage for their flavour and antimicrobial properties.

Wine manufacture depends on the fermentation of grape juice by wild yeasts present on the skin of grapes. The final concentration of alcohol is 8–15%, high enough to kill the yeasts, and the wine is often left to mature over a number of years. Some unconverted sugar may remain.

Other common drinks prepared by fermentation are cider from apple juice and the Japanese saké made from rice.

Industrial alcohol can be prepared from carbohydrate-containing waste substances such as molasses.

Cheese manufacture

Cheese manufacture usually depends on the combined activities of bacteria and fungi. During the process the milk sugar lactose is fermented to lactic acid which causes the milk protein, casein, to curdle. The solid curds, containing protein and fat, are separated from the liquid whey and acted on by bacteria and/or fungi. Inoculation with different microbes produces different varieties of cheese, such as cheddar from *Lactobacillus* species. Some famous cheeses are ripened with *Penicillium* species, such as roquefort (*P. roqueforti*), camembert (*P. camemberti*), Danish blue and Italian gorgonzola.

The souring of cream during butter manufacture and the flavour of butter are caused by lactic acid streptococci. *Lactobacillus* species are also used in the production of sauerkraut (from cabbages), silage and pickles. A useful article on cheese manufacture was published in *Scientific American* of May 1985, pp. 66–73.

Baking

Another important fermentation industry which utilises yeast is baking. Strains of *S. cerevisiae* selected for their high production of carbon dioxide, the raising agent, are used by bakeries. The alcohol produced at the same time is driven off as a vapour by the heat of baking.

Antibiotics

Since the 1930s a great deal of research has been devoted to isolating from bacteria and fungi natural chemicals that have antibiotic properties, that is which inhibit the growth of, or kill, other micro-organisms. They have found application in medicine, veterinary science, agriculture, industry and pure research. Soil-dwelling organisms are a particularly rich source of antibiotics since in the micro-ecosystems in which they exist there is much competition, and antibiotics form part of the natural 'armoury' for establishing ecological niches. Soil samples from all over the world are continually being screened for potentially useful new antibiotics.

The first antibiotic to be exploited was **penicillin**, which is produced by several species of the fungus *Penicillium*, notably *P. notatum* and *P. chrysogenum*, the latter being the current commercial source. The impact of its introduction during the 1940s was enormous because it was active against all staphylococcal infections, a wide range of Gram positive bacteria and yet virtually non-toxic to the patient. It is still the most important antibiotic, and new synthetic derivatives of it are continually being introduced to improve its effectiveness, the starting material still being penicillin prepared from large-scale culture (fermentation) of the fungus.

Griseofulvin is another antibiotic obtained from *Penicillium* species (particularly *P.griseofulvum*). It has antifungal properties and is especially effective against athlete's foot and ringworm when taken orally. **Fumagillin** is a particular type of antibiotic obtained from *Aspergillus fumigatus*. It is frequently used against amoebic dysentery.

One of the most productive sources of antibiotics has been the genus *Streptomyces*, a bacterium resembling a miniature fungus, of which there are many species and from which over 500 antibiotics have been identified. More than 50 of these have found practical applications, including **streptomycin**, **chloramphenicol** and the **tetracyclines**.

Streptomycin was discovered soon after penicillin. It proved almost as dramatically successful, increasing the range of pathogens that could be treated. For example, unlike penicillin, it is active against the tuberculosis bacillus. *Bacillus* species have also proved fruitful, such as *Bacillus brevis* which produces **gramicidin**.

Single cell protein (SCP) and new food sources

A new food source of recent years is 'single cell protein' (SCP), a term which refers to protein derived from the large-scale growth of micro-organisms such as bacteria, yeasts and other fungi, and algae. The protein may be used for human consumption or animal feed. It may also be a useful source of minerals and vitamins (as well as containing fat and carbohydrate). There are several advantages in using micro-organisms as a food source: they occupy less room than conventional crops and animals, grow much more rapidly and can grow on a wide variety of cheap or waste products of agriculture or industry. Examples of these are petroleum products, methane, methanol, ethanol, sugars, molasses, cheese whey and waste from pulp and paper mills.

One of the major products is **Pruteen**, whose production is described in section 11.3.10. Early hopes for SCP have been dampened for various reasons.

(*a*) Agricultural surpluses, particularly of the high protein products like grain and dairy products, are now common in developed countries such as the USA and Europe.

(*b*) Developing countries where protein is scarce cannot afford the investment in equipment, and lack relevant expertise.

(*c*) There has been an increase in production, and reduction in price, of competitive animal feed additives like soyabean, fishmeal and gluten from maize (the latter being a by-product of biotechnological fuel programmes).

(*d*) There was a rise in oil prices in the late 1970s; SCP production is energy-intensive.

ICI and Rank, Hovis, McDougall now produce a mycoprotein from the fungus *Fusarium*. This protein is unusual in being used for human consumption, and ICI may produce it on a larger scale using the expertise gained from Pruteen manufacture.

New energy sources: biogas and gasohol

Many methods are being explored for exploiting living organisms and biological processes for energy supplies. Artificial photosynthesis generating hydrogen gas (a fuel) from water, is a long-term possibility. Another basic strategy is to exploit the energy trapped in biomass. This usually means upgrading the raw material (biomass) to a higher fuel value by fermentation. Although this may result in a *net loss* of energy from the original biomass, the product has a higher energy content per unit mass and is

therefore more easily stored and transported. Among the raw materials currently being investigated are waste materials such as animal manures, sewage sludge, domestic wastes, food wastes, paper wastes, spoilt crops, sugar cane tops and molasses. Various crops (such as maize, sugarcane, sugarbeet) and water plants (such as kelps and water hyacinth) might also be used. Two processes currently dominate, namely production of **biogas** (methane) by anaerobic digestion and of ethanol by yeast fermentation.

Biogas
Overall equation:
$$C_6H_{12}O_6 \longrightarrow 3CH_4 + 3CO_2$$
glucose   methane   carbon dioxide

energy value: 16 kJ per g   56 kJ per g

Biogas is about 54–70% methane. Most of the rest is carbon dioxide, with traces of nitrogen, hydrogen and other gases. (Natural gas is about 80% methane.) A mixture of micro-organisms is used in the fermentation, including a group of bacteria called **methanogens** which can produce methane from carbon dioxide and hydrogen. These are **archaebacteria** (see fig 2.2). A wide range of waste materials or plant products can be used for fermenting (see above). In the USA the water hyacinth, a vigorous plant which can block canals and water ways, has been used. Most operations are small-scale and for local fuel use, as is common in India and China.

The manure from one cow in one year can be converted to methane equivalent of over 227 litres of petrol; 0.5 kg of cow manure could generate enough gas to cook a family's meals for a day.

In China, over 18 million family-scale digesters have been built. The gas is typically used for cooking, lighting, tractor or car fuel and for running electricity generators.

On a larger scale, the gas can be a by-product of landfill, sewage or factory waste (such as sugar factories, distilleries). It can be used to drive electricity generators in sewage works and waste treatment plants. In Britain, rubbish could be a major source of methane (up to 20 litres of gas per kilogram of refuse). At the moment the gas is collected from landfill sites by sinking pipes into the compacted rubbish and sucking out the gas.

Ethanol
Overall equation:
$$C_6H_{12}O_6 \longrightarrow 2C_2H_5OH + 2CO_2$$
glucose   ethanol

energy value: 16 kJ per g   30 kJ per g

Ethanol, or 'power alcohol', has been produced successfully in Brazil in the '**Proalcool**' programme. Sugarcane juice (with some molasses) is the starting material. Ethanol is distilled from the fermented product. Over 11 000 million litres were produced in 1985 and cars in Brazil are adapted to run on ethanol (its main use). In the USA the product is known as **Gasohol** and the initial biomass is starch from maize. Over 2280 million litres per year were being produced in the mid-1980s. As a motor fuel it can be used either pure or blended with petrol.

Toxic chemicals and xenobiotics

A xenobiotic is a chemical compound synthesised by humans which is not naturally found in living organisms and cannot normally be metabolised (broken down) by them. It is not surprising, therefore, that many toxic chemicals are xenobiotics. About 300 million tonnes of hazardous waste are produced per year in the USA and, in 1979, 2 million tonnes of pesticide were used in the western hemisphere. One approach to the cleaning up of all of the toxic waste is to 'design' or discover micro-organisms with the relevant enzymes to break down the chemicals. Such organisms can be cultured in fermenters with the waste. Some plasmids (see section 2.2.4) carry genes coding for suitable enzymes. These may be brought together in bacteria for the purposes of degrading specific chemicals. For example, a strain of the bacterium *Pseudomonas putida* has been created which breaks down octane, xylene and camphor. Genetic engineering may be used more extensively in the future.

Waste disposal

See above for sewage disposal (technically a type of biotechnology), new energy sources, toxic chemicals and xenobiotics, and SCP (new foods). Treatment of waste water from domestic and industrial sources uses biological processes similar to those in sewage works.

Extraction of metals from minerals and solutions

Bacteria can be used to chemically extract ('leach') copper and other metals from rocks containing low concentrations of metals. These are often the waste from mines. Micro-organisms may similarly be used to remove toxic heavy metal pollution from waste waters of mining operations and other industrial processes. Thus both economic and environmental benefits are possible. Bacterial leaching is currently used on a commercial scale for copper and uranium extraction. One possible important future use of environmental significance is the leaching out of inorganic sulphur from coal. If other bacteria could be used to remove organic sulphur, the burning of desulphurised coal would not release sulphur dioxide, thought to be one of the main contributors to acid rain.

Enzyme technology and biosensors

Using the versatility of the enzymes of micro-organisms, new methods of making many industrially important chemicals are being introduced. In section 6.9 a detailed account of the procedures involved and the use of enzymes is given, together with an account of the related technology of biosensors and biochips.

Genetic engineering

Genetic engineering is the manipulation of genes. The process is variously referred to as **genetic manipulation**, **gene cloning** (since a clone of genetically identical organisms can be grown from the original altered cell), or **recombinant DNA technology**. In genetic engineering, pieces of DNA (genes) are introduced into a host by means of a carrier (vector) system. The foreign DNA becomes a permanent feature of the host, being replicated and passed on to daughter cells along with the rest of its DNA. The donated DNA could come from another organism or might be an artificially synthesised gene.

In fig 2.25 an outline of the normal procedure is given, but it should be noted that some knowledge of DNA and genetics will be necessary to understand this (see chapters 22 and 23). In fig 2.25 the vector is a plasmid (see section 2.2.4), but bacteriophages or other viruses are sometimes used, the latter mainly for animal cells. **Restriction endonucleases** are enzymes which cut (cleave) DNA. There are many different types, each of which cuts at a different specific base sequence. Some leave overlapping ends of DNA called '**sticky ends**' because their base sequences are complementary.

Only a very small proportion of treated bacterial cells will be successfully transformed, and in order to recognise these cells the vector used usually contains a '**marker gene**'. This is often a gene for resistance to a particular antibiotic, for example tetracycline resistance. At the end of the process, any transformed bacterial cells can therefore be identified by growing the bacteria on a medium containing tetracycline. Two early examples of the commercial application of this technique are the production of human insulin and human growth hormone.

Insulin production. Ever since the cause of diabetes mellitus was established as a shortage of the hormone insulin, sufferers have been provided with insulin derived from the pancreases of slaughtered sheep and cattle. Due to minor differences in the chemical composition of insulin from different species, some patients showed damaging side-effects as a result of the injections. Insulin is a protein, and it was argued that if the genes for this protein could be inserted into a bacterium, then it might be possible to culture the bacterium on a commercial scale. One of the potential problems in transferring genes between eukaryotes and prokaryotes is that different regulatory mechanisms exist, even if the genes themselves can be transferred. But successful transfer of human insulin genes into bacteria has now been achieved and the bacteria are grown by the fermentation methods described in section 11.3.10. The process developed by Eli Lilly is described here; their product is known as **Humulin**.

The procedure is complicated by the fact that insulin has two polypeptide chains (fig 5.32), the A chain and the B chain. The amino acid sequences of these chains are known and the first procedure developed involves synthesising two

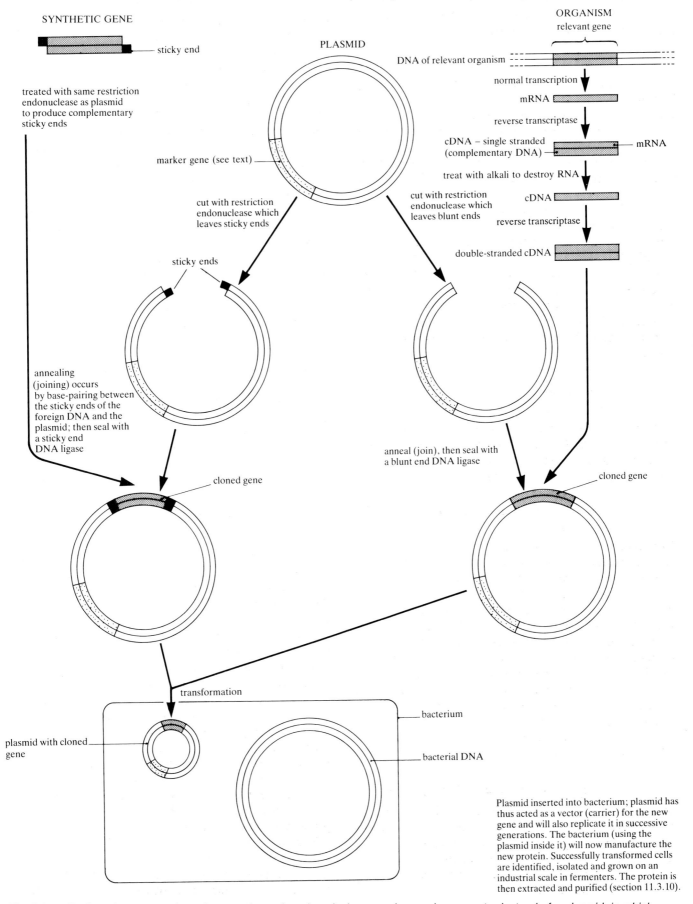

NB A double-stranded DNA molecule, ⟨⟩⟨⟩⟨⟩⟨⟩ is represented as: ═══╪═══

complementary strands of DNA

SYNTHETIC GENE

sticky end

treated with same restriction endonuclease as plasmid to produce complementary sticky ends

PLASMID

marker gene (see text)

cut with restriction endonuclease which leaves sticky ends

cut with restriction endonuclease which leaves blunt ends

ORGANISM
relevant gene

DNA of relevant organism

normal transcription

mRNA

reverse transcriptase

cDNA – single stranded (complementary DNA) — mRNA

treat with alkali to destroy RNA

cDNA

reverse transcriptase

double-stranded cDNA

sticky ends

annealing (joining) occurs by base-pairing between the sticky ends of the foreign DNA and the plasmid; then seal with a sticky end DNA ligase

anneal (join), then seal with a blunt end DNA ligase

cloned gene

cloned gene

transformation

bacterium

plasmid with cloned gene

bacterial DNA

Plasmid inserted into bacterium; plasmid has thus acted as a vector (carrier) for the new gene and will also replicate it in successive generations. The bacterium (using the plasmid inside it) will now manufacture the new protein. Successfully transformed cells are identified, isolated and grown on an industrial scale in fermenters. The protein is then extracted and purified (section 11.3.10).

Fig 2.25 *Outline of two procedures for genetic engineering. A virus may be used as a vector instead of a plasmid, in which case the final stage is 'transduction', not 'transformation'*

artificial DNA molecules (genes) which code for the two appropriate amino acid chains. The synthesised DNA molecules are then introduced into separate bacteria by the process for synthetic genes shown in fig 2.25. After growing large quantities of the bacteria in fermenters, the bacteria are lysed (split open) and the chains purified. The two chains are then chemically combined in an oxidation reaction, and have been shown to function exactly as natural human insulin. More recently, an alternative procedure, using one synthetic gene which mimics the normal human gene and codes for a molecule called **proinsulin**, has been developed. Once purified, proinsulin is converted into insulin by digesting away part of the molecule.

Insulin was the first genetically engineered protein to be used in humans and it is now used in routine clinical practice throughout the world.

Growth hormone (somatotrophin). The role and importance of growth hormone (GH) is discussed in section 21.8.1. Unlike insulin, GH from animals other than humans is not effective in humans. In roughly one child in 5 000 too little GH is secreted during childhood, and in extreme cases the child suffers from pituitary dwarfism (IQ is unaffected). Injections of GH can overcome this problem, but the hormone is in short supply and is very expensive because it can only be obtained from pituitary glands of human cadavers. Trials of GH produced by genetic engineering began first in the USA and then followed in Britain in the late 1980s. Although very expensive, hopefully all requirements will eventually be met at reasonable cost.

Other proteins. Some success has been achieved with a group of anti-viral proteins called **interferons** which can be used to help fight viral diseases. The genes have been cloned in *E. coli*, yeasts and animal cells, with the best yields coming from yeast cells. One advantage of using yeast in genetic engineering is that it is a eukaryote and therefore gene expression is easier to achieve if the foreign genes are also from eukaryotes. Another important application of genetic engineering is in the development of vaccines. Vaccines contain antigens, which are normally proteins. Genetically engineered vaccines are now available for hepatitis B and foot-and-mouth disease, both viral diseases. There is currently intensive research to find vaccines for AIDS and malaria, where one of the main problems is the isolation of a suitable antigen.

2.6   Harmful micro-organisms

2.6.1   Deterioration of food and materials

Saprotrophic bacteria and fungi attack and decompose organic materials and can therefore present many problems to humans, despite their vital role as decomposers in the biosphere. Foods such as grain and fruit which are stored in large quantities must be protected, and food spoilage generally is a constant problem. There are many different and economically costly ways of preserving food. Natural fabrics, leather and other consumer goods manufactured from raw materials are also subject to attack, particularly by fungi. Fungi which live on cellulose cause rots in damp timber and fabrics. Much money is spent on preservation of all these materials.

2.6.2   Bacteria, viruses and fungi as agents of disease

Organisms which cause disease are called **pathogens**. They are parasitic on their hosts. All viruses are parasitic and therefore usually cause symptoms of disease in their hosts. Some bacteria and fungi are also pathogenic. Pathogenic bacteria more commonly affect animals than plants, whereas the reverse is true of fungi. Viruses are more indiscriminate. Human disease and disease of our domestic animals and crops is the inevitable consequence.

Bacterial and viral diseases

Some important animal diseases that are caused by viruses are foot-and-mouth disease of cattle, swine fever, fowl pest and myxomatosis of rabbits. Virus infections of plants commonly cause a yellow mottling of leaves called **leaf mosaic**, and crinkly or dwarfed leaves. They also cause stunting of growth with consequent reductions in yields. Some important crop diseases are turnip yellow mosaic virus (TYMV), tobacco mosaic virus (TMV), tomato bushy stunt and tomato spotted wilt virus. The striped appearance of some tulip varieties is caused by a virus and these tulips are sold as a specific variety by horticulturists. Plant viruses are apparently always RNA viruses.

Important bacterial diseases of animals include *Salmonella* food poisoning of pigs and poultry. Bacterial diseases of plants include crown gall of fruit trees and fire blight of apples and pears (*Agrobacterium tumefaciens* and *Erwinia amylovorum* respectively).

Table 2.6 describes some common viral diseases of humans and can be compared with table 2.7 which gives similar information for bacterial diseases. Further information is contained in fig 2.6.

Methods of transmission

It is convenient to discuss the transmission of viral and bacterial diseases together since the principal methods are the same. Examples of all the methods described below are given in tables 2.6 and 2.7.

Droplet infection

Respiratory infections in particular are usually spread by droplet infection. Sneezing and coughing result in a violent expulsion of millions of tiny droplets of liquid (mucus and saliva) which can carry living microbes and be inhaled by other people, particularly in crowded and poorly ventilated places. Standard hygiene should include appropriate use of handkerchiefs or tissues and ventilation of rooms.

Some microbes, like the smallpox virus and the tuberculosis bacterium, are quite resistant to drying out, so may be carried in dust containing the dried remains of droplets. Even talking can result in microscopic droplets of saliva being released, so such infections are difficult to prevent if the microbe is virulent.

Contagion (direct physical contact)

Relatively few diseases are spread by direct physical contact with an infected person or animal. They include the **venereal**, that is sexually transmitted, diseases such as gonorrhoea and syphilis. A disease closely related to syphilis, namely yaws, is common in tropical countries and requires direct skin contact. Contagious viral diseases include trachoma, a common eye disease of tropical countries, the common wart and herpes simplex which causes 'cold sores'. Leprosy and tuberculosis are caused by *Mycobacterium* species and are contagious bacterial diseases.

Vectors

A **vector** is an organism that transmits a pathogen. It picks up the infection from an organism called the **reservoir**. For example the flea is the vector of the two bacterial diseases endemic typhus and plague (bubonic plague or Black Death) and the reservoir is the rat. In the case of rabies, a viral disease, the vector and reservoir are the same, for example dog or bat.

2.7 What are (a) the vectors and (b) the reservoirs for (i) epidemic typhus and (ii) yellow fever? (See tables 2.6 and 2.7.)

Table 2.6 Some common viral diseases of humans.

Name of disease	Caused by	Parts of body affected	Method of spread	Type of vaccination*
Influenza	A myxovirus (DNA virus) three types, A, B, and C, of varying severity	Respiratory passages: epithelial lining of trachea and bronchi	Droplet infection	Killed virus: must be of right strain
Common cold	Large variety of viruses, most commonly rhino-virus (RNA virus)	Respiratory passages: usually upper passages only	Droplet infection	Living or inactivated virus given as intramuscular injection; not very effective because so many different strains of rhinovirus
Smallpox**	Variola virus (DNA virus) a pox virus	Respiratory passages, then skin	Droplet infection (contagion possible via wounds in skin)	Living attenuated virus applied by scratching skin; no longer carried out
Mumps	A paramyxovirus (RNA virus)	Respiratory passages, then generalised infection throughout body via blood, particularly salivary glands; also testes in adult males	Droplet infection (or contagion via infected saliva to mouth)	Living attenuated virus
Measles	A paramyxovirus (RNA virus)	Respiratory passages (mouth to bronchi) spreading to skin and intestines	Droplet infection	Living attenuated virus
German measles (Rubella)	Rubella virus	Respiratory passages, lymph nodes in neck, eyes and skin	Droplet infection	Living attenuated virus; more essential for girls because disease causes complication in pregnancy
Poliomyelitis ('polio')	Poliovirus (a picornavirus) (RNA virus), three strains exist	Pharynx and intestines, then blood; occasionally motor neurones in spinal cord, when paralysis may occur	Droplet infection or via human faeces (see cholera, table 2.8)	Living attenuated virus given orally, usually on sugar lump
Yellow fever	An arbovirus, that is arthropod-borne virus (RNA virus)	Lining of blood vessels and liver	Vector – arthropods, e.g. ticks, mosquitoes	Living attenuated virus (control of vectors also important)
AIDS (see section 14.14.3)	Retrovirus (RNA virus)	Skin cancer (Kaposi's sarcoma), blood (septicaemia), brain (dementia)	Sexual intercourse – homo- and heterosexuals	Not available

* types of vaccination – see table 2.8
** last recorded natural case in Somalia, October 1977; disease extinct, though virus kept in a few laboratories

Table 2.7 Some common bacterial diseases of humans.

Name of disease	Caused by	Parts of body affected	Method of spread	Type of vaccination* or antibiotic
Diphtheria	*Corynebacterium diphtheriae* (rod-shaped, Gram +)	Upper respiratory tract, mainly throat. Harmful toxin spread by blood to all parts of body. Toxin affects heart.	Droplet infection	Toxoid
Tuberculosis (TB)	*Mycobacterium tuberculosis* (rod-shaped, member of Actinomycetes)	Mainly lungs	Droplet infection. Drinking milk from infected cattle.	BCG living attenuated bacteria. Must test first to see if already immune. Antibiotics, e.g. streptomycin.
Whooping cough	*Bordetella pertussis* (rod-shaped, Gram −)	Upper respiratory tract, inducing violent coughing	Droplet infection	Killed bacteria
Gonorrhoea	*Neisseria gonorrhoeae* (coccus, Gram −)	Reproductive organs: mainly mucous membranes of urino-genital tract. Newborn infants may acquire serious eye infections if they pass through infected birth canal.	Contagion by sexual contact	Antibiotics, e.g. penicillin, streptomycin
Syphilis	*Treponema pallidum* (a spirochaete)	Reproductive organs, then eyes, bones, joints, central nervous system, heart and skin	Contagion by sexual contact	Antibiotics, e.g. penicillin
Typhus	*Rickettsia*	'Epidemic typhus' more serious than 'endemic typhus'. Similar to typhoid. Linings of blood vessels causing clots. Skin rash.	Epidemic typhus: vector – louse. Endemic typhus: vector – rat flea. From rat to rat by flea and lice.	Killed bacteria or living non-virulent strain. Antibiotics, e.g. tetracyclines, chloramphenicol (control of vectors also important).
Tetanus	*Clostridium tetani* (rod-shaped, Gram +)	Blood. Toxin produced which affects motor nerves of spinal cord and hence muscles, causing lockjaw and spreading to the muscles. Often fatal.	Wound infection	Toxoid
Cholera	*Vibrio cholerae* (comma-shaped, Gram −)	Alimentary canal: mainly small intestine. Toxin affects lining of intestine.	Faecal contamina-tion: (a) food- or water-borne of material contaminated with faeces from infected person; (b) handling of con-taminated objects; (c) vector, e.g. flies moving from human faeces to food.	Killed bacteria: short-lived protection and not always effective. Antibiotics, e.g. tetracyclines, chloramphenicol.
Typhoid fever	*Salmonella typhi* (= *S. typhosa*) (rod-shaped, Gram −)	Alimentary canal, then spreading to lymph and blood, lungs, bone marrow, spleen	As cholera	Killed bacteria (TAB vaccine)
Bacterial dysentery (bacillary dysentery)	*Shigella dysenteriae* (rod-shaped, Gram −)	Alimentary canal, mainly ileum and colon	As cholera	No vaccine. Antibiotic, e.g. tetracyclines.
Bacterial food poisoning (gastro-enteritis or salmonellosis)	*Salmonella* spp. (rod-shaped, Gram −)	Alimentary canal	Mainly foodborne-meat from infected animals, mainly poultry and pigs. Also via faecal con-tamination as cholera.	No vaccine. Antibiotics, e.g. tetracyclines; usually not necessary and not very effective.

* types of vaccination – see table 2.8

32

Table 2.8 Types of vaccine.

(1) **Attenuated living micro-organism**
An **attenuated** micro-organism is one whose virulence has been greatly reduced by some laboratory procedure, e.g. growing it in a high temperature. It may be a mutant variety with the same antigens, but lacking virulence.
(2) **Killed micro-organism**
Killed by some laboratory procedure, e.g. exposure to 75% alcohol (TAB vaccine). Antigens still present.
(3) **Toxoids**
A toxoid is an inactivated toxin that retains its antigenic properties. It is inactivated by laboratory treatment, e.g. treatment with formaldehyde.
(4) **Mild strain of virus**
Closely related but non-pathogenic strain, e.g. formerly smallpox/cowpox.

In the cases mentioned the vector acts as a second host in which the pathogen can multiply. Insects can also carry pathogens on the outsides of their bodies. For example, houseflies walking and feeding on faeces from a person infected with a gut disease, such as cholera, typhoid or dysentery, might transmit the pathogen to food likely to be consumed by humans.

Faecal contamination

With diseases that affect the alimentary canal the pathogens leave the body in the faeces. Three common means of transmitting disease arise from this.

Waterborne. The classic waterborne diseases are cholera, typhoid (both caused by flagellated bacteria) and dysentery. If insanitary conditions prevail, faeces of infected individuals are often deposited in or near water which may be used for drinking. In this way disease spreads rapidly through a population.

Foodborne. Food may be contaminated by traces of faecal matter by washing it in contaminated water, touching it with unwashed hands or when a vector such as a housefly touches it.

Contamination of objects. Various objects may become contaminated with faecal matter either directly or via handling. Subsequent handling by another person may lead to hand-to-mouth passage of the disease.

'True' foodborne

Salmonella food poisoning is commonly spread in the meat of infected animals if it is undercooked. *Clostridium botulinum* (fig 2.6) is a bacterium which causes **botulism**, an often fatal form of food poisoning because the toxin it produces is one of the most toxic substances known (the lethal dose for mice is $5 \times 10^{-5}\,\mu g$). It can grow in protein-rich foods, particularly tinned meats.

Contamination of wounds

Ignoring the bites of animal vectors, there are certain diseases associated with contamination of wounds. Gas gangrene and tetanus are both infections of deep wounds caused by *Clostridium* species, usually picked up from the soil. More superficial wounds and burns are easily infected with staphylococci and streptococci.

Fungal diseases

Some important and familiar diseases are shown in table 2.9. The best-known obligate parasites, namely mildews, rusts and smuts, are included in the table. Although obligate parasites do not kill their hosts, they cause yield losses and make them more vulnerable to other diseases and adverse conditions. They are of great economic importance when they attack crop plants. For example, powdery mildews can cause total yield losses of up to 10% in cereals such as barley. A large industry has grown up to produce fungicides which can be used to protect crops.

Parts of plants infected include underground organs, as in wart disease of potatoes; leaves by rusts, powdery mildews, downy mildews and black spot; flowers, by smuts and ergot; ripe fruit, by soft rots and moulds.

2.7　Practical work

The following practical work is designed to cover some of the basic microbiological techniques associated with bacteriology, using milk as a relatively safe source of bacteria. Milk is a useful food source for bacteria as well as mammals, and certain bacteria are characteristically associated with it.

2.7.1　Bacterial content of milk

Bacteria inevitably enter milk during milking and handling, even under the most hygienic conditions. Milking is normally followed immediately by cooling to retard bacterial growth. The untreated (raw) milk is pasteurised, a heat process intended to kill pathogenic bacteria, though many non-pathogenic bacteria survive. Bacteria present are:

15–30 °C *Streptococcus lactis* (Gram +) dominates, together with many other streptococci (Gram +) and coryneform bacteria (for example *Microbacterium*, *Brevibacterium*) which resemble

Table 2.9 Some common fungal diseases. Also included is the phylum Oomycota, now classified in the kingdom Protoctista, not Fungi.

Phylum	Disease	Host	Fungus	Notes
Oomycota	Potato blight	Potato	*Phytophthora infestans*	Caused Irish potato famine of 1845, with many people emigrating to America as a result.
	Downy mildews	Vine	*Plasmopara viticola*	One of the most destructive diseases of vineyards. Accidentally introduced into Europe from America in nineteenth century.
		Onion, tobacco, cabbage, wallflower	*Peronospora* spp.	Not normally serious diseases.
	(Water-)mould	Fish and fish eggs	*Saprolegnia* spp.	Many do significant damage in fish hatcheries.
Zygomycota	Soft rot	Apples and other fruit in storage	*Rhizopus stolonifer*	Not normally a serious disease. Other fungi cause more serious soft rots.
Ascomycota	Powdery mildews	Hop, cereals, apple, rose and others	*Erysiphe* and other genera *E. graminis* attacks cereals	Serious obligate parasites. Economically important, particularly with cereals.
	Dutch elm disease	Elm	*Ceratocystis ulmi*	Has devastated elms in many parts of the world, including America, Britain and the rest of Europe.
	Brown rot	Stone fruits, e.g. peach, plum	*Monilinia fructigena*	Serious disease worldwide.
	Apple scab	Apple	*Venturia inaequalis*	One of the most important parasites of apples. Infects leaves, twigs, young fruit, weakening host and reducing fruit quality.
	Ergot	Rye	*Claviceps purpurea*	Produces sclerotia* known as ergots in place of host ovaries. Ergots contain alkaloids related to the hallucinogen LSD which can be fatal if eaten.
	Black spot	Rose	*Diplocarpon rosae*	Common infection of roses.
	Aspergillosis (farmer's lung)	Humans, birds	*Aspergillus fumigatus*	Disease of lungs similar to tuberculosis. Rare in humans. Sometimes associated with mouldy hay.
Basidiomycota	Rusts	Many, e.g. bean, cereals, coffee, carnation	Numerous, e.g. *Puccinia graminis* (black stem rust of wheat) *Albugo* (white blister rusts)	Economically important. Patches of spores formed at surface, often rust-coloured.
	Smuts	Many, e.g. onion, cereals	Numerous, e.g. *Ustilago avenae* (loose smut of oats)	Produce black, sooty masses of spores. Economically important, mainly in cereals. Grain itself may be infected, making it useless.
Fungi Imperfecti	Ringworm, Athlete's foot	Humans	*Trichophyton* spp. and others	Ringworm more common in children, athlete's foot in adults. Skin infections.
	Vascular wilt	Potato, flax, tomato, banana, palm	*Fusarium* spp.	Infects vascular (conducting) tissue, causing wilting.

* sclerotia (sing. sclerotium) – hard-walled, resistant, resting body produced by some fungi, often as a means of overwintering.

Lactobacilli but may have swollen ends to the rods (coryneform means club-shaped).

Streptococcus lactis grows well at 10 °C but growth ceases at > 40 °C.

30–40 °C *Lactobacillus* (Gram +) and coliform (gut-living) bacilli (Gram −) dominate, such as *E. coli*.

Streptococcus lactis and *Lactobacillus* are lactic acid bacteria. They produce lactic acid during fermentation (anaerobic respiration) of lactose (milk sugar) and the accumulating acid causes souring of milk. Colonies of *S. lactis* and *Lactobacillus* are relatively small (maximum diameter of a few millimetres on a culture) and never pigmented, appearing chalky white. *S. lactis* forms smooth-textured colonies with entire edges. If finely divided calcium carbonate is included in the nutrient agar, streptococci show clear zones around each colony where lactic acid dissolves the calcium carbonate. Streptococci are responsible for the normal souring of milk. They have the usual appearance under the microscope (fig 2.26). Lactobacilli are rods which tend to stick together in long chains (fig 2.26). Colonies may have a rough surface texture with irregular edges.

A number of other bacteria may be found in milk, including the gut-living rod *Alcaligenes* (Gram −) found singly or in chains. It may be recognised on MacConkey's agar by a yellowish (alkaline) zone around each colony.

2.7.2 Bacteriology experiments

The following three experiments are exercises in the use of microbiological techniques. The second and third experiments are extensions of the first. The first experiment is to culture milk bacteria. The second is to stain bacteria for examination with a light microscope. The third involves the counting of bacterial colonies using the technique of serial dilution.

Experiment 2.1: To investigate the bacterial content of fresh and stale milk

The aims of the experiment are to determine the effect of leaving milk unrefrigerated for 24 h and why milk becomes stale. Milk is almost a complete food for humans and the experiments show that it is also a good culture medium for certain bacteria.

Materials

4 sterile nutrient agar plates	fresh pasteurised milk
inoculating loop	stale milk (milk left at room temperature for 24 h)
Bunsen burner	incubator set at 35 °C
indelible marker or wax pencil	

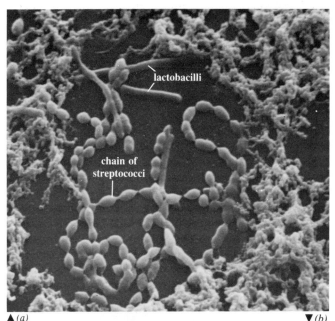

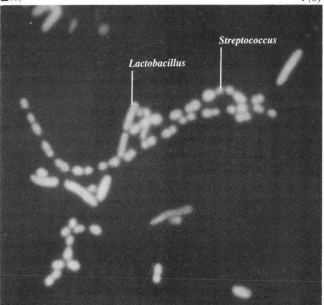

Fig 2.26 (a) Scanning electron micrograph of bacteria recovered from milk. (b) Bacteria recovered from milk by membrane filtration. Chains of streptococci and rods are clearly visible

Method

(1) Place the inoculating loop in the Bunsen burner flame until the loop is red-hot (fig 2.27a).

(2) Allow the loop to cool and then dip it into a sample of fresh, well-shaken milk.

(3) Lift the lid of a sterile agar plate slightly with the other hand and lightly spread the contents of the inoculating loop over the surface of the agar as described in fig 2.27b.

(4) Close the lid of the plate and return the loop to the Bunsen burner flame until red-hot.

(5) Label the base of the plate with an indelible marker (or wax pencil).
(6) Repeat with a second plate and another sample of fresh milk.
(7) Flame the loop again and having allowed it to cool, dip it into a sample of stale milk.

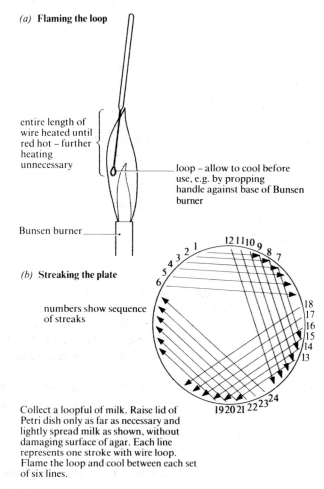

(a) **Flaming the loop**

entire length of wire heated until red hot – further heating unnecessary

loop – allow to cool before use, e.g. by propping handle against base of Bunsen burner

Bunsen burner

(b) **Streaking the plate**

numbers show sequence of streaks

Collect a loopful of milk. Raise lid of Petri dish only as far as necessary and lightly spread milk as shown, without damaging surface of agar. Each line represents one stroke with wire loop. Flame the loop and cool between each set of six lines.

Fig 2.27 *Flaming a wire loop and inoculating nutrient agar with milk bacteria using the streaking technique*

(8) Spread the contents of the loop over the surface of a third plate and then close the lid.
(9) Label the base of the plate with an indelible marker.
(10) Repeat with a fourth plate and a second sample of stale milk.
(11) Place the four plates in an incubator at 35 °C for about three days. They should be placed upside down to prevent condensation falling onto the cultures. After incubation, the two halves of each plate should be taped together for safety reasons.
(12) Record the appearance of the colonies and compare with the description in section 2.7.1.

Notes

(1) Students may pour their own plates, if McCartney tubes of sterile, molten, nutrient agar are supplied.

(2) The particular streaking technique used progressively reduces the number of bacteria in each streak. It is suitable in situations where large numbers of bacteria are present, as in milk, and is normally used to isolate pure colonies of bacteria from mixed cultures.
(3) The plates can be placed in a refrigerator after incubation until required. This prevents further bacterial growth.
(4) When the plates are no longer required they should be placed in a disposable autoclave bag and autoclaved for 15 min before final disposal.
(5) Other experiments using milk could be performed. The experiment above is the simplest. The effect of refrigeration could be studied. Also if samples of raw (unpasteurised) milk could be obtained (for instance direct from a dairy farm) the effect of the process of pasteurisation on the bacterial content of milk could be studied. Pasteurisation of milk can be accomplished by placing raw milk in a sterile test-tube plugged with cotton wool and heating at 63 °C for 35 min in a water bath. A third variation would be to incubate some plates at 10 °C instead of 35 °C. This lower temperature favours growth of *Streptococcus lactis* compared with *Lactobacillus*.

Experiment 2.2: To stain bacteria for examination with a light microscope

Although direct microscopic examination of living bacteria is possible using a phase contrast microscope, it is more common to kill and stain bacteria before examination.

One stain which is important in the identification of bacteria is the Gram stain. It was first developed by a Danish physician, Christian Gram, in 1884. Before staining, all bacteria are colourless. Afterwards **Gram positive** bacteria are stained **violet** and **Gram negative** bacteria stained **red**. The difference between the two types of bacteria is described in section 2.2.2 (cell walls).

Materials

basic stain = crystal violet (0.5% aqueous)
mordant = Lugol's iodine
decolouriser = acetone–alcohol (50:50 acetone: absolute alcohol)
counterstain = safranin (1% aqueous)
wire loop
Bunsen burner
glass slides scrupulously clean (wipe with alcohol)
forceps
staining rack set up over sink or dish
distilled water in a wash bottle
blotting paper
immersion oil and microscope with oil immersion lens

Method

(Stages 1 to 6 should take about 5 min.) (Based on *Bacteriology*, Humphries, J., John Murray, 1974.)

(1) **Prepare a smear** of bacteria on the slide as follows. Flame a wire loop and cool. Place a loopful or two of tap water on the centre of a clean slide. Touch the wire loop lightly on a selected bacterial colony from the experiment above, opening the lid of the plate a minimal amount for safety reasons. Transfer the bacteria to the slide and gently mix with the water. Spread the bacteria over the slide, using the loop, to cover an area about 3 × 1 cm. Flame the loop again. It is important to achieve the correct thickness of the smear. It should appear only faintly opalescent and is more usually too thick than too thin. It should also be of even thickness. Allow the smear to become perfectly dry in air (a few minutes).

(2) **Fix the smear.** Holding the slide with forceps, pass it horizontally just over a yellow bunsen flame three times. It is important that it is not overheated and should feel comfortable against the skin after each passage over the flame. Fixing kills the bacteria by coagulating the cytoplasm and also makes them stick to the slide.

(3) Staining is likely to soil the bench so should be done on a rack over a sink or dish. A rack can be made by arranging two glass or metal rods across the sink or dish 5 cm apart and absolutely horizontal. If supported on plasticine they are easily adjusted. Flood the slide with crystal violet stain. Leave for 30 s.

(4) Wash off with Lugol's iodine; flood with Lugol's iodine and leave for 30 s. Wash off the iodine with distilled water from a wash bottle.

(5) Flood the slide with acetone–alcohol until no more colour is seen to come off (about 3 s); *immediately* wash with water to prevent excessive decolourisation. Repeat if necessary (only experience will show how much washing is needed).

(6) Flood the slide with safranin and leave for 1 min. Wash off the stain with water. Gently dry the slide between sheets of clean blotting paper and allow to dry finally in air.

(7) Apply a drop of immersion oil and examine under the oil immersion lens (section A2.3.2).

Results

Are your observations in agreement with the description given in section 2.7.1 of the bacterial content of milk?

Experiment 2.3: To compare the numbers of bacteria present in fresh and stale milk

If a single bacterium is placed on nutrient agar it will grow to form a colony which is easily seen with the naked eye, unlike the original bacterium. This can be made use of when counting bacteria.

After sterilising the apparatus, the first part of the experiment involves the technique of serial dilution. The numbers of bacteria in milk are vast, so counting can be made more manageable by diluting by a known factor and taking a small sample of known volume. A series of dilutions is prepared. In the second part of the experiment samples of each dilution are cultured and the one giving the most suitable number of colonies (a reasonably large number but with no overlap of colonies) when grown on agar is used to calculate the number of bacteria in a given volume of milk.

Materials

6 sterile nutrient agar plates	indelible marker
8 1 cm³ graduated pipettes	Bunsen burner
	100 cm³ distilled water
1 10 cm³ graduated pipette	fresh milk
	stale milk
6 test-tubes and test-tube rack	70% alcohol
	aluminium foil
cotton wool	glass spreader

Sterilisation of apparatus

(1) Place cotton wool plugs in each of six test-tubes and cover the plugs loosely with aluminium foil.

(2) Place a small piece of cotton wool in the top of each of eight 1 cm³ graduated pipettes and one 10 cm³ graduated pipette and wrap each pipette separately in aluminium foil.

(3) Place the test-tubes and pipettes in a hot air oven at 160 °C for 60 min (bottles of media and water should not be sterilised in an oven).

(4) Allow all the apparatus to cool before use.

Serial dilution of milk and inoculation of agar plates

(1) Label the six sterile plugged test-tubes F1, F2, F3, S1, S2 and S3, and remove the aluminium foil covers from the plugs.

(2) Label the base of each of six sterile nutrient agar plates F1, F2, F3, and S1, S2, S3.

(3) Transfer 9.9 cm³ of sterile distilled water to each of the six test-tubes using the following technique.

   (*a*) Remove the cotton wool plug from the flask containing sterile distilled water using the little finger and fourth finger of one hand.

   (*b*) Whilst holding the plug, draw up 9.9 cm³ of water using the sterile 10 cm³ graduated pipette held in the other hand.

   (*c*) Replace the plug.

   (*d*) Remove the plug from the first test-tube using the same method as in (*a*).

   (*e*) Transfer 9.9 cm³ of water to the test-tube.

   (*f*) Replace the plug.

   (*g*) Repeat for the five remaining test-tubes.

(4) Shake the sample of fresh milk and transfer 0.1 cm^3 of this milk using a sterile 1 cm^3 pipette to tube F1, removing and replacing the plug as before. This gives a ×100 dilution.

(5) Shake the tube gently to ensure thorough mixing.

(6) Using a fresh pipette, transfer 0.1 cm^3 from tube F1 to the sterile plate labelled F1, lifting the lid by a minimal amount.

(7) Dip a glass spreader in 70% alcohol, allow excess alcohol to drip off and then hold the spreader vertically in a Bunsen burner flame.

(8) Cool the spreader and spread the sample of milk over the surface of the plate.

(9) Re-sterilise the spreader.

(10) Using the same pipette as in point (6), transfer 0.1 cm^3 from tube F1 to tube F2, removing and replacing the bungs as before.

(11) Shake the tube F2 to ensure thorough mixing. This gives a ×10 000 dilution.

(12) Repeat the procedure from (6)–(9), substituting F2 for F1.

(13) Repeat from (10)–(11), using F3 for F2. This gives a × 1 000 000 dilution. Repeat (6)–(9) using F3 for F1.

(14) Repeat the serial dilution technique using the sample of stale milk and prepare plates S1, S2 and S3.

(15) Incubate the six plates upside down at 35 °C for about three days.

(16) The lids of the plates should then be taped down to avoid the risk of pathogens being spread.

(17) Examine the plates for bacterial growth. Count the numbers of individual colonies where practical. Record results in the form of a table and use them to calculate the number of bacteria in 1 cm^3 of undiluted milk.

Notes

See notes (3) and (4) at the end of experiment 2.1.

2.7.3 Practical work with fungi

The methods for handling fungi are in many cases the same as those for bacteria, being standard microbiological techniques. Many saprotrophic fungi, like bacteria, can be cultured on nutrient agar and, if pure cultures are required, the sterile techniques described in section 2.7.2 should be used. Common fungi suitable for culture in this way are *Mucor*, *Rhizopus*, *Penicillium* and *Aspergillus*, and a suitable culture medium is a 2% malt agar prepared in petri dishes. Selected fungi can be isolated from mixed cultures grown by chance contamination of substrates such as bread, fruit, and other moist foods. Spores can be transferred and added to the culture medium by a sterile mounted needle. Cultures can be conveniently examined with low power stereoscopic microscopes.

Chapter Three

Variety of life–protoctists and plants

In chapter 2 it was stated that all cellular organisms seem to fall into two natural groups, prokaryotes and eukaryotes. The eukaryotes have certain important features in common, summarised in table 2.1; one of these is the presence of nuclei in their cells.

There is evidence that a key event in the evolution of eukaryotes was the invasion of a primitive ancestor by aerobic prokaryotes that developed into mitochondria (the **endosymbiont theory** – see section 9.3.1). Another important event is thought to have been a subsequent invasion by photosynthetic prokaryotes (blue-green bacteria) that developed into chloroplasts. Cells from this line of evolution are believed to have given rise to the ancestors of plants. A plant is a **photosynthetic eukaryote** or an **autotrophic eukaryote**, autotrophic meaning that it uses carbon dioxide as a source of carbon. Eukaryotes that lack chloroplasts are non-photosynthetic and described as **heterotrophic** (using an organic source of carbon). Among the early eukaryotes are thought to have been the algae, fungi, slime moulds and protozoa. The basic stock of

eukaryotes from which these groups arose were probably simple unicellular organisms which moved by beating flagella. Many of these organisms exhibited both animal- and plant-like characteristics. Evolutionary relationships among these primitive groups are still not clear, and it is therefore difficult to divide the groups into kingdoms. Margulis and Schwartz* proposed a system in 1982 which, like any other system, cannot be regarded as perfect, but which has been widely adopted and is currently recommended by the Institute of Biology. Five kingdoms are proposed, one prokaryote kingdom, the Prokaryotae and four eukaryote kingdoms (see fig 3.1).

The most controversial group is the **Protoctista (protoctists)** because it is probably an unnatural group. It contains eukaryotes that are generally regarded as identical or similar to the ancestors of modern plants, animals and fungi. The Protoctista includes two groups, the algae and the protozoans, which formerly had taxonomic status, and

* Margulis, L. & Schwartz, K.V. (1982), *Five kingdoms: an illustrated guide to the phyla of life on Earth*, W.H. Freeman & Co.

Table 3.1 Differences between plants and animals.

	Typical animal	Typical plant
Nutrition	Heterotrophic (see chapter 10)	Autotrophic (see chapter 9)
Locomotion	Motile – essential for finding food	Non-motile
Sensitivity	Controlled by hormones and nervous system – latter allows rapid reactions and is essential for movement and locomotion	Controlled by hormones only – no nervous system. Responds slowly to stimuli, usually by growth.
Excretion	Special excretory structures in most multi-cellular animals, particularly for nitrogenous excretion	Few excretory products and no special excretory organs
Osmoregulation	Special structures to carry out osmoregulation	Active osmoregulation not required owing to presence of cell walls
Growth	Occurs throughout body	Restricted to certain regions called meristems (in multicellular plants)
Surface area to volume ratio	Compact body for ease of movement	Large surface area to volume ratio for efficient trapping of light and exchange of materials. Branching often occurs
Cell structure	No rigid cell wall	Rigid cell wall containing cellulose
	Vacuoles small and temporary	Large permanent vacuole containing cell sap
	No chloroplasts or other plastids	Chloroplasts (containing chlorophyll) or other plastids present
	Stores carbohydrate as glycogen	Stores carbohydrate as starch
	Centrioles present	Centrioles absent

which are now regarded as containing organisms too widely different to be placed in one phylum. In addition, it includes one group of organisms which were previously placed in the fungi, the Oomycota or oomycetes, but which are now regarded as ancestral to fungi. The slime moulds, a group of organisms which are motile but which produce spores in sporangia, are also included in the Protoctista.

Further discussion on classification is given later in this chapter (see algae) and at the beginning of chapter 4. Plants are distinguished from animals according to the features shown in table 3.1.

Kingdom Protoctista
(Gk. *protos*, very first; *ktistos*, to establish)

3.1   Characteristics and classification of protoctists

Protoctists include all eukaryotic organisms which are no longer classified as animals, plants or fungi. Many are unicellular organisms or collections of similar cells, and the kingdom includes all algae, protozoa, slime moulds and the Oomycota. Algae and protozoa are adapted for aquatic habitats. The classification and characteristics of the Protoctista are summarised in fig 3.1 and given in more detail later in fig 3.2 (algae), table 3.4 (Oomycota) and table 3.5 (protozoa).

3.2   Algae

3.2.1   Characteristics and classification of algae

The algae form a large group of protoctistans of great biological importance and significance to humans (section 3.2.7). Their bodies lack true stems, roots and leaves. Originally therefore, they were classified with fungi in the division Thallophyta (see footnote page 17). However, as more has been discovered about the group it has become obvious that there is great variety among the algae. They are best thought of as oxygen-producing photosynthetic eukaryotes that evolved in, and have exploited, an aquatic environment. It is true that some have escaped on to the land, but the worldwide productivity of

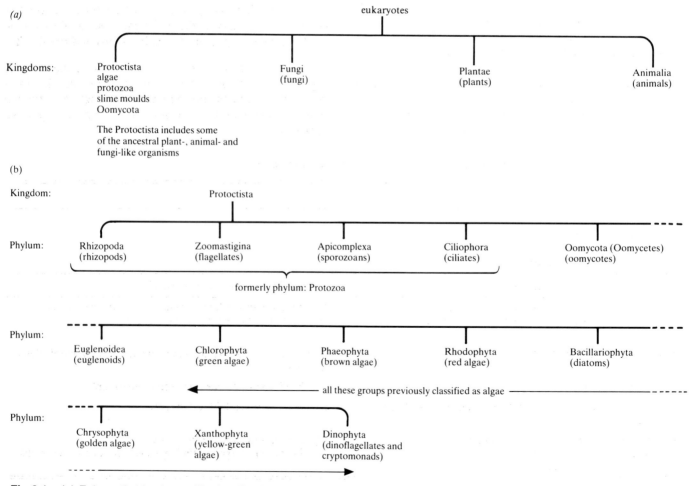

Fig 3.1 (a) *Eukaryotic kingdoms*, (b) *classification of Protoctista*

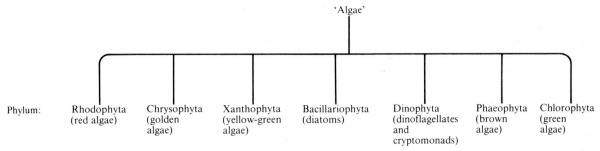

'Algae'

| Phylum: | Rhodophyta (red algae) | Chrysophyta (golden algae) | Xanthophyta (yellow-green algae) | Bacillariophyta (diatoms) | Dinophyta (dinoflagellates and cryptomonads) | Phaeophyta (brown algae) | Chlorophyta (green algae) |

Fig 3.2 *The phyla of eukaryotic algae*

these coastal and land forms is insignificant compared with those in the oceans and fresh water.

If algae are thought of in this way, the blue-green bacteria (cyanobacteria), formerly known as the blue-green algae, have to be excluded from the group, since they are prokaryotes. There remains the important point, however, that blue-green bacteria produce oxygen in photosynthesis, whilst other photosynthetic prokaryotes do not. This requires the presence of chlorophyll *a* and photosystem II (section 9.4.2) so that water can be split into hydrogen and oxygen, an important advance over other photosynthetic bacteria. Little is known about how this advance was achieved, though some types intermediate between blue-green bacteria and other bacteria are being discovered. Our understanding of the link between blue-green bacteria, algae and plants is improved by the evidence of the endosymbiont theory which suggests that chloroplasts originated as blue-green bacteria (section 9.3.1).

Fortunately, the eukaryotic algae seem to fall naturally into distinct groups, chiefly on the basis of their photosynthetic pigments. These groups are given the status of phyla in modern classifications. The relationship between the phyla is still the subject of research which is fundamental to our knowledge of the origins of plants and the link between prokaryotes and eukaryotes.

The phyla are shown in fig 3.2 and a current view of their relationships in fig 3.3. Characteristics of the algae and of some of the main phyla are shown in table 3.2.

3.2.2 Asexual reproduction in the algae

Algae show both asexual and sexual reproduction. Below is a summary of the asexual types found, which range from simple to complex.

Vegetative reproduction. In some colonial forms, the colonies fragment to produce separate, smaller colonies. In the larger, thalloid algae, such as *Fucus*, new thalli may develop from the main thallus and break off.

Fragmentation. This occurs in filamentous algae, such as *Spirogyra*. The filament breaks in a controlled manner somewhere along its length to form two filaments and could be regarded as a form of vegetative reproduction.

Table 3.2 Classification and characteristics of two of the main groups of algae.

Algae

| *General characteristics* |
| Body is a photosynthetic thallus |
| Almost all are specialised for an aquatic existence |
| Great range of size and form |

Phylum Chlorophyta ('green algae')	*Phylum Phaeophyta* ('brown algae')
Dominant photosynthetic pigment is chlorophyll; therefore green in appearance. Chlorophylls *a* and *b* present (as in plants)	*Dominant photosynthetic pigment is brown and called fucoxanthin. Chlorophylls *a* and *c* present
Store carbohydrate as starch (insoluble)	*Store carbohydrate as soluble laminarin and mannitol. Also store fat
Mostly freshwater	Nearly all marine (three freshwater genera only).
Large range of types, e.g. unicellular, filamentous, colonial, thalloid	Filamentous or thalloid, often large
e.g. *Chlamydomonas*, a unicellular, motile alga *Spirogyra*, a filamentous alga *Ulva*, a thalloid, marine alga	e.g. *Fucus*, a thalloid, marine alga *Laminaria*, large thalloid, marine alga; one of the kelps

* a diagnostic feature.

41

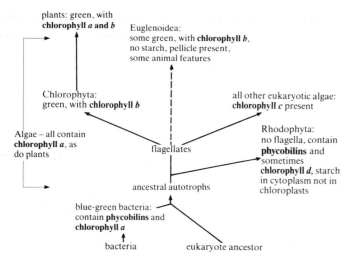

Fig 3.3 *Possible relationships between different groups of algae, bacteria, euglenoids and plants. For further discussion, see B. S. Rushton,* School Science Review, *62, no. 221, 648–54, June 1981*

Binary fission. In this process a unicellular organism divides into two equal halves, the nucleus dividing by mitosis.

Zoospores. These are motile, flagellate spores produced by many algae, for example *Chlamydomonas*.

Aplanospores. These are non-motile spores produced, for example, by some brown algae.

3.2.3 Sexual reproduction in the algae

Sexual reproduction involves the combination of genetic material from two individuals of the same species. The commonest method in the algae is by fusion of two morphologically (structurally) identical gametes. The process is called **isogamy**, and the gametes **isogametes**. *Spirogyra* and some species of *Chlamydomonas* are isogamous.

Sometimes, one of the gametes is less motile, or larger, and the process is then called **anisogamy**. In *Spirogyra* the gametes are structurally identical, but one moves while the other is stationary. This could be regarded as physiological anisogamy. A third variation occurs when one gamete is large and stationary while the other is small and motile. The gametes are known as female and male respectively and the process is called **oogamy**. *Fucus* and some species of *Chlamydomonas* are oogamous. The female gamete is larger because it contains food reserves for the developing zygote after fertilisation.

The three types of sexual reproduction are associated with an increase in complexity of body structure because, although some simple algae such as *Chlamydomonas* show

oogamy, it is more widespread in the complex algae, such as the Phaeophyta. Oogamy is the only type occurring in plants.

Unfortunately, a confusing variety of terms is associated with gametes and organs of sexual reproduction in the algae and plants, especially algae. The main terms are explained below.

The gametes of the fungi, algae, mosses and ferns are produced in structures called **gametangia**. The male gametangium is called an antheridium and the female gametangium an oogonium or archegonium.

An **oogonium*** is a simple female gametangium found in many algae and fungi, and the female gamete or gametes it contains are called **oospheres**. A fertilised oosphere is called an **oospore** and this typically develops into a thick-walled resting spore capable of surviving adverse conditions. A general term for a female gamete is an **ovum** or **egg-cell**, although the term oosphere is sometimes used more loosely to mean ovum.

An **archegonium** is a more complex female gametangium, characteristic of bryophytes, ferns and many conifers, and is described later in the chapter.

An **antheridium** produces male gametes called **antherozoids** or **spermatozoids**. They are motile because they possess one or many flagella. They are characteristic of the

---

* The term oogonium also refers to a cell of an animal that divides to produce oocytes in the ovary (see chapter 20).

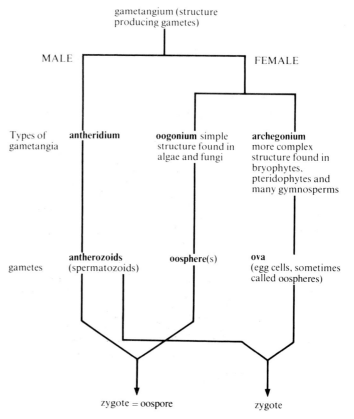

Fig 3.4 *Types of gametangia and gametes*

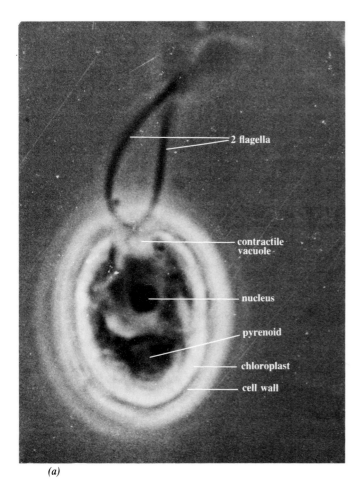

(a)

2 flagella

contractile vacuole

nucleus

pyrenoid

chloroplast

cell wall

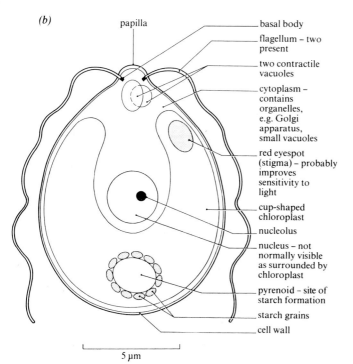

(b)

papilla

basal body

flagellum – two present

two contractile vacuoles

cytoplasm – contains organelles, e.g. Golgi apparatus, small vacuoles

red eyespot (stigma) – probably improves sensitivity to light

cup-shaped chloroplast

nucleolus

nucleus – not normally visible as surrounded by chloroplast

pyrenoid – site of starch formation

starch grains

cell wall

5 μm

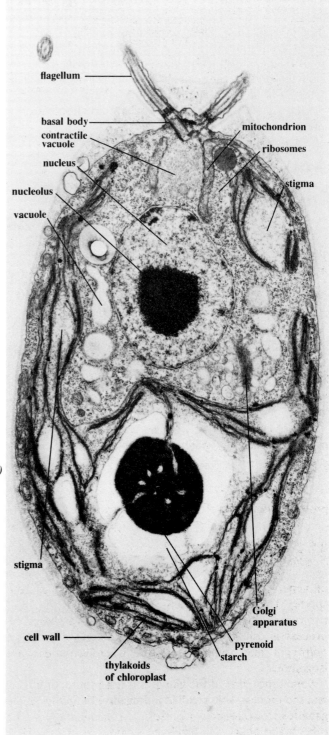

(c)

flagellum

basal body

contractile vacuole

nucleus

nucleolus

vacuole

mitochondrion

ribosomes

stigma

stigma

cell wall

thylakoids of chloroplast

starch

pyrenoid

Golgi apparatus

Fig 3.5 *(a)* Chlamydomonas *as seen with the light microscope (× 600). (b) Diagram of structure of* Chlamydomonas. *(c) Electron micrograph of* Chlamydomonas reinhardtii *(× 1400)*

43

fungi, algae, mosses, ferns and some conifers. Male gametes produced by animals are called **spermatozoa** (**sperm** or **sperms**). These terms are summarised in fig 3.4.

For the purposes of this chapter there is little point in preserving the distinction between the different names for gametes of the same sex, so all male gametes will be referred to as sperm and all female gametes as ova.

Like fungi, some algae show heterothallism (section 2.4.2).

3.2.4   Phylum Chlorophyta

Characteristics of the Chlorophyta are summarised in table 3.2.

Chlamydomonas is a unicellular motile alga living mainly in stagnant water, such as in ponds and ditches, particularly where it is rich in soluble nitrogenous compounds, as in farmyards. The cells often occur in numbers large enough to impart a green colour to the water. A few species are marine, or live in brackish coastal waters.

Structure

Chlamydomonas is motile and possesses contractile vacuoles. Its structure is illustrated in fig 3.5. The electron micrograph reveals the presence of eukaryote organelles in

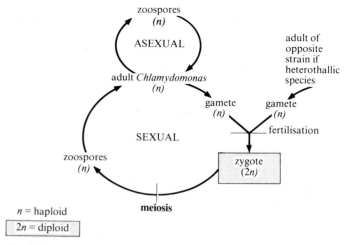

Fig 3.6 *Diagrammatic life cycle of* Chlamydomonas

the cytoplasm, such as Golgi apparatus, mitochondria, ribosomes and small vacuoles. The **pyrenoid** is a structure found in the chloroplasts of most algae. It is a protein body and probably consists mainly of the carbon-dioxide-fixing enzyme ribulose bisphosphate carboxylase. It is associated with the storage of carbohydrates such as starch. The **red eye spot** detects changes in light intensity and the cell responds by moving towards, or staying in, light of

Fig 3.7 *Reproduction in* Chlamydomonas. *(a) Asexual reproduction. (b) Sexual reproduction in an isogamous species*

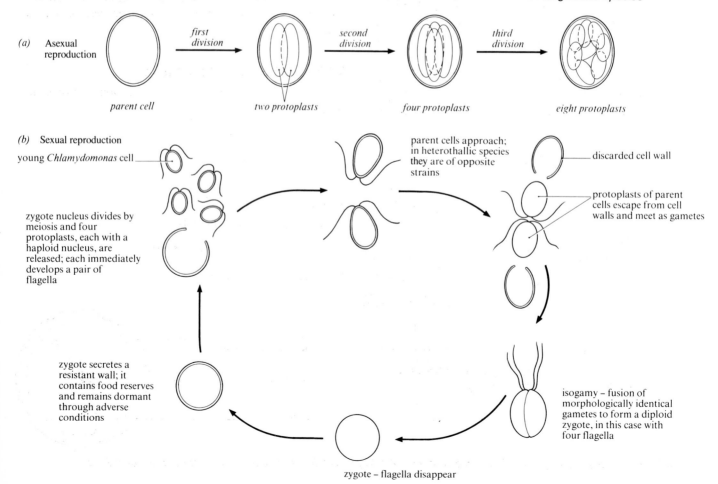

44

optimum intensity for photosynthesis. Such a response to light is called **phototaxis** (section 15.1.2). The cell moves by beating its two flagella and has a corkscrew motion because it rotates as it moves forward.

Life cycle

The life cycle of *Chlamydomonas* is summarised in fig 3.6. The adult is haploid.

Asexual reproduction

Asexual reproduction is by **zoospores**. The parent cell withdraws its flagella and the protoplast inside the cell wall divides into 2–16 daughter protoplasts (usually four). During this process the nuclear division is by mitosis and the chloroplast also divides. The daughter protoplasts develop new cell walls, eye spots and flagella. Centrioles (basal bodies) are involved in the formation of new flagella. The daughter cells, now called zoospores, are liberated by gelatinisation of the parent cell wall. Each grows to a full-sized *Chlamydomonas* cell. The process is illustrated in fig 3.7*a*.

Sexual reproduction

Some *Chlamydomonas* species are homothallic and some are heterothallic, and different species may be isogamous, anisogamous or oogamous. Reproduction of an isogamous species is illustrated in fig 3.7*b*. At germination, the first division of the zygote nucleus is by meiosis, resulting in a return to the haploid condition of the adult. The young *Chlamydomonas* cells released may be called zoospores until they reach maturity.

Spirogyra is a non-branching filamentous alga, living in ponds and other bodies of still, fresh water. The majority of species are floating and they are characteristically slimy.

Structure

The cylindrical walls are joined end to end to form a filament as shown in fig 3.8. Each cell is identical, so there is no division of labour. It has a narrow peripheral layer of cytoplasm and a large vacuole across which run strands of cytoplasm. The nucleus is suspended by these strands in a central position. The peripheral cytoplasm contains one or more chloroplasts that are spiral in form.

Growth and reproduction

Growth of the filament is intercalary, that is can occur by division and subsequent growth of any cell along the length of the filament (contrast most plants where primary growth is confined to apical regions). The nucleus of a given cell divides by mitosis, which is followed by division of the whole cell as a new cell wall grows from the edges of the filament inwards. The two daughter cells thus formed grow to normal size, causing the filament to grow in length.

Asexual reproduction is by fragmentation, as mentioned earlier (section 3.2.2).

Sexual reproduction is by a method confined to the filamentous algae in which two filaments line up alongside each other and adjacent cells become connected by short tubular outgrowths of the cells. The whole cell contents behave like gametes and the process can be regarded as *anisogamous* because, although the gametes are morphologically identical, one is motile and crosses through the connecting tube to the other. The process is called **conjugation**.

3.2.5 Phylum Phaeophyta

Characteristics of the Phaeophyta are summarised in table 3.2.

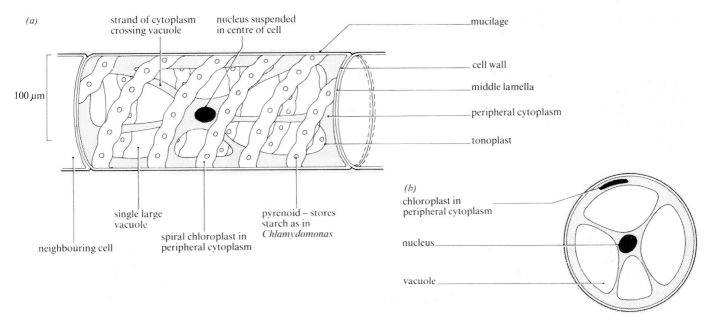

Fig 3.8 *Structure of* Spirogyra. *(a) Diagram of side view. (b) Diagram of TS cell in region of nucleus showing cylindrical nature of cells*

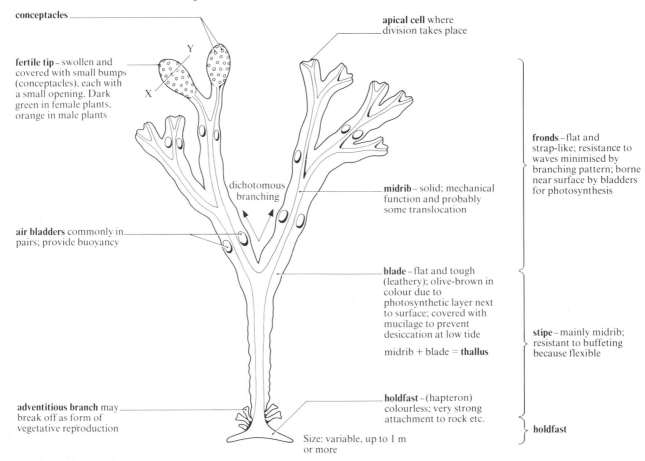

X–Y see fig 3.10

conceptacles

apical cell where division takes place

fertile tip – swollen and covered with small bumps (conceptacles), each with a small opening. Dark green in female plants, orange in male plants

fronds – flat and strap-like; resistance to waves minimised by branching pattern; borne near surface by bladders for photosynthesis

dichotomous branching

midrib – solid; mechanical function and probably some translocation

air bladders commonly in pairs; provide buoyancy

blade – flat and tough (leathery); olive-brown in colour due to photosynthetic layer next to surface; covered with mucilage to prevent desiccation at low tide

midrib + blade = **thallus**

stipe – mainly midrib; resistant to buffeting because flexible

adventitious branch may break off as form of vegetative reproduction

holdfast – (hapteron) colourless; very strong attachment to rock etc.

holdfast

Size: variable, up to 1 m or more

Fig 3.9 *External features of* Fucus vesiculosus, *with notes on structure, particularly adaptations to environment*

Species of the genus *Fucus* are common on rocky shores off the British coast. They are well adapted to the relatively harsh conditions of the littoral zone, the zone alternately exposed and covered by the tides.

There are three common species and these are often found at three different levels, or zones, on the shore, a phenomenon called **zonation**. They are principally zoned according to their ability to withstand exposure to air. Their chief recognition features and positions on the shore are noted below.

F.spiralis (flat wrack) – towards high tide mark. If suspended, the thallus adopts a slight spiral twist.

F.serratus (common, serrated or saw wrack) – middle zone. Edge of the thallus is serrated.

F.vesiculosus (bladder wrack) – towards low tide mark. Possesses air bladders for buoyancy.

The external features of *F. vesiculosus* are shown in fig 3.9 and some of the main features of internal structure in fig 3.10.

The body, or thallus, shows some division of labour between different tissues. This trend is carried further in the Phaeophyta than in any other algal group. Adaptations to environment are discussed below.

Reproductive structures

Sexual reproduction is oogamous. *F.vesiculosus* and *F.serratus* are dioecious, that is they have separate male and female organisms. *F. spiralis* is hermaphrodite, having male and female organs on the same organism and in the same conceptacles. The sex organs develop inside conceptacles on the 'fertile' tips of some fronds. Each conceptacle has a pore (**ostiole**) for later release of the sex organs. Their structures are shown in fig 3.10.

The adult is diploid and the gametes are produced by meiosis.

Adaptations to environment

Before discussing the adaptations of *Fucus* to its environment, some mention must be made of the nature of this environment, which is relatively hostile. Being intertidal, the different species are subjected to varying degrees of exposure to air when the tide recedes. Therefore they must be protected against drying out. Temperatures may change rapidly, as when a cold sea advances into a hot rock pool. Salinity is another factor to which the organism has adapted, and this may increase in an evaporating rock pool, or decrease during rain. The surge and tug of the tide, and the pounding of waves, are additional factors which demand mechanical strength if they are to be withstood. Large waves can pick up stones and cause great damage as they crash down.

Morphological adaptations (overall structure)

The thallus is firmly anchored by a **holdfast** (fig 3.9). This forms an intimate association with its substrate, usually rock, and is extremely difficult to dislodge. In fact, the rock often breaks before the holdfast.

The thallus is dissected owing to its dichotomous branching in one plane, and this minimises resistance to water. It is also tough but non-rigid. The midrib of the thallus is strong and flexible.

F. vesiculosus possesses air bladders for buoyancy, thus holding its fronds up near the surface for maximum interception of light for photosynthesis.

Chloroplasts are mainly located in the surface layers for maximum exposure to light for photosynthesis.

Physiological adaptations

The dominant photosynthetic pigment is the brown pigment **fucoxanthin**. This is an adaptation to photosynthesising under water because fucoxanthin strongly absorbs blue light, which penetrates water much further than longer wavelengths such as red light.

The thallus secretes large quantities of mucilage which fills spaces within the body and exudes on to its surface. This helps to prevent desiccation by retaining water.

The solute potential of the cells is lower than that of sea water, so water is not lost by osmosis.

Reproductive adaptations

Release of gametes is synchronised with the tides. At low tide the thallus dries and squeezes the sex organs, which are protected by mucilage, out of the conceptacles. As the tide advances, the walls of the sex organs dissolve and release the gametes.

The male gametes are motile and chemotactic, attracted by a chemical secretion of the female gametes.

The zygote develops immediately after fertilisation, minimising the risk of being swept out to sea.

3.2.6 Trends in the algae

Even among the few examples described, it can be seen that there is a wide range of algal types, ranging from unicellular organisms such as *Chlamydomonas* to relatively large seaweeds, such as *Fucus*, with differentiated bodies showing some division of labour. Some large brown algae even possess conducting tissues, although none possess true vascular tissue (xylem and phloem).

Within the algae there is also a trend in sexual reproduction from simple isogamy and anisogamy to oogamy. Caution has to be exercised in trying to use these trends to establish evolutionary relationships between algal groups. The relationships are still not clear and the group from which land plants are thought to have evolved, the Chlorophyta (green algae), contains simple unicellular types as well as complex types, together with a range of sexual reproduction from isogamy to oogamy.

3.2.7 Importance of the algae

The role of algae in the biosphere

Modern estimates are that at least half the world's productivity, that is carbon fixation, comes from the oceans. This is contributed by the algae, the only vegetation in the sea. Considering their large surface area, the oceans might be expected to contribute a larger fraction, but photosynthesis is confined to the surface layers where light is available and here nutrient availabil-

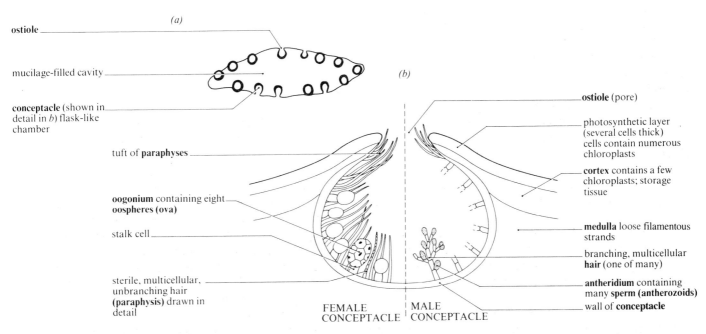

Fig 3.10 *Reproductive organs of* Fucus vesiculosus. *(a) VS fertile tip (low power). (b) VS conceptacle (high power)*

ity, particularly nitrogen and phosphorus, is a limiting factor.

Algae are vital as primary producers (chapter 12), being at the start of most aquatic food chains, including freshwater as well as virtually all ocean food chains. These chains lead through zooplankton*, crustaceans, and so on to fish. Many of the algae are microscopic unicells and these are the chief components of phytoplankton.*

Carbon fixation is not the only consequence of photosynthesis (section 9.2). The oxygen in the atmosphere is maintained by photosynthesis and at least half of this must therefore come from the algae, more than is contributed by forests on land.

Alginic acid, agar and carrageenin

A number of useful substances, including alginic acid, agar and carrageenin, are extracted from algae. **Alginic acid** and its derivatives (alginates) are polysaccharides extracted from the middle lamellae and cell walls of brown algae, such as *Laminaria*, *Ascophyllum* and *Macrocystis*. The fronds of the algae (seaweeds) are harvested in large quantities from shallow coastal waters, such as *Macrocystis* off the coast of California. The purified alginates are non-toxic and readily form gels. They are used as thickeners and gelling agents in a wide variety of industrial products; for example hand creams in cosmetics, emulsifiers in ice cream, polishes, medicines and paints, gelling agents in confectionery and glazes in ceramics.

Agar is a polysaccharide extracted from red algae. Like alginates, it forms gels and is probably most familiar as a convenient medium on which to culture bacteria and fungi. Here it is prepared as a dilute solution, mixed with nutrients and allowed to set as a jelly. It is also used for much the same purposes as alginates.

Carrageenin (**carragheen**) is another cell wall polysaccharide and is extracted mainly from the red alga *Chondrus crispus*. It is chemically very similar to agar and is used for the same purposes.

Diatomite (kieselguhr)

Algae of the phylum Bacillariophyta are mainly unicellular and are called **diatoms**. They have a characteristic cell wall structure containing silica. When they die they sediment, so that on the sea bed or lake bottom extensive deposits can be built up over long periods of time. The resulting 'diatomaceous earth' has a high proportion of silica (up to 90%) and when purified can be used as an inert filtering material, as in sugar refining and brewing, as a filler in paints and paper, and can be used in insulation materials that have to withstand extremes of temperature.

Fertiliser

A traditional but small-scale use of the larger seaweeds (red and brown algae) has been as a fertiliser on coastal farms. They are richer in potassium, but poorer in nitrogen and phosphorus, than farm manure, and are of limited success.

Food

Some algae are used directly for human consumption, particularly in the Far East. *Porphyra*, a delicate red seaweed, and *Laminaria*, a large brown seaweed, are commonly used either raw or prepared in some way. *Porphyra* is made into laver bread in South Wales, a traditional dish in which the boiled alga is mixed with oatmeal and cooked in butter. In the search for new foods, much attention has been paid to mass culture of algae. Few have provided greater success in this field. However, the blue-green bacterium *Spirulina* shows promise as a food source.

Sewage disposal

Algae contribute to some extent to the microbial life of sewage works, the sewage providing nutrients for microscopic green algae as well as for bacteria, fungi and protozoa. Algae are particularly useful to humans in open 'oxidation ponds', which are used especially in tropical and subtropical countries. Ponds between 1 m and 1.5 m deep receive raw sewage, and oxygen provided by algal photosynthesis is vital for the other aerobic microorganisms that utilise the sewage. The algae can also be harvested occasionally and processed for animal fodder.

Research

Unicellular algae show characteristics typical of plants and often make ideal research material since they can be grown in large numbers under precisely controlled conditions without occupying a great deal of space. For example, the use of *Chlorella* in research on photosynthesis has been rewarding, as described in section 9.4.3. Algae are also used in ion uptake experiments and were important in early work on the structure of cell walls and flagella.

Harmful effects

Under certain conditions algae produce 'blooms'; that is dense masses of material. This is particularly true in relatively warm conditions when there is high nutrient availability. The latter may be artificially induced by human activity, as when sewage is added to water, or inorganic fertilisers run off from agricultural land into rivers and lakes. As a result, an explosive growth of primary producers (algae) occurs and far more than usual therefore die before being eaten. The subsequent process of decomposition is carried out by aerobic bacteria which, in turn, multiply and deplete the water of oxygen. This sequence may be rapid and the lack of oxygen may lead to the death of fish and other animals and plants. The increase of nutrients which starts the process is called **eutrophication**, and if rapid constitutes a form of pollution.

Toxins produced by algal blooms (and by blue-green bacteria) can also increase mortality. They can be a

---

* Plankton are minute algae (phytoplankton) and animals (zooplankton) floating in the surfaces of the oceans and lakes. They are of great economic and ecological importance.

serious problem in lakes, including those of fish farms, especially where intensive addition of fertilisers to farmlands adds to the eutrophication problems. Similar problems may occur as a result of algal blooms in the oceans. In addition, toxins may be stored by shellfish feeding on the algae and be passed on to humans causing, for example, paralytic shellfish poisoning.

Algae also cause problems in water storage reservoirs, where their products may taint the water and where they can grow on and block the beds of sand used as filters.

> **3.1** The problems just mentioned are greater in lowland reservoirs. Why should this be so?
>
> **3.2** Algae are not associated with disease, unlike many fungi and bacteria. What is the reason for this?

3.3 Phylum Euglenoidea (euglenoids)

Characteristics of the Euglenoidea are summarised below. The phylum shows a mixture of features that makes these organisms difficult to classify. In the past it has been included in both the plant and animal kingdoms.

Euglena is a common unicellular organism of freshwater ponds, ditches and any other water that is rich in soluble organic matter. Like *Chlamydomonas* it may be present in numbers sufficient to impart a green colour to the water, chlorophyll being the dominant pigment. Its structure is shown in fig 3.11, together with notes on some of its features.

Euglena lacks a cell wall. The outer layer of the cell is the cell surface membrane, immediately below which is the proteinaceous **pellicle**. This is flexible and permits the organism to assume various shapes. As the pellicle surrounds the cytoplasm, it can be regarded as a form of **exoskeleton**. It consists of a number of thickened longitudinal strips and microfibrils articulating with each other. When minute fibrils within the cytoplasm, called **myo-**

Table 3.3 Characteristics of Euglenoidea.

Phylum Euglenoidea ('euglenoids')
Dominant photosynthetic pigment is chlorophyll; therefore green in appearance.
Chlorophylls *a* and *b* present
Store carbohydrate as paramylum (similar to starch)
Mostly freshwater
Unicellular and motile
*Pellicle present instead of cellulose cell wall
Eye spot and contractile vacuoles present
e.g. *Euglena*

* a diagnostic feature.

nemes, contract, they cause the strips of the pellicle to slide over one another and effect a change in body shape. This is called **euglenoid movement**. Details of its more usual method of locomotion by means of its long flagellum are included in fig 3.11 (see eye spot, photoreceptor and long flagellum) and in section 17.6.3.

Asexual reproduction is by longitudinal binary fission. No sexual reproduction occurs.

Nutrition

Green species of *Euglena* are autotrophic, synthesising their own food from carbon dioxide, water and mineral salts. However, they are dependent on an external supply of vitamins B_1 and B_{12} which, like animals, they cannot synthesise for themselves. Although *Euglena* resembles animals in this respect, a large number of protoctists in other groups also share this inability.

A few *Euglena* species lack chloroplasts and are therefore colourless and non-photosynthetic (heterotrophic). They have a saprotrophic mode of nutrition, carrying out extracellular digestion, and grow most abundantly where putrefaction is taking place since decaying material is rich in organic compounds. Other colourless forms may be capable of ingesting small food particles at the base of the gullet, where the pellicle is absent, and carrying out intracellular digestion (holozoic nutrition, section 10.1.1). Food may be driven into the gullet by the action of the flagella. These species resemble the protozoan *Peranema* (section 3.5.2).

If green species of *Euglena* are kept in darkness for a prolonged period they lose their chloroplasts and become colourless. If the medium contains organic nutrients, they survive saprotrophically. Chloroplasts return when the organisms are returned to light.

3.4 Phylum Oomycota (oomycetes)

Characteristics of the Oomycota are given in table 3.4. In the phylum are a number of pathogenic organisms, including the downy mildews (see table 2.10). One of these, *Phytophthora infestans*, will be studied as an example of a parasite.

Phytophthora infestans is a pathogen of economic importance because it parasitises potato crops, causing a potentially devastating disease known as potato blight. It is similar in its structure and mode of attack to another member of the Oomycota, *Peronospora*, which is a common, but less serious, disease of wall-flower, cabbage and other members of the plant family Cruciferae.

Blight is usually first noticed in the leaves in August, though infection normally starts in the spring when the organism grows up to the leaves from the tubers in which the mycelium has over-wintered.

A mycelium of branched, aseptate hyphae ramifies through the intercellular spaces of the leaves, giving off **branched haustoria** which push into the mesophyll cells and

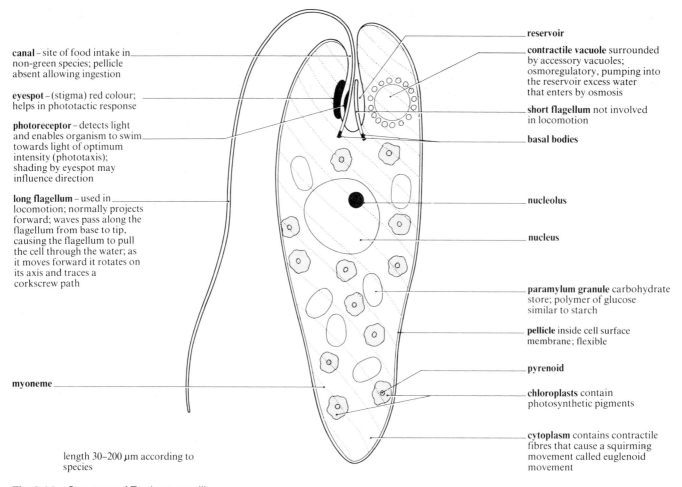

canal – site of food intake in non-green species; pellicle absent allowing ingestion

eyespot – (stigma) red colour; helps in phototactic response

photoreceptor – detects light and enables organism to swim towards light of optimum intensity (phototaxis); shading by eyespot may influence direction

long flagellum – used in locomotion; normally projects forward; waves pass along the flagellum from base to tip, causing the flagellum to pull the cell through the water; as it moves forward it rotates on its axis and traces a corkscrew path

myoneme

length 30–200 μm according to species

reservoir

contractile vacuole surrounded by accessory vacuoles; osmoregulatory, pumping into the reservoir excess water that enters by osmosis

short flagellum not involved in locomotion

basal bodies

nucleolus

nucleus

paramylum granule carbohydrate store; polymer of glucose similar to starch

pellicle inside cell surface membrane; flexible

pyrenoid

chloroplasts contain photosynthetic pigments

cytoplasm contains contractile fibres that cause a squirming movement called euglenoid movement

Fig 3.11 *Structure of* Euglena gracilis

absorb nutrients from them (fig 3.12). In warm, humid conditions the mycelium produces long, slender structures called **sporangiophores** which emerge from the lower surface of the leaf through stomata or wounds. These branch and give rise to **sporangia** (fig 3.12). In warm conditions sporangia may behave as spores, being blown or splashed by raindrops on to other plants, where further infection takes place. A hypha emerges from the sporangium and penetrates the plant via a stoma, lenticel or wound. In cool conditions, the sporangium contents may divide to form motile zoospores (a primitive feature) which, when released, swim in surface films of moisture.

Table 3.4 Characteristics of Oomycota.

Phylum Oomycota

Sexual reproduction by oogamy, involving fusion of an oosphere (female gamete) with a male gamete to produce an oospore

Asexual reproduction by means of biflagellate zoospores produced in sporangia

Non-septate hyphae

Cell walls contain cellulose

e.g. *Phytophthora infestans*, facultative parasite causing potato blight
Pythium, many important facultative parasites, some cause damping-off of seedlings
Peronospora, obligate parasites, causing downy mildews of crucifers, e.g. cabbages

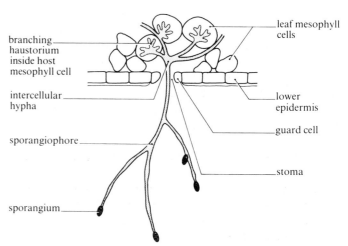

branching haustorium inside host mesophyll cell

intercellular hypha

sporangiophore

sporangium

leaf mesophyll cells

lower epidermis

guard cell

stoma

Fig 3.12 Phytophthora infestans *growing in a diseased potato leaf, with sporangiophores emerging from the underside of the leaf*

They may encyst until conditions are suitable once more for hyphal growth, then produce new infections.

Diseased plants show individual leaflets with small, brown, dead, 'blighted' areas. Inspection of the lower surface of an infected leaflet reveals a fringe of white sporangiophores around the dead area. In warm, humid conditions, the dead area spreads rapidly through the whole leaf and into the stem. Some sporangia may fall to the ground and infect potato tubers where infection spreads very rapidly causing a form of dry rot in which the tissues are discoloured a rusty brown in an irregular manner from the skin to the centre of the tuber.

First the base and then the rest of the plant becomes a putrid mass as the dead areas become secondarily infected with decomposing bacteria (saprotrophs). *Phytophthora* thus kills the whole plant, unlike its close relative *Peronospora* which is an obligate parasite. In this respect, *Phytophthora* is not a typical obligate parasite and it is sometimes described as facultative, though the distinction is perhaps not worth stressing here.

The organism normally overwinters as a dormant mycelium within lightly infected potato tubers. Except where the potato is native (Mexico, Central and South America) it is thought that the organism rarely reproduces sexually, unlike *Peronospora*, but under laboratory conditions it can be induced to do so. Like *Peronospora*, it produces a resistant resting spore. It is the result of fusion between an antheridium and an oogonium, and a thick-walled oospore is produced. This can remain dormant in the soil over winter and cause infection in the following year.

In the past, *Phytophthora* epidemics have had serious consequences. The disease is thought to have been accidentally introduced into Europe from America in the late 1830s and caused a series of epidemics that totally destroyed the potato crop in Ireland in 1845 and in subsequent years. Widespread famine resulted and many starved to death, victims as much of complex economic and political influences as of the disease. Many Irish families emigrated to North America as a result.

The disease is also of interest because in 1845 Berkeley provided the first clear demonstration that micro-organisms cause disease by showing that the organism associated with potato blight *caused* the disease, rather than being a by-product of decay.

Knowledge of the life cycle of potato blight has since led to methods of controlling the disease. These are summarised below.

(1) Care must be taken to ensure that no infected tubers are planted.
(2) New plantings must not be made in soil known to have carried the disease a year previously, since the organism can survive up to one year in the soil. Crop rotation may therefore help.
(3) All diseased parts of infected plants should be destroyed before lifting tubers, for example by burning or spraying with a corrosive solution such as sulphuric acid. This is because tubers can be infected from decaying haulms (stems) and aerial parts.
(4) Since the organism can overwinter in unlifted tubers, care must be taken to ensure that all tubers are lifted in an infected field.
(5) The organism can be attacked with copper-containing fungicides, such as Bordeaux mixture. Spraying must be carried out at the correct time to prevent an attack, since infected plants cannot be saved. It is usual to spray at fortnightly intervals, from the time that the plants are a few centimetres high until they are well matured. Tubers intended as seed potatoes can be sterilised externally by immersion in a dilute mercury(II) chloride solution.
(6) Accurate monitoring of meteorological conditions, coupled with an early warning system for farmers, can help to decide when spraying should be carried out.
(7) Breeding for resistance to the blight has been carried out for some years. The wild potato, *Solanum demissum*, is known to show high resistance and has been used in breeding experiments. One great obstruction to producing the required immunity lies in the fact that the organism exists in many strains and no potato has been found to be resistant to all of them. New strains of the organism may appear as new strains of potato are introduced. This is a familiar problem in plant pathology and emphasises the need for conservation of the wild ancestors of our modern crop plants as sources of genes for disease resistance.

3.5 Phyla Rhizopoda, Zoomastigina, Apicomplexa, Ciliophora

At the beginning of chapter 3 it was stated that the fungi, algae, slime moulds and protozoans might have evolved from an earlier group of eukaryote ancestors which moved by means of flagella (see fig 3.14). These flagellate organisms, represented by present-day forms belonging to the Kingdom Protoctista, probably displayed both plant- and animal-like characteristics. Some may have possessed chlorophyll and carried out photosynthesis only; some may have lost their chlorophyll, probably by mutation and fed heterotrophically exclusively; whilst some may have photosynthesised and fed heterotrophically. This suggests that there were few real differences between plant and animal cells at an early stage in evolution.

The earliest 'animal-like' unicellular organisms were probably similar in their basic features to present-day protozoans, as represented by organisms belonging to the phyla Rhizopoda, Zoomastigina, Apicomplexa and Ciliophora.

The earliest 'plant-like' organisms were probably similar in their basic features to present-day euglenoids (phylum Euglenophyta, section 3.3), which display a mixture of plant and animal features.

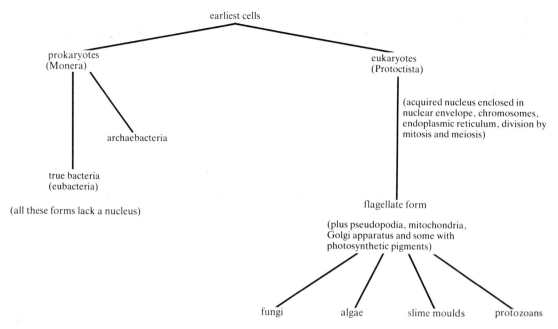

earliest cells

prokaryotes
(Monera)

archaebacteria

true bacteria
(eubacteria)

(all these forms lack a nucleus)

eukaryotes
(Protoctista)

(acquired nucleus enclosed in
nuclear envelope, chromosomes,
endoplasmic reticulum, division by
mitosis and meiosis)

flagellate form

(plus pseudopodia, mitochondria,
Golgi apparatus and some with
photosynthetic pigments)

fungi   algae   slime moulds   protozoans

Fig 3.13 *Possible early cell evolution*

3.5.1 Protozoans

The protozoa (*protos*, first; *zoon*, animal) are a diverse group of organisms. There are over 50 000 known species, and they are found in all environments where water is present. Each protozoan functions as an independent unit and is able to perform effectively all the activities necessary for life (table 3.5).

Within the protozoans there exists a range of different levels of **cellular organisation**. In the simplest forms (such as *Amoeba proteus)* the cell is relatively undifferentiated and there are few organelles which can be related to any specific activity. In the more differentiated protozoa (for example *Paramecium caudatum*) organisation has become highly complex. Here, elaboration of the living material into numerous highly efficient **organelles** has enabled these organisms to perform particular activities much more effectively.

It is a matter of debate whether a protozoan should be regarded as **unicellular** or **non-cellular** (acellular). The term unicellular will be used in this book. Non-cellular implies an equivalence between the whole body of the protozoan, which exhibits all the attributes of life, and the whole body of a multicellular animal. If protozoans are to be called unicellular, then comparison must be made between the whole protozoan body and a single cell of the body of a multicellular animal.

When the body of a multicellular animal is compared with that of a protozoan, it is obvious that structurally the protozoan is much simpler. However, if individual cells from multicellular animals are compared with protozoans the story is quite the opposite. Because the protozoan has to accomplish all of life's processes it is not surprising that it exhibits a much more complex organisation than that of multicellular animal cells. Cells of these animals are usually

designed to perform only one specific function. Consequently these cells exhibit a diminished level of structural complexity but an increased efficiency in performing their specific functions.

3.5.2 Phylum Zoomastigina

Peranema (fig 3.14) is an example of a flagellate, and its resemblance to some species of *Euglena* is mentioned in section 3.3.

Peranema is predatory and feeds on organisms such as *Euglena*. A '**rod organelle**' located near the cytostome touches the prey, protrudes and becomes attached to its surface. The anterior end then dilates and envelops the victim, the rods helping to push it through the cytostome.

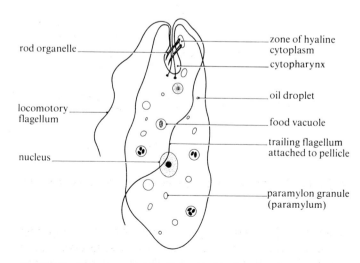

rod organelle

locomotory flagellum

nucleus

zone of hyaline cytoplasm

cytopharynx

oil droplet

food vacuole

trailing flagellum attached to pellicle

paramylon granule (paramylum)

Fig 3.14 Peranema trichophorum *showing the main structures visible under the light microscope*

Table 3.5 Classification of the protozoa.

Characteristic features
Unicellular
No tissues
Specialised parts of living material form organelles
Exhibit all forms of heterotrophism
Reproduction by fission; gametic fission when whole animal breaks up into gametes

NB Compound protozoans – these have more elaborate organelles and their structure foreshadows the multicellular animal phyla
(*a*) every cell is capable of reproduction
(*b*) all cells are structurally and physiologically similar and not organised into primary germ layers

Phylum Zoomastigina	*Phylum Rhizopoda*	*Phylum Ciliophora*	*Phylum Apicomplexa*
Some possess chromatophores, others do not	No chromatophores	No chromatophores	No chromatophores
Semi-rigid cell Covering is a pellicle	Some secrete tests, shells or skeleton, others possess no specific outer covering	Pellicle	Pellicle
Definite shape Adult movement by one or several flagella	Variable shape Adult movement by pseudopodia	Definite shape Adult movement by numerous cilia arranged in tracts	Definite shape Most exhibit no external structures for locomotion, any movement is limited
One nucleus	One nucleus	Macronucleus and micronucleus	One nucleus
Asexual reproduction by longitudinal binary fission	Asexual reproduction by binary fission	Asexual reproduction by transverse binary fission	Asexual reproduction by spores and schizogony (growth and reproduction)
		Sexual reproduction by conjugation	Sexual reproduction occurs in the life history
Multiple fission in cyst	May sporulate	Rarely sporulate	Large numbers of resistant spores after syngamy
e.g. *Euglena* *Peranema* *Trypanosoma*	e.g. *Amoeba* *Arcella* *Polystomella*	e.g. *Paramecium* *Vorticella* *Stentor*	e.g. *Monocystis* *Plasmodium*

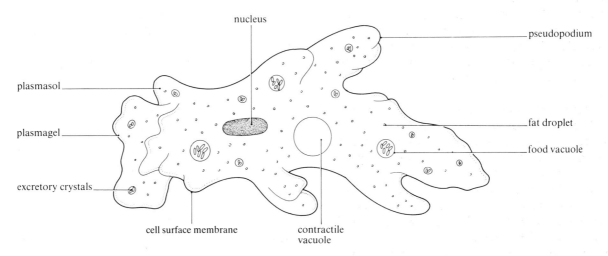

Fig 3.15 Amoeba proteus – *structures visible under the light microscope*

They may also shred the prey to some extent, so exposing its inner parts to enzyme action. Swallowing takes about eight minutes to complete, and the prey is then digested within food vacuoles.

3.5.3 Phylum Rhizopoda

Amoeba (fig 3.15) is a well-known example of the Rhizopoda. When compared with ciliates or flagellates

Amoeba shows little real differentiation either in its surface layer or within the cell. This may, in part, be due to the nature of its locomotory process which requires considerable cytoplasmic mobility. Nevertheless, its apparent simple construction is deceptive and it should be remembered that this animal is sufficiently well organised to carry out effectively all of the processes for life within its own minute mass of living material.

Amoeba is a free-living microscopic animal found living on the mud at the bottom of shallow freshwater ponds and streams where there is some movement of water. It is **omnivorous** feeding on a varied diet of algae, ciliates and flagellates. It measures approximately 0.1 mm in diameter and is subdivided into a nucleus and cytoplasm surrounded by a delicate cell surface membrane. The nucleus is embedded in the cytoplasm but occupies no fixed position. It is concerned with the organisation and integration of the animal's metabolic and reproductive processes.

The cytoplasm is differentiated into an outer layer of clear **plasmagel** called **ectoplasm**, and an inner mass of granular **plasmasol**, or **endoplasm**. The endoplasm contains fat droplets and a variety of vacuoles containing food matter in various stages of digestion, indigestible remains of food and crystals of excretory materials. It possesses a variable number of transient contractile vacuoles, which regularly become filled with water from the cytoplasm and then ultimately expel it into the surrounding pond water. The vacuoles perform an osmoregulatory function for the animal.

The animal constantly changes shape. This is brought about by temporary outpushings of the cytoplasm, called **pseudopodia**, which are regularly formed at any part of the animal's surface. These lobose pseudopodia are used for locomotion and feeding.

Amoeba does not possess any specific sensory organelles, however it is able to respond to a variety of stimuli. For example, it can discriminate between different types of food, it moves away from bright light, strong chemical solutions and persistent mechanical irritation. Violent disturbances cause it to withdraw all pseudopodia and remain stationary for some time.

Asexual reproduction takes place by binary fission. This form of mitotic cell division is triggered off in response to limits imposed either by the surface area to volume ratio, and/or the ratio between cytoplasmic volume and nuclear volume. The nucleus, which contains between 500–600 very small chromosomes, divides first. This is followed by a constriction and elongation of the cytoplasm which forces the daughter chromosomes apart and towards their respective poles. Ultimately two daughter amoebae of roughly equal proportions separate from each other. Under ideal conditions the whole process can be completed within 30 min. Each daughter *Amoeba* then proceeds to feed and grow to maximum size.

This particular form of reproduction is thought to be the only one that occurs in *Amoeba proteus*. Earlier reports of sporulation and cyst formation are now disregarded.

However both of these activities do take place in other species of *Amoeba*.

3.5.4 Phylum Ciliophora

Paramecium (fig 3.16) is a well-known ciliate. It demonstrates a very high degree of **cellular differentiation**, exhibiting many complex organelles which have been designed to perform specific functions for the organism. Not only does its structure show specialisation, but its reproductive activity is also complex.

Paramecium lives in stagnant water, or slow-flowing fresh water containing decaying organic matter. It possesses a constant elongate body shape with a blunt anterior end and a tapered posterior. Its whole body is covered by a thin, flexible **pellicle**. The pellicle has a uniform appearance, being composed of a lattice of hexagonally shaped pits each perforated by a pair of **cilia**. The whole body of the organism is covered with cilia, which are generally arranged in longitudinal rows, diagonally aligned along its length (fig 3.17). Perforating the ridges of each pit are holes which provide the exits for flask-shaped structures called **trichocysts**. When stimulated, fine, sharply tipped threads are discharged from them. They serve as a means of anchorage during feeding activity (fig 3.18).

Beneath the pellicle is a clear layer of firm plasmagel, the ectoplasm, which exhibits considerable complexity. **Kinetosomes**, the structures from which cilia are formed, are found here. A single fibril (kinetodesmal fibril) arises from each kinetosome and extends forwards and obliquely to its right. It joins other fibrils from adjacent kinetosomes to form a longitudinal bundle of striated fibrils, called a **kinetodesma**. The kinetosomes and fibrils of a particular row are collectively known as a **kinety**. There is also a dense network of fibrils situated in the endoplasm near the cytostome. This is the **motorium**. It has branches which interconnect with the fibrils of the ectoplasm, and the whole fibrillar system is thought to be the controlling centre for ciliary activity.

At the boundary between the ectoplasm and the more granular endoplasm are large bundles of microfilaments called **M fibres** (myonemes). These are contractile in their activity and promote a change of shape in *Paramecium* to enable it to squeeze through narrow spaces.

Near the anterior end of the animal, and on its ventral surface, is a permanent ciliated shallow depression called the **oral groove**. It extends backwards, and tapers into a much narrower tube-like gullet at the end of which is a portion of naked endoplasm, the **cytostome** (mouth). Within the gullet are rows of closely packed cilia arranged into sheets (undulating membranes). When feeding, the cilia of the oral groove suck a current of water, containing bacteria and other suspended particles, into the gullet. The gullet cilia drive the food into the cytostome. Here the food particles, together with a drop of water are enclosed in a food vacuole and ingested into the endoplasm. These vacuoles move away from the cytostome and begin to follow a distinct pathway through the endoplasm as a result of cyclosis (the circulation of organelles within the cell cytoplasm). Any indigestible material is egested at a fixed point, the **cytoproct**, by active vacuolar activity (exocytosis).

Two fixed contractile vacuoles are present in the endoplasm. Both are dorsally situated, one at the anterior end, the other at the posterior end. Around each contractile vacuole are a number of radial canals which fill with fluid before emptying into the main vacuole. The posterior contractile vacuole empties and fills at a faster rate than the anterior one, because of greater endosmosis (intake of water) in the region of the gullet.

Dorsal to the gullet, and towards the centre of the body, lie the **two nuclei**. The larger, bean-shaped **macronucleus** is polyploid; it controls metabolism and differentiation within the animal. The **micronucleus** is diploid; it controls reproductive activity and gives rise to new macronuclei. It is always active when nuclear reorganisation takes place during the life history of the animal.

Paramecium swims by the rhythmic beating of its cilia. Each cilium beats a little in advance of the one immediately behind so that waves of ciliary activity pass over the animal (metachronal rhythm). The direction of each wave is slightly oblique, causing the animal to swim in a spiral manner and at the same time to rotate about its longitudinal axis.

Detection of external stimuli probably takes place through the cilia, especially the stiff, non-locomotory ones at the posterior end. *Paramecium* is sensitive to touch, different concentrations of chemicals, oxygen and carbon dioxide levels, and changes in light intensity. If it encounters unfavourable conditions, or meets an obstruction, *Paramecium* is able to stop its cilia beating, reverse

(a)

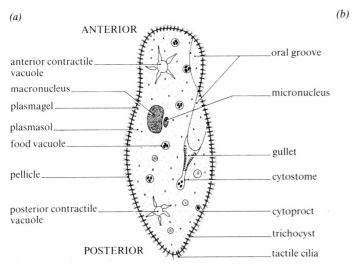

(b)

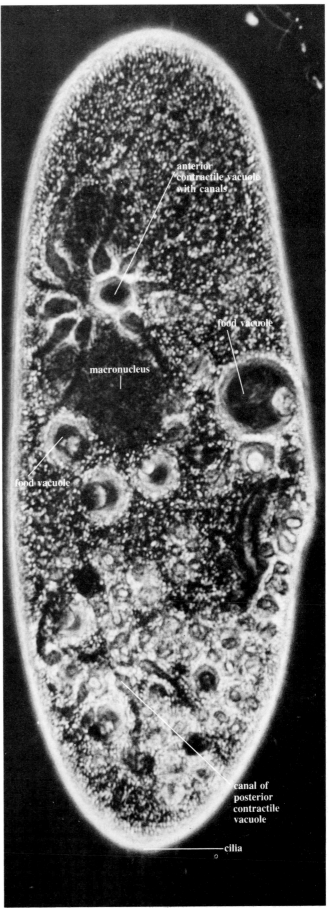

Fig 3.16 *(a) Paramecium caudatum – structures visible under the light microscope. (b) Micrograph of Paramecium caudatum showing structural details (× 832)*

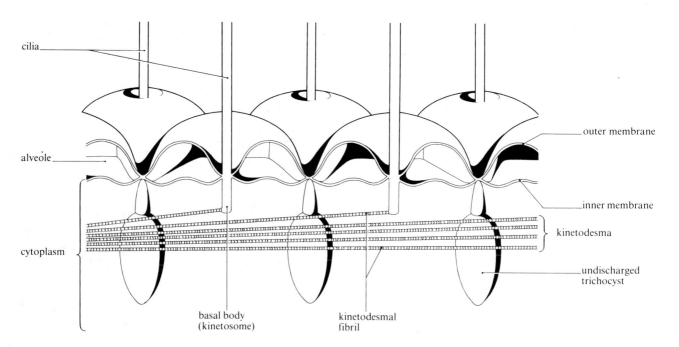

cilia

alveole

cytoplasm

outer membrane

inner membrane

kinetodesma

undischarged
trichocyst

basal body
(kinetosome)

kinetodesmal
fibril

Fig 3.17 *Pellicle and infraciliature of* Paramecium caudatum, *taken from electron micrographs and based upon the work of Grell, Ehret and Powers*

their beat, and then swim forward at an angle to its original path. This is continued until a clear route, or favourable conditions, are restored. It is an example of trial and error behaviour and tends to keep the animal in an optimum environment. The position of the motorium near the mouth, and its connections with the kinetodesmata make it ideally placed to receive and respond in advance of conditions ahead of the animal (fig 3.19).

Most species of *Paramecium* divide asexually by means of transverse binary fission. Both nuclei increase in size elongate and pull themselves apart. The macronucleus divides amitotically, randomly distributing its chromosomes between the two newly formed macronuclei. The micronucleus undergoes mitosis. A spindle forms within its nuclear envelope and very small chromosomes are shared between the two daughter micronuclei in the normal manner. A cytoplasmic constriction forms around the middle of the animal which finally ruptures, producing two daughter paramecia. By the time separation is complete, both daughters have a full complement of organelles.

At times of food shortage, a form of sexual reproduction called **conjugation** occurs (fig 3.20). It only takes place between compatible mating types of the same species. Meiosis and nuclear exchange take place resulting in the production of offspring with a wide variety of genotypes. The process is as shown in fig 3.20.

(1) Two different compatible mating types (**conjugants**) adhere to each other at their oral grooves.
(2) The pellicle breaks down and a **cytoplasmic bridge** is

established between them. Attachment can last several hours. Respective macronuclei disintegrate. Each micronucleus divides meiotically to form four daughter micronuclei.
(3) Three micronuclei disintegrate and disappear. Which shall disappear and which survive depends on their relative positions in the cytoplasm.
(4) The remaining nucleus in each conjugant divides once mitotically to form two identical gametic nuclei. One gamete remains stationary ('female' nucleus), whilst the other ('male' nucleus) migrates via the cytoplasmic bridge into the opposite conjugant.
(5) The male and female nuclei fuse to form a **zygotic nucleus** (synkaryon). Exchange of genetic material is now complete.
(6) Conjugants separate and are called ex-conjugants. The zygotic nucleus of each divides mitotically to form eight daughter nuclei.
(7) Four become macronuclei and four micronuclei; then three micronuclei degenerate.
(8) Binary fission of each ex-conjugant takes place. Two macronuclei enter each new cell and each micronucleus undergoes mitosis.
(9) Further binary fission results in separation of macronuclei and another mitotic division of the micronucleus. The end result is four daughter paramecia formed from each ex-conjugant.

Continued amitotic division of the macronucleus during binary fission leads to unequal distribution of its chromosomes. This upsets the normal coordinated activity of *Paramecium* and leads to '**depression**'. The situation can be rectified by a process called **autogamy**, a modified form of sexual reproduction which occurs every 3–4 weeks (fig 3.21). It leads to the production of new macronuclei containing the 'normal' number of chromosomes, and is a

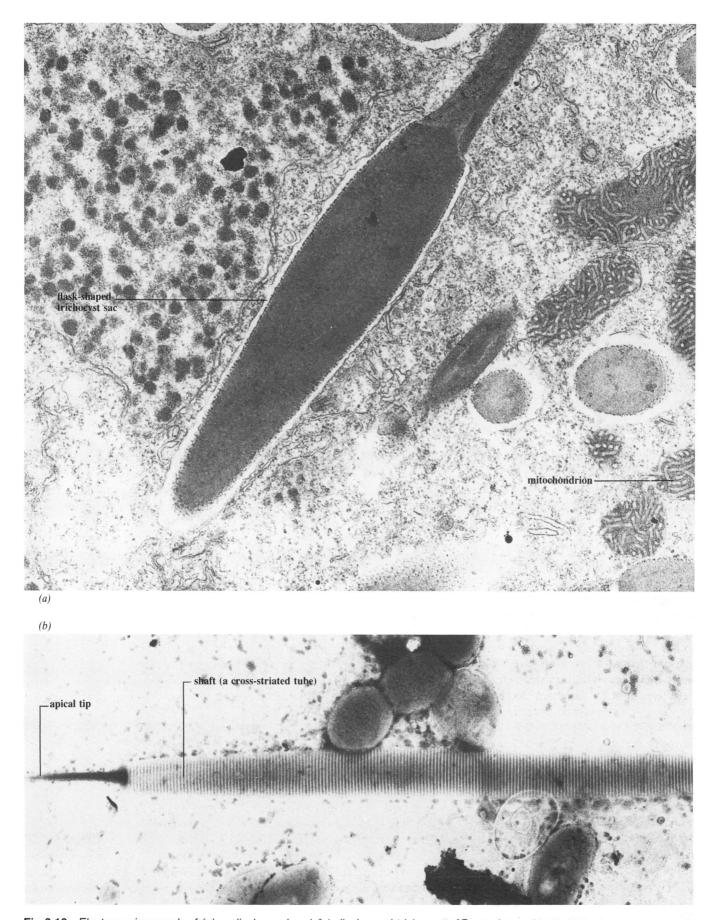

(a)

flask-shaped
trichocyst sac

mitochondrion

(b)

shaft (a cross-striated tube)

apical tip

Fig 3.18 *Electron micrograph of (a) undischarged and (b) discharged trichocyst of* P. caudatum *(× 50 000)*

type of self-fertilisation occurring within a single individual. It occurs as shown in fig 3.21.

(1) The micronucleus divides into eight haploid nuclei. Six of the nuclei break down. The macronucleus degenerates.
(2) The two remaining nuclei fuse to form a zygotic nucleus.

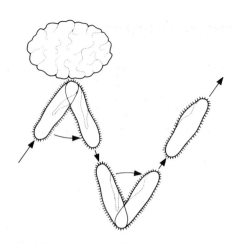

Fig 3.19 *Avoiding reaction of* Paramecium

(3) The zygotic nucleus divides twice to form four nuclei, two of which become macronuclei and two micronuclei.
(4) Binary fission occurs to produce two daughter paramecia. Normal nuclear condition of the macronucleus has thus been restored.

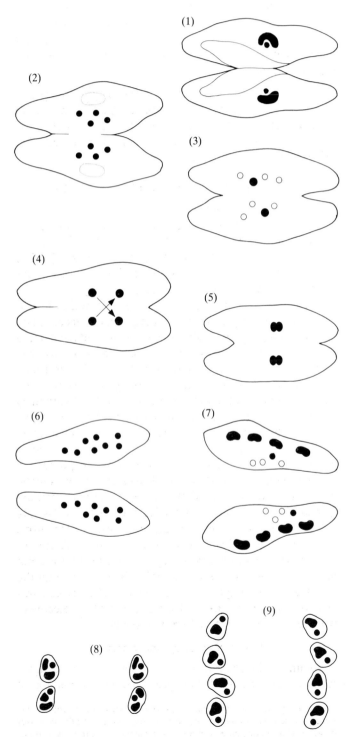

Fig 3.20 *Conjugation in* Paramecium caudatum

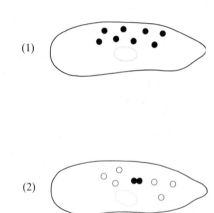

Fig 3.21 *Autogamy in* Paramecium caudatum

Kingdom Plantae

Plants are described as **autotrophic eukaryotes** at the beginning of this chapter. This definition excludes fungi since fungi are heterotrophic. For the purposes of this book, fungi have been classified as a separate kingdom. The only other autotrophic eukaryotes, apart from plants, are algae.

The *International Code of Botanical Nomenclature* recommends that each major group of the plant kingdom should be called a division, rather than a phylum as in the animal kingdom. Strictly speaking the terms phylum and division are not equivalent, but for the sake of simplicity phylum is used rather than division in this book.

3.6 Phylum Bryophyta – liverworts and mosses

There are fossil records of blue-green bacteria living 3000 million years ago and eukaryotic organisms have existed for more than 1000 million years. However, the first organisms to colonise the land, primitive plants, did not do so until about 420 million years ago. Probably the greatest single problem to overcome in making the transition from water to land is that of desiccation. Any plant not protected in some way, for example by a waxy cuticle, will tend to dry out and die very rapidly. Even if this difficulty is overcome, there remain other problems, notably that of successfully achieving sexual reproduction. In the algae this usually involves a male gamete which must swim in water to reach the female gamete.

The first plants to colonise the land are generally thought to have evolved from the green algae (fig 3.3), a few advanced members of which evolved reproductive organs, namely archegonia (female) and antheridia (male), that enclosed and thus protected the gametes within. This, and certain other factors that helped to prevent desiccation, enabled some of them to invade the land.

One of the main themes that will be stressed when considering the different groups of land plants will be their gradually increasing independence from water.

The main problems associated with the transition from an aquatic to a terrestrial environment are summarised below.

Desiccation. Air is a drying medium and water is essential for life for many reasons (section 5.1.2). Means of obtaining water and conserving it are required.

Reproduction. Delicate sex cells must be protected and motile male gametes (sperm) require water if they are to reach the female gametes.

Support. Air, unlike water, offers no support to the plant body.

Nutrition. Plants require light and carbon dioxide for photosynthesis, so at least part of the body must be above ground. Minerals and water, however, are at ground level or below ground, and to make efficient use of these, part of the plant must grow below ground in darkness.

Gaseous exchange. For photosynthesis and respiration, carbon dioxide and oxygen must be exchanged with the atmosphere rather than a surrounding solution.

Environmental variables. Water, particularly large bodies of water like lakes and oceans, provides a very constant environment. A terrestrial environment, however, is much more subject to changes in important factors such as temperature, light intensity, ionic concentration and pH.

It will be seen in the remainder of this chapter that plants have successfully exploited the land by gradual changes in structure and function. It is the main changes that the student should try to understand rather than the detailed differences between plants.

3.6.1 Classification and characteristics of Bryophyta

The simplest group of land plants is the phylum Bryophyta, which includes two main classes, the Hepaticae (liverworts) and the Musci (mosses). The classification and characteristics of the Bryophyta are summarised in table 3.6.

The bryophytes are relatively poorly adapted to life on land, so are mainly confined to damp, shady places. They are small simple plants, with strengthening and conducting tissues absent or poorly developed. There is no true vascular tissue (xylem or phloem). They lack true roots, being anchored by thin filamentous outgrowths of the stem called **rhizoids**. Water and mineral salts can be absorbed by the whole surface of the plant, including the rhizoids, so that the latter are mainly for anchorage, unlike true roots. (*True* roots also possess vascular tissue, as do *true* stems and leaves.) Thus the stems and leaves of bryophytes are not homologous with stems and leaves of vascular plants, where they are part of a diploid sporophyte not a haploid gametophyte. The plant surface lacks a cuticle, or has only a delicate one, and so has no barrier against loss (or entry) of water. Nevertheless, most bryophytes have adapted to survive periods of dryness using mechanisms that are not fully understood. For example, it has been shown that the well-known xerophytic moss *Grimmia pulvinata* can survive total dryness for longer than a year at 20 °C. Recovery is rapid as soon as water becomes available.

Alternation of generations

In common with all land plants* and some advanced algae, such as *Laminaria*, bryophytes exhibit **alternation of**

---

* All plant groups are terrestrial, although a few species have returned to water as a secondary adaptation, such as aquatic ferns and the flowering plant *Zostera*. 'Land plants' will refer to all plants. (Note that algae are no longer classified as plants.)

Table 3.6 Classification and characteristics of the Bryophyta.

Phylum Bryophyta

General characteristics
Alternation of generations in which the gametophyte generation is dominant
No vascular tissue, that is no xylem or phloem
Body is a thallus, or differentiated into simple 'leaves' and 'stems'
No true roots, stems or leaves: the gametophyte is anchored by filamentous rhizoids
Sporophyte is attached to, and is dependent upon, the gametophyte for its nutrition
Spores are produced by the sporophyte in a spore capsule on the end of a slender stalk above the gametophyte
Live mainly in damp, shady places

Class Hepaticae (or *Hepaticopsida*) (liverworts)	*Class Musci* (or *Bryopsida*) (mosses)
Gametophyte is a flattened structure that varies from being a thallus(rare) to 'leafy' with a stem(majority), with intermediate lobed types	Gametophyte 'leafy' with a stem and generally more differentiated than liverworts
'Leaves' (of leafy types) in three ranks along the stem	'Leaves' spirally arranged
Rhizoids unicellular	Rhizoids multicellular
Capsule of sporophyte splits into four valves for spore dispersal: elaters aid dispersal	Capsule of sporophyte has an elaborate mechanism of spore dispersal, dependent on dry conditions and involving teeth or pores; elaters absent

e.g. *Pellia*, a thallose liverwort
   Marchantia, a thallose liverwort, with antheridia and archegonia on stalked structures above the thallus
   Lophocolea, a leafy liverwort, common on rotting wood
	e.g. *Funaria*
	   Mnium, a common woodland moss similar in appearance to *Funaria*
	   Sphagnum, bog-moss: forms peat in wet acid habitats (bogs)

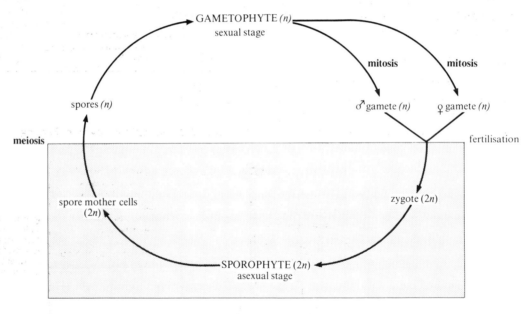

Fig 3.22 *Generalised life cycle of a plant showing alternation of generations. Note the haploid stages (n) and diploid stages (2n). The gametophyte is always haploid and always produces gametes by mitosis. The sporophyte is always diploid and always produces spores by meiosis*

generations. Two types of organism, a **haploid gametophyte** generation and a **diploid sporophyte** generation, alternate in the life cycle (the cycle from the zygote of one generation to the zygote of the next generation). The cycle is sum-marised in fig 3.22. The haploid generation is called the gametophyte (*gameto*, gamete; *phyton*, plant) because it undergoes sexual reproduction to produce gametes. Production of gametes involves mitosis, so the gametes are also haploid. The gametes fuse to form a diploid zygote which grows into the next generation, the diploid sporophyte generation. It is called the sporophyte because it undergoes asexual reproduction to produce spores. Production of spores involves meiosis, so that there is a return to the haploid condition. The haploid spores give rise to the gametophyte generation.

One of the two generations is always more conspicuous and occupies a greater proportion of the life cycle; this is said to be the **dominant generation**. In the bryophytes, the gametophyte generation is dominant. In all other land plants the sporophyte generation is dominant. It is customary to place the dominant generation in the top half of the life cycle diagram.

Fig 3.22 should be studied carefully because it summarises the life cycle of all land plants, including the flowering plants, which are the most advanced. One point that must be remembered is that gamete production involves mitosis, not meiosis as in animals; meiosis occurs in the production of spores.

3.6.2 Class Hepaticae – liverworts

Characteristics of the Hepaticae are summarised in table 3.6. They are more simple in structure than mosses and, on the whole, more confined to damp and shady habitats. They are found on the banks of streams, on damp rocks and in wet vegetation. Most liverworts show regular lobes or definite 'stems' with small, simple 'leaves'. The simplest of all though are the thalloid liverworts where the body is a flat thallus with no stem or leaves. One of these, *Pellia*, is used as an example.

Pellia is a liverwort common throughout Britain. The plant is a dull green with flat branches about 1 cm wide. Its external features are shown in fig 3.23.

3.6.3 Class Musci – mosses

Characteristics of the mosses are summarised in table 3.6. They have a more differentiated structure than liverworts but, like liverworts, are small and found mainly in damp habitats. They often form dense cushions.

Funaria is a common moss of fields, open woodland and disturbed ground, being one of the early colonisers of such ground. It is especially associated with freshly burned areas, for example after heath fires. It is also a common weed in greenhouses and gardens. Its external features are illustrated in fig 3.26.

Life cycle

The sex organs, male antheridia and female archegonia, are found at the tips of separate shoots on the gametophyte generation (see fig 3.25). Within them male and female gametes are produced from gamete mother cells whose nuclei divide by mitosis. A diagrammatic summary of the life cycle is shown in fig 3.24. Note that the sex organs protect the developing gametes from desiccation. The

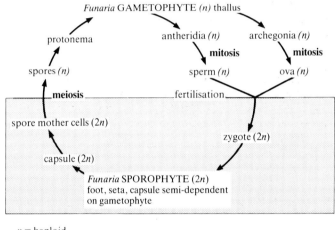

n = haploid

$2n$ = diploid

Fig 3.24 *Diagrammatic life cycle of* Funaria

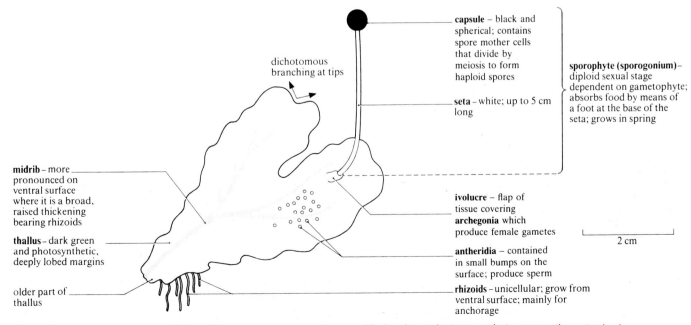

Fig 3.23 *External features of* Pellia. *The gametophyte is shown with the dependent sporophyte generation attached*

61

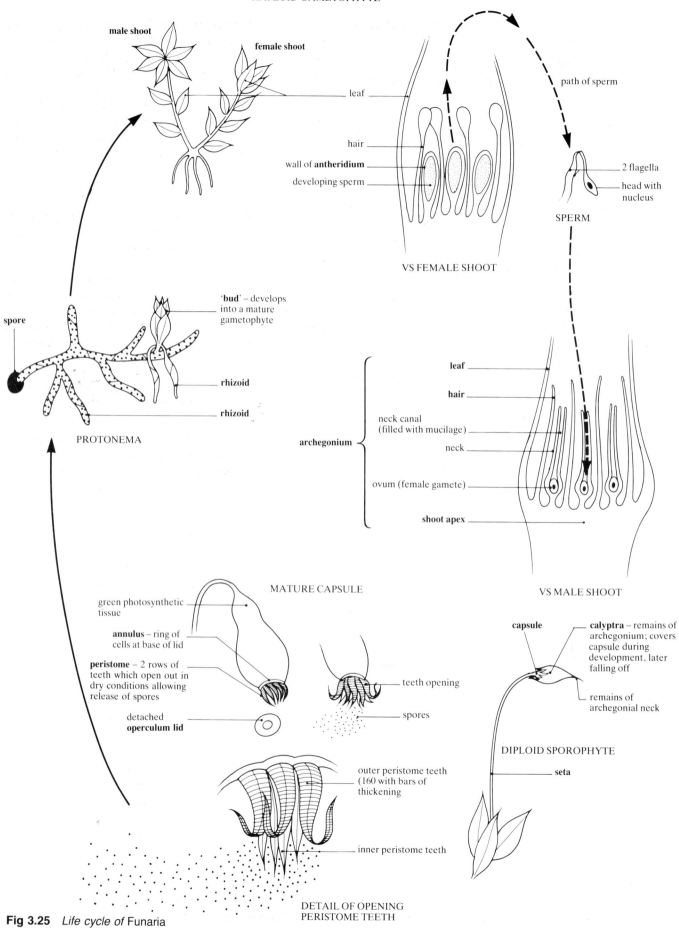

HAPLOID GAMETOPHYTE

male shoot

female shoot

leaf

path of sperm

hair

wall of **antheridium**

developing sperm

2 flagella

head with nucleus

SPERM

VS FEMALE SHOOT

spore

'bud' – develops into a mature gametophyte

rhizoid

rhizoid

PROTONEMA

leaf

hair

neck canal (filled with mucilage)

neck

archegonium

ovum (female gamete)

shoot apex

VS MALE SHOOT

MATURE CAPSULE

green photosynthetic tissue

annulus – ring of cells at base of lid

peristome – 2 rows of teeth which open out in dry conditions allowing release of spores

detached **operculum lid**

teeth opening

spores

capsule

calyptra – remains of archegonium; covers capsule during development, later falling off

remains of archegonial neck

DIPLOID SPOROPHYTE

seta

outer peristome teeth (160 with bars of thickening

inner peristome teeth

DETAIL OF OPENING PERISTOME TEETH

Fig 3.25 *Life cycle of* Funaria

ovum is also protected by mucilage found in the neck canal of the archegonium.

Fertilisation.   Water is essential for fertilisation. When the surface of the plant is wet, mature antheridia absorb water and burst, releasing the male gametes (antherozoids or sperms) on to the surface. The sperms are biflagellate (have two flagella) and are produced in such large numbers that the fluid covering the plant has a milky appearance. They swim towards the archegonia, attracted by sucrose secreted by the archegonial necks. Such movement is an example of chemotaxis (section 15.1.2). Sperms swim down the neck of each archegonium to the venter at its base, which contains the female gamete or ovum. Fertilisation, that is fusion of the sperm nucleus with the ovum nucleus, takes place in the venter and the product is a diploid zygote.

Development of the zygote.   Usually only one zygote develops. The zygote grows into a sporophyte, consisting of a foot, seta (stalk), and capsule. The foot grows back into the gametophyte and acts as an absorptive organ through which it obtains nutrients (fig 3.26). The developing sporophyte contains chloroplasts and therefore appears green and is capable of producing some of its own food requirements by photosynthesis. It is therefore, only semi-dependent on the gametophyte. As the sporophyte grows, the archegonium enlarges at first to contain it and is called the **calyptra**. Later, the sporophyte seta elongates and the calyptra ruptures and remains like a cap on the capsule until the latter is almost mature (fig 3.25).

Asexual reproduction.   The sporophyte capsule contains spore mother cells that divide by meiosis to produce haploid spores. When mature the thin-walled cells of the annulus swell with water and force the lid off the capsule, exposing an inner and an outer ring of teeth (inner and outer peristome). Each ring contains 16 teeth. The outer teeth have specially thickened (lignified) walls which result in the teeth curling inwards and closing over the end of the capsule in damp conditions. In dry conditions they curl back and the inner teeth part, so allowing escape of spores (fig 3.25). Dry conditions favour dispersal (rain would quickly carry spores to the ground). The fact that the capsule is borne about 3 cm above the gametophyte by the seta also increases the chances of wind currents catching and dispersing the very light spores.

Germination.   The spore germinates on finding a suitable damp habitat and grows to form a filamentous structure called a **protonema**, resembling a filamentous green alga. It produces several 'buds', each of which develops into a moss gametophyte. The life cycle of *Funaria* is summarised in fig 3.25.

Success of adaptation to land

Mosses are well adapted to a terrestrial environment in their mode of spore dispersal, which depends on the drying out of the capsule and the dispersal of small, light spores by wind. However, they still show a great reliance on water for the following reasons.

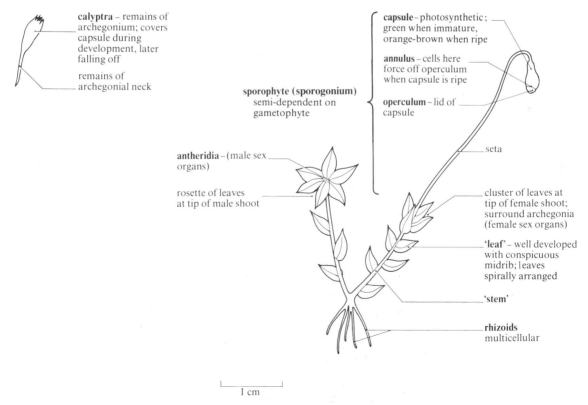

calyptra – remains of archegonium; covers capsule during development, later falling off

remains of archegonial neck

capsule – photosynthetic; green when immature, orange-brown when ripe

annulus – cells here force off operculum when capsule is ripe

operculum – lid of capsule

sporophyte (sporogonium) semi-dependent on gametophyte

seta

antheridia – (male sex organs)

rosette of leaves at tip of male shoot

cluster of leaves at tip of female shoot; surround archegonia (female sex organs)

'leaf' – well developed with conspicuous midrib; leaves spirally arranged

'stem'

rhizoids multicellular

1 cm

Fig 3.26   *Structure of* Funaria. *The gametophyte is shown with the semi-dependent sporophyte generation attached*

(1) They are still dependent on water for reproduction because sperms must swim to the archegonia. They are adapted to release their sperms when water is available since only then do the antheridia burst. They are partly adapted to land because the gametes develop in protective structures, the antheridia and archegonia.

(2) There are no special supportive structures, so the plants are restricted in upward growth.

(3) They are dependent on availability of water and mineral salts close to or at the surface of the soil, because they have no roots to penetrate the substrate. However, rhizoids are present for anchorage, an adaptation to a solid substratum.

> **3.3** Liverworts and mosses have sometimes been described as the amphibians of the plant world. Briefly explain why this should be so.

3.7 Phyla Lycopodophyta (clubmosses), Sphenophyta (horsetails) and Filicinophyta (ferns)

The oldest known of these plants are fossils from the end of the Silurian period, 380 million years old. Whether these plants evolved from bryophytes or independently from algae is not known, but they are the earliest known vascular plants. **Vascular plants** are those containing **vascular tissue**, that is the conducting tissues of xylem and phloem. In order to emphasise what a major advance vascular tissue represents compared with the simple conducting cells of some bryophytes and algae, all vascular plants were formerly included in one division, the **Tracheophyta**, with clubmosses, horsetails and ferns and the more advanced group, the seed-bearing plants, being classified as subdivisions.

Vascular tissue is a feature of the sporophyte generation, that generation which in the bryophytes is small and dependent on the gametophyte. Its occurrence in the sporophyte and not the gametophyte generation is one reason why the sporophyte generation becomes conspicuous in all vascular plants.

Vascular tissue has two important properties of relevance here. Firstly, it forms a **transport system**, conducting food and water around the multicellular body, thus allowing development of large, complex bodies. Secondly, these bodies can be **supported** because xylem, apart from being a conducting tissue, contains lignified cells of great strength and rigidity. In some extinct ferns, xylem developed extensively as a result of secondary growth to form wood, which is the major supporting tissue of trees and shrubs. Another lignified tissue, sclerenchyma, also develops in vascular plants and supplements the mechanical role of xylem (section 8.2.1). The vascular tissue of ferns shows certain basic features compared with flowering plants. The xylem contains tracheids rather than vessels and the phloem contains sieve cells rather than sieve tubes (section 8.2.2).

The earliest known vascular plants, the Psilopsida, a group which is now almost extinct, lacked roots, but these appeared later. Roots penetrate the soil with the result that water can be obtained more easily, the xylem conducting it to other parts of the plant. From the early groups of rooted plants the clubmosses, horsetails and ferns have survived to the present.

Once plant bodies could achieve support above the ground, there must have been competition for light and a tendency for taller forms to evolve. The period following the Silurian, the Devonian, is marked by the appearance of 'trees' up to 3 m tall and sometimes with woody trunks 2 m thick. By the next period, the Carboniferous, great swampy forests of giant clubmosses and horsetails were widespread, eventually giving rise to the coal seams of today. In these forests insects and amphibians first became abundant. Elsewhere, ferns and tree-ferns (not supported by any wood) also occurred, and these were the dominant vegetation for about 70 million years, from the Devonian to the Permian, when conifers and later flowering plants largely replaced them (see the geological time scale in appendix 5).

Despite these advances in adapting to a land environment, which are associated with the sporophyte generation, there remains the major problem of the gametophyte. This is even smaller and more susceptible to desiccation than the bryophyte gametophyte and is called a **prothallus**, dying as soon as it has reproduced to form the sporophyte. It produces sperms which must swim to reach the female gametes.

Heterospory

In some clubmosses and ferns, the gametophyte is protected by remaining in the spores of the previous sporophyte generation. In such cases there are two types of spore and the plants are therefore described as **heterosporous**. Plants producing one type of spore, like the bryophytes, are described as **homosporous**.

Heterosporous plants produce large spores called **megaspores** and small spores called **microspores**. Certain technical terms are applied to the structures associated with spore production and for convenience these are summarised in table 3.7 and fig 3.27.

Megaspores give rise to female gametophytes (prothalli) that bear archegonia, and microspores give rise to male gametophytes (prothalli) that bear antheridia. Sperms produced by the antheridia then travel to a female prothallus. Both male and female prothalli remain protected inside their respective spores. The microspore is small, can be produced in large numbers, and is dispersed by wind from the parent sporophyte; the male prothallus that the microspore contains is therefore dispersed with it.

Table 3.7 Glossary of terms associated with spore production.

strobilus or cone – a collection of sporophylls
sporophyll – a leaf which produces sporangia (*phyllon*, leaf)
megasporophyll – a leaf which produces a megasporangium
microsporophyll – a leaf which produces a microsporangium
sporangium – a structure in which spores are produced by plants; it is associated with asexual reproduction
megasporangium – a sporangium which produces megaspores
microsporangium – a sporangium which produces microspores
megaspore – a relatively large spore which grows to produce a female gametophyte
microspore – a relatively small spore which grows to produce a male gametophyte
homosporous – producing only one type of spore, e.g. *Pellia, Funaria, Dryopteris*
heterosporous – producing two types of spore, megaspores and microspores, e.g. *Selaginella* and all spermatophytes

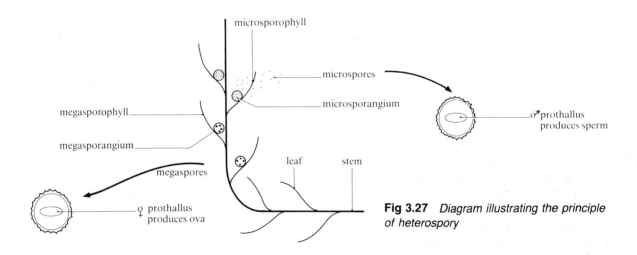

Fig 3.27 *Diagram illustrating the principle of heterospory*

Table 3.8 Classification and characteristics of the ferns, clubmosses and horsetails.

General characteristics
Alternation of generations in which the sporophyte generation is dominant
Gametophyte is reduced to a small, simple prothallus
Sporophyte has true roots, stems and leaves with vascular tissues

Phylum Lycopodophyta (clubmosses)	*Phylum Sphenophyta* (horsetails)	*Phylum Filicinophyta* (ferns)
Leaves relatively small (microphyllous*) and spirally arranged around the stem	Leaves relatively small (microphyllous*) and arranged in whorls around the stem	Leaves relatively large (macrophyllous*) and called fronds; spirally arranged around the stem
Homosporous and heterosporous forms	Homosporous	Homosporous (mostly)
Sporangia usually in strobili (cones)	Sporangia in strobili (cones) on distinctive sporangiophores	Sporangia usually in clusters (sori)
e.g. *Selaginella*, a heterosporous clubmoss *Lycopodium*, a homosporous clubmoss	e.g. *Equisetum* (only surviving genus)	e.g. *Dryopteris filix-mas* (male fern) *Pteridium* (bracken)

* microphyllous – having a single mid-vein. Usually small.   macrophyllous – having branching veins. Large.

The evolution of heterospory is an important step in the evolution of seed-bearing plants, as will be shown later.

3.7.1 Phylum Filicinophyta – ferns

Characteristics of the Filicinophyta are summarised in table 3.8. They are usually restricted to damp, shady habitats. Few ferns are capable of growing in full sunlight, although bracken (*Pteridium*) is a common exception. Ferns are common in tropical rain forests, where temperature, light and humidity are favourable.

The male fern (*Dryopteris filix-mas*) is probably the most common British fern and is found in damp woods, hedgerows and other shady places throughout the country.

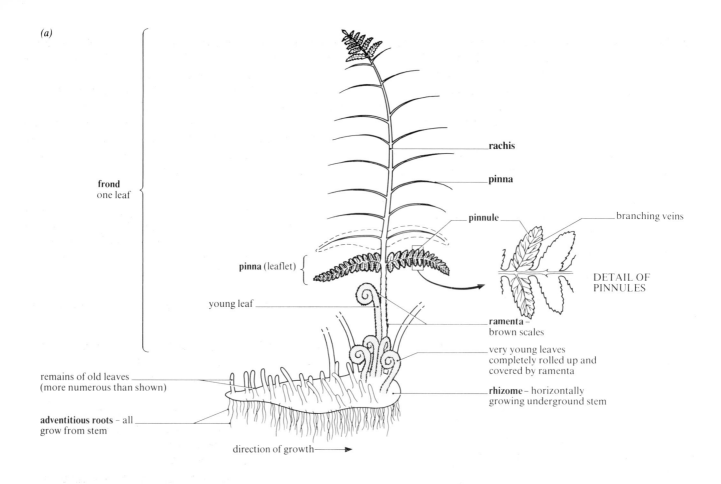

(a)

frond
one leaf

rachis

pinna

pinnule

branching veins

pinna (leaflet)

DETAIL OF
PINNULES

young leaf

ramenta –
brown scales

very young leaves
completely rolled up and
covered by ramenta

remains of old leaves
(more numerous than shown)

rhizome – horizontally
growing underground stem

adventitious roots – all
grow from stem

direction of growth

(b)

(c)

rachis

pinna

pinnule

placenta
sorus

indusium

Fig 3.28 *External features of the sporophyte generation of Dryopteris filix-mas, the male fern. (a) Diagram with details of one pair of pinnae; others have the same structure. (b) The fronds. (c) Underside of frond showing sori (some covered with indusium)*

The **fronds** (leaves) of the sporophyte may reach a metre or more in height and grow from a thick horizontal stem, or **rhizome**. This bears **adventitious roots**. Branches from the main stem may eventually break away and give rise to separate plants, a form of vegetative reproduction. The bases of the fronds are covered with dry brown scales called **ramenta** that protect the young leaves from frost or drought. The young leaves show a characteristic tightly rolled structure. The ramenta gradually become smaller and less dense up the main axis of the frond. This axis is called the **rachis**, and the leaflets either side the **pinnae**. The small rounded subdivisions of the pinnae are called **pinnules**. The external features of the sporophyte of *Dryopteris filix-mas* are shown in fig 3.28.

Life cycle

A diagrammatic summary of the life cycle of *Dryopteris* is shown in fig 3.29.

Asexual reproduction. Spores are produced during late summer in structures called **sporangia**. Sporangia develop in clusters called **sori** on the undersides of pinnules (fig 3.29*a*). Each sorus has a protective covering called an **indusium**. Inside each sporangium diploid spore mother cells divide by meiosis to produce haploid spores. All the spores are identical, so *Dryopteris* is homosporous. When mature, the indusium shrivels and drops off, and the exposed sporangium walls begin to dry out. In each wall is a conspicuous strip of cells, the **annulus**, with thickenings on their inner and radial walls (fig 3.29*b*). The annulus extends only part of the way around the sporangium, and completing the circuit are thin-walled cells forming a region called the **stomium**. As the cells of the annulus dry, their thin outer walls are pulled inwards by the shrinking cytoplasm. The tension thus caused across the whole strip makes the cells of the adjacent stomium suddenly rupture and the annulus curl back. At the moment of rupture spores are catapulted from the sporangium. Eventually the cytoplasm pulls away from the annulus walls altogether, suddenly releasing the tension across the annulus so that it returns violently to its original position, throwing out the remaining spores.

Germination. The spores can remain dormant for a short period and when suitable moist conditions are present germinate to form the gametophyte generation. The gametophyte is a thin heart-shaped plate of cells about 1 cm in diameter (fig 3.29*a*). It is green and photosynthetic and is anchored by unicellular rhizoids to the soil. This delicate prothallus lacks a cuticle and is prone to drying out, so can only survive in damp conditions.

Sexual reproduction. The gametophyte (prothallus) produces simple antheridia and archegonia on its lower surface (fig 3.29*a*). These sex organs protect the gametes within them. Gametes are produced by mitosis of gamete mother cells, the antheridia producing sperm and

each archegonium an ovum, as in the bryophytes. Each sperm has a tuft of flagella. When ripe, and conditions are wet, each antheridium releases its sperm, which swim through a film of water towards the archegonia. This is a chemotactic response to malic acid (2-hydroxybutanedioic acid) secreted by the necks of the archegonia. Cross-fertilisation usually occurs because the antheridia mature before the archegonia. The product of fertilisation is a diploid zygote. Note that fertilisation is still dependent on water as in the bryophytes.

Development of the zygote. The zygote grows into the sporophyte generation. The young embryo develops a foot with which to absorb nutrients from the gametophyte until its own roots and leaves can take over the role of nutrition. The gametophyte soon withers and dies.

The life cycle is summarised non-pictorially in fig 3.30.

3.4 How are ferns better adapted to life on land than liverworts or mosses?

3.5 Which of the following are nutritionally self-supporting?
(*a*) Mature liverwort and moss gametophytes
(*b*) Mature liverwort and moss sporophytes
(*c*) Mature fern gametophytes
(*d*) Mature fern sporophytes

3.6 In what main respects are mosses, liverworts and ferns poorly adapted to life on land?

3.7 How can ferns spread?

3.8 How is the zygote of liverworts (or mosses) and ferns (*a*) protected, (*b*) supplied with food?

3.7.2 Phylum Lycopodophyta – clubmosses

Characteristics of the Lycopodophyta are summarised in table 3.8. Note that, despite their superficial resemblance to mosses, these plants are structurally more advanced than true mosses which are bryophytes. The Lycopodophyta were once far more extensive than they are today, with many tree species as mentioned on page 64. They are intermediate between ferns and seed plants in their adaptation to land.

Selaginella is a mainly tropical genus, with one British species, *S. selaginoides*. It is quite common in mountainous areas of north-west Britain, favouring moist conditions on wet rocks and pastures and being common near streams. It has a creeping stem, that is one that usually lies flat on the ground, with short erect branches. The external features of a common greenhouse species of *Selaginella, S. kraussiana*, are shown in fig 3.31. It has small leaves produced in four rows and arranged in opposite pairs, each pair having one larger (lower) and one smaller (upper) leaf. Each leaf has a **ligule**, a small, membranous outgrowth near its base.

Fig 3.29 *(a) Life cycle of* Dryopteris filix-mas

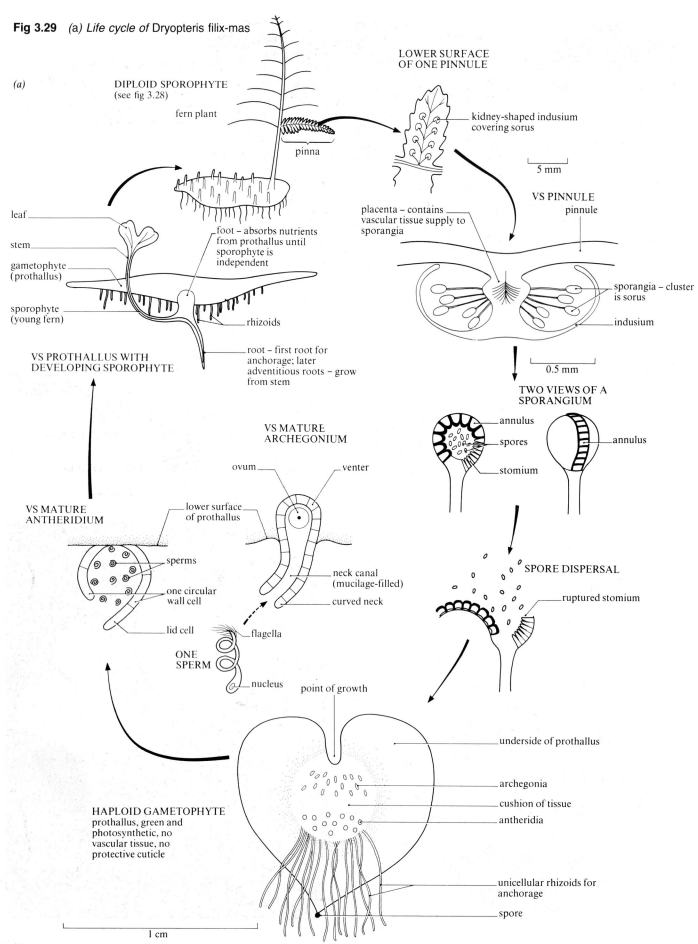

(a)

DIPLOID SPOROPHYTE
(see fig 3.28)

fern plant

pinna

LOWER SURFACE
OF ONE PINNULE

kidney-shaped indusium
covering sorus

5 mm

leaf

stem

gametophyte
(prothallus)

sporophyte
(young fern)

foot – absorbs nutrients
from prothallus until
sporophyte is
independent

rhizoids

root – first root for
anchorage; later
adventitious roots – grow
from stem

VS PROTHALLUS WITH
DEVELOPING SPOROPHYTE

VS PINNULE
pinnule

placenta – contains
vascular tissue supply to
sporangia

sporangia – cluster
is sorus

indusium

0.5 mm

TWO VIEWS OF A
SPORANGIUM

annulus

spores

stomium

annulus

VS MATURE
ARCHEGONIUM

ovum

venter

lower surface
of prothallus

neck canal
(mucilage-filled)

curved neck

VS MATURE
ANTHERIDIUM

sperms

one circular
wall cell

lid cell

ONE
SPERM

flagella

nucleus

SPORE DISPERSAL

ruptured stomium

point of growth

underside of prothallus

archegonia

cushion of tissue

antheridia

HAPLOID GAMETOPHYTE
prothallus, green and
photosynthetic, no
vascular tissue, no
protective cuticle

unicellular rhizoids for
anchorage

spore

1 cm

68

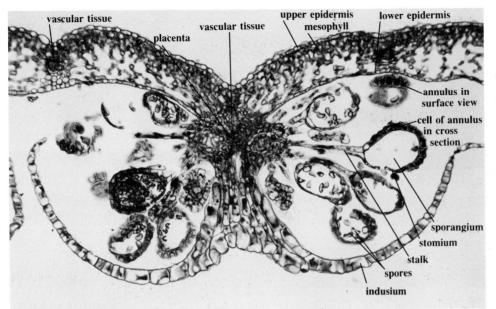

vascular tissue

placenta

vascular tissue

upper epidermis
mesophyll

lower epidermis

annulus in
surface view

cell of annulus
in cross
section

sporangium

stomium

stalk

spores

indusium

(b)

Fig 3.29 (cont.) (b)–(e)
photomicrographs. (b) LS of a sorus.
(c) LS antheridia. (d) LS
archegonium. (e) prothallus with
first frond emerging.

(c)

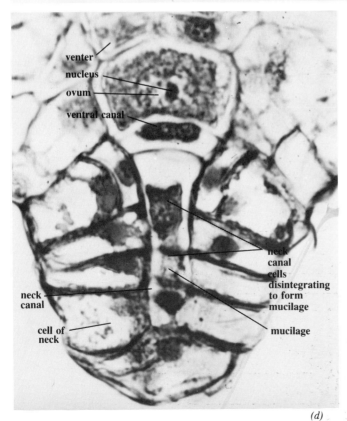

cell of prothallus

stalk cell of antheridium

developing sperms

antheridium

lid cell

sperms

venter

nucleus

ovum

ventral canal

neck
canal
cells
disintegrating
to form
mucilage

neck
canal

cell of
neck

mucilage

1mm

first
sporophyte leaf

sporophyte stem

gametophyte
prothallus

apical notch
of prothallus

(d)

(e)

69

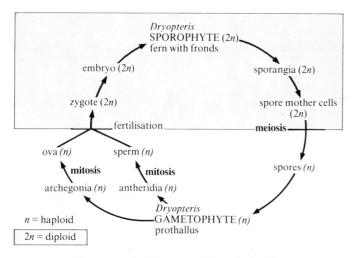

Fig 3.30 *Diagrammatic life cycle of* Dryopteris filix-mas

Root-like structures called **rhizophores** grow down from the stem and branch into adventitious roots at their free ends.

Reproduction involves the production of vertical branches called **strobili**, or **cones**, differing in structure from the rest of the plant. They consist of four vertical rows of leaves of equal size that produce sporangia on their dorsal surfaces, and are therefore **sporophylls**.

Life cycle

A diagrammatic summary of the life cycle of *Selaginella* is shown in fig 3.32. For the meanings of some of the terms used below, consult table 3.7.

Asexual reproduction. *Selaginella* produces strobili or cones as described above. The lower leaves are megasporophylls producing megasporangia, while the upper leaves are microsporophylls producing microsporangia (fig 3.33). Each megasporangium produces four megaspores and each microsporangium produces many microspores; in both cases the spore mother cells undergo meiosis. Since there are two types of spore, *Selaginella* is described as heterosporous.

Spore development and sexual reproduction. The microspores develop into male gametophytes. During development the microspores are released and they may be dispersed or sift down the strobilus to the megasporophylls. The contents of each microspore become a male prothallus, consisting of one vegetative cell and a single antheridium, inside which flagellate sperm are produced by mitosis. The prothallus, a reduced gametophyte generation, is non-photosynthetic and entirely dependent on food stored within the microspore. This food can therefore be traced back to the sporophyte generation.

The megaspores develop into female gametophytes. Again development begins before the spores are shed and the contents of each megaspore become a female prothallus, a reduced gametophyte generation. The top of the prothallus is revealed by the spore splitting. It develops rhizoids and becomes partially green and photosynthetic. However, most of its food, like that of the male gametophyte, comes from a food store in the spore and is derived from the previous sporophyte generation. The female prothallus produces archegonia at its surface, inside each of which an ovum is produced by mitosis.

Note that the mature gametophytes are *not* independent plants, unlike all the previous examples of land plants studied in this chapter. This is an important evolutionary

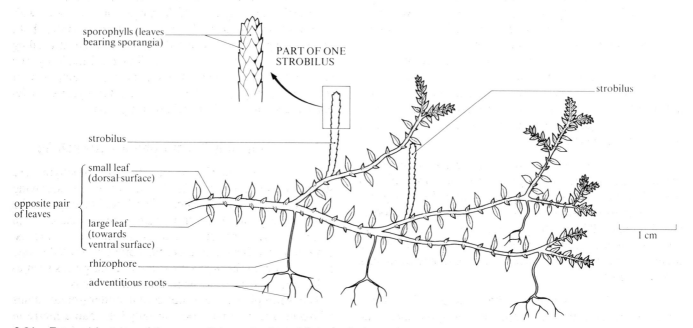

3.31 *External features of the sporophyte generation of* Selaginella kraussiana, *a common greenhouse species. In* Selaginella selaginoides, *the only naturally occurring British species, all leaves are the same size. Also, it lacks rhizophores, having adventitious roots only*

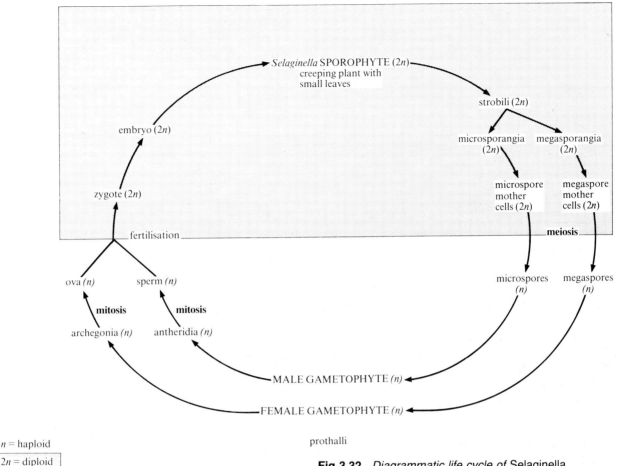

prothalli

Fig 3.32 *Diagrammatic life cycle of* Selaginella

advance, an adaptation to life on land that results in the gametophyte generation, previously vulnerable to desiccation, now being partly protected by the spore. No longer free-living, it is supplied with food stored in the spore by the preceding sporophyte generation.

This advantage is accompanied by the disadvantage that sperm have to travel from the male prothallus, inside the microspore, to the female prothallus inside the megaspore. Self-fertilisation of the gametophyte is no longer possible, and the spores (and hence gametophytes) may be widely separated if dispersed.

Fertilisation. The walls of the microspores rupture and the sperm escape. Moist conditions are still needed for this to occur, and sperm swim to the archegonia of the female prothallus. The latter is still in the megaspore, and this may still be on the parent sporophyte or may have been released. Sperm swim down the neck of the archegonium and one will fuse with the ovum to produce a diploid zygote.

Development of the zygote. The zygote develops into an embryo sporophyte. The upper part of the embryo becomes an elongated structure, the **suspensor**, that pushes the embryo down into the food store of the gametophyte and megaspore. The embryo develops a root,

stem and leaves, obtaining food through an absorptive foot until it is an independent, photosynthetic plant.

Note that the larger size of the megaspore compared with the microspore is due to the food it contains, enabling both the female gametophyte and the embryo of the succeeding sporophyte generation to grow. Thus food made by one sporophyte generation is used in the early development of the next sporophyte generation. The life cycle of *Selaginella* is summarised pictorially in fig 3.33.

3.7.3 Phylum Sphenophyta – horsetails

Characteristics of the Sphenophyta are summarised in table 3.8. *Equisetum* is the only surviving genus and has about 25 species distributed throughout the world (except in Australia). Many are associated with damp or wet habitats, such as ponds and marshes. However, *Equisetum arvense*, the common or field horsetail, is common throughout Britain in drier places such as fields and roadsides, wasteground and gardens.

The sporophytes have horizontal underground stems (rhizomes) and aerial shoots usually less than a metre in height. The shoots have characteristic whorls of small pointed scale leaves at each node. They may be 'sterile' vegetative shoots or 'fertile' shoots bearing strobili (cones).

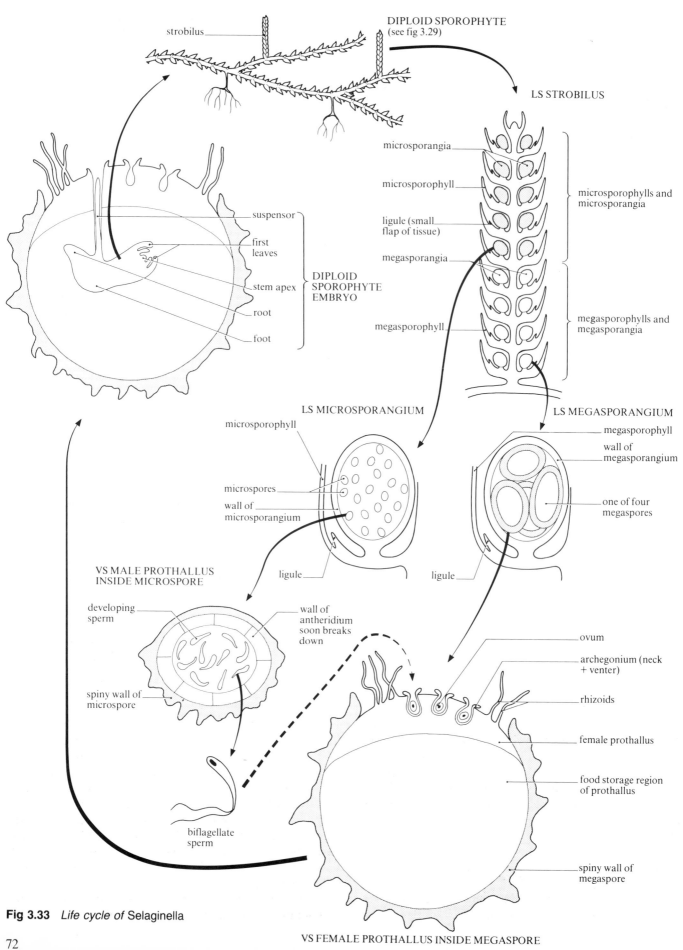

strobilus

DIPLOID SPOROPHYTE
(see fig 3.29)

LS STROBILUS

microsporangia

microsporophyll

ligule (small flap of tissue)

megasporangia

megasporophyll

microsporophylls and microsporangia

megasporophylls and megasporangia

suspensor

first leaves

stem apex

root

foot

DIPLOID SPOROPHYTE EMBRYO

LS MICROSPORANGIUM

microsporophyll

microspores

wall of microsporangium

ligule

LS MEGASPORANGIUM

megasporophyll

wall of megasporangium

one of four megaspores

ligule

VS MALE PROTHALLUS INSIDE MICROSPORE

developing sperm

wall of antheridium soon breaks down

spiny wall of microspore

biflagellate sperm

ovum

archegonium (neck + venter)

rhizoids

female prothallus

food storage region of prothallus

spiny wall of megaspore

Fig 3.33 *Life cycle of* Selaginella

VS FEMALE PROTHALLUS INSIDE MEGASPORE

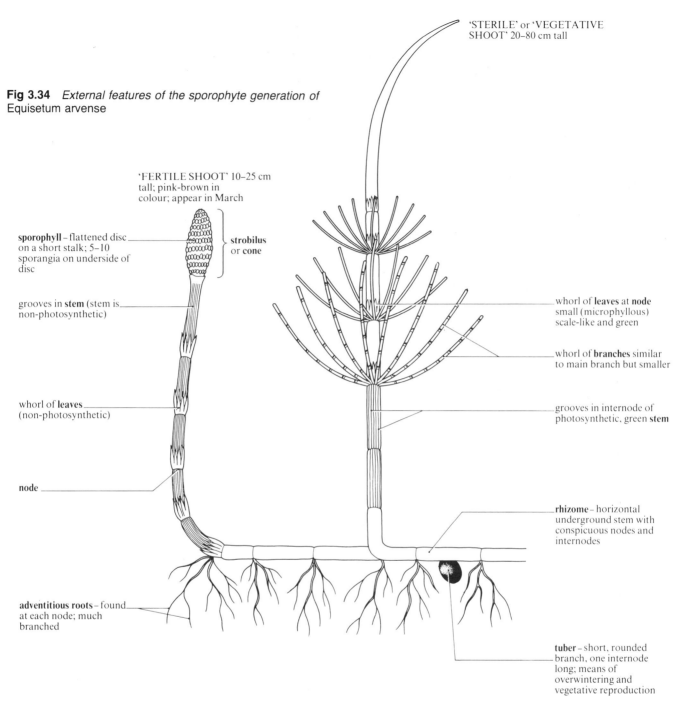

Fig 3.34 *External features of the sporophyte generation of Equisetum arvense*

'STERILE' or 'VEGETATIVE SHOOT' 20–80 cm tall

'FERTILE SHOOT' 10–25 cm tall; pink-brown in colour; appear in March

sporophyll – flattened disc on a short stalk; 5–10 sporangia on underside of disc

strobilus or **cone**

grooves in **stem** (stem is non-photosynthetic)

whorl of **leaves** (non-photosynthetic)

node

adventitious roots – found at each node; much branched

whorl of **leaves** at **node** small (microphyllous) scale-like and green

whorl of **branches** similar to main branch but smaller

grooves in internode of photosynthetic, green **stem**

rhizome – horizontal underground stem with conspicuous nodes and internodes

tuber – short, rounded branch, one internode long; means of overwintering and vegetative reproduction

Sterile shoots are green and have a whorl of narrow branches, as well as scale leaves, at each node. Fertile shoots of most species are pale brown to colourless, unbranched, and with a single strobilus at the apex; in some species they are green and branching. Internodes (the regions between nodes) of all parts have a number of longitudinally running grooves and the main branches are hollow. The external features of *E. arvense* are illustrated in fig 3.34.

3.8   Seed-bearing plants

The most successful group of land plants have seeds. In this section stress will be placed on the adapta-tions which made them successful and they will be compared with the less advanced groups already studied.

The seed-bearing plants probably have their origin among extinct seed-producing members of the clubmosses, horsetails and ferns. If the clubmoss *Selaginella* has been studied it will be noted that its life cycle is essentially the same as that of seed plants, except that in *Selaginella* the megaspore is released whereas in seed plants it is retained. However, it will be assumed in the following discussion that *Selaginella* has not been studied and the life cycle of seed plants will be compared with that of homosporous plants like the ferns.

One of the main problems for plants living on land is the vulnerability of the gametophyte generation. For example, in ferns the gametophyte is a delicate prothallus and it produces male gametes, or sperm, dependent on water for swimming. In seed plants, however, the gametophyte generation is protected and very much reduced. It is only by comparing the life cycle of seed plants with those of more primitive plants that it becomes obvious that alternation of generations still occurs in seed plants. Three important advances have been made by seed plants, first the development of heterospory, secondly the development of seeds, and thirdly the development of non-swimming male gametes.

Heterospory

An important advance towards the seed-plant life cycle came with the evolution of plants that produced two types of spore, microspores and megaspores. Such plants are termed **heterosporous** and are discussed in the introduction (section 3.7). Table 3.7 contains a glossary of terms relating to spore production in the life cycle of heterosporous types (see also fig 3.27). All seed plants are heterosporous.

The contents of a microspore become a male gametophyte, and those of a megaspore become a female gametophyte. In both cases the mature gametophyte is very reduced and is not released from the spore, unlike the free-living independent gametophytes of homosporous plants such as *Dryopteris*. By being retained within the spores, the gametophytes are protected from desiccation, an important adaptation to life on land. They are non-photosynthetic and dependent on food stored in the spores by the preceding sporophyte generation. The most extreme reduction of gametophytes takes place in the flowering plants, as will be discussed later.

Megaspores are produced in megasporangia on megasporophylls and microspores in microsporangia on microsporophylls. The equivalent structure to a megasporangium in a seed plant is called an **ovule**. Within an ovule only

one megaspore or female gametophyte develops, and this is called an **embryo sac**. The equivalent structure to a microsporangium is called a **pollen sac**. Within the pollen sac many microspores develop which are called **pollen grains**.

Evolution of seeds

In seed plants, the megaspores are not released from the sporophyte, unlike the situation in more ancestral heterosporous plants. Instead they are retained in the ovules (megasporangia) still attached to the sporophyte. Inside each megaspore, the female gametophyte (embryo sac) develops and produces one or more female gametes, or ova. Once the female gamete is fertilised, the ovule is called a **seed**. Thus a seed is a fertilised ovule. The ovule, later the seed, has the following main advantages associated with it.

(1) The female gametophyte is protected by the ovule. It is totally dependent upon the parent sporophyte and is not susceptible to desiccation as would be a free-living gametophyte.

(2) After fertilisation it develops a food store, supplied by the parent sporophyte plant to which it is still attached. The food will be used by the developing zygote (the next sporophyte generation) at germination.

(3) The seed is specialised to resist adverse conditions and can remain dormant until conditions are suitable for germination.

(4) The seed may be modified to facilitate dispersal from the parent gametophyte.

The seed is a complex structure because it contains cells from three generations, a parent sporophyte, a female gametophyte and the embryo of the next sporophyte

Fig 3.35 *Relationships between the gametophyte and sporophyte generations in different groups of plants*

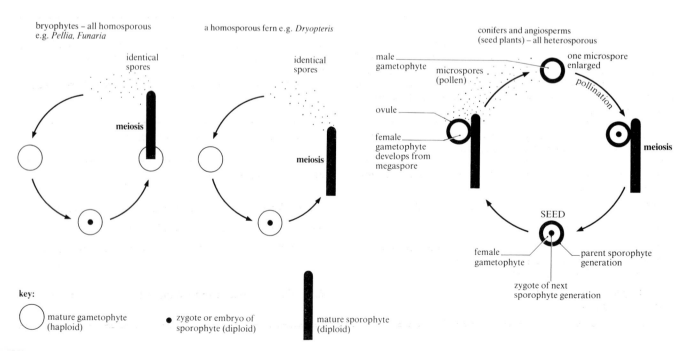

bryophytes – all homosporous
e.g. *Pellia, Funaria*

identical spores

meiosis

a homosporous fern e.g. *Dryopteris*

identical spores

meiosis

conifers and angiosperms
(seed plants) – all heterosporous

male gametophyte

microspores (pollen)

one microspore enlarged

pollination

ovule

female gametophyte develops from megaspore

meiosis

SEED

female gametophyte

parent sporophyte generation

zygote of next sporophyte generation

key:

○ mature gametophyte (haploid)

● zygote or embryo of sporophyte (diploid)

▮ mature sporophyte (diploid)

generation. This is summarised in fig 3.35. All the essentials for life are supplied by the parent sporophyte and it is not until the seed is mature, containing a food store and an embryo sporophyte, that it is dispersed from the parent sporophyte.

Evolution of non-swimming male gametes, and fertilisation independent of water

In the groups previously studied, sexual reproduction has been achieved by sperm swimming in surface moisture to the ova. With seed plants certain problems arise. Male gametes must reach the female gametes for fertilisation to be achieved and it has already been noted that male and female gametophytes develop separately, the latter being retained in the ovules of the sporophyte. Male gametes are produced by the male gametophytes inside the microspores, or pollen grains. Instead of developing as swimming sperm, the male gametes remain non-motile and are carried by pollen grains from the pollen sacs (microsporangia) to the vicinity of the ovules. This transfer is called **pollination**. The final stage of transfer involves growth of a **pollen tube** towards the ovule, down which the non-motile male gametes can pass to achieve fertilisation. At no stage then is water necessary for sperm. Only in a few primitive seed plants, such as the cycads, are sperm released from the pollen tubes, an indication of a link with non-seed-bearing plants. Fig 3.35 compares the life cycle of seed plants with representative non-seed-bearing plants in a way that emphasises the nature of seeds and the relationships between sporophyte and gametophyte generations.

The requirement for pollination is a possible disadvantage since the process is likely to be haphazard and difficult to achieve, and the production of large amounts of pollen is biologically expensive. Originally pollination is thought to have been achieved by wind. However, early in the evolution of seed plants flying insects appeared (in the Carboniferous era about 300 million years ago) bringing the possibility of more efficient pollination by insects. One group of seed plants, the flowering plants, have exploited this method to a high degree.

> **3.9** The chances of survival and development of wind-blown pollen grains (microspores) are much less than those of spores of *Dryopteris*. Why?
>
> **3.10** Account for the fact that megaspores are large and microspores are small.

3.8.1 Classification and characteristics of seed-bearing plants

Classification and characteristics of the seed-bearing plants are summarised in table 3.9.

Table 3.9 shows the two groups of seed-bearing plants, the **conifers** and **angiosperms**. The latter are commonly known as the flowering plants. In conifers ovules, later seeds, are located on the surfaces of specialised leaves called megasporophylls or **ovuliferous scales**. These are arranged in cones. In angiosperms seeds are enclosed, affording even more protection to the gametophyte and subsequent zygote. The structures enclosing the seeds are **carpels** and are thought to be equivalent to megasporophylls (leaves) which are folded up to enclose the ovules (megasporangia). One or more carpels may be present.

The hollow base of a carpel, or group of fused carpels, is

Table 3.9 Classification and characteristics of the seed-bearing plants.

Seed-bearing plants	
General characteristics	
Heterosporous, i.e. two types of spore: microspores and megaspores; microspore = pollen grain, megaspore = embryo sac	
The embryo sac (megaspore) remains completely enclosed in the ovule (megasporangium); a fertilised ovule is a seed.	
Sporophyte is the dominant generation; gametophyte generation is severely reduced	
Water is not needed for sexual reproduction because male gametes do not swim (except in a few primitive members); they are conveyed to the ovum by a pollen tube to effect fertilisation	
Complex vascular tissues in roots, stems and leaves	
Phylum Coniferophyta (mainly conifers; yews, cycads, ginkgos and others belong to different phyla (see below))	*Phylum Angiospermophyta* (flowering plants)
'Naked' seeds: this means that the seeds are exposed, i.e. not enclosed in an ovary	Seeds are enclosed in an ovary
Usually cones on which sporangia and spores develop	Produce flowers in which sporangia and spores develop
No fruit because no ovary	After fertilisation, the ovary develops into a fruit
No vessels in xylem, only tracheids; no companion cells in phloem, only albuminous cells (similar in function to companion cells, but different in origin)	Xylem contains vessels; phloem contains companion cells
Phylum Cycadophyta – cycads *Phylum* Ginkgophyta – ginkgos	*Classes* Dicotyledoneae and Monocotyledoneae (see table 3.9)

called an **ovary**. The ovary encloses the ovules. After fertilisation the ovary is called a **fruit** and the ovules are called **seeds**. Either the fruit or the seed (sometimes both) is modified to assist in dispersal.

Fig 3.36 compares, by means of simple diagrams, the different spore-bearing structures of vascular plants and enables a comparison to be made of some of the terms which have been used.

3.8.2 Phylum Coniferophyta

Characteristics of the Coniferophyta are summarised in table 3.9.

Conifers are a successful group of plants of worldwide distribution, accounting for about one-third of the world's forests. They are trees or shrubs, mostly evergreen, with needle-like leaves. Most of the species are found at higher altitudes and further north than any other trees. Conifers are commercially important as 'softwoods', being used not only for timber but for resins, turpentine and wood pulp. They include pines, larches (which are deciduous), firs, spruces and cedars. A typical conifer *Pinus sylvestris*, the Scots pine, is described below.

Pinus sylvestris is found throughout central and northern Europe, Russia and North America. It is native to Scotland, though it has been introduced elsewhere in Britain. It is planted for ornament and timber, being a stately, attractive tree up to 36 m in height with a characteristic pink to orange-brown flaking bark. It grows most commonly on sandy or poor mountain soils and consequently the root system is often shallow and spreading. Its external features are illustrated in fig 3.37.

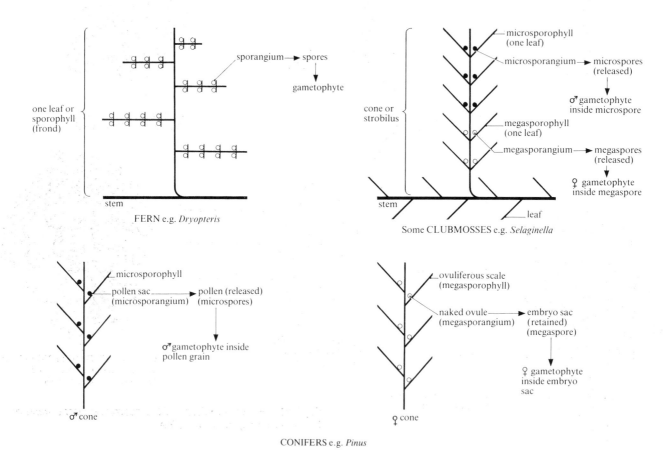

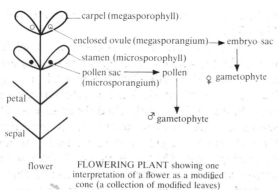

stamen = microsporophyll

pollen sac = microsporangium

pollen = microspores

carpel = megasporophyll

ovule = megasporangium

embryo sac = megaspore

Fig 3.36 *Diagrammatic comparison of spore-bearing structures in vascular plants*

76

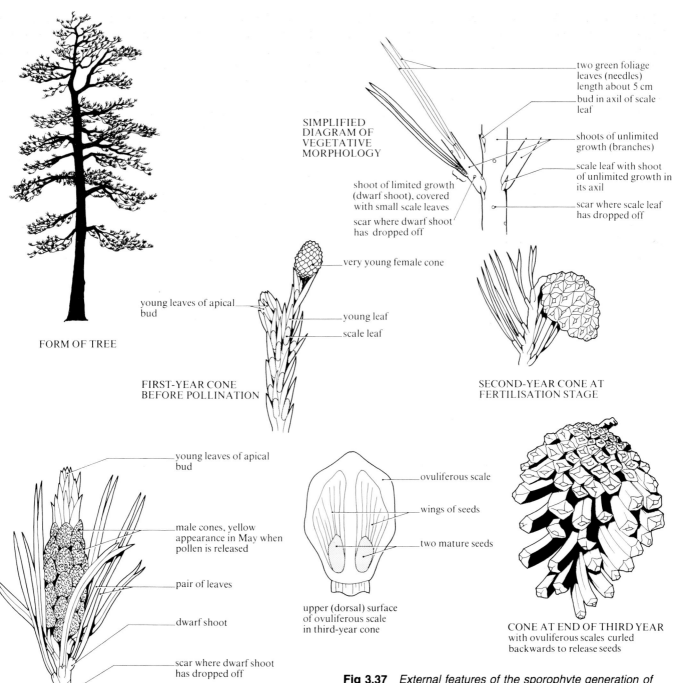

FORM OF TREE

SIMPLIFIED DIAGRAM OF VEGETATIVE MORPHOLOGY

two green foliage leaves (needles) length about 5 cm

bud in axil of scale leaf

shoots of unlimited growth (branches)

scale leaf with shoot of unlimited growth in its axil

scar where scale leaf has dropped off

shoot of limited growth (dwarf shoot), covered with small scale leaves

scar where dwarf shoot has dropped off

very young female cone

young leaves of apical bud

young leaf

scale leaf

FIRST-YEAR CONE BEFORE POLLINATION

SECOND-YEAR CONE AT FERTILISATION STAGE

young leaves of apical bud

male cones, yellow appearance in May when pollen is released

pair of leaves

dwarf shoot

scar where dwarf shoot has dropped off

ovuliferous scale

wings of seeds

two mature seeds

upper (dorsal) surface of ovuliferous scale in third-year cone

CONE AT END OF THIRD YEAR with ovuliferous scales curled backwards to release seeds

GROUP OF MALE CONES

Fig 3.37 *External features of the sporophyte generation of* Pinus sylvestris, *the Scots pine*

Each year a whorl of lateral buds around the stem grows out into a whorl of branches. The roughly conical appearance of *Pinus* and other conifers is due to the transition from whorls of shorter (younger) branches at the tops to longer (older) branches lower down. The latter usually die and drop off as the tree grows, leaving the mature trees bare for some distance up their trunks (fig 3.37).

The main branches and trunk continue growth from year to year by the activity of an apical bud. They are said to show **unlimited growth**. They have spirally arranged scale leaves, in the axils of which are buds that develop into very short branches (2–3 mm) called **dwarf shoots**. These are shoots of **limited growth** and at their tips grow two leaves. Once the shoot has grown, the scale leaf at its base drops off leaving a scar. The leaves are needle-like, reducing the surface area available for the loss of water. They are also covered with a thick, waxy cuticle and have sunken stomata, further adaptations for conserving water. These xeromorphic features ensure that the tree does not lose too much water from its evergreen leaves during cold seasons, when water may be frozen or difficult to absorb from the soil. After two to three years the dwarf shoots and leaves drop off together, leaving a further scar.

The tree is the sporophyte generation and is heterosporous. In spring, male and female cones are produced on the same tree. The male cones are about 0.5 cm in diameter, rounded and found in clusters behind the apical buds at the bases of new shoots. They develop in the axils of scale leaves in the place of dwarf shoots. Female cones arise in the axils of scale leaves at the tips of new strong shoots, at some distance from the male cones and in a more scattered arrangement. Since they take three years to complete growth and development, they are of various sizes, ranging from about 0.5–6 cm on a given tree. They are green when young, becoming brown or reddish-brown in their second year. Both male and female cones consist of spirally arranged, closely packed sporophylls around a central axis (fig 3.37).

Each sporophyll of a male cone has two microsporangia or pollen sacs on its lower surface. Inside each pollen sac, pollen mother cells divide by meiosis to form pollen grains or microspores. Each grain has two large air sacs to aid in wind dispersal. During May the cones become yellow in appearance as they release clouds of pollen. At the end of the summer they wither and drop off.

Each sporophyll of a female cone consists of a lower bract scale and a larger upper ovuliferous scale. On its upper surface are two ovules side by side, inside of which one megaspore mother cell divides by meiosis to produce four megaspores. Only one of these develops. Pollination takes place during the first year of the cone's development, but fertilisation does not take place until the pollen tubes grow during the following spring. The fertilised ovules become winged seeds. They continue to mature during the second year and are dispersed during the third year. By this time the cone is relatively large and woody and the scales bend outwards to expose the seeds prior to wind dispersal.

3.8.3 Phylum Angiospermophyta – flowering plants

Characteristics of the Angiospermophyta are summarised in table 3.9.

Angiosperms are better adapted to life on land than any other plants. After their appearance during the Cretaceous period, 135 million years ago, they rapidly took over from conifers as the dominant land vegetation on a world scale, and proliferated as different habitats were successfully exploited. Some angiosperms even returned to fresh water, and a few to salt water.

One of the most characteristic features of angiosperms, apart from the enclosed seeds already mentioned, is the presence of flowers instead of cones. This has enabled many of them to utilise insects, and occasionally birds or even bats, as agents of pollination. In order to attract these animals, flowers are usually brightly coloured, scented and offer pollen or nectar as food. In some cases the flowers have become indispensable to the insects. The result is that, in some cases, the evolution of insects and flowering plants has become closely linked and there are many highly specialised, mutually dependent, relationships. The flower generally becomes adapted to maximise the chances of pollen transfer by the insect and the process is therefore more reliable than wind pollination. Insect-pollinated plants need not, therefore, produce as much pollen as wind-pollinated plants. Nevertheless, many flowers are specialised for wind pollination.

Life cycle

The life cycle of a typical flowering plant is summarised in fig 3.38.

The purpose of fig 3.38 is to emphasise the links with the life cycles of more primitive plants, a detailed description of the life cycle being reserved for section 20.2. It is still the same, in essence, as that shown in fig 3.22. Note particularly the stages at which meiosis and mitosis occur. Gametes are still produced by mitosis, and spores by meiosis, as in other plants with alternation of generations. Strictly speaking, the flower is an organ of both asexual and sexual reproduction since it produces spores (asexual reproduction) within which gametes are produced (sexual reproduction). Note that a pollen grain is a spore, not a male gamete, as it *contains* male gametes. As noted before, the pollen grain carries the male gametes to the female parts, thus avoiding the need for swimming sperm.

The process of endosperm development is also shown in fig 3.38. The endosperm develops into a food store and the manner of its formation is unique to angiosperms.

Dicotyledons and monocotyledons

The angiosperms are divided into two major groups that are given the status of classes. The most commonly used names for the two groups are the monocotyledons and dicotyledons, usually abbreviated to **monocots** and **dicots**. A summary of the ways in which they differ is given in table 3.10. Few of the differences are diagnostic if used alone, since exceptions may exist, but a combination of several characters would lead to positive identification. The modern view is that monocots probably evolved from ancestral dicots.

Angiosperms may be **herbaceous** (non-woody) or **woody**. Woody plants become shrubs or trees. They grow large amounts of secondary xylem (wood) that offers support, as well as being a conducting tissue, and is produced as a result of the activity of the vascular cambium. Herbaceous plants, or herbs, rely on turgidity and smaller quantities of mechanical tissues such as collenchyma, sclerenchyma and xylem for support, and they are consequently smaller plants. They either lack a vascular cambium or, if present, it shows restricted activity. Many herbaceous plants are **annuals**, completing their life

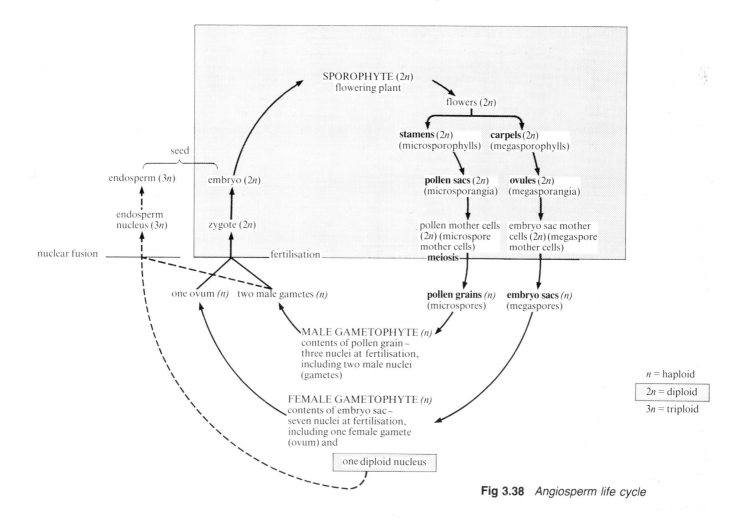

Fig 3.38 *Angiosperm life cycle*

n = haploid
2n = diploid
3n = triploid

Table 3.10 Major differences between dicotyledons and monocotyledons.

	Class Dicotyledoneae	*Class Monocotyledoneae*
Leaf morphology	Reticulate venation (net-like pattern of veins)	Parallel venation (veins are parallel)
	Lamina (blade) and petiole (leaf stalk)	Lanceolate (elongate)
	Dorso-ventral (dorsal and ventral surfaces differ)	Identical dorsal and ventral surfaces
Stem anatomy	Ring of vascular bundles	Vascular bundles scattered
	Vascular cambium usually present, giving rise to secondary growth	Vascular cambium usually absent, so no secondary growth (exceptions occur, e.g. palms)
Root morphology	Primary root persists as a tap root that develops lateral roots (secondary roots)	Adventitious roots from the base of the stem take over from the primary root, giving rise to a fibrous root system
Root anatomy	Few groups of xylem (2–8) (see section 14.5)	Many groups of xylem (commonly up to 30)
	Vascular cambium often present, giving rise to secondary growth	Vascular cambium usually absent, so no secondary growth
Seed morphology	Embryo has two cotyledons (seed leaves)	Embryo has one cotyledon
Flowers	Parts mainly in fours and fives	Parts usually in threes
	Perianth segments usually differ forming a calyx and corolla.	Perianth segments identical, with no distinct calyx and corolla
	Often insect pollinated	Often wind pollinated
Examples	Pea, rose, buttercup, dandelion	Grasses, iris, orchids, lilies

cycles from germination to seed production in one year. Some produce organs of perennation such as bulbs, corms and tubers by means of which they overwinter or survive periods of adverse conditions such as drought (section 20.1.1). They may then be **biennial** or **perennial**, that is they produce their seeds and die in their second year or they survive from year to year. Shrubs and trees are perennial, and may be **evergreen**, producing and shedding leaves all year round so that leaves are always present, or **deciduous**, shedding leaves in seasons of cold or drought.

The morphology (structure) of representative angiosperms will be described in figs 3.40–43 to illustrate their diversity.

(a)

(b)

Fig 3.39 *Structure of a (a) monocot and (b) a dicot leaf*

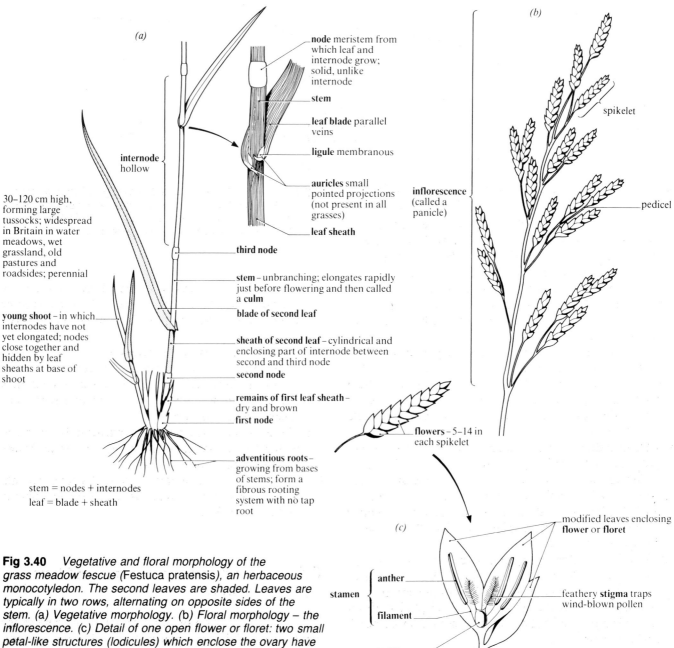

Fig 3.40 *Vegetative and floral morphology of the grass meadow fescue (*Festuca pratensis*), an herbaceous monocotyledon. The second leaves are shaded. Leaves are typically in two rows, alternating on opposite sides of the stem. (a) Vegetative morphology. (b) Floral morphology – the inflorescence. (c) Detail of one open flower or floret: two small petal-like structures (lodicules) which enclose the ovary have been omitted*

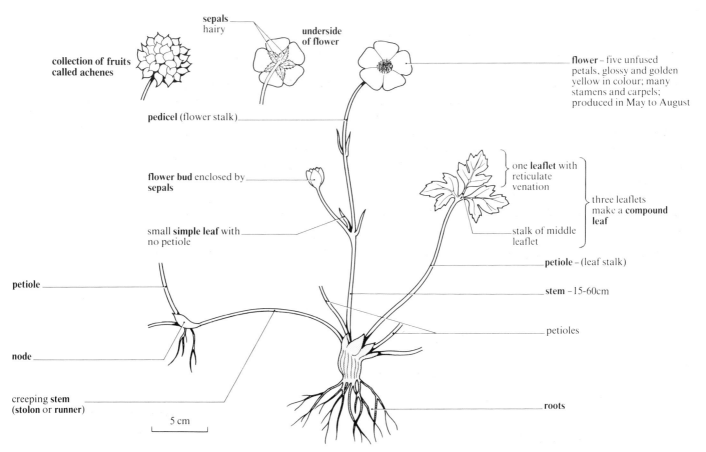

collection of fruits called achenes

sepals hairy

underside of flower

flower – five unfused petals, glossy and golden yellow in colour; many stamens and carpels; produced in May to August

pedicel (flower stalk)

flower bud enclosed by sepals

small **simple leaf** with no petiole

one **leaflet** with reticulate venation

three leaflets make a **compound leaf**

stalk of middle leaflet

petiole – (leaf stalk)

stem – 15-60cm

petiole

petioles

node

creeping **stem** (**stolon** or **runner**)

roots

5 cm

3.8.4 Summary of adaptations of conifers and angiosperms to life on land

In section 3.6 the problems associated with the transition from an aquatic to a terrestrial environment were discussed and summarised. Having studied representatives of the major groups of land plants we can return to consider why the conifers and angiosperms are so well adapted to life on land. Their major advantage over other plants is related to their reproduction. Here they are better adapted in three important ways.

(1) The gametophyte generation is very reduced. It is always protected inside sporophyte tissue, on which it is totally dependent. In mosses and liverworts, where the gametophyte is conspicuous, and in ferns where it is a free-living prothallus, the gametophyte is susceptible to drying out.

(2) Fertilisation is not dependent on water as it is in other plant groups, where sperm swim to the ova. The male gametes of seed plants are non-motile and are carried within pollen grains that are suited for dispersal by wind or insect. Final transfer of the male gametes after pollination is by means of pollen tubes, the ova being enclosed within ovules.

(3) Conifers and flowering plants produce seeds. Development of seeds is made possible by the retention of ovules and their contents on the parent sporophyte.

Other ways in which spermatophytes are adapted to life on land are summarised below and discussed in more detail elsewhere in the book.

(a) Xylem and sclerenchyma are lignified tissues provid-

Fig 3.41 *Vegetative and floral morphology of the creeping buttercup (Ranunculus repens), an herbaceous dicotyledon. It is a common perennial plant throughout Britain, found in wet fields, woods, gardens and on waste ground*

ing support in all vascular plants. Many of these show secondary growth with deposition of large amounts of wood (secondary xylem). Such plants become trees or shrubs.

(b) True roots, also associated with vascular plants, enable water in the soil to be exploited efficiently.

(c) The plant is protected from desiccation by an epidermis with a waterproof cuticle, or by cork after secondary thickening has taken place.

(d) The epidermis of aerial parts, particularly leaves, is perforated by stomata, allowing gaseous exchange between plant and atmosphere.

(e) Plants show other adaptations to hot dry environments (xeromorphic adaptations) as described in sections 18.2.3 and 19.3.2.

3.9 Comparative summary of land plants

Fig 3.44 provides a summary of some of the key features of the land plants discussed in this chapter, with particular emphasis on life cycles.

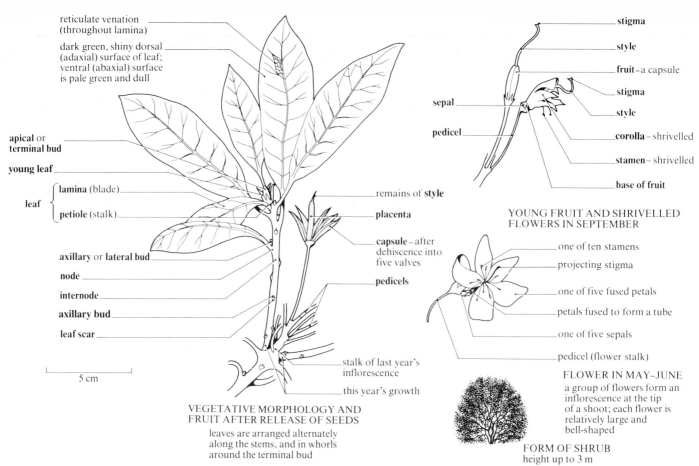

reticulate venation (throughout lamina)

dark green, shiny dorsal (adaxial) surface of leaf; ventral (abaxial) surface is pale green and dull

apical or terminal bud

young leaf

leaf { lamina (blade) / petiole (stalk) }

axillary or lateral bud

node

internode

axillary bud

leaf scar

5 cm

remains of **style**

placenta

capsule – after dehiscence into five valves

pedicels

stalk of last year's inflorescence

this year's growth

VEGETATIVE MORPHOLOGY AND FRUIT AFTER RELEASE OF SEEDS
leaves are arranged alternately along the stems, and in whorls around the terminal bud

stigma

style

fruit – a capsule

stigma

style

sepal

pedicel

corolla – shrivelled

stamen – shrivelled

base of fruit

YOUNG FRUIT AND SHRIVELLED FLOWERS IN SEPTEMBER

one of ten stamens

projecting stigma

one of five fused petals

petals fused to form a tube

one of five sepals

pedicel (flower stalk)

FLOWER IN MAY–JUNE
a group of flowers form an inflorescence at the tip of a shoot; each flower is relatively large and bell-shaped

FORM OF SHRUB
height up to 3 m

Fig 3.42 *Vegetative and floral morphology of the wild* Rhododendron, R. ponticum, *an evergreen dicotyledonous shrub. It is commonly planted in woods and gardens. Originally introduced, it has become naturalised, favouring acid soils (sandy or peaty) on heaths and in woods*

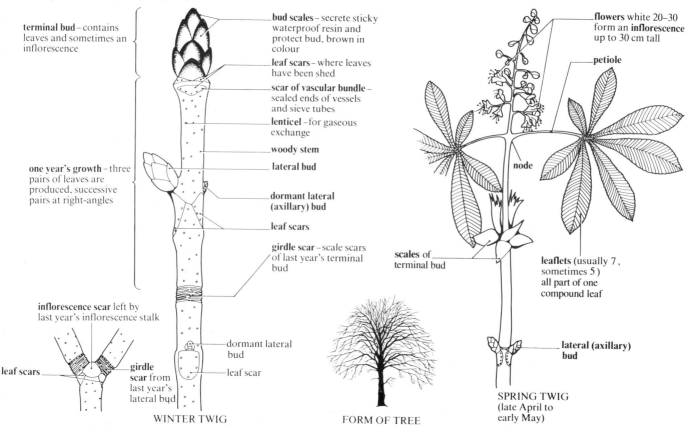

terminal bud – contains leaves and sometimes an inflorescence

one year's growth – three pairs of leaves are produced, successive pairs at right-angles

inflorescence scar left by last year's inflorescence stalk

leaf scars

girdle scar from last year's lateral bud

WINTER TWIG

bud scales – secrete sticky waterproof resin and protect bud, brown in colour

leaf scars – where leaves have been shed

scar of vascular bundle – sealed ends of vessels and sieve tubes

lenticel – for gaseous exchange

woody stem

lateral bud

dormant lateral (axillary) bud

leaf scars

girdle scar – scale scars of last year's terminal bud

dormant lateral bud

leaf scar

FORM OF TREE

flowers white 20–30 form an **inflorescence** up to 30 cm tall

petiole

node

scales of terminal bud

leaflets (usually 7, sometimes 5) all part of one compound leaf

lateral (axillary) bud

SPRING TWIG
(late April to early May)

Fig 3.43 *Vegetative and floral morphology of the horse chestnut* (Aesculus hippocastanum), *a deciduous dicotyledonous tree. The tree may reach 30 m or more in height*

	Phylum BRYOPHYTA (liverworts and mosses) *Pellia*, a liverwort *Funaria*, a moss	Phyla LYCOPODOPHYTA, SPHENOPHYTA and FILICINOPHYTA (clubmosses, horsetails, ferns and others) *Dryopteris*, a fern	*Selaginella*, a clubmoss	Phyla CONIFEROPHYTA and ANGIOSPERMOPHYTA (seed plants) Angiospermophyta (flowering plants)
Presence or absence of vascular tissue (xylem and phloem)	non-vascular	vascular	vascular	vascular
Relative importance of sporophyte and gametophyte generations in life cycle	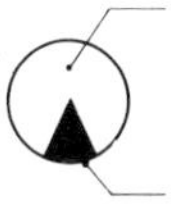 gametophyte *(n)* gametophyte conspicuous sporophyte *(2n)*	 sporophyte conspicuous	 sporophyte conspicuous	 sporophyte conspicuous
Mature gametophyte	thallus; lobed or with simple 'leaves', free-living and photosynthetic	prothallus; free-living and photosynthetic	♀ prothallus retained in megaspore; semi-dependent on sporophyte food; megaspore released from sporophyte ♂ prothallus retained in microspore; dependent on sporophyte food; microspore released from sporophyte	♀ only eight nuclei; retained in megaspore (embryo sac); dependent on sporophyte food; megaspore released from sporophyte ♂ only three nuclei; retained in microspore (pollen grain); dependent on sporophyte food; microspore released from sporophyte
Male gametes	free-swimming sperm; therefore water needed	free-swimming sperm; therefore water needed	free-swimming sperm; therefore water needed	non-motile nuclei inside pollen; pollination followed by growth of pollen tube towards ♀ gamete ensures fertilisation; water not needed
Mature sporophyte	dependent (*Pellia*) or semi-dependent (*Funaria*) on gametophyte; no stem, leaves or roots; only capsule, seta and foot	fern plant with true leaves (fronds), stem and roots	clubmoss with true leaves, stem and roots	flowering plant with true leaves, stem and roots
Homosporous or heterosporous	homosporous	homosporous	heterosporous	heterosporous
Habitat	moist, shady places	often moist and shady	moist	wide range, including very dry habitats (dependent on species)
Life cycle	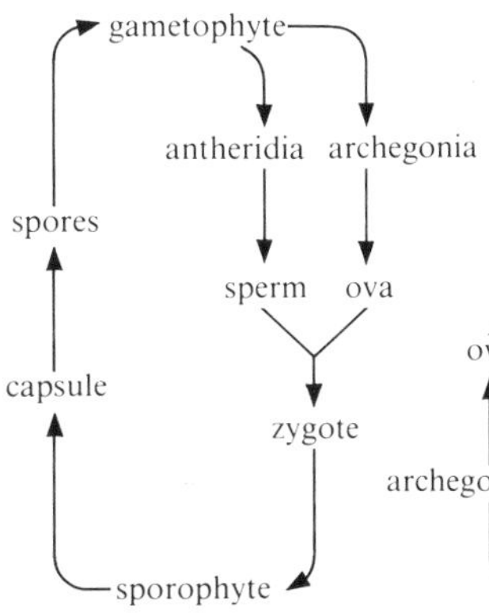 spores produced in a capsule	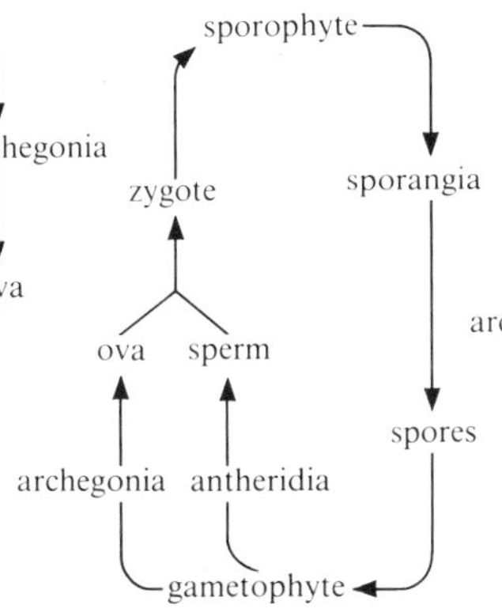 spores produced on unmodified leaves	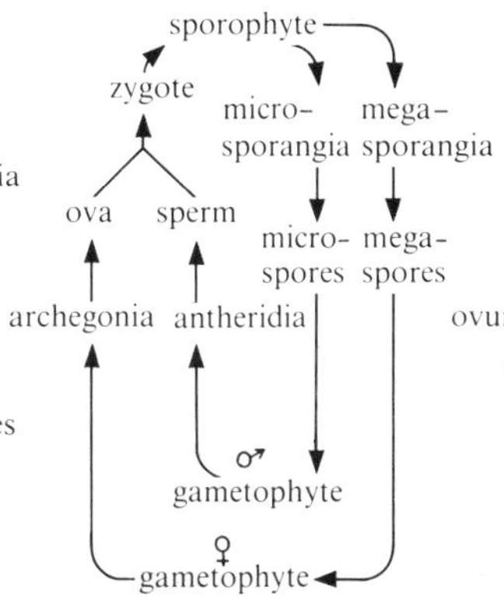 spores produced in cones (strobili)	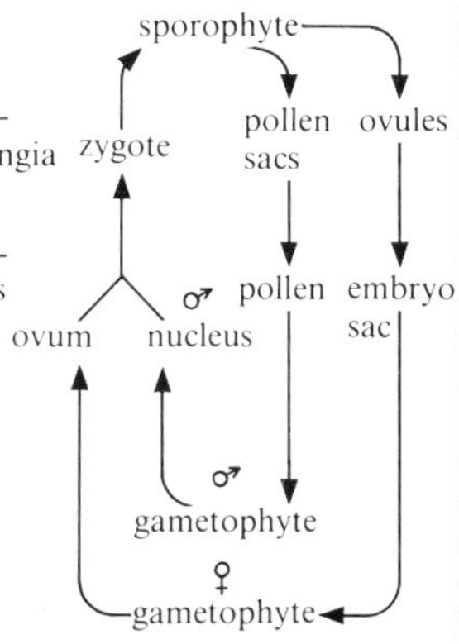 spores (pollen and embryo sac) produced in flowers; no antheridia or archegonia

Fig 3.44 *Comparison of the major groups within the kingdom Plantae*

Chapter Four

Variety of life – animals

4.1   Kingdom Animalia

As discussed at the beginning of chapter 2, it is believed that at some point in time two distinct lines evolved from the very earliest cell forms. They were a group without a nuclear envelope enclosing the nuclear material, the prokaryotes, and a group with a nuclear envelope enclosing a true nucleus, the eukaryotes.

4.2   Origins and trends

All animal phyla are composed of multicellular, heterotrophic organisms. One group, the sponges (phylum Porifera), do not form true tissues (table 4.1), but in all other animals within the multicellular body similar cells operate collectively and become specialised functionally to form **tissues**. Many different tissues can be formed, each performing different functions. This is called differentiation or **division of labour** within the organism and it may be extensive. The advantage of this is that tissues generally perform specific tasks more effectively than individual cells.

Cellular activity in a tissue is coordinated so that the cells collectively function as a unit. A number of tissues may work together as an **organ**, and a group of organs working together forms an **organ system**. Just as cells are unable to act independently within a tissue, so organs and organ systems are subordinate to some means of coordination by the body. The net effect is an overall cooperation between the various systems which enables the organism to live as an effective, unique individual entity. The development of specific tissues, organs and organ systems is a feature of multicellular organisation and will be discussed in greater detail throughout the course of this chapter.

4.2.1   Phylogenetic origins of the kingdom Animalia

It is most probable that multicellular animals originated from the Protoctista, but it is much less certain which of the protoctistan groups, or how many of them, might have been ancestral. Two main hypotheses have been discussed in recent years. The first was put forward by

Table 4.1 Classification of phylum Porifera.

Phylum Porifera (pore-bearing) – sponges

Characteristic features
Some cellular differentiation, but no tissue organisation
Two layers of cells – outer pinacoderm and inner choanoderm (of collared flagellate cells)
Adults sessile
All marine
Body frequently lacks symmetry
Single body cavity
Numerous pores in body wall
Usually a skeleton of calcareous or siliceous spicules, or horny fibres
No differentiated nervous system
Asexual reproduction by budding
All are hermaphrodite, most protandrous
Embryonic development includes blastula and larval stages
Great regenerative power
'Dead-end' phylum – it has not given rise to any other group of organisms

Class Calcarea (calcareous spicules) e.g. *Leucosolenia*, *Sycon*

Class Hexactinellida (siliceous six-rayed spicules) e.g. *Euplectella* Venus flower basket) (fig 4.1)

Class Demospongiae (siliceous spicules, not six-rayed; or spongin fibres; or without skeletal elements), bath sponges e.g. *Halichondria*

Haeckel in 1866. He suggested that certain protoctistans, probably protozoans, divided repeatedly into daughter cells which failed to separate. Some anatomical and functional differences arose between the collection of cells, leading to specialisation. This produced a multicellular organism showing limited division of labour: in effect a forerunner of the cnidarians. It is probably true that the Porifera arose in this way. Some poriferan cells are almost identical to a family of flagellates called choanoflagellates, and it is likely that the sponges evolved from a colonial form of them (**colonial theory**).

The second hypothesis was put forward by Hadzi in 1944, who suggested that the nucleus of a protozoan divided repeatedly to give a multinucleate protozoan. This condition is seen today in the ciliated opalinids and in the Cnidosporidia. Subsequent internal division produced a multicellular condition (**syncitial theory**). This hypothesis has been widely accepted as a means of explaining how all multicellular animals other than the sponges have originated. What is more, Hadzi proposed that the turbellarian platyhelminths were more primitive than the Cnidaria. He suggested that the platyhelminths evolved as a result of internal division of multinucleate protozoa, and that the Cnidaria arose from the Turbellaria after some of them had adopted a sessile mode of life. To support his line of argument Hadzi pointed out that bilateral symmetry already existed in many of the protoctistans and that the cnidarians may not be strictly diploblastic (two-layered), as cells are often found in the mesogloea (fig 4.5). The covering of cilia over the turbellarian body also suggests possible relationships with ciliates.

Nevertheless, because the cnidarian type studied in this book possesses a construction that is simpler than in the platyhelminths mentioned, it has been decided to place it before the platyhelminths in the study of the various animal phyla. However, this must not be judged to be an indication of its actual point of origin. This is still a point of great debate amongst zoologists.

4.2.2   Trends in the kingdom Animalia

The onset of multicellularity and the increasing size of the organisms produced many physiological and anatomical problems which it was necessary for the animals to solve if they were to be successful. Some of the more important ones, and the ways they have been overcome, are summarised as follows.

(1) Large animals with many cells require much more food than the unicellular protozoa.

(2) Animals have become entirely heterotrophic, and in most cases holozoic (section 10.1.1).

(3) Development of an **alimentary canal** has enabled the animals to ingest large food particles, digest them and absorb the soluble products. This is followed by the egestion of insoluble food remains.

(4) A variety of feeding habits has been developed, incorporating carnivorous, herbivorous and omnivorous modes of life. Some are parasitic.

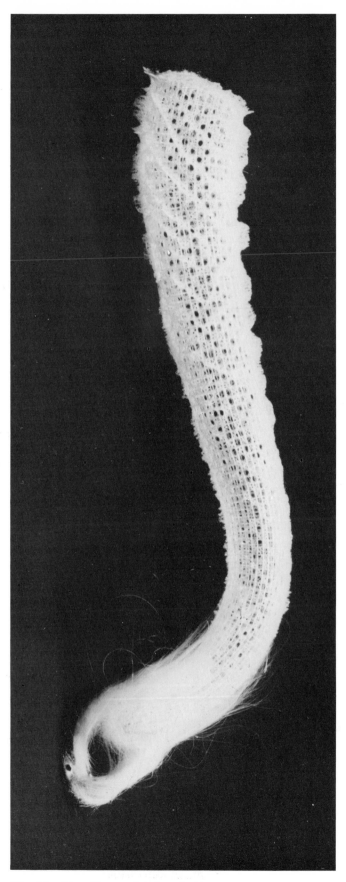

Fig 4.1 *The siliceous 'skeleton' of Venus flower basket (Euplectella). This is a deep sea member of the subkingdom Parazoa – the sponges*

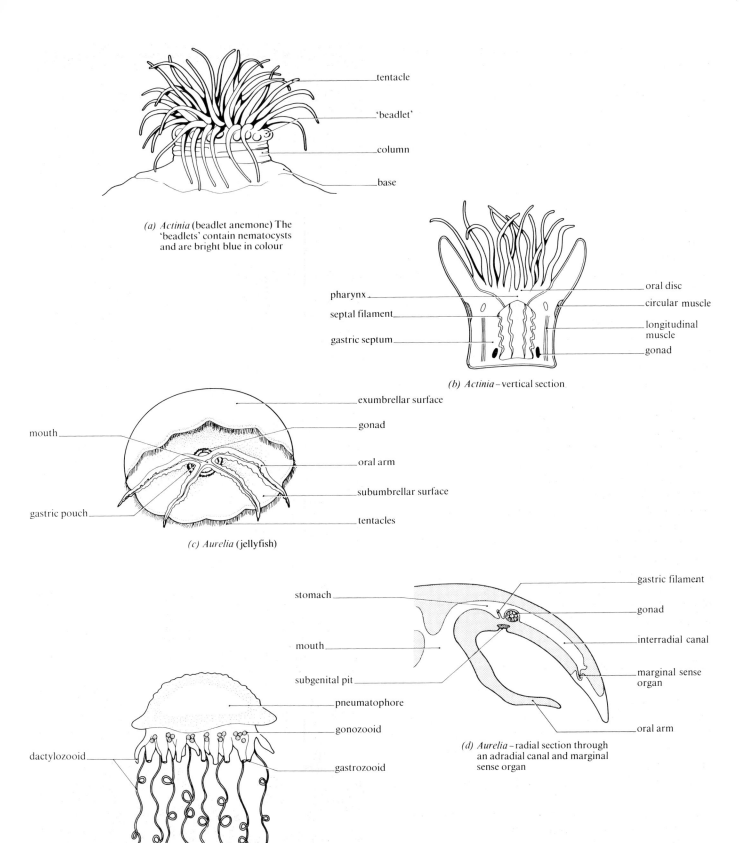

tentacle

'beadlet'

column

base

(a) *Actinia* (beadlet anemone) The 'beadlets' contain nematocysts and are bright blue in colour

pharynx

septal filament

gastric septum

oral disc

circular muscle

longitudinal muscle

gonad

(b) *Actinia* – vertical section

mouth

gastric pouch

exumbrellar surface

gonad

oral arm

subumbrellar surface

tentacles

(c) *Aurelia* (jellyfish)

stomach

mouth

subgenital pit

pneumatophore

gonozooid

gastrozooid

dactylozooid

gastric filament

gonad

interradial canal

marginal sense organ

oral arm

(d) *Aurelia* – radial section through an adradial canal and marginal sense organ

(e) *Physalia* diagrammatic

Fig 4.2   *A variety of cnidarians*

(5) To meet increased demands for food, and to enable organisms to search for it, an efficient means of locomotion has been devised.

(6) **Muscle** and **skeletal systems** (exo- or endoskeletons) have been developed to (a) aid maintenance of general body shape, (b) protect and provide support for inner structures, and (c) provide propulsive forces that will enable the organism to move.

(7) Most animals have adopted a **bilaterally symmetrical shape**. This gives a compact, generally elongated form which offers least resistance to movement. It also confers anterior, posterior, dorsal, ventral, right and left aspects to the animals. These areas may undergo further specialisation in different organisms. (NB Cnidaria and echinoderms are notable exceptions. They are sessile or slow moving. The former exhibit radial symmetry whilst the latter exhibit pentamerous symmetry. This enables them to encounter environmental changes from all directions.)

(8) Development of a central nervous system has occurred to coordinate all body activities. Sense organs receive stimuli, the central nervous system processes the information, and effectors produce an appropriate response.

(9) The major sense organs and nerve centres become situated at the anterior end of the body. Here they are ideally placed to be the first structures to encounter any environmental changes ahead. This process is called **cephalisation** and results in the formation of a definite head region.

(10) The nervous system is complemented by the parallel development of another coordinating system, the **endocrine system** (section 16.6). Together they serve to maintain the animal's steady state.

(11) Increase in size causes the spatial problem of the separation of central tissue from the body wall and the environment. Hence there has been the development of a **transport system**. This consists of a fluid tissue, usually **blood**, pumped around the body in vessels by a muscular heart or contractile vessels.

(12) The transport system provides a means by which oxygen, carbon dioxide, soluble food and excretory materials are transported throughout the body. At various points they may be taken up and utilised by the tissues, or expelled from the body.

(13) The relatively impermeable outer covering of an animal means that there are few areas which can be used for exchange of materials between the body and its surrounding environment. This adds to the need for an efficient transport system.

(14) The development of a multicellular animal from a single-celled zygote is often a long, complex process. It is therefore necessary for a period of **embryonic development** to take place, quite often followed by a larval phase and **metamorphosis** before the adult form is achieved.

4.3   Phylum Cnidaria (Coelenterata)

4.3.1   Class Hydrozoa

The majority of animals in this class are marine, but the genus *Hydra* is found in fresh water.

Obelia (fig 4.3) is a marine form and lives in shallow coastal waters attached to rocks, shells, seaweeds or wooden piles. It exists in two distinctly different forms during its life history. There is a sessile form, comprising of branching colonies of numerous minute **hydroid polyps**. Growing in the angles of the lower branches are **blastostyles** which give rise, by budding, to small swimming **medusae**. The medusa is the active free-living form of the animal. It provides a means of dispersal, and is the only form that is able to reproduce sexually. There are no special osmoregulatory mechanisms in either form as their cell contents are isosmotic with sea water. Gaseous exchange is by diffusion over the whole surface.

The colonial form

The colony consists of many hydroid individuals, each interconnected by a thin hollow tube, the **coenosarc**. Characteristically the coenosarc consists of a continuous cavity, the **enteron**, running throughout the colony, surrounded by a wall composed of outer **ectodermis**, **mesogloea** and inner **endodermis**. At the end of each branch the coenosarc continues into a cup-shaped structure, the **hydranth** (hydroid polyp). The coenosarc secretes outside itself a thin protective exoskeleton of chitin, the **perisarc**. This expands around the hydranth to form a **hydrotheca**. All hydranths are feeding structures and their collective activity supplies the whole colony with an adequate means of nourishment. Much of the structural organisation of the hydranths can be directly related to their function. Each hydranth is sac-like in shape and possesses an oral aperture at its apex. Surrounding the oral aperture is a circle of approximately 24 **tentacles** which are used to capture food which is generally small crustacea. Each tentacle consists of an internal layer of endodermal cells surrounded by an external covering of ectodermal cells. Being sessile, *Obelia* has to rely for its food source on water currents carrying prey, or prey swimming towards it. The radial symmetry of the hydranths and the branched nature of the whole colony enable the tentacles to sweep through large quantities of water and to encounter food from all directions. Each tentacle possesses batteries of **nematoblasts**, cells which each contain a **nematocyst**. There are three main types of nematocysts: penetrants, volvants and glutinants; they operate as the food-collecting structures of *Obelia*. When the projecting **cnidocils** are touched the nematocyst contents are automatically discharged, usually in large numbers (fig 4.4). They penetrate, hold and generally kill the prey, depending on the type of nematocyst. The highly developed longitudinal muscle tails of the tentacles holding the prey contract and succeed in carrying the prey to the mouth where it is ingested and passed into the enteron.

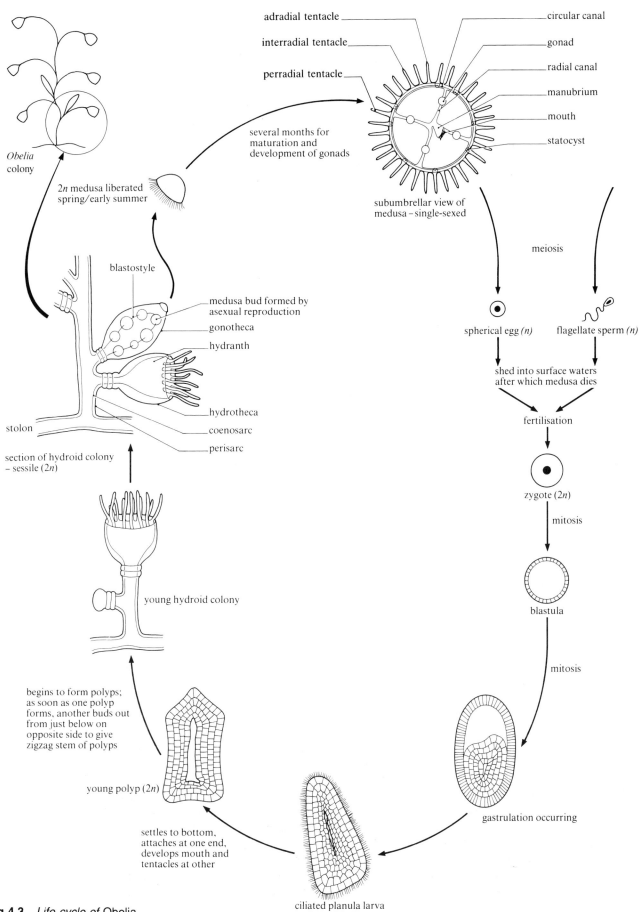

adradial tentacle
interradial tentacle
perradial tentacle

circular canal
gonad
radial canal
manubrium
mouth
statocyst

subumbrellar view of
medusa – single-sexed

several months for
maturation and
development of gonads

Obelia
colony

2*n* medusa liberated
spring/early summer

blastostyle

medusa bud formed by
asexual reproduction

gonotheca

hydranth

hydrotheca

stolon

coenosarc

perisarc

section of hydroid colony
– sessile (2*n*)

meiosis

spherical egg *(n)*

flagellate sperm *(n)*

shed into surface waters
after which medusa dies

fertilisation

zygote (2*n*)

mitosis

blastula

mitosis

young hydroid colony

begins to form polyps;
as soon as one polyp
forms, another buds out
from just below on
opposite side to give
zigzag stem of polyps

young polyp (2*n*)

gastrulation occurring

settles to bottom,
attaches at one end,
develops mouth and
tentacles at other

ciliated planula larva

Fig 4.3 *Life cycle of* Obelia

Table 4.2 Classification of phylum Cnidaria.

Phylum Cnidaria (Coelenterata)

Characteristic features
Diploblastic animals: body wall composed of two layers of cells, an outer ectoderm and an inner endoderm; these layers are separated by a structureless, gelatinous layer of mesogloea which may contain cells that have migrated from the other layers
Tissue level of organisation achieved
Single cavity, corresponds to the coelenteron; primarily inhalent, secondarily exhalent
Single opening for ingestion and egestion
Radial symmetry exhibited
Sedentary polyp forms which may be solitary or colonial; medusoid forms, free swimming and solitary
Nervous system is a collection of cells forming an irregular net or plexus
Asexual reproduction by budding or strobilation
Sexual reproduction produces characteristic planula larva
Polymorphism exhibited, but most individuals reducible either to hydroid or medusoid type

Class Hydrozoa	Class Scyphozoa	Class Anthozoa
Polyp dominant	Polyp present	Polyp only
Medusa simple	Large medusa dominant	No medusa
No mesenteries	Mesenteries present only in young polyp	Large mesenteries normally present
No gullet	No gullet	Gullet lined by ectoderm
Gonads ectodermal	Gonads endodermal	Gonads endodermal
Polyps solitary or colonial	Polyp is the hydratuba	Polyps solitary or colonial (in corals)
Nematocysts	Nematocysts	Nematocysts
e.g. *Hydra*	e.g. *Aurelia* (jellyfish)	e.g. *Actinia* (anemone)
Obelia		*Madrepora* (coral)

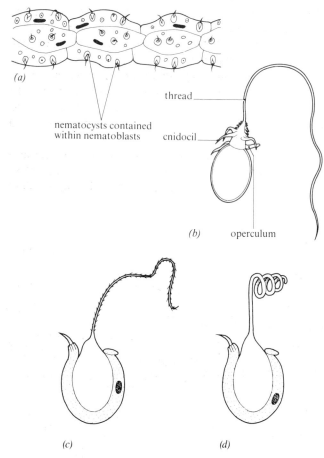

(a)

thread

nematocysts contained within nematoblasts

cnidocil

(b)   operculum

(c)   (d)

Fig 4.4 *(a) Portion of tentacle of hydranth showing batteries of nematoblasts. (b) Discharged penetrant nematocyst. (c) Volvant nematocyst. (d) Glutinant nematocyst*

Glandular cells (fig 4.5) in the endodermis secrete proteolytic enzymes into the enteron which begin extracellular digestion of the food. **Flagellate cells**, also of the endodermis, assist in the circulation of food particles and fluid in the enteron, whilst **pseudopodial cells** ingest food particles by phagocytic activity. Intracellular digestion within the food vacuoles completes the digestive process.

Many of the endodermal cells possess muscle tails containing contractile proteinaceous fibres embedded in the mesogloea. The muscle tails of the endodermis are arranged horizontally to the long axis of the body. Contraction of these enables the hydranth to become longer and thinner. The ectodermis of the hydranth also possesses **musculo-epithelial** cells, but their muscle tails are arranged parallel to the long axis of the body and allow the hydranth to become shorter and fatter and to retract more or less completely into its protective hydrotheca if it is irritated in any way. **Sensory cells** and undifferentiated **interstitial** cells are also present in the ectodermis.

Flagellate cells found in the endodermis circulate food particles throughout the coenosarc to all parts of the colony. At any point endodermal cells can engulf food for their own needs. After digestion has been completed in the endodermis, soluble food diffuses to the ectodermis through the mesogloea (fig 4.5). Any unwanted food remains are egested via the oral aperture.

On either side of the mesogloea is a **nerve net** composed of numerous **multipolar nerve cells**. The network is denser in the region of the oral disc and the tentacles, and is in contact with the sensory cells of the ectodermis and endodermis.

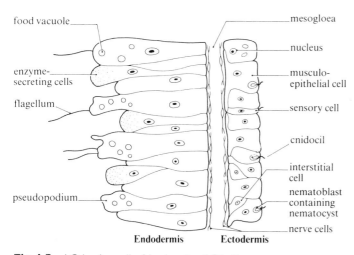

food vacuole

mesogloea

nucleus

enzyme-secreting cells

musculo-epithelial cell

flagellum

sensory cell

cnidocil

interstitial cell

pseudopodium

nematoblast containing nematocyst

nerve cells

Endodermis　　**Ectodermis**

Fig 4.5 *LS body wall of hydranth of* Obelia

Each blastostyle is a hollow extension of the coenosarc. It has no mouth or tentacles and is surrounded by a portion of the perisarc called the **gonotheca** which is open at its free end. Medusae are budded off and eventually expelled via the gonothecal opening into the sea.

The medusa

This is the **pelagic** sexual stage in the life history of *Obelia*. It is a bell-shaped, inverted polyp and possesses a convex exumbrellar and a concave subumbrellar surface. Hanging down from the centre of the subumbrellar surface is the **manubrium** at the end of which is a four-lobed oral aperture.

The whole of the outer surface is covered by ectodermal cells. The oral aperture provides the opening for the gullet which leads into the enteron. From here, four radial canals pass outwards to the edge of the bell and link up with a circular canal. The gullet, enteron and canals are lined with endodermis possessing numerous flagellate cells.

Initially about 24 tentacles hang vertically downwards from the edge of the bell, but more are added as the medusa matures. Their structure is essentially the same as in the hydranth, but in addition they each possess a swelling at their base where interstitial cells have accumulated. The interstitial cells are used to replace lost or damaged tentacular nematoblasts. Ingestion is aided by the tentacles, which bend over towards the mouth by infolding of the medusa margin. Digestion and distribution of food is accomplished in a similar way to that of the hydranth.

Halfway along each radial canal, and protruding from the subumbrellar surface, are the **gonads**. They consist of an outer layer of ectodermal cells and a core of endodermal cells derived from the radial canal itself. Germ cells originate in the ectoderm of the manubrium and migrate to the gonads. It is here that meiosis occurs. Each medusa is unisexual and will produce either **flagellate sperms** or **spherical eggs**. Liberation of the gametes occurs when the sacs burst. After this the medusa dies.

At the base of each adradial tentacle is a fluid-filled sac lined with ectodermal cells called a **statocyst**. Each sac possesses a **statolith** of calcium carbonate (fig 4.6). Alongside, and attached to, the statolith is a series of sensory protoplasmic processes. The whole apparatus enables the medusa to detect changes in its orientation during swimming.

The medusa swims actively by contracting the margin of its body inwards towards the manubrium. This forces a jet of water out and backwards from the subumbrellar surface and propels the animal forward. The whole process is effected by contraction of musculo-epithelial cells of the ectodermis, in particular a ring of well-developed, striated circular muscle fibres at the edge of the bell on the subumbrellar side. There are also muscle tails arranged radially on the subumbrellar surface. The normal shape of the medusa is regained by the elasticity of the thick layer of mesogloea that is present between the exumbrellar and subumbrellar surfaces of the animal.

In order to coordinate the overall activity of the medusa the nervous system has necessarily to be more complex than that of the polyp. As well as the characteristic nerve net, nerve cells have been concentrated into two rings, one external and one internal to the circular canal. Primarily the inner ring controls the subumbrellar muscle fibres whilst the outer ring receives impulses from the statocysts. However, they also interconnect with each other, the sensory cells and with processes innervating the tentacles.

The medusa normally swims through the water with the margins of its bell horizontally aligned, however if the bell is tilted, angular displacement of the statoliths will cause impulses to be produced by the sensory processes which are passed to the outer nerve ring. They are then relayed to the inner ring and the subumbrellar musculature will be stimulated to take the appropriate **reflex corrective action** to restore equilibrium.

In *Obelia* it can be seen that the organisation of cells into tissues has permitted different regions of the animal to specialise in performing particular tasks well. This is called

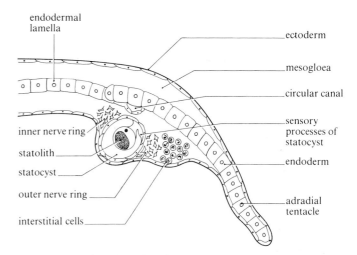

endodermal lamella

ectoderm

mesogloea

circular canal

inner nerve ring

sensory processes of statocyst

statolith

endoderm

statocyst

outer nerve ring

adradial tentacle

interstitial cells

Fig 4.6 *VS statocyst of* Obelia *medusa at base of an adradial tentacle*

differentiation and may lead to **division of labour**. However, there has to be cooperation between different areas of the colony. In a differentiated organism each cell is no longer able to perform all those processes necessary to keep itself alive and must depend to some extent on the activities of neighbouring tissues. For example, the tentacles, equipped with nematocysts, adequate musculature and innervation, capture prey and pass it to the enteron. Here the endodermis is specialised to digest the food. The soluble food materials are then passed back to the tentacles, among other areas, to provide them with the energy and raw materials they need to capture the next meal, and so on. This mutual cooperation between cells and tissues, as seen in *Obelia*, has become an established characteristic of all multicellular animals.

4.3.2 Polymorphism

The occurrence of structurally and functionally different types of individual within the same organism during its life history is called polymorphism. In *Obelia* there are feeding individuals (**gastrozooids**), individuals only capable of asexual reproductive activity (**gonozooids**), and free-living sexually reproductive zooids (medusae).

4.3.3 Alternation of generations and metagenesis

Whilst it is clear that *Obelia* undergoes an alternation of asexual and sexual phases during its life history, this should not be confused with alternation of generations in plants. In the majority of plants there is a regular alternation between a haploid gametophyte and a diploid sporophyte (section 3.6.1), whereas in *Obelia* both asexual and sexual phases are diploid. The only haploid cells in the life cycle of *Obelia* are the gametes. Since the colonial form does not produce gametes it may be regarded as the juvenile stage and the term **metagenesis** would be more appropriate to describe the life cycle of *Obelia*. Metagenesis implies deferment of the sexually reproductive phases rather than alternating phases of mitosis and meiosis.

4.4 Phylum Platyhelminthes

4.4.1 The triploblastic condition

This is the embryological situation where a third layer, the **mesoderm**, has developed which separates the ectoderm from the endoderm. The presence of mesoderm in the body is significant in several respects. It allows triploblastic organisms to increase in size and this results in considerable separation of the alimentary canal from the body wall. This poses problems of transport of materials between the endodermal and ectodermal layers. In animals where the mesoderm completely fills the space between the endoderm and ectoderm (**acoelomate condi-**

tion), transport problems are overcome by a dorso-ventral flattening of the body and maintenance of a large surface area in relation to volume. Thus diffusion of materials between environment and tissues is adequate to satisfy metabolic requirements. In animals where a space (the **coelom**) develops within the mesoderm (**coelomate condition**) transport systems are developed which carry materials from one part of the body to another. The presence of the mesoderm layer has been utilised to form a variety of organs, which may combine together and contribute towards an organ system level of organisation. Examples of such systems include the central nervous system and digestive, excretory and reproductive systems. The muscular activity of triploblastic organisms is also much improved. This is necessary as their increased size renders the ciliary mode of locomotion inadequate.

The platyhelminths are designed on the triploblastic body plan and are the most ancestral group of organisms to utilise mesoderm. They are the earliest animals to have developed organs and organ systems from the mesoderm. Much of the mesoderm, though, remains undifferentiated and forms a packing tissue, the **mesenchyme** or parenchyma, which supports and protects the organs of the body.

The phylum is divided into three classes; two of these are completely parasitic, whereas the other class, the most typical, contains free-living forms. The platyhelminths possess a clearly differentiated 'head' situated anteriorly, and a distinct posterior end. There are clearly defined dorsal and ventral surfaces. Many structures (such as eyes) are symmetrically arranged on the right- and left-hand sides of the body. Such organisation, where the right side is approximately the mirror image of the left and where there is a distinct anterior end, is said to constitute **bilateral symmetry**.

No transport system has developed, hence in the basic body structure all parts are in close proximity to food and oxygen supplies. All platyhelminths are thin and flat to provide a large surface area to volume ratio for gaseous exchange, and many forms possess a much-branched gut ramifying throughout the body to facilitate digestion and absorption of food materials. In addition, excretory material is collected from all parts by a branched system of excretory tubes ending in **flame cells** (see chapter 19).

4.4.2 Class Turbellaria

Planaria is a free-living, carnivorous flatworm found in freshwater streams and ponds. It remains under stones during the day, emerging only at night to feed. It is black in colour and can measure up to 15 mm in length. It has an elongated, extremely flattened body, with a relatively broad anterior 'head' possessing a pair of dorsal eyes, and a posterior end that is clearly tapered. *Planaria* is bilaterally symmetrical, a body design associated with an active mode of life (fig 4.7).

There is a single gut opening, the mouth, which is located on its ventral surface towards the posterior end of the body.

Table 4.3 Classification of phylum Platyhelminthes.

Phylum Platyhelminthes

Characteristic features
Triploblastic
Bilaterally symmetrical
Unsegmented
Acoelomate
Central nervous system anteriorly placed; very simple network; ganglia
Excretory system of branching tubes ending in flame cells
Flattened dorsoventrally
Mouth but no anus
Complex hermaphroditic reproductive system
Larval form usually present

Class Turbellaria	*Class Trematoda*	*Class Cestoda*
Free living; aquatic	Endoparasitic	Endoparasitic
Delicate, soft, leaf-like body	Leaf-like	Elongated body divided into proglottides which are able to break off
Suckers rarely present	Usually ventral sucker in addition to sucker on proscolex	Suckers and hooks on proscolex
Ciliated cellular outer covering; cuticle absent	Thick cuticle; no cilia in adult	Thick cuticle; no cilia in adult
Enteron present	Enteron	No enteron
Sense organs in adult	Sense organs only in free-living stages	Sense organs only in free-living stages
Simple life history	Complex life history	Complex life history
e.g. *Planaria*	e.g. *Fasciola* (liver fluke)	e.g. *Taenia* (tapeworm)

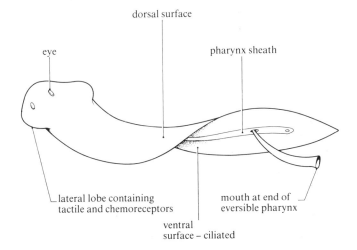

Fig 4.7 Planaria lugubris *showing external features*

Planaria has a complex body wall which contributes towards the mechanisms of locomotion, protection and capture of prey (fig 4.8). The epidermis consists of columnar cells which are ciliated at the sides and on the ventral surface of the organism. Interspersed between the ventral ciliated cells are tracts of glandular cells which secrete slime. The epidermal cells also possess **rhabdites**. These are secreted during food capture and help entangle the prey.

Below the epidermis are several layers of muscle cells. There is an outer circular, a middle diagonal and an inner longitudinal layer. There are also dorso-ventral muscle tracts. Each layer consists of collections of individual muscle fibres, not just muscle tails of epithelial cells as in the cnidarians. The complex pattern of musculature enables the animal to perform all kinds of agile movements in the water, and therefore contributes to its complex behavioural activities.

Planaria feeds on small worms, crustacea and on the dead bodies of larger organisms. Anteriorly the body wall possesses sense cells which enable the animal to detect prey from some distance away. *Planaria* takes up a position on top of the prey and pins it down by means of muscular contractions of its body. Rhabdites and slime are exuded from the worm's ventral surface and the sticky fluid helps entangle the prey. An **eversible pharynx** protrudes and engulfs the prey. If the prey is large, enzymes are secreted on to it to begin extracellular digestion, and pumping activity of the pharynx breaks the food into smaller particles. These are then ingested into the intestine. Endodermal cells line the three main branches of the intestine, one branch leading to the front and two to the rear of the animal (fig 4.9). Numerous blind-ending lateral caecae arise from each of these branches. The intestine thus possesses a large surface area for digestion and absorption. The large size of the intestine penetrates most parts of the body and facilitates diffusion of materials to and from the body cells. Additional enzymes are secreted by the gland cells of the intestine and continue the process of extracellular digestion. Small particles of food are finally engulfed by phagocytic cells. Digestion is completed intracellularly. From here the soluble food passes to the

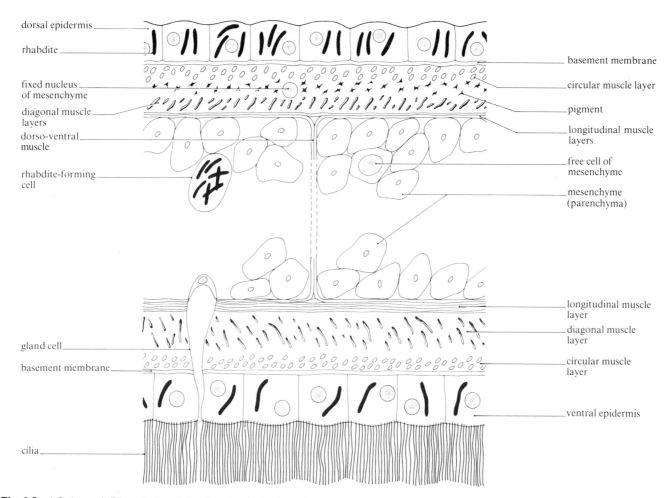

Fig 4.8 *LS through* Planaria lugubris *showing both dorsal and ventral body walls*

Labels for Fig 4.8 (left side, top to bottom): dorsal epidermis, rhabdite, fixed nucleus of mesenchyme, diagonal muscle layers, dorso-ventral muscle, rhabdite-forming cell, gland cell, basement membrane, cilia.

Labels for Fig 4.8 (right side, top to bottom): basement membrane, circular muscle layer, pigment, longitudinal muscle layers, free cell of mesenchyme, mesenchyme (parenchyma), longitudinal muscle layer, diagonal muscle layer, circular muscle layer, ventral epidermis.

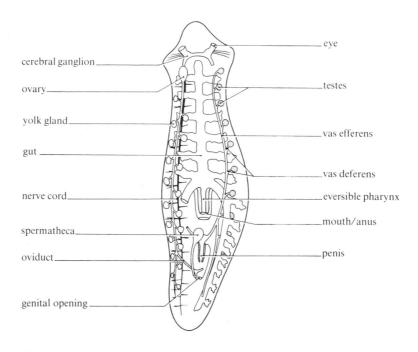

Fig 4.9 *Anatomy of* Planaria lugubris

Labels for Fig 4.9 (left side, top to bottom): cerebral ganglion, ovary, yolk gland, gut, nerve cord, spermatheca, oviduct, genital opening.

Labels for Fig 4.9 (right side, top to bottom): eye, testes, vas efferens, vas deferens, eversible pharynx, mouth/anus, penis.

rest of the body by diffusion via the mesenchyme and by amoeboid cells. Undigested food remains are egested through the mouth.

A distinct excretory system has differentiated within the mesodermal layer. It consists of two longitudinal excretory canals which open on to the dorsal surface via a number of pores. Each canal has many branches which end in flame cells (section 19.4.3). Excretory substances are actively secreted into the flame cells and ultimately passed out of the body via ducts. The flame cell system also provides an efficient osmoregulatory mechanism for the planarian.

Asexual reproduction occurs by **transverse fission**. The posterior end of the flatworm adheres to the substrate by the secretion of slime, while the anterior half pulls away from it until the worm splits in two. The split takes place just behind the pharynx. Both portions then proceed to regenerate all missing parts.

Planarians are hermaphrodite and possess a complex reproductive system. The arrangement is shown in fig 4.9. Sperm are produced by the germinal epithelium of numerous **testes** and pass into **vasa efferentia**, which in turn empty their contents into two longitudinally arranged **vasa deferentia**. Mature sperm are stored in **seminal vesicles** until required. Each seminal vesicle leads to a muscular, protrusible **penis** contained in the **genital atrium**.

Ova are produced by the paired **ovaries** which are situated laterally within the mesenchyme. Each ovary is connected, via a short portion of the **oviduct**, to a **receptaculum seminis**. This is where fertilisation will take place. Each oviduct passes to the rear of the body. Along its length it receives lateral ducts from **yolk** and **shell glands**. Both oviducts finally join together in the **genital atrium**. Another structure, the **copulatory sac** (or spermatheca) also opens into the genital atrium.

Copulation takes place prior to fertilisation. Two worms adhere to each other by their ventral surfaces. Sperms from one worm are transferred to the copulatory sac of the other when its penis is inserted into the genital atrium of its partner. After this, the worms separate. Within each organism the sperm swim from the copulatory sac, up the oviducts to each receptaculum seminis where fertilisation occurs. The fertilised eggs then pass down the oviducts and become coated with yolk cells and shell substances. **Cocoons** containing several eggs and many yolk cells are formed in the genital atrium and eventually expelled from the body. Within a few weeks the eggs hatch and small worms emerge.

4.4.3 Class Trematoda

Fasciola hepatica (figs 4.10 and 4.11) belongs to the class Trematoda, which is one of the major groups of parasites in the animal kingdom. It is endoparasitic, living in the bile ducts of sheep, its most important, or **primary**, **host**. Other primary hosts are cattle and, occasionally, humans. Many differences exist between *Fasciola* and the free-living *Planaria lugubris*. These differences can be

attributed to the adaptations that *Fasciola* has evolved in order to survive as an endoparasite. Associated with its parasitic mode of life is a complex life history, involving three larval stages (the miracidium, redia and cercaria), and opportunities for increasing their numbers during the life cycle. Within some of the larval forms are germinal cells, as distinct from somatic cells, which undergo normal mitotic division to produce even more individuals. Such a process is called **polyembryony**, the cells of each new individual being products of divisions of the original zygotic cell. The large numbers of offspring produced in this way help to offset the high mortality rate that inevitably occurs during infection of new hosts. For a part of its life history *Fasciola* infests a **secondary host**, the freshwater snail *Limnea truncatula*, in which some of its larval stages are able to live and multiply.

Each stage in the life history of *Fasciola* exhibits structural, physiological and reproductive adaptations suited to its mode of life. Some of these are listed below.

Adult fluke. The body is thin and flat and pressed against the side of the bile duct so that it does not interrupt the bile flow. It is attached to the wall of the bile duct by oral and ventral suckers which enable it to maintain its position in the duct, and spines on the body wall, which point backwards, prevent it from being flushed down the duct in the bile flow. The body wall protects the worm against the host's enzymes. The gland cells situated here secrete material which protects the parasite against the host's antitoxins (fig 4.11). The body wall is also the area of nitrogenous excretion (primarily ammonia) and the site of gaseous exchange for the parasite. Respiration is thought to be largely anaerobic (but oxygen is utilised if present). However, no relationship has been established between oxygen uptake and carbon dioxide output. The muscular pharynx has a pumping action which enables the ingestion of viscous materials such as blood, other tissues and mucus.

A complex, hermaphrodite reproductive system ensures that fertilisation (either self- or cross-fertilisation) can occur.

Miracidium. This is the first of the larval stages of *Fasciola* (fig 4.12). It has a ciliated epidermis which provides means for swimming in water or in moisture on vegetation. The miracidium is attracted to its molluscan secondary host by chemotaxis. An apical papilla attaches it to the snail's foot and an apical gland secretes proteolytic enzymes on to the surface of the snail to assist in the penetration of the host's tissues. Penetration is further aided by circular and longitudinal muscle cells which help the larva to wriggle through the tissues of the host. In this way it migrates to the digestive glands of the secondary host via the lymph channels. There are germ cells present which give rise to subsequent larval forms.

Sporocyst. This is an immobile, closed germinal sac containing germinal cells. The cells proliferate by

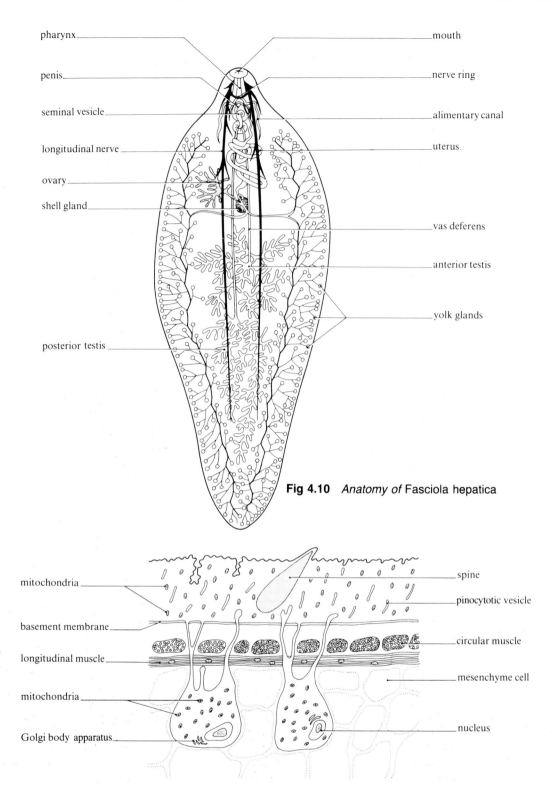

pharynx
penis
seminal vesicle
longitudinal nerve
ovary
shell gland
posterior testis

mouth
nerve ring
alimentary canal
uterus
vas deferens
anterior testis
yolk glands

Fig 4.10 *Anatomy of* Fasciola hepatica

mitochondria
basement membrane
longitudinal muscle
mitochondria
Golgi body apparatus

spine
pinocytotic vesicle
circular muscle
mesenchyme cell
nucleus

Fig 4.11 *Section through body wall of* Fasciola hepatica *showing ultrastructure*

polyembryony into many rediae. This is therefore a multiplicative phase in the life history of *Fasciola*.

Redia. This stage has a muscular pharynx to suck in fluids and tissues from its host. Circular and longitudinal muscle cells aid locomotion of the larva; two posterior lateral flaps provide purchase at the posterior end, whilst an anterior collar provides 'grip' at the other end.

Germinal cells proliferate into more rediae, or into cercariae, by polyembryony; so this is also a multiplicative phase. There is a birth pore for the escape of the second generation of rediae or the cercariae.

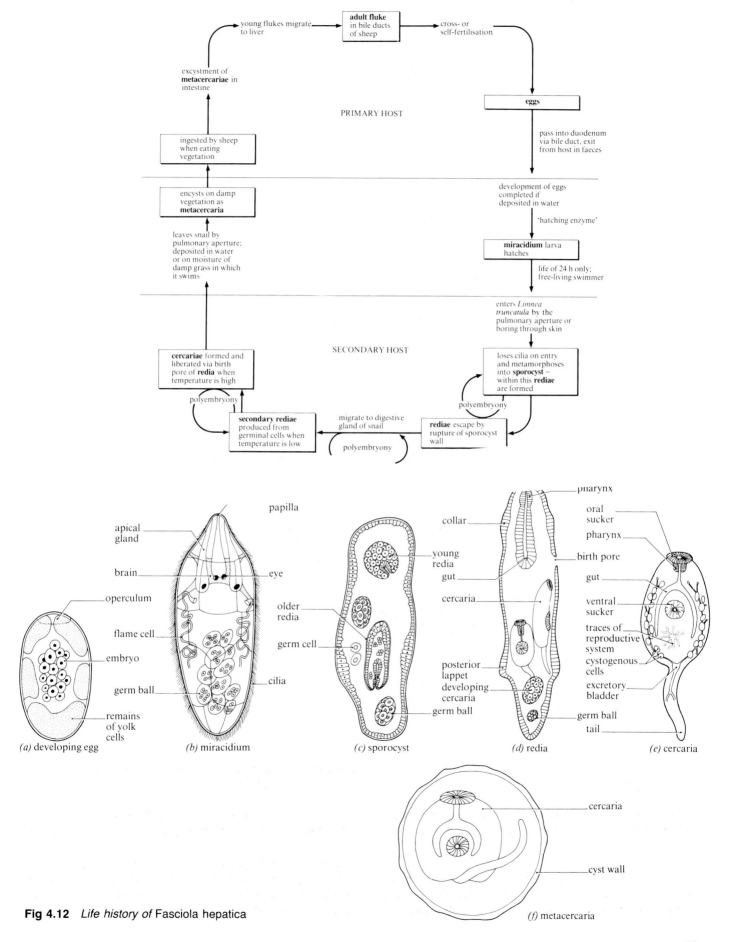

Fig 4.12 *Life history of* Fasciola hepatica

Cercaria. This bears many features in common with the adult fluke, which include oral and ventral suckers for anchorage to suitable substrates, such as grass. There is also a tail to assist in locomotion through water or moisture on vegetation. Cystogenous glands are present which secrete a cyst wall to form a metacercaria.

Metacercaria. No further development of this stage occurs until it is swallowed by sheep. It has considerable powers of resistance to low temperatures, but is susceptible to desiccation (fig 4.12).

Limnea truncatula is an amphibious snail inhabiting ponds, muddy tracks and damp vegetation. It is able to withstand adverse conditions. Therefore the sporocyst and redia stages of *Fasciola's* life history, which develop within the snail, are themselves directly protected from such unfavourable conditions. Indeed, in conditions of low temperature, rediae produce daughter rediae instead of cercariae. The rediae remain within the snail and can overwinter within the host, only producing cercariae when warmer weather returns in the spring. *Limnea* is also a very rapid breeder. It has been estimated that one snail may produce up to 160 000 offspring in 12 weeks. If all of these offspring contain developmental stages of *Fasciola*, then the chances of cercariae escaping from the snails and entering new, uninfected primary hosts will be considerably increased. The amphibious mode of life of *Limnea* ensures that when the cercariae escape there is water available in which to disperse.

The release of young adult flukes from the metacercaria (excystment) takes place in the gut of the sheep or cow. The process is initiated in the stomach by high carbon dioxide levels and a temperature of around 39 °C. Under these conditions the parasite releases proteolytic enzymes which digest a hole in the cyst wall at one point. Emergence of young flukes is triggered off by the presence of bile in the digestive juices of the small intestine.

The young flukes burrow through the intestinal wall and migrate to the liver via the coelom. For a time they feed on liver tissue, but about six weeks after infection they become permanently attached in the bile ducts.

Fasciola can have several effects on its host. A heavy infection can cause death. Liver metabolism of the host is interfered with when cercariae migrate through it. Cells are destroyed and bile ducts may be blocked; large-scale erosion of the liver (liver rot) will cause dropsy. Little, or absence of, bile in the gut can affect digestion, and the excretory wastes of *Fasciola* can have a toxic effect on the host.

The following measures can be taken against *Fasciola*. Drainage of the pasture land and introduction of snail-eating geese and ducks to the pastures (a method of biological control) will help to remove the secondary host *Limnea*. The filling in of ponds and use of elevated drinking troughs will also help to achieve this. Use of lime on the land will help to prevent the hatching of the eggs of the parasite, as they will not hatch in water with a pH of more than 7.5. For sheep which are already infected, the administration of carbon tetrachloride kills some of the fluke stages in the liver.

4.5   Body cavities

It has been seen in the platyhelminths that the mesoderm completely fills the space between the ectodermal and endodermal layers and forms a solid middle layer. The only cavity present is the **archenteron**. Since its lumen is, in reality, in contact with the exterior environment, the archenteron cannot be regarded as a true internal body cavity (coelom). Such a condition, without a coelom, as illustrated by the platyhelminths is said to be **acoelomate** (fig 4.13*a*).

In most recent groups, an extensive internal space or body cavity is developed, called a **coelom**, which separates

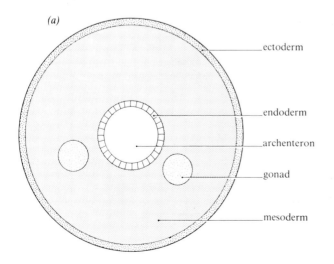

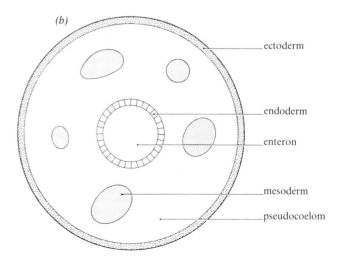

Fig 4.13 *The (a) acoelomate and (b) pseudocoelomate condition*

the body wall from the alimentary tract. The space is generally filled with fluid and may perform the following functions:

(1) may act as a hydrostatic skeleton;
(2) enable activities of the body wall and alimentary canal to operate independently of each other;
(3) permit animals to become much larger;
(4) the fluid of the cavity may act as a circulatory medium for the transport of food, waste materials and gases;
(5) waste materials and excess fluids may be temporarily stored here;
(6) provides space for the enlargement of internal organs;
(7) may play a part in the osmoregulatory activity of organisms.

4.5.1   Types of body cavity

The **pseudocoelom**, which is found in nematodes and rotifers (fig 4.13b), is derived from the hollow space situated in the blastula at an early stage in embryological development. It is bounded on the outside by ectoderm and on the inside by the endodermal wall of the alimentary canal. The internal organs remain free within the pseudocoelom, and the spaces between them are filled with large, vacuolated mesodermal cells. Thus the organs are separated and can operate independently of each other, and the mesodermal cells, which are easily deformed, enable the organism to change its body shape readily. The pseudocoelom appears to be the forerunner of another type of body cavity, the coelom.

The **coelom** is preceded by the blastocoel during embryological development, hence the coelom should be regarded as the **secondary body cavity**. When it develops it reduces the blastocoel to a series of blood-filled spaces bound by mesodermally derived walls. The coelom arises when a split occurs within the embryonic mesoderm. The cavity of the coelom is thus bounded by a lining of mesodermal cells called the **peritoneum**. Vertical portions of the peritoneum are called **mesenteries** and they help in the suspension of the alimentary canal from the body wall. The coelomic cavity is filled with **coelomic fluid**.

Examination of the embryological development of the coelom in various organisms indicates that it may arise in more than one way. Opinion is divided as to which method is the most ancestral. Three methods of coelom formation are shown in fig 4.14.

The result of coelom formation is that a layer of mesoderm (**somatic**) is applied to the ectoderm and

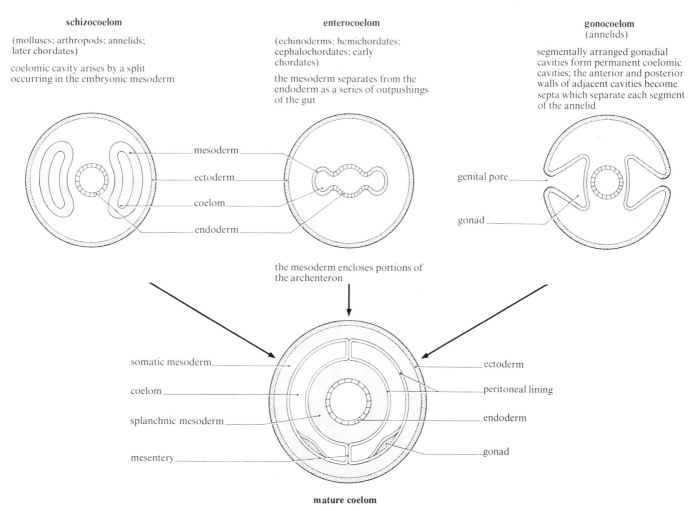

Fig 4.14 *Methods of coelom formation*

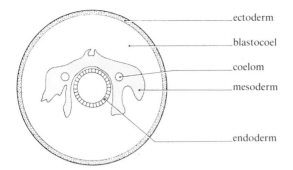

ectoderm

blastocoel

coelom

mesoderm

endoderm

Fig 4.15 *The haemocoel condition*

becomes part of the body wall, and a layer of mesoderm (**splanchnic**) associates with the endoderm of the alimentary tract to form the musculature of the gut. The relatively stationary coelomic fluid separates the body wall from the alimentary canal, and any organs which protrude into the cavity are bounded by peritoneum.

In the case of the **haemocoel**, which is found in arthropods and molluscs, the coelom has been almost completely obliterated by a greatly enlarged blastocoel (fig 4.15). The blastocoel consists of sinuses filled with blood. The blood is generally circulated in the haemocoel. Gonads are never differentiated from the haemocoel wall, and the coelom is confined to cavities of excretory organs and the gonoducts.

4.6   Phylum Nematoda

Ascaris lumbricoides is a common nematode which parasitises humans and pigs. The worms possess slender, elongated bodies, tapering at each end. The male is smaller than the female and is more curved at its posterior end. Covering the body of each worm is a cuticle made of three layers of diagonally arranged collagen fibres. This allows the body some degree of contraction and extension. Beneath the cuticle the epidermis is thickened into four longitudinal internal ridges, one dorsal, one ventral and two lateral. The epidermis forms a **syncitium**, since it is a mass of cytoplasm containing many nuclei and is bounded by a cell surface membrane. Below the epidermis

Table 4.4 Characteristics of phylum Nematoda.

Phylum Nematoda

Characteristic features
Triploblastic, pseudocoelomate
Bilaterally symmetrical
Unsegmented
Elongated, round 'worms' with pointed ends
Alimentary canal with mouth and anus
Sexes separate (dioecious)
Some free living, many important plant and animal parasites
Anterior end shows a degree of cephalisation

is a single layer of longitudinal, obliquely striated muscle fibres. The layer is divided into four bands, each band occupying the space between two ridges of epidermis. Each muscle cell has its contractile elements located peripherally. Its other end is bulbous and possesses a slender cytoplasmic extension which connects with a longitudinal nerve cord. (This contrasts with other organisms where nerve fibres pass to the muscles.)

The mouth is surrounded by three 'lips' which possess sensory papillae. It leads into a muscular pharynx lined by cuticle and epidermis. The pharynx pumps some of its host's food into the straight intestine of the worm. The gut is provided with valves at both ends which prevent regurgitation of the food. Microvilli are present in the intestine which is secretory and absorptive in its anterior and posterior regions respectively. Unwanted material is egested via the anus.

The pseudocoel which separates the alimentary canal from the body wall consists of a small number of vacuolated cells filled with a protein-rich fluid. The so-called excretory system (it may be entirely osmoregulatory in function) consists of two longitudinally orientated lateral canals which unite anteriorly to form a single canal leading to the exterior by a ventral excretory pore. There is a nerve ring around the pharynx which is associated with a number of ganglia. From it, distinct dorsal, ventral and lateral cords run the length of the worm. Anteriorly, six nerves pass from the ring to the head sense organs.

Locomotion is achieved by undulating waves of contraction and relaxation of the muscle bands acting against the turgidity of the pseudocoel. Absence of circular muscle permits bending in only the dorso-ventral plane.

The female has two ovaries which pass their contents via oviducts to the uteri where the eggs are gathered. The uteri lead to a vagina which opens on the ventral surface by the female genital pore. The single testis of the male is a long tube which enlarges into a sperm duct and opens to the surface near the posterior end. A seminal vesicle is present which joins the rectum just before the cloaca. During copulation the male inserts its copulatory spicules into the genital atrium of the female and distends the vagina. The amoeboid sperm are ejected into the vagina and pass to the uteri where fertilisation occurs. Details of the life cycle of *A. lumbricoides* can be seen in fig 4.16.

4.7   Phylum Annelida

4.7.1   The coelomate body plan and metameric segmentation

Whereas the nematodes possess a pseudocoel, later groups, from the annelids onwards, have developed a fluid-filled coelom. Coelomic fluid separates the body wall from the alimentary tract. The majority of the mesoderm which lines the coelom develops into muscle; that of the body wall aids locomotion of the whole animal, whilst that of the gut causes **peristalsis** of food. Transport of

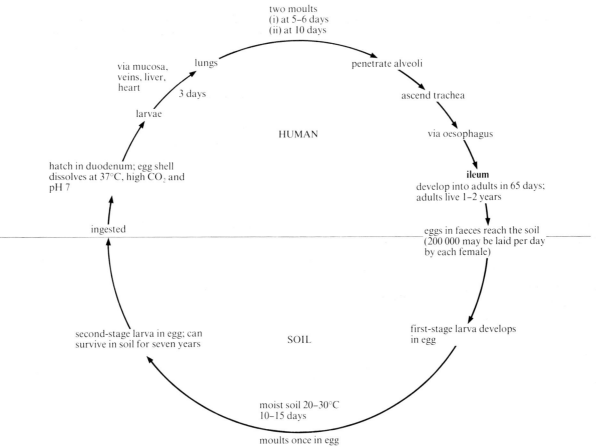

two moults
(i) at 5–6 days
(ii) at 10 days

lungs   penetrate alveoli

via mucosa,   ascend trachea
veins, liver,
heart   HUMAN   via oesophagus

3 days

larvae   **ileum**
                                            develop into adults in 65 days;
hatch in duodenum; egg shell   adults live 1–2 years
dissolves at 37°C, high CO₂ and
pH 7

ingested   eggs in faeces reach the soil
                                       (200 000 may be laid per day
                                        by each female)

second-stage larva in egg; can   SOIL   first-stage larva develops
survive in soil for seven years   in egg

moist soil 20–30°C
10–15 days

moults once in egg

Fig 4.16 *Life cycle of* Ascaris lumbricoides

Table 4.5 Classification of phylum Annelida.

Phylum Annelida

Characteristic features
Triploblastic, coelomate
Bilaterally symmetrical
Metamerically segmented
Perivisceral coelom
Pre-oral prostomium
Central nervous system of paired supra-oesophageal ganglia connected to ventral nerve cord by commissures
Solid, ventral nerve cord, usually double with segmented nerves
Excretory organs are segmental, ectodermal in origin, ciliated, and called nephridia
Definite cuticle secreted by ectoderm
Chaetae of chitin arranged segmentally (except leeches)
Larva typically a trochophore

Class Polychaeta	*Class Oligochaeta*	*Class Hirudinea*
Marine	Inhabit freshwater or damp earth	Ectoparasitic with suckers anterior and posterior
Distinct head	No distinct head	No distinct head
Chaetae numerous on parapodia	Few chaetae – in pairs or single, no parapodia	Small fixed number of segments, no chaetae or parapodia
Dioecious	Hermaphrodite	Hermaphrodite
Gonads not localised but extending throughout whole body	Gonads localised in few segments	Gonads localised in small number of segments
Fertilisation is external	Copulation and cross-fertilisation	Cross-fertilisation
No cocoon	Clitellum with eggs laid in cocoon	Eggs laid in cocoon
Free-swimming trochophore larva	No larval stage, development direct	No larval stage
e.g. *Nereis* (ragworm)	e.g. *Lumbricus* (earthworm)	e.g. *Hirudo* (leech)
Arenicola (lugworm)		

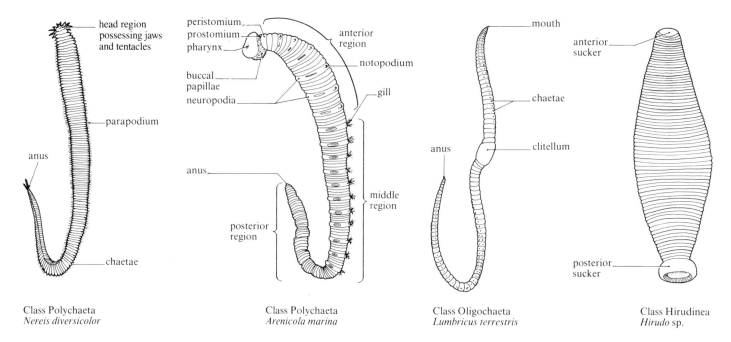

Labels on figures (left to right):

Class Polychaeta
Nereis diversicolor
- head region possessing jaws and tentacles
- parapodium
- anus
- chaetae

Class Polychaeta
Arenicola marina
- peristomium
- prostomium
- pharynx
- buccal papillae
- neuropodia
- anterior region
- notopodium
- gill
- middle region
- posterior region
- anus

Class Oligochaeta
Lumbricus terrestris
- mouth
- chaetae
- clitellum
- anus

Class Hirudinea
Hirudo sp.
- anterior sucker
- posterior sucker

Fig 4.17 *A variety of annelids*

materials between the gut wall and the body wall (and vice versa) is achieved by a well-developed **blood vascular system**.

Another evolutionary advance which took place amongst the coelomates was that of **metameric segmentation**. It is a phenomenon which originates in the mesoderm but usually affects both mesodermal and ectodermal regions of the body. As a result the body becomes divided transversely into a number of similar parts or segments. In the annelids, where it is clearly seen, the subdivisions may be indicated externally by constrictions of the body surface. Internally the segments are separated from each other by septa extending across the coelom. However the segments are not entirely independent, as a number of organ systems run the length of the body, penetrating each individual segment in turn.

Once segmentation has been established, individual or small groups of segments may become further modified and specialised in many ways to perform a variety of different functions. Such differences may occur by elaboration of organs within a segment, by fusion or even loss of segments.

It is thought that the coelom arose in the immediate ancestors of the annelids and was exploited as an adaptation for the burrowing habit. Burrowing would have given the annelids a two-fold advantage over their competitors: protection against predators, and exploitation of new ecological niches. The coelom of the annelid provided a form of **hydrostatic skeleton** against which its muscles could act during locomotion and burrowing. Contraction of the circular muscles produces a pressure in the coelomic fluid that forces the body to elongate. Similarly contraction of the longitudinal muscles would produce a pressure in the coelomic fluid that would cause the body to widen. It would be a further advantage if the action of the circular and longitudinal muscles could be localised to certain regions as this would facilitate burrowing. Metameric segmentation, resulting in a subdivision of the muscle layers, provided this mechanism. Segmentation of the nervous system to coordinate muscle activity, and of the blood and excretory systems to accommodate the needs of the muscles is then thought to have followed.

4.7.2 Class Polychaeta

Nereis is an elongated, cylindrical bristle-worm. It lives in estuaries under stones or in mud burrows. The segmented nature of its body is clearly visible externally. All segments, apart from those most anterior and posterior, are very similar to each other. On either side of each segment is a lateral projection, the **parapodium**. It is **biremous** and consists of an upper **notopodium** and a lower **neuropodium** (fig 4.18). Rods called **acicula** support both these processes, and two tufts of **chaetae** emanate from each structure forming fan-like bodies. Two additional outgrowths of the parapodia are noticeable: a dorsal and a ventral **cirrus**.

The body wall of the worm consists of a thin cuticle secreted by a single layer of columnar epithelium. There is a thin outer layer of circular muscle below which is a thicker layer of longitudinal muscle. The longitudinal muscle is split into two dorso-lateral and two ventro-lateral bundles which run lengthways in the body. Oblique muscles extend from the midline to mid-lateral regions and join with circular muscles. All are obliquely striated.

Septa, composed of a double layer of peritoneum, separate the coelom into individual segments. Coelomic fluid is present which contains amoeboid cells and a variety of dissolved materials. It bathes all organs and aids both excretory and reproductive processes.

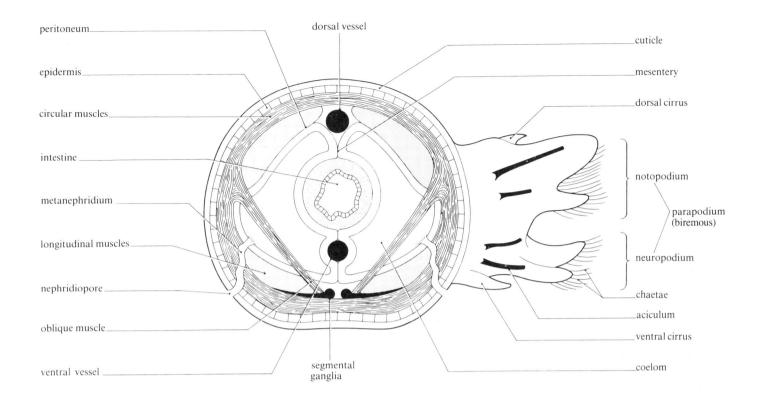

Fig 4.18 *Cross-section of trunk segment of* Nereis diversicolor

The alimentary canal runs from mouth to anus and is more complex anteriorly. Prey is captured by two horny jaws at the end of an eversible pharynx. It is swallowed when the pharynx is retracted.

Nereis possesses a clearly differentiated head which displays a considerable degree of **cephalisation** (fig 4.19). The head consists of an anterior **prostomium** and a **posterior peristomium**. On the prostomium is a pair of dorsal sensory **tentacles** and two pairs of eyes, whilst a pair of fleshy palps extend from its ventro-lateral regions. The mouth is situated between the two head parts, and on the peristomium are four pairs of long, flexible tactile **cirri**.

Internally, there is an increase of nerve cells and nerve tissue at the anterior end compared with the platyhelminths and nematodes. This concentration of nerve tissue is composed of a pair of relatively large fused **cerebral ganglia** which supply the prostomium via the prostomial nerves. The cerebral ganglia are connected to a **double ventral nerve cord** by a pair of circumoesophageal commissures.

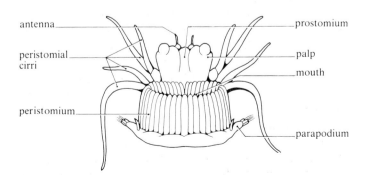

Fig 4.19 *Ventral view of head of* Nereis diversicolor

The double ventral nerve cord runs throughout the length of the worm. In each segment the ventral cord bears a pair of ganglia from which lateral nerves extend. These nerves are mixed, containing sensory and motor nerve components.

The excretory organs are **nephridia**. A pair is found in all but the first and the last segments. Nitrogenous waste is principally ammonia. The blood and tissue fluids of the animal can remain isotonic with the marine environment over a wide range of salinities. Thus there are few problems of osmoregulation.

There is an efficient blood vascular system with blood confined to closed vessels. Blood flows forwards in a dorsal longitudinal vessel and passes into a ventral vessel in each segment via two pairs of lateral segmental vessels. These run into the parapodia where they branch into capillaries and rejoin before joining the ventral vessel. Circulation is maintained by contractile activity of the major vessels and by waves of muscular contraction running along the body wall which squeeze the blood vessels, forcing blood along.

The parapodia are extensively vascularised and function as the animal's gaseous exchange surface. Haemoglobin, dissolved in the plasma, increases the blood's oxygen carrying capacity.

Nereis crawls by using its parapodia in an oar-like manner, and swims by the coordinated activity of the parapodia and lateral flexing of the body brought about by contraction and relaxation of the body wall musculature.

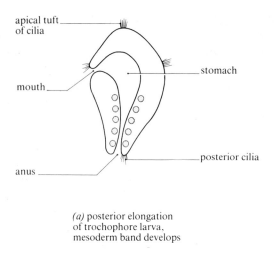

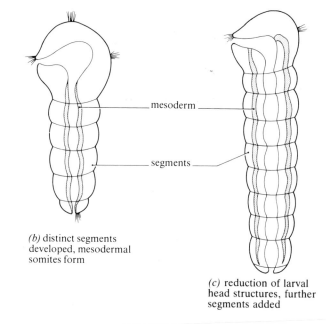

(a) posterior elongation of trochophore larva, mesoderm band develops

(b) distinct segments developed, mesodermal somites form

(c) reduction of larval head structures, further segments added

Fig 4.20 *Development of segmentation in polychaetes during metamorphosis of the trochophore larva*

The sexes are separate in nereids. Gametes are formed in most segments from germinal cells in the peritoneum. Prior to fertilisation, worms leave their burrows and swim near the surface of the water. Spawning occurs at a definite time of the year, generally in early spring. Males shed their sperm into the water, and segments of the female burst to release eggs, millions of gametes being present in the sea at the same time. After discharge of the gametes the adult worms die.

Fertilisation is external. The zygote develops into a ciliated **trochophore larva** (fig 4.20). This later metamorphoses; its lower region elongates and develops several segments. Its ciliated bands disappear, and the trochophore settles on the sea bed where it develops into the adult form.

4.7.3 Class Oligochaeta

Lumbricus, an earthworm, is an elongated, cylindrical organism, approximately 12–18 cm in length. The anterior end of the body is tapered, whilst the posterior end is dorso-ventrally flattened. Despite being a terrestrial animal, it has not fully overcome all the problems associated with life on land. In order to protect itself from desiccation it lives underground in burrows in damp soil, and emerges only at night to feed and reproduce. The differences in body form exhibited by *Lumbricus* as compared with *Nereis* are the result of its adaptation to a subterranean life.

The body is streamlined with no projecting structures which might impede its passage through the soil. The prostomium is a small, rounded structure without sensory appendages overlying the mouth. Each segment, except

the first and last, possesses four pairs of chaetae, two positioned ventrally and two ventro-laterally. The chaetae protrude from chaetigerous sacs located in the body wall and are able to be protracted or retracted by the action of specialised muscle blocks (fig 4.21). They are used during locomotory activity. Longer chaetae are present on segments 10–15, 26 and 32–37, and are used during copulation. Another reproductive structure, the **clitellum**, is situated on segments 32–37 (fig 4.22). Here the epidermis is dorsally and laterally swollen with gland cells that form a very noticeable saddle. The clitellum aids in the processes of copulation and cocoon formation.

The structure of the body wall is similar to that of *Nereis* and is shown in fig 4.21. There is a terminal mouth and anus. Food is ingested by the muscular action of a non-eversible pharynx. The gut is straight, and its digestive and absorptive surface is increased by the presence of a **typhlosole** (a dorsal longitudinal fold on the intestine which protrudes into the gut lumen). *Lumbricus* is a detritus

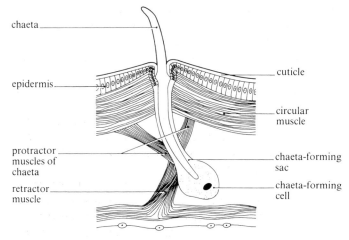

Fig 4.21 *VS body wall of* Lumbricus terrestris *through chaeta*

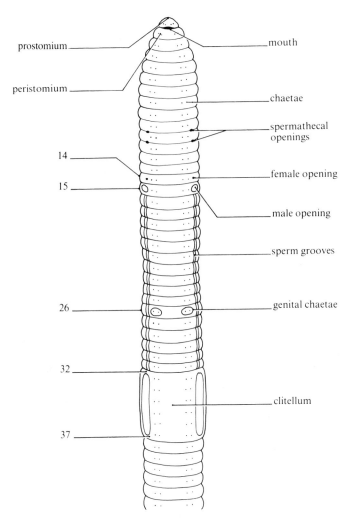

prostomium

peristomium

mouth

chaetae

spermathecal openings

14

15

female opening

male opening

sperm grooves

genital chaetae

26

32

37

clitellum

Fig 4.22 *Ventral view of anterior region of* Lumbricus terrestris

feeder, digesting organic materials from the soil it swallows. Food is absorbed into blood capillaries lining the intestinal wall. The majority of the soil passes straight through the worm, much of it eventually being deposited as castings onto the surface of the ground.

Secretions of coelomic fluid via dorsal pores, and mucus from epidermal mucous glands, keep the worm's thin cuticle moist. It is here that gaseous exchange occurs by diffusion, a process that is helped by the presence of networks of looped blood capillaries in the epidermal layer.

There is a pair of excretory and osmoregulatory nephridia in every segment except the first three and the last one. They open on to the surface of the worm in front of the ventro-lateral chaetae via nephridiopores. **Chloragogenous cells** found around the gut also aid in excretion.

Blood, collected from the segments, flows forwards in the dorsal contractile vessel. It is passed to the median ventral vessel by five pairs of muscular, lateral **pseudohearts** in segments 7–11. Valves present in the hearts and the dorsal vessel prevent backflow. The ventral

vessel distributes blood to all segments via lateral branches.

Though there is no noticeable aggregation of sensory structures at its anterior end, *Lumbricus* possesses sensory cells which respond to touch, chemicals and light. These are distributed throughout the epidermis. The central nervous system is similar to that of *Nereis*. Giant fibres present in the ventral cord enable the worm to contract its whole body in response to particularly irritating stimuli, whilst the general design of the nervous system permits the coordinated activity of the muscle layers necessary for the normal burrowing and locomotory activity of the worm.

The reproductive system and behaviour of earthworms is very complex. This can be associated with their terrestrial mode of life and the necessity to avoid desiccation of gametes and fertilised eggs. *Lumbricus* is hermaphrodite (fig 4.23). This is an adaptation to a relatively sedentary existence. Contact between worms is infrequent, but when it does occur, because they are hermaphrodite, any two worms of the same species will be able to copulate. This involves reciprocal transfer of male gametes and leads to mutual fertilisation.

The sex organs are grouped at the anterior end of each worm. The exact location of the reproductive organs in specific segments is shown in fig 4.23. Mating and subsequent laying of fertilised eggs in **cocoons** is a complicated process which can be summarised as follows.

During the spring and summer months, on warm, moist nights, worms protrude from their burrows, rarely leaving them completely, and pair with one of their neighbours. The ventral surfaces of two worms press against each other, the head of one worm pointing to the tail of the other. Such a position ensures that segments 9–11 of one worm are opposite the clitellum of the other and vice versa.

The long chaetae of the clitellar region and segments 10–15 and 26 are thrust into the body of each mating partner to maintain close contact during copulation.

The epidermis of each worm secretes a **mucus sheath** around itself from segments 11–31; this keeps the sperms of each partner separate during copulation and provides a closed channel for their passage.

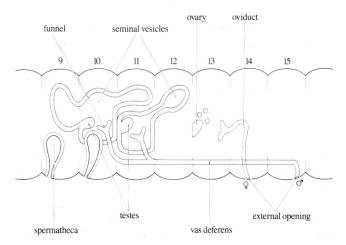

funnel

seminal vesicles

ovary

oviduct

9   10   11   12   13   14   15

spermatheca

testes

vas deferens

external opening

Fig 4.23 *Lateral view of position of reproductive organs*

105

In the clitellar regions a tube common to both partners is secreted which binds them tightly together.

Sperms from the seminal vesicles leave each worm via the openings of the vasa deferentia on segment 15 and are propelled backwards along the ventral seminal grooves of each worm. Their movement is facilitated by contraction of arch-shaped muscles found among the longitudinal muscle layer of segments 15–32. When the sperms reach segments 9 and 10 of their partner they pass into its **spermathecae**.

Once sperm have been exchanged (the process takes 3–4 h), the worms separate. Two days later **cocoon** formation begins.

The glandular epithelium secretes a tough chitinous tube around itself. This becomes the sheath of the cocoon. **Albumen** is secreted into the cocoon by the clitellum. This later nourishes the embryo. Expansion of the segments behind the cocoon force it towards the anterior of the worm. As this is happening 10–12 eggs are passed into it from the oviducal openings of segment 14. When the cocoon passes the spermathecal openings on segments 9 and 10, sperm are deposited in it and the eggs are fertilised. The cocoon is eventually forced clear of the worm. Its ends seal quickly, thus preventing desiccation. Initially the cocoon is yellow, but later dries and darkens in colour.

Cocoons are formed every 3–4 days until all sperms have been used. This can continue for a whole year without further pairing being necessary.

Development is direct, there being no free-swimming larval stage. Usually only one embryo develops per cocoon, and a young worm hatches 2–12 weeks after laying, depending on environmental conditions.

Agricultural importance of earthworms

Burrowing activity permits greater penetration of air into the soil, and improves the drainage capacity of the soil. It also enables roots to grow downwards through the soil more easily. Mixing and churning of the soil is brought about when earth which contains inorganic particles is brought up to the surface from lower regions.

Worms do not swallow particles greater than 2 mm in diameter. Thus when soil is deposited on the surface as casts, it is stone-free and provides a good medium for seed germination. Earthworm activity at the soil surface may cover seeds and promote more effective germination.

Leaves may be pulled underground by worms and partially digested. The remainder of the leaves will add to the organic content of the soil, as will the excretory wastes and secretions of worms and the bodies of dead worms.

The pH of worm casts is approximately 7. This has the advantage of preventing soils becoming excessively acid or alkaline.

4.8   Phylum Mollusca

The phylum Mollusca consists of a diverse group of organisms which include slow-moving snails and slugs, relatively sedentary bivalves, such as clams, and highly active cephalopods (fig 4.24). With over 80 000 living species and 35 000 fossil species it is second only in size to the Arthropoda. One of the molluscs, the giant squid, is the largest non-vertebrate animal, weighing several tonnes and measuring 16 m in length.

The formation of a protective shell, possession of external or internal fertilisation mechanisms, and use of gills or lungs for gaseous exchange has enabled molluscs to colonise aquatic and terrestrial environments and thus occupy a wide range of ecological niches. However a shell can be a handicap to locomotion, and some of the more active molluscs show a reduction or loss of the shell.

There is strong evidence that molluscs may have evolved from an ancestral annelid-like ancestor. For instance, molluscs and annelid polychaetes exhibit spiral cleavage during embryological development, and form almost identical trochophore larvae. Also the discovery of a molluscan 'living fossil' *Neopilina*, which possesses segmentally arranged gills, gonads, excretory organs and shell muscle, suggests that the early molluscs were built on a segmental plan, as are the annelids.

However, there is equally as much evidence to suggest that the Mollusca may have had an ancestor among the early platyhelminth turbellarians. The ancestry of the molluscs, therefore, is still far from clear.

4.9   Phylum Arthropoda

The phylum Arthropoda contains more species than any other phylum. Arthropods have exploited every type of habitat on land and in water and exist at all latitudes. Within each class there is tremendous **adaptive radiation** (section 24.7.6). The arthropod body design can be regarded as an elaboration of the segmented body plan of annelids. Ancestral arthropods possessed a series of similar simple appendages along the length of their bodies, which probably served a variety of purposes such as gaseous exchange, food gathering, locomotion and detection of stimuli. The success of the arthropods is said to be

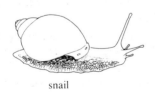
snail

clam

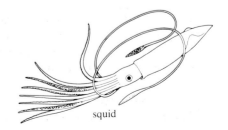
squid

Fig 4.24 *A variety of molluscs*

Table 4.6 Classification of phylum Mollusca.

Phylum Mollusca

Characteristic features
Unsegmented, triploblastic coelomates
Usually bilaterally symmetrical
Body divided into a head, ventral muscular foot and dorsal visceral hump
Skin soft, and over the hump it forms a mantle which secretes a calcareous shell
Heart and open haemocoelic system
Respiratory pigment usually haemocyanin
Nervous system consisting of circumoesophageal ring with cerebral and pleural ganglia, pedal cords and visceral loops
Basically oviparous with a trochophore larva

There are six classes of molluscs but only the three major classes are described here.

Class Gastropoda	Class Pelycopoda (Bivalvia)	Class Cephalopoda
Terrestrial, marine and freshwater	Aquatic	Aquatic
Asymmetrical	Bilateral symmetry	Bilateral symmetry, long axis of body dorso-ventral
At some stage in their development they show torsion of the visceral mass	No torsion of visceral mass	No torsion of visceral mass
Anus is anterior		
Shell of one piece, usually coiled	Body laterally compressed and is enclosed by two valves (hence the term 'bivalve')	Chambered shell often reduced and internal or wholly absent
Head, eyes and sensory tentacles	Head greatly reduced in size, tentacles absent	Head highly developed. Tentacles and well-developed eyes
Land forms lost gills and converted mantle cavity into a lung	Large plate-like gills	Gills
Radula	Filter feeder	Radula and horny beak
Internal fertilisation	External fertilisation	Internal fertilisation
e.g. *Helix aspersa* (land snail) *Patella* (limpet) *Buccinum* (whelk) *Limax* (slug)	e.g. *Mytilus edulis* (marine mussel) *Ostrea* (oyster)	e.g. *Sepia officinalis* (cuttlefish) *Loligo* (squid) *Octopus vulgaris* (octopus)

the result of a process called '**arthropodisation**', which is the exploitation of potentialities latent within the annelid body plan. Many factors have contributed to this success and some of the major ones are listed below.

(1) The evolution of a firm **exoskeleton** (cuticle) which is resistant to changes of shape. This has been used to form a sytem of levers. **Joints** have developed between many of them leading to the formation of serially arranged, **jointed appendages**. Each segment is attached to its neighbour by means of a modified portion of cuticle which is thin and flexible. This allows each segment or lever to be moved independently of its adjacent component (fig 4.25). Constituents of the insect exoskeleton are as follows.

The **epicuticle** is composed of an outer cement layer of lipoprotein, two waterproof wax layers and a cuticulum layer associated with polyphenols. It is 3–6 μm thick and is the main waterproofing layer. It is almost impermeable and affords protection against entry of micro-organisms.

The **procuticle** is composed of chitin, arthropodin and resilin. Chitin is an amino polysaccharide which gives the cuticle a degree of flexibility. Arthropodin is a protein which complexes with the chitin. Its degree of hardness is increased if it is tanned, that is the arthropodin/chitin complex reacts with phenols, and during this reaction its molecular structure becomes much firmer due to the formation of many additional cross linkages.

$$\text{Arthropodin (soft)} \xrightarrow[\text{process}]{\text{tanning}} \text{Sclerotonin (hard)}$$

Resilin is an elastic protein. It is a natural rubber made up of amino acid chains running in all directions and randomly joined together.

(2) Portions of the exoskeleton, including many appendages, have been modified to form a variety of structures serving many different purposes. In addition, groups of adjacent appendages may carry out similar functions. This further increases the efficiency and complexity of the activity (fig 4.26).

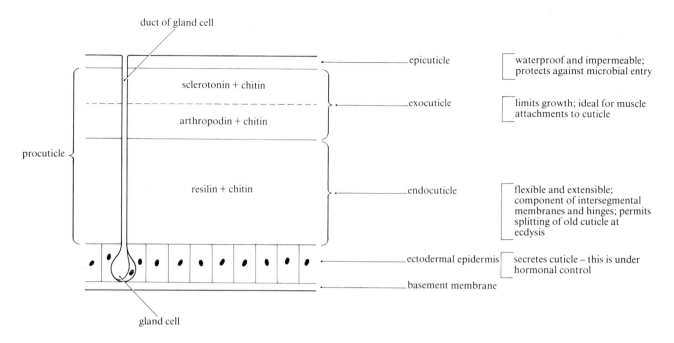

duct of gland cell

sclerotonin + chitin

arthropodin + chitin

procuticle

resilin + chitin

gland cell

epicuticle — waterproof and impermeable; protects against microbial entry

exocuticle — limits growth; ideal for muscle attachments to cuticle

endocuticle — flexible and extensible; component of intersegmental membranes and hinges; permits splitting of old cuticle at ecdysis

ectodermal epidermis — secretes cuticle – this is under hormonal control

basement membrane

Fig 4.25 *VS body wall of insect to show layers of exoskeleton*

(3) The division of labour which occurs in the arthropods has contributed to the development of distinct regions of the body, namely the **head**, and in many cases a **thorax** and an **abdomen**. The head incorporates sensory receptors (such as eyes, antennae and statocysts) and feeding appendages. The brain is much larger than in annelids, and **cephalisation** is more pronounced.

(4) The cuticle is waterproof, enabling some arthropod species, notably the insects, to exploit terrestrial habitats.

(5) A firm surface for muscle attachment is provided by the inner surface of the exoskeleton. The continuous muscle layers of the annelids are no longer apparent in arthropods. Instead, **antagonistic** pairs of muscles are present which facilitate the separate movement of individual appendages or segments (fig 4.27).

(6) Arthropod muscle has become **striated**. This increases the speed of muscle contraction and hence the animal's speed of response.

(7) A hard exoskeleton has imposed a size limitation on the arthropods, by virtue of its weight. Growth is difficult and can only take place if the exoskeleton is periodically shed; hence **ecdysis** (**moulting**) has evolved. However the arthropod is vulnerable to attack by predators during this period, and generally seeks the protection of shelter before undergoing the process.

(8) A coelom is not present as the main body cavity. Instead a **haemocoel** has developed. This is used to distend the body during moulting so that the old cuticle can be split open and cast off.

4.9.1 General adult insect morphology

External anatomy

The body of an adult insect is usually divided into three distinct regions, the head, thorax and abdomen (fig 4.29). The head bears one pair of jointed **antennae** whose form may vary considerably, eyes which may be of two types (**compound** and **simple**), and movable mouthparts. The mouthparts of insects are very diverse in form and function. Indeed insects may be classified into two groups on the basis of the construction of their mouthparts: chewing insects or mandibulate, and sucking insects or haustellate.

Whilst the two types of mouthparts differ considerably in appearance, their components are homologous (parts of different species that have the same evolutionary origin but serve different purposes because the organisms possessing them have undergone adaptive radiation). The mandibulate type is the more ancestral. Three pairs of appendages make up the mouthparts of an adult insect. They are the **mandibles**, **maxillae** and **segmental palps**, and the second maxillae which are usually divided into an upper lip (**labrum**) and a lower lip (**labium**).

The thorax is subdivided into three regions, the **pro-**, **meso-**, and **metathoracic segments**. A pair of **spiracles** is present on the meso- and metathoracic segments. The thorax also bears three pairs of **legs**. Again there is tremendous variation in the construction of the legs and the functions they perform. The legs may be modified for walking, running, leaping, swimming, grasping or even the production of sound. Most insects possess wings, but some orders such as Thysanura and Collembola, are entirely wingless. Usually there are two pairs of **wings**, one pair on

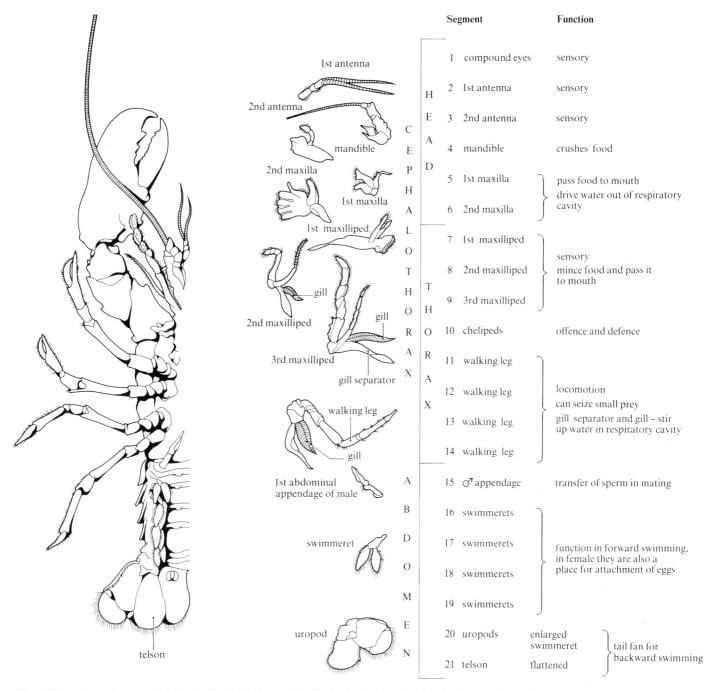

	Segment		Function
H E A D	1	compound eyes	sensory
	2	1st antenna	sensory
	3	2nd antenna	sensory
	4	mandible	crushes food
	5	1st maxilla	pass food to mouth
	6	2nd maxilla	drive water out of respiratory cavity
T H O R A X	7	1st maxilliped	sensory
	8	2nd maxilliped	mince food and pass it to mouth
	9	3rd maxilliped	
	10	chelipeds	offence and defence
	11	walking leg	locomotion
	12	walking leg	can seize small prey
	13	walking leg	gill separator and gill – stir up water in respiratory cavity
	14	walking leg	
A B D O M E N	15	♂ appendage	transfer of sperm in mating
	16	swimmerets	function in forward swimming, in female they are also a place for attachment of eggs
	17	swimmerets	
	18	swimmerets	
	19	swimmerets	
	20	uropods	enlarged swimmeret — tail fan for backward swimming
	21	telson	flattened

Fig 4.26 *Appendages of lobster (after Buchsbaum) to illustrate their variety in structure and function*

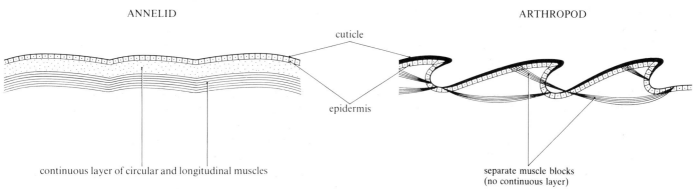

ANNELID   ARTHROPOD

cuticle

epidermis

continuous layer of circular and longitudinal muscles

separate muscle blocks
(no continuous layer)

Fig 4.27 *Comparison of the body wall of an annelid and an arthropod*

109

Table 4.7 Classfication of phylum Arthropoda.

Phylum Arthropoda

Characteristic features
Triploblastic, coelomate
Segmented, bilaterally symmetrical
Coelom much reduced, perivisceral cavity a haemocoel
Central nervous system of paired pre-oral ganglia connected by commissures to a ventral nerve cord; the ventral nerve cord is double, solid with segmental ganglia and nerves
No nephridia
Secreted exoskeleton of chitin and sometimes calcareous matter
Each segment typically bears a pair of jointed appendages used for locomotion or feeding or sensory purposes
Cilia completely lacking externally
Dorsal heart with open vascular system
Many larval forms found within the phylum

Superclass Crustacea*	*Class Chilopoda*	*Class Insecta*	*Class Arachnida*
Mainly aquatic	Mainly terrestrial	Mainly terrestrial	Terrestrial
Ill-defined cephalothorax	Clearly defined head	Well-defined head, thorax, abdomen	Divisions into prosoma and opisthosoma
Head of six segments 2 pairs of antennae	Head of six segments 1 pair of antennae	Head of six segments 1 pair of antennae	Prosoma of six segments not in any way homologous with the head of other arthropods – no antennae
Pair of compound eyes raised on stalks	Eyes simple, compound or absent	Pair of compound eyes and simple eyes	Simple eyes
At least three pairs of mouthparts (gnathites)	One pair of gnathites	Usually three pairs of gnathites	No true gnathites
	Numerous segments each bearing one pair of similar appendages	Three thoracic segments each with a pair of legs. Second and third thoracic segments usually have one pair of wings each	Segments 4–7 each possess a pair of walking legs
Abdomen typically 11 segments	Body segments all similar	Abdomen typically 11 segments	Abdomen typically 13 segments not all externally visible in some cases
Genital apertures in thoracic segments	Median genital opening	Genital apertures near anus on abdomen	Genital apertures on second abdominal segment
Straight gut	Straight gut	Gut may be coiled	Highly specialised gut to deal with liquid food
Hepatic caecae open into mesenteron	No hepatic caecae	No hepatic caecae	Hepatic caecae open into mesenteron
Typically nauplius or other larval form. May be direct development	No larval form	Commonly a complicated meta-morphosis. Development may be direct with a nymphal stage or indirect with larval stages	No larval form
Typically gaseous exchange by gills – outgrowths of the body wall or limbs	Gaseous exchange by tracheae	No gills in adult. Gaseous exchange by tracheae	Gaseous exchange by internal air spaces 'lungs' or 'gill' books or tracheae
e.g. *Daphnia* (water-flea) *Astacus* (crayfish)	e.g. *Lithobius* (centipede) *Iulus* (millipede)	e.g. *Periplaneta* (cockroach) *Apis* (bee) *Pieris* (white butterfly)	e.g. *Scorpio* (scorpion) *Epeira* (web-spinning spider)

*This superclass contains many classes e.g. Malacostraca

the mesothorax and the other pair on the metathorax. The most primitive forms of wing are the membranous wings as seen in the Hymenoptera and Diptera. Other forms evolving from these include the hairy wings of the Trichoptera, scaly wings of the Lepidoptera, the leathery covers (tegmina) of the Orthoptera, and the horny wing covers (elytra) of the Coleoptera. In the Diptera the second pair of wings has been replaced by a pair of balancing organs, the **halteres**.

The number of abdominal segments varies from between 3–11. Most segments are without any appendages, but segments 8 and 9 may possess reproductive appendages. In

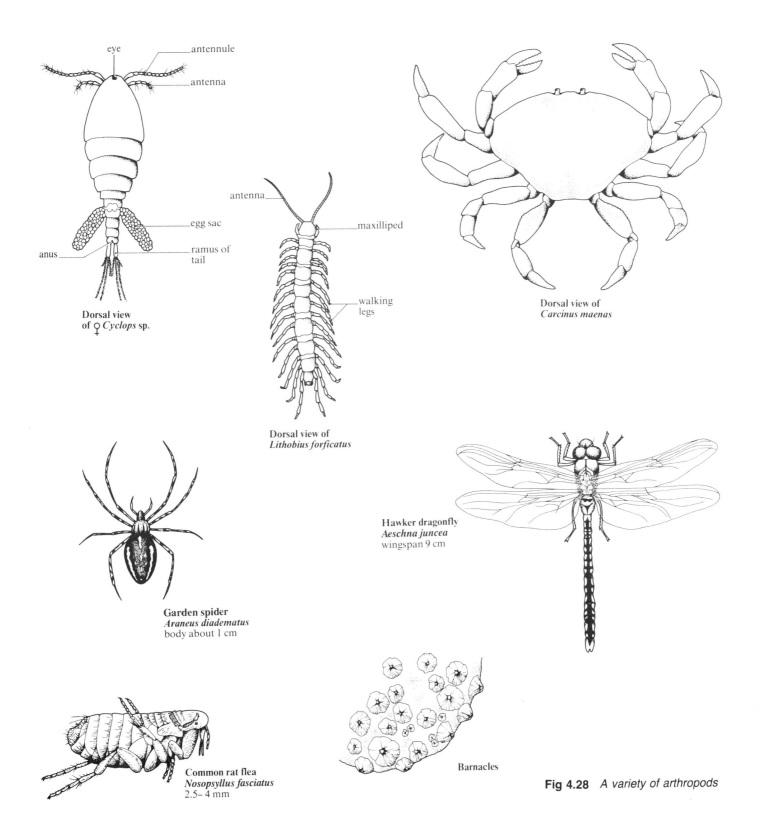

eye

antennule

antenna

egg sac

ramus of tail

anus

Dorsal view of ♀ *Cyclops* sp.

antenna

maxilliped

walking legs

Dorsal view of *Lithobius forficatus*

Dorsal view of *Carcinus maenas*

Garden spider *Araneus diadematus* body about 1 cm

Hawker dragonfly *Aeschna juncea* wingspan 9 cm

Common rat flea *Nosopsyllus fasciatus* 2.5–4 mm

Barnacles

Fig 4.28 *A variety of arthropods*

some species the 11th segment bears **cerci**. Spiracles are usually present on abdominal segments 1–8.

Internal anatomy

The alimentary canal generally consists of a mouth, pharynx, oesophagus, crop, gizzard and intestine (fig 4.30).

The size of the organs varies according to the nature of the diet of the insect.

The blood of an insect does not circulate through its body in arteries and veins; instead it flows through a haemocoel. The dorsal heart is a tube closed posteriorly and open at its anterior. It is perforated at points along its length by pairs of lateral holes called **ostia**. Contractions of the heart,

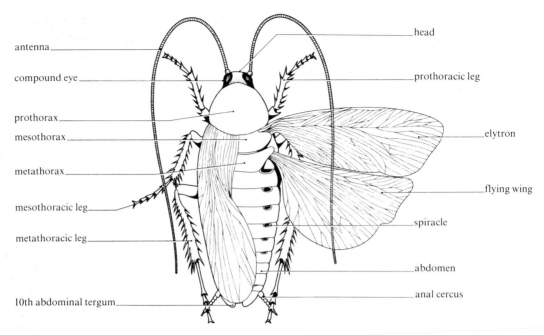

antenna

compound eye

prothorax

mesothorax

metathorax

mesothoracic leg

metathoracic leg

10th abdominal tergum

head

prothoracic leg

elytron

flying wing

spiracle

abdomen

anal cercus

Fig 4.29 *Dorsal view of a cockroach (male)*

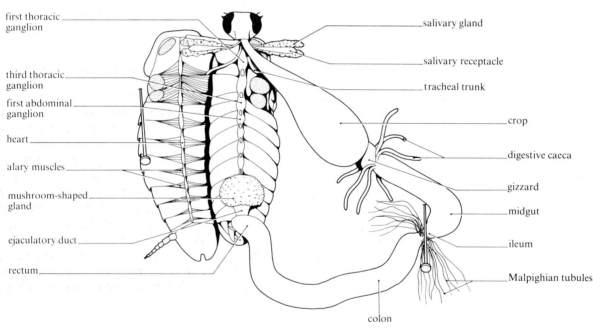

first thoracic ganglion

third thoracic ganglion

first abdominal ganglion

heart

alary muscles

mushroom-shaped gland

ejaculatory duct

rectum

salivary gland

salivary receptacle

tracheal trunk

crop

digestive caeca

gizzard

midgut

ileum

Malpighian tubules

colon

Fig 4.30 *Internal anatomy of male cockroach*

beginning posteriorly and moving forwards, force the blood towards the anterior end of the insect. The blood is pumped out into the haemocoel where it bathes all tissues. It later percolates back into the heart.

Commissures connect the dorsal cerebral ganglion to a ventral suboesophageal ganglion. The ventral nerve cord consists of segmentally arranged ganglia joined by connectives. It runs the length of the insect body. Cephalisation is considerable, this being correlated with the highly developed antennae and eyes.

Gaseous exhange is effected by a **tracheal system** consisting of a number of air tubes which pass to all parts of the body. The tubes open out onto the surface of the insect via apertures called **spiracles** (fig 4.29). The spiracles may be opened and closed, the mechanism being closely related to the levels of oxygen and carbon dioxide in the blood.

Malpighian tubules are the chief organs of nitrogenous excretion in most insects. They excrete **uric acid**. The uric acid is mixed with faeces in the hindgut and eventually expelled via the rectum. If necessary, water can be reabsorbed by the rectum.

The sexes are separate in insects. Typically the female reproductive system consists of two ovaries and two lateral oviducts which unite to form a common oviduct. This leads to the vagina. Accessory glands and a spermatheca are also present. The male possesses a pair of testes and a pair of lateral sperm ducts (vasa deferentia). The lower part of each duct is enlarged to form a seminal vesicle in which sperm are stored. The sperm ducts unite to form a common ejaculatory duct which leads into an extensible or eversible penis. Accessory glands which secrete seminal fluid are also present (fig 4.30).

4.9.2   Insect life histories

Life histories of insects are very variable and often highly complex. In many, a process called **metamorphosis** (*meta*, change; *morphe*, form) occurs. This is an abrupt change of form or structure of the animal during the course of its life cycle.

In the more primitive insect groups the larval stages often resemble the adult (**imago**) during development. Each successive larval form (called a **nymph** or **instar**) usually looks more and more like the adult. This form of development is termed **hemimetabolous** metamorphosis. It may be further subdivided into **gradual** metamorphosis, where the nymph lives in the same habitat as the adult and eats the same food, or **incomplete** metamorphosis, where the nymph possesses adaptive features which enable it to live in a different habitat and eat different food from that of the adult. This avoids competition for food between juvenile and adult.

In later groups, the larval stages are morphologically quite distinct from the adult. The final **larval** moult produces a sedentary **pupa**, inside which the drastic metamorphosis produces the adult tissues, using components from the degenerating larval tissues. This is called **holometabolous** or **complete** metamorphosis. Metamorphosis is under hormonal control, and is discussed more fully in section 21.7.

4.9.3   Classification of insects based on types of metamorphosis used in the life history

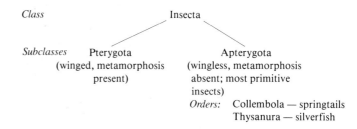

Hemimetabolous metamorphosis	Holometabolous metamorphosis
Direct development	Indirect development
Wings develop externally	Wings develop internally
Miniature stages – nymphs resemble adults	Immature stages – larvae differ in structure and function from adults
	Last pre-adult stage a pupa
Orders:	*Orders:*
Ephemeroptera – mayflies	Diptera – blowflies, mosquitoes
Dictyoptera – cockroaches	Lepidoptera – butterflies,
Orthoptera – locusts	moths

4.9.4   Implications of metamorphosis

Metamorphosis enables the juvenile and adult forms to live in different habitats and exploit different sources of food, that is to occupy different ecological niches. This reduces competition between juveniles and adults. For instance, dragonfly nymphs (naiads) prey upon aquatic insects and exchange gases via gills, whereas the adults attack terrestrial insects, live in air and exchange gases via tracheae. Also, lepidopteran larvae generally feed on foliage and possess chewing mouthparts, whereas the adults drink nectar and have sucking mouthparts.

Since adults seldom grow after the last moult, metamorphosis also allows the immature stages to provide the feeding and growing periods of the insect's life history.

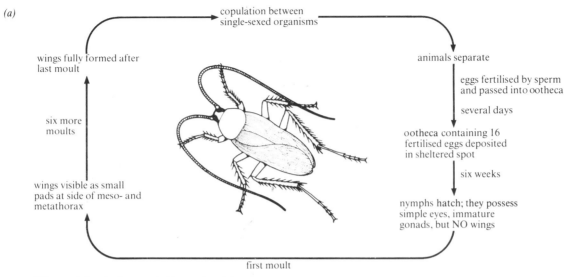

(a)

copulation between single-sexed organisms

wings fully formed after last moult

animals separate

eggs fertilised by sperm and passed into ootheca

several days

six more moults

ootheca containing 16 fertilised eggs deposited in sheltered spot

six weeks

wings visible as small pads at side of meso- and metathorax

nymphs hatch; they possess simple eyes, immature gonads, but NO wings

first moult

NB nymph lives in the same habitat as the adult and eats the same food – gradual metamorphosis

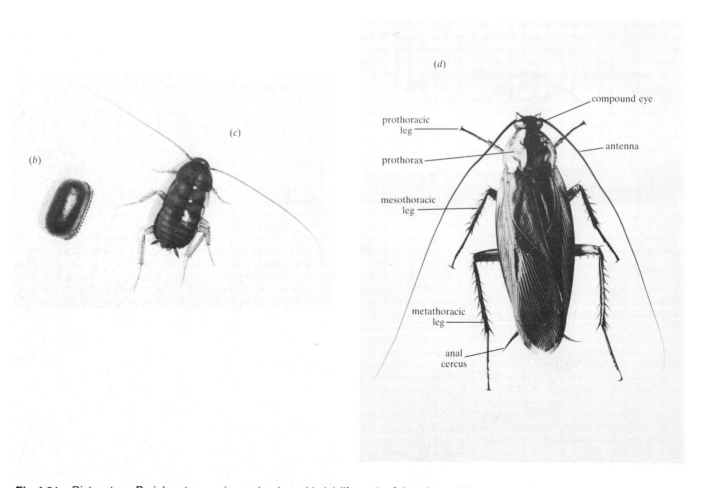

(d)

(c)

(b)

compound eye

prothoracic leg

antenna

prothorax

mesothoracic leg

metathoracic leg

anal cercus

Fig 4.31 *Dictyoptera,* Periplaneta americana *(cockroach): (a) life cycle (b) ootheca, (c) nymph, (d) imago*

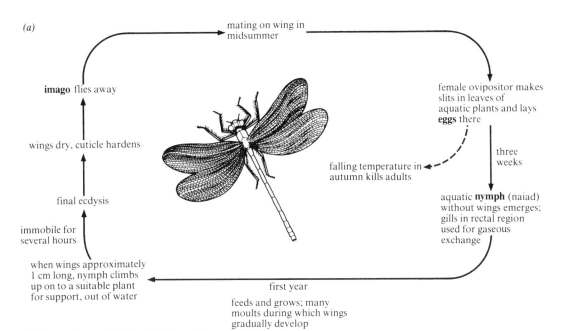

(a)

mating on wing in midsummer

imago flies away

wings dry, cuticle hardens

final ecdysis

immobile for several hours

when wings approximately 1 cm long, nymph climbs up on to a suitable plant for support, out of water

female ovipositor makes slits in leaves of aquatic plants and lays **eggs** there

falling temperature in autumn kills adults

three weeks

aquatic **nymph** (naiad) without wings emerges; gills in rectal region used for gaseous exchange

first year

feeds and grows; many moults during which wings gradually develop

NB nymphs resemble the adult (imago) in general body form but possess adaptations which fit them to live in water, a habitat quite different to the terrestrial environment of the imago – **incomplete metamorphosis**

(b)

folded mask

wings developing

(c)

Fig 4.32 *Odonata*, Aeschna juncea *(dragonfly): (a) life cycle, (b) aquatic nymph, (c) imago*

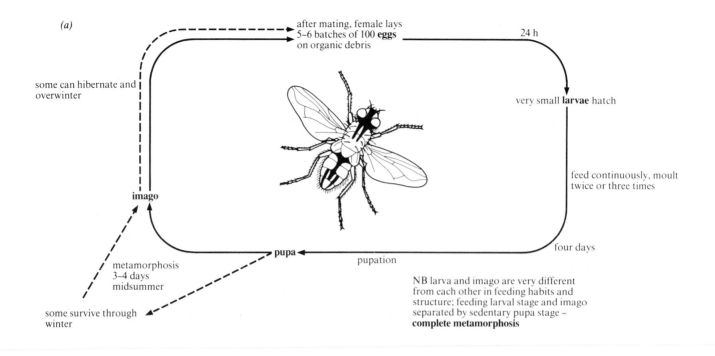

(a)

after mating, female lays
5–6 batches of 100 **eggs**
on organic debris

24 h

some can hibernate and
overwinter

very small **larvae** hatch

feed continuously, moult
twice or three times

imago

pupa

pupation

four days

metamorphosis
3–4 days
midsummer

NB larva and imago are very different
from each other in feeding habits and
structure; feeding larval stage and imago
separated by sedentary pupa stage –
complete metamorphosis

some survive through
winter

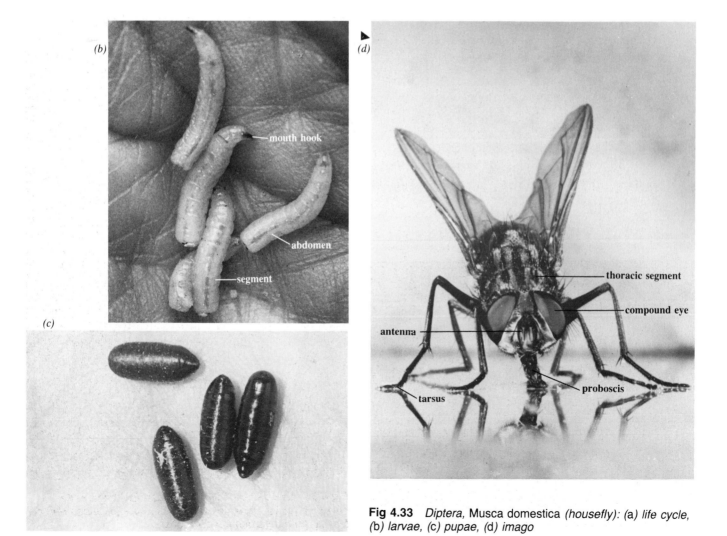

(b)

mouth hook

abdomen

segment

(c)

(d)

thoracic segment

compound eye

antenna

proboscis

tarsus

Fig 4.33 *Diptera*, Musca domestica *(housefly): (a) life cycle,*
(b) larvae, (c) pupae, (d) imago

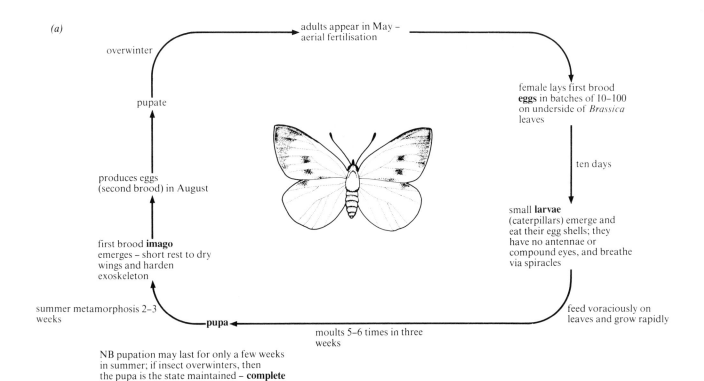

(a)

overwinter → adults appear in May – aerial fertilisation

pupate

female lays first brood **eggs** in batches of 10–100 on underside of *Brassica* leaves

ten days

produces eggs (second brood) in August

small **larvae** (caterpillars) emerge and eat their egg shells; they have no antennae or compound eyes, and breathe via spiracles

first brood **imago** emerges – short rest to dry wings and harden exoskeleton

summer metamorphosis 2–3 weeks

feed voraciously on leaves and grow rapidly

pupa

moults 5–6 times in three weeks

NB pupation may last for only a few weeks in summer; if insect overwinters, then the pupa is the state maintained – **complete metamorphosis**

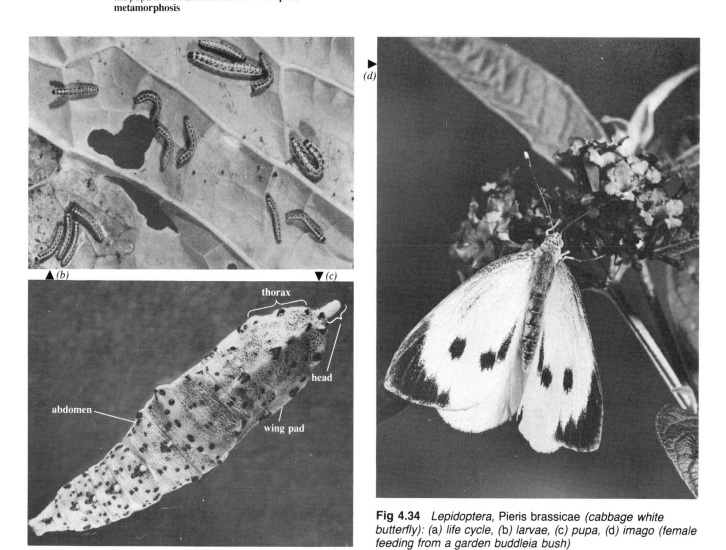

▲ *(b)*

▼ *(c)*

▶ *(d)*

thorax

head

wing pad

abdomen

Fig 4.34 *Lepidoptera,* Pieris brassicae *(cabbage white butterfly): (a) life cycle, (b) larvae, (c) pupa, (d) imago (female feeding from a garden buddleia bush)*

Table 4.8 Classification of phylum Echinodermata.

Phylum Echinodermata

Characteristic features
Triploblastic, coelomate
All marine
Water vascular system is part of the coelom
Tube feet
Calcareous exoskeleton
No special excretory organs present
Sexes are separate
Basic larval stage called dipleurula, possesses ciliated band and is the main dispersive phase – pelagic
Larva bilaterally symmetrical, adult shows pentamerous symmetry

Class Stelleroidea	*Class Echinoidea*	*Class Crinoidea*	*Class Holothuroidea*	*Subclass Ophiuroidea*
Free-living	Free-living	Attached during part or all of its life by aboral stalk	Free-living	Free-living
Star-shaped, flattened	Globular	Star-shaped	Cucumber-shaped	Star-shaped
Arms not sharply demarcated from disc	Does not possess arms	Arms	Body not drawn into arms	Very long arms sharply demarcated from central disc
Few calcareous plates in body wall; movable spines	Numerous plates in body wall	No spines	No external spines	Spines and calcareous plates
e.g. *Asterias* (starfish)	e.g. *Echinocardium* (sea urchin)	e.g. *Antedon* (feather star)	e.g. *Holothuria* (sea cucumber)	e.g. *Ophiothrix* (brittle star)

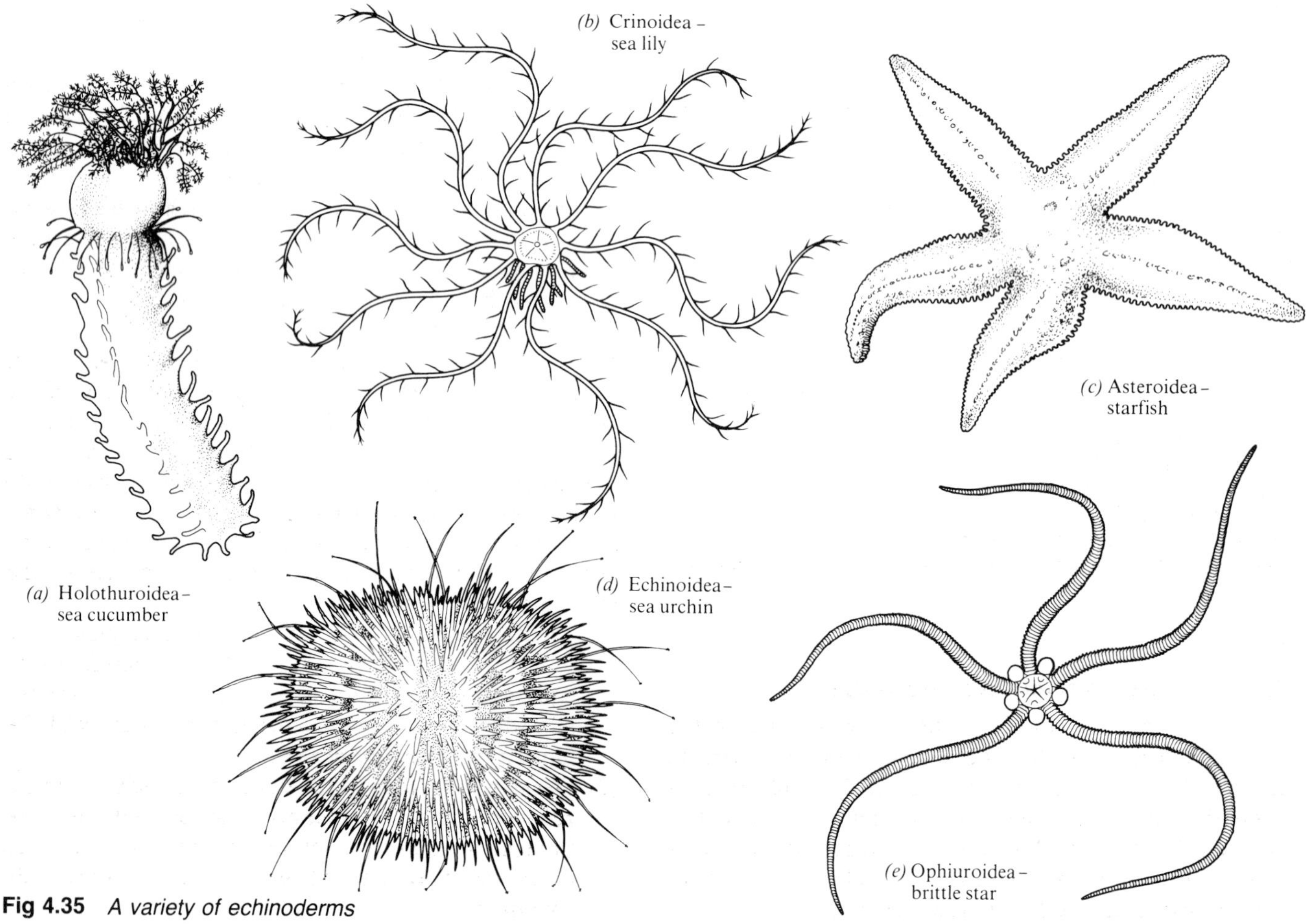

(a) Holothuroidea – sea cucumber

(b) Crinoidea – sea lily

(c) Asteroidea – starfish

(d) Echinoidea – sea urchin

(e) Ophiuroidea – brittle star

Fig 4.35 *A variety of echinoderms*

4.10 Phylum Echinodermata

There are over 5 000 known species of echinoderms. They are all marine and are largely bottom-dwellers inhabiting shorelines and shallow seas. The adult forms exhibit **pentamerous symmetry** (a modified form of radial symmetry), this being secondarily developed from a bilateral ancestor (fig 4.35). Their most unique characteristic is the possession of a **water vascular system**, a complex of tubes surrounding the mouth and passing into the arms and tube feet.

There are a number of striking resemblances between the echinoderms and chordates that lend support to the suggested affinities between them. For example, **radial cleavage** occurs during the development of echinoderm and chordate embryos. This contrasts with the **spiral cleavage** of annelids, molluscs and arthropods (fig 4.36). The **blastopore** (the opening of the blastocoel) forms the anus in echinoderms and chordates whereas it becomes the mouth in the annelids, molluscs and arthropods. On the basis of the fate of the blastopore, the former organisms are classed as **deuterostomes**, whilst the latter are termed **protostomes**.

4.11 Phylum Chordata

Table 4.9 Characteristic features of chordates.

Phylum Chordata

Characteristic features
Notochord present at some stage in the life history. This is a flexible rod of tightly packed, vacuolated cells held together within a firm sheath
Triploblastic, coelomate
Bilateral symmetry
Pharyngeal (visceral) clefts present
Dorsal, hollow nerve cord
Segmental muscle blocks (myotomes) on either side of the body
Post-anal tail
Closed blood system
Blood flows forwards ventrally, backwards dorsally
Ventral vessel connected to dorsal vessel by blood vessels located in the visceral arches
Limbs formed from more than one body segment

4.11.1 Non-vertebrate chordates

A brief review of the major features of the three non-vertebrate classes Hemichordata, Urochordata and Cephalochordata shows a gradual trend towards possession of all chordate features throughout the entire life cycle. It establishes links between the protostome (non-vertebrate) phyla, the echinoderms and the vertebrates and suggests a course for chordate evolution.

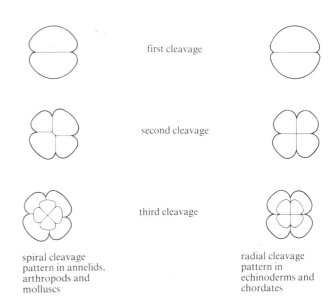

first cleavage

second cleavage

third cleavage

spiral cleavage pattern in annelids, arthropods and molluscs

radial cleavage pattern in echinoderms and chordates

Fig 4.36 *Comparison of methods of cleavage in animals*

4.11.2 Chordate phylogeny

It has been proposed that the urochordate **ascidian tadpole** evolved into a pelagic fish-like chordate by **neoteny**, a process whereby the organism becomes sexually mature and reproduces whilst retaining the body form of the larval stage (fig 4.38).

It is thought that the ascidian larva shared a common ancestor with the echinoderm larva (fig 4.39). The ciliated circumoral band of the echinoderm larva, in moving to a dorsal position and rolling inwards with its associated nervous tissue, gave rise to the **dorsal hollow nerve cord**. The tail muscle and notochord evolved and increased the locomotory power and internal support of the larva, hence increasing the organism's size and activity.

During the late Devonian and lower Carboniferous periods land generally rose whilst the sea level was lowered. Consequently there was a redistribution of the aquatic medium. Professor Romer has suggested that it was this increasing lack of water and drying up of large areas that forced the crossopterygians (lobe-finned fish) onto land to seek new aquatic habitats (fig 4.40). As a result they began to spend more time on land. In order to exploit the terrestrial environment vertebrates had to overcome the following major problems.

(1) Breathing gaseous oxygen – crossopterygians possessed well-developed lungs and were therefore well equipped at the outset. Even so, they still possessed gills which were their main respiratory organs, with the lungs as a secondary means.

(2) Desiccation – there is evidence that early amphibia retained their fish ancestor's scales and they are thought never to have ventured far from water. It was during the Permian that they evolved resilient body coverings.

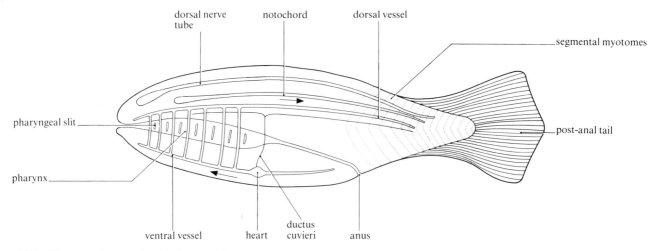

Fig 4.37 *Diagram showing basic chordate features*

Labels: dorsal nerve tube, notochord, dorsal vessel, segmental myotomes, pharyngeal slit, pharynx, ventral vessel, heart, ductus cuvieri, anus, post-anal tail

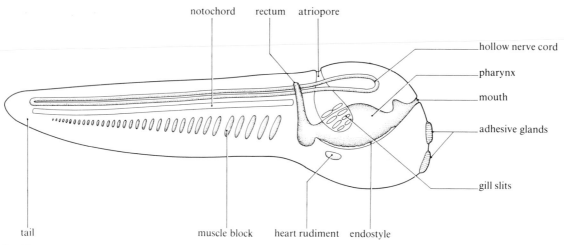

Fig 4.38 *Ascidian tadpole*

Labels: notochord, rectum, atriopore, hollow nerve cord, pharynx, mouth, adhesive glands, gill slits, tail, muscle block, heart rudiment, endostyle

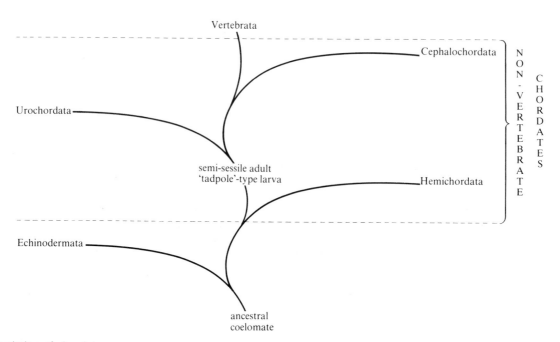

Fig 4.39 *Evolution of chordates*

Labels: Vertebrata, Cephalochordata, NON-VERTEBRATE CHORDATES, Urochordata, semi-sessile adult 'tadpole'-type larva, Hemichordata, Echinodermata, ancestral coelomate

Table 4.10 Classification of non-vertebrate chordates.

	Class Hemichordata	*Class Urochordata*	*Class Cephalochordata*
Non-chordate features	Terminal anus Blood flows forward in dorsal blood vessel Pelagic larva similar to holothurian echinoderm larva	No trace of notochord in adult No nerve cord in adult Adult a sessile filter feeder, structurally nothing like a chordate	
Chordate features	Tripartite body plan of preoral proboscis, collar and trunk Pharyngeal slits, may have arisen initially to dispose of excess water created by feeding mechanism. Latterly developed into food-collecting device	Gill slits in adult multiplied to form large filter-feeding pharynx Larva an ascidian tadpole, possesses the following features: notochord, pharyngeal slits, dorsal tubular nerve cord, segmental myotomes, post-anal tail	Fish-like animals showing all recognisable chordate features Notochord extends length of body in larval and adult stages Large pharynx with clefts forms feeding mechanism Ciliated gill bars Pharyngeal slits open into atrium Segmental myotomes No head or limbs
	e.g. *Saccoglossus*	e.g. *Ciona intestinalis*	e.g. *Amphioxus lanceolatus*

Table 4.11 Classification of subphylum Vertebrata.

Subphylum Vertebrata

Characteristic features
Well-developed central nervous system including brain
Internal skeleton
Pharyngeal clefts few in number
Kidneys for nitrogenous excretion and osmoregulation
Muscular ventral heart
Two pairs of limbs

Class Agnatha (Cyclostomata)	Class Chondrichthyes	Class Osteichthyes	Class Amphibia	Class Reptilia	Class Aves	Class Mammalia
Slimy skin	Skin with placoid scales	Skin with cycloid scales	Soft skin	Skin dry with horny scales and bony plates	Skin bears feathers, legs have scales	Skin bears hair with two types of glands, sebaceous or sudoriparous
	Cartilaginous skeleton	Bony skeleton				
Paired limbs present	Paired, fleshy pectoral and pelvic fins	Paired pectoral and pelvic fins supported by rays	Paired pentadactyl limbs	Paired pentadactyl limbs usually present	Paired pentadactyl limbs, front pair form wings	Paired pentadactyl limbs
	Visceral clefts present as separate gill openings	Visceral clefts present as separate gill openings, but covered by a bony flap (operculum)	Visceral clefts present in tadpole only, lungs in adult	Visceral clefts never develop gills	Visceral clefts never develop gills	Visceral clefts never develop gills
	Lateral line system well developed	Lateral line system well developed	Lateral line system in tadpole only	No lateral line	No lateral line	No lateral line
	Inner ear, no middle or external ear	Inner ear, no middle or external ear	Inner and middle ear, no external ear	Inner and middle ear, no external ear	Inner and middle ear, no external ear	External, middle and inner ear, middle ear develops 3 ear ossicles
	No larval stage	Larval stage	Larval stage	No larval stage	No larval stage	No larval stage
	Eggs produced, internal fertilisation	Eggs produced, external fertilisation	Eggs produced, external fertilisation	Oviparous eggs laid, or eggs retained until hatching (ovoviviparous), internal fertilisation	Yolky eggs in calcareous shells, oviparous, internal fertilisation	Eggs develop within mother (except two genera), viviparous, internal fertilisation
						Muscular diaphragm between thorax and abdomen
e.g. *Myxine* (hag fish) *Lampetra* (lamprey)	e.g. *Scyliorhinus* (dogfish)	e.g. *Clupea* (herring)	e.g. *Rana* (frog) *Bufo* (toad)	e.g. *Natrix* (grass snake) *Crocodylus* (crocodile)	e.g. *Columba* (pigeon) *Aquila* (eagle)	e.g. *Homo* (human) *Canis* (dog)

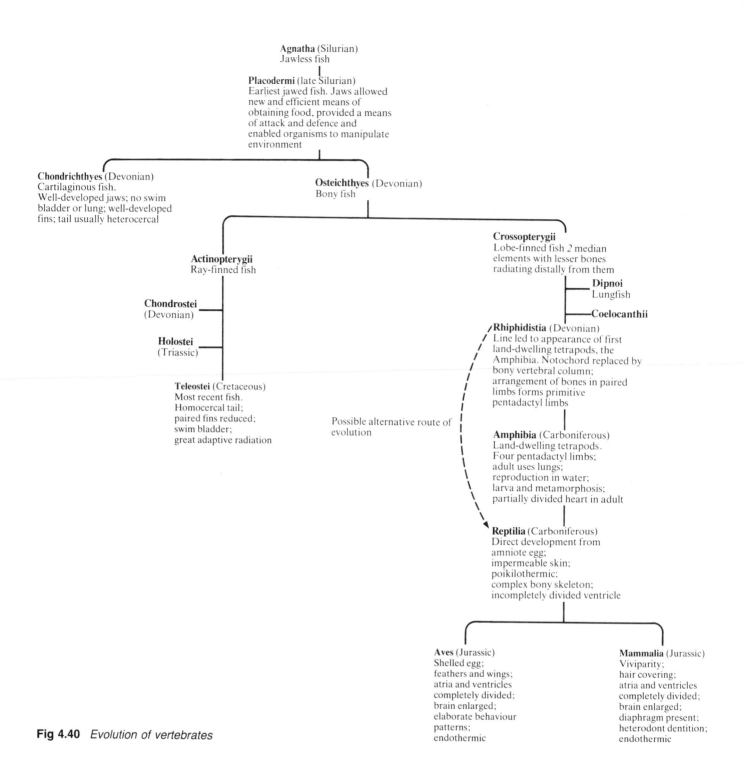

Agnatha (Silurian)
Jawless fish

Placodermi (late Silurian)
Earliest jawed fish. Jaws allowed
new and efficient means of
obtaining food, provided a means
of attack and defence and
enabled organisms to manipulate
environment

Chondrichthyes (Devonian)
Cartilaginous fish.
Well-developed jaws; no swim
bladder or lung; well-developed
fins; tail usually heterocercal

Osteichthyes (Devonian)
Bony fish

Actinopterygii
Ray-finned fish

Crossopterygii
Lobe-finned fish 2 median
elements with lesser bones
radiating distally from them

Chondrostei
(Devonian)

Dipnoi
Lungfish

Coelocanthii

Holostei
(Triassic)

Rhiphidistia (Devonian)
Line led to appearance of first
land-dwelling tetrapods, the
Amphibia. Notochord replaced by
bony vertebral column;
arrangement of bones in paired
limbs forms primitive
pentadactyl limbs

Teleostei (Cretaceous)
Most recent fish.
Homocercal tail;
paired fins reduced;
swim bladder;
great adaptive radiation

Possible alternative route of
evolution

Amphibia (Carboniferous)
Land-dwelling tetrapods.
Four pentadactyl limbs;
adult uses lungs;
reproduction in water;
larva and metamorphosis;
partially divided heart in adult

Reptilia (Carboniferous)
Direct development from
amniote egg;
impermeable skin;
poikilothermic;
complex bony skeleton;
incompletely divided ventricle

Aves (Jurassic)
Shelled egg;
feathers and wings;
atria and ventricles
completely divided;
brain enlarged;
elaborate behaviour
patterns;
endothermic

Mammalia (Jurassic)
Viviparity;
hair covering;
atria and ventricles
completely divided;
brain enlarged;
diaphragm present;
heterodont dentition;
endothermic

Fig 4.40 *Evolution of vertebrates*

(3) Increased effect of gravity – the apparent increase in body weight in air meant new stresses on the vertebral column. It changed from being a compression strut to being a girder. Limbs and girdles were developed.

(4) Change in locomotion – paired appendages became the main locomotory structures with the tail used for balance. In fish, locomotion is effected by the body and tail with the paired fins being used for balance.

(5) Reproduction – tetrapods must either develop methods for protecting eggs from desiccation or return

to water to reproduce. Amphibia have not solved this problem and adopt the latter alternative.

(6) Irritability – changes in the sensory receptors had to be made to cope with the new stimuli present in the terrestrial environment.

4.1 Using the key provided in fig 4.41 attempt to classify as far as you can the range of animals provided by your teacher.

Fig 4.41 *Key for the classification of some non-vertebrate animals found in moist terrestrial or freshwater habitats*

1. Organism unicellular, or colony of similar cells .... PROTOCTISTA
   Organism multicellular, some or all cells specialised for different functions 2

2. Animal forming whitish or greenish encrusting growth on stones or branches etc.; pierced by small holes; texture spongy .... PORIFERA (Spongillidae)

   Animals not forming an encrusting growth 3

3. Animal attached, with tubular body surmounted by tentacles; flexible and retracting on disturbance .... CNIDARIA (Hydrozoa)

   Animal microscopic, usually under 2 mm with conspicuous crown of cilia. The body usually with a capsule or cuticle often of definite plates; often a 'foot' ending in two terminal small processes .... ROTIFERA
   Animal not as above 4

4. Animal under 2 mm, elongate, body forked posteriorly; cuticle usually bearing spines and scales. Cilia at front and on parts of surface .... GASTROTRICHA
   Animal very small, with one or two suckers; with or without forked tail (cercaria stage) .... TREMATODA
   Animal unlike either of above 5

5. Animal colonial, with a number of tentaculate heads, the tentacles bearing cilia. Usually attached but capable in few cases of slow creeping .... POLYZOA
   Animal not colonial 6

6. Animals with a hard inflexible shell (MOLLUSCA), from which a soft, unsegmented, body projects 7
   Animals without a hard inflexible shell 8

7. Snails, with single coiled shell, or limpet-like .... GASTROPODA
   Bivalves, with two valves or shells joined by a hinge ..... PELYCOPODA

8. Segmented or unsegmented worm-like animals, never with jointed limbs, where segmented with more than 14 segments 9
   Segmented animals with a hard jointed integument and usually with jointed limbs (ARTHROPODA) 18

9. Unsegmented animals without bristles or suckers 10
   Segmented animals either with bristles on the segments or a sucker at front and hind ends 12

10. Round-sectioned, elongate, with body enclosed in cuticle; no cilia, ends of body usually pointed .... NEMATODA
    Animal variable shape, often flat, not very elongate. No cuticle; body capable of contraction. Ciliated 11

11. Reddish-yellow worm about 1 cm long with protrusible proboscis above the mouth (usually protruded when animal dropped into spirit) NEMERTINA
    Flat, often with proboscis bearing the mouth; movement characteristically gliding over substratum or surface of water. Colour usually black, yellowish, brown, or whitish .... PLATYHELMINTHES: TURBELLARIA

12. Animals bearing a sucker at both ends; segmentation well marked; never bearing bristles (leeches) .... HIRUDINEA
    Animals never with sucker at both ends, (and only in one small group of parasites, on crayfish with any suckers). Usually with bristles grouped into bundles, one on each side ventrally and one or more on each side dorsally per segment (OLIGOCHAETA) 13

13. Lacking bristles, sucker posteriorly .... BRANCHIOBDELLIDAE
    Bristles present, free-living, no suckers 14

14. Rather small, transparent worms with long chaetae .. NAIDIDAE
    Larger worms, or if small, without prominent chaetae 15

15. Usually more than two chaetae per bundle 16
    Never more than two chaetae per bundle 17

16. Small (to 36 mm) whitish, reddish, or yellowish worms with straight or S-shaped, pointed bristles ... ENCHYTRAEIDAE
    Small to large (to 200 mm) worms, some of the bristles forked at the tip .... TUBIFICIDAE

17. The families Lumbriculidae, Phreoryctidae, Criodrilidae and Lumbricidae are

all here and are best separated on characters of the reproductive system

18. Arthropods: small animals with globular or pear shaped, usually transparent bodies; three or four pairs of limbs, usually with hairs or bristles and often branched .... IMMATURE CRUSTACEA

Arthropods also with globular but usually opaque bodies, often bright red and mostly lacking visible segments; four pairs of limbs when adult; three when immature .... ARACHNIDA
19

Arthropods with elongate bodies, bearing 3 pairs of legs or lacking legs; head bearing antennae (1 pair) and mouth parts, thorax bearing legs, abdomen usually legless except for cerci or gills .... INSECTA

Arthropods with more than six legs, segmentation well marked, head bearing antennae (2 pairs), usually with abdominal appendages .... CRUSTACEA
21

19. Animal a typical spider, hairy body, unwetted by water and building air-filled 'bell' chamber .... *Argyronecta aquatica*

Animal smaller, without attached air bubble   20

20. Animals small (1 mm), elongate, with short, stumpy legs, living usually among wet moss, etc .... TARDIGRADA

Animals with rounded body, elongate legs, palps with 4 or 5 joints, often brightly coloured (red) or patterned brown. Usually above 1 mm, adults often free swimming, sometimes crawling .... ACARINA (mites)

21. Small crustacea, less than 3 mm, variable number of legs, usually transparent and free swimming in surface waters (zooplankton); often with carapace and with branched appendages bearing bristles .... MICROCRUSTACEA

Large crustacea, more than 3 mm, bottom living or swimming above the bottom 22

22. More than 5 pairs legs 23
    5 pairs legs only 24

23. Animal flattened laterally from side to side, swimming on side, abdominal appendages of two kinds, the longest at the back ...... AMPHIPODA (Gammaridae)

Animal dorso-ventrally flattened, all limbs of equal size, resembling woodlouse .... ISOPODA (Asellotidae)

24. Body crab-like .... *Eriocheir sinensis* (mitten crab)

Body longer, lobster-like ..... *Astacus pallipes* (crayfish)

Chapter Five

Chemicals of life

5.1 Introduction to biochemistry

The study of the chemicals of living organisms, or biochemistry, has been closely associated with the great expansion in biological knowledge that has taken place during this century. Originally a supporting subject, particularly for medicine, biochemistry has grown into a discipline in its own right, studied to degree level in most centres of higher education. Its importance lies in the fundamental understanding it gives us of physiology, that is the way in which biological systems work. This in turn finds application in fields like agriculture (development of pesticides, herbicides and so on); medicine (including the whole pharmaceutical industry); fermentation industries with their vast range of useful products, including alcoholic drinks; food and nutrition, including dietetics, food production and preservation; and some more recent applications like enzyme technology and production of new types of food and fuel.

Biochemistry is also one of the great unifying themes in biology. At this level, what is often striking about living organisms is not so much their differences as their similarities.

5.1.1 Elements found in living organisms

The Earth's crust contains approximately 100 chemical elements and yet only 16 of these are essential for life. These 16 are listed in table 5.1. The four most common elements in living organisms are, in order, hydrogen, carbon, oxygen and nitrogen. These account for more than 99% of the mass and numbers of atoms found in all living organisms. The four most common elements in the Earth's crust, however, are oxygen, silicon, aluminium, and

sodium. The biological importance of hydrogen, oxygen, nitrogen and carbon is largely due to their having valencies of 1, 2, 3 and 4 respectively and their ability to form more stable covalent bonds than any other elements with these valencies (see appendix A1.1.3).

The importance of carbon

Carbon exhibits many unique features in its chemistry that are fundamental to life. A whole branch of chemistry, organic chemistry, is devoted to the study of carbon and its compounds. What are these unique features? Carbon has an atomic number of six because it has six electrons orbiting a nucleus containing six protons (fig A1.1). The nucleus also contains six neutrons, the protons and neutrons combining to give it a mass number of 12. In its chemistry it acquires a full (stable) outer shell of eight electrons by sharing four electrons. Thus, it is covalent (shares electrons) and has a valency of four (it shares four electrons).

A simple example of this sharing is shown in fig A1.2d, for the compound methane, whose **molecular formula** is CH_4 and whose **structural formula** is also shown in fig A1.2d.

> **5.1** From what you have read, what is the difference between molecular and structural formulae?

When carbon is joined to four atoms or groups, the four bonds are arranged symmetrically in a tetrahedron (fig 5.1). If the three-dimensional arrangement of atoms is important, the convention shown in fig 5.1b can be used. Another convention commonly used is to omit carbon

Table 5.1 The elements found in living organisms.

Chief elements of organic molecules	Ions		Trace elements			
H hydrogen	Na⁺	sodium	Mn	manganese	B	boron
C carbon	Mg²⁺	magnesium	Fe	iron	Al	aluminium
N nitrogen	Cl⁻	chlorine	Co	cobalt	Si	silicon
O oxygen	K⁺	potassium	Cu	copper	V	vanadium
P phosphorus	Ca²⁺	calcium	Zn	zinc	Mo	molybdenum
S sulphur					I	iodine

Elements in each column are arranged in order of atomic mass, not abundance. Those in the first three columns are found in all organisms.
(Based on A. L. Lehninger, *Biochemistry*, Worth. N.Y. 1970)

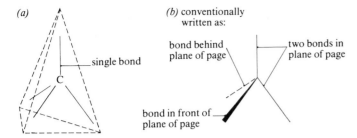

(a) single bond

(b) conventionally written as:

bond behind plane of page

two bonds in plane of page

bond in front of plane of page

Fig 5.1 *Tetrahedral arrangement of carbon bonds*

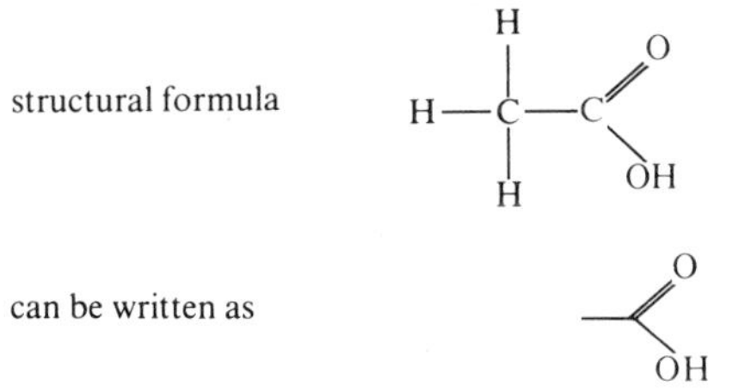

structural formula

can be written as

Fig 5.2 *Two ways of representing the structural formula of ethanoic acid (acetic acid), CH₃COOH*

atoms, and any hydrogen atoms joined to carbon atoms, from the structural formula. A simple example is ethanoic acid (acetic acid), shown in fig 5.2. You will see also from this figure why its molecular formula can be represented as $CH_3.COOH$.

Knowing that the valency of carbon is 4, it is possible to deduce the location of missing hydrogen atoms. The advantage of this convention is two-fold: it simplifies diagrams of structural formulae and allows stress to be placed on the more important chemical groups.

The importance of carbon, then, lies in the way it forms strong, stable covalent bonds. This it can do with other carbon atoms, and with other types of atoms.

Carbon can form covalent bonds with other carbon atoms to form stable chains or rings, a property not shown to such an extent by any other element (fig 5.3). This ability is largely responsible for the vast variety of organic compounds; C—C bonds can be regarded as the skeletons of organic molecules.

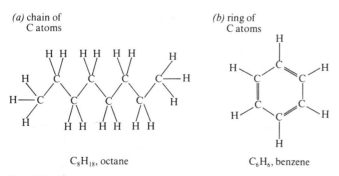

(a) chain of C atoms

C_8H_{18}, octane

(b) ring of C atoms

C_6H_6, benzene

Fig 5.3 *Examples of chain and ring structures formed by C–C bonds*

5.2 Draw the structural formulae of (a) octane and (b) benzene, using the convention described in fig 5.2.

Carbon atoms commonly form covalent bonds with H, N, O, P, and S. Combination with these and other elements contributes to the large variety of organic compounds.

Multiple bonds. A further important property of carbon is its ability to form strong multiple bonds, a property shared with oxygen and nitrogen. The multiple bonds are:

double bonds: $>C=C<$ $>C=O$ $>C=N-$

triple bonds: $-C\equiv C-$ $-C\equiv N$
(rare in nature)

Compounds containing double $=$ or triple $\equiv$ carbon–carbon bonds are called unsaturated. In a saturated carbon compound, all carbon–carbon bonds are single.

5.3 Draw the structural formula for the unsaturated organic compound ethene (ethylene), C_2H_4.

Summary. The important chemical properties of carbon are
(1) it is a relatively small atom with a low mass,
(2) it has the ability to form four strong, stable covalent bonds,
(3) it has the ability to form carbon–carbon bonds, thus building up large carbon skeletons with ring and/or chain structures,
(4) it has the ability to form multiple covalent bonds with other carbon atoms, oxygen and nitrogen.

This unique combination of features is responsible for the enormous variety of organic molecules. Variation occurs in three major ways: **size**, determined by the carbon skeleton, **chemistry**, determined by the associated elements and chemical groups and how saturated the carbon skeleton is, and **shape**, determined by geometry, that is angles of the bonds.

5.1.2 Simple biological molecules

Having seen which elements are found in organisms, it is necessary to examine which compounds they form. Again, there is a fundamental similarity between all living organisms. Water is the most abundant compound, typically constituting between 60–95% of the fresh mass of an organism. Certain simple organic molecules are also universally found; these act as building blocks for larger molecules and are listed in table 5.2. They are discussed more fully later.

Thus relatively few types of molecule give rise to the larger molecules and structures of living cells. They are the kinds of molecules which biologists speculate could have

Table 5.2 Chemical 'building blocks' of organic compounds.

Small molecules ('building blocks')	Constituent of
amino acids	proteins
sugars (monosaccharides)	polysaccharides and nucleic acids
fatty acids, glycerol and choline	lipids
aromatic bases	nucleic acids

been synthesised in the 'primeval soup' of chemicals which is thought to have existed in the early history of the planet, before life itself appeared (section 24.1). These simple molecules are made in turn from even simpler, inorganic molecules, notably carbon dioxide, nitrogen and water.

Importance of water

Without water, life could not exist on this planet. It is doubly important to living organisms because it is both a vital chemical constituent of living cells and, for many, a habitat. It is worth while, then, looking at some of its chemical and physical properties.

These are rather unusual and due mostly to its small size,

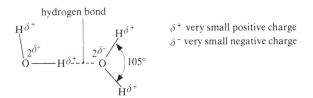

Fig 5.4 *Two water molecules, showing polarity of the molecules and formation of a hydrogen bond between them*

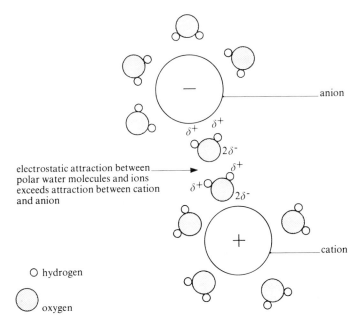

its polarity and to hydrogen-bonding between its molecules. Polarity is an uneven charge distribution over a molecule. In water, one end of the molecule is slightly positive and the other slightly negative. This is known as a **dipole**. The more electronegative oxygen atom tends to attract the single electrons of the hydrogen atoms. Water molecules therefore have an electrostatic attraction for each other, opposite charges coming together and causing them to behave as if they were 'sticky' (fig 5.4). These attractions are not as strong as normal ionic bonds and are called **hydrogen bonds**. With these features in mind, some of the biologically significant properties of water can be examined.

Biological significance of water

Solvent properties. Water is an excellent solvent for polar substances. These include ionic substances like salts, whose charged particles (ions) dissociate (separate) in water when the substance dissolves as described in fig 5.5, and some non-ionic substances like sugars and simple alcohols which contain charged (polar) groups within the molecules, such as the hydroxyl (—OH) groups of sugars and alcohols.

Once a substance is in solution its molecules or ions can move about freely, thus making it more chemically reactive than if it were solid. Thus the majority of the cell's chemical reactions take place in aqueous solutions. Non-polar substances, such as lipids, are immiscible with water and can serve to separate aqueous solutions into compartments, as with membranes. Non-polar parts of molecules are repelled by water and usually group together in its presence, as when oil droplets coalesce in water to form larger oil droplets, that is non-polar molecules are **hydrophobic** (water-hating). Such hydrophobic interactions are important in maintaining the stability of membranes, many protein molecules, nucleic acids and other subcellular structures.

Water's solvent properties also mean that it acts as a transport medium, as in the blood, lymphatic and excretory systems, the alimentary canal and in xylem and phloem.

High heat capacity. The specific heat capacity of water is the amount of heat, measured in joules, required to raise the temperature of 1 kg of water by 1 °C. Water has a high heat capacity. This means that a large increase in heat energy results in a relatively small rise in temperature. This is because much of the energy is used in breaking the hydrogen bonds (overcoming the 'stickiness') which restrict the mobility of the molecules.

Fig 5.5 (left) *Distribution of water molecules around an anion (−) and a cation (+). Note that the more negatively charged oxygen atom of water faces inwards to the cation but outwards from the anion. This occurs when ionic substances dissolve in water. Due to their polarity, water molecules weaken the attraction between ions of opposite charge and then surround the ions, keeping them apart. The ions are said to be hydrated*

Temperature changes within water are minimised as a result of its high heat capacity. Biochemical processes therefore operate over a smaller temperature range, proceeding at more constant rates and are less likely to be inhibited by extremes of temperature. Water also provides a very constant external environment for many cells and organisms.

High heat of vaporisation. Latent heat of vaporisation (or relative latent heat of vaporisation) is a measure of the heat energy required to vaporise a liquid, that is to overcome the attractive forces between its molecules so that they can escape as a gas. A relatively large amount of energy is needed to vaporise water. This is due to the hydrogen bonding. As a result, water has an unusually high boiling point for such a small molecule.

The energy imparted to water molecules to vaporise them thus results in loss of energy from their surroundings, that is a cooling effect occurs. This is made use of in sweating and panting of mammals, the opening of the mouth of some reptiles in sunshine, such as crocodiles, and may be important in cooling transpiring leaves. The high heat of vaporisation means that a large amount of heat can be lost with minimal loss of water from the body.

High heat of fusion. Latent heat of fusion (or relative latent heat of fusion) is a measure of the heat energy required to melt a solid, in this case ice. With its high heat capacity, water requires relatively large amounts of heat energy to thaw it. Conversely, liquid water must lose a relatively large amount of heat energy to freeze. Contents of cells and their environments are therefore less likely to freeze. Ice crystals are particularly damaging if they develop inside cells.

Density and freezing properties. The density of water decreases below 4 °C and ice therefore tends to float. It is the only substance whose solid form is less dense than its liquid form.

Since ice floats, it forms at the surface first and the bottom last. If ponds froze from the bottom upwards, freshwater life could not exist in temperate or arctic climates. Ice insulates the water below it, thus increasing the chances of survival of organisms in the water. This is important in cold climates and cold seasons, and must have been particularly so in the past, such as during Ice Ages. Also, the ice thaws more rapidly by being at the surface. The fact that water below 4 °C tends to rise also helps to maintain circulation in large bodies of water. This may result in nutrient cycling and colonisation of water to greater depths.

High surface tension and cohesion. Cohesion is the force whereby individual molecules stick together. At the surface of a liquid, a force called surface tension exists between the molecules as a result of inwardly acting cohesive forces between the molecules. These cause the surface of the liquid to occupy the least possible surface area (ideally a sphere). Water has a higher surface tension than any other liquid. The high cohesion of water molecules is important in cells and in translocation of water through xylem in plants (section 14.4). At a less fundamental level, many small organisms rely on surface tension to settle on water or to skate over its surface.

Water as a reagent. Water is biologically significant as an essential metabolite, that is it participates in the chemical reactions of metabolism. In particular, it is used as a source of hydrogen in photosynthesis (section 9.4.2) and is used in hydrolysis reactions.

Water and evolutionary change. The importance of water to living organisms is reflected in the fact that its shortage appears to have been a major selection pressure in the development of species. This is a recurrent theme in chapters 3 and 4 where, for example, the restrictions placed on certain plants by their motile gametes are discussed. All terrestrial organisms are adapted to obtain and conserve water, and the extreme adaptations of xerophytes, desert animals and so on, provide some fascinating examples of biological design.

Some of the biologically important functions of water are summarised in table 5.3.

Table 5.3 Some biologically important functions of water.

All organisms
Structure – high water content of protoplasm
Solvent and medium for diffusion
Reagent in hydrolysis
Support for aquatic organisms
Fertilisation by swimming gametes
Dispersal of seeds, gametes and larval stages of aquatic organisms, and seeds of some terrestrial species e.g. coconut

Plants
Osmosis and *turgidity* (important in many ways, such as growth (cell enlargement), support, guard cell mechanism)
Reagent in photosynthesis
Transpiration and *translocation* of inorganic ions and organic compounds
Germination of seeds – swelling and breaking open of the testa and further development

Animals
Transport
Osmoregulation
Cooling by evaporation, such as sweating, panting
Lubrication, as in joints
Support – hydrostatic skeleton
Protection, for example lachrymal fluid, mucus
Migration in ocean currents

5.1.3 Macromolecules

The simpler organic molecules associate to form larger molecules. A **macromolecule** is a giant molecule made from many repeating units; it is therefore a **polymer** and the unit molecules are called **monomers**. There are three types of macromolecule, namely polysaccharides, proteins and nucleic acids and their constituent monomers are monosaccharides, amino acids and nucleotides respectively.

Macromolecules account for over 90% of the dry mass of cells. Table 5.4 summarises the important properties of macromolecules.

Differences between macromolecules are discussed in detail later, but one key point is that nucleic acids and proteins can be regarded as 'informational' molecules, whereas polysaccharides are 'non-informational'. This means that the *sequence* of subunits is important in proteins and nucleic acids and is much more variable than in polysaccharides, where only one or two different subunits are normally used. The reasons for this will become clear later. In the rest of this chapter, we shall be studying the three classes of macromolecules and their subunits in detail. In addition, lipids, although generally much smaller molecules (average M_r 750–2 500) will be included since they generally associate with each other into much larger groups of molecules.

Table 5.4 Characteristics of macromolecules.

Property	Polysaccharides	Proteins	Nucleic acids
M_r (relative formula mass or molecular mass)	10^4–10^6 (typical)	10^4–10^6 (typical)	10^4–10^{10} (typical)
Subunits	**monosaccharides** (Many types, though few commonly used. Usually only 1 type per molecule. Sometimes 2 types alternate.)	**amino acids** (20 common types. All may be used in 1 molecule.)	**nucleotides** (5 types, 4 used in DNA, 4 in RNA.)
Branching or non-branching	May be branched	No branching	No branching
Type of bond joining subunits	Glycosidic bond – 2 types	Peptide bond	Phosphodiester bond (Sugar–phosphate bond)

Other properties common to all three types of macromolecule: (i) bonds between subunits are formed by elimination of water (**condensation**), (ii) formation of bonds requires energy, (iii) bonds between subunits are broken by addition of water (**hydrolysis**).

Table 5.5 Some common chemical groups found in organic molecules.

Aldehyde group	$-C{\overset{H}{\underset{O}{\lesseqgtr}}}$ or $-CHO$ $\xrightarrow[\text{oxidation}]{O}$ $-COOH$ carboxylic acid $\xrightarrow[\text{reduction}]{2H}$ $-CH_2OH$ primary alcohol	
Keto group (compound containing this group is called a ketone)	$\overset{C}{\underset{C}{>}}C=O$ $\xrightarrow[\text{reduction}]{2H}$ $-CHOH$ secondary alcohol	
Hydroxyl group	$-OH$	
Primary alcohol group	$-CH_2OH$ or $-C{\overset{H}{\underset{OH}{-}}}H$	
Secondary alcohol group	$CHOH$ or $>C{\overset{H}{\underset{OH}{<}}}$	
Carboxyl group	$>COOH$ or $-C{\overset{O}{\underset{OH}{<}}}$	
Carbonyl group	$>C=O$ note that this group is present in aldehydes, ketones and carboxylic acids	

5.2   Carbohydrates (saccharides)

Carbohydrates are substances with the general formula $C_x(H_2O)_y$, where x and y are variable numbers; their name (hydrate of carbon) is derived from the fact that hydrogen and oxygen are present in the same proportions as in water. All carbohydrates are aldehydes or ketones and all contain several hydroxyl groups. Their chemistry is determined by these groups. For example, aldehydes are very easily oxidised and hence are powerful reducing agents. The structures of these and some related groups are shown in table 5.5, together with some typical chemical reactions of aldehydes and ketones.

Carbohydrates are divided into three main classes, mono-, di- and polysaccharides, as shown in fig 5.6.

5.2.1   Monosaccharides

Monosaccharides are single sugar units. Their general formula and some of their properties are shown in fig 5.6. They are classified according to the number of carbon atoms as trioses (3C), tetroses (4C), pentoses (5C), hexoses (6C) and heptoses (7C). Of these, pentoses and hexoses are the most common.

5.4   What would be the molecular formula of each of these types of sugar?

The chief functions of monosaccharides are summarised in table 5.6.

It will be seen from table 5.6 that monosaccharides are important as energy sources and as building blocks for the synthesis of larger molecules. They are suitable for the latter role because they are chemically reactive molecules and show a wide variety of structures, including the variation in number of carbon atoms already mentioned. Some other important features which contribute to their variety are discussed below.

Aldoses and ketoses

In monosaccharides, all the carbon atoms except one have a hydroxyl group attached. The remaining carbon atom is either part of an aldehyde group, in which case the monosaccharide is called an **aldose** or **aldo sugar**, or is part of a keto group, when it is called a **ketose** or **keto sugar**. Thus all monosaccharides are aldoses or ketoses. The two simplest monosaccharides are the trioses glyceraldehyde and dihydroxyacetone. Glyceraldehyde has an aldehyde

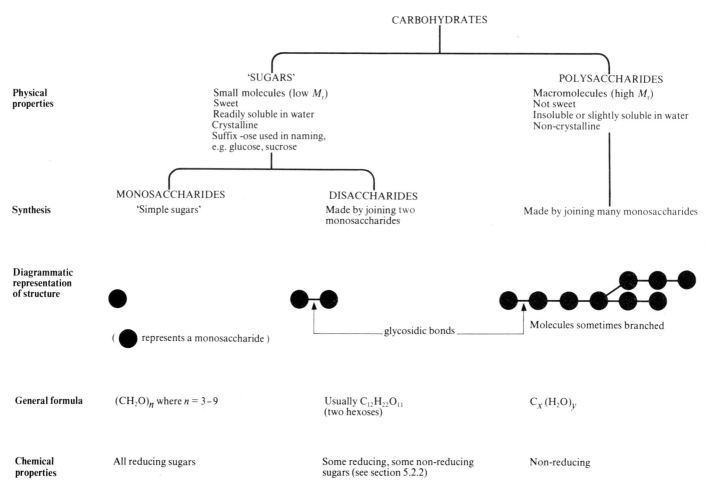

Fig 5.6 *Classification of carbohydrates. Note that the term 'sugar' is superfluous but convenient because mono- and disaccharides share certain properties, such as sweetness of taste*

Table 5.6 Chief functions of monosaccharides.

Trioses $C_3H_6O_3$ e.g. glyceraldehyde, dihydroxyacetone
Intermediates in respiration (see glycolysis), photosynthesis (see dark reactions) and other branches of carbohydrate metabolism
Glyceraldehyde→glycerol→triglyceride(lipid)

Tetroses $C_4H_8O_4$
Rare in nature, occurring mainly in bacteria (not discussed further in this chapter)

Pentoses $C_5H_{10}O_5$ e.g. ribose, ribulose
Synthesis of nucleic acids; ribose is a constituent of RNA, deoxyribose of DNA
Synthesis of some coenzymes, e.g. NAD, NADP, coenzyme A, FAD, FMN
Synthesis of AMP, ADP, and ATP
Synthesis of the polysaccharides called pentosans (see polysaccharides)
Ribulose bisphosphate is the CO_2 acceptor in photosynthesis

Hexoses $C_6H_{12}O_6$ e.g. glucose, fructose, galactose, mannose
Source of energy when oxidised in respiration; glucose is the most common respiratory substrate and the most common monosaccharide
Synthesis of disaccharides; monosaccharide units can link together to form larger molecules; combinations of 2–10 monosaccharides are termed oligosaccharides, the most commonly occurring oligosaccharides are the disaccharides, made up of two monosaccharides
Synthesis of the polysaccharides called hexosans (see polysaccharides), glucose is particularly important in this role

Derivatives of Monosaccharides
Some important derivatives of monosaccharides are sugar alcohols, sugar acids, deoxy sugars and amino sugars

(i) Sugar $\xrightarrow[\text{reduction}]{2H}$ **sugar alcohol**, e.g. glycerol, used in lipid synthesis. Sugar alcohols sometimes act as storage carbohydrates, e.g. mannitol in *Fucus* and some fruits.

—CHO (aldose) $\xrightarrow{2H}$ —CH_2OH
>C=O (ketose) $\xrightarrow{2H}$ >CHOH

(ii) Sugar $\xrightarrow[\text{oxidation}]{O}$ **sugar acid** Sugar acids are important intermediates in carbohydrate metabolism. Some, e.g. glucuronic acid, are constituents of the polysaccharides used in gums, mucilages and cell walls. Vitamin C (ascorbic acid) is a sugar acid derived from a hexose.

—CHO (aldose) $\xrightarrow{O}$ —COOH
—CH_2OH (aldose $\xrightarrow{-2H}$ CHO $\xrightarrow{O}$ —COOH
or ketose)

(iii) Sugar $\xrightarrow[\text{lost by}]{\text{oxygen atom}}$ **deoxy sugar** Most important is deoxyribose, formed by deoxygenation of ribose, and used in DNA synthesis.
replacing —OH group
with—H

(iv) Sugar $\xrightarrow[\text{(amino group)}]{\text{add —NH}_2}$ **amino sugar** e.g. glucosamine, used in synthesis of chitin and formed in many polysaccharides of vertebrates. Galactosamine used in synthesis of cartilage.
—OH of carbon 2→ —NH_2

131

group and dihydroxyacetone a keto group, and they can be regarded as the parent compounds of the aldoses and ketoses respectively (fig 5.7).

5.5 If you have little experience of chemistry, it might be useful to answer the following with reference to fig 5.7.
(a) What is the valency of each element?
(b) What is the total number of each type of atom? Does it conform with the molecular formula?
(c) How many hydroxyl groups does each molecule contain? Could this have been predicted knowing they were trioses?
(d) Can you recognise any other chemical groups not labelled in this diagram?

Fig 5.8 shows some other common aldoses and ketoses. In general, aldoses, such as ribose and glucose, are more common than ketoses, such as ribulose and fructose.

5.6 Which sugars in fig 5.8 are pentoses and which are hexoses?

Optical isomerism

Another important structural characteristic of monosaccharides is the occurrence of isomerism. If two different compounds have the same molecular formula, they are said to be isomers of each other. Two types of isomerism occur, structural and stereoisomerism. **Structural isomerism** is due to different linkings of the atoms or groups within the molecules. Thus all hexoses are structural isomers of each other (fig 5.8 – compare glucose, mannose, galactose and fructose, all with the same molecular formula, $C_6H_{12}O_6$).

Stereoisomerism occurs when the same atoms or groups are joined together but are arranged differently in space. Two kinds of stereoisomerism occur, geometric and optical isomerism. We are not concerned with geometric isomerism, which involves certain compounds containing double bonds. Optical isomerism, however, is an important, biologically significant feature of monosaccharides and amino acids.

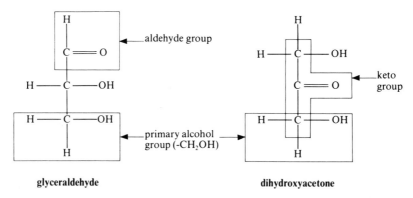

Fig 5.7 *Structures of glyceraldehyde and dihydroxyacetone. Note carefully the positions of the aldehyde and keto groups. Aldehyde groups are always at the end of the chain of C atoms*

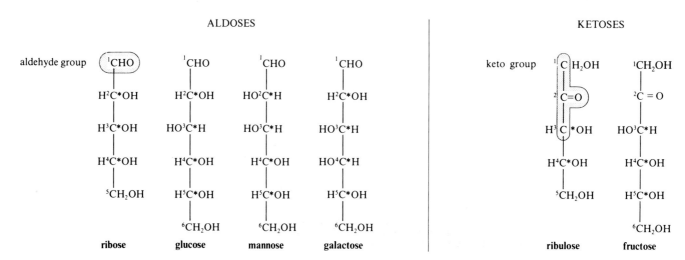

Fig 5.8 *Some common aldoses and ketoses. C atoms are numbered according to convention. * asymmetric C atoms – see optical isomerism*

Certain solid compounds when dissolved to form a solution (and certain liquid compounds) possess the power to rotate the plane of vibration of plane-polarised light, and are said to be **optically active**. Plane-polarised light is light vibrating in one plane only, whereas light normally vibrates in all planes perpendicular to its direction of transmission. It is easier to understand this by studying fig 5.9, which shows normal light being artificially converted to plane-polarised light by a **polariser**, and then having its plane of polarisation changed by an optically active substance.

If the substance rotates the plane of polarisation to the right, it is said to be **dextro-rotatory** and if to the left **laevo-rotatory**. The degree of rotation can be measured by an instrument called a polarimeter.

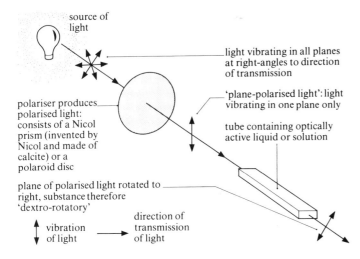

Fig 5.9 *Principles of optical isomerism*

Optical isomerism is a property of any compound which can exist in two forms whose structures are mirror images. Like right- and left-handed gloves, such structures cannot be superimposed on each other. In organic compounds this occurs when a carbon atom has four different atoms or groups attached to it. Such a carbon atom is called an **asymmetric carbon atom**. The principle is illustrated in fig 5.10a where it can be seen that the tetrahedral arrangement of bonds about the central, asymmetric carbon atom means that there are two possible arrangements of the groups in space, forming two mirror images. Glyceraldehyde provides a simple example of a monosaccharide showing optical isomerism. It possesses one asymmetric carbon atom and exists in two forms as shown in fig 5.10b.

Dextro-rotatory compounds are given the prefix 'd'- or more recently (+) and laevo-rotatory compounds 'l'- or more recently (−). The two isomers of glyceraldehyde are called the D-isomer (dextro-rotatory) and the L-isomer (laevo-rotatory). (Note the use of *small* capitals.) All optical isomers of monosaccharides can be structurally related to one of these two forms of glyceraldehyde. By convention, if the asymmetric carbon atom furthest from the aldehyde or ketone group has its hydroxyl group in the same position as in D-glyceraldehyde, the isomer is called the D-isomer, and if in the same position as in L-glyceraldehyde, it is called the L-isomer. **This is irrespective of the direction in which they rotate plane-polarised light.** All the isomers shown in fig 5.8 are D-isomers, as are virtually all naturally occurring monosaccharides. However, the direction in which they rotate plane-polarised light varies. D-glucose for example rotates it to the right, whereas D-fructose rotates it to the left. The D- and L-isomers of glucose are shown in fig 5.11 and it will be seen that they are mirror images.

Although D- and L-isomers of the same substance have the same chemical and physical properties, and are therefore given the same chemical name, such as D- and L-glyceraldehyde, their three-dimensional differences have biological significance in one important respect, namely that enzymes, which depend on recognition by shape, can distinguish between the mirror images. It is possible that early in evolution an arbitrary bias to accept the D-isomers of sugars was established since their L-isomers are rare in nature. Naturally occurring amino acids, however, are all L-isomers.

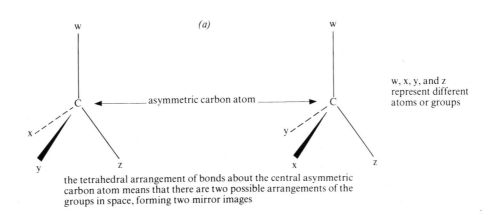

the tetrahedral arrangement of bonds about the central asymmetric carbon atom means that there are two possible arrangements of the groups in space, forming two mirror images

Fig 5.10 *Optical isomerism (a) Diagram illustrating the principle of asymmetry of carbon atoms.*

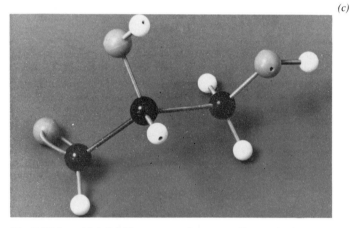

(b)

D(+)-glyceraldehyde

L(−)-glyceraldehyde

* asymmetric carbon atom

(c)

Fig 5.10 (cont.) *(b) Two ways of representing optical isomers (mirror images) of glyceraldeyde and (c) molecular models of the two isomers*

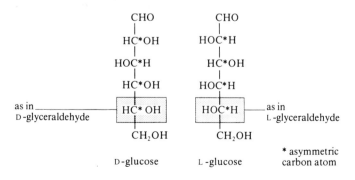

as in D -glyceraldehyde

as in L -glyceraldehyde

D -glucose

L -glucose

* asymmetric carbon atom

Fig 5.11 *D- and L-isomers of glucose. Note that they are mirror images*

Ring structures

Fig 5.8 shows pentoses and hexoses represented as straight chain molecules. Because of the bond angles between carbon atoms, however, it is possible for sugars with five and six carbon atoms to form stable ring structures. In pentoses, the first carbon atom joins with the oxygen atom on the fourth carbon atom to give a five-membered ring called a **furanose ring**, as shown in fig 5.12. In hexoses which are aldoses, for example glucose, the first carbon atom combines with the oxygen atom on carbon five to give a six-membered ring, as shown in fig 5.13. This structure is known as a **pyranose ring**. In hexoses which are ketoses, such as fructose, the *second* carbon atom combines with the oxygen atom on carbon five to give a furanose ring.

The ring structures of pentoses and hexoses are the usual forms, with only a small proportion of the molecules existing in the 'open chain' form at any one time. The ring structure is the form incorporated into disaccharides and polysaccharides.

H
|
^{1}C = O
|
H——^{2}C——OH
|
H——^{3}C——OH
|
H——^{4}C——OH
|
^{5}CH$_2$OH

D-ribose

carbon 1 shown to right of ring by convention

^{5}CH$_2$OH   O   H
^{4}C $_1$C
H   H   H   OH
$_3$C————$_2$C
OH   OH

D-ribose with 5-membered furanose ring

conventionally written as:

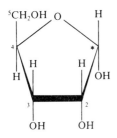

The ring is at right-angles to the plane of the paper. The front portion is represented by thicker lines. The side-groups are located above and below the ring.

* see α - and β - isomers

Fig 5.12 *Three conventional ways of representing the structure of ribose*

H
|
^{1}C = O
|
H——^{2}C——OH
|
HO——^{3}C——H
|
H——^{4}C——OH
|
H——^{5}C——OH
|
^{6}CH$_2$OH

D-glucose

^{6}CH$_2$OH
$_5$ O
H   H   H
$_4$ OH   H $_1$
OH   OH
$_3$ H $_2$ OH

D-glucose with 6-membered pyranose ring
* see α - and β - isomers

Fig 5.13 *Two conventional ways of representing the structure of glucose*

The ring structures have biological significance because their formation results in another carbon atom becoming asymmetric. This carbon atom is asterisked in figs 5.12 and 5.13 and you will notice that each has a hydroxyl group (–OH) attached to it. This may be below the plane of the ring, as shown in figs 5.12 and 5.13, or above the plane of the ring. The former is the α-isomer, the latter the β-isomer. The existence of α- and β-isomers leads to greater chemical variety and is of importance in, for example, the formation of starch and cellulose (see polysaccharides).

5.7 (*a*) Draw side by side, for comparison, the α- and β-isomers of D-ribose and D-glucose. (*b*) What type of isomerism is this?

5.2.2 Disaccharides

Fig 5.6 summarises some of the properties of disaccharides. They are formed by condensation reactions between two monosaccharides, usually hexoses, as shown in fig 5.14.

The bond formed between two monosaccharides is called a **glycosidic bond** and it normally forms between carbon atoms 1 and 4 of neighbouring units (a 1,4 bond). The process can be repeated indefinitely to build up the giant molecules of polysaccharides (fig 5.14). The monosaccharide units are called **residues** once they have been linked. Thus a maltose molecule contains two glucose residues.

The most common disaccharides are maltose, lactose and sucrose:

glucose + glucose   = maltose,
glucose + galactose = lactose,
glucose + fructose = sucrose.

Maltose is formed by the action of amylases (enzymes) on starch during digestion, for example in animals or germinating seeds. It is converted to glucose by the action of a maltase. Lactose, or milk sugar, is found exclusively in milk. Sucrose, or cane sugar, is most abundant in plants, where it is translocated in large quantities through phloem tissue. It is sometimes stored because it is relatively inert metabolically. It is obtained commercially from sugar cane and sugar beet and is the 'sugar' we normally buy in shops.

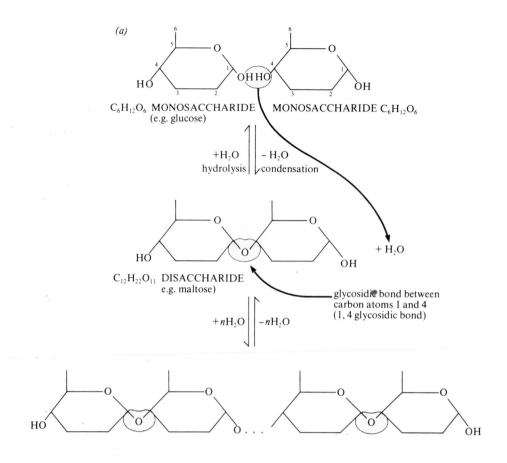

(a)

$C_6H_{12}O_6$ MONOSACCHARIDE
(e.g. glucose)

MONOSACCHARIDE $C_6H_{12}O_6$

$+H_2O$ hydrolysis $\quad -H_2O$ condensation

$C_{12}H_{22}O_{11}$ DISACCHARIDE
e.g. maltose)

$+ H_2O$

glycosidic bond between carbon atoms 1 and 4
(1, 4 glycosidic bond)

$+nH_2O \quad -nH_2O$

$(C_6H_{10}O_5)_n$ POLYSACCHARIDE
(e.g. starch)

(b)

cellulose

Fig 5.14 (a) Formation of a disaccharide and a polysaccharide. NB Only the relevant parts of the molecules are shown. A complete glucose molecule is shown in fig 5.13. (b) Structure of cellulose. In some polysaccharides, units may be spun through 180° at each successive condensation, as in cellulose

Reducing sugars

All monosaccharides and some disaccharides, including maltose and lactose, are reducing sugars. Sucrose is a non-reducing sugar. The chemistry of reduction is related to the activity of the aldehyde group of aldo sugars and the combined ketone and primary alcohol groups of keto sugars. In non-reducing sugars, these groups are not available to participate in a reaction since they are linked together in the glycosidic bond. Two common tests for reducing sugars, Benedict's test and Fehling's test (section 5.8) make use of the ability of these sugars to reduce copper from a valency of 2 to a valency of 1. Both tests involve use of an alkaline solution of copper(II) sulphate ($CuSO_4$) which is reduced to insoluble copper(I) oxide (Cu_2O).

Ionic equation: $Cu^{2+} + e^- \longrightarrow Cu^+$
blue solution   brick-red precipitate

5.2.3   Polysaccharides

Fig 5.6 summarises some of the properties of polysaccharides. They function chiefly as food and energy stores (for example starch and glycogen) and as structural materials (for example cellulose). They are convenient storage molecules for several reasons: their large size makes them more or less insoluble in water, so they exert no osmotic or chemical influence in the cell; they fold into compact shapes (see below) and they are easily converted to sugars by hydrolysis when required.

As we have already seen, polysaccharides are polymers of monosaccharides; polymers of pentoses are called **pentosans** and polymers of hexoses are called **hexosans**. Polymers of glucose are **glucosans**.

Starch

Starch is a polymer of glucose. It is a major fuel store in plants, but is absent from animals where the equivalent is glycogen (see below). Starch has two components, amylose and amylopectin. Amylose has a straight chain structure consisting of several thousand glucose residues, though the chain coils helically into a more compact shape. Amylopectin is also compact as it has many branches, formed by 1,6 glycosidic bonds (fig 5.15). It has up to twice as many glucose residues as amylose.

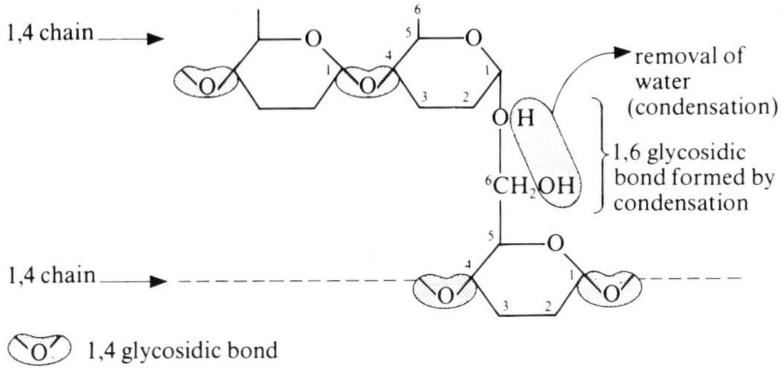

Fig 5.15 *Structure of amylopectin showing formation of one branch*

A suspension of amylose in water gives a blue-black colour with iodine–potassium iodide solution, whereas a suspension of amylopectin gives a red-violet colour. This forms the basis of the test for starch (section 5.8). Starch molecules accumulate to form starch grains. These are visible in many plant cells, notably in the chloroplasts of leaves (fig 9.6), in storage organs such as the potato tuber, and in seeds of cereals and legumes. The grains appear to be made of layers of starch and are usually of a characteristic size and shape for a given plant species.

5.8 With reference to your answer to question 5.7, which isomer, α- or β-glucose, is used in making starch?

Glycogen

Glycogen is the animal equivalent of starch, being a storage polysaccharide made from glucose; many fungi also store it. In vertebrates, glycogen is stored chiefly in the liver and muscles, both centres of high metabolic activity where it provides a useful source of glucose for use in respiration. It is very similar in structure to amylopectin, but shows more branching. It forms tiny granules inside cells which are usually associated with smooth endoplasmic reticulum (fig 7.5). The metabolism of glycogen is described in chapter 11.

Cellulose

Cellulose is another polymer of glucose.

5.9 Study the structure of cellulose in fig 5.14b. Is it made of α- or β- glucose residues? (This should reveal why successive glucose residues are rotated at 180° in the cellulose molecule.)

About 50% of the carbon found in plants is in cellulose and it is the most abundant organic molecule on Earth. It is virtually confined to plants, although it is found in some non-vertebrates and ancestral fungi. Its abundance is a result of its being a structural component of all plant cell walls, constituting about 20–40% of the wall on average. The structure of the molecule reveals its suitability for this role. It consists of long chains of glucose residues with about 10 000 residues per chain (fig 5.14b). Hydroxyl groups (—OH) project outwards from each chain in all directions and form hydrogen bonds with neighbouring chains, thus establishing a rigid cross-linking between chains. The chains associate in groups to form microfibrils, which are arranged in larger bundles to form macrofibrils. These have tremendous tensile strength (some idea of this strength can be obtained by testing cotton, which is almost pure cellulose) and their arrangement in layers in a cementing matrix of other polysaccharides is described in section 7.3.1. Despite their combined strength, the layers are fully permeable to water and solutes, an important property in the functioning of plant cells. Apart from being a structural compound, cellulose is an important food source for some animals, bacteria and fungi. The enzyme cellulase, which catalyses the digestion of cellulose to glucose, is relatively rare in nature and most animals, including humans, cannot utilise cellulose despite its being an abundant and potentially valuable source of glucose. Ruminant mammals like the cow, however, have bacteria living symbiotically in their guts which digest cellulose. The abundance of cellulose and its relatively slow rate of breakdown in nature have ecological implications because it means that substantial quantities of carbon are 'locked up' in this substance, and carbon is one of the chief materials required by living organisms. Commercially, cellulose is extremely important. It is used, for example, to make cotton goods and is a constituent of paper.

Callose

Callose is an amorphous polymer of glucose found in a wide variety of locations in plants and often formed in response to wounding or stress. It is particularly important in phloem sieve tubes (chapter 14). It has 1,3 glycosidic linkages.

Inulin

Inulin is an unusual polysaccharide, being a polymer of fructose. It is used as a food store, particularly in roots and tubers of the family Compositae, for example *Dahlia* tubers.

5.2.4 Compounds closely related to polysaccharides

Brief mention has already been made of amino sugars and sugar acids in table 5.6. These molecules often participate in the same ways as simple sugars in the formation of polysaccharides, and their products are generally called mucopolysaccharides. They are of great biological significance.

Chitin

Chitin is closely related to cellulose in structure and function, being a structural polysaccharide. It occurs in some fungi, where its fibrous nature contributes to cell wall structure, and in some animal groups, particularly the arthropods where it forms an essential part of the exoskeleton. Structurally it is identical to cellulose except that the hydroxyl (—OH) group at carbon atom 2 is replaced by —NH.CO.CH$_3$. This is a result of the amino sugar glucosamine combining with an acetyl group (CH$_3$.CO—). Chitin is therefore a polymer of acetylglucosamine (fig 5.16). It forms bundles of long parallel chains like cellulose.

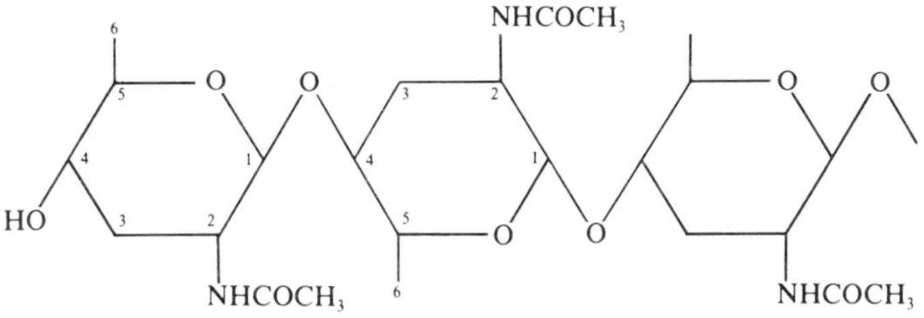

Fig 5.16 *Structure of chitin*

Glycoproteins and glycolipids

Glycoproteins and glycolipids are important molecules containing a polysaccharide unit, and the structure and biological functions of glycolipids are described further in section 5.3.8. Table 5.7 gives further examples of the many roles played by polysaccharides and closely related compounds. The chief point to note is the existence of a great variety of molecules with different structures which, in turn, leads to a wide range of functions.

> **5.10** Summarise the structures and functions of starch, glycogen, cellulose and chitin in a table similar to table 5.7.
>
> **5.11** What structural features of carbohydrates account for the wide variety of polysaccharides?

5.3 Lipids

Lipids are sometimes classified loosely as those water-insoluble organic substances which can be extracted from cells by organic solvents such as ether, chloroform and benzene. They cannot be defined precisely because their chemistry is so variable, but we could say that true lipids are esters of fatty acids and an alcohol.

Esters are organic compounds formed by a reaction between an acid and an alcohol:

$$\text{acid} + \text{alcohol} \xrightarrow{\text{'esterification'}} \text{ester} + \text{water}$$

e.g. CH$_3$.COOH + C$_2$H$_5$OH → CH$_3$COOC$_2$H$_5$ + H$_2$O
ethanoic acid (acetic acid) + ethanol (ethyl alcohol) → ethyl ethanoate + water

—COO— is an ester linkage. Note that an ester linkage is formed by a condensation reaction.

> **5.12** What is a condensation process?

5.3.1 Constitutents of lipids

Fatty acids

Fatty acids contain the acidic group —COOH (the carboxyl group) and are so named because some of the larger molecules in the series occur in fats. They have the general formula R.COOH where R is hydrogen or an alkyl group such as —CH$_3$, —C$_2$H$_5$, and so on (increasing by —CH$_2$ for each subsequent member of the series). R usually has many carbon atoms in lipids. Most fatty acids have an even number of carbon atoms between 14 and 22 (most commonly 16 or 18). The most common fatty acids are shown in fig 5.17. Note the characteristically long chain of carbon and hydrogen atoms forming a hydrocarbon tail. Many of the properties of lipids are determined by these tails, including their insolubility in water. The tails are hydrophobic (*hydro*, water; *phobos*, fear).

Fatty acids sometimes contain one or more double bonds (C=C), such as oleic acid (fig 5.17). In this case they are said to be **unsaturated**, as are lipids containing them. Fatty acids and lipids lacking double bonds are said to be **saturated**. Unsaturated fatty acids melt at much lower temperatures than saturated fatty acids. Oleic acid, for example, is the chief constituent of olive oil and is liquid at normal temperatures (M.pt. 13.4 °C), whereas palmitic and stearic acids (M.pts. 63.1 °C and 69.6 °C respectively) are solid at temperatures to which living organisms are normally exposed.

Table 5.7 Further examples of polysaccharides and closely related compounds.

	Structure	Name of 'polysaccharide'	Units (residues)	Function
Structural	Cell wall matrix in plants	Pectins	Galactose (a hexose) and galacturonic acid (the acid sugar of galactose). Pectic acid is polygalacturonic acid.	Often form gels (commercial jelling agents).
		Hemicelluloses	Very mixed. Sugar residues (mainly pentoses) and sugar acids.	Further information on cell wall matrix in chapter 7.
	Bacterial cell walls	Murein	A polysaccharide cross-linked with amino acids. Polysaccharide is alternating units of 2 amino sugars (acetyglucosamine and another similar nitrogen-containing monosaccharide – compare chitin).	A structural component equivalent to cellulose of plant cell walls. Unique to prokaryotes.
		Other polysaccharides		Some are *antigens* in Gram positive bacteria.
	Outer coats of animal cells	Glycoprotein, glycolipids and other polysaccharides, e.g. hyaluronic acids		Important in ability of cells to 'recognise' each other and in antigenic properties. Intercellular lubrication.
	Connective tissue	Hyaluronic acid	Alternating sugar acid and amino sugar residues.	Part of ground substance of vertebrate connective tissue. Important lubricant – found in synovial fluid in joints and vitreous humour of eye. Forms very viscous solution. Major component of cartilage, bone and other connective tissues. Also cornea.
	'mucopolysaccharides' – repeating pairs of units, one of which is always an amino sugar, e.g. glucosamine	Chondroitin sulphate	Similar to hyaluronic acid.	
Protective		Heparin	Related to chondroitin.	Anticoagulant in mammalian blood and connective tissue. Secreted by most cells.
		Gums and mucilages	Sugars (arabinose, galactose, xylose and rhamnose), and sugar acids (glucuronic and galacturonic acids).	Swell in water; gums form gels or sticky solutions, mucilages form looser gels or a slimy mass. Large, open, flexible molecules, often complex and highly branched. Formed as a result of injury as hard, glossy exudates, e.g. gum arabic from *Acacia*. May also retain water for drought resistance.
Food storage		Mannan, arabinan	Mannose, arabinose.	In some plants.
		Hemicelluloses	See above.	Some seeds, e.g. dates.

NB The suffix -an is used for polymers, e.g. glucose→glucans (e.g. starch); mannose→mannans etc.

stearic acid, $C_{17}H_{35}COOH$, a saturated fatty acid; in palmitic acid, $C_{15}H_{31}COOH$, the tail is two carbon atoms shorter

oleic acid, $C_{17}H_{33}COOH$, an unsaturated fatty acid

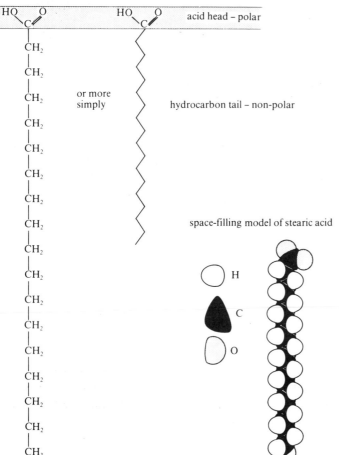

Fig 5.17 *Some examples of common fatty acids*

5.13 Cells of poikilothermic ('cold-blooded') animals usually have a higher proportion of unsaturated fatty acids than homeothermic ('warm-blooded') animals. Can you account for this?

Alcohols

Most lipids are esters of the alcohol **glycerol** (fig 5.18), and are therefore called **glycerides**.

5.3.2 Formation of a lipid

Glycerol has three hydroxyl (—OH) groups, all of which can condense with a fatty acid to form an ester. Usually all three undergo condensation reactions as shown in fig 5.18, and the lipid formed is therefore called a **triglyceride**.

5.3.3 Properties and functions of triglycerides

Triglycerides are the commonest lipids in nature and are further classified as fats or oils, according to whether they are solid (fats) or liquid (oils) at 20 °C. The higher the proportion of unsaturated fatty acids, the lower their melting points.

5.14 Tristearin and triolein are both lipids. Which is more likely to be an oil?

Triglycerides are non-polar and therefore relatively insoluble in water. They are less dense than water and therefore float.

A major function of lipids is to act as energy stores. They have a higher calorific value than carbohydrates, that is a given mass of lipid will yield more energy on oxidation than an equal mass of carbohydrate (see also chapter 11). This is because lipids have a higher proportion of hydrogen and an almost insignificant proportion of oxygen compared with carbohydrates.

Animals store extra fat when hibernating, and fat is also found below the dermis of the skin of vertebrates where it serves as an insulator. Here it is extensive in mammals living in cold climates, particularly in the form of blubber in aquatic mammals such as whales, where it also contributes to buoyancy. Plants usually store oils rather than fats. Seeds, fruits and chloroplasts are often rich in oils and some seeds are commercial sources of oils, for example the coconut, castor bean, soyabean and sunflower seed. When fats are oxidised, water is a product. This metabolic water can be very useful to some desert animals, such as the kangaroo rat, which stores fat for this purpose (section 19.3.4).

5.15 A camel stores fat in the hump primarily as a water source rather than as an energy source. (a) By what metabolic process would water be made available from fat? (b) Carbohydrate could also be used as a water source in the same process. What advantage does fat have over carbohydrate?

5.3.4 Waxes

Waxes are esters of fatty acids with long-chain alcohols. Their functions are summarised in table 5.8.

5.3.5 Phospholipids

Phospholipids are lipids containing a phosphate group. The commonest type, phosphoglycerides, are formed when one of the primary alcohol groups ($-CH_2OH$) of glycerol forms an ester with phosphoric acid (H_3PO_4) instead of a fatty acid (fig 5.19).

The molecule consists of a phosphate head (circled in fig 5.19) with two hydrocarbon tails (the fatty acids).

5.3.6 Steroids and terpenes

Steroids and terpenes can be classified as lipids from the substances involved in their synthesis, although they do not contain fatty acids. They are derived from 5-carbon hydrocarbon building blocks called isoprene units, C_5H_8. Steroids all contain a nucleus composed of 17 carbon atoms, methyl groups ($-CH_3$) are usually attached at positions 18 and 19 as shown in fig 5.20 and a side-chain generally occupies position 17.

In humans, the steroid present in largest amounts is cholesterol (fig 5.20), a key intermediate in the synthesis of related steroids, as well as being an important constituent of membranes. It is a steroid alcohol, or sterol and is made in the liver.

Steroids are abundant in plants and animals and have many important biochemical and physiological roles, as table 5.8 reveals. It is interesting to note that some plant steroids can be converted into animal hormones in the presence of relevant enzymes.

Terpenes also have a wide range of physiological roles, particularly in plants (see table 5.8).

5.3.7 Lipoproteins

Lipoproteins are associations of lipids with proteins. Their functions are summarised in table 5.8.

5.3.8 Glycolipids

Glycolipids are associations of lipids with carbohydrates. The carbohydrate forms a polar head to the molecule, and glycolipids, like phospholipids, are found in membranes (table 5.8).

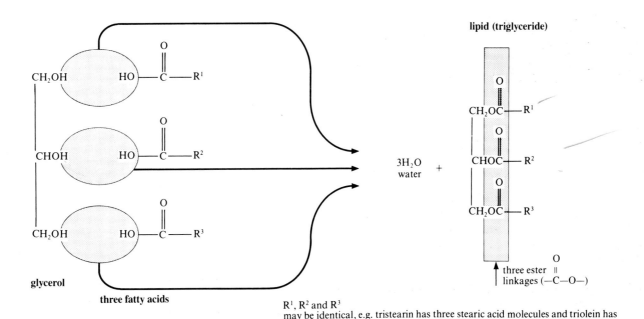

Fig 5.18 *Formation of a lipid from fatty acids and glycerol by condensation reactions*

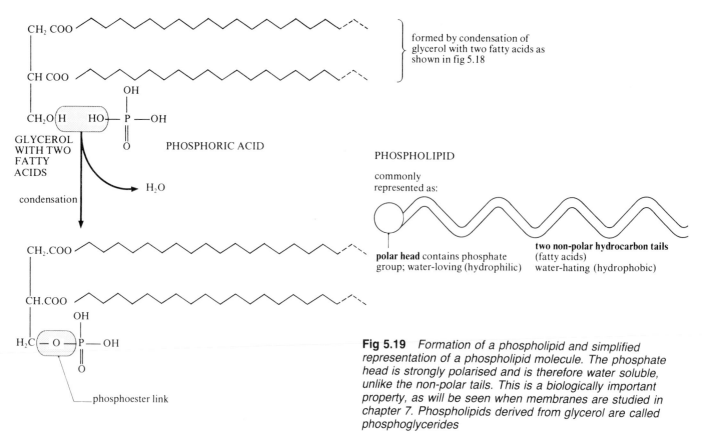

formed by condensation of
glycerol with two fatty acids as
shown in fig 5.18

PHOSPHOLIPID

commonly
represented as:

polar head contains phosphate
group; water-loving (hydrophilic)

two non-polar hydrocarbon tails
(fatty acids)
water-hating (hydrophobic)

GLYCEROL
WITH TWO
FATTY
ACIDS

condensation

PHOSPHORIC ACID

H_2O

phosphoester link

Fig 5.19 *Formation of a phospholipid and simplified
representation of a phospholipid molecule. The phosphate
head is strongly polarised and is therefore water soluble,
unlike the non-polar tails. This is a biologically important
property, as will be seen when membranes are studied in
chapter 7. Phospholipids derived from glycerol are called
phosphoglycerides*

Table 5.8 Functions of lipids other than fats and oils.

Waxes
Mainly used as waterproofing material by plants and animals as in:
additional protective layer on cuticle of epidermis of some plant organs, e.g. leaves, fruits, seeds (particularly xerophytes),
skin, fur and feathers of animals,
exoskeleton of insects (see chitin).
Beeswax is a constituent of the honeycomb of bees.

Phospholipids
Constituents of membranes.

Steroids
Bile acids, e.g. cholic acid. Constituents of bile, forming part of bile salts which emulsify and solubilise lipids during digestion (chapter 10).
Sex hormones, e.g. oestrogen, progesterone, testosterone (chapter 20).
Cholesterol (absent from plants (see text)).
Vitamin D – rickets occurs if deficient.
Cardiac poisons e.g. digitalis, used for heart therapy.
Adrenocortical hormones (corticosteroids) e.g. aldosterone, corticosterone, cortisone (chapter 16).

Terpenes
Scents and flavours in 'essential oils' of plants e.g. menthol in mint, camphor (2, 3, or 4 isoprene units).
Gibberellins – plant growth substances with 4 isoprene units (see chapter 15).
Phytol – component of chlorophyll (chapter 9) and vitamins A, E, and K, has 4 isoprene units.
Cholesterol – derived from terpenes with 6 isoprene units.
Carotenoids – photosynthetic pigments with 8 isoprene units (chapter 9).
Natural rubber – thousands of isoprene units in regular linear arrangements.

Lipoproteins
Membranes are lipoprotein structures.
Form in which lipids are transported in blood plasma and lymph.

Glycolipids
Components of cell membranes, particularly in myelin of nerve cells and on outer surfaces of nerve cells; chloroplast membranes.

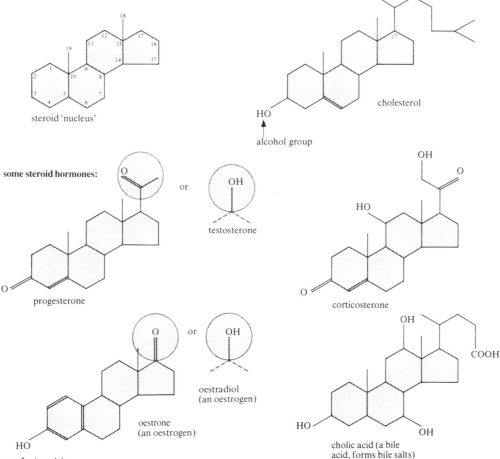

steroid 'nucleus'

cholesterol

alcohol group

some steroid hormones:

O or OH

testosterone

progesterone

O

corticosterone

HO

O or OH

oestradiol
(an oestrogen)

oestrone
(an oestrogen)

HO

cholic acid (a bile
acid, forms bile salts)

Fig 5.20 *Structures of steroids*

5.4 Amino acids

Over 170 amino acids are currently known to occur in cells and tissues. Of these, 26 are constituents of proteins, 20 occurring commonly in protein structure (table 5.9).

Plants are able to make all the amino acids they require from simpler substances. However, animals are unable to synthesise all that they need, and therefore must obtain some 'ready-made' amino acids directly from their diet. These are termed **essential amino acids**. It must be emphasised, however, that when considered as components of proteins within the animal body, the essential amino acids are no more important than those that the animals synthesise themselves. They are simply 'essential' because they cannot be synthesised by them.

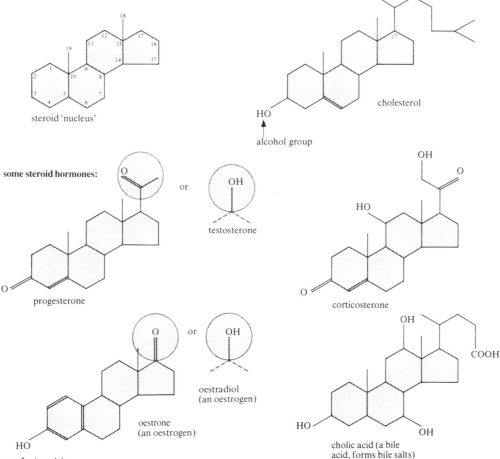

Fig 5.21 *General formula of an amino acid*

5.4.1 Structure and range of amino acids

With the exceptions of proline and hydroxyproline which are imino acids (table 5.9), all amino acids are α amino acids, that is the amino (—NH₂) group is attached to the α carbon of the related carboxylic (—COOH) group. The general formula of an amino acid can be seen in fig 5.21.

The majority of amino acids possess one acidic carboxylic group and one basic amino group and are termed 'neutral' amino acids. However, in some cases there may be more than one amino group present, giving rise to **basic amino acids**, or more than one carboxylic group giving rise to **acidic amino acids**. The R group represents the residual part of the molecule. Its composition varies considerably in different amino acids and is responsible for the unique properties they display.

The simplest amino acid, glycine (fig 5.22), is formed when R is substituted by H. When the R group is anything but H, the groups attached to the α carbon will all be different. The carbon is therefore asymmetric. This means that the amino acid will possess two optically active forms. In fact, all amino acids but glycine are optically active and may exist in either the D or L forms. In nature amino acids are generally found in their L form. Optical isomerism is discussed earlier in this chapter.

143

Table 5.9 Common amino acids in proteins.

Amino acid	Abbreviation	R group	Notes
Alanine	Ala	$-CH_3$	
Arginine	Arg	$-[CH_2]_3-NH-C(=NH)-NH_2$	Basic amino acid, essential for children
Asparagine	Asn	$-CH_2-CO-NH_2$	R contains amide group
Aspartic acid	Asp	$-CH_2-COOH$	Acidic amino acid
Cysteine	Cys	$-CH_2-SH$	R contains sulphur
Glutamine	Gln	$-[CH_2]_2-CO-NH_2$	R contains amide group
Glutamic acid	Glu	$-[CH_2]_2-COOH$	Acidic amino acid
Glycine	Gly	$-H$	
Histidine	His	$-CH_2-$ (imidazole ring: N, NH, H)	Basic amino acid, essential for children
Isoleucine	Ile	$-C(H)(CH_3)-C_2H_5$	Essential amino acid
Leucine	Leu	$-CH_2-CH(CH_3)_2$	Essential amino acid
Lysine	Lys	$-[CH_2]_4-NH_2$	Basic amino acid, essential
Methionine	Met	$-[CH_2]_2-S-CH_3$	Essential amino acid, R contains sulphur
Phenylalanine	Phe	$-CH_2-$ (aromatic ring)	Essential amino acid, R contains aromatic ring
Proline	Pro	$CH_2^{*}-CH_2$ / CH_2 \ NH — H, COOH (structure of whole molecule)	= NH instead of $-NH_2$, called an **imino** acid; *$-C(H)(OH)-$ in hydroxyproline.
Serine	Ser	$-CH_2-OH$	
Threonine	Thr	$-C(OH)(H)-CH_3$	Essential amino acid
Tryptophan	Trp	$-CH_2-$ (indole ring with NH)	R contains aromatic ring, essential amino acid
Tyrosine	Tyr	$-CH_2-$ (aromatic ring) $-OH$	R contains aromatic ring
Valine	Val	$-CH(CH_3)(CH_2)$	Essential amino acid

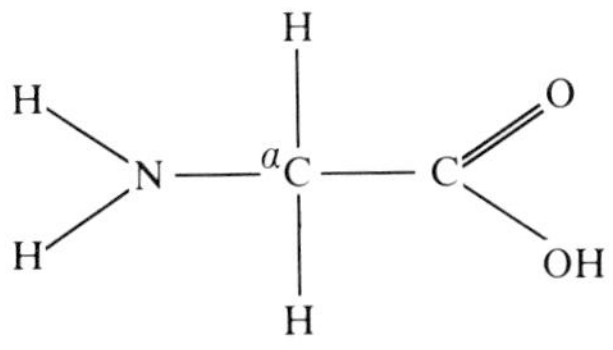

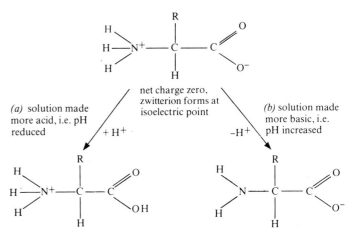

Fig 5.22 Glycine

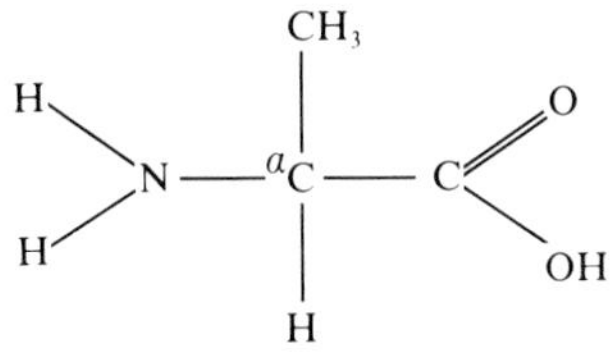

Fig 5.23 Alanine

When R is substituted by —CH$_3$, the amino acid alanine is formed (fig 5.23).

Table 5.9 shows the names, three-letter abbreviations and R groups of the commonly occurring amino acids.

Rare amino acids

A small number of rare amino acids occur in organisms. They are derivatives of some of the common amino acids. For example, hydroxyproline is a derivative of proline, and is found in collagen; hydroxylysine is a derivative of lysine, and is also found in collagen.

There is no DNA triplet code for the rare amino acids, and they are derived by the modification of their parent amino acids after they have been incorporated into a polypeptide chain.

Non-protein amino acids

Over 150 of these are known to occur, either free or in a combined form in cells, but never in proteins. For example, ornithine and citrulline are important metabolic intermediates in the synthesis of arginine. Also GABA (γ-amino butyric acid) is virtually unique to the nervous system. It is an inhibitory neurotransmitter, important in the brain.

5.4.2   Properties of amino acids

Amino acids are colourless, crystalline solids. They are generally soluble in water, but insoluble in organic solvents. In neutral aqueous solutions they exist as dipolar ions (**zwitterions**) and are **amphoteric**, possessing both basic and acidic properties.

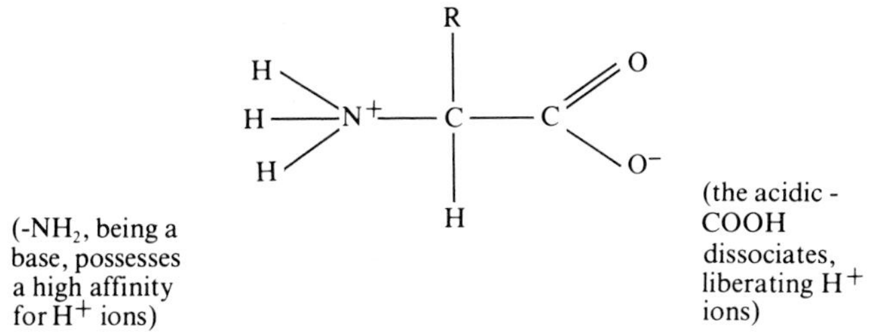

(-NH$_2$, being a base, possesses a high affinity for H$^+$ ions)

(the acidic - COOH dissociates, liberating H$^+$ ions)

Fig 5.24 Neutral zwitterion form of an amino acid

Each amino acid has its own specific pH at which it will exist in its neutral zwitterion form and will be strongly dipolar (fig 5.24). If it is placed in an electric field at this pH, the amino acid will migrate neither to the cathode nor to the anode. The pH causing this electrical neutrality is called the **isoelectric point** of the amino acid. Thus each amino acid possesses its own specific isoelectric point.

The amphoteric nature of amino acids is useful biologically as it means that they can act as buffers in solutions, resisting changes in pH. They do this by donating H$^+$ ions as pH increases and accepting H$^+$ ions as pH decreases. Fig 5.25 demonstrates what happens when an amino acid at the pH of its isoelectric point (*a*) has an acid added to it, and (*b*) has a base added to it.

net charge zero, zwitterion forms at isoelectric point

(a) solution made more acid, i.e. pH reduced   + H$^+$

(b) solution made more basic, i.e. pH increased   –H$^+$

H$^+$ions accepted. The amino acid becomes positively charged and will migrate to the negative electrode (cathode) if placed in an electric field. Net charge +

H$^+$ ions donated. The amino acid becomes negatively charged and will migrate to the positive electrode (anode) if placed in an electric field. Net charge –

Fig 5.25 Effect of (a) acid and (b) alkali on the isoelectric point of an amino acid

5.4.3   Linkages

Amino acids are able to form a variety of chemical bonds with other reactive groups. These will later be shown to be of great significance in protein structure and function.

Peptide bond

This is formed when a water molecule is eliminated during interaction between the amino group of one amino acid and the carboxylic group of another. Elimination of water is known as **condensation** and the linkage formed is a covalent carbon–nitrogen bond, called a **peptide bond** (fig 5.26). The compound formed is a **dipeptide**. It possesses a free amino group at one end, and a free carboxylic group at the other. This enables further combination between the dipeptide and other amino acids. If many amino acids are joined together in this way, a **polypeptide** is formed (fig 5.27).

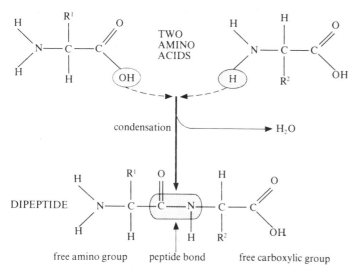

Fig 5.26 *Formation of a dipeptide by condensation of two amino acids*

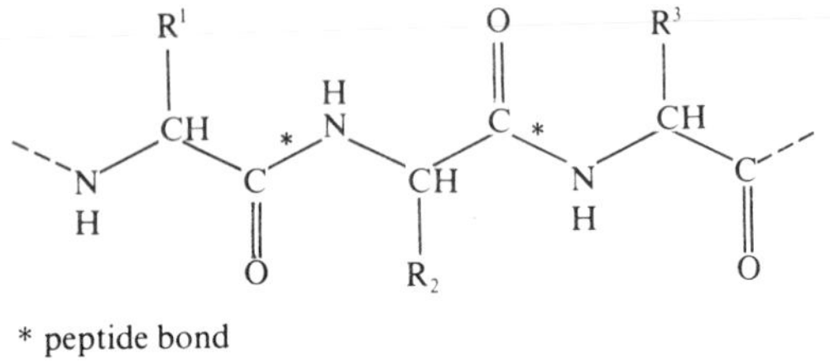

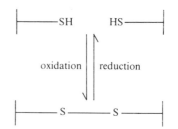

* peptide bond

Fig 5.27 *Part of a polypeptide showing the joining of three amino acids*

5.16 Write down the structural formula of the tripeptide formed by alanine, glycine and serine joined together in that order.

Ionic bond

At a suitable pH an interaction may occur between ionised amino and carboxylic groups. The result is the formation of an ionic bond (fig 5.28). In an aqueous environment this bond is much weaker than a covalent bond and can be broken by changing the pH of the medium.

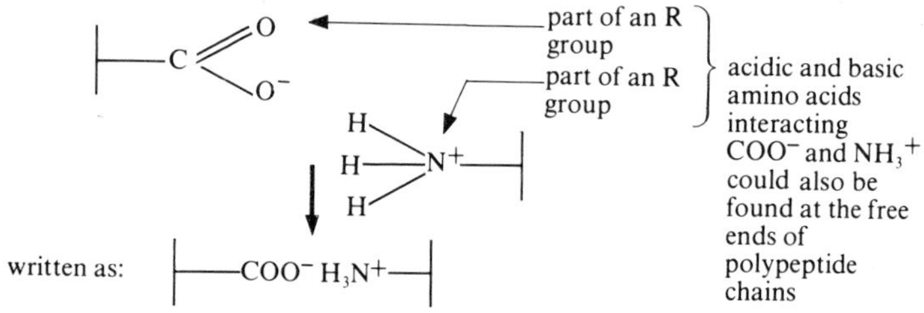

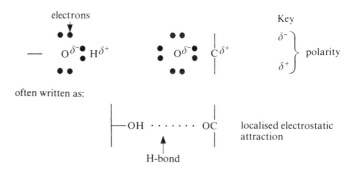

Fig 5.28 *Ionic bond formation*

Disulphide bond

When two molecules of cysteine combine, neighbouring cysteine sulphydryl (—SH) groups are oxidised and subsequently form a disulphide bond (fig 5.29). Interchain or intrachain disulphide bonds may be formed. This is significant in protein structure (figs 5.32 and 5.33).

Fig 5.29 *Formation of a disulphide bond*

Hydrogen bond

Electropositive hydrogen atoms attached to oxygen or nitrogen in —OH or —NH groups have a tendency to share the electrons of a neighbouring electronegative oxygen atom such as the O of a =CO group (fig 5.30). The hydrogen bond is weak, but as its occurrence is frequent, the total effect makes a considerable contribution towards molecular stability, as in the structure of silk (fig 5.35*a*).

Fig 5.30 *Formation of a hydrogen bond*

5.5   Proteins

Proteins are complex organic compounds always containing the elements carbon, hydrogen, oxygen and nitrogen, and in some cases sulphur. Some proteins form complexes with other molecules containing phosphorus, iron, zinc and copper. Proteins are macromolecules of high M_r (relative formula mass or molecular mass), between several thousands and several millions, consisting of chains of amino acids. Twenty different amino acids are commonly found in naturally occurring proteins. The potential variety of proteins is unlimited because the sequence of amino acids in each protein is specific for that protein (section 22.6) and is genetically controlled by the DNA of the cell in which it is manufactured. Proteins are the most abundant organic molecules to be found in cells and comprise over 50% of their total dry mass. They are an essential component of the diet of animals and may be converted to both fat and carbohydrate by the cells. Their structural diversity enables them to display a great range of structural and metabolic activities within the organism.

5.5.1 Size of protein molecules

Simple peptides containing two, three or four amino acid residues are called di-, tri- and tetrapeptides respectively. Polypeptides are chains of many amino acid residues (up to several thousand – table 5.10). A protein may possess one or more polypeptide chains.

5.5.2 Classification of proteins

Because of the complexity of protein molecules and their diversity of function, it is very difficult to classify them in a single, well-defined fashion. Three alternative methods are given in tables 5.11, 5.12 and 5.13.

5.5.3 Structure of proteins

Each protein possesses a characteristic three-dimensional shape, its **conformation**. In describing the three-dimensional structure of proteins it is usual to refer to four separate levels of organisation as follows.

Primary structure

The primary structure is the number and sequence of amino acids held together by peptide bonds in a polypeptide chain (fig 5.31). It was the double Nobel prizewinner F. Sanger, working in Cambridge, who pioneered work on elucidating the amino acid sequence of proteins. He worked specifically with insulin (fig 5.32), a hormone, and this was the first protein to have its amino acid sequence determined. It took him ten years (1944–54) to discover the sequence. Insulin is a compound of 51 amino acids, with a M_r of 5 733. The protein is composed of two polypeptide chains held together by disulphide bridges.

Today, much of the amino acid sequencing is accomplished by machine and an estimated 8 000 protein primary structures are known. Another example, lysozyme is shown in fig 5.33.

There are over 10 000 proteins in the human body, all composed of different arrangements of the 20 fundamental amino acids. The sequence of amino acids of a protein dictates its biological function. In turn, this sequence is strictly controlled by the sequence of bases in DNA (section 22.6). Substitution of just a single amino acid can cause a major alteration in a protein's function, as in the condition of sickle cell anaemia (section 23.9). Analysis of amino acid sequences of homologous proteins from different species is of interest, because it offers evidence about the possible taxonomic relationships between different species. This is dealt with in chapter 24.

> **5.17** (*a*) Let the letters A and B represent two different amino acids. Write down the sequences of all the possible tripeptides that could be made with just these two amino acids.
> (*b*) From your answer to (*a*), what is the formula for calculating the number of different tripeptides that can be formed from two different amino acids?
> (*c*) How many polypeptides, 100 amino acids in length, could be formed from two different amino acids?
> (*d*) How many polypeptides, 100 amino acids in length (a modest length for a protein), could be made using all 20 common amino acids?
> (*e*) How many peptides/polypeptides could be made (any length) from all 20 common amino acids?

Table 5.10 Sizes of some proteins.

Protein	M_r (molecular mass)	Number of amino acid residues	Number of polypeptide chains
Ribonuclease	12 640	124	1
Lysozyme	13 930	129	1
Myoglobin	16 890	153	1
Haemoglobin	64 500	574	4
α amylase	97 600	Not known	2
TMV (Tobacco mosaic virus)	≈ 40 000 000	≈ 336 500	2 130

The largest protein complexes are found in viruses where M_rs of over 40 000 000 are commonly found.

Table 5.11 Classification of proteins according to structure.

Type	Nature	Function
Fibrous 	Secondary structure most important (little or no tertiary structure) Insoluble in water Physically tough Long parallel polypeptide chains cross-linked at intervals forming long fibres or sheets	Perform structural functions in cells and organisms, e.g. components of connective tissue, collagen (tendons, bone matrix), myosin (in muscle sarcomere), silk (spiders' webs), keratin (hair, horn, nails, feathers)
Globular 	Tertiary structure most important Polypeptide chains tightly folded to form spherical shape Easily soluble to form a colloidal suspension	Globulins of blood serum – important in immunology Form enzymes, antibodies and some hormones, e.g. insulin Important in protoplasm as they hold water and other substances and serve to maintain molecular organisation
Intermediate	Fibrous but soluble	e.g. fibrinogen – forms insoluble fibrin when blood clots

Table 5.12 Classification of proteins according to composition.

Proteins

(i) *Simple*
Only amino acids form their structure

(ii) *Conjugated*
Complex compounds consisting of globular proteins and non-proteinaceous material; the non-proteinaceous material is called a **prosthetic** group

(i) *Simple proteins*

Name	Properties	Location
Albumins	Neutral Soluble in water Soluble in dilute salt solution	Egg albumen Serum albumin of blood
Globulins	Neutral Insoluble in water Soluble in dilute salt solution	Antibodies in blood Blood fibrinogen
Histones	Basic Soluble in water Insoluble in dilute ammonia solution	Associated with nucleic acids, in nucleo-proteins of cell
Scleroproteins (only in animal kingdom)	Insoluble in water and most other solvents	Keratin of hair, skin, feathers; collagen of bone matrix and tendon; elastin of ligament

(ii) *Conjugated proteins*

Name	Prosthetic group	Location
Phosphoprotein	Phosphoric acid	Casein of milk Vitellin of egg yolk
Glycoprotein	Carbohydrate	Blood plasma Mucin (component of saliva)
Nucleoprotein	Nucleic acid	Component of viruses Chromosomes Ribosome structure
Chromoprotein	Pigment	Haemoglobin – haem (iron-containing pigment) Phytochrome (plant pigment) Cytochrome (respiratory pigment)
Lipoprotein	Lipid	Membrane structure Lipid transported in blood as lipoprotein
Flavoprotein	FAD (flavin adenine dinucleotide, see section 11.3.6)	Important in electron transport chain in respiration
Metal proteins	Metal	E.g. nitrate reductase, the enzyme in plants which converts nitrate to nitrite

Table 5.13 Protein classification according to function. Proteins are also important in membranes where they function as enzymes, receptor sites and transport sites.

Type	Examples	Occurrence/function
Structural	Collagen	Component of connective tissue, bone, tendons, cartilage
	Sclerotin	Exoskeleton of insects
	α-keratin	Skin, feathers, nails, hair, horn
	Elastin	Elastic connective tissue (ligaments)
	Mucoproteins	Synovial fluid, mucous secretions
	Viral coat proteins	'Wraps up' nucleic acid of virus
Enzymes	Trypsin	Catalyses hydrolysis of protein
	Ribulose bisphosphate carboxylase	Catalyses carboxylation (addition of CO_2) of ribulose bisphosphate in photosynthesis
	Glutamine synthetase	Catalyses synthesis of the amino acid glutamine from glutamic acid + ammonia
Hormones	Insulin $\}$ Glucagon	Help to regulate glucose metabolism
	ACTH	Stimulates growth and activity of the adrenal cortex
Transport	Haemoglobin	Transports O_2 in vertebrate blood
	Haemocyanin	Transports O_2 in some non-vertebrate blood
	Myoglobin	Transports O_2 in muscles
	Serum albumin	Transport in blood, e.g. fatty acids, lipids
Protective	Antibodies	Form complexes with foreign proteins
	Fibrinogen	Precursor of fibrin in blood clotting
	Thrombin	Involved in clotting mechanism
Contractile	Myosin	Moving filaments in myofibril of sarcomere
	Actin	Stationary filaments in myofibril of sarcomere
Storage	Ovalbumin	Egg white protein
	Casein	Milk protein
Toxins	Snake venom	Enzymes
	Diphtheria toxin	Toxin made by diphtheria bacteria

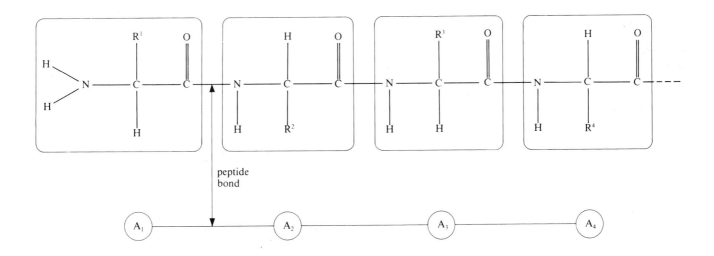

Fig 5.31 *Portion of a polypeptide chain to show primary structure. A_1, A_2, A_3 and A_4 represent different amino acids*

Fig 5.32 (below) *Primary structure (sequence of amino acids) of insulin. The molecule consists of two polypeptide chains held together by two disulphide bridges*

Fig 5.33 (right) *The primary structure of lysozyme. Lysozyme is an enzyme that is found in many tissues and secretions of the human body, in plants, and in the whites of eggs. Its function is to catalyse the breakdown of the cell walls of bacteria. The molecule consists of a single polypeptide chain of 129 amino acid residues. There are four intrachain disulphide bridges*

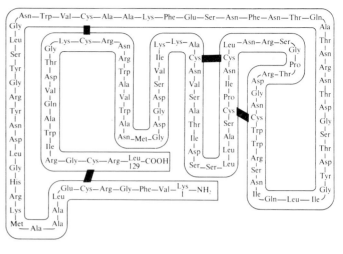

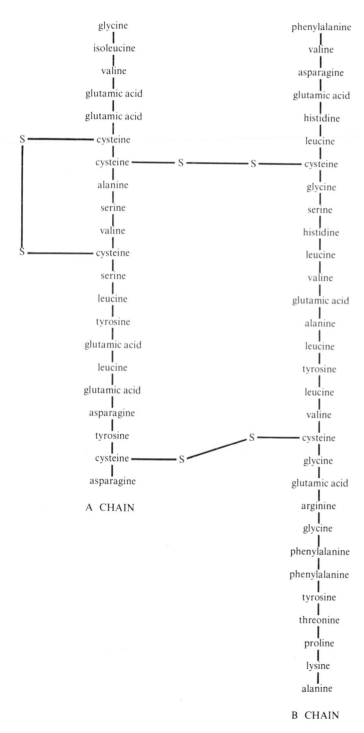

A CHAIN

B CHAIN

Secondary structure

In addition to the primary structure there is a specific secondary structure. It usually takes the form of an extended spiral spring, the α-helix, whose structure is maintained by many hydrogen bonds which are formed between adjacent CO and NH groups. The H atom of the NH group of one amino acid is bonded to the O atom of the CO group three amino acids away (fig 5.34). X-ray diffraction analysis data indicate that the α-helix makes one complete turn for every 3.6 amino acids.

A protein which is entirely α-helical, and hence fibrous, is keratin. It is the structural protein of hair, wool, nails, claws, beaks, feathers and horn, as well as being found in vertebrate skin. Its hardness and stretchability vary with the degree of cross-linking by disulphide bridges between neighbouring chains.

Theoretically, all CO and NH groups can participate in hydrogen bonding as described, so the α-helix is a very stable, and hence a common, structure. In spite of this, most proteins are globular molecules in which there are also regions of β-sheet (see below) and irregular structure. This is due mainly to interference in hydrogen bonding by certain R groups, the occurrence of disulphide bridges between different parts of the same chain and the inability of the amino acid proline to make hydrogen bonds.

Another type of secondary structure is the β-pleated sheet. Silk fibroin, the protein used by silkworms when spinning their cocoon threads, is entirely in this form. This protein comprises a number of adjacent chains which are more extended than the α-helices. They are arranged in a parallel fashion but running in opposite directions to each other. They are joined together by hydrogen bonds formed between the C=O and NH groups of one chain and the NH and C=O groups of adjacent chains. Again, all NH and C=O groups are involved in hydrogen bonding, so the structure is very stable. This is called the β-**configuration**, and the whole structure is known as a β-**pleated sheet** (fig 5.35). The sheet of fibroin has a high tensile strength and cannot be stretched, but the arrangement of the polypeptides makes the silk very supple. In globular proteins a single polypeptide chain may fold back on itself and form regions of β-pleated sheet.

Yet another arrangement is seen in the fibrous protein collagen. Here three polypeptide chains are wound around each other to form a triple helix. There are about 1 000 amino acid residues in each chain, and the complete triple helix compound is called **tropocollagen** (fig 5.36). Again the protein cannot be stretched and this is an essential part of its functioning, for example in tendons, bone and other connective tissue. Proteins which exist entirely in the form of helical coils, such as keratin and collagen, are exceptional.

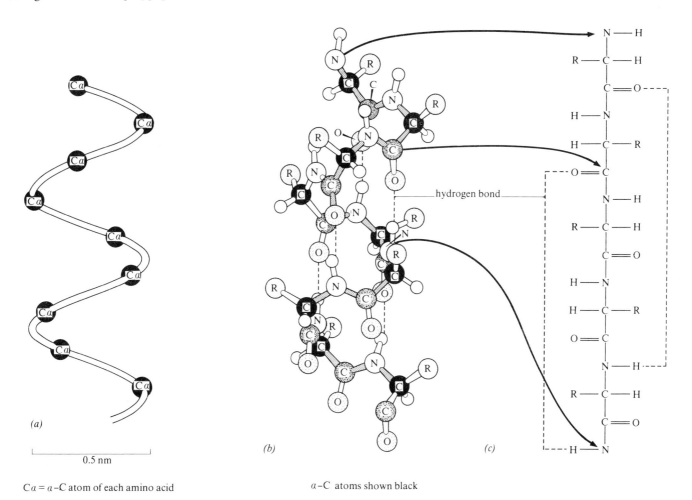

0.5 nm

Cα = α–C atom of each amino acid

α–C atoms shown black

Fig 5.34 Structure of the α-helix. (a) The α-C atoms are shown. A line joining them describes an α-helix. (b) the entire α-helix. (c) Part of the α-helix straightened out. Hydrogen bonds hold the helix in place

Fig 5.35 (below) Beta-pleated sheet. The chains are held parallel to each other by the hydrogen bonds that form between the NH and CO groups. The side-groups (R) are above and below the plane of the sheet. (a) Two antiparallel polypeptide chains. (b) Drawing of three parallel chains to show pleating of structure between R groups

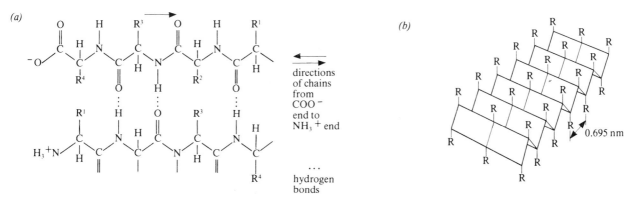

directions of chains from COO⁻ end to NH₃⁺ end

... hydrogen bonds

0.695 nm

Fig 5.36 *Collagen triple-helix structure*

Tertiary structure

Usually the polypeptide chain bends and folds extensively, forming a precise, compact 'globular' shape. This is the protein's tertiary conformation and it is maintained by the interaction of the three types of bond already discussed, namely ionic, hydrogen and disulphide bonds as well as hydrophobic interactions (fig 5.37). The latter are quantitatively the most important and occur when the protein folds so as to shield hydrophobic side-groups from the aqueous surroundings, at the same time exposing hydrophilic side-chains.

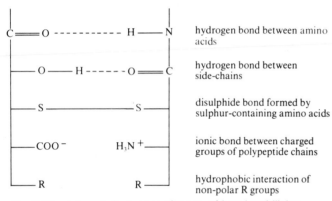

hydrogen bond between amino acids

hydrogen bond between side-chains

disulphide bond formed by sulphur-containing amino acids

ionic bond between charged groups of polypeptide chains

hydrophobic interaction of non-polar R groups

Fig 5.37 *(above) Summary of types of bond stabilising secondary and tertiary structures of proteins. Hydrophobic interactions (associations of non-polar molecules or parts of molecules) to exclude water molecules in the aqueous environment of the cell are particularly important in maintaining structure, as in membranes*

The tertiary structure of a protein can be determined by X-ray crystallography. By early 1963, and after many years work, Kendrew and Perutz had elucidated the secondary and tertiary structures of myoglobin using this technique (fig 5.38):

primary structure – single polypeptide chain of 153 amino acids, the sequence was elucidated in the early 1960s;

secondary structure – about 75% of the chain is α-helical (eight helical sections);

tertiary structure – non-uniform folding of the α-helical chain into a compact shape;

prosthetic group – haem group (contains iron).

Further information about the functions of myoglobin can be found in chapter 14. The elucidation of tertiary structure is still very time-consuming and there are still only about 300 proteins whose tertiary structure is known. Use of computers and other techniques to predict tertiary structure, based on knowledge of primary and secondary structures, is a fast-growing area of molecular biology. There would follow the possibility of designing proteins with particular shapes for particular functions, with important applications in industry and medicine.

Quaternary structure

Many highly complex proteins consist of an aggregation of polypeptide chains held together by hydrophobic interactions and hydrogen and ionic bonds. Their precise arrangement is the quaternary structure. Haemoglobin exhibits such a structure. It consists of four separate polypeptide chains of two types, namely two α-chains and two β-chains. The two α-chains each contain 141 amino acids, while the two β-chains each contain 146 amino acids.

H₂N— Val — Leu — Ser — Glu — Gly — Glu — Trp — Gln — Leu — Val(10) — Leu — His — Val — Tyr — Ala — Lys — Val —

Glu — Ala — Asp(20) — Val — Ala — Gly — His — Gly — Gln — Asp — Ile — Leu — Ile(30) — Arg — Leu — Phe — Lys —

Ser — His — Pro — Glu(40) — Thr — Leu — Glu — Lys — Phe — Asp — Arg — Phe — Lys — His — Leu — Lys(50) — Thr —

Glu — Ala — Glu — Met — Lys — Ala — Ser — Glu — Asp(60) — Leu — Lys — Gly — His — His — Glu — Ala — Glu —

Leu — Thr(70) — Ala — Leu — Gly — Ala — Ile — Leu — Lys — Lys — Gly(80) — His — His — Glu — Ala — Glu —

Leu — Lys — Pro — Leu(90) — Ala — Gln — Ser — His — Ala — Thr — Lys — His — Lys — Ile — Pro(100) — Ile — Lys —

Tyr — Leu — Glu — Phe — Ile(110) — Ser — Glu — Ala — Ile — Ile — His — Val — Leu — His — Ser — Arg — His —

Pro(120) — Gly — Asn — Phe — Gly — Ala — Asp — Ala — Gln — Gly — Ala(130) — Met — Asn — Lys — Ala — Leu — Glu —

Leu — Phe — Arg — Lys(140) — Asp — Ile — Ala — Ala — Lys — Tyr — Lys — Glu — Leu — Gly(150) — Tyr — Gln — Gly —COOH

Fig 5.38 *(above and opposite)   (a) Primary structure of myoglobin. (b) X-ray diffraction pattern of myoglobin (sperm whale). The regular array of spots is a result of scattered (diffracted) beams of X-rays striking the photographic film. The photograph is a two-dimensional section through a three dimensional array of spots. The pattern and intensity of the spots are used to determine the arrangement of atoms in the molecule. From J. C. Kendrew,* Scientific American, *December 1961. (c) Conformation of myoglobin deduced from high resolution X-ray data. (d) Structure of myoglobin*

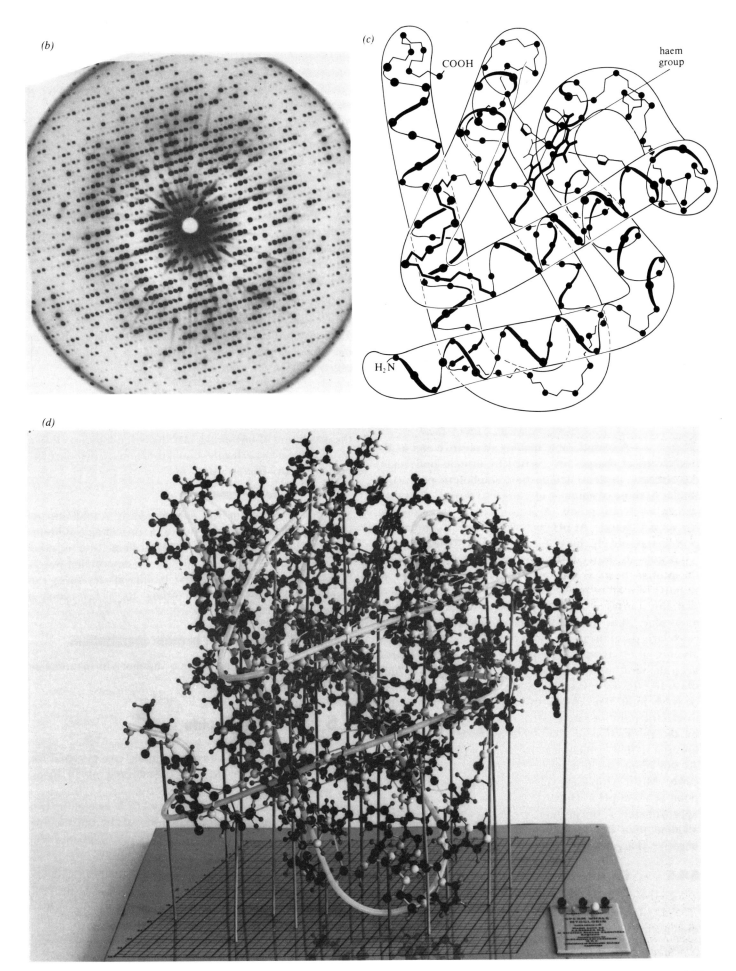

(b)

(c)

COOH

haem group

H₂N

(d)

153

Fig 5.39 *Structure of haemoglobin. The molecule consists of four chains: two alpha chains and two beta chains. Each chain carries a haem to which one molecule of oxygen binds. The assembly of a protein from separate subunits is an example of quaternary structure*

The complete structure of haemoglobin was worked out by Kendrew and Perutz and is illustrated in fig 5.39.

The protein coats of some viruses, such as the tobacco mosaic virus, are composed of many polypeptide chains arranged in a highly ordered fashion (fig 2.14).

5.5.4 Electrical properties of proteins

A considerable number of positive and negative electrical charges are carried by protein molecules. Accordingly, proteins demonstrate amphoteric properties similar to those of amino acids, and also possess their own specific isoelectric points. At its isoelectric point a protein has no net charge. At pHs below its isoelectric point the protein is positively charged, and above its isoelectric point it is negatively charged. In either case the net charge which the protein bears is the same (either all positive or all negative) for all its molecules. The net electrostatic effect that this causes is repulsion between adjacent protein molecules, thus preventing their aggregation. At its isoelectric point the protein is at its least soluble due to the absence of electrostatic repulsion which would otherwise keep the molecules apart. The process of souring of milk illustrates this well. Casein is a soluble protein component of fresh milk. The pH of fresh milk is well above the isoelectric point of casein. However, when milk is soured by the production of lactic acid by bacteria, its pH is lowered to that of the isoelectric point of casein (pH 4.7). At this point casein is precipitated in the form of white curds. Most cytoplasmic proteins possess an isoelectric point at about pH 6. However since the cytoplasmic pH is approximately 7, the proteins are in an environment more alkaline than their isoelectric points, and hence carry negative charges.

5.5.5 Denaturation and renaturation of proteins

Denaturation is the loss of the specific three-dimensional conformation of a protein molecule. The change may be temporary or permanent, but the amino acid sequence of the protein remains unaffected. If denaturation occurs, the molecule unfolds and can no longer perform its normal biological function. A number of agents may cause denaturation as follows.

Heat or radiation. e.g. infra-red or ultra-violet light. Kinetic energy is supplied to the protein causing its atoms to vibrate violently, so disrupting the weak hydrogen and ionic bonds. Coagulation of the protein then occurs.

Strong acids and alkalis and high concentrations of salts. Ionic bonds are disrupted and the protein is coagulated. Breakage of peptide bonds may occur if the protein is allowed to remain mixed with the reagent for a long period of time.

Heavy metals. Cations form strong bonds with carboxylate anions and often disrupt ionic bonds. They also reduce the protein's electrical polarity and thus increase its insolubility. This causes the protein to precipitate out of solution.

Organic solvents and detergents. These reagents disrupt hydrophobic interactions and form bonds with hydrophobic (non-polar) groups. This in turn causes the disruption of intramolecular hydrogen bonding. When alcohol is used as a disinfectant it functions to denature the protein of any bacteria present.

Renaturation

Sometimes a protein will spontaneously refold into its original structure after denaturation, providing conditions are suitable. This is called **renaturation**, and is good evidence that tertiary structure can be determined purely by primary structure and that biological structures can spontaneously assemble according to a few general principles.

5.5.6 Mammalian protein metabolism

Fig 5.40 provides a summary of mammalian protein metabolism.

5.6 Nucleic acids

Nucleic acids, like proteins, are essential for life. They constitute the genetic material of all living organisms, including the simplest virus.

The elucidation of the structure of DNA, one of the two types of nucleic acid, represents one of the outstanding milestones in biology because it finally solved the problem of how living cells, and therefore organisms, accurately replicate themselves and encode the information needed to control their activities. Table 5.4 shows that nucleic acids are made up of units called **nucleotides**. These are arranged to form extremely long molecules known as **polynucleotides**. Thus, to understand their structure, it is necessary first to study the structure of the nucleotide.

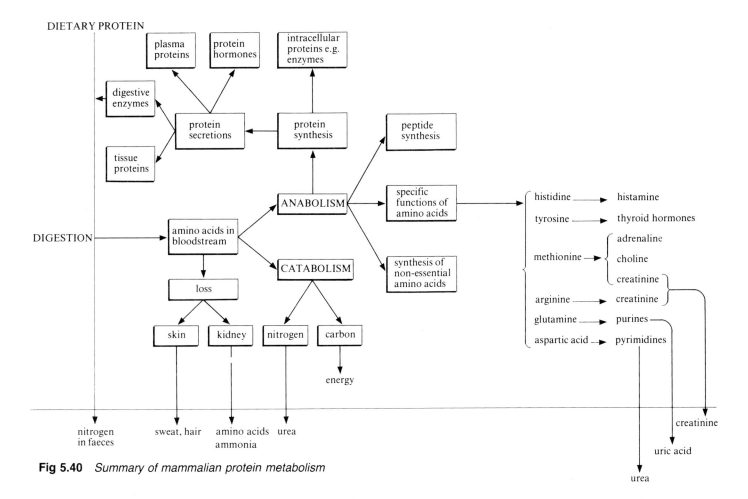

Fig 5.40 *Summary of mammalian protein metabolism*

5.6.1 Structure of nucleotides

A nucleotide has three components, a 5-carbon sugar, a nitrogenous base and phosphoric acid.

Sugar. The sugar has five carbon atoms, therefore it is a pentose. There are two types of nucleic acids, depending on the pentose they contain. Those containing ribose are called ribonucleic acids (RNA) and those containing deoxyribose (ribose with an oxygen atom removed from carbon atom 2) deoxyribonucleic acids (DNA) (fig 5.41).

Bases. Each nucleic acid contains four different bases, two derived from purine and two from pyrimidine. The nitrogen in the rings gives the molecules their basic nature. The bases are the purines, adenine (A) and guanine (G), and the pyrimidines, thymine (T) in DNA or uracil (U) in RNA and cytosine (C). Purines have two rings and pyrimidines one ring in their structure.

Note that RNA contains uracil in place of thymine in DNA. Thymine is chemically very similar to uracil (it is 5-methyl uracil, that is uracil with a methyl group on carbon atom 5). The bases are commonly represented by their initial letters A, G, T, U and C.

Phosphoric acid (fig 5.41). This gives nucleic acids their acid character. Fig 5.42 shows how the sugar, base and phosphoric acid combine to form a nucleotide.

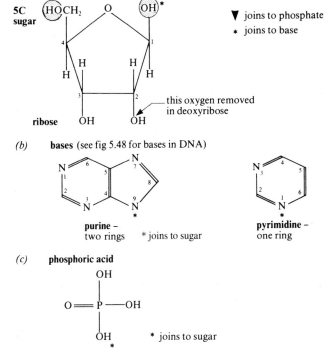

Fig 5.41 *Components of nucleotides*

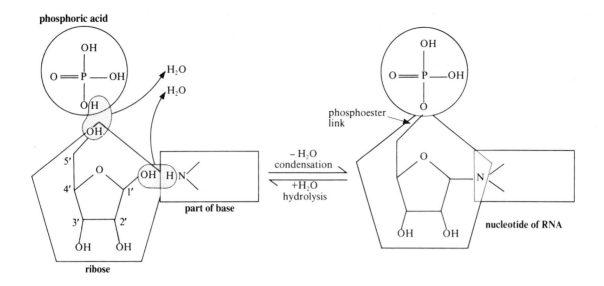

phosphoric acid

OH

O=P—OH

OH

OH

5'

4'

O

3'   2'

1'

OH   OH

ribose

H₂O

H₂O

OH H N

part of base

− H₂O
condensation

+H₂O
hydrolysis

OH

O=P—OH

OH

phosphoester
link

O

O

N

OH   OH

nucleotide of RNA

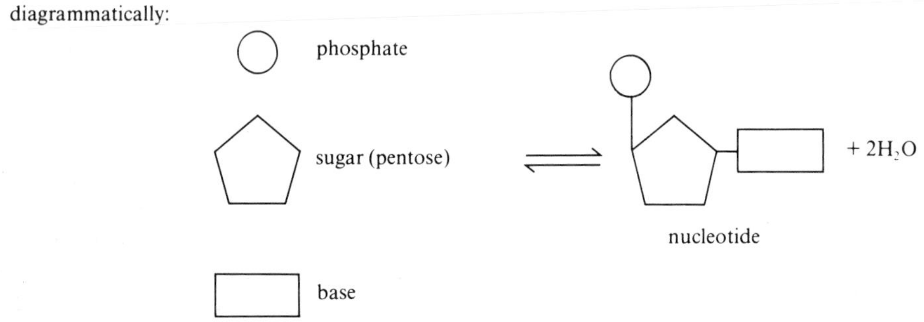

diagrammatically:

◯ phosphate

⬠ sugar (pentose)

▭ base

+ 2H₂O

nucleotide

Fig 5.42 *Formation of a nucleotide. Carbon atoms of ribose are, by convention, numbered 1' to 5' to avoid confusion with C atoms 1 to 9 of bases*

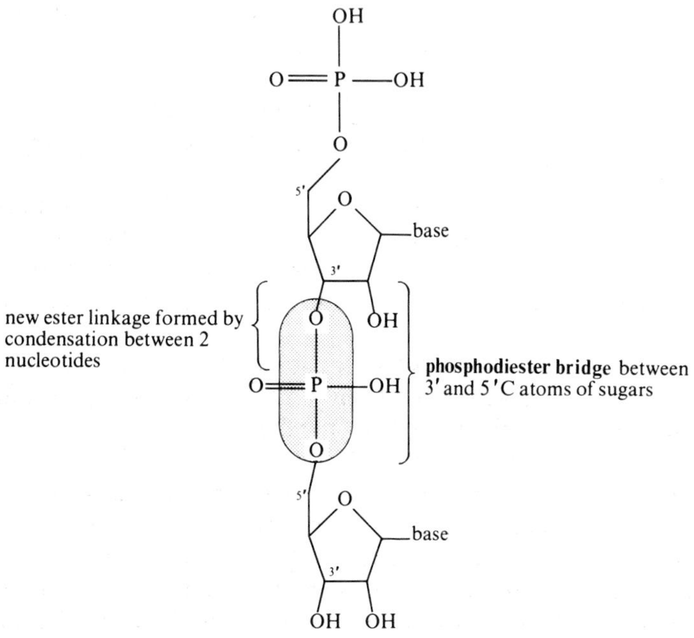

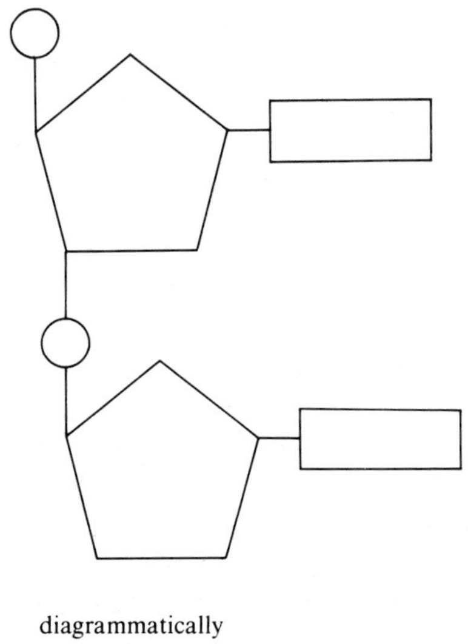

OH

O=P—OH

O

5'   O

base

3'

new ester linkage formed by
condensation between 2
nucleotides

O   OH

O=P—OH

O

phosphodiester bridge between
3' and 5' C atoms of sugars

5'   O

base

3'

OH   OH

diagrammatically

Fig 5.43 *Structure of a dinucleotide*

156

The combination of a sugar with a base forms a compound called a **nucleoside**. This occurs with the elimination of water and therefore is a condensation reaction. A nucleotide is formed by further condensation between the nucleoside and phosphoric acid forming a phosphoester link.

Different nucleotides are formed according to the sugars and bases used. Nucleotides are not only used as building blocks for nucleic acids, but they and their derivatives form several important coenzymes, including adenosine monophosphate (AMP), diphosphate (ADP) and triphosphate (ATP), cyclic AMP, coenzyme A, nicotinamide adenine dinucleotide (NAD) and its phosphate NADP, and flavin adenine dinucleotide (FAD) (chapter 6).

5.6.2 Structure of dinucleotides and polynucleotides

Two nucleotides join to form a **dinucleotide** by condensation between the phosphate group of one with the sugar of the other to form a phosphodiester bridge, as shown in fig 5.43. The process is repeated up to several million times to make a **polynucleotide**. An unbranched sugar–phosphate backbone is formed by phosphodiester bridges between the 3′ and 5′ carbon atoms of the sugars as shown in fig 5.44.

Phosphodiester linkages are formed from strong covalent bonds and this confers strength and stability on the polynucleotide chain. This is an important point in preventing breakage of the 'chain' during DNA replication (chapter 22).

5.6.3 Structure of DNA

Like proteins, polynucleotides can be regarded as having a primary structure, which is the sequence of nucleotides, and a three-dimensional structure. Interest in the structure of DNA intensified when it was realised in the early part of this century that it might be the genetic material. Evidence for this is presented in section 22.4.

By the early 1950s the Nobel prize-winning chemist Linus Pauling of the USA had worked out the α-helical structure which is common to many fibrous proteins, and was applying himself to the problem of the structure of DNA, which evidence suggested was also a fibrous molecule. At the same time Maurice Wilkins and Rosalind Franklin of King's College, London were tackling the same problem using the technique of X-ray crystallography. This involved the difficult and time-consuming process of preparing pure fibres of the salt of DNA from which they managed to get complex X-ray diffraction patterns. These reveal the gross structure of the molecule but are not as detailed as those from pure crystals of proteins (fig 5.45). Meanwhile, James Watson and Francis Crick of the Cavendish Laboratory in Cambridge had chosen what was to prove the successful approach. Using all the chemical and physical information they could gather, they began

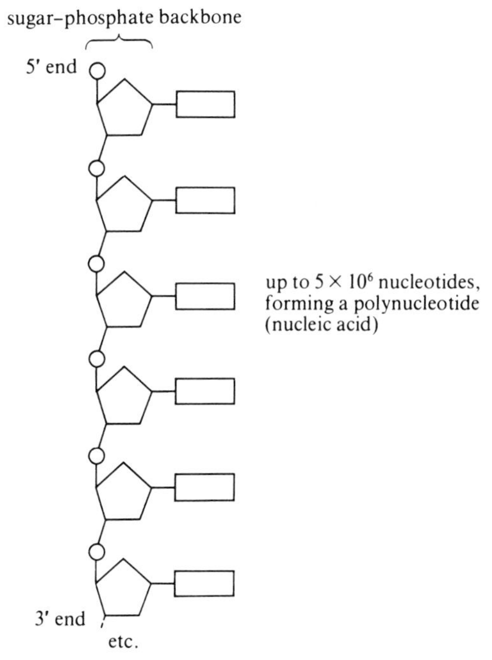

Fig 5.44 *Formation of a polynucleotide*

building scale models of polynucleotides in the hope that a convincing structure would emerge. Watson's book *The Double Helix* provides a fascinating insight into their work.

Two lines of evidence proved crucial. Firstly, they were in regular communication with Wilkins and had access to the X-ray diffraction data, against which they were able to test their models. These data strongly suggested a helical structure (fig 5.45) with regularity at a spacing of 0.34 nm along its axis. Secondly, they realised the significance of some evidence published by Erwin Chargaff in 1951 concerning the ratio of the different bases found in DNA. Although important, this evidence had generally been overlooked. Table 5.14 shows some of Chargaff's data, and supporting data obtained since.

5.18 Examine the table. What does it reveal about the ratios of the different bases?

Watson and Crick had been exploring the idea that there may be two helical chains of polynucleotides in DNA, held together by pairing of bases between neighbouring chains. The bases would be held together by hydrogen bonds.

5.19 If this model were correct, can you predict from Chargaff's data which base combines with which in each pair?

Fig 5.46 shows how the base pairs are joined by hydrogen bonds. Adenine pairs with thymine, and guanine with cytosine; the adenine–thymine pair has two hydrogen bonds and the guanine–cytosine pair has three hydrogen bonds. Watson tried pairing the bases in this way, and recalls 'my morale skyrocketed, for I suspected that we

Table 5.14 Relative amounts of bases in DNA from various organisms.

Source of DNA	Adenine	Guanine	Thymine	Cytosine
Man	30.9	19.9	29.4	19.8
Sheep	29.3	21.4	28.3	21.0
Hen	28.8	20.5	29.2	21.5
Turtle	29.7	22.0	27.9	21.3
Salmon	29.7	20.8	29.1	20.4
Sea urchin	32.8	17.7	32.1	17.3
Locust	29.3	20.5	29.3	20.7
Wheat	27.3	22.7	27.1	22.8
Yeast	31.3	18.7	32.9	17.1
Escherichia coli (a bacterium)	24.7	26.0	23.6	25.7
ΦX174 bacteriophage (a virus)	24.6	24.1	32.7	18.5

Amounts are in molar proportions on a percentage basis.

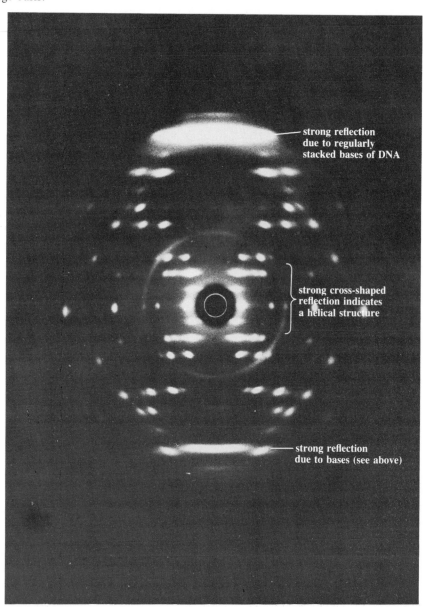

strong reflection due to regularly stacked bases of DNA

strong cross-shaped reflection indicates a helical structure

strong reflection due to bases (see above)

Fig 5.45 *X-ray diffraction photograph of a fibre of DNA. This is the kind of pattern from which the double helical structure was originally deduced (photograph by courtesy of Dr J. M. Squire)*

now had the answer to the riddle of why the number of purine residues exactly equalled the number of pyrimidine residues'.* He noticed the neat way in which the bases fit and that the overall size and shape of the base pairs was identical, both being three rings wide (fig 5.46). Hydrogen bonding between other combinations of bases, while possible, is much weaker. The way was finally open to building the correct model of DNA, whose structure is summarised in figs 5.47–49.

Features of the DNA molecule

Watson and Crick showed that DNA consists of two polynucleotide chains. Each chain forms a right-handed helical spiral and the two chains coil around each other to form a double helix (fig 5.47). The chains run in opposite directions, that is are **antiparallel**, the so-called 3′ end of one being opposite the 5′ end of the other (remember the 3′,5′ phosphodiester linkages). Each chain has a sugar–phosphate backbone with bases which project at right-angles and hydrogen bond with the bases of the opposite chain across the double helix (fig 5.48). The sugar–phosphate backbones are clearly seen in a space-filling model of DNA (fig 5.49). The width between the two backbones is constant and equal to the width of a base pair, that is the width of a purine plus a pyrimidine. Two purines would be too large, and two pyrimidines too small, to span the gap between the two chains. Along the axis of the molecule the base pairs are 0.34 nm apart, accounting for the regularity indicated by X-ray diffraction. A complete turn of the double helix comprises 3.4 nm, or ten base pairs. There is no restriction on the sequence of bases in one chain, but because of the rules of base pairing, the sequence in one chain determines that in the other. The two chains are thus said to be **complementary**.

* From *The Double Helix*, James D. Watson, Weidenfeld & Nicolson, 1968.

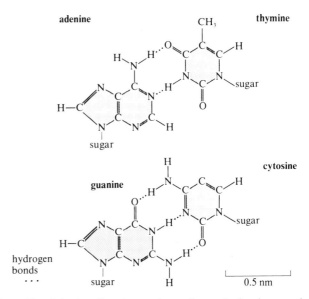

Fig 5.46 *Adenine–thymine and guanine–cytosine base pairs*

Watson and Crick published their model in 1953 in the journal *Nature*, and together with Maurice Wilkins were awarded the Nobel Prize for their work in 1962, the same year that Kendrew and Perutz received Nobel prizes for their work on the three-dimensional structure of proteins, also based on X-ray crystallography.

To act as genetic material, the structure had to be capable of carrying coded information and of accurate replication. Its suitability for this was not overlooked by Watson and Crick who, with masterly understatement near the end of their paper said 'It has not escaped our notice that the specific pairing we have postulated immediately suggests a possible copying mechanism for the genetic material.'* In a second paper that year they discussed the

* Watson, J. D. & Crick, F.H.C. (1953) *Nature* **171**, 737.

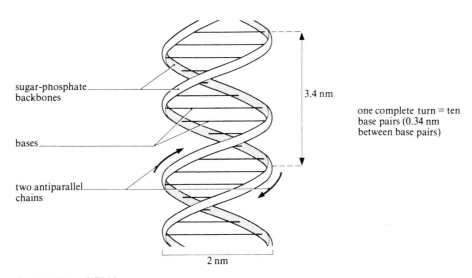

one complete turn = ten base pairs (0.34 nm between base pairs)

Fig 5.47 *Diagrammatic structure of DNA*

159

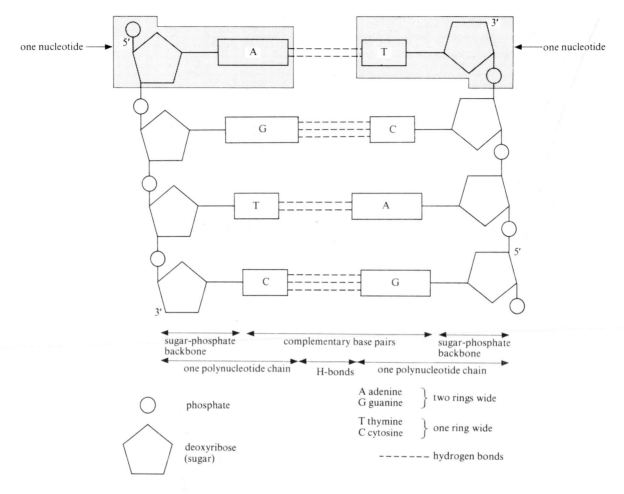

one nucleotide →

one nucleotide ←

sugar-phosphate backbone ↔ complementary base pairs ↔ sugar-phosphate backbone

one polynucleotide chain ↔ H-bonds ↔ one polynucleotide chain

○   phosphate

⬠   deoxyribose (sugar)

A adenine
G guanine } two rings wide

T thymine
C cytosine } one ring wide

- - - - - - hydrogen bonds

Fig 5.48 *DNA – diagrammatic structure of straightened chains*

genetic implications of the structure, and these are dealt with in chapter 22. This discovery, in which structure was shown to be so clearly related to function even at the molecular level, gave great impetus to the science of molecular biology.

5.6.4   Structure of RNA

RNA is normally single stranded, unlike DNA. Certain forms of RNA do assume complex structures, notably transfer RNA (tRNA) and ribosomal RNA (rRNA). Another form is messenger RNA (mRNA). These are involved in protein synthesis and are discussed in chapter 22.

5.7   Other biochemically important molecules

Apart from the well-defined classes of organic molecule already discussed, many other complex organic compounds occur in living cells. Of particular note are the vitamins and a group of substances which assist enzymes in their functioning, the cofactors or coenzymes. They contain miscellaneous chemical structures, some being nucleotides or their derivatives. Vitamins are discussed in section 10.3.10 and cofactors in section 6.2.

Inorganic ions and molecules of biochemical importance are discussed in section 9.12 with mineral nutrition.

5.8   Identification of biochemicals

It is recommended that you first familiarise yourself with the following tests by using pure samples of the chemicals being tested. Once the techniques have been mastered and familiarity with the colour changes obtained, various tissues can be studied.

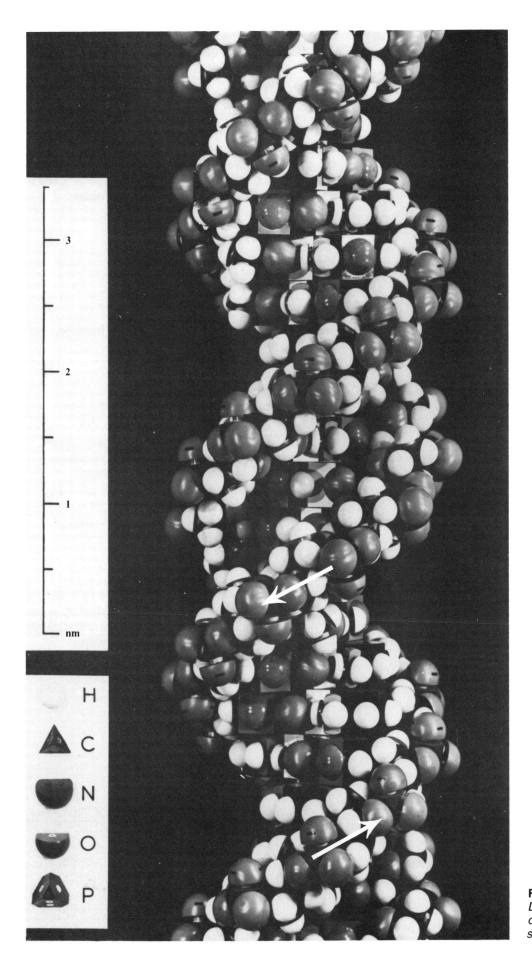

3

2

1

nm

H

C

N

O

P

Fig 5.49 *Space-filling model of DNA. Arrows indicate the directions of the two antiparallel sugar–phosphate backbones*

Experiment 5.1: Identification of biochemicals in pure form

N.B. Any heating that has to be done in the following tests should be carried out in a water bath at the boiling point of water. Direct heating of test-tubes should not take place.

Materials

pH paper
test-tubes
test-tube rack
Bunsen burner
teat pipettes
spatula
1 cm^3 syringe
iodine/potassium iodide solution
Benedict's reagent

dilute sulphuric acid
sodium hydrogencarbonate
Sudan III
Millon's reagent
5% potassium hydroxide solution
1% copper sulphate solution
DCPIP (dichlorophenolindophenol) solution
1% starch solution (cornflour is a recommended source)
1% glucose solution
1% sucrose solution (Analar sucrose must be used to avoid contamination with a reducing sugar)
olive oil or corn oil
absolute ethanol
egg albumen
1% lactose solution
1% fructose solution

Carbohydrates

Reducing sugars. The reducing sugars include all monosaccharides, such as glucose and fructose, and some disaccharides, such as maltose. Use 0.1–1% sugar solutions.

Test	Observation	Basis of test
Benedict's test		
Add 2 cm^3 of a solution of the reducing sugar to a test-tube. Add an equal volume of Benedict's solution. Shake and bring gently to the boil, shaking continuously to minimise spitting.	The initial blue colouration of the mixture turns green, then yellowish and may finally form a brick-red precipitate.	Benedict's solution contains copper sulphate. Reducing sugars reduce soluble blue copper sulphate, containing copper(II) ions (Cu^{2+}) to insoluble red-brown copper oxide containing copper(I). The latter is seen as a precipitate.

Additional information
The mixture is likely to bump violently during heating and extra care should therefore be taken. The test is **semi-quantitative**, that is a rough estimation of the amount of reducing sugar present will be possible. The final precipitate will appear green to yellow to orange to red-brown with increasing amounts of reducing sugar. (The initial yellow colour blends with the blue of the copper sulphate solution to give the green colouration.)

Test	Observation	Basis of test
Fehling's test		
Add 2 cm^3 of a solution of the reducing sugar to a test-tube. Add 1 cm^3 of Fehling's solution A and 1 cm^3 of Fehling's solution B. Shake and bring to the boil.	The initial blue colouration of the mixture turns green to yellow and finally a brick-red precipitate is formed.	As Benedict's test.

Additional information
Not as convenient as Benedict's test because Fehling's solutions A and B have to be kept separate until the test. Also it is not as sensitive.

Non-reducing sugars. The most common non-reducing sugar is sucrose, a disaccharide. If reducing sugars have been shown to be absent (negative result in above test) a brick-red precipitate in the test below indicates the presence of a non-reducing sugar. If reducing sugars have been shown to be present, a heavier precipitate will be observed in the following test than with the reducing test if non-reducing sugar is also present.

Test	Observation	Basis of test
Add 2 cm^3 of sucrose solution to a test-tube. Add 1 cm^3 dilute hydrochloric acid. Boil for one minute. Carefully neutralise with sodium hydrogencarbonate (check with pH paper) – care is required because effervescence occurs. Carry out Benedict's test.	As Benedict's test.	A disaccharide can be hydrolysed to its monosaccharide constitutents by boiling with dilute hydrochloric acid. Sucrose is hydrolysed to glucose and fructose, both of which are reducing sugars and give the reducing sugar result with the Benedict's test.

Starch. This is only slightly soluble in water, in which it forms a colloidal suspension. It can be tested in suspension or as a solid.

Test	Observation	Basis of test
Iodine/potassium iodide test		
Add 2 cm^3 1% starch solution to a test-tube. Add a few drops of I$_2$/KI solution. Alternatively add the latter to the solid form of starch.	A blue-black colouration.	A polyiodide complex is formed with starch.

Cellulose and lignin. See appendix A2.4.2 (staining).

Lipids

Lipids include oils (such as corn oil and olive oil), fats and waxes.

Test	Observation	Basis of test
Sudan III		
Sudan III is a red dye. Add 2 cm^3 oil to 2 cm^3 of water in a test-tube. Add a few drops of Sudan III and shake.	A red-stained oil layer separates on the surface of the water, which remains uncoloured.	Fat globules are stained red and are less dense than water.
Emulsion test		
Add 2 cm^3 fat or oil to a test-tube containing 2 cm^3 of absolute ethanol. Dissolve the lipid by shaking vigorously. Add an equal volume of cold water.	A cloudy white suspension	Lipids are immiscible with water. Adding water to a solution of the lipid in alcohol results in an emulsion of tiny lipid droplets in the water which reflect light and give a white, opalescent appearance.

Proteins

A suitable protein for these tests is egg albumen.

Test	Observation	Basis of test
Millon's Test		
Add 2 cm³ protein solution or suspension to a test-tube. Add 1 cm³ Millon's reagent and boil. NB Millon's reagent is poisonous: take care!	A white precipitate forms which coagulates on heating and turns red or salmon pink.	Millon's reagent contains mercury acidified with nitric acid, giving mercury(II) nitrate and nitrite. The amino acid tyrosine contains a phenol group which reacts to give a red mercury(II) complex. This is a reaction given by all phenolics and is not specific for proteins. Protein usually coagulates on boiling, thus appearing solid. The only common protein lacking tyrosine likely to be used is gelatin.
Biuret Test		
Add 2 cm³ protein solution to a test-tube. Add an equal volume of 5% potassium hydroxide solution and mix. Add 2 drops of 1% copper sulphate solution and mix. No heating is required.	A mauve or purple colour develops slowly.	A test for peptide bonds. In the presence of dilute copper sulphate in alkaline solution, nitrogen atoms in the peptide chain form a purple complex with copper(II) ions (Cu^{2+}). Biuret is a compound derived from urea which also contains the —CONH— group and gives a positive result.

Vitamin C (ascorbic acid)

This test can be conducted on a quantitative basis if required, in which case the volumes given below must be measured accurately. A suitable source of vitamin C is a 50/50 mix of fresh orange or lemon juice with distilled water. Vitamin C tablets may also be purchased.

Test	Observation	Basis of test
Using 0.1% ascorbic acid solution as a standard. Add 1 cm³ of DCPIP solution to a test-tube. Fill a 1 cm³ syringe with 0.1% ascorbic acid. Add the acid to the DCPIP drop by drop, stirring gently with the syringe needle. Do not shake.* Add until the blue colour of the dye just disappears. Note the volume of ascorbic acid solution used.	Blue colour of dye disappears to leave a colourless solution.	DCPIP is a blue dye which is reduced to a colourless compound by ascorbic acid, a strong reducing agent.

* Shaking the solution would result in oxidation of the ascorbic acid by oxygen in the air. The effects of shaking and of boiling could be investigated.

5.20 How could you determine the concentration of ascorbic acid in an unknown sample?

5.21 You are provided with three sugar solutions. One contains glucose, one a mixture of glucose and sucrose, and one sucrose.
(a) How could you identify each solution?
(b) Supposing that the apparatus were available, and time permitted, briefly discuss any further experiments you could perform to confirm your results.

5.22 How would you make 100 cm³ of a 10% glucose solution?

5.23 Starting with stock solutions of 10% glucose and 2% sucrose how would you make 100 cm³ of a mixture of final concentration 1% sucrose and 1% glucose?

Experiment 5.2: Identification of biochemicals in tissues

A biochemist is often faced with the problem of wanting to identify chemicals (qualitative analysis) or to measure their amounts (quantitative analysis) in living tissue. Sometimes the chemical can be tested for directly, but often some kind of extraction and purification process must first be embarked upon.

A convenient exercise is to take a range of common foods and plant material and to test for the range of biochemicals listed in experiment 5.1 above. An extraction procedure is designed where possible to give a clear, colourless solution for testing, and you should note the rationale behind the procedures so that you could design your own if necessary.

Materials

As for experiment 5.1 up to DCPIP solution
pestle and mortar
microscope
slides and cover-slips
razor blade
watch glass
Schultz's solution
phloroglucinol + conc. hydrochloric acid
potato tuber
apple
cotton wool
woody stem
seeds/nuts
soaked peas
beans

Microscopic examination of thin sections of tissue

Suitable for: Visible storage products, particularly starch grains, such as potato tuber.

As above with appropriate staining or other chemical testing

Suitable for: Reducing sugars – Mount in a few drops of Benedict's reagent, heat gently to boiling; add water if necessary to prevent drying.

Starch – Mount section in dilute iodine/potassium iodide solution.

Protein – Mount in a few drops of Millon's reagent, heat gently to boiling; add water if necessary to prevent drying.

Oil and fat – Stain material, such as seed, with Sudan III and wash with water and/or 70% ethanol. Then section and mount.

Cellulose, lignin, etc. – see appendix A2.4.2 for staining.

Testing a clear, aqueous solution

Decolourise tissue if necessary: Pigments may interfere with colour tests but can usually be removed with an organic solvent such as 80% ethanol or 80% propanone. *Care must be taken to avoid naked flames.* However, remember these solvents may also remove lipids and soluble sugars.
Suitable for: Removing chlorophyll from leaves.

Homogenise (grind) material: Sugars and proteins – Small pieces of solid material can be ground with a small quantity of water using a pestle and mortar or a food mixer. The ground material should be squeezed through several layers of pre-moistened fine muslin or nylon and/or filtered or centrifuged to remove solid material. This may be unnecessary if a fairly colourless, fine suspension is obtained. The clear solution can be tested as usual, with further dilution if necessary. The solid residue may also be tested if appropriate.
Lipids – Grind material, transfer to a test-tube and boil. Lipids will escape as oil droplets. Perform the Sudan III test. Alternatively take thin shavings of nuts or other foods, including coloured foods, and do the emulsion test.
Suitable for:

Fruit, such as apple, orange	(vitamin C, sugars)
Nuts	(oils)
Castor oil seed	(oil)
Pea seed	(protein)
Pine kernels	(protein and oil)
Potato	(starch, vitamin C)
Egg	(protein)

Subdivision of the above materials, such as into seeds, flesh, skin and juice, may be possible.

Chapter Six

Enzymes

Enzymes are protein molecules produced by living cells. Each cell contains several hundred enzymes. They are used to promote a vast number of rapid chemical reactions between temperature limits suitable for the particular organism, that is approximately 5–40 °C. In order to achieve the same speeds of reaction outside the organism, high temperatures would be necessary, as well as marked changes in other conditions. These would be lethal to a living cell, which operates in such a way as to prevent any marked change in its normal working conditions. Thus enzymes can be defined as biological **catalysts**, that is they speed up reactions. They are vitally important because in their absence reactions in the cell would be too slow to sustain life.

Enzyme reactions may be either **anabolic** (involved in synthesis) or **catabolic** (involved in breakdown). The sum total of all these reactions in a living cell or organism constitutes its metabolism. Therefore metabolism consists of anabolism and catabolism. An example of an enzyme involved in anabolism is glutamine synthetase:

$$\text{glutamic acid + ammonia + ATP} \xrightarrow[\text{synthetase}]{\text{glutamine}} \text{glutamine + water + ADP + P}_i$$

(ATP is adenosine triphosphate, ADP is adenosine diphosphate and P_i is inorganic phosphate.) An example of an enzyme involved in catabolism is maltase:

$$\text{starch + water} \xrightarrow{\text{amylase}} \text{maltose}$$

Commonly, a number of enzymes are used in sequence to convert one substance into one or several products via a series of intermediate compounds. The chain of reactions is referred to as a **metabolic pathway**. Many such pathways can proceed simultaneously in the cell. The reactions proceed in an integrated and controlled manner and this can be attributed to the **specific** nature of enzymes. A single enzyme generally will catalyse only a single reaction. Thus enzymes serve to control the chemical reactions that occur within cells and ensure that they proceed at an efficient rate.

6.1 Catalysis and energy of activation

Biological catalysts (that is enzymes) possess the following major properties: all are globular proteins; they increase the rate of a reaction without themselves

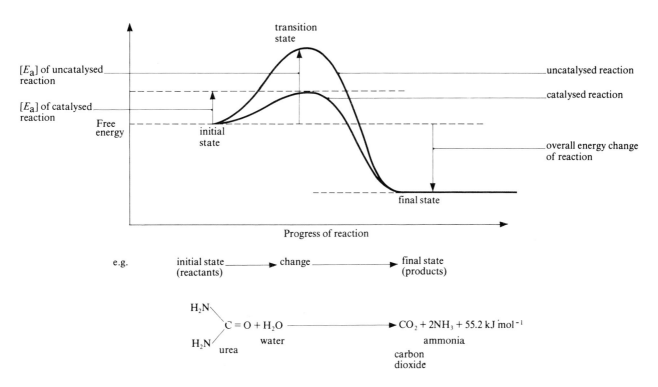

Fig 6.1 *Energy diagram showing a catalysed and uncatalysed chemical reaction (see also appendix 1)*

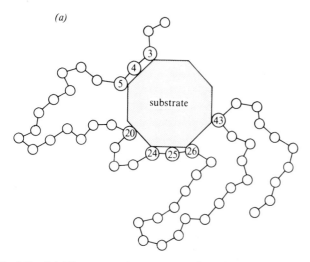

(a)

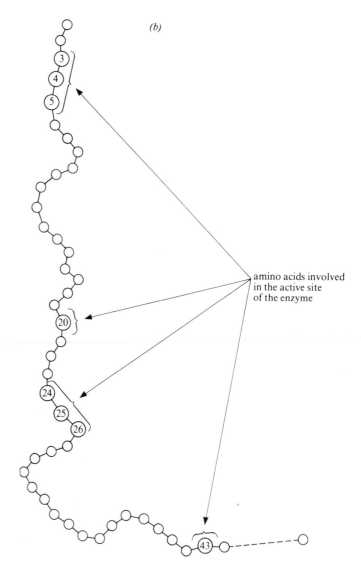

(b)

amino acids involved
in the active site
of the enzyme

Fig 6.2 *(a) Diagrammatic representation of substrate complexing with an enzyme active site. (b) Position of the amino acids of the active site shown on the primary structure of the enzyme*

being used up; their presence does not alter the nature or properties of the end product(s) of the reaction; a very small amount of catalyst effects the change of a large amount of substrate; their activity varies with pH, temperature, pressure and substrate and enzyme concentrations; the catalysed reaction is reversible; they are specific, that is an enzyme will generally catalyse only a single reaction.

Consider a mixture of petrol and oxygen maintained at room temperature. Although a reaction between the two substances is thermodynamically possible, it does not occur unless energy is applied to it, such as a simple spark. The energy required to make substrates react is called **activation energy** [E_a]. The greater the amount of activation energy required, the slower will be the rate of reaction at a given temperature. Enzymes, by functioning as catalysts, serve to reduce the activation energy required for a chemical reaction to take place (fig 6.1). They speed up the overall rate without altering, to any great extent, the temperature at which it occurs.

An **enzyme** combines with its **substrate** to form a short-lived enzyme/substrate complex (fig 6.2). Within this complex the chances of reactions occurring are greatly enhanced. Once a reaction has occurred, the complex breaks up into **products** and enzyme. The enzyme remains unchanged at the end of the reaction and is free to interact again with more substrate.

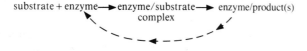

6.1.1 Mechanism of enzyme action

Detailed study has revealed that most enzymes are far larger molecules than the substrates they act on and that only a very small portion of the enzyme, between 3–12 amino acids, comes into direct contact with the substrate in the enzyme/substrate complex. This region is called the **active site** of the enzyme. It is here that binding of the substrate or substrates occurs (fig 6.2). The remaining amino acids, which comprise the bulk of the enzyme, function to maintain the correct globular shape of the molecule which, as will be explained below, is important if the active site is to function at the maximum rate (fig 6.3).

Enzymes are very specific and it was suggested by Fischer in 1890 that this was because the enzyme had a particular shape into which the substrate or substrates fit exactly. This is often referred to as the 'lock and key' hypothesis, where the substrate is the **key** whose shape is complementary to the enzyme or **lock** (fig 6.4).

When an enzyme/substrate complex is formed it is 'activated' into forming the products of the reaction. Once formed, the products no longer fit into the active site and escape into the surrounding medium, leaving the active site free to receive further substrate molecules.

In 1959 Koshland suggested a modification to the 'lock and key' analogy. Working from evidence that suggested that enzymes and their active sites were physically rather

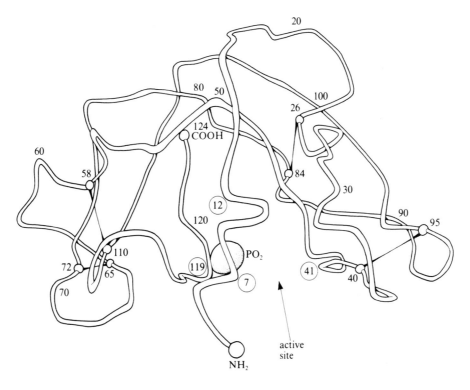

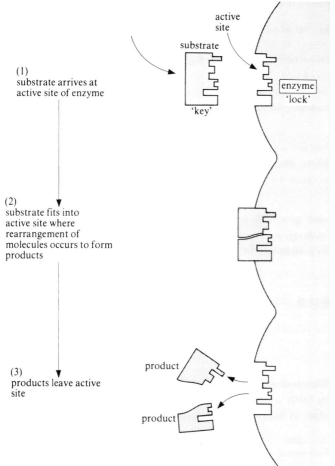

active
site

(1)
substrate arrives at
active site of enzyme

(2)
substrate fits into
active site where
rearrangement of
molecules occurs to form
products

(3)
products leave active
site

Fig 6.4 *Fischer's 'lock and key' hypothesis (1890).
Sequence of events when the union of a substrate with its
enzyme occurs*

Fig 6.3 *Tertiary structure of ribonuclease. The amino acids
involved at the active site are numbers 12 and 119 (histidines)
and 7 and 41 (lysines). Ribonuclease hydrolyses ribonucleic
acids to nucleotides (From Kartha, Bello & Harker (1967)
Nature, 213, 864.)*

more flexible structures than hitherto described, he
envisaged a dynamic interaction occurring between enzyme and substrate. He argued that when a substrate
combines with an enzyme, it induces changes in the enzyme
structure. The amino acids which constitute the active site
are moulded into a precise formation which enables the
enzyme to perform its catalytic function most effectively
(fig 6.5). This is called the '**induced-fit**' hypothesis. A
suitable analogy would be that of a hand changing the
shape of a glove as the glove is put on. Further refinements
to the hypothesis have been made as details of individual
reactions became known. In some cases, for example, the
substrate molecule changes shape slightly before binding.

6.2 Enzyme cofactors

Many enzymes require non-protein compo-
nents called **cofactors** for their efficient activity. They were
discovered as substances that had to be present for enzyme
activity, even though, unlike enzymes, they were stable at
relatively high temperatures. Cofactors may vary from
simple inorganic ions to complex organic molecules, and
may either remain unchanged at the end of a reaction or be
regenerated by a later process. The **enzyme/cofactor**
complex is called a **holoenzyme**, whilst the enzyme portion,
without its cofactor, is called an **apoenzyme**. There are
three recognised types of cofactor: inorganic ions, prosthe-

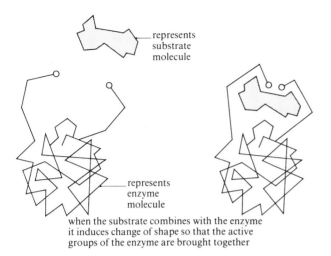

represents substrate molecule

represents enzyme molecule

when the substrate combines with the enzyme it induces change of shape so that the active groups of the enzyme are brought together

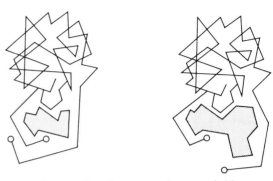

larger and smaller compounds are unsuitable for reacting with the enzyme

Fig 6.5 *Diagram to show Koshland's 'induced-fit' hypothesis. (From J. C. Marsden & C. F. Stoneman (1977)* Enzymes and equilibria, *Heinemann Educational Books)*

tic groups and coenzymes. Many organic molecules, some related to vitamins, act as cofactors. The molecule may be tightly bound to the enzyme (as in prosthetic groups) or only loosely associated with it (as in coenzymes). In both cases the molecules act as carriers of groups of atoms, single atoms or electrons that are being transferred from one place to another in an overall metabolic pathway.

6.2.1 Inorganic ions (alternatively known as enzyme activators)

These are thought to mould either the enzyme or the substrate into a shape such that an enzyme/substrate complex can be formed, hence increasing the chances of a reaction occurring between them and therefore increasing the rate of reaction catalysed by that particular enzyme. For example, salivary amylase activity is increased in the presence of chloride ions.

6.2.2 Prosthetic groups (for example FAD, FMN, biotin, haem)

The organic molecule is integrated in such a way that it effectively assists the catalytic function of its enzyme, as in flavin adenine dinucleotide (FAD). This contains riboflavin (vitamin B_2) which is the hydrogen-accepting part of FAD (fig 6.6). It is concerned with cell oxidation pathways such as part of the respiratory chain in respiration (chapter 11).

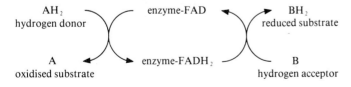

AH_2 hydrogen donor — enzyme-FAD — BH_2 reduced substrate

A oxidised substrate — enzyme-FADH$_2$ — B hydrogen acceptor

Net effect: 2H transferred from A to B. One holoenzyme acts as a link between A and B.

Haem

Haem is an iron-containing prosthetic group. It has the shape of a flat ring (a '**porphyrin ring**' as is found in

chlorophyll) with an iron atom at its centre. It has a number of biologically important functions.

Electron carrier. Haem is the prosthetic group of cytochromes (see respiratory chain, chapter 11), where it acts as an electron carrier. In accepting electrons the iron is reduced to Fe(II); in handing on electrons it is oxidised to Fe(III). In other words it takes part in oxidation/reduction reactions by reversible changes in the valency of the iron.

Oxygen carrier. Haemoglobin and myoglobin are oxygen-carrying proteins that contain haem groups. Here the iron remains in the reduced, Fe(II) form (see section 14.13.1).

Other enzymes. Haem is found in catalases and peroxidases, which catalyse the decomposition of hydrogen peroxide into water and oxygen. It is also found in a number of other enzymes.

6.2.3 Coenzymes (for example NAD, NADP, coenzyme A, ATP)

Nicotinamide adenine dinucleotide (NAD) (fig 6.7)

This is derived from the vitamin nicotinic acid and can exist in both a reduced and an oxidised form. In the oxidised state it functions in catalysis as a hydrogen acceptor

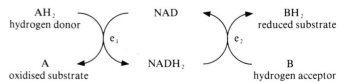

AH_2 hydrogen donor — NAD — BH_2 reduced substrate
e_1 — e_2
A oxidised substrate — NADH$_2$ — B hydrogen acceptor

where e_1 and e_2 are two different dehydrogenase enzymes.

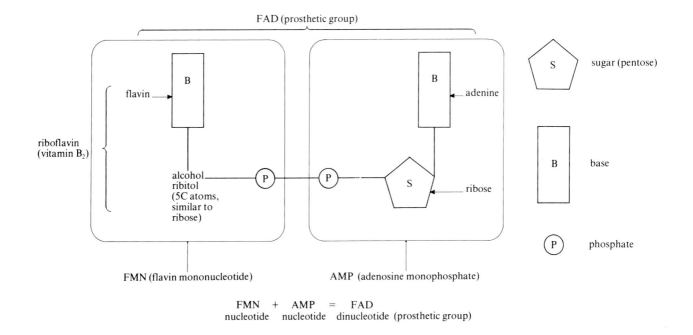

Fig 6.6 *Vitamin and prosthetic group interrelationship. The structure of FAD (flavin adenine dinucleotide) is shown*

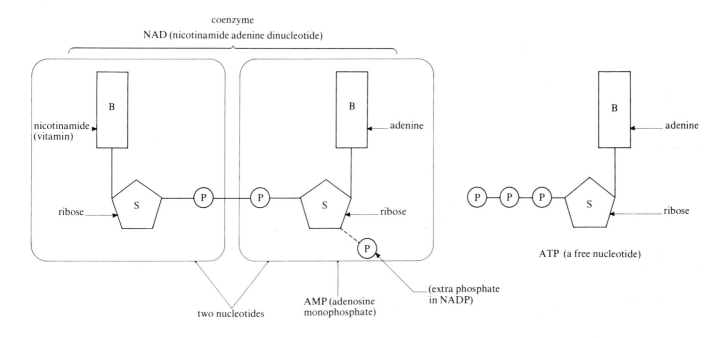

Fig 6.7 *Vitamin and coenzyme interrelationship. The structures of NAD, NADP and ATP are shown*

Net effect: 2H transferred from A to B. Here the coenzyme acts as a link between two different enzyme systems e_1 and e_2.

6.3 The rate of enzyme reactions

The rate of an enzyme reaction is measured by the amount of substrate changed or amount of product formed, during a period of time.

The rate is determined by measuring the slope of the tangent to the curve in the initial stage of the reaction (shown as (*a*) in fig 6.8). The steeper the slope, the greater is the rate. If activity is measured over a period of time, the rate of reaction usually falls, most commonly as a result of a fall in substrate concentration (see next section).

6.4 Factors affecting the rate of enzyme reactions

When investigating the effect of a given factor on the rate of an enzyme-controlled reaction, all other factors should be kept **constant** and at **optimum levels** wherever possible. Initial rates only should be measured, as explained above.

Computer program. ENZYME: The program simulates enzyme-controlled reactions. The user can generate four types of graph: free energy vs. reaction coordinate; kinetic energy vs. temperature; amount of product or reaction rate vs. time, substrate concentration, enzyme concentration, pH or temperature.

6.4.1 Enzyme concentration

Provided that the substrate concentration is maintained at a high level, and other conditions such as pH and temperature are kept constant, the rate of reaction is proportional to the enzyme concentration (fig 6.9). Invariably reactions are catalysed by enzyme concentrations which are much lower than substrate concentrations. Thus as the enzyme concentration is increased, so will be the rate of the enzymatic reaction.

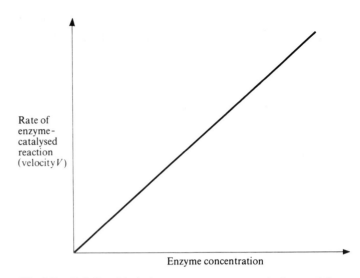

Fig 6.9 *Relationship between enzyme concentration and the rate of an enzyme-controlled reaction*

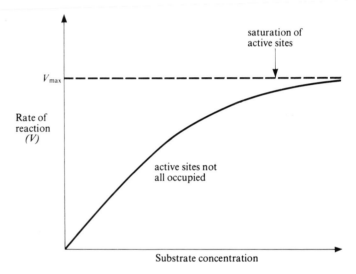

Fig 6.10 *Effect of substrate concentration on the rate of an enzyme-controlled reaction*

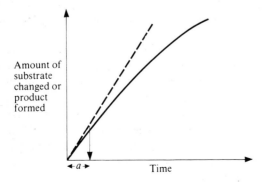

Fig 6.8 *The rate of an enzyme-controlled reaction*

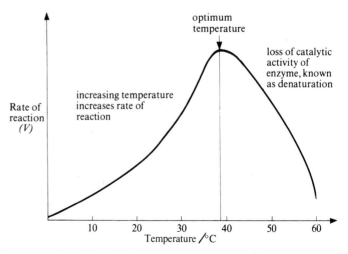

Fig 6.11 *The effect of temperature on the activity of an enzyme such as salivary amylase*

172

Experiment 6.1: To determine the effect of enzyme concentration on the hydrolysis of sucrose by sucrase (invertase)

Materials

2% sucrose solution
1%, 0.75%, 0.5% sucrase (invertase) solutions
Benedict's reagent
12 test-tubes and rack
water baths at 38 °C and 100 °C
glass rods
stopclock
distilled water
labels
Bunsen burner

Method

(1) Add 2 cm³ of clear blue Benedict's reagent to 2 cm³ of clear colourless 1% sucrase solution. Heat the mixture in the water bath maintained at 100 °C for 5 min (Benedict's test).
(2) Repeat (1) using 2 cm³ of clear colourless 2% sucrose solution and then 2 cm³ of distilled water.
(3) Boil 5 cm³ 1% sucrase solution.
(4) Take 8 clean, dry test-tubes, label 1–8, and add 1 cm³ Benedict's reagent to each.
(5) Add 5 cm³ of 2% sucrose solution to a test-tube labelled S and place in the water bath maintained at 38 °C throughout the experiment.
(6) Add 5 cm³ of 1% sucrase solution to a test-tube labelled E and place in the water bath at 38 °C.
(7) Leave both test-tubes and contents in the water bath for 5 min to allow them to equilibrate with their surroundings.
(8) Add the enzyme solution to the sucrose solution, invert the test-tube to thoroughly mix the two solutions.
(9) Immediately start the stopclock and replace the tube containing the reaction mixture in the water bath.
(10) Throughout the experiment agitate the mixture continuously to ensure thorough mixing.
(11) After 30 s of incubation remove 1 cm³ of mixture and place in test-tube 1.
(12) Repeat this procedure every 30 s placing the samples in tubes 2–8 in turn.
(13) Heat tubes 1–8 in the water bath at 100 °C for 5 min. Note the time when the first positive reducing sugar test is obtained indicated by a brick-red precipitate.
(14) Repeat the experiment using the boiled enzyme from (3).
(15) Repeat the entire sequence/experiment twice using the 0.75% and 0.5% sucrase solutions.
(16) Record your observations and comment on your results.

6.4.2 Substrate concentration

For a given enzyme concentration, the rate of an enzymatic reaction increases with increasing substrate concentration (fig 6.10). The theoretical maximum rate (V_{max}) is never quite obtained, but there comes a point when any further increase in substrate concentration produces no significant change in reaction rate. This is because at high substrate concentrations the active sites of the enzyme molecules at any given moment are virtually saturated with substrate. Thus any extra substrate has to wait until the enzyme/substrate complex has dissociated into products and free enzyme before it may itself complex with the enzyme. Therefore at high substrate levels, both enzyme concentration, and the time it takes for dissociation of the enzyme/substrate molecule, limit the rate of the reaction.

6.4.3 Temperature

The effect of temperature on the rate of a reaction can be expressed as the temperature coefficient, Q_{10}.

$$Q_{10} = \frac{\text{rate of reaction at } (x + 10)°C}{\text{rate of reaction at } x \text{ °C}}$$

Over a range of 0–40 °C, Q_{10} for an enzyme-controlled reaction is 2. In other words, the rate of an enzyme-controlled reaction is doubled for every rise of 10 °C. Heat increases molecular motion, thus the reactants move more quickly and chances of their bumping into each other are increased. As a result there is a greater probability of a reaction being caused. The temperature that promotes maximum activity is referred to as the optimum temperature. If the temperature is increased above this level, then a decrease in the rate of the reaction occurs despite the increasing frequency of collisions. This is because the secondary and tertiary structures of the enzyme have been disrupted, and the enzyme is said to be **denatured** (fig 6.11).

> **6.1** Explain how denaturing an enzyme may affect its efficiency as a catalyst.

If temperature is reduced to near or below freezing point, enzymes are **inactivated**, not denatured. They will regain their catalytic influence when higher temperatures are restored.

Today techniques of quick-freezing food are in widespread use as a means of preserving food for extensive periods. This not only prevents growth and multiplication of micro-organisms, but also deactivates their digestive enzymes thus making it impossible for them to decompose food. The natural enzymes in the food itself are also inactivated. However, once frozen, it is necessary to keep the food at subzero temperatures until it is to be prepared for consumption.

Experiment 6.2: To investigate the distribution of catalase in a soaked pea, and to determine the effect of different temperatures on its activity

Catalase is an enzyme which catalyses the decomposition of hydrogen peroxide, liberating oxygen gas as shown by effervescence:

$$2H_2O_2 \xrightarrow{\text{catalase}} 2H_2O + O_2$$

Hydrogen peroxide is a toxic by-product of metabolism in certain plant and animal cells, and is efficiently removed by catalase, which is one of the fastest acting enzymes known (at 0 °C one molecule of catalase can decompose 40 000 molecules of hydrogen peroxide per second). Catalase is found in microbodies (see chapter 7) and peroxisomes (see chapters 7 and 9).

Materials

a supply of soaked peas
hydrogen peroxide solution
test-tubes and rack
water baths at 40 °C, 60 °C, 70 °C, 80 °C and 100 °C
clock
thermometer
scalpels, scissors and forceps
test-tube holder
glass rod
white tile

Method

(1) Test for the presence of catalase by crushing a soaked pea and adding a few drops of hydrogen peroxide solution.
(2) Remove the seed coats from three soaked peas and test separately for catalase activity in both the seed coats and the cotyledons.
(3) Place two test-tubes containing distilled water in a water bath at 40 °C.
(4) Boil three whole peas in a test-tube and then place the boiled peas in one of the tubes in the water-bath.
(5) Place three whole unboiled peas in the other test-tube in the water bath.
(6) Allow enough time for the peas to reach the temperature of the water bath (at least 10 min).
(7) Test each pea for catalase activity.
(8) Repeat the experiment at 50, 60, 70, 80 and 100 °C.
(9) Record your observations and comment on your results.

> **6.2** Study fig 6.12 carefully and comment on the shapes of the curves given for the enzymatic reaction at different temperatures.

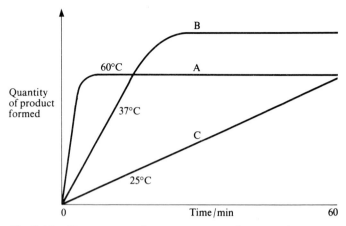

Fig 6.12 *Time course of an enzyme reaction at various temperatures*

6.4.4 pH

Under conditions of constant temperature, every enzyme functions most efficiently over a narrow pH range. The optimum pH is that at which the maximum rate of reaction occurs (fig 6.13 and table 6.1). When the pH is altered above or below this value, the rate of enzyme activity diminishes. Changes in pH alter the ionic charge of the acidic and basic groups that help to maintain the specific shape of the enzyme (section 5.5.4). The pH change leads to an alteration in enzyme shape, particularly at its active site. If extremes of pH are encountered by an enzyme, then it will be denatured. The optimum pH of an enzyme is not always the same as the pH of its immediate intracellular environment. This suggests that the enzyme's surroundings may be exerting some form of control over its activity.

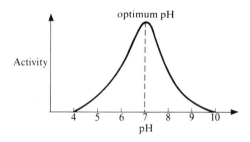

Fig 6.13 *Activity/pH curve for an enzyme*

Table 6.1 Optimum pH values for some enzymes.

Enzyme	Optimum pH
Pepsin	2.00
Sucrase	4.50
Enterokinase	5.50
Salivary amylase	6.80
Catalase	7.60
Chymotrypsin	7.00–8.00
Pancreatic lipase	9.00
Arginase	9.70

Experiment 6.3: To investigate the effect of different pH values on enzyme activity

Materials

Benedict's reagent
buffer solutions at pH 3,5,7,9,11
1% starch solution
water bath at 38 °C
Bunsen burner
asbestos mat
test-tube holder, test-tubes and rack
5 cm³ graduated pipettes
thermometer
stopclock
distilled water
stock solution of salivary
amylase (such as contained in saliva)

Method

(1) Rinse out the mouth with 5 cm³ of distilled water and spit this out.

(2) Swill 10 cm³ of distilled water round the mouth for 1 min and then collect this liquid.

(3) Make up the volume of salivary amylase to 40 cm³ with distilled water.

(4) Test the salivary amylase, starch and buffer solutions for the presence of reducing sugar using Benedict's reagent.

(5) Label a test-tube pH3 and add 2 cm³ of starch solution.

(6) Add 2 cm³ of buffer solution pH3 to the same test-tube and mix the two solutions thoroughly.

(7) Boil at least 4 cm³ of enzyme solution and place 4 cm³ in a labelled test-tube.

(8) Add 4 cm³ of unboiled enzyme solution to another labelled test-tube and place all three test-tubes in the water bath and allow the solutions to reach 38 °C (approximately 1 min).

(9) Place a small quantity of Benedict's reagent in each of 11 test-tubes and label them 1–11.

The following three stages must be carried out very quickly:

(10) When the solutions in the water bath have equilibrated, add the buffered starch solution to the unboiled enzyme solution.

(11) Mix the two solutions thoroughly by inverting the test-tube and replace the tube in the water bath.

(12) Start the stopclock and immediately remove a small quantity of reaction mixture (approximately the same volume as the Benedict's reagent) and place it in the test-tube labelled 1.

(13) Throughout the experiment the mixture must be shaken vigorously.

(14) After one minute of incubation remove a second, approximately equal volume, of the mixture and place it in test-tube 2.

(15) Repeat the removal of similar-sized samples of mixture at minute intervals for a further 9 min and place in test-tubes 3–11.

(16) Perform Benedict's tests on test-tubes 1–11 and note the time of incubation at which a positive result (a brick-red precipitate) is first achieved.

(17) Repeat the experiment using the boiled enzyme solution from (7).

(18) Repeat the entire experiment using each of the other buffer solutions.

(19) Plot a graph of time taken for hydrolysis to occur against pH and comment on your results.

6.3 (a) In fig 6.14, what is the optimum pH for the activity of enzyme B?

(b) Give an example of an enzyme which could be represented by (i) activity curve A, (ii) activity curve B.

(c) Why does the enzyme activity of C decrease at pH values between 8 and 9?

(d) Why is pH control important *in vivo*?

(e) 1 cm³ of a catalase solution was added to hydrogen peroxide solution at different pH values and the time taken to collect 10 cm³ of oxygen was measured. The results are given below.

pH of solution	Time (min) to collect gas
4.00	20.00
5.00	12.50
6.00	10.00
7.00	13.60
8.00	17.40

Draw a graph of these results and comment on them.

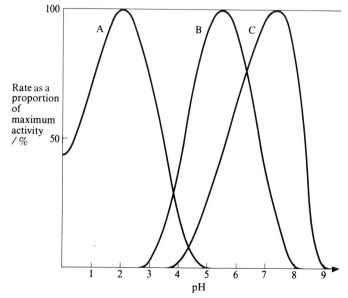

Fig 6.14 *The effect of pH on the activity of three enzymes, A, B and C*

6.5   Enzyme inhibition

A variety of small molecules exist which can reduce the rate of an enzyme-controlled reaction. They are called **enzyme inhibitors**. Inhibition may be reversible or irreversible.

6.5.1   Reversible inhibition

The inhibitor may be easily removed from the enzyme under certain conditions.

Competitive reversible inhibition

Here a compound, structurally similar to that of the usual substrate, associates with the enzyme's active site, but is unable to react with it. While it remains there it prevents access of any molecules of true substrate. As the genuine substrate and inhibitor **compete** for position in the active site, this form of inhibition is called competitive inhibition. It is able to be reversed, for if the substrate concentration is increased, the rate of reaction will be increased.

> **6.4** Why should the rate of reaction increase under these conditions?

An example of competitive inhibition is illustrated in fig 6.15.

The knowledge of competitive inhibition has been utilised in **chemotherapy**. This is the use of chemicals to destroy infectious micro-organisms without damaging host tissues. During the Second World War, **sulphonamides**, chemical derivatives of sulphanilamide, were used extensively to prevent the spread of microbial infection. The sulphonamides are similar in structure to para-aminobenzoate (PAB), a substance essential to the growth of many pathogenic bacteria. The bacteria require PAB for the production of folic acid, an important enzyme cofactor. Sulphonamides act by interfering with the synthesis of folic acid from PAB.

Animal cells are insensitive to sulphonamides even though they require folic acid for some reactions. This is because they use pre-formed folic acid and do not possess the necessary metabolic pathway for making it.

(a)

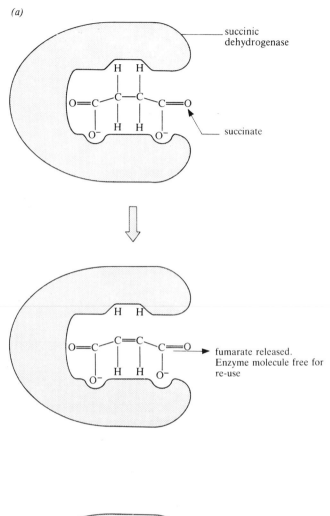

(b)

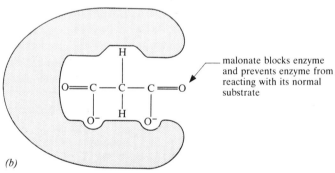

Fig 6.15 *An example of competitive inhibition. (a) The action of the enzyme succinic dehydrogenase on succinate. (b) Competitive inhibition of the enzyme by malonate*

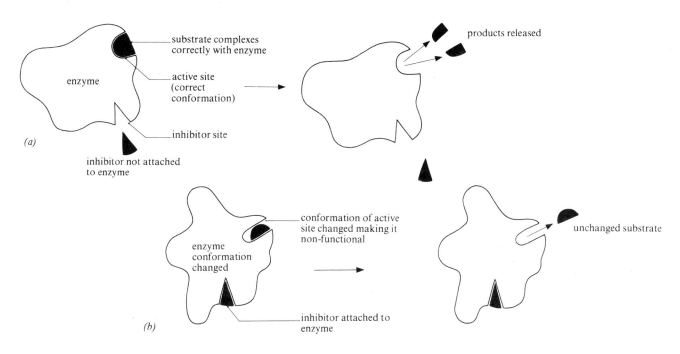

Fig 6.16 *The principle of non-competitive inhibition. (a) Normal reaction. (b) Non-competitive inhibition*

Non-competitive reversible inhibition

This type of inhibitor has no real structural similarity to the substrate and forms an enzyme/inhibitor complex at a point on the enzyme other than its active site (fig 6.16). It has the effect of altering the globular structure of the enzyme, so that even though the genuine substrate may be able to bind with the enzyme, catalysis is unable to take place. For example, cyanide combines with metallic ions (acting as prosthetic groups) of some enzymes (such as copper ions of cytochrome oxidase) and inhibits their activity. The rate of reaction will continue to decrease with increasing inhibitor concentration. When inhibitor saturation is reached, the rate of the reaction will be almost nil.

6.5.2 Irreversible inhibition (fig 6.17)

Very small concentrations of chemical reagents such as the heavy metal ions mercury (Hg^{2+}), silver (Ag^+) and arsenic (As^+), or iodoacetic acid, completely inhibit some enzymes. They combine permanently with sulphydryl ($-SH$) groups and cause the protein of the enzyme molecule to precipitate. If these are components of the active site then the enzyme is inhibited.

> **6.5** What would be the effect on the rate of reaction between an inhibitor of this kind and the substrate if substrate concentration is increased?

Diisopropylfluorophosphate (DFP), a nerve gas used in warfare, forms an enzyme/inhibitor complex with the amino acid serine at the active site of the enzyme acetylcholinesterase. This enzyme deactivates the chemical transmitter substance acetylcholine. One of the functions of acetylcholine is to aid the passage of a nerve impulse from one neurone to another across a synaptic gap (section 16.1). When the impulse has been transmitted acetylcholinesterase functions to deactivate acetylcholine almost immediately by hydrolysing it into choline and ethanoate. Once this has been completed the neurone is free to pass on another impulse. If acetylcholinesterase is inhibited, acetylcholine accumulates and nerve impulses are constantly propagated, causing prolonged muscle contraction. Paralysis or death is the end result. Some insecticides currently in use (such as parathion) have a similar effect on insects.

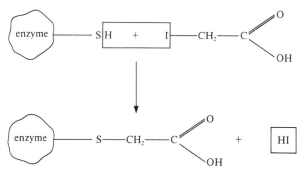

Fig 6.17 *Irreversible inhibition of an enzyme by iodoacetic acid*

6.6 Allosteric enzymes

The activity of these enzymes is regulated by compounds which are not their substrates and which bind to the enzyme at specific sites well away from the active site. They modify enzyme activity by causing a reversible change in the structure of the enzyme's active site. Compounds of this nature are called **allosteric effectors** and they may speed up (**allosteric activators**) or slow down (**allosteric inhibitors**) the reaction rate of an allosteric enzyme by increasing or decreasing the affinity of the enzyme for its substrate. An example of this is provided by the enzyme phosphofructokinase which catalyses the phosphorylation of fructose-6-phosphate to fructose-1-6-diphosphate. This reaction occurs during the glycolysis section of the respiratory pathway. When ATP is at a high concentration, it inhibits the enzyme phosphofructokinase allosterically. However, when cell metabolism increases and more ATP is used up, the overall concentration of ATP decreases and the pathway once again comes into operation (figs 6.18 and 6.19).

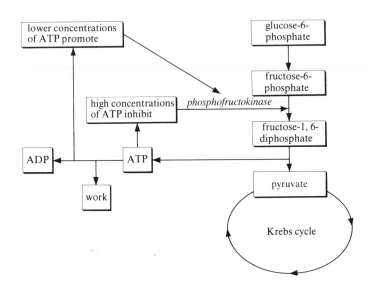

Fig 6.18 *Possible mechanism of the allosteric effect of ATP on phosphofructokinase. (After D. Harrison (1975)* Patterns in Biology, *Arnold)*

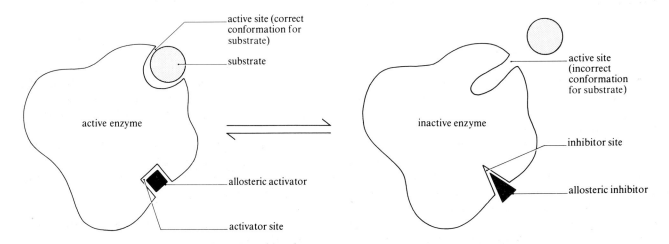

Fig 6.19 *Schematic representation of allosteric enzyme activity*

6.6.1 End-product inhibition (negative feedback inhibition)

When the end product of a metabolic pathway begins to accumulate, it may act as an allosteric inhibitor on the enzyme controlling the first step of the pathway. The affinity of the enzyme for its substrate would therefore be lowered, and further production of the end product decreased or prevented. This is called end-product inhibition and is an example of a negative feedback mechanism (section 18.1) serving to control an aspect of metabolic activity (fig 6.20).

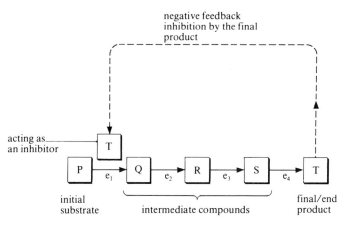

Fig 6.20 *Final/end-product inhibition. $e_1 - e_4$ are specific enzymes of a metabolic pathway*

6.7 Control of metabolism

There are over 500 enzymes present in a typical cell. Their activity and concentration will fluctuate continuously. How then, is control and integration of metabolism achieved? The answer lies in the specificity of action of enzymes, their spatial organisation and their functional interaction with other cellular components. Two distinct types of metabolic pathway exist in a cell which clearly demonstrate these features; these are the linear and branched metabolic pathways.

6.7.1 The linear metabolic pathway

A number of enzymes are arranged together in an organised fashion as a **multi-enzyme complex**. They are usually membrane-bound (fig 6.21a). The linear order of enzymes permits **self-regulation** by negative feedback inhibition, the rate of the pathway being controlled by the concentration of the end product. Such close-knit organisation also serves to reduce to a minimum interference from other reactions. Each enzyme is interdependent with the ones adjacent to it, and molecules are passed as products from one enzyme, to become the substrate of the next enzyme in the chain until the specific end product is formed.

6.7.2 The branched metabolic pathway.

In this pathway, a number of different end products may be formed. Which one would depend on the conditions prevailing in the cell at the time (fig 6.21b). Control of end-product formation would be influenced by feedback inhibition. Here again, a multi-enzyme system is in operation, but the enzymes are in solution and in no way closely associated with each other.

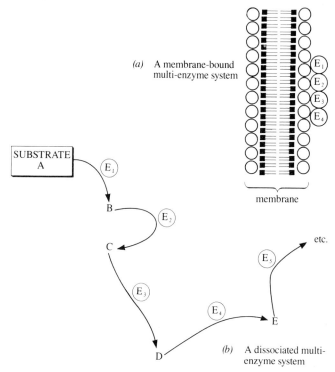

Fig 6.21 *Multi-enzyme systems. (a) A membrane-bound multi-enzyme system. (b) A dissociated multi-enzyme system, where B, C, D and E could be products depending upon the conditions in the cell*

6.6 Consider the multi-enzyme system shown below:

$$A \xrightarrow{e_1} B \xrightarrow{e_2} C \xrightarrow{e_3} D \xrightarrow{e_4} X$$
$$A \xrightarrow{e_5} P \xrightarrow{e_6} Q \xrightarrow{e_7} R \xrightarrow{e_8} S$$

(a) If e_1 is specific for A, and the end-product X inhibits e_1, what does this tell you about the binding sites of A and X on the enzyme?

(b) How might an excess of X regulate the metabolic pathway?

(c) What is the name given to the type of control system operating here?

(d) Why are several enzymes needed in any metabolic pathway?

6.7 Summarise the characteristic properties of enzymes.

Table 6.2 Enzyme classification.

Group	Reaction catalysed	Examples
Oxidoreductase	Transfer of H or O atoms, or electrons, from one molecule to another	Two types, dehydrogenases and oxidases $AH_2 + B \underset{}{\overset{dehydrogenase}{\rightleftharpoons}} A + BH_2$ transfer of H_2 e.g. $CH_3CHO + NADH_2 \underset{dehydrogenase}{\overset{alcohol}{\rightleftharpoons}} CH_3CH_2OH + NAD$ (p. 330) ethanal alcohol (ethanol) $AH_2 + O \overset{oxidase}{\rightleftharpoons} A + H_2O$ e.g. reduced cytochrome $+ \frac{1}{2}O_2 \underset{oxidase}{\overset{cytochrome}{\rightleftharpoons}}$ cytochrome $+ H_2O$ (p. 328)
Transferase	Transfer of a specific group from one molecule to another. The group may be methyl-, acyl-, amino- or phosphate	$AB + C \rightleftharpoons A + BC$ e.g. transaminases (transfer amino groups) (pp. 268 & 340) phosphorylases (add inorganic phosphate) glycogen $+ P_i \underset{phosphorylase}{\overset{glycogen}{\rightleftharpoons}}$ glucose-1-phosphate in respiration
Hydrolase	Formation of two products from a substrate by hydrolysis (splitting molecule with water)	$AB + H_2O \rightleftharpoons AOH + BH$ e.g. lipase, amylase, peptidases, other digestive enzymes (section 10.4.9)
Lyase	Non-hydrolytic addition or removal of groups from substrates. C–C, C–N, C–O or C–S bonds may be split	e.g. decarboxylases (remove CO_2) $\underset{\substack{\|\\OH}}{R-C}-\overset{O}{\overset{\|\!\!\|}{C}} \overset{decarboxylase}{\rightleftharpoons} R-\underset{H}{\overset{O}{\overset{\|\!\!\|}{C}}} + CO_2$ $CH_3COCOOH \underset{decarboxylase}{\overset{pyruvate}{\rightleftharpoons}} CH_3CHO + CO_2$ (p. 330) pyruvic acid ethanal e.g. carboxylases (add CO_2) $RuBP + H_2O + CO_2 \overset{RuBP\ carboxylase}{\rightleftharpoons} 2GP$ (p.265)
Isomerase	Intramolecular rearrangement – one isomer converted into another	$AB \rightleftharpoons BA$ e.g. glucose-1-phosphate $\overset{phosphoglucomutase}{\rightleftharpoons}$ glucose-6-phosphate in respiration glucose-6-phosphate $\underset{isomerase}{\overset{phosphoglucomutase}{\rightleftharpoons}}$ fructose-6-phosphate (p. 327)
Ligase	Join together two molecules by synthesis of new C–C, C–N, C–O or C–S bonds using energy from ATP	$X + Y + ATP \rightleftharpoons XY + ADP + P_i$ e.g. synthetases e.g. aminoacyl tRNA synthetases (p. 822)

NB This classification is often not followed in the common names of enzymes, e.g. DNA **polymerase** (p. 814).

6.8   Enzyme classification

In 1961 a systematic nomenclature for enzymes was recommended by a commission of the International Union of Biochemistry. The enzymes were placed into six groups according to the general type of reaction which they catalyse. Each enzyme was given a systematic name, accurately describing the reaction it catalyses. However, since many of these names were very long and complicated, each enzyme was allocated a 'trivial' name for everyday use. This consists of (i) the name of the substrate acted upon by the enzyme, (ii) the type of reaction catalysed, and (iii) the suffix -ase. For example, ribulose bisphosphate carboxylase; substrate: ribulose bisphosphate ($+CO_2$); type of reaction: carboxylation (addition of CO_2). The enzymes are classified as in table 6.2.

6.9   Enzyme technology

Use of enzymes in industry is growing and, with the versatility of micro-organisms in producing

Table 6.3 Summary of some common industrial uses of enzymes.

Application	Enzymes used	Uses	Problems
Biological detergents	Primarily proteases, produced in an extracellular form from bacteria	Used for pre-soak conditions and direct liquid applications	Allergic response of process workers; now overcome by encapsulation techniques
	Amylase enzymes	Detergents for machine dishwashing to remove resistant starch residues	
Baking industry	Fungal alpha-amylase enzymes; normally inactivated about 50 °C, destroyed during baking process	Catalyse breakdown of starch in the flour to sugar. Yeast action on sugar produces carbon dioxide. Used in production of white bread, buns, rolls	
	Protease enzymes	Biscuit manufacture to lower the protein level of the flour	
Baby foods	Trypsin	To pre-digest baby foods	
Brewing industry	Enzymes produced from barley during mashing stage of beer production	Degrade starch and proteins to produce simple sugars, amino acids and peptides used by the yeasts to enhance alcohol production	
	Industrially produced enzymes: amylases, glucanases, proteases betaglucanase amyloglucosidase proteases	Now widely used in the brewing process: split polysaccharides and proteins in the malt improve filtration characteristics low-calorie beer remove cloudiness during storage of beers	
Fruit juices	Cellulases, pectinases	Clarify fruit juices	
Dairy industry	Rennin, derived from the stomachs of young ruminant animals (calves, lambs, kids)	Manufacture of cheese, used to split protein	Older animals cannot be used as with increasing age rennin production decreases and is replaced by another protease, pepsin, which is not suitable for cheese production. In recent years the great increase in cheese consumption together with increased beef production has resulted in increasing shortage of rennin and escalating prices
	Microbially produced enzyme	Now finding increasing use in the dairy industry	
	Lipases	Enhance ripening of blue-mould cheeses (Danish blue, Roquefort)	
	Lactases	Break down lactose to glucose and galactose	
Starch industry	Amylases, amyloglucosidases and glucoamylases	Converts starch into glucose and various syrups	
	Glucose isomerase	Converts glucose into fructose (high-fructose syrups derived from starchy materials have enhanced sweetening properties and lower calorific values)	
	Immobilised enzymes	Production of high fructose syrups	Widely used in USA and Japan but EEC restrictive practices to protect sugar beet farmers prohibits use
Rubber industry	Catalase	To generate oxygen from peroxide to convert latex to foam rubber	
Paper industry	Amylases	Degrade starch to lower viscosity product needed for sizing and coating paper	
Photographic industry	Protease (ficin)	Dissolve gelatin off scrap film allowing recovery of silver present	

Based on Table 5.2, *Biotechnology*, 2nd ed., John E. Smith, New Studies In Biology (1988), Edward Arnold and *Enzymes, their nature and role*, Wiseman & Gould, Hutchinson Educational.

enzymes, new methods of making many industrially important chemicals are possible. High cost, purified enzymes are also being increasingly used in medicine, notably for diagnostic purposes in blood and urine tests. Enzymes can accomplish reactions at normal temperatures and pressures which would otherwise require expensive, energy-demanding high temperatures and/or pressures, or might not be possible at all.

Enzymes may be extracted from cells and purified before use, used within whole cells or, in the near future, it is anticipated that 'designer enzymes' will be 'tailor made' for specific tasks (**protein engineering**). The latter depends on being able to predict the three-dimensional shape into which a given primary structure will fold. Once a protein has been designed with the desired shape, an artificial gene could be made which would code for the protein and the gene could be inserted, say, into the bacterium *E. coli* for mass production. Table 6.3 summarises some of the current uses of enzymes in industry.

6.9.1 Immobilisation of enzymes

Apart from the problem of obtaining the enzyme in a suitable form in the first place (usually from a micro-organism), other major problems in enzyme technology include how to keep the organism functioning for a long period of time (**stabilisation**) and, in particular, how to immobilise the enzyme. **Immobilisation** is the process of attaching the enzyme to, or trapping it in, an inert solid support or carrier. This enables the reactants to be passed over the enzyme in a **continuous** process, or for the enzyme to be used in a **batch reactor** (a reactor that deals with one batch at a time), the important point being that in both cases the enzyme can be recovered at the end of the reaction for re-use. The alternative of mixing the enzyme in solution with the reactants is technically simpler, but once

the reaction is over this will involve either wasteful loss of the enzyme or potentially expensive recovery techniques. Therefore, where possible, it is used only for cheaper enzymes such as amylases and proteases.

Insoluble polymers, in the form of membranes or particles, are typically used as supports for the enzyme. Immobilised whole microbial cells are also sometimes used, for example inside polyacrylanide beads. Glucose isomerase (see starch industry, table 6.3) is currently the enzyme most commonly used in immobilised form, and can operate continuously for 1 000 hours at 60 °C. Class experiments using alginate beads to immobilise enzymes, for example urease, can provide a demonstration of this principle (consult, for example, the London Centre for Biotechnology, or NCSB Reading).

6.9.2 Biosensors

In industry, medicine, agriculture and environmental science, it is sometimes useful to be able to monitor the presence of specific chemicals both accurately and rapidly. For example, continuous monitoring of an industrial fermentation process would allow conditions such as pH, temperature, and substrate concentration to be maintained precisely at optimum levels; or in medicine, rapid diagnosis might be possible, based on detection of chemicals such as sodium, potassium or glucose in blood or urine samples. Biosensors are a relatively new and important innovation in this field. Our taste buds and olfactory (smell) areas may be regarded as biosensors. A **biosensor** uses an **immobilised** biological molecule (usually an enzyme or an antibody) or a whole microbial cell to detect or 'sense' a particular substance. The biosensor does this by reacting specifically with the substance to be detected (hence the use of enzymes or antibodies) to give a product which is used to generate an electrical signal by means of a

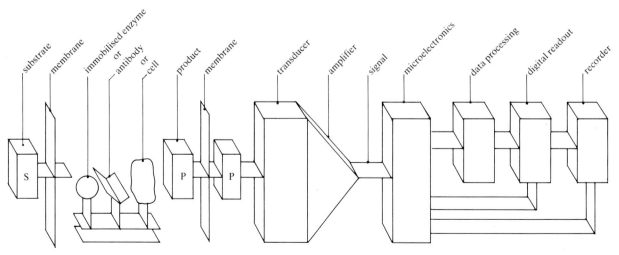

Fig 6.22 *Schematic outline of a biosensor. The substance to be measured (substrate) passes through a thin membrane and then encounters the biological sensing agent – usually an enzyme, an antibody or a whole microbial cell. The substrate and sensor interact to give a product which may be an electrical current, heat, a gas, or a soluble chemical. The product then passes through another membrane to the transducer which detects and measures the product, producing an electrical signal which is amplified and used to give an immediate read-out*

182

device called a **transducer**. The product of the reaction could be heat, a gas, a soluble chemical or an electric current generated by electron or proton flow during the reaction. The nature of the transducer varies with the nature of the product it has to detect, but it always produces an electrical signal in response to the product. This electrical signal is amplified and processed to give an instant read-out (fig 6.22).

Details of an experiment to detect urea using the enzyme urease are available from the London Centre for Biotechnology.

A number of sensing devices have already been introduced, such as glucose monitors for medical purposes and nerve gas sensors for military use. Enzymes are being used increasingly in medicine for routine automatic analysis of body fluids, for example for metabolites, drugs and hormones. A glucose oxidase 'electrode' or biosensor is one of the most developed products. This can measure the amount of glucose in a blood sample. It is hoped that in the relatively near future it will be possible to implant such devices in blood vessels in the skin of diabetics, allowing them to monitor more accurately their insulin requirements. Ultimately, it may be possible to link the biosensor to a minipump, so that insulin is automatically released when needed, thus in effect providing the diabetic with an automatic pancreas. This fine control would reduce the common secondary effects of diabetes, such as eye and kidney damage, suffered by some diabetics as a result of relatively crude treatment by occasional injections.

One of the major future developments in biosensor research will be towards making more-sensitive miniature sensors. This could be the result of developing '**biochips**'. Just as large computers have been reduced in size by the introduction of silicon microchips containing the complex circuits required, so further size reduction may be possible by using **semiconducting organic molecules**, such as proteins in place of silicon. Electrical signals will then pass along these molecules; electrical circuits could be just one molecule wide. Biochips would be small enough to implant in the human body. Interfaced with the relevant devices, numerous medical applications become possible, such as artificial sense organs or artificial regulation of heartbeat.

Another important development in biosensor research will be multifunctional biosensors, each of which it is anticipated will be sensitive to at least ten different stimuli.

Chapter Seven

Cells

The basic unit of structure and function in living organisms is the **cell**. This concept, known as the **cell theory**, evolved gradually during the nineteenth century as a result of microscopy. The study of cells by microscopy came to be known as **cytology**. Later in the nineteenth century, and during this century, much more experimentation was brought into the study of cells, and there is now a large branch of biology called **cell biology** in which many techniques are combined to gain an understanding of living organisms at the cellular level. Like biochemists, cell biologists often deal with fundamental processes which are common to most or all cells, so that cell biology, like biochemistry, is a unifying theme in biology. Table 7.1 summarises some of the historically important events in the development of cell biology.

There is good reason why life should have a cellular basis. The cell, in essence, is a self-perpetuating chemical system. In order to maintain the concentration of chemicals required the system has to be physically separate from its environment, yet capable of exchange with its environment so that chemicals which are raw materials can be acquired and those which are waste products can be removed. In this way, by doing work, the system can maintain stability (homeostasis, chapter 18). In all cases the barrier between the chemical system and its environment is the cell surface membrane; this helps to control exchanges between the two and so forms the boundary of the cell.

Cells always contain cytoplasm, as well as genetic material in the form of DNA. The DNA controls the activities of the cell and can replicate itself so that new cells are formed. The concept that new cells only come from pre-existing cells also dates from the nineteenth century (table 7.1) and is an essential part of the cell theory.

Table 7.1 Some historically important events in cell biology.

1590	Jansen invented the **compound microscope**, which combines two lenses for greater magnification.
1665	Robert Hooke, using an improved compound microscope, examined cork and used the term 'cell' to describe its basic units. He thought the cells were empty and the walls were the living material.
1650–1700	Antony van Leeuwenhoeck, using a good quality simple lens (mag. ×200), observed nuclei and unicellular organisms, including bacteria. In 1676, bacteria were described for the first time as '**animalcules**'.
1700–1800	Further descriptions and drawings published, mainly of plant tissues, although the microscope was generally used as a toy.
1827	Dolland dramatically improved the quality of lenses. This was followed by a rapid spread of interest in microscopy.
1831–3*	Robert Brown described the nucleus as a characteristic spherical body in plant cells.
1838–9*	Schleiden (a botanist) and Schwann (a zoologist) produced the '**cell theory**' which unified the ideas of the time by stating that **the basic unit of structure and function in living organisms is the cell**.
1840*	Purkyne gave the name **protoplasm** to the contents of cells, realising that the latter were the living material, not the cells walls. Later the term **cytoplasm** was introduced (cytoplasm + nucleus = protoplasm).
1855*	Virchow showed that all cells arise from pre-existing cells by cell division.
1866	Haeckel established that the nucleus was responsible for storing and transmitting hereditary characters.
1866–88	Cell division studied in detail and chromosomes described.
1880–3	Plastids, e.g. chloroplasts, discovered.
1890	Mitochondria discovered.
1898	Golgi apparatus discovered.
1887–1900	Improvements in microscopes, fixatives, stains and sectioning. Cytology† started to become experimental. Embryology was studied to establish how cells interact during growth of a multicellular organism. Cytogenetics‡, with its emphasis on the functioning of the nucleus in heredity, became a branch of cytology.
1900	Mendel's work, forgotten since 1865, was rediscovered giving an impetus to cytogenetics. Light microscopy had almost reached the theoretical limits of resolution, thus slowing down the rate of progress.
1930s	Electron microscope developed, enabling much improved resolution.
1946 to present	Electron microscope became widely used in biology, revealing much more detailed structure in cells. This 'fine' structure is called **ultrastructure**.

* significant events in the origin and development of the cell concept.

† **cytology** – the study of cells, especially by microscopy.

‡ **cytogenetics** – the linking of cytology with genetics, mainly relating structure and behaviour of chromosomes during cell division to results from breeding experiments.

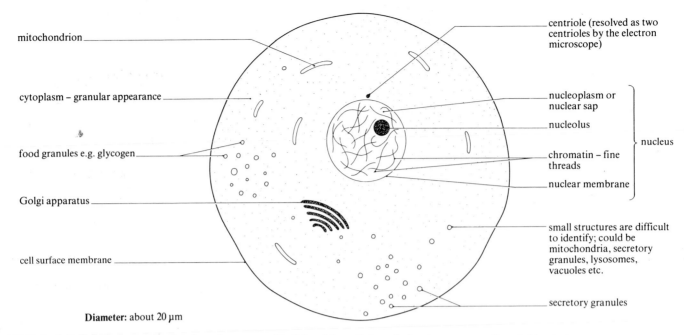

mitochondrion

cytoplasm – granular appearance

food granules e.g. glycogen

Golgi apparatus

cell surface membrane

Diameter: about 20 µm

centriole (resolved as two centrioles by the electron microscope)

nucleoplasm or nuclear sap

nucleolus

chromatin – fine threads

nuclear membrane

nucleus

small structures are difficult to identify; could be mitochondria, secretory granules, lysosomes, vacuoles etc.

secretory granules

Fig 7.1 (above) *Typical animal cell such as an epithelial cell from lining of cheek as seen with a light microscope*

Fig 7.2 (below) *Ultrastructure of a generalised animal cell as seen with the electron microscope. NB for simplicity, only some of the rough endoplasmic reticulum is shown covered with ribosomes. Similarly only some of the free ribosomes are shown*

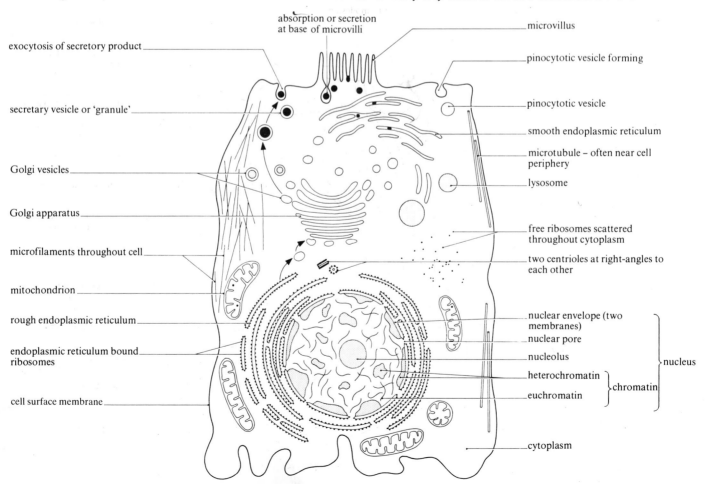

exocytosis of secretory product

absorption or secretion at base of microvilli

microvillus

pinocytotic vesicle forming

secretary vesicle or 'granule'

pinocytotic vesicle

smooth endoplasmic reticulum

Golgi vesicles

microtubule – often near cell periphery

lysosome

Golgi apparatus

microfilaments throughout cell

free ribosomes scattered throughout cytoplasm

two centrioles at right-angles to each other

mitochondrion

rough endoplasmic reticulum

nuclear envelope (two membranes)

nuclear pore

endoplasmic reticulum bound ribosomes

nucleolus

heterochromatin

chromatin

nucleus

cell surface membrane

euchromatin

cytoplasm

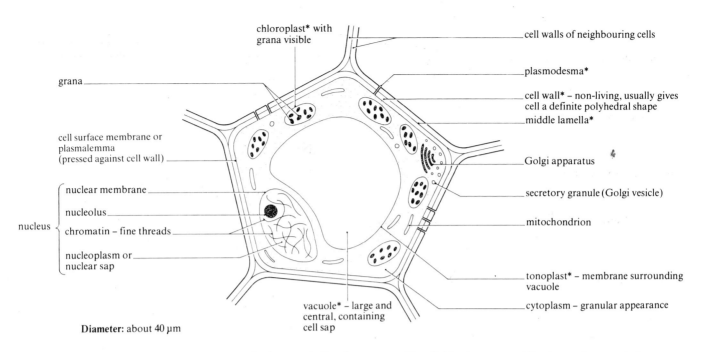

Fig 7.3 (above) *Typical plant cell such as a leaf mesophyll cell as seen with a light microscope. *Features characteristic of plant cells but not animal cells*

Fig 7.4 (below) *Ultrastructure of a generalised plant cell as seen with the electron microscope*

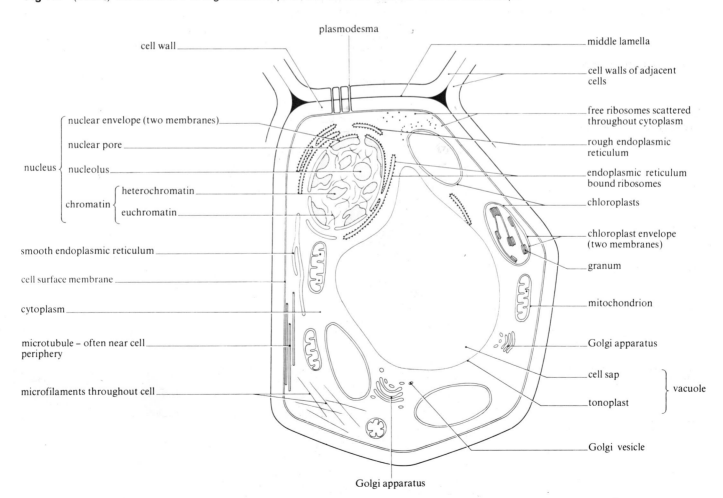

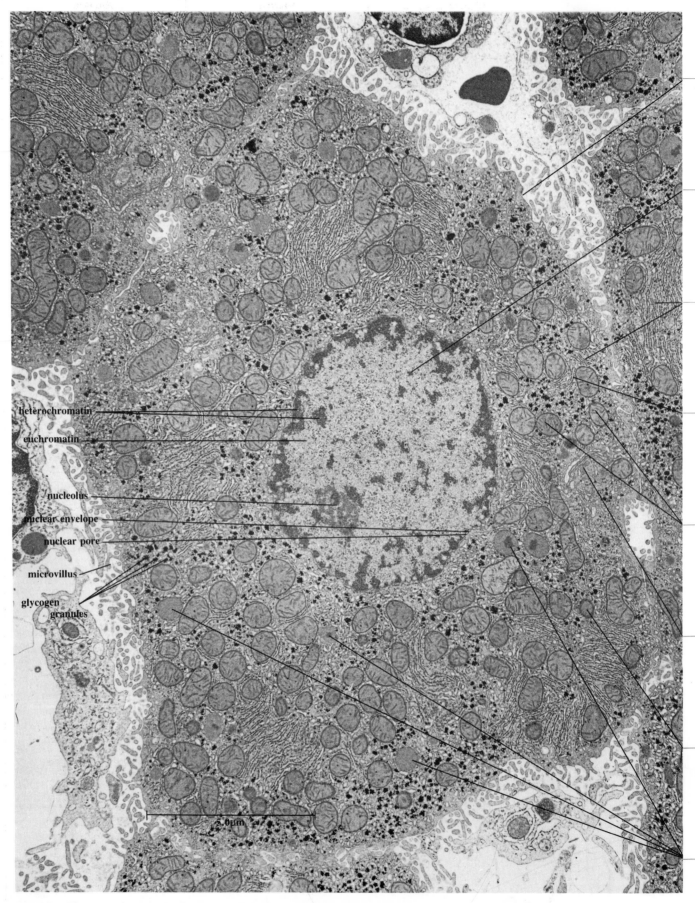

heterochromatin

euchromatin

nucleolus

nuclear envelope

nuclear pore

microvillus

glycogen
granules

Fig 7.5 *Electron micrograph of a thin section of a representative animal cell, a rat liver cell (hepatocyte × 9 600)*

	Diagram	Structure	Functions
cell surface membrane	**Cell surface membrane (plasmalemma)** protein lipid bilayer protein	Two layers of lipid (bilayer) sandwiched between two protein layers	A partially permeable barrier controlling exchange between the cell and its environment
nucleus	**Nucleus** nuclear envelope (two membranes) nuclear pore heterochromatin euchromatin } chromatin nucleolus nucleoplasm	Largest cell organelle, enclosed by an **envelope** of two membranes that is perforated by **nuclear pores**. It contains **chromatin** which is the extended form taken by chromosomes during interphase. It also contains a **nucleolus**.	Chromosomes contain DNA, the molecule of inheritance. DNA is organised into genes which control all the activities of the cell. Nuclear division is the basis of cell replication, and hence reproduction. The nucleolus manufactures ribosomes.
endoplasmic reticulum	**Endoplasmic reticulum (ER)** ribosomes cisterna	A system of flattened, membrane-bounded sacs called **cisternae**, forming tubes and sheets. It is continuous with the outer membrane of the nuclear envelope.	If ribosomes are found on its surface it is called **rough ER**, and transports proteins made by the ribosomes through the cisternae. **Smooth ER,** (no ribosomes) is a site of lipid and steroid synthesis.
ribosomes	**Ribosomes** large subunit small subunit	Very small organelles consisting of a large and a small subunit. They are made of roughly equal parts of protein and RNA. Slightly smaller ribosomes are found in mitochondria (and chloroplasts in plants).	Sites of protein synthesis, holding in place the various interacting molecules involved. They are either bound to the ER or lie free in the cytoplasm. They may form **polysomes** (polyribosomes), collections of ribosomes strung along messenger RNA.
mitochondria	**Mitochondria** (sing. **mitochondrion**) phosphate granule ribosome matrix crista envelope (two membranes) circular DNA	Surrounded by an envelope of two membranes, the inner being folded to form **cristae**. Contains a **matrix** with a few ribosomes, a circular DNA molecule and phosphate granules.	In aerobic respiration cristae are the sites of oxidative phosphorylation and electron transport, and the matrix is the site of Krebs cycle enzymes and fatty acid oxidation.
Golgi apparatus	**Golgi apparatus** Golgi vesicles dictyosome or Golgi body	A stack of flattened, membrane-bounded sacs, called **cisternae**, continuously being formed at one end of the stack and budded off as vesicles at the other. Stacks may form discrete dictyosomes as in plant cells, or an extensive network as in many animal cells.	Processing in cisternae and transport in vesicles of many cell materials, such as enzymes from the ER. Often involved in secretion and lysosome formation.
lysosome	**Lysosomes**	A simple spherical sac bounded by a single membrane and containing digestive (hydrolytic) enzymes. Contents appear homogeneous.	Many functions, all concerned with breakdown of structures or molecules. See text for role in autophagy, autolysis, endocytosis and exocytosis.
microbodies	**Microbodies**	A roughly spherical organelle bounded by a single membrane. Its contents appear finely granular except for occasional striking crystalloid or filamentous deposits.	All contain catalase, an enzyme that breaks down hydrogen peroxide. All are associated with oxidation reactions. In plants, are the site of the glyoxylate cycle.

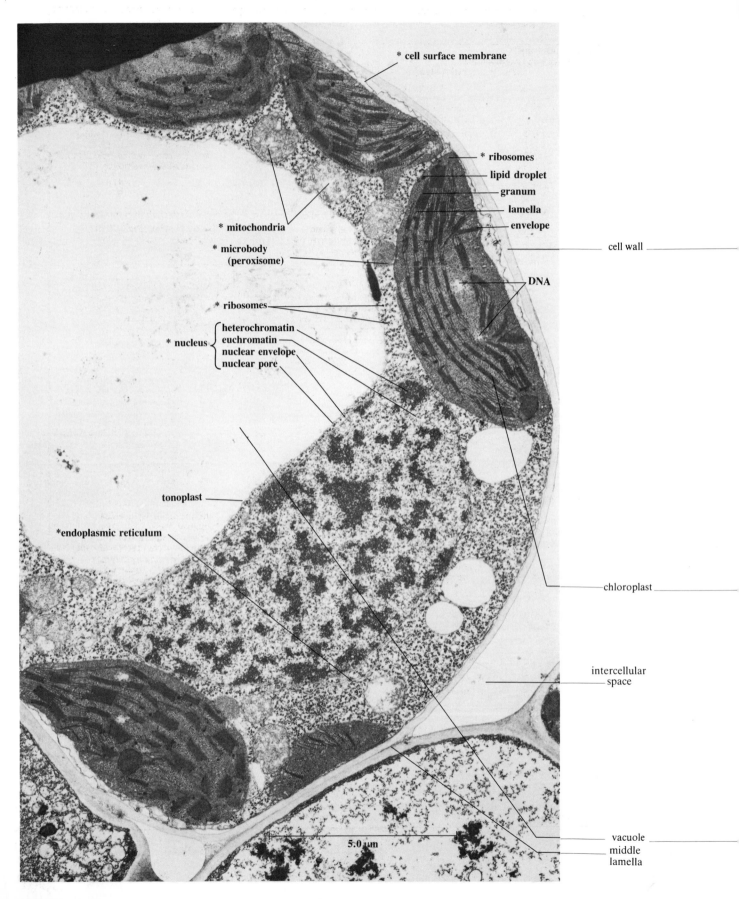

* cell surface membrane

* ribosomes
lipid droplet
granum
lamella
envelope

cell wall

* mitochondria

* microbody
(peroxisome)

DNA

* ribosomes

* nucleus {
heterochromatin
euchromatin
nuclear envelope
nuclear pore
}

tonoplast

*endoplasmic reticulum

chloroplast

intercellular
space

5.0 μm

vacuole

middle
lamella

Fig 7.6 *Electron micrograph of a thin section of a representative plant cell, a leaf mesophyll cell (× 15 000)*
Notes provided on fig 7.5

190

Diagram	Structure	Functions

Cell wall, middle lamella, plasmodesmata (sing. plasmodesma)

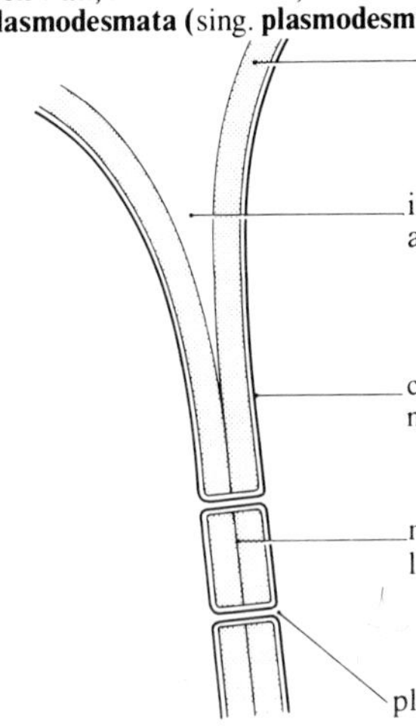

cell wall

intercellular air space

cell surface membrane

middle lamella

plasmodesma

Detail of plasmodesma

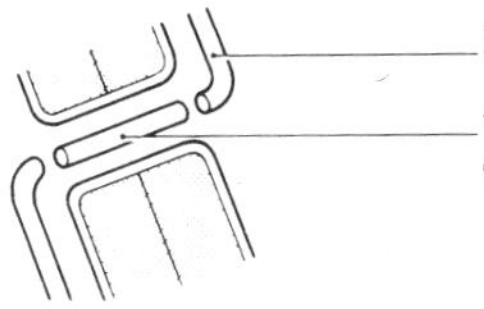

ER

tubular core

	A rigid cell wall surrounding the cell, consisting of cellulose microfibrils running through a matrix of other complex polysaccharides, namely hemicelluloses and pectic substances. May be secondarily thickened in some cells.	Provides mechanical support and protection. It allows a pressure potential to be developed which aids in support. It prevents osmotic bursting of the cell. It is a pathway for movement of water and mineral salts. Various modifications, such as lignification, for specialised functions.
	Thin layer of pectic substances (calcium and magnesium pectates).	Cements neighbouring cells together.
	A fine cytoplasmic thread linking the cytoplasm of two neighbouring cells through a fine pore in the cell walls. The pore is lined with the cell surface membrane and has a central tubular core, often associated at each end with ER.	Enables a continuous system of cytoplasm, the **symplast**, to be formed between neighbouring cells for transport of substances between cells.

Chloroplast

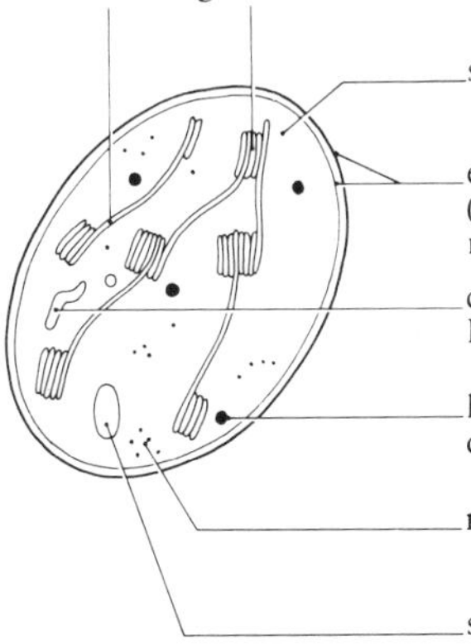

photosynthetic membranes with chlorophyll

lamella   granum

stroma

envelope (two membranes)

circular DNA

lipid droplet

ribosomes

starch grain

	Large plastid containing chlorophyll and carrying out photosynthesis. It is surrounded by an envelope of two membranes and contains a gel-like **stroma** through which runs a system of membranes that are stacked in places to form **grana**. It may store starch. The stroma also contains ribosomes, a circular DNA molecule and lipid droplets.	It is the organelle in which photosynthesis takes place, producing sugars and other substances from carbon dioxide and water using light energy trapped by chlorophyll. Light energy is converted to chemical energy.

Large central vacuole

(Smaller vacuoles may occur in plant and animal cells such as food vacuoles, contractile vacuoles.)

	A sac bounded by a single membrane called the **tonoplast**. It contains **cell sap**, a concentrated solution of various substances, such as mineral salts, sugars, pigments, organic acids and enzymes. Typically large in mature cells.	Storage of various substances including waste products. It makes an important contribution to the osmotic properties of the cell. Sometimes it functions as a lysosome.

Protoplasm is the nineteenth-century term given to the living contents of cells, which were observed at the time as little more than a fluid in which the processes of life took place. It is now known, particularly as a result of electron microscopy, that division of labour takes place in the protoplasm, with minute structures having particular functions. These definite structures are called organelles, that is 'small organs'. All eukaryotic cells are built on the same basic plan with a limited number of organelles carrying out the different biochemical activities in separate compartments. **This 'compartmentation' is the key to understanding cell organisation**.

The first organelle to be discovered was the nucleus, described by Robert Brown in 1831 (table 7.1). The smallest organelles are ribosomes, and these are found in all cells. Some organelles are found only in specialised cells, such as chloroplasts which are found in photosynthetic cells.

In section A2.3 details of the use of a light microscope are given and differences between light microscopes and electron microscopes are described, together with some of the techniques used with microscopy which will be referred to in this chapter. In this chapter the structure of the eukaryote cell will be described. Chapter 2 should be referred to for a comparison with the prokaryotic cell. Eukaryotes comprise protoctists, fungi, plants and animals.

7.1   Typical animal and plant cells

Figs 7.1 and 7.2 show the appearance of typical animal and plant cells as seen with the light microscope at the maximum resolution of × 1 500. Figs 7.3 and 7.4 show the increased detail that can be seen with the aid of the electron microscope.

> **7.1** With reference to figs 7.1–7.4, what additional structures are revealed by the electron microscope compared with the light microscope?
>
> **7.2** With reference to figs 7.1–7.4, what structures are found (a) in plant cells but not in animal cells, and
> (b) in animal cells but not in plant cells?

Figs 7.5 and 7.6 show actual electron micrographs of representative animal and plant cells, together with summaries of the structures and functions of the different parts seen.

7.2   Structures common to animal and plant cells

7.2.1   Cell membranes

Cell membranes are important for a number of reasons. They separate the contents of cells from their external environments, controlling exchange between the two, and they enable separate compartments to be formed inside cells in which specialised metabolic pathways can take place. Chemical reactions, such as the light reactions of photosynthesis in chloroplasts and oxidative phosphorylation of respiration in mitochondria, sometimes take place on the membranes themselves. They also act as receptor sites for recognising external stimuli such as hormones and other chemicals, either from the external environment or from other parts of the organism. An understanding of their properties is essential to an understanding of cell function.

It has been known since the turn of the century that cell membranes do not behave simply like semi-permeable membranes that allow only the passage of water and other small molecules such as gases. Instead they are better described as **partially permeable**, since other substances such as glucose, amino acids, fatty acids, glycerol and ions can diffuse slowly through them, and they also exert a measure of active control over what substances they allow through.

Early work showed that organic solvents, such as alcohol, ether and chloroform, penetrate membranes even more rapidly than water. This suggested that membranes have non-polar portions; in other words that they contain lipids. This was later confirmed by chemical analysis which showed that membranes are comprised almost entirely of proteins and lipids. The proteins are discussed later. The lipids are mainly phospholipids, glycolipids and sterols.

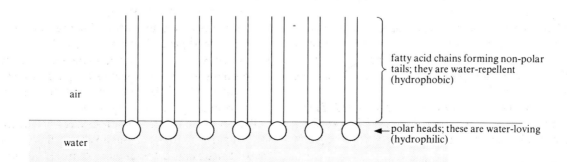

air

water

fatty acid chains forming non-polar tails; they are water-repellent (hydrophobic)

polar heads; these are water-loving (hydrophilic)

Fig 7.7 *Monolayer of polar lipid molecules, such as phospholipids, at the surface of water*

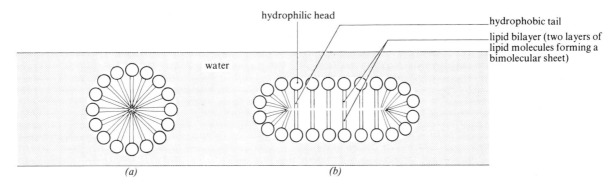

hydrophilic head

hydrophobic tail

lipid bilayer (two layers of lipid molecules forming a bimolecular sheet)

water

(a)   (b)

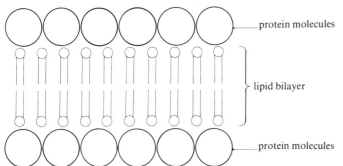

protein molecules

lipid bilayer

protein molecules

Fig 7.8 (above) *Sections through (a) a spherical micelle and (b) a rod-shaped micelle formed by polar lipids in water*

Fig 7.9 (left)   *Davson–Danielli model of membrane structure*

Fig. 7.10 (below) *Electron micrograph showing the surface membrane of a red blood cell (× 250 000). The arrows indicate the three-layered structure of the membrane (dense-light-dense). The stain used contained osmium which is taken up by the hydrophilic regions of proteins and lipids*

Phospholipids (containing a phosphate group) have a polar* head and two non-polar tails (fig 5.19). **Glycolipids** are lipids combined with carbohydrate. Like phospholipids, glycolipids have polar heads and non-polar tails. **Sterols** are steroid alcohols. The most abundant sterol is cholesterol (fig 5.20). Unlike phospholipids and glycolipids, cholesterol is completely non-polar.

If a thin layer of polar lipids, such as phospholipids, is spread over the surface of water, the molecules orientate themselves into a single monomolecular layer, a **monolayer**, as shown in fig 7.7. The non-polar hydrophobic tails project out of the water, whilst the polar hydrophilic heads lie in the surface of the water.

If the polar lipid is present in large enough amounts to more than cover the surface of the water, or if it is shaken up with the water, particles known as **micelles** are formed, in which hydrophobic tails project inwards away from the water as shown in fig 7.8.

Fig 7.8 shows a type of micelle in which two layers of lipid molecules occur, known as a **lipid bilayer**. Phospholipid bilayers like this have many of the properties of living cell

membranes. Davson and Danielli proposed, in 1935, that such a structure, coated with protein molecules on both surfaces, might occur in cell membranes. Their membrane model is summarised in fig 7.9. With the introduction of the electron microscope membranes could be clearly seen for the first time, and surface membranes of both animal and plant cells showed a characteristic three-layered (**trilaminar**) appearance. An example of this is shown in fig 7.10.

In 1959 Robertson combined the available evidence and put forward the '**unit membrane' hypothesis** which proposed that all biological membranes shared the same basic structure:

(*a*) they are about 7.5 nm wide;

(*b*) they have a characteristic trilaminar appearance when viewed with the electron microscope;

(*c*) the three layers are a result of the same arrangement of proteins and polar lipids as proposed by Davson and Danielli (fig 7.9) and represent two protein layers surrounding a central lipid layer.

The unit membrane hypothesis has since been modified in the light of evidence from a variety of sources, notably freeze fracturing, a technique described in section A2.5 that is very important in the investigation of membrane structure. The technique allows membranes to be split and

* Remember that polar groups or molecules are charged and have an affinity for water (hydrophilic); non-polar groups or molecules do not mix with water (hydrophobic) (section 5.1.2).

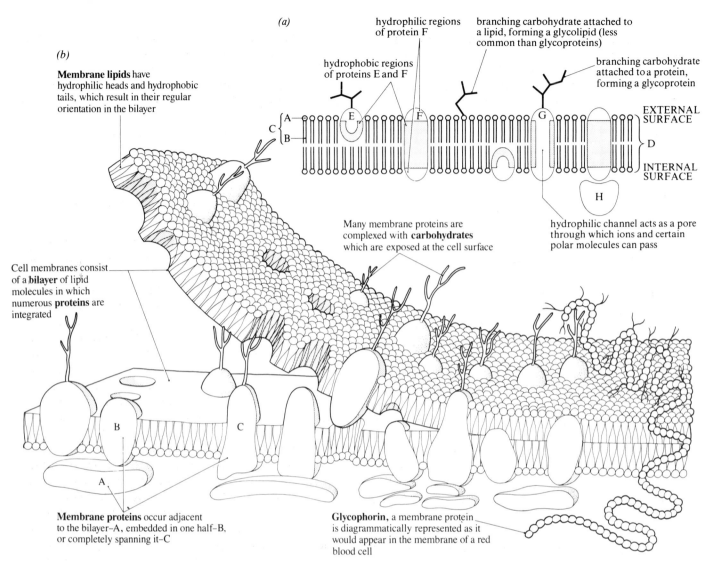

(a)

(b)

Membrane lipids have hydrophilic heads and hydrophobic tails, which result in their regular orientation in the bilayer

hydrophilic regions of protein F

hydrophobic regions of proteins E and F

branching carbohydrate attached to a lipid, forming a glycolipid (less common than glycoproteins)

branching carbohydrate attached to a protein, forming a glycoprotein

EXTERNAL SURFACE

INTERNAL SURFACE

Many membrane proteins are complexed with **carbohydrates** which are exposed at the cell surface

hydrophilic channel acts as a pore through which ions and certain polar molecules can pass

Cell membranes consist of a **bilayer** of lipid molecules in which numerous **proteins** are integrated

Membrane proteins occur adjacent to the bilayer–A, embedded in one half–B, or completely spanning it–C

Glycophorin, a membrane protein is diagrammatically represented as it would appear in the membrane of a red blood cell

Fig 7.11 (a) *Modern fluid mosaic model of membrane structure. Glycoproteins and glycolipids are associated only with the external surfaces of membranes. (b) Three-dimensional model of membrane structure*

7.3 (*a*) What are the structures represented by the labels A, B, C and D in fig 7.11 (*a*)? (*b*) What common component of structure D has been omitted?

the surfaces inside to be examined. It has the advantage that the membranes are preserved in a life-like state by instant freezing, rather than being subjected to chemical fixation which might alter the arrangement of the components. Freeze fracturing reveals the presence of particles (mainly proteins) which penetrate into, and sometimes right through, the lipid bilayer. In general, the more metabolically active the membrane, the more protein particles that are found; chloroplast membranes (75% protein) have many particles (fig 9.13), whereas the metabolically inert myelin sheath (18% protein) has none. The inner and outer faces of membranes also differ in their particle distribution.

In 1972, Singer and Nicolson put forward the '**fluid mosaic**' model of membrane structure in which a mosaic of protein molecules floats in a fluid lipid layer. This model is shown in its modern form in fig 7.11.

In this model the lipid bilayer remains unchallenged as the unit membrane, but it is regarded as a dynamic structure in which proteins can float in the lipid like islands, some moving about freely while others are fixed in position, sometimes by microfilaments running into the cytoplasm. Lipids also move about.

Proteins

Some proteins penetrate only part of the way into the membrane while others penetrate all the way through. Usually they have hydrophobic portions which interact with the lipids, with hydrophilic portions facing the aqueous contents of the cell at the membrane surface. In all there are thousands of different proteins which can occur in cell membranes. They may be purely structural or have some additional function. Some, for example, act as **carrier** molecules, transporting specific substances through the membrane. The carrier may be part of an active pump mechanism (discussed later). It is believed that hydrophilic **channels** or **pores** sometimes occur within a protein, or between adjacent protein molecules. The pore spans the membrane, allowing the passage through the membrane of polar molecules that would otherwise be excluded by the lipid region. Such a protein-lined pore is shown in fig 7.11a.

Other membrane proteins may act as enzymes, specific receptor molecules, electron carriers and energy transducers in photosynthesis and respiration, and so on. Also present in membranes are glycoproteins. These have branching carbohydrate portions resembling antennae on their free surfaces, as shown in fig 7.11. The 'antennae' are made up of a number of sugar residues and may be of many different, but precisely defined, patterns owing to the diversity of linkages between sugars and the existence of α- and β-isomers as described in chapter 5. They are important as recognition features in a number of ways. For example, sugar-recognition sites of two neighbouring cells may bind to each other causing cell-to-cell adhesion. This may enable cells to orientate themselves and to form tissues, such as during cell differentiation. Recognition is also the basis of various control systems and of the immune response, where glycoproteins act as antigens. Certain molecules in solution may bear recognition sites which enable them to be taken up specifically by cells with complementary recognition sites. The addition of sugar residues (glycosylation) to proteins by the Golgi apparatus for this purpose is discussed later (section 7.2.7). Sugars can therefore function as informational molecules and in this sense are comparable with proteins and nucleic acids.

Lipids

Variations in lipid composition affect such properties as fluidity and permeability, the usual consistency of the lipids being similar to that of olive oil. Unsaturated lipids have kinks in their fatty acid tails (section 5.3.1 and fig 5.17). These prevent close packing of the molecules and make the membrane structure more open and fluid. Fluidity also increases with decreasing length of fatty acid tails and the lipid cholesterol is important in regulating fluidity within certain limits. Fluidity affects membrane activity, such as the ease with which membranes fuse with each other, and the activity of membrane-bound enzymes and transport proteins.

Glycolipids contribute to recognition sites in the same way as glycoproteins.

Summary of cell membranes

A summary of the features of biological membranes is given below.
(1) Different types of membranes differ in thickness but most fall within the range 5–10 nm, for example cell surface membranes are 7.5 nm wide.
(2) Membranes are lipoprotein structures (lipid + protein), with carbohydrate (sugar) portions attached to the *external* surfaces of some lipid and protein molecules. Typically, 2–10% of the membrane is carbohydrate.

(3) The lipids spontaneously form a bilayer owing to their polar heads and non-polar tails.
(4) The proteins are variable in function.
(5) The sugars are involved in recognition mechanisms.
(6) The two sides of a membrane may differ in composition and properties.
(7) Both lipids and proteins show rapid lateral diffusion in the plane of the membrane unless anchored or restricted in some way.

7.2.2 Transport across the cell surface membrane

In chapter 14 the problems of long-distance transport within the bodies of multicellular plants and animals are discussed. Living organisms are also faced with the problem of short-distance transport across cell membranes which, although only 5–10 nm wide, present barriers to the movement of ions and molecules, particularly polar molecules such as glucose and amino acids that are repelled by the non-polar lipids of membranes. Transport across membranes is vital for a number of reasons, for example to maintain a suitable pH and ionic concentration within the cell for enzyme activity, to obtain certain food supplies for energy and raw materials, to excrete toxic substances or secrete useful substances and to generate the ionic gradients essential for nervous and muscular activity. In the following account, movement across the cell surface membrane will be discussed, although similar movements occur across the membranes of cell organelles within cells. There are four basic methods of entry into, or exit from, cells, namely diffusion, osmosis, active transport and endocytosis or exocytosis. The first two processes are passive, that is they do not require the expenditure of energy by the cell; the latter two are active, energy-consuming processes.

Diffusion and facilitated diffusion

Gases, like the respiratory gases oxygen and carbon dioxide, diffuse rapidly in solution through membranes, moving from regions of high concentration to regions of low concentration down diffusion gradients. Ions and small polar molecules such as glucose, amino acids, fatty acids and glycerol normally diffuse slowly through membranes. Uncharged and fat soluble (lipophilic) molecules pass through membranes much more readily, as already noted.

A modified form of diffusion known as **facilitated diffusion** exists in which the substance is allowed through the membrane by a specific molecule. This molecule may possess a specific channel that admits only one type of substance. An example is the movement of glucose into red blood cells, which is not inhibited by respiratory inhibitors and is therefore not an active process.

Osmosis

Water diffuses through membranes, a process called osmosis (see section A1.5).

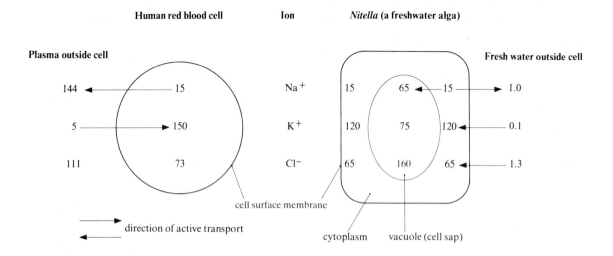

Human red blood cell   Ion   *Nitella* (a freshwater alga)

Plasma outside cell   Fresh water outside cell

cell surface membrane

direction of active transport

cytoplasm   vacuole (cell sap)

Active transport

Active transport is the energy-consuming transport of molecules or ions across a membrane against a concentration gradient. Energy is required because the substance must be moved against its natural tendency to diffuse in the opposite direction. Movement is usually unidirectional, unlike diffusion which is reversible.

When movement of ions is considered, two factors will influence the direction in which they diffuse. One is concentration, the other is electrical charge. An ion will usually diffuse from a region of its high concentration to a region of its low concentration. It will also generally be attracted towards a region of opposite charge, and move away from a region of similar charge. Thus ions are said to move down **electrochemical gradients**, which are the combined effects of both electrical and concentration gradients. Strictly speaking then, active transport of ions is their movement against an electrochemical gradient. It has been shown that cells maintain a potential difference, that is a charge, across their cell surface membranes and that, for almost all cells studied, the inside of the cell is negative with respect to the outside medium. Thus cations (positively charged ions) are usually electrically attracted into the cells and anions repulsed. However, their relative concentrations inside and outside of the cell also helps to determine in which direction they actually diffuse.

The major ions of extracellular and intracellular fluids are sodium (Na^+), potassium (K^+) and chloride (Cl^-) ions. Study fig 7.12 which shows the concentrations of certain ions in the cytoplasm and cell sap of a plant cell, and the cytoplasm of an animal cell.

The data show that both of these types of cell have ionic compositions very different from their external solutions. For example, like most cells, they have a much higher potassium content inside than outside. Another typical feature is the higher concentration of potassium inside relative to sodium.

If respiration of the red blood cells is specifically inhibited, for example with cyanide, the ionic composition

Fig 7.12 *Concentrations (mM) of Na^+, K^+ and Cl^- ions in two types of cell and their environments*

of the cells gradually changes until it comes into equilibrium with the plasma. This suggests that the ions can diffuse passively through the cell surface membrane of the red blood cells, but that normally respiration supplies the energy for active transport to maintain the concentrations shown in fig 7.12. In the case of both types of cell shown in fig 7.12, sodium is actively pumped out of the cell and potassium is actively pumped in. It is possible to calculate that there is no net tendency for chloride ions to enter red blood cells from the plasma, despite the higher concentration of chloride ions in the plasma. This is due to the strictly negative charge of the cell contents relative to the outside, which tends to repulse chloride ions; in other words it is the **electrochemical gradient** that determines the movement of the ions, as already explained, and this is true of all cells. A careful study of the figures provided for *Nitella* in fig 7.12 will show similar examples of how movement of ions is not solely determined by concentration. The potential difference across the cell surface membrane of red blood cells is −10 mV, and of *Nitella* is −140 mV.

In recent years it has been shown that the cell surface membranes of most cells possess **sodium pumps** that actively pump sodium ions out of the cell. Usually, though not always, the sodium pump is coupled with a potassium pump which actively accumulates potassium ions from the external medium and passes them into the cell. The combined pump is called the **sodium–potassium pump** (Na^+–K^+ pump).

Since this pump is a common feature of cells and has a number of important functions, it provides a good example of active transport.

It has been studied in animal cells and has been shown to be driven by ATP. Its physiological importance is revealed by the observation that more than a third of the ATP consumed by a resting animal is used to pump sodium and potassium. This is essential in controlling cell volume (osmoregulation), in maintaining electrical activity in nerve and muscle cells and in driving active transport of some other substances such as sugars and amino acids. Also, high cell concentrations of potassium are needed for protein synthesis, glycolysis, photosynthesis and other vital processes.

The pump is essentially a protein which spans the membrane from one side to the other. On the inside it accepts sodium and ATP, while on the outside it accepts potassium. The transfer of sodium and potassium across the membrane is thought to be brought about by conformational changes in the protein. The protein also acts as an ATPase, catalysing the hydrolysis of ATP with release of energy to drive the pump. A possible sequence of events is summarised in fig 7.13. Note that for every $2K^+$ taken into the cell, $3Na^+$ are removed. Thus a negative potential is built up inside the cell and a potential difference across the membrane.

As sodium is pumped out, it generally diffuses back in passively. However, the membrane is relatively impermeable to sodium so that back diffusion is very slow. Membranes are usually about 100 times more permeable to potassium ions than to sodium, so potassium diffuses in much more rapidly.

7.4 Try to explain the following observations.
(a) When K^+ ions are removed from the medium surrounding red blood cells, sodium influx into the cells and potassium efflux (outflux) increase dramatically.
(b) If ATP is introduced into cells, Na^+ efflux is stimulated.

Active transport is carried out by all cells but it assumes particular significance in certain physiological processes. The process is particularly associated with epithelial cells as in the gut lining and kidney tubules, because these are active in secretion and absorption.

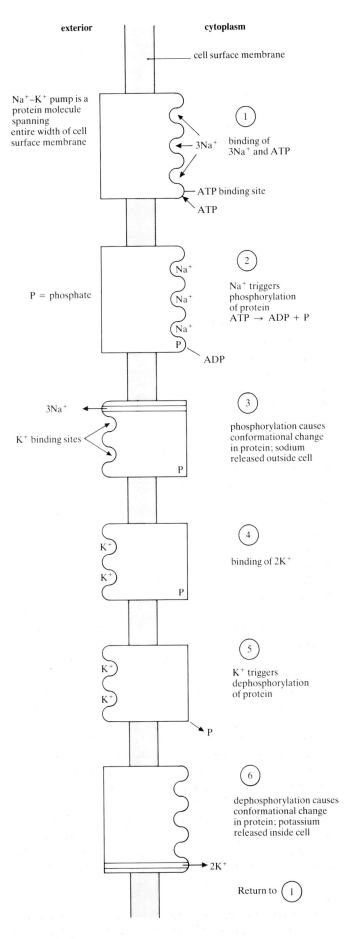

Fig 7.13 *Outline of a possible scheme for the operation of a sodium–potassium pump in red blood cells. Each event in the cycle is a consequence of the previous event. Given a supply of sodium, potassium and ATP the pump will continue to run. Changes in the conformation of the protein are caused by addition or removal of phosphate (phosphorylation or dephosphorylation respectively)*

197

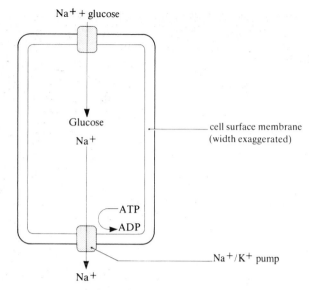

Na⁺ + glucose

Glucose

Na⁺

cell surface membrane
(width exaggerated)

ATP

ADP

Na⁺/K⁺ pump

Na⁺

Fig 7.14 *Active transport of glucose through the cell surface membrane of an intestinal cell or kidney cell. (Based on Fig 36–12, L. Stryer (1981) Biochemistry, 2nd ed., Freeman)*

Active transport in the intestine. When the products of digestion are absorbed in the small intestine they must pass through the epithelial cells lining the gut wall. Glucose, amino acids and salts then pass through the cells of the blood capillary walls, into the blood and thence to the liver. Soon after feeding, relatively high concentrations of digested foods are found in the gut and absorption is partly a result of diffusion. However, this is very slow and must be supplemented by active transport. Such active transport is coupled to a sodium–potassium pump as shown in fig 7.14.

As sodium is pumped out by the sodium–potassium pump, so it tends to diffuse back in. Situated in the membrane is a protein which requires both sodium and glucose to function. These are transported together **passively** into the cell. Sodium thus 'pulls' the glucose into the cell. A similar sodium–amino-acid carrier protein operates in the active transport of amino acids into cells, the active part of the process being the pumping back of sodium ions. In the absence of a sodium gradient the carriers may still act, providing that the external concentration of glucose or amino acids is greater than the internal concentration, that is facilitated diffusion can occur.

Active transport in nerve cells and muscle cells. In nerve cells and muscle cells a sodium–potassium pump is responsible for the development of a potential difference, called the **resting potential**, across the cell surface membrane (see conduction of nervous impulses, section 16.1, and muscle contraction, section 17.4). A pump similar to the Na⁺–K⁺ pump occurs in the membranes of the sarcoplasmic reticulum in muscle cells, where calcium is actively pumped into the sarcoplasmic reticulum at the expense of ATP (chapter 17).

Active transport in the kidney. Active transport of glucose and sodium occurs from the proximal convoluted tubules of the kidney (fig 19.27) and the kidney cortex actively transports sodium. These processes are described more fully in chapter 19.

Endocytosis and exocytosis

Endocytosis and exocytosis are active processes involving the bulk transport of materials through membranes, either into cells (endocytosis) or out of cells (exocytosis).

Endocytosis occurs by an infolding or extension of the cell surface membrane to form a vesicle* or vacuole*. It is of two types.

(1) Phagocytosis ('cell eating') – material taken up is in solid form. Cells specialising in the process are called **phagocytes** and are said to be **phagocytic**; for example some white blood cells. The sac formed during uptake is called a **phagocytic vacuole**. (See section 7.2.8.)

(2) Pinocytosis ('cell drinking') – material taken up is in liquid form (a solution, colloid or fine suspension). Vesicles formed are often extremely small, in which case the process is known as **micropinocytosis** and the vesicles as **micropinocytotic**.

Pinocytosis is particularly associated with amoeboid protozoans and many other, often amoeboid, cells, such as leucocytes, embryo cells, liver cells, and certain kidney cells involved in fluid exchange. It can also occur in plant cells.

Exocytosis is the reverse process of endocytosis by which materials are removed from cells, such as solid, undigested remains from food vacuoles or reverse pinocytosis in secretion (see section 7.2.7).

7.2.3 The nucleus

Nuclei are found in all eukaryotic cells, the only common exceptions being mature phloem sieve tube elements and mature red blood cells of mammals. In some protozoa, such as *Paramecium*, two nuclei exist, a micronucleus and a meganucleus. Normally, however, cells contain only one nucleus. Nuclei are conspicuous because they are the largest of cell organelles, and they were the first to be described by light microscopists. They are typically spherical to ovoid in shape and about 10 μm in diameter by 20 μm in length.

The nucleus is vitally important because it controls the cell's activities. This is because it contains the genetic (hereditary) information in the form of DNA. Not only this, but the DNA is capable of replication and this can be followed by nuclear division thus ensuring that the daughter nuclei also contain DNA. Nuclear division precedes cell division, and all daughter cells possess nuclei. The nucleus is surrounded by a nuclear envelope and contains chromatin, one or more nucleoli and nucleoplasm.

The single membrane (**nuclear membrane**) surrounding

* Vacuole – fluid-filled, membrane-bound sac. Vesicle – small vacuole.

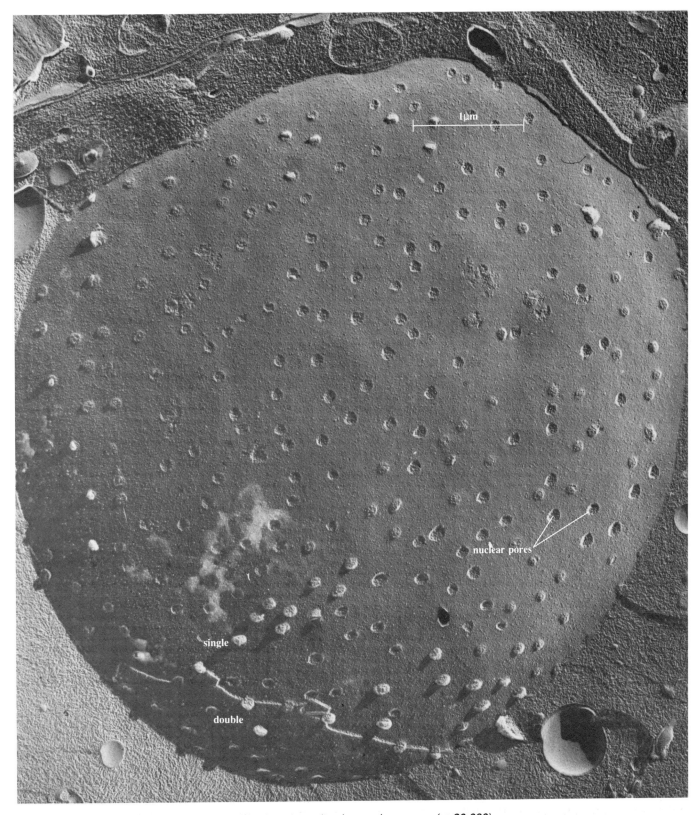

Fig 7.15 *Electron micrograph of freeze-etched nucleus showing nuclear pores (× 30 000)*

the nucleus, as shown by the light microscope, is actually a **nuclear envelope** composed of two membranes. The outer membrane is continuous with the endoplasmic reticulum (ER) as shown in figs 7.3 and 7.4, and like the ER may be covered with ribosomes engaged in protein synthesis. The nuclear envelope is perforated by **nuclear pores** (fig 7.5) and these are particularly well revealed by freeze etching as shown in fig 7.15. Nuclear pores allow exchange of substances between the nucleus and the cytoplasm, for example the exit of messenger RNA (mRNA) and of ribosomal subunits and the entry of ribosomal proteins, nucleotides and molecules that regulate the activity of DNA. The pore has a definite structure formed by fusion of the outer and inner membranes of the envelope. This controls the passage of molecules through the pore.

Within the nucleus is a gel-like matrix called **nucleoplasm** (or nuclear sap) which contains chromatin and one or more nucleoli. Nucleoplasm contains a variety of chemical substances such as ions, proteins (including enzymes) and nucleotides, either in true or colloidal solution.

Chromatin is composed mainly of coils of DNA bound to basic proteins called **histones**. The organisation of histones and DNA into bead-like structures called nucleosomes, and the packing of the nucleosomes in the chromatin, are described in section 22.4.

The term chromatin means 'coloured material' and refers to the fact that this material is easily stained for viewing with the microscope. During nuclear division chromatin stains more intensely and hence becomes more conspicuous because it condenses into more tightly coiled threads called **chromosomes**. During interphase (the period between nuclear divisions) it becomes more dispersed. However, some remains tightly coiled and continues to stain intensely. This is called **heterochromatin** and is seen as characteristic dark patches usually occurring near the nuclear envelope (figs 7.3–6). The remaining, loosely coiled chromatin is located towards the centre of the nucleus and is called **euchromatin**. These individual fibres are too dispersed to be visible under the light microscope and they are thought to contain the DNA which is genetically active during interphase. Thus cells in which a wide variety of genes are being expressed, as in liver cells, will show more euchromatin and less heterochromatin than cells in which few genes are being expressed, such as mucus-producing cells.

Until recently, it was imagined that interphase chromosomes lie randomly, like spaghetti, inside the nucleus. Recent work, using serial sectioning through nuclei, suggests that chromosomes are held in precise domains. The relative positions of chromosomes and their genes may influence such events as chromosome mutation and gene expression, and may change during development.

The **nucleolus** is a conspicuous rounded structure within the nucleus, whose function is the manufacture of ribosomal RNA (fig 7.5). One or more nucleoli may be present. It stains intensely because of the large amounts of DNA and RNA it contains. It has a dense fibrillar region of DNA of one or several different chromosomes, called the **nucleolar organiser**. This contains many copies of the genes that code for ribosomal RNA. The nucleoli disperse and are no longer visible during prophase (the early stage of cell division) and the organisers re-organise the nucleoli during telophase (at the end of nuclear division).

Around the central region of the nucleolus is a less dense, peripheral region containing granules where ribosomal RNA is beginning to be folded and assembled with proteins into ribosomes. The partly assembled ribosomes move out through the nuclear pores into the cytoplasm, where assembly is completed.

7.2.4 Cytoplasm

In the introduction to this chapter it was pointed out that the living contents of eukaryote cells are divided into nucleus and cytoplasm, the two together forming the protoplasm. Cytoplasm consists of an aqueous ground substance containing a variety of cell organelles and other inclusions such as insoluble waste or storage products.

The cytosol or ground substance

The **cytosol** is the soluble part of the cytoplasm. It forms the ground substance or 'background material' of the cytoplasm and is located between the cell organelles. It contains a system of microfilaments (see section 7.2.10) but otherwise appears transparent and structureless in the electron microscope. It is about 90% water and forms a solution which contains all the fundamental biochemicals of life. Some of these are ions and small molecules forming true solutions, such as salts, sugars, amino acids, fatty acids, nucleotides, vitamins and dissolved gases. Others are large molecules which form colloidal solutions (section A1.4), notably proteins and to a lesser extent RNA. A colloidal solution may be a sol (non-viscous) or a gel (viscous); often the outer regions of cytoplasm are more gel-like, as with the ectoplasm of *Amoeba* (section 17.6).

Apart from acting as a store of vital chemicals, the ground substance is the site of certain metabolic pathways, an important example being glycolysis. Synthesis of fatty acids, nucleotides and some amino acids also takes place.

The most common view of cytoplasm is the static one of cells that have been killed and prepared for microscopy. However, when *living* cytoplasm is examined, great activity is usually seen as cell organelles move about and occasional '**cytoplasmic streaming**' occurs. This is an active mass movement of cytoplasm which may be particularly prominent in certain cells such as young sieve tube elements.

7.2.5 Endoplasmic reticulum (ER)

One of the most important discoveries to be made when the electron microscope was introduced was the occurrence of a complex system of membranes running through the cytoplasm of all eukaryotic cells. This network, or reticulum, of membranes was named the **endoplasmic reticulum** and although it is often extensive, it is below the limits of resolution of the light microscope. The membranes were seen to be covered with small particles which later became known as ribosomes. At about the same time a cell fraction was isolated by differential centrifugation

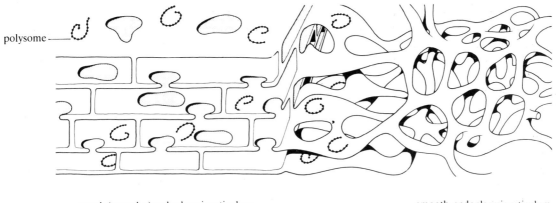

polysome

rough (granular) endoplasmic reticulum   smooth endoplasmic reticulum

Fig 7.16 *Three-dimensional model of endoplasmic reticulum*

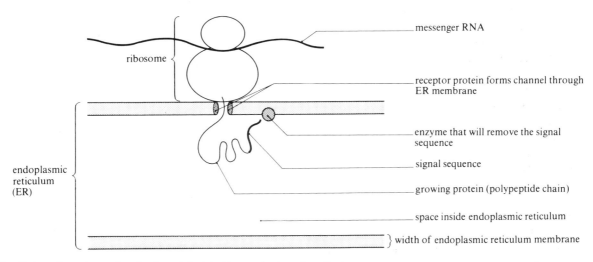

ribosome

messenger RNA

receptor protein forms channel through ER membrane

enzyme that will remove the signal sequence

signal sequence

growing protein (polypeptide chain)

space inside endoplasmic reticulum

} width of endoplasmic reticulum membrane

endoplasmic reticulum (ER)

Fig 7.17 *Entry of newly synthesised protein into the endoplasmic reticulum*

which was shown to be capable of protein synthesis. Examination of this fraction in the electron microscope revealed many small membrane-bound sacs (vesicles), each covered externally with ribosomes. These sacs were called **microsomes**. The **microsomal fraction** is now known to be formed during the homogenisation procedure. When the ER is broken up into small pieces it reseals into vesicles. Thus microsomes do not exist as such in intact cells.

Typically, the ER appears in thin sections as pairs of parallel lines (membranes) running through the cytoplasm, as shown in figs 7.3–6. Occasionally though, a section will glance through the surface of these membranes and show that, in three dimensions, the ER is usually sheet-like rather than tubular. A possible three-dimensional structure is shown in fig 7.16. The ER consists of flattened, membrane-bound sacs called **cisternae**. These may be covered with ribosomes, forming **rough ER**, or ribosomes may be absent, forming **smooth ER**, which is usually more tubular. Both types are concerned with the synthesis and transport of substances.

Rough ER is concerned with the transport of proteins which are made by ribosomes on its surface. Details of protein synthesis are given in chapter 22. For the present it is sufficient to know that the growing protein, which consists of a chain of amino acids called a polypeptide chain, is bound to the ribosome until its synthesis is complete. At the beginning of protein synthesis, the first part of the growing chain may consist of a 'signal sequence' which fits a specific receptor in the ER membrane, thus binding the ribosome to the ER. The receptor forms a channel through which the protein can pass into the ER cisternae (fig 7.17). Once inside, the signal sequence is removed and the protein folds up into its tertiary structure, thus trapping it inside the ER.

> **7.5** A high proportion of amino acid residues in signal sequences are non-polar. Suggest a reason for this.

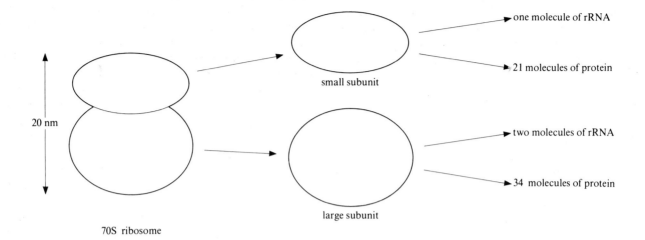

70S ribosome

small subunit → one molecule of rRNA

→ 21 molecules of protein

large subunit → two molecules of rRNA

→ 34 molecules of protein

20 nm

Fig 7.18 *Structure of a 70S ribosome. (The subunits of 80S ribosomes possess more proteins and the large subunit possesses three rRNA molecules)*

The protein is now transported through the cisternae, usually being extensively modified en route. For example, it may be phosphorylated or converted into a glycoprotein. A common route for the protein is via smooth ER to the Golgi apparatus from whence it can be secreted from the cell or passed on to other organelles in the same cell, such as storage bodies or lysosomes.

If a protein is made which does not possess a signal sequence, the ribosome making it remains free in the cytoplasm and the protein is released into the cytosol for use in the cell.

One of the chief functions of smooth ER is lipid synthesis. For example, in the epithelium of the intestine the smooth ER makes lipids from fatty acids and glycerol absorbed from the gut and passes them on to the Golgi apparatus for export. Steroids are a type of lipid and smooth ER is extensive in cells which secrete steroid hormones, such as the adrenal cortex and the interstitial cells of the testis. In the liver both rough and smooth ER are involved in detoxication. In muscle cells a specialised form of smooth ER, called sarcoplasmic reticulum, is present (section 17.4).

7.2.6 Ribosomes

Ribosomes are minute organelles, about 20 nm in diameter, found in large numbers throughout the cytoplasm of living cells, both prokaryotic and eukaryotic. A typical bacterial cell contains about 10 000 ribosomes, while eukaryotic cells possess many times more than this number. They are the sites of protein synthesis.

Each ribosome consists of two subunits, one large and one small as shown in fig 7.18. Being so small they are the last organelles to be sedimented in a centrifuge, requiring a force of 100 000× gravity for 1–2 h. Sedimentation has revealed two basic types of ribosome, called 70S* and 80S ribosomes. The 70S ribosomes are found in prokaryotes

* S=Svedberg unit. This is related to the rate of sedimentation in a centrifuge, the greater the S number, the greater the rate of sedimentation.

and the slightly larger 80S ribosomes occur in the cytoplasm of eukaryotes. It is interesting to note that chloroplasts and mitochondria contain 70S ribosomes, suggesting that these eukaryotic organelles are related in some way to prokaryotes (section 9.3.1).

Ribosomes are made of roughly equal amounts by mass of RNA and protein (hence they are ribonucleoprotein particles). The RNA is termed **ribosomal RNA (rRNA)** and is made in nucleoli. The distribution of rRNA molecules and protein molecules is given in fig 7.18. Together these molecules form a complex three-dimensional structure which is capable of spontaneous self-assembly.

During protein synthesis at ribosomes, amino acids are joined together one by one to form polypeptide chains. The process is described in detail in chapter 22. The ribosome acts as a binding site where the molecules involved can be precisely positioned relative to each other. These molecules include messenger RNA (mRNA), which carries the genetic instructions from the nucleus, transfer RNA (tRNA), which brings the required amino acids to the ribosome, and the growing polypeptide chain. In addition there are chain initiation, elongation and termination factors to be accommodated. The process is so complex that it could not occur efficiently, if at all, without the ribosome.

Two populations of ribosomes can be seen in eukaryotic cells, namely free and ER-bound ribosomes (figs 7.3, 7.5 and 7.16). All of the ribosomes have an identical structure but some are bound to the ER by the proteins that they are making, as explained in the previous section. Such proteins are usually secreted. An example of a protein made by free ribosomes is haemoglobin in young red blood cells.

During protein synthesis, the ribosome moves along the thread-like mRNA molecule. Rather than one ribosome at a time passing along the RNA, the process is carried out more efficiently by a number of ribosomes moving simultaneously along the mRNA, like beads on a string. The resulting chains of ribosomes are called **polyribosomes** or **polysomes**. They form characteristic whorled patterns on the ER as shown in fig 7.16 and they can be isolated intact by centrifugation.

7.2.7 Golgi apparatus

The Golgi apparatus was discovered by Camillo Golgi in 1898, using special staining techniques. However, its structure was only revealed by electron microscopy. It is found in virtually all eukaryotic cells and consists of a stack of flattened, membrane-bound sacs called **cisternae**, together with a system of associated vesicles called **Golgi vesicles**. In plant cells a number of separate stacks called **dictyosomes** are found (fig 7.6). In animal cells a single larger stack is thought to be more usual. It is difficult to build up a three-dimensional picture of the Golgi apparatus from thin sections but it is believed from such evidence as negative staining that a complex system of interconnected tubules is formed around the central stack, as shown in fig 7.19.

At one end of the stack new cisternae are constantly being formed by fusion of vesicles which are probably derived from buds of the smooth ER. This 'outer' or 'forming' face is convex, whilst the other end is the concave 'inner' or 'maturing' face where the cisternae break up into vesicles once more. The whole stack consists of a number of cisternae thought to be moving from the outer to the inner face.

The function of the Golgi apparatus is to transport and chemically modify the materials contained within it. It is

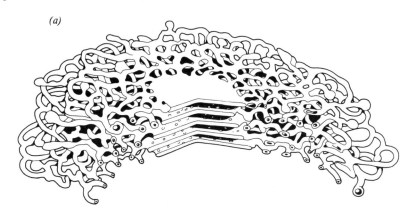

(a)

Fig 7.19 (a) The three-dimensional structure of the Golgi apparatus. (b) Transmission electron micrograph showing two Golgi apparatuses. The left-hand one shows a dictyosome in vertical section. The right-hand one shows the topmost cisternum viewed from above (× 50 000)

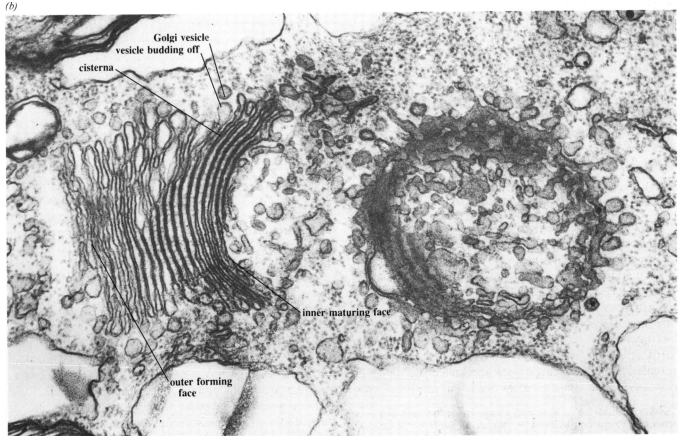

(b)

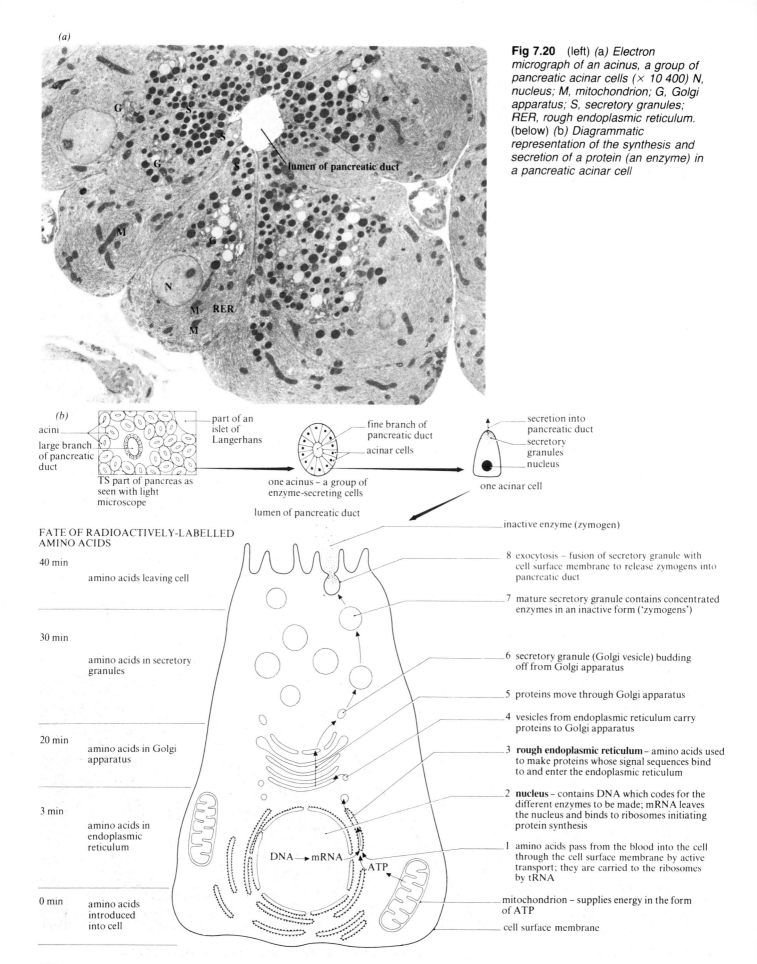

(a)

Fig 7.20 (left) *(a) Electron micrograph of an acinus, a group of pancreatic acinar cells (× 10 400) N, nucleus; M, mitochondrion; G, Golgi apparatus; S, secretory granules; RER, rough endoplasmic reticulum. (below) (b) Diagrammatic representation of the synthesis and secretion of a protein (an enzyme) in a pancreatic acinar cell*

G

S

S

G

S

lumen of pancreatic duct

M

N

M   **RER**

M

(b)

acini

large branch of pancreatic duct

part of an islet of Langerhans

TS part of pancreas as seen with light microscope

fine branch of pancreatic duct

acinar cells

one acinus – a group of enzyme-secreting cells

lumen of pancreatic duct

secretion into pancreatic duct

secretory granules

nucleus

one acinar cell

FATE OF RADIOACTIVELY-LABELLED AMINO ACIDS

40 min — amino acids leaving cell

30 min — amino acids in secretory granules

20 min — amino acids in Golgi apparatus

3 min — amino acids in endoplasmic reticulum

0 min — amino acids introduced into cell

inactive enzyme (zymogen)

8 exocytosis – fusion of secretory granule with cell surface membrane to release zymogens into pancreatic duct

7 mature secretory granule contains concentrated enzymes in an inactive form ('zymogens')

6 secretory granule (Golgi vesicle) budding off from Golgi apparatus

5 proteins move through Golgi apparatus

4 vesicles from endoplasmic reticulum carry proteins to Golgi apparatus

3 **rough endoplasmic reticulum** – amino acids used to make proteins whose signal sequences bind to and enter the endoplasmic reticulum

2 **nucleus** – contains DNA which codes for the different enzymes to be made; mRNA leaves the nucleus and binds to ribosomes initiating protein synthesis

1 amino acids pass from the blood into the cell through the cell surface membrane by active transport; they are carried to the ribosomes by tRNA

DNA → mRNA

ATP

mitochondrion – supplies energy in the form of ATP

cell surface membrane

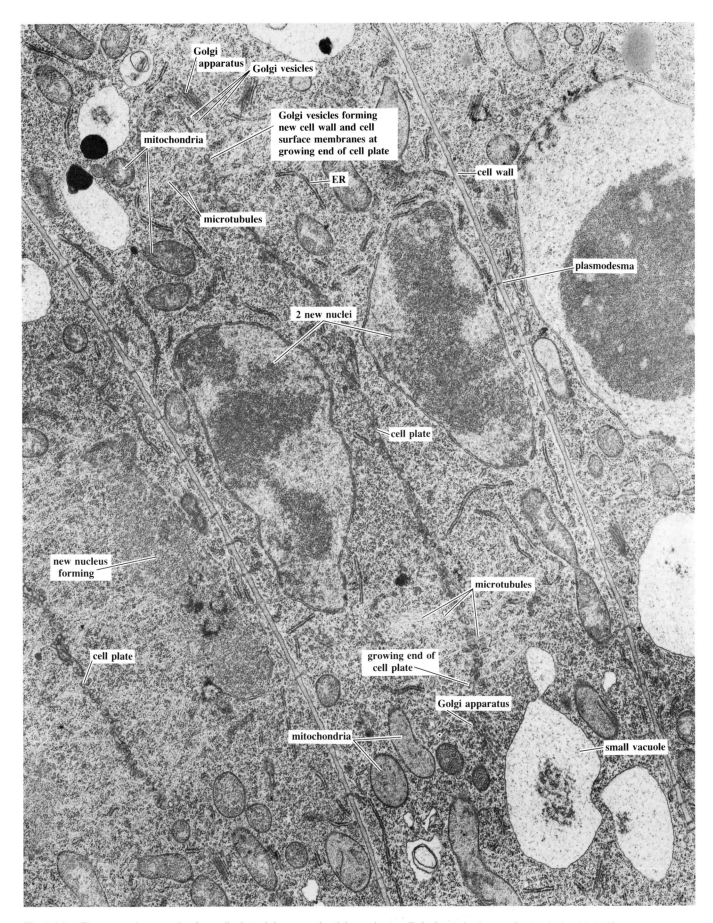

Fig 7.21 *Electron micrograph of a cell plate (phragmoplast) in a plant cell during telophase of mitosis (× 15 000)*

particularly important and prominent in secretory cells, a good example being provided by the acinar cells of the pancreas. These cells secrete the digestive enzymes of the pancreatic juice into the pancreatic duct, along which they pass to the duodenum. Fig 7.20a is an electron micrograph of such a cell, and fig 7.20b a diagrammatic representation of the secretion pathway.

Details of the pathway have been confirmed by using radioactively labelled amino acids and following their incorporation into protein and subsequent passage through different cell organelles. This can be done by homogenising samples of tissue at different times after supplying the amino acids, separating the cell organelles by centrifugation and finding which organelles contain the highest proportion of the amino acids. After concentration in the Golgi apparatus, the protein is carried in Golgi vesicles to the cell surface membrane. The final stage in the pathway is secretion of the inactive enzyme by reverse pinocytosis. The digestive enzymes secreted by the pancreas are synthesised in an inactive form so that they do not attack and destroy the cells that make them. An inactive enzyme is called a **proenzyme** or **zymogen**. An example is trypsinogen which is converted to active trypsin in the duodenum.

In general, proteins received by the Golgi apparatus from the ER have had short carbohydrate chains added to become glycoproteins (like the membrane proteins shown in fig 7.11). These carbohydrate 'antennae' can be remodelled in the Golgi apparatus, possibly to become markers that direct the proteins to their correct destinations. However, the exact details of how the Golgi apparatus sorts and directs molecules are unknown. The process of combining carbohydrates with proteins to form glycoproteins is called **glycosylation** and occurs during the production of many proteins.

The Golgi apparatus is also sometimes involved in the secretion of carbohydrates, an example being provided by the synthesis of new cell walls by plants. Fig 7.21 shows the intense activity which goes on at the 'cell plate', the region between two newly formed daughter nuclei where the new cell wall is laid down after nuclear division (mitosis or meiosis).

Golgi vesicles are steered into position at the cell plate by microtubules (described later) and fuse. Their membranes become the new cell surface membranes of the daughter cells, while their contents contribute to the middle lamella and new cell walls. Radioactively labelled glucose fed to dividing plant cells has been shown by autoradiography to appear in the Golgi apparatus and subsequently to be incorporated into cell wall polysaccharides within Golgi vesicles. These are polysaccharides of the cell wall matrix rather than cellulose, whose synthesis does not occur in Golgi vesicles.

Secretion by the pancreatic acinar cell and the formation of new plant cell walls are examples of the way in which many cell organelles can combine to perform one function.

An important glycoprotein secreted by the Golgi apparatus is **mucin**, which forms **mucus** in solution. It is secreted by goblet cells of the respiratory and intestinal epithelia. The root cap cells of plants contain Golgi apparatus which secretes a mucous polysaccharide, helping to lubricate the tip of the root as it penetrates the soil. The Golgi apparatus in leaf glands of the insectivorous plants *Drosera* (sundews) and *Pinguicula* (butterworts) secretes a sticky slime and enzymes which trap and digest insects. The slime, wax, gum and mucilage secretions of many cells are released by Golgi apparatus.

The Golgi apparatus is also sometimes involved in lipid transport. When digested, lipids are absorbed as fatty acids and glycerol in the small intestine. They are resynthesised to lipids in the smooth ER, coated in protein and then transported through the Golgi apparatus to the plasma membrane where they leave the cell, mainly to enter the lymphatic system.

A second important function of the Golgi apparatus, in addition to the secretion of proteins, glycoproteins, carbohydrates and lipids, is the formation of lysosomes, described below.

7.2.8 Lysosomes

Lysosomes (*lysis*, splitting; *soma*, body) are found in most eukaryotic cells, but are particularly abundant in animal cells exhibiting phagocytic activity. They are bounded by a single membrane and are simply sacs that contain hydrolytic (digestive) enzymes, such as proteases, nucleases, lipases and acid phosphatases. The contents of the lysosome are acidic and the enzymes have a low optimum pH. The enzymes have to be kept apart from the rest of the cell or they would destroy it. In animal cells, lysosomes are usually spherical and 0.2–0.5 μm in diameter. They have a characteristically homogeneous appearance in electron micrographs (fig 7.22).

In plant cells the large central vacuoles may act as lysosomes, although bodies similar to the lysosomes of animal cells are sometimes seen in the cytoplasm, particularly in dying cells. Most of the work on lysosomes has been done with animal cells.

The enzymes contained within lysosomes are synthesised on rough ER and transported to the Golgi apparatus. Golgi vesicles containing the processed enzymes later bud off and are called **primary lysosomes**. These have a number of functions, mostly involving digestive processes within the cell, but sometimes involving secretion of digestive enzymes. Their functions are summarised below and in fig 7.23.

Digestion of material taken in by endocytosis

The process of endocytosis is explained in section 7.2.2. Primary lysosomes may fuse with the vesicles or vacuoles formed by endocytosis to form **secondary lysosomes** in

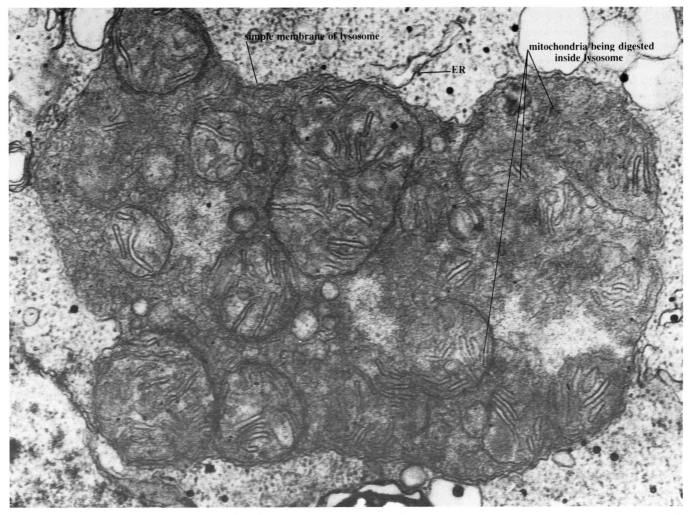

Fig 7.22 *Electron micrograph of a secondary lysosome (× 90 750)*

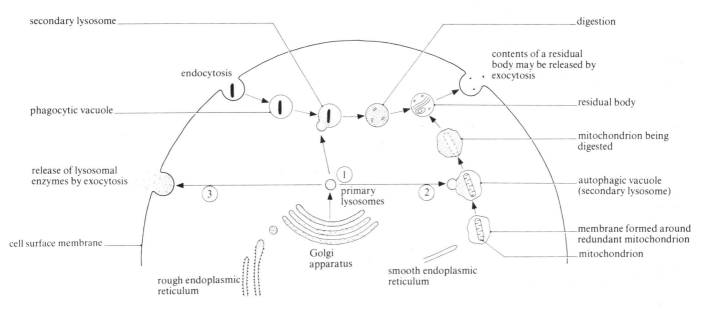

Fig 7.23 (above) *Three possible uses of a primary lysosome. The numbers 1, 2 and 3 refer to the order in which these pathways are discussed in the text*

which the material taken in by endocytosis is digested. This material might be taken in for food, as in some protozoans such as *Amoeba*, or for defensive purposes, as is the case when phagocytic white blood cells and macrophages ingest bacteria. The secondary lysosome may also be called a **food vacuole**. The products of digestion are absorbed and assimilated by the cytoplasm of the cell leaving undigested remains. The secondary lysosome is now termed a **residual body**. These usually migrate to the cell surface membrane and egest their contents (exocytosis). In certain cells, such as heart muscle and liver cells (hepatocytes), the residual bodies are stored.

An interesting example of the role of lysosomes occurs in the thyroid gland, where cells are stimulated by thyroid-stimulating hormone (TSH) to take up thyroglobulin by pinocytosis. The pinocytic vesicles so formed fuse with primary lysosomes and the thyroglobulin is partially hydrolysed to produce the active hormone thyroxine before the lysosome fuses with the cell surface membrane, thus secreting the hormone into the blood.

Autophagy

Autophagy is the process by which unwanted structures within the cell are removed. They are first enclosed by a single membrane, usually derived from smooth ER, and this structure then fuses with a primary lysosome to form a secondary lysosome, the **autophagic vacuole**, in which the unwanted material is digested. This is part of the normal turnover of cytoplasmic organelles, old ones being replaced by new ones. It becomes more frequent in cells undergoing reorganisation during differentiation.

Release of enzymes outside the cell (exocytosis)

Sometimes the enzymes of primary lysosomes are released from the cell. This occurs during the replacement of cartilage by bone during development. Similarly the matrix of bone may be broken down during the remodelling of bone that can occur in response to injury, new stresses and so on. In this case the enzymes are secreted from the lysosomes of cells known as **osteoclasts**.

Autolysis

Autolysis is the self-destruction of a cell by release of the contents of lysosomes within the cell. In such circumstances lysosomes have sometimes been aptly named 'suicide bags'. Autolysis is a normal event in some differentiation processes and may occur throughout a tissue, as when a tadpole tail is resorbed during metamorphosis. It also occurs after cells die. Sometimes it occurs as a result of certain lysosomal diseases or after cell damage.

7.2.9 Peroxisomes or microbodies

Peroxisomes or microbodies are common organelles of eukaryotic cells (fig 7.5). They are spherical, 0.3–1.5 μm in diameter (slightly smaller on average than mitochondria) and bounded by a single membrane. Their contents are finely granular, sometimes with a distinctive crystalline core which is a crystallised protein (enzyme) and they are derived from the ER, with which they often remain in close association.

Their most distinctive feature is the presence of the enzyme **catalase**, which catalyses the decomposition of hydrogen peroxide to water and oxygen (hence the name peroxisome). Hydrogen peroxide is a by-product of certain cell oxidations and is also very toxic, so must be eliminated immediately. Catalase is the fastest-acting enzyme known and its activity can be demonstrated by dropping a piece of fresh, preferably ground, liver into hydrogen peroxide, when rapid evolution of oxygen is observed. The cells of liver contain large numbers of peroxisomes (fig 7.5). Animal peroxisomes participate in a number of metabolic processes involving oxidation, but more details are known about plant peroxisomes. They can be divided into three types. **Glyoxysomes**, so called because they metabolise a compound called glyoxylate, are concerned with conversion of lipids to sucrose in lipid-rich seeds, such as in the endosperm of castor oil seeds (see question 21.7 and glyoxylate cycle, section 11.5). **Leaf peroxisomes** are important in the process of photorespiration in which they are intimately associated with chloroplasts and mitochondria, the three organelles often being found in close proximity as shown in fig 9.28. Hydrogen peroxide is produced during the photorespiratory pathway as shown in fig 9.28. A third group of **non-specialised peroxisomes** is found in other tissues.

7.2.10 The cytoskeleton

With the advent of electron microscopy it quickly became obvious that the cytoplasm of cells had a great deal more organisation than had previously been realised, and that extensive division of labour occurred between membrane-bound cell organelles and small organelles like ribosomes and centrioles. More recently structure has been revealed at an even finer level in the apparently structureless 'background' or ground substance of the cytoplasm. Complex networks of fibrous protein structures have been shown to exist in all eukaryotic cells. Collectively known as the **cytoskeleton**, these fibres are of at least three types: **microtubules**, **microfilaments** and **intermediate filaments**. They are concerned with movement, either by or within cells, and with the ability of cells to maintain their shapes.

Microtubules

Nearly all eukaryotic cells contain unbranched, hollow cylindrical organelles called **microtubules**. They are very fine tubes, having an external diameter of about 24 nm and with walls about 5 nm thick made up of helically arranged globular subunits of a protein called **tubulin**, as shown in fig 7.24. Their typical appearance in electron micrographs is shown in fig 7.21. They may extend for several micrometres in length. At intervals, cross-bridges (arms) sometimes project from their walls and these are probably involved in

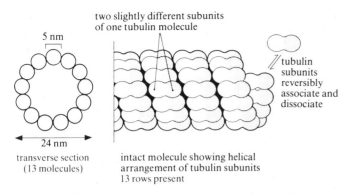

two slightly different subunits of one tubulin molecule

5 nm

24 nm

transverse section (13 molecules)

intact molecule showing helical arrangement of tubulin subunits 13 rows present

tubulin subunits reversibly associate and dissociate

Fig 7.24 (above) *Probable arrangement of tubulin subunits in a microtubule*

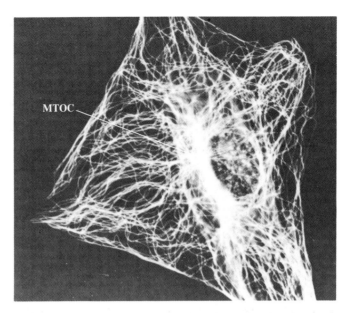

MTOC

Fig 7.25 *Immunofluorescence micrograph showing the distribution of microtubules in a cell (fibroblast) by reaction with a fluorescent antibody. The microtubules radiate from the microtubule-organising centre (MTOC) lying just outside the nucleus. The MTOC contains the centriole at its centre*

linking with adjacent microtubules, as occurs in cilia and flagella. Growth of microtubules occurs at one end by addition of tubulin subunits. It is inhibited by a number of chemicals, such as **colchicine**, which have been used to investigate the functions of microtubules. Growth apparently requires a template to start and certain very small ring-like structures that have been isolated from cells, and which consist of tubulin subunits, appear to serve this function. In intact animal cells, centrioles probably also serve this function and are therefore sometimes known as microtubule-organising centres, or MTOCs. Centrioles contain short microtubules as shown in fig 22.3.

The network of microtubules in cells has been strikingly revealed by a technique known as **immunofluorescence microscopy** in which fluorescent markers are tagged to antibody molecules that bind specifically to the protein whose distribution is being investigated. If an antibody to tubulin is used, a distribution like that shown in fig 7.25 is revealed by the light microscope. The microtubules radiate from the centrosphere around the centrioles. Satellite proteins around the centrioles act as MTOCs.

Microtubules are involved in a number of cell processes, some of which are listed below.

Centrioles, basal bodies, cilia and flagella. Centrioles are small hollow cylinders (about 0.3–0.5 μm long and about 0.2 μm in diameter) that occur in pairs in most animal and lower plant cells in a distinctly staining region of the cytoplasm known as the **centrosome** or **centrosphere**. Each contains nine triplets of microtubules as shown in fig 22.3. At the beginning of nuclear division the centrioles replicate themselves and the two new pairs migrate to opposite poles of the spindle, the structure on which the chromosomes become aligned (section 22.2). The spindle itself is made of microtubules ('spindle fibres') presumably synthesised using centrioles as MTOCs. The microtubules control separation of chromatids or chromosomes as described in chapter 22. Cells of higher plants lack centrioles, although they do produce spindles during nuclear division. The cells are thought to contain smaller MTOCs that are not easily visible even with the electron microscope. Another possible function of

centrioles as MTOCs is discussed in intracellular transport below.

Identical in structure to centrioles are **basal bodies**, formerly known as **kinetosomes** or **blepharoplasts**. They are always found at the base of cilia and flagella and probably originate from replication of centrioles. They also seem to act as MTOCs because cilia and flagella contain a characteristic '9 + 2' arrangement of microtubules (section 17.6 and figure 17.31).

In spindles, as well as in cilia and flagella, microtubules undergo sliding motions which in the former case move chromosomes or chromatids and in the latter are responsible for beating movements. Further details of these activities are given in chapters 17 and 22.

Intracellular transport. Microtubules have also been implicated in the movements of other cell organelles such as Golgi vesicles, an example being the guiding of Golgi vesicles to the cell plate shown in fig 7.21. There is a constant traffic in cells of Golgi vesicles and of vesicles from the ER to the Golgi apparatus, and time-lapse photography reveals regular movements of larger organelles, such as lysosomes and mitochondria, in many cells. Such movements may be both random and non-random and are believed to be typical of most cell organelles. They are suspended if the microtubule system is disrupted.

Cytoskeleton. Microtubules also have a passive architectural role in cells, their long, fairly rigid, tube-like structure acting in a skeletal fashion to form a '**cytoskeleton**'. They help to determine the shape of cells during development and to maintain the shape of differentiated cells, often being found in a zone just beneath the

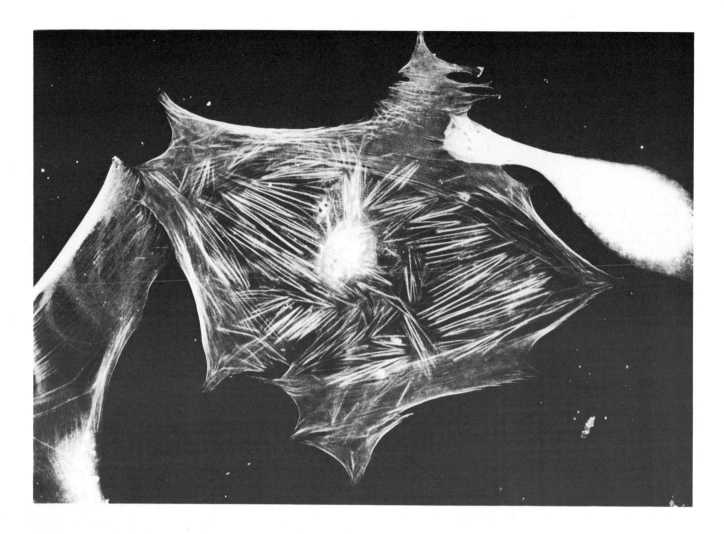

Fig 7.26 *Immunoflorescence micrograph showing the distribution of actin microfilaments in a cell (fibroblast) by reaction with a fluorescent antibody. Long stress fibres are revealed. Fluorescence staining of tropomyosin indicates that actin and tropomysin are found in the same fibre bundles.*

cell surface membrane. The axons of nerve cells, for example, contain longitudinally running bundles of microtubules (possibly involved in transport along the axons). Animal cells in which microtubules are disrupted revert to a spherical shape. In plant cells the alignment of microtubules exactly corresponds with the alignment of cellulose fibres during deposition of the cell wall, thus indirectly establishing cell shape.

Microfilaments

Microfilaments are very fine protein filaments about 7 nm in diameter. Recently it has been shown that they are abundant in all eukaryotic cells and consist of the protein **actin** which is found in muscle. In fact, 10–15% of the total protein of all cells so far examined has proved to be actin and immunofluorescence microscopy shows a cytoskeleton of actin similar to that of microtubules (fig 7.26).

Microfilaments often occur in sheets or bundles (stress fibres) just below the cell surface membrane and at the interface between stationary and moving cytoplasm where cytoplasmic streaming is taking place. They are probably involved in endocytosis and exocytosis. A much smaller proportion of myosin filaments is also found in cells, myosin being the other major protein of muscle. Interac-

tions between actin and myosin are the cause of muscle contraction, as described in section 17.4. This and other evidence suggests that microfilaments are involved in cell motility (whether of the whole cell or within the cell), although this is not controlled in exactly the same way as in muscle. Actin filaments may operate alone in some cases, and with myosin in others. A good example of the latter is in microvilli (section 7.2.11). Microfilaments are constantly being assembled and disassembled in cells that show motility (see, for example, amoeboid movement, section 17.6). A final example of the use of microfilaments is during cleavage of animal cells which is brought about by constriction of a ring of microfilaments after nuclear division.

Intermediate filaments

A third distinct type of filament of intermediate size (8–10 nm in diameter) also has certain cytoskeletal and motility roles.

7.2.11 Microvilli

One of the best understood cytoskeletal systems is that of microvilli. Microvilli are finger-like extensions of the cell surface membrane of some animal cells. They increase the surface area by as much as 25 times and are particularly numerous on cells specialised for absorption, such as on intestinal epithelium and kidney tubule epithelium. The fringe of the microvilli can just be seen with a light microscope and is called a **brush border**. More irregular and transitory extensions of the cell surface membrane also occur, as shown in figs 7.3 and 7.5, and these play roles in exocytosis and endocytosis.

Each microvillus contains bundles of about 40 cross-linked actin filaments which are associated with myosin filaments at the base of the microvillus in a region called the **terminal web**. The terminal web contains actin and intermediate filaments and extends across the whole cell just below the microvilli. The system as a whole ensures that the microvilli remain upright and retain their shape, while still probably allowing backward and forward movement through the interactions of actin and myosin (compare muscle contraction). The increase in surface area provided by the microvilli not only improves the efficiency of absorption, but also, in the gut, of digestion because certain digestive enzymes are associated with their surface (see section 10.4.9).

Plant cells lack microvilli because their rigid cell walls impose restrictions on extensions of the cell surface membrane. However, it is interesting to note the comparable increases in membrane surface area achieved by transfer cells for purposes of transport (fig 14.27 and section 14.8.4).

7.2.12 Mitochondria

Mitochondria are found in all aerobic eukaryotic cells and their structure and function are briefly summarised in figs 7.3–6. Their chief function is aerobic respiration and they are described in detail in section 11.5.

7.3 Structures characteristic of plant cells

As noted already, the cells of higher plants contain all the organelles found in animal cells with the exception of centrioles. They also possess extra structures which are the theme of this section.

7.3.1 Cell walls

Plant cells, like those of prokaryotes and fungi, are surrounded by a relatively rigid wall which is secreted by the living cell (the protoplast) within. Plant cell walls differ in chemical composition from those of the prokaryotes and the fungi, as reference to table 2.1 will show. However, they share some of the same functions, such as protection and support, and impose the same physical restraints on cell movement. The wall laid down during cell division of plants is called the **primary wall**. This may later be thickened to become a **secondary wall**. Formation of the primary wall is described in this section and a micrograph of an early stage in wall formation is shown in fig 7.21.

Structure of the cell wall

The primary wall consists of cellulose microfibrils running through a **matrix** of complex polysaccharides. Cellulose is a polysaccharide whose chemical structure is described in section 5.2.3. Of particular relevance to its role in cell walls is its fibrous nature and high tensile strength, which approaches that of steel. Individual molecules of cellulose are long chains cross-linked by hydrogen bonds to other molecules to form strong bundles called **microfibrils**. Microfibrils form the framework of the cell wall within the cell wall matrix. The matrix consists of polysaccharides which are usually divided for convenience into **pectins** and **hemicelluloses** according to their solubility in a number of solvents used in extraction procedures. **Pectins**, or **pectic substances**, are usually extracted first, having relatively high solubility. They are a mixed group of acidic polysaccharides (built up from the sugars arabinose and galactose, the sugar acid galacturonic acid, and methanol). They form long branching or straight molecules. The **middle lamella** that holds neighbouring cell walls together is composed of sticky, gel-like magnesium and calcium pectates. In the walls of ripening fruit certain insoluble pectic substances are converted back to soluble pectins. These form gels when sugar is added and are therefore used as commercial gelling agents.

Hemicelluloses are a mixed group of alkali-soluble polysaccharides (including polymers of the sugars xylose, galactose, mannose, glucose and glucomannose). Like cellulose they form chain-like molecules, but the chains are less organised, shorter and more branched. Cell walls are hydrated and 60–70% of their mass is usually water. Water can move freely through free space in the cell wall and also contributes to the chemical and physical properties of the cell wall polysaccharides.

Mechanically strong materials, like cell walls, in which more than one component is present are known as **composite materials** and they are generally stronger than any of their components in isolation. Fibre–matrix systems are used widely in engineering and a study of their properties is an important branch of both modern engineering and biology. The matrix transfers stress to the fibres, which have a high tensile strength. The matrix also improves resistance to compression and shear, spreads out the fibres and protects them from abrasion and possible chemical attack. An example of a matrix traditionally used

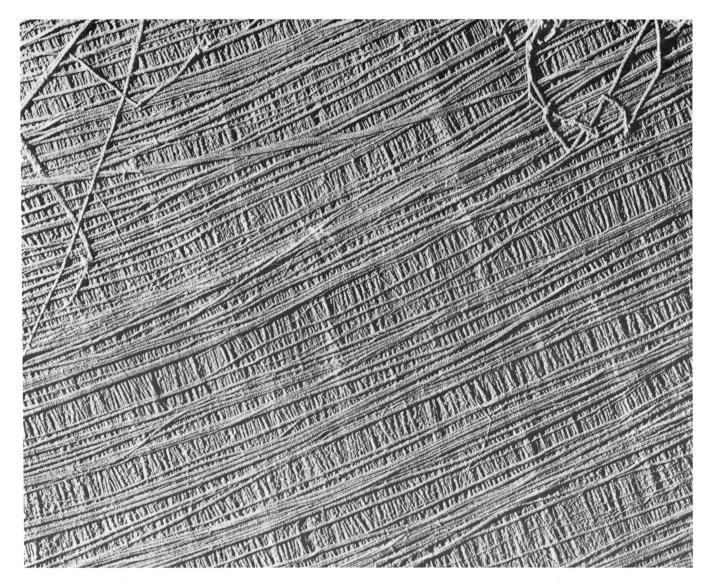

in engineering is concrete which may be reinforced in various ways, such as by steel rods. A more modern and lighter structural material is a glass or carbon fibre reinforced plastic in which the plastic acts as the matrix. Wood is a composite whose strength is due to its cell walls. Other rigid biological composites include bone, cartilage and arthropod cuticles. Pliant composites also exist such as connective tissue and skin.

In some cells, such as leaf mesophyll cells, the primary wall remains the only wall. In most, however, extra layers of cellulose are laid down on the inside surface of the primary wall (the outside surface of the cell surface membrane), thus building up a secondary wall. This usually occurs after the cell has reached a maximum size; but a few cells, such as those of collenchyma, continue to grow during this phase. Secondary thickening of plant cell walls should not be confused with secondary thickening (secondary growth) of the whole plant, which is an increase in girth resulting from the addition of new cells.

The cellulose fibres of a given layer of secondary thickening are usually orientated at the same angle, but

Fig 7.27 *Electron micrograph of layers from the wall of the green seaweed* Chaetomorpha melagonium *showing cellulose microfibrils some 20 nm wide; the contrast is due to shadowing with a platinum/gold alloy*

different layers are orientated at different angles, forming an even stronger cross-ply structure. This is shown in fig 7.27.

Some cells such as xylem vessel elements and sclerenchyma undergo extensive **lignification** whereby lignin, a complex polymer (not a polysaccharide), is deposited in all the cellulose layers (often a primary layer and three secondary layers). In some cells, such as protoxylem, the lignin is laid down in annular, spiral or reticulate patterns as shown in fig 8.11. In others it is complete, apart from pits which represent areas of the primary wall originally occupied by groups of plasmodesmata and known as pit fields (section 8.1.3 and fig 8.7). Lignin cements and anchors cellulose fibres together. It acts as a very hard and rigid matrix giving the cell wall extra tensile and particularly compressional strength (prevents buckling). It further protects the cells from physical and chemical damage. Together with the cellulose, which remains in the wall, it is responsible for the unique characteristics of wood as a construction material.

Functions of the cell wall

The main functions of plant cell walls are summarised below.

(1) Mechanical strength and skeletal support is provided for individual cells and for the plant as a whole. Extensive lignification increases strength in some walls (small amounts are present in most walls).

(2) Cell walls are fairly rigid and resistant to expansion and therefore allow development of turgidity when water enters the cell by osmosis. This contributes to the support of all plants and is the main source of support in herbaceous plants and organs such as leaves which do not undergo secondary growth. The cell wall prevents the cell from bursting when exposed to a dilute solution.

(3) Orientation of cellulose microfibrils limits and helps to control cell growth and shape because the cell's ability to stretch is determined by their arrangement. If, for example, cellulose microfibrils form hoops in a transverse direction around the cell, the cell will stretch, as it fills with water by osmosis, in a longitudinal direction.

(4) The system of interconnected cell walls (the **apoplast**) is a major pathway of movement for water and dissolved mineral salts. The walls are held together by middle lamellae. The cell walls also possess minute pores through which structures called **plasmodesmata** can pass, forming living connections between cells, and allowing all the protoplasts to be linked in a system called the **symplast**.

(5) Cell walls develop a coating of waxy cutin, the cuticle, on exposed epidermal surfaces reducing water loss and risk of infection. Cork cell walls undergo impregnation with suberin which serves a similar function after secondary growth.

(6) The walls of xylem vessels, tracheids and sieve tubes (with their sieve plates) are adapted for long-distance translocation of materials through the cells, as explained in chapters 8 and 14.

(7) The cell walls of root endodermal cells are impregnated with suberin forming a barrier to water movement (section 14.1.7).

(8) Some cell walls are modified as food reserves, as in storage of hemicelluloses in some seeds.

(9) The cell walls of transfer cells develop an increased surface area and the consequent increase in surface area of the cell surface membrane increases the efficiency of transfer by active transport (section 14.8.4).

7.3.2 Plasmodesmata

Plasmodesmata are living connections that pass between neighbouring plant cells through very fine pores in adjacent walls. Fig 7.6 summarises what little is known about their structure and function. They are sometimes found in groups known as primary pit fields as described in section 8.1.3. Sieve plate pores of phloem sieve tubes are derived from plasmodesmata.

7.3.3 Vacuoles

A vacuole is a fluid-filled sac bounded by a single membrane. Animal cells contain relatively small vacuoles, such as phagocytic vacuoles, food vacuoles, autophagic vacuoles and contractile vacuoles. However, plant cells, notably mature parenchyma and collenchyma cells, have a large central vacuole surrounded by a membrane called the **tonoplast** (fig 7.4). The fluid they contain is called **cell sap**. It is a concentrated solution of mineral salts, sugars, organic acids, oxygen, carbon dioxide, pigments and some waste and 'secondary' products of metabolism. The functions of vacuoles are summarised below.

(1) Water generally enters the concentrated cell sap by osmosis through the differentially permeable tonoplast. As a result a pressure potential builds up within the cell and the cytoplasm is pushed against the cell wall. Osmotic uptake of water is important in cell expansion during cell growth, as well as in the normal water relations of plants.

(2) The vacuole sometimes contains pigments in solution called **anthocyans**. These include **anthocyanins**, which are red, blue and purple, and other related compounds which are shades of yellow and ivory. They are largely responsible for the colours in flowers (for example in roses, violets and *Dahlia*), fruits, buds and leaves. In the latter case they contribute to autumn shades, together with the photosynthetic pigments of chloroplasts. They are important in attracting insects, birds and other animals for pollination and seed dispersal.

(3) Plant vacuoles sometimes contain hydrolytic enzymes and act as lysosomes during life. After cell death the tonoplast, like all membranes, loses its partial permeability and the enzymes escape causing autolysis.

(4) Waste products and certain secondary products of plant metabolism may accumulate in vacuoles. For example, crystals of waste calcium oxalate are sometimes observed. The role of secondary products is not always clear, as in the case of alkaloids which may be stored in vacuoles. They, like tannins (which are astringent to the taste), may offer protection from consumption by herbivores. Tannins are particularly common in vacuoles (as well as in the cytoplasm and cell walls) of leaves, wood, bark, unripe fruits and seed coats. Latex may accumulate in vacuoles, usually in a milky emulsion, as in dandelion stems. Certain cells, known as **laticifers**, are specialised for this function. The latex of the rubber tree *Helvea brasiliensis* contains the enzymes and intermediates needed for rubber synthesis, and the latex of the opium poppy contains alkaloids.

(5) Some of the dissolved substances act as food reserves, which can be utilised by the cytoplasm when necessary, for example sucrose, mineral salts and inulin.

7.3.4 Plastids

Plastids are organelles found only in plant cells and in higher plants develop from small bodies called **proplastids** found in meristematic regions. They are surrounded by two membranes (the envelope). Proplastids can develop into several types of plastid depending on where they are found in the plant. There are various ways of classifying the different types and a simple scheme is given below.

Chloroplasts. These are plastids that contain chlorophyll and carotenoid pigments and carry out photosynthesis. They are found mainly in leaves and are described in section 9.3.1 in the context of photosynthesis.

Chromoplasts. These are non-photosynthetic coloured plastids containing mainly red, orange or yellow pigments (carotenoids). They are particularly associated with fruits (such as the tomato and red pepper) and flowers in which their bright colours serve to attract insects, birds and other animals for pollination and seed dispersal. The orange pigment of carrot roots is also contained in chromoplasts.

Leucoplasts. These are colourless plastids lacking pigments. They are usually modified for food storage, and are particularly abundant in storage organs such as roots, seeds and young leaves. They are further classified according to food stored; for example **amyloplasts** store starch (see fig 15.16), **lipidoplasts** (elaioplasts or oleoplasts) store lipids either as oil or fat, as in oily nuts and sunflower seeds, and **proteoplasts** store protein, as in some seeds.

Chapter Eight

Histology

All multicellular organisms possess groups of cells of similar structure and function assembled together to form tissues. The study of tissues is called **histology**. A **tissue** can be defined as a group of physically linked cells and associated intercellular substances that is specialised for a particular function or functions. This specialisation, whilst leading to an increased efficiency of action by the organism as a whole, means that the collective activity of different tissues must be coordinated and integrated if the organism is to be viable.

Different tissues are often grouped together into larger functional units called **organs**. Internal organs are more obvious in animals than in plants, where they do not exist as such except perhaps for the vascular bundles. In animals, organs form the parts of the even larger functional units known as **systems**, for example the digestive system (pancreas, liver, stomach, duodenum and so on) and the vascular system (heart and blood vessels).

The cells of a tissue may be all of one type, for example parenchyma, collenchyma and cork in plants and squamous epithelium in animals. Alternatively, the tissues may contain a mixture of different cell types, as in xylem and phloem in plants and areolar connective tissue in animals. Generally the cells of a tissue share a common embryological origin.

The study of tissue structure and function relies heavily on light microscopy and the associated techniques of preserving, staining and sectioning material. These techniques are described in section A2.4.

In this chapter the histology of flowering plants is studied at the level of detail which can be seen with the light microscope. In some cases, though, reference is made to structure as revealed by the scanning electron microscope in order to provide greater clarification. In relating structure to function in tissues it is important to bear in mind the three-dimensional structures of the cells and their relationship to one another. This kind of information is usually 'pieced together' by examining material in thin section, most commonly in transverse section (TS) and longitudinal section (LS). Neither type of section alone can give all the information required, but a combination of the two can often reveal the necessary information. Some cells, such as xylem vessels and tracheids in plants, can easily be examined whole by macerating the tissues. This involves the breakdown of soft tissues leaving behind the harder, lignified xylem vessels, tracheids and fibres.*

Plant tissues can be divided into two groups, either consisting of one type of cell or of more than one type. Animal tissues are divided into four groups: epithelial, connective, muscle and nervous tissue. Table 8.1 shows the characteristic features, functions and distribution of plant tissues.

8.1 Simple plant tissues – tissues consisting of one type of cell

8.1.1 Parenchyma

Structure

The structure of parenchyma is shown in fig 8.1. The cells are usually roughly spherical (isodiametric) though they may be elongated.

Functions and distribution

The cells are unspecialised and act as **packing tissue** between more specialised tissues, as in the pith, cortex and medullary rays. They form a large part of the bulk of various organs, such as the stem and root, and they also occur among the xylem vessels (xylem parenchyma) and phloem cells (phloem parenchyma).

The osmotic properties of parenchyma cells are important because when turgid they become tightly packed and provide support for the organs in which they are found. This is particularly important in the stems of herbaceous plants, where they form the main means of support. During periods of water shortage the cells of such plants lose water and this results in the plants wilting.

Although structurally unspecialised, the cells are **metabolically active** and are the sites of many of the vital activities of the plant body.

A system of air spaces runs between the cells through which gaseous exchange can take place between living cells and the external environment through stomata or lenticels. Oxygen for respiration and carbon dioxide for photosynthesis can thus diffuse through the spaces, such as in the spongy mesophyll layer of the leaf.

---

* The structure of some plant tissues is dealt with elsewhere in this book. More detailed structure of phloem is given in chapter 14 where its structure is related to its function in translocation. Development of plant tissues from meristematic cells is discussed in chapter 21, together with secondary growth and the structure of wood (secondary xylem) and cork.

Table 8.1 Characteristic features, functions and distribution of plant tissues.*

Tissue	Living or dead	Wall material	Cell shape	Main functions	Distribution
Parenchyma	Living	Cellulose, pectins and hemicelluloses	Usually isodiametric, sometimes elongated	Packing tissue. Support in herbaceous plants. Metabolically active. Intercellular air spaces allow gaseous exchange. Food storage. Transport of materials through cells or cell walls.	Cortex, pith, medullary rays and packing tissue in xylem and phloem
Modified parenchyma (a) epidermis	Living	Cellulose, pectins and hemicelluloses, and covering of cutin	Elongated and flattened	Protection from desiccation and infection. Hairs and glands may have additional functions.	Single layer of cells covering entire primary plant body
(b) mesophyll	Living	Cellulose, pectins and hemicelluloses	Isodiametric, irregular or column-shaped depending on location	Photosynthesis (contains chloroplasts). Storage of starch.	Between the upper and lower epidermis of leaves
(c) endodermis	Living	Cellulose, pectins and hemicelluloses, and deposits of suberin	As epidermis	Selective barrier to movement of water and mineral salts (between cortex and xylem) in roots. Starch sheath with possible role in geotropic response in stems.	Around vascular tissue (innermost layer of cortex)
(d) pericycle	Living	Cellulose, pectins and hemicelluloses	As parenchyma	In roots it retains meristematic activity producing lateral roots and contributing to secondary growth if this occurs.	In roots between central vascular tissue and endodermis

NB The pericycle in the stem is made of sclerenchyma and has a different origin.

Tissue	Living or dead	Wall material	Cell shape	Main functions	Distribution
Collenchyma	Living	Cellulose, pectins and hemicelluloses	Elongated and polygonal with tapering ends	Support (a mechanical function)	Outer regions of cortex, e.g. angles of stems, midrib of leaves
Sclerenchyma (a) fibres	Dead	Mainly lignin. Cellulose, pectins and hemicelluloses also present.	Elongated and polygonal with tapering interlocking ends	Support (purely mechanical)	Outer regions of cortex, pericycle of stems, xylem and phloem
(b) sclereids	Dead	As fibres	Roughly isodiametric, though variations occur	Support or mechanical protection	Cortex, pith, phloem, shells and stones of fruits, seed coats

Xylem — Mixture of living and dead cells. Xylem also contains fibres and parenchyma which are as previously described.

Tissue	Living or dead	Wall material	Cell shape	Main functions	Distribution
tracheids and vessels	Dead	Mainly lignin. Cellulose, pectins and hemicelluloses also present.	Elongated and tubular	Translocation of water and mineral salts. Support.	Vascular system

Phloem — Mixture of living and dead cells. Phloem also contains fibres and sclereids which are as previously described.

Tissue	Living or dead	Wall material	Cell shape	Main functions	Distribution
(a) sieve tubes	Living	Cellulose, pectins and hemicelluloses	Elongated and tubular	Translocation of organic solutes (food)	Vascular system
(b) companion cells	Living	Cellulose, pectins and hemicelluloses	Elongated and narrow	Work in association with sieve tubes	Vascular system

* Tissues associated with secondary growth, such as wood and cork, are described in chapter 21.

(a)

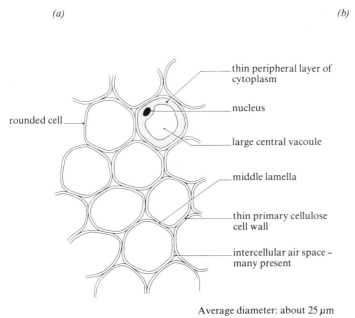

rounded cell —

thin peripheral layer of cytoplasm

nucleus

large central vacoule

middle lamella

thin primary cellulose cell wall

intercellular air space – many present

Average diameter: about 25 μm

Fig 8.1 *Structure of parenchyma cells. (a) TS, cells are usually roughly isodiametric (spherical), though may be elongated. (b) TS* Helianthus *stem pith*

(b)

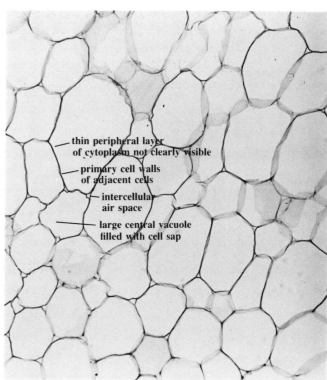

thin peripheral layer of cytoplasm not clearly visible

primary cell walls of adjacent cells

intercellular air space

large central vacuole filled with cell sap

Parenchyma cells are often sites for food storage, most notably in storage organs, such as potato tubers where the parenchyma cortex stores starch in **amyloplasts**. Food is also stored in medullary rays. Rare examples occur of parenchyma cells storing food in thickened cell walls, for example the hemicelluloses of date seed endosperm.

The walls of parenchyma cells are important pathways of water and mineral salt transport through the plant (part of the 'apoplast pathway' described in chapter 14). Substances may also move through cytoplasmic routes between neighbouring cells.

Parenchyma cells may become modified and more specialised in certain parts of the plant. Some examples of tissues that can be regarded as modified parenchyma are discussed below.

Epidermis. This is the layer, one cell thick, that covers the whole of the primary plant body. Its basic function is to protect the plant from desiccation and infection. During secondary growth it may be ruptured and replaced by a cork layer as described in section 21.6.6. The structure of typical epidermal cells is shown in fig 8.2.

The epidermal cells secrete a waxy substance called **cutin** which forms a layer of variable thickness called the **cuticle** within and on the outer surface of the cell walls. This helps to reduce water loss by evaporation from the plant surface as well as helping to prevent the entry of pathogens.

If the surfaces of leaves are examined in a light microscope it can be seen that the epidermal cells of dicotyledonous leaves are irregularly arranged and often have wavy margins, while those of monocotyledons tend to be more regular and rectangular in shape. At intervals,

specialised epidermal cells called **guard cells** occur in pairs side by side, with a pore between them called a **stoma**. These features are shown in figs 8.2b and c. Guard cells have a distinctive shape and are the only epidermal cells that contain chloroplasts, the rest being colourless. The size of the stoma is adjusted by the turgidity of the guard cells as described in chapter 14. The stomata allow gaseous exchange to occur during photosynthesis and respiration and are most numerous in the leaf epidermis, though they are also found in the stem. Water vapour also escapes through the stomata, and this is part of the process called transpiration.

Sometimes epidermal cells grow hair-like extensions which may be unicellular or multicellular and serve a wide variety of functions. In roots, unicellular hairs grow from a region just behind the root tip and increase the surface area for absorption of water and mineral salts (fig 14.16). In climbing plants, such as goosegrass (*Galium aparine*), hooked hairs often occur and function to prevent the stems from slipping from their supports.

More often epidermal hairs are an additional protective feature. They may assist the cuticle in reducing water loss by trapping a layer of moist air next to the plant, as well as reflecting radiation. Some hairs are water absorbing, notably on xerophytic plants. Others may have a mechanical protective function as with short, stiff bristles. The hairs of the stinging nettle (*Urtica dioica*) are hard with a bulbous tip and as they knock against an animal's body their fragile tip breaks off and the jagged end pierces the skin. The cell contents at their bases enter the wound, acting as an irritant poison. Hairs may form barriers around the nectaries of flowers preventing access to crawling

217

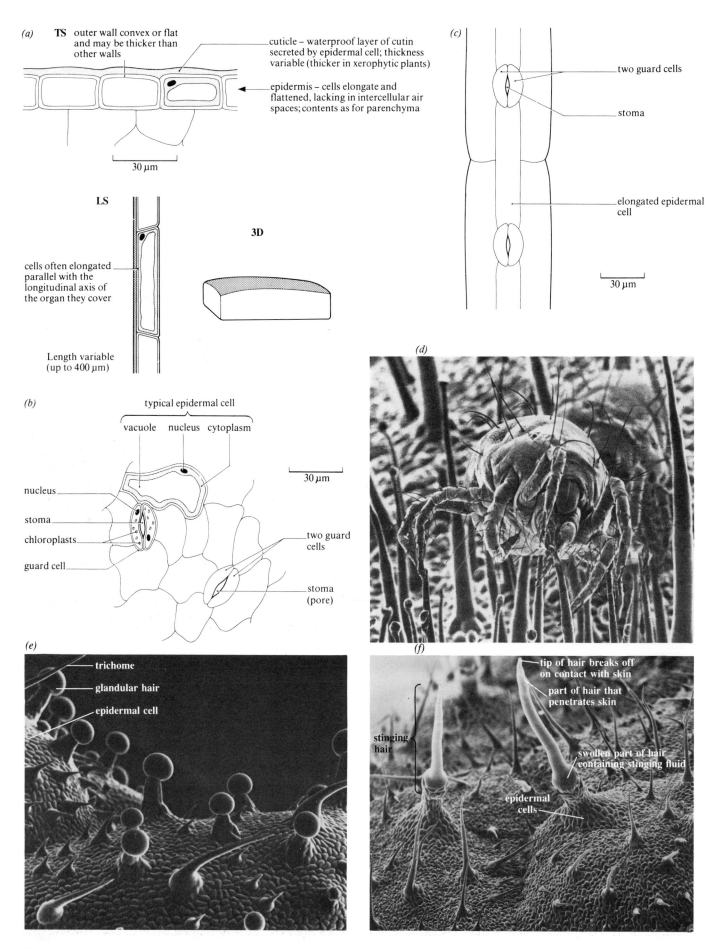

(a) **TS** outer wall convex or flat and may be thicker than other walls

cuticle – waterproof layer of cutin secreted by epidermal cell; thickness variable (thicker in xerophytic plants)

epidermis – cells elongate and flattened, lacking in intercellular air spaces; contents as for parenchyma

30 μm

LS

cells often elongated parallel with the longitudinal axis of the organ they cover

3D

Length variable (up to 400 μm)

(c)

two guard cells

stoma

elongated epidermal cell

30 μm

(b)

typical epidermal cell

vacuole   nucleus   cytoplasm

30 μm

nucleus

stoma

chloroplasts

guard cell

two guard cells

stoma (pore)

(d)

(e)

trichome

glandular hair

epidermal cell

(f)

tip of hair breaks off on contact with skin

part of hair that penetrates skin

stinging hair

swollen part of hair containing stinging fluid

epidermal cells

218

Fig. 8.2 (opposite) *Structure of epidermal cells. (a) Epidermal cells seen in TS, LS and three-dimensions. (b) Surface view of dicotyledon leaf epidermis (for TS stoma see fig 14.16). (c) Surface view of monocotyledon leaf epidermis. (d) Spider mite trapped and killed by the hair glands of a potato leaf. An enzyme capable of digesting animal matter has been found in one type of glandular hair in the potato, so the potato could be regarded as a carnivorous plant. Many other plants not normally thought of as carnivorous may have similar abilities. (e) Young leaf of* Cannabis sativa *with adaxial glands and trichomes. (f) Leaf surface of* Urtica dioica *(stinging nettle)*

(a)

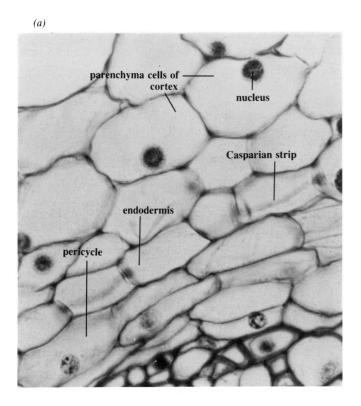

(b)

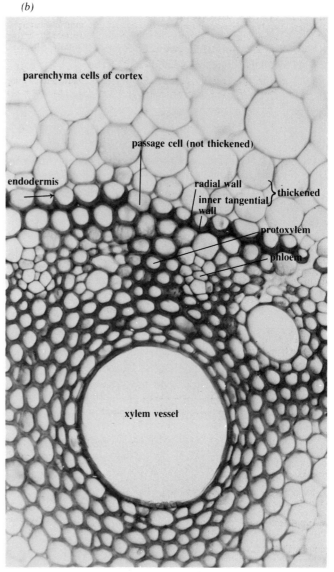

Fig 8.3 *Structure of root endodermis. (left) (a) TS young endodermis with Casparian band (HP). (above) (b) TS old dicotyledonous root showing endodermis*

insects and helping to promote cross-pollination by larger flying insects (see white dead-nettle, fig 20.17).

Glandular cells are also a common feature of the epidermis and these may be hair-like. They may secrete a sticky substance that traps and kills insects, either for protection or, if the exudate contains enzymes, for digestion and subsequent absorption of food. Such plants may be regarded as carnivorous (fig 8.2d). Glandular hairs are sometimes responsible for the scents given off by plants, such as on the leaves of lavender (*Lavendula*).

Mesophyll (see also figs 9.3 and 9.4). This is the packing tissue found between the two epidermal layers of leaves and consists of parenchyma modified to carry out photosynthesis. Photosynthetic parenchyma is sometimes called **chlorenchyma**. The cytoplasm of such cells contains numerous chloroplasts where the reactions of photosynthesis occur. In dicotyledons there are two distinct layers of

mesophyll, an upper layer consisting of column-shaped cells forming the **palisade mesophyll**, and a lower layer of more irregularly shaped cells, containing fewer chloroplasts, called **spongy mesophyll**. Most photosynthesis is carried out in the palisade mesophyll, while large intercellular air spaces between spongy mesophyll cells allow efficient gaseous exchange.

Endodermis (see also fig 14.17). This is the layer of cells surrounding the vascular tissue of plants and can be regarded as the innermost layer of the cortex. The cortex is usually parenchymatous, but the endodermis may be modified in various ways, both physiologically and structurally. It is more conspicuous in roots, where it is one cell thick, than in stems because in roots each cell develops a **Casparian strip**, a band of **suberin** (a fatty substance) that runs round the cell (fig 8.3). At a later stage further thickenings of the wall may take place. The structure and

219

function of root endodermis are shown in fig 14.17.

In the stems of dicotyledons the vascular bundles form a ring and the endodermis is the layer, one to several cells thick, immediately outside of this ring (fig 14.15). In this situation the endodermis often appears no different from the rest of the cortex, but may store starch grains and form a **starch sheath** which becomes visible when stained with iodine solution. These starch grains may sediment inside the cells in response to gravity, making the endodermis important in the geotropic response in the same way as root cap cells (section 15.2.2). In monocotyledonous stems the vascular bundles are scattered and no endodermis can be distinguished around them.

Pericycle. Roots possess a layer of parenchyma, one to several cells thick, called the **pericycle**, between the central vascular tissue and the endodermis (fig 14.17). It retains its meristematic capacity and produces lateral roots. It also contributes to secondary growth if this occurs. In stems there is usually no equivalent layer.

Companion cells. These are specialised parenchyma cells found adjacent to sieve tubes and vital for the functioning of the latter. They are very active metabolically and have a denser cytoplasm with smaller vacuoles than normal parenchyma cells. Their origin, structure and function are described later in this chapter (section 8.2.2).

8.1.2   Collenchyma

Collenchyma consists, like parenchyma, of living cells but is modified to give support and mechanical strength.

Structure

The structure of collenchyma is shown in fig 8.4. It shows many of the features of parenchyma but is characterised by the deposition of extra cellulose at the corners of the cells. The deposition occurs after the formation of the primary cell wall. The cells also elongate parallel to the longitudinal axis of the organ in which they are found.

Function and distribution

Collenchyma is a mechanical tissue, providing support for those organs in which it is found. It is particularly important in young plants, herbaceous plants and in organs such as leaves where secondary growth does not occur. In these situations it is an important strengthening tissue supplementing the effects of turgid parenchyma. It is the first of the strengthening tissues to develop in the primary plant body and, because it is living, can grow and stretch without imposing limitations on the growth of other cells around it.

In stems and petioles its value in support is increased by its location towards the periphery of the organ. It is often found just below the epidermis in the outer region of the cortex and gradually merges into parenchyma towards the inside, thus forming a hollow cylinder in three dimensions. Alternatively strengthening ridges may be formed, as along the fleshy petioles of celery (*Apium graveolus*) and the angular stems of plants such as dead-nettle (*Lamium*). In dicotyledonous leaves it appears as solid masses running the length of the midrib, providing support for the vascular bundles.

8.1.3   Sclerenchyma

The sole function of **sclerenchyma** is to assist in providing support and mechanical strength for the plant. Its distribution within the plant is related to the stresses to which different organs are subjected. Unlike collenchyma, the mature cells are dead and incapable of elongation so they do not mature until elongation of the living cells around them is complete.

Structure

There are two types of sclerenchyma cell, namely **fibres**, which are elongated cells, and **sclereids** or **stone cells**, which are usually roughly spherical, although both may vary considerably in size and shape. Their structures are shown in figs 8.5 and 8.6 respectively. In both cases the primary cell wall is heavily thickened with deposits of **lignin**, a hard substance with great tensile and compressional strength. A high tensile strength means that it does not break easily on stretching, and a high compressional strength means that it does not buckle easily.

Deposition of lignin takes place in and on the primary cellulose cell wall, and as the walls thicken, the living contents of the cells are lost with the result that the mature cells are dead. In both fibres and sclereids structures called **simple pits** appear in the walls as they thicken. These represent areas where lignin is not deposited on the primary wall owing to the presence of groups of **plasmodesmata** (strands of cytoplasm that connect neighbouring cells through minute pores in the adjacent cell walls). Each group of plasmodesmata forms one pit. The pits are described as simple because they are tubes of constant width. Their development is best explained diagrammatically as shown in fig 8.7.

Function and distribution of fibres

Individual sclerenchyma fibres are strong owing to their lignified walls. Collectively their strength is enhanced by their arrangement into strands or sheets of tissue that extend for considerable distances in a longitudinal direction. In addition, the ends of the cells interlock with one another, increasing their combined strength.

Fibres are found in the pericycle of stems, forming a solid

Fig 8.4 *Structure of collenchyma cells. (a) TS, cells are polygonal in outline. (b) LS, cells are elongated (up to 1 mm in length). (c) TS collenchyma from Helianthus stem. (d) LS collenchyma from Helianthus stem*

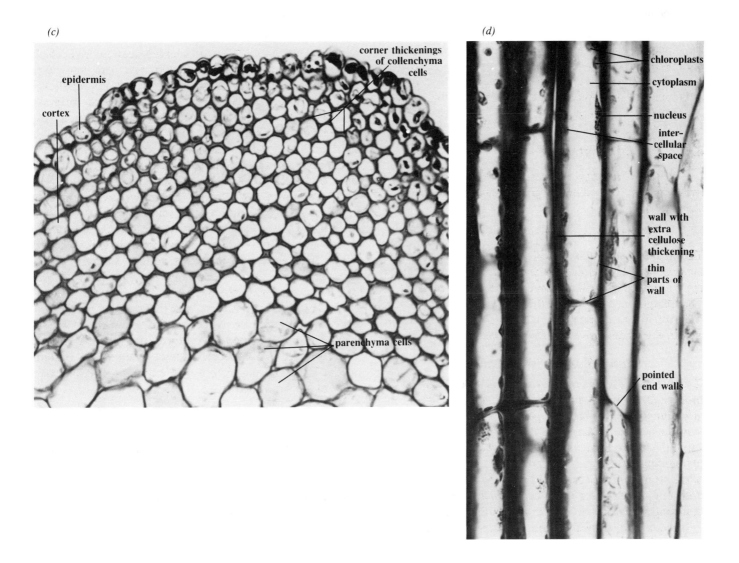

(a) **TS** Cells are polygonal in outline

- vacuole
- nucleus
- cytoplasm
} normal living cell contents

intercellular air spaces extremely small or non-existent

middle lamella

20 µm

thinner side walls

thickenings of extra cellulose at each corner

(b) **LS** Cells are elongated (up to 1 mm in length)

end walls often pointed

thick part of wall – thickenings run length of cell

thin part of wall

(c)

epidermis

cortex

corner thickenings of collenchyma cells

parenchyma cells

(d)

chloroplasts

cytoplasm

nucleus

inter-cellular space

wall with extra cellulose thickening

thin parts of wall

pointed end walls

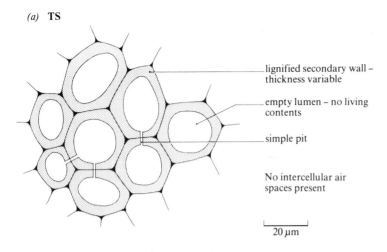

(a) **TS**

lignified secondary wall –
thickness variable

empty lumen – no living
contents

simple pit

No intercellular air
spaces present

20 μm

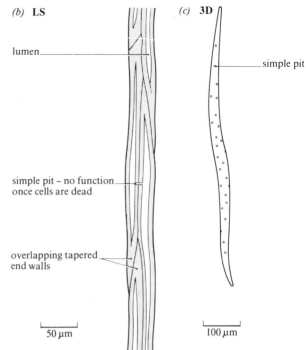

(b) **LS**

lumen

simple pit – no function
once cells are dead

overlapping tapered
end walls

50 μm

(c) **3D**

simple pit

100 μm

Fig 8.5 *Structure of sclerenchyma cells. (a) TS, cells are
polygonal in outline. (b) LS, cells are elongated (length very
variable, commonly > 1 mm, up to 250 mm reported).
(c) Three-dimensional appearance. (d) TS sclerenchyma from
Helianthus stem. (e) LS sclerenchyma from Helianthus stem*

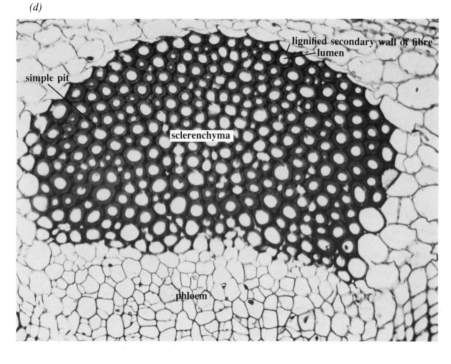

(d)

lignified secondary wall of fibre
lumen

simple pit

sclerenchyma

phloem

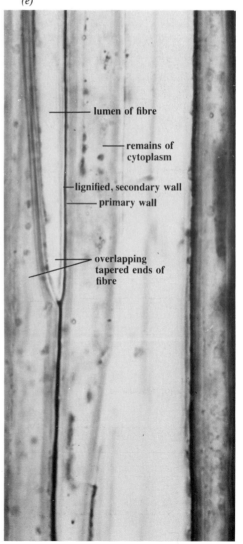

(e)

lumen of fibre

remains of
cytoplasm

lignified, secondary wall
primary wall

overlapping
tapered ends of
fibre

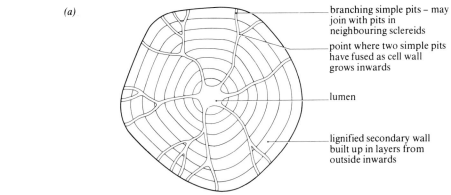

(a)

branching simple pits – may join with pits in neighbouring sclereids

point where two simple pits have fused as cell wall grows inwards

lumen

lignified secondary wall built up in layers from outside inwards

(b)

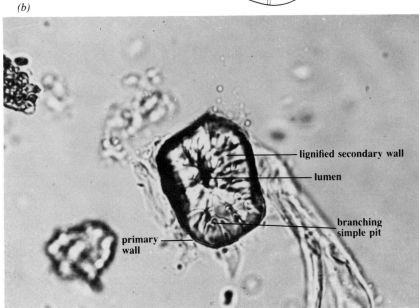

lignified secondary wall

lumen

branching simple pit

primary wall

Fig 8.6 *Structure of sclerenchyma sclereids. (a) TS or LS, cells are isodiametric. (b) Entire sclereid from macerated flesh of pear fruit (× 400)*

Fig 8.7 (below) *Development of simple pits in sclerenchyma fibres and sclereids*

TS before lignification

close group of plasmodesmata ('primary pit field') – primary cell walls are thinner in this area as a result of slower growth

adjacent primary cell walls of two cells

middle lamella

plasmodesma

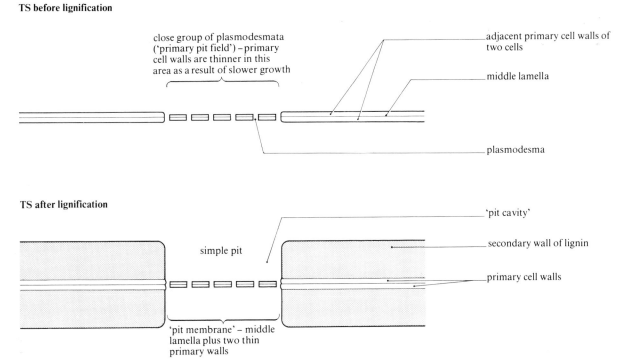

TS after lignification

simple pit

'pit cavity'

secondary wall of lignin

primary cell walls

'pit membrane' – middle lamella plus two thin primary walls

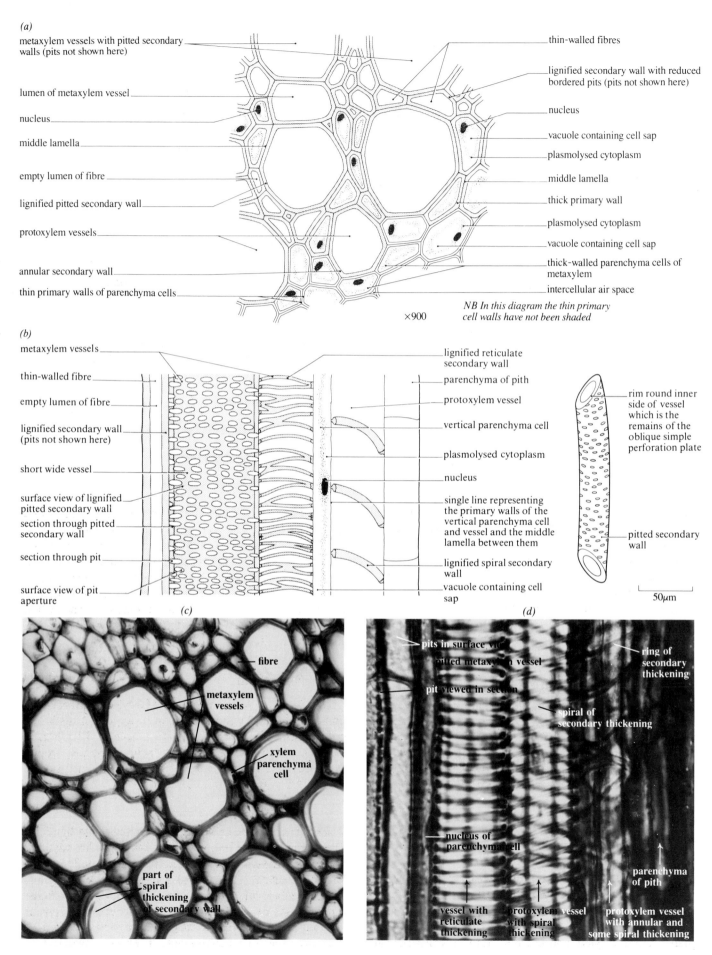

(a)

metaxylem vessels with pitted secondary walls (pits not shown here)

lumen of metaxylem vessel

nucleus

middle lamella

empty lumen of fibre

lignified pitted secondary wall

protoxylem vessels

annular secondary wall

thin primary walls of parenchyma cells

thin-walled fibres

lignified secondary wall with reduced bordered pits (pits not shown here)

nucleus

vacuole containing cell sap

plasmolysed cytoplasm

middle lamella

thick primary wall

plasmolysed cytoplasm

vacuole containing cell sap

thick-walled parenchyma cells of metaxylem

intercellular air space

×900

NB In this diagram the thin primary cell walls have not been shaded

(b)

metaxylem vessels

thin-walled fibre

empty lumen of fibre

lignified secondary wall (pits not shown here)

short wide vessel

surface view of lignified pitted secondary wall

section through pitted secondary wall

section through pit

surface view of pit aperture

lignified reticulate secondary wall

parenchyma of pith

protoxylem vessel

vertical parenchyma cell

plasmolysed cytoplasm

nucleus

single line representing the primary walls of the vertical parenchyma cell and vessel and the middle lamella between them

lignified spiral secondary wall

vacuole containing cell sap

rim round inner side of vessel which is the remains of the oblique simple perforation plate

pitted secondary wall

50μm

(c)

fibre

metaxylem vessels

xylem parenchyma cell

part of spiral thickening of secondary wall

(d)

pits in surface view pitted metaxylem vessel

pit viewed in section

ring of secondary thickening

spiral of secondary thickening

nucleus of parenchyma cell

vessel with reticulate thickening

protoxylem vessel with spiral thickening

protoxylem vessel with annular and some spiral thickening

parenchyma of pith

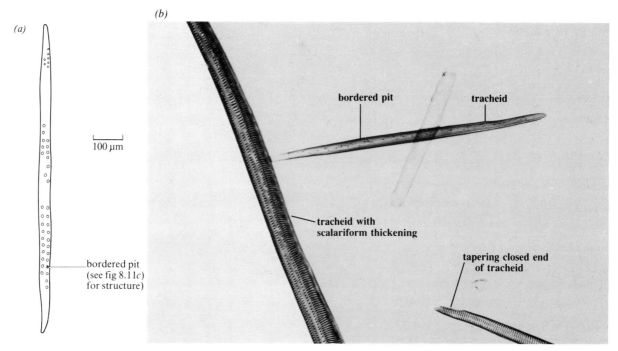

(a)

100 µm

bordered pit
(see fig 8.11c)
for structure)

(b)

bordered pit   tracheid

tracheid with
scalariform thickening

tapering closed end
of tracheid

Fig 8.8 (opposite) *Structure of primary xylem. (a) TS.* *(b) LS. (c) TS primary xylem from Helianthus stem.* *(d) LS primary xylem from Helianthus stem*

Fig 8.9 (above) *Structure of tracheids. (a) Tracheid with bordered pits (tracheids may also have annular, spiral, scalariform and reticulate thickening, like vessels, see fig 8.11g). (b) Tracheids from macerated wood of* Pinus *(× 120)*

rod of tissue 'capping' the vascular bundles of dicotyledons, and a hollow cylinder around the vascular bundles of monocotyledons (see fig 14.15). They often form a layer in the cortex below the epidermis of stems or roots, in the same way as collenchyma, forming a hollow cylinder that contains the rest of the cortex and vascular tissue. Fibres also occur in both xylem and phloem, either individually or in groups, as described in section 8.2.

Function and distribution of sclereids

Sclereids are generally scattered singly or in groups almost anywhere in the plant body, but are most common in the cortex, pith, phloem and in fruits and seeds.

Depending on numbers and position, they confer firmness or rigidity on those structures in which they are found. In the flesh of pear fruits they occur in small groups and are responsible for the 'grittiness' of these fruits when eaten. In some cases they form very resilient, solid layers, as in the shells of nuts, and the stones (endocarp) of stone fruits. In seeds they commonly toughen the testa (seed coat).

8.2   Plant tissues consisting of more than one type of cell

There are two types of conducting tissue in plants, namely **xylem** and **phloem**, both of which contain more than one type of cell. Together they constitute the **vascular tissue** whose function in translocation is described in chapter 14. Xylem conducts mainly water and mineral salts from the roots up to other parts of the plant, while phloem conducts mainly organic food from the leaves both up and down the plant. Both tissues may be increased in amount as a result of secondary growth as described in chapter 21. Secondary xylem may become extensive, when it is known as **wood**. The structure of wood is shown in figs 21.25 and 21.26.

8.2.1   Xylem

Xylem has two major functions, the conduction of water and mineral salts, and support. Thus it has both a physiological and a structural role in the plant. It consists of four cell types, namely tracheids, vessel elements, parenchyma and fibres. These are illustrated in TS and LS in fig 8.8.

Tracheids

Tracheids are single cells that are elongated and lignified. They have tapering end walls that overlap with adjacent tracheids in the same way as sclerenchyma fibres. Thus they have mechanical strength and give support to the plant. They are dead with empty lumens when mature. Tracheids represent the original, primitive water-conducting cells of vascular plants and are the only cells found in the xylem of the more ancestral vascular plants. They have given rise, in

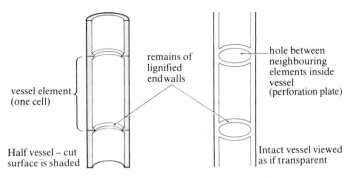

Fig 8.10 (above) *Fusion of vessel elements to form a vessel*

vessel element (one cell)

remains of lignified end walls

hole between neighbouring elements inside vessel (perforation plate)

Half vessel – cut surface is shaded

Intact vessel viewed as if transparent

Fig 8.11 (below) *Structure of protoxylem and metaxylem vessels. (a) Protoxylem vessels. (b) Micrograph of annular and spiral protoxylem vessels. (c) Micrograph of metaxylem reticulate vessels from macerated wood.*

other plants, to xylem fibres and vessels which are described later. Despite their ancestral nature, they obviously function efficiently because conifers, most of which are trees, rely exclusively on tracheids to conduct water from the roots to the aerial parts. Water can pass through the empty lumens without being obstructed by living contents. It passes from tracheid to tracheid through the pits via the 'pit membranes', formed as described in fig 8.7, or through unlignified portions of the cell walls. The pattern of lignification of the walls resembles that of vessels which are described below. Fig 8.9 illustrates the structure of tracheids. Angiosperms have relatively fewer tracheids than vessels, and vessels are thought to be more effective transporting structures, possibly necessary owing to the larger leaves and higher transpiration rates of this group.

Vessels

Vessels are the characteristic conducting units of angiosperm xylem. They are very long, tubular structures formed by the fusion of several cells end to end in a row.

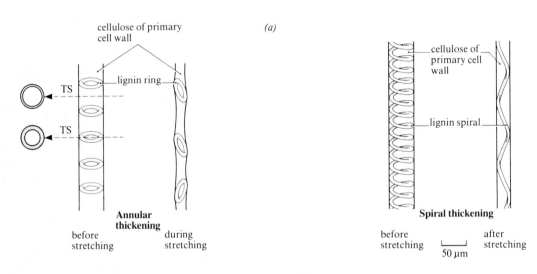

(a)

cellulose of primary cell wall

lignin ring

TS

TS

Annular thickening

before stretching

during stretching

cellulose of primary cell wall

lignin spiral

Spiral thickening

before stretching

after stretching

50 μm

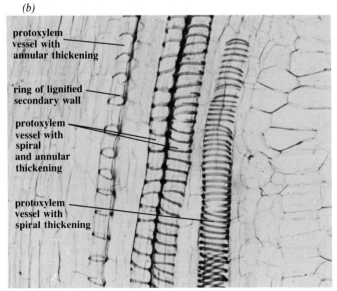

(b)

protoxylem vessel with annular thickening

ring of lignified secondary wall

protoxylem vessel with spiral and annular thickening

protoxylem vessel with spiral thickening

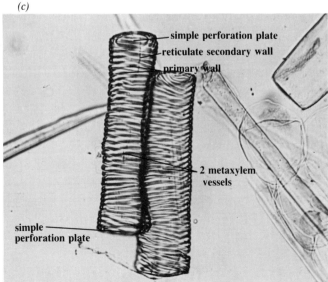

(c)

simple perforation plate

reticulate secondary wall

primary wall

2 metaxylem vessels

simple perforation plate

Fig 8.11 (cont.) *(d) Pitted and reticulate metaxylem vessels. (e) Micrograph of metaxylem pitted vessel from macerated wood. (f) Scanning electron micrograph of metaxylem vessels (× 18 000). Appearance of these vessels in TS will vary according to which part of the vessel is sectioned as indicated in the diagram of extreme left vessel in part (a). (g) TS bordered pit to show structure*

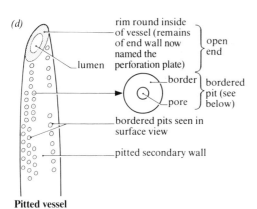

(d)

rim round inside of vessel (remains of end wall now named the perforation plate) — open end

lumen

border — bordered pit (see below)
pore

bordered pits seen in surface view

pitted secondary wall

Pitted vessel

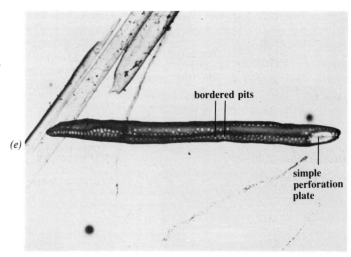

(e)

bordered pits

simple perforation plate

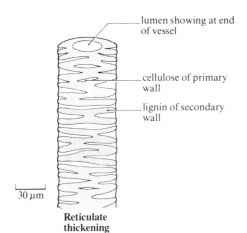

lumen showing at end of vessel

cellulose of primary wall

lignin of secondary wall

30 μm

Reticulate thickening

Size is very variable; the longest are several metres in length, though commonly several centimetres long.

Scalariform thickening is similar to reticulate but with fewer interconnections between the bars of thickening. It is less commonly seen. It usually grades into reticulate thickening by progressive lignification

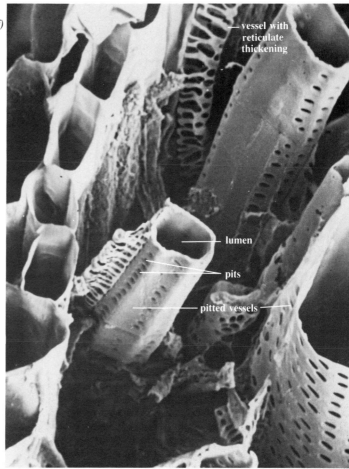

(f)

vessel with reticulate thickening

lumen

pits

pitted vessels

(g)

TS bordered pit to show structure

border   pore   border

pit

lignin of secondary wall

primary cell walls

lignin arches over pit forming border

middle lamella

torus – lignified thickening which can block pore, acting as a valve

Each of the cells forming a xylem vessel is equivalent to a tracheid and is called a **vessel element**. However, vessel elements are shorter and wider than tracheids. The first xylem to appear in the growing plant is called **primary xylem** and develops in the root and shoot apices. Differentiated xylem vessel elements appear in rows at the edges of the procambial strands (shown in figs 21.18 and 21.20). A vessel is formed when the neighbouring vessel elements of a given row fuse as a result of their end walls breaking down. A series of rims is left around the inner side of the vessel marking the remains of the end walls. The fusion of elements is shown in fig 8.10.

Protoxylem and metaxylem

The first vessels form the **protoxylem**, located in the part of the apex, just behind the apical meristem, where elongation of surrounding cells is still occurring. Mature protoxylem vessels can be stretched as surrounding cells elongate because lignin is not deposited over the entire cellulose wall, but only in rings or in spirals as shown in fig 8.11. These act as reinforcement for the tubes during elongation of the stem or root. As growth proceeds more xylem vessels develop and these undergo more extensive lignification completing their development in the mature regions of the organ and forming **metaxylem**. Meanwhile, the earliest protoxylem vessels have stretched and collapsed. Mature metaxylem vessels cannot stretch or grow because they are dead, rigid, fully lignified tubes and were they to develop before the living cells around them had finished elongating they would impose severe restraints on elongation.

Metaxylem vessels show three basic patterns of lignification, namely scalariform, reticulate and pitted, as shown in fig 8.11.

The long, empty tubes of xylem provide an ideal system for translocating large quantities of water over long distances with minimal obstruction to flow. As with tracheids, water can pass from vessel to vessel through pits or through unlignified portions of the cell wall. The walls also have high tensile strength, being lignified, another important feature because it prevents tubes collapsing when conducting water under tension (section 14.4).

The second main function of xylem, namely support, is also fulfilled by the collection of lignified tubes. In the primary plant body the distribution of xylem in the roots is central, helping to withstand the tugging strains of the aerial parts as they bend or lean over. In the stems the vascular bundles are arranged either peripherally in a ring, as in dicotyledons, or scattered, as in monocotyledons, so that in both cases separate rods of xylem run through the stem and provide some support. The supporting function becomes much more important if secondary growth takes place. During this process extensive growth of secondary xylem occurs which supports the large structure of trees and shrubs, taking over from collenchyma and sclerenchyma as the chief mechanical tissue. The nature and extent of the thickness is modified to some extent by the stresses received by the growing plant, so that reinforcement growth can occur and give maximum support.

Xylem parenchyma

Xylem parenchyma occurs in both primary and secondary xylem but it is more extensive and assumes greater importance in the latter. It has thin cellulose cell walls and living contents, as is typical of parenchyma.

Two systems of parenchyma exist in secondary xylem, derived from meristematic cells called ray initials and fusiform initials, as described in section 21.6.6 and fig 21.21. The ray parenchyma is the more extensive (fig 21.24). It forms radial sheets of tissue called medullary rays which maintain a living link through the wood between the pith and cortex. Its functions include food storage, deposition of tannins, crystals and so on, radial transport of food and water, and gaseous exchange through the intercellular spaces.

Fusiform initials normally give rise to xylem vessels or phloem sieve tubes and companion cells but occasionally they give rise to parenchyma cells. These form vertical rows of parenchyma in the secondary xylem.

Xylem fibres

Xylem fibres, like xylem vessels, are thought to have originated from tracheids. They are shorter and narrower than tracheids and have much thicker walls, but they have pits similar to those in tracheids and are often difficult to distinguish from them in section because intermediate cell types occur. Xylem fibres closely resemble the sclerenchyma fibres already described, having overlapping end walls. Since they do not conduct water they can have much thicker walls and narrower lumens than xylem vessels and are therefore stronger and confer additional mechanical strength to the xylem.

8.2.2 Phloem

Phloem resembles xylem in possessing tubular structures modified for translocation. However, the tubes are composed of living cells with cytoplasm and have no mechanical function. There are five cell types in the phloem, namely sieve tube elements, companion cells, parenchyma, fibres and sclereids.

Sieve tubes and companion cells

Sieve tubes are the long tube-like structures that translocate solutions of organic solutes like sucrose throughout the plant. They are formed by the end-to-end fusion of cells called **sieve tube elements** or **sieve elements**. Rows of these cells can be seen developing from the procambial strands of apical meristems where primary phloem develops, together with primary xylem, in vascular bundles.

The first phloem formed is called **protophloem** and, like protoxylem, it is produced in the zone of elongation of the growing root or stem (figs 21.18 and 21.20). As the tissues around it grow and elongate it becomes stretched and much of it eventually collapses and becomes non-functional.

Fig 8.12 *Structure of phloem. (a) Diagram of TS.
(b) Micrograph of TS of primary phloem of* Helianthus *stem
(× 450). (c) Diagram of LS. (d) Micrograph of LS of
primary phloem of* Cucurbita *stem (× 432)*

(a) **TS**

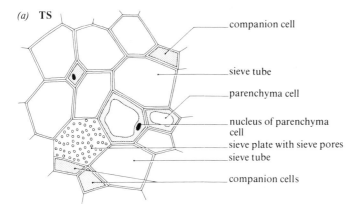

- companion cell
- sieve tube
- parenchyma cell
- nucleus of parenchyma cell
- sieve plate with sieve pores
- sieve tube
- companion cells

(b)

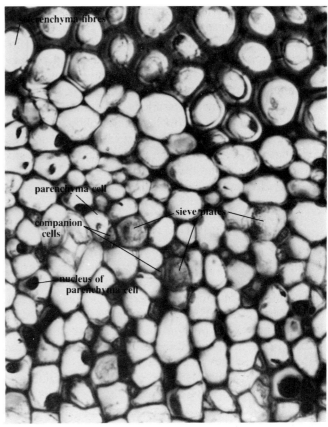

- sclerenchyma fibres
- parenchyma cell
- companion cells
- sieve plate
- nucleus of parenchyma cell

(c) **LS**

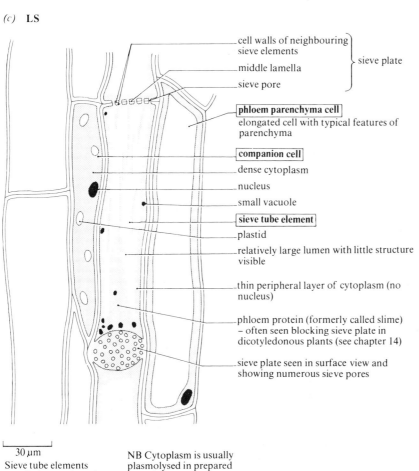

- cell walls of neighbouring sieve elements
- middle lamella } sieve plate
- sieve pore

phloem parenchyma cell
elongated cell with typical features of parenchyma

companion cell
- dense cytoplasm
- nucleus
- small vacuole

sieve tube element
- plastid
- relatively large lumen with little structure visible
- thin peripheral layer of cytoplasm (no nucleus)
- phloem protein (formerly called slime) – often seen blocking sieve plate in dicotyledonous plants (see chapter 14)
- sieve plate seen in surface view and showing numerous sieve pores

├──30 μm──┤
Sieve tube elements usually longer than shown

NB Cytoplasm is usually plasmolysed in prepared material

(d)

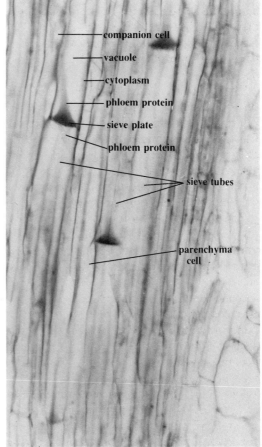

- companion cell
- vacuole
- cytoplasm
- phloem protein
- sieve plate
- phloem protein
- sieve tubes
- parenchyma cell

229

Meanwhile, however, more phloem continues to be produced and the phloem that matures after elongation has ceased is called **metaphloem**.

Sieve tube elements have a very distinctive structure. Their walls are made of cellulose and pectic substances, like parenchyma cells, but their nuclei degenerate and are lost as they mature and the cytoplasm becomes confined to a thin layer around the periphery of the cell. Although they lack nuclei, the sieve elements remain living but are dependent on the adjacent companion cells which develop from the same original meristematic cell. The two cells together form a functional unit, the companion cell having dense, very active cytoplasm. The detailed structure of the cells is revealed by the electron microscope and is described in chapter 14 (see figs 14.22 and 14.23 and section 14.2.2).

A conspicuous and characteristic feature of sieve tubes that is visible in the light microscope is the **sieve plate**. This is derived from the two adjoining end walls of neighbouring sieve elements. Originally plasmodesmata run through the walls but the canals enlarge to form pores, making the walls look like a sieve and allowing a flow of solution from one element to the next. Thus sieve tubes are spanned at intervals by sieve plates that mark successive sieve elements. The structure of sieve tubes, companion cells and phloem parenchyma as seen with the electron microscope is shown in fig 8.12.

Secondary phloem, which develops from the vascular cambium like secondary xylem, appears similar in structure to primary phloem except that it is crossed by bands of lignified fibres and medullary rays of parenchyma as shown in figs 21.25 and 21.26. It is much less extensive than secondary xylem and is constantly being replaced as described in section 21.6.

Phloem parenchyma, fibres and sclereids

Phloem parenchyma and fibres are found in dicotyledons but not in monocotyledons. Phloem parenchyma has the same structure as parenchyma elsewhere, though the cells are generally elongated. In secondary phloem, parenchyma occurs in medullary rays and vertical strands as already described for xylem parenchyma. Phloem parenchyma and xylem parenchyma have the same functions.

Phloem fibres are exactly similar to the sclerenchyma fibres already described. They occur occasionally in the primary phloem, but more frequently in the secondary phloem of dicotyledons. In secondary phloem they form vertically running bands of cells. Since the secondary phloem is subject to stretching as growth continues, the sclerenchyma probably helps to resist this pressure.

Sclereids occur frequently in phloem, especially in older phloem.

8.3 Animal epithelial tissue

Epithelial tissue is arranged in single or multilayered sheets and covers the internal and external surfaces of the body of an organism. True epithelial tissue arises embryonically from either the ectoderm, which provides epithelium for the skin, nervous system and parts of the fore- and hindgut, or the endoderm, which provides epithelium for the remainder of the alimentary canal, the liver and pancreas. It should be mentioned here that the inner lining of blood vessels, called endothelium, is not true

Table 8.2 Classification of epithelial tissues.

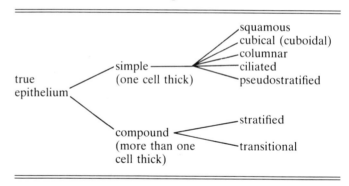

Fig 8.13 *Simple squamous epithelium: (a) diagram; (b) photomicrograph (small blood vessel)*

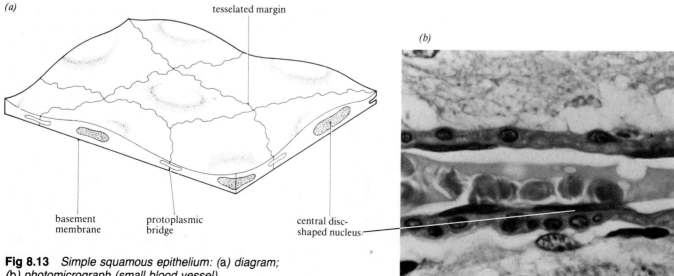

epithelium as it is derived from embryonic mesoderm.

Epithelial cells are held together by small amounts of cementing substance containing a carbohydrate derivative called hyaluronic acid. The bottom layer of cells rests on a **basement membrane** composed of a network of collagenous fibres usually secreted by underlying tissues. As epithelial cells are not supplied with blood vessels, they rely on diffusion of oxygen and nutrients from lymph vessels which ramify adjacent intercellular spaces. Nerve endings may penetrate the epithelium.

Epithelial tissue functions to protect underlying structures from injury through abrasion or pressure, and from infection. Stress is combated by the tissue becoming thickened and keratinised, and where cells are sloughed off due to constant friction the epithelium shows a very rapid rate of cell division so that lost cells are speedily replaced. The free surface of the epithelium is often highly differentiated and may be absorptive, secretory or excretory in function, or bear sensory cells and nerve endings specialised for stimulus reception.

Epithelial tissues are classified into the following types indicated in table 8.2 according to the number of cell layers and the shape of the individual cells. In many areas of the body the different cell types intermix and the epithelia cannot be classified into distinct types.

(a)

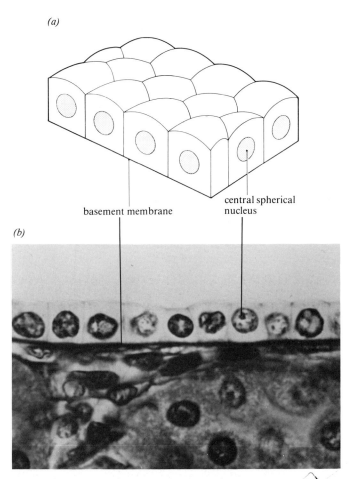

(b)

Fig 8.14 *Cubical (cuboidal) epithelium: (a) diagram; (b) photomicrograph (kidney)*

8.3.1 Simple epithelia

Squamous epithelium

The cells are thin, flattened and contain little cytoplasm enclosing a centrally placed disc-shaped nucleus (fig 8.13). The margins of squamous cells are tesselated (irregular) and provide a mosaic outline in surface view. There are often protoplasmic connections between adjacent cells which help to bind them firmly together. Squamous epithelium occurs in areas such as the Bowman's capsules of the kidney, the alveolar lining of the lungs and the blood capillary walls where its thinness permits diffusion of materials through it. It also provides smooth linings to hollow structures such as blood vessels and the chambers of the heart where it allows the relatively friction-free passage of fluids through them.

Cubical epithelium

This is the least specialised of all epithelia and, as the name implies, the cells are cube-shaped and possess a central spherical nucleus (fig 8.14). When viewed from the surface the cells are either pentagonal or hexagonal in outline. They form the lining of many ducts such as the salivary, pancreatic and collecting ducts of the kidney where they are non-secretory. Cubical epithelium in other parts of the body is secretory and is found in many glands such as the salivary, mucus, sweat and thyroid glands.

Columnar epithelium

These cells are tall and quite narrow, thus providing more cytoplasm per unit area of epithelium (fig 8.15). Each cell possesses a nucleus situated at its basal end. Secretory goblet cells are often interspersed among the epithelial cells and the epithelium may be secretory and/or absorptive in function. There is frequently a conspicuous striated border of **microvilli** at the free surface end of each cell which increases the surface area of the cell for absorption and secretion. Columnar epithelium lines the stomach, where mucus secreted by goblet cells protects the stomach lining from the acidic contents of the stomach and from digestion by enzymes. It also lines the intestine where mucus again protects it from self-digestion and at the same time lubricates the passage of food. In the small intestine digested food is actually absorbed through the epithelium into the bloodstream. Columnar epithelium lines and protects many kidney ducts, and is a component of the thyroid gland and gall bladder.

Ciliated epithelium

Cells of this tissue are usually columnar in shape but bear numerous cilia at their free surfaces (fig 8.16). They are always associated with mucus-secreting goblet cells producing fluids in which the cilia set up currents. Ciliated epithelium lines the oviducts, ventricles of the brain, the spinal canal and the respiratory passages, where it serves to move materials from one location to another.

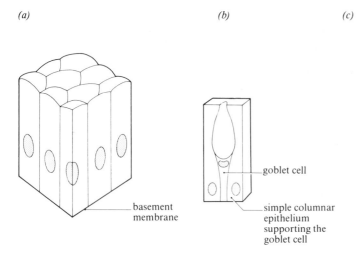

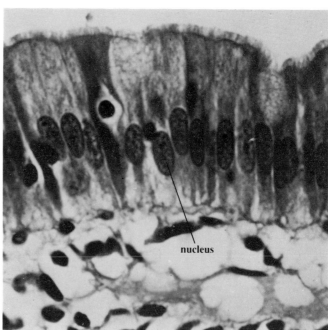

Fig 8.15 *(a) Columnar epithelium, (b) showing goblet cell, (c) photomicrograph of columnar epithelium (trachea)*

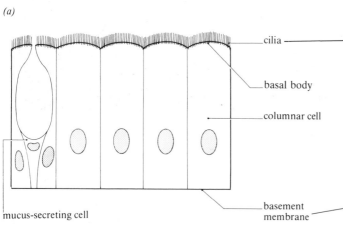

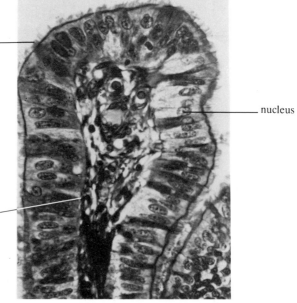

Fig 8.16 *Simple ciliated epithelium: (a) diagram; (b photomicrograph (oviduct)*

Pseudostratified epithelium

When viewed in section the nuclei of this type of epithelium appear to be at several different levels because all the cells do not reach the free surface (fig 8.17). Nevertheless the epithelium is still only one layer of cells thick with each cell attached to the basement membrane. This epithelium is found lining the urinary tract, the trachea (as pseudostratified columnar), other respiratory passages as (pseudostratified columnar ciliated) and as a component of the olfactory mucosa.

8.3.2 Compound epithelia

Stratified epithelium

This tissue comprises a number of layers of cells, is correspondingly thicker than simple epithelium and forms a relatively tough, impervious barrier. The cells are formed by mitotic division of the germinal layer which rests on the basement membrane (fig 8.18). The first-formed cells are cuboid in shape, but as they are pushed outwards towards the free surface of the tissue they become flattened. In this condition the cells are called **squames**. They may remain uncornified, as in the oesophagus, where the epithelium protects the underlying tissues against mechanical damage

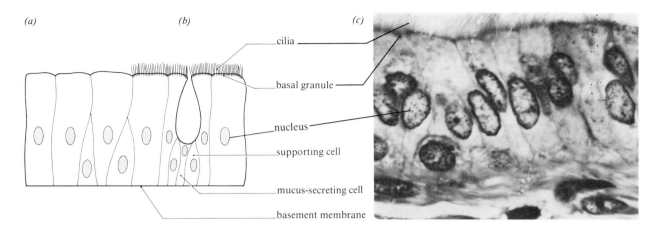

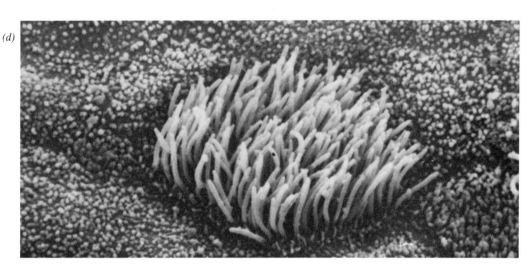

Fig 8.17 *Pseudostratified epithelium: (a) columnar; (b) ciliated; (c) photomicrograph of respiratory, ciliated epithelium; (d) scanning electron micrograph of cilia*

cilia

basal granule

nucleus

supporting cell

mucus-secreting cell

basement membrane

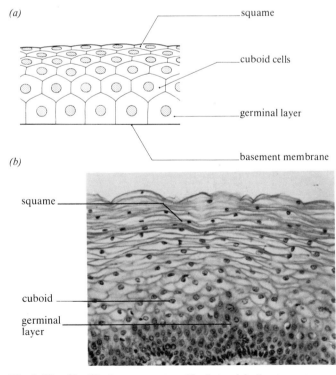

squame

cuboid cells

germinal layer

basement membrane

squame

cuboid

germinal layer

Fig 8.18 *Stratified squamous epithelium: (a) diagram; (b) photomicrograph (vagina)*

by friction with food just swallowed. In other areas of the body the squames may be transformed into a dead horny layer of **keratin** which ultimately flakes away. In this condition the epithelium is said to be **cornified**, and is found in particular abundance on external skin surfaces, lining the buccal cavity and the vagina, where it affords protection against abrasion.

According to the shape of the cells which make up the stratified epithelium, it may be termed stratified squamous (located in parts of the oesophagus), stratified cuboidal (in the sweat gland ducts), stratified columnar (in the mammary gland ducts), and stratified transitional (in the bladder).

Transitional epithelium

This is often regarded as a modified type of stratified epithelium. It consists of 3–4 layers of cells all of similar size and shape except at the free surface where they are more flattened (fig 8.19). The superficial cells do not slough off, and all cells are able to modify their shape when placed under differing conditions. This property is important in locations where structures are subjected to considerable distention such as the urinary bladder, ureter and the pelvic region of the kidney. The thickness of the tissue also prevents urine escaping into the surrounding tissues.

233

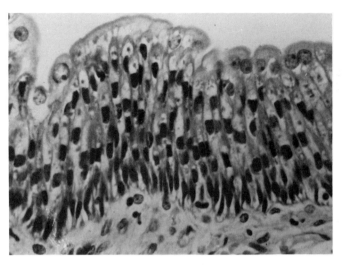

Fig 8.19 *Transitional epithelium (bladder)*

surface via ducts (table 8.3). **Endocrine** are glands where the secretion is passed directly into the bloodstream. Endocrine glands possess no ducts and are alternatively termed **ductless glands** (section 16.6 and fig 8.20).

The secretions produced by glandular cells are released in three different ways. In **merocrine glands** the secretion, produced within the cells, is simply passed through the cell membrane at the cell's free surface. No cytoplasm is lost. This occurs in simple goblet cells, the sweat glands and the exocrine regions of the vertebrate pancreas. In **apocrine glands** a portion of the cell's distal cytoplasm is lost as the secretion is released, as in the secretions of the mammary glands. In **holocrine glands** the whole cell breaks down to release its secretory product and is extruded from the epithelial layer. Sebaceous glands show this mode of secretion.

Sometimes a cell may secrete different materials each by

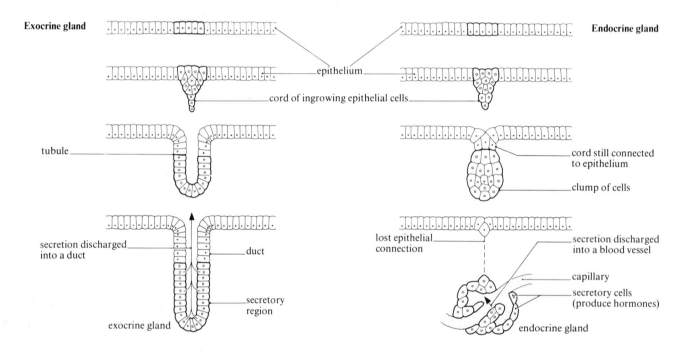

Fig 8.20 *The development of exocrine and endocrine glands (From Freeman & Bracegirdle (1975) An atlas of histology, Heinemann Education Books Ltd., London.)*

8.3.3   Glandular epithelia

Amongst the epithelial cells there may be individual glandular cells, such as the **goblet cells**, or aggregates of glandular cells forming a **multicellular gland**. An epithelium containing many goblet cells is called a mucous membrane.

Two types of glandular cell exist, called exocrine and endocrine. **Exocrine** are those where the secretion is delivered to the free surface of the epithelium (fig 8.20). Multicellular exocrine glands pass their products to the

a different method. Such an example can be found in the mammary gland where a lipid product is secreted by the apocrine mechanism and a protein secretion is released in a merocrine fashion.

If the glandular cells secrete a viscous mucous product they are called **mucous cells** or **mucocytes**, whereas if the secretion is clear, watery in consistency and contains enzymes they are termed **serous cells** or **serocytes**. If both kinds of secretion are produced within the same gland then the gland is called a **mixed gland**.

Multicellular exocrine glands exist in a number of forms of variable complexity as shown in table 8.3.

8.4   Animal connective tissue

Connective tissue is the major supporting tissue of the body. It includes the skeletal tissue, bone and

Type of gland	Structure	Location	Type of gland	Structure	Location
Simple tubular	secretory portion tubular in design	Crypts of Lieberkühn in the ileum of higher vertebrates Fundic region of stomach	Simple alveolar	secretory portion sac-like in construction	Mucus glands in skin of frog
Simple coiled tubular		Sweat glands in Man	Simple branched alveolar		Sebaceous glands in mammalian skin
Simple branched tubular		Fundic region of stomach Brunner's glands in mammalian small intestine	Compound alveolar		Exocrine parts of pancreas Mammary gland
Compound tubular		Brunner's glands in mammal Salivary glands	Compound tubular-alveolar	many branched ducts possessing a mixture of tubular and alveolar secretory portions	Submaxillary glands Mammary glands Salivary glands

Table 8.3 Different forms of multicellular exocrine glands.

Table 8.4 Types of connective tissue.

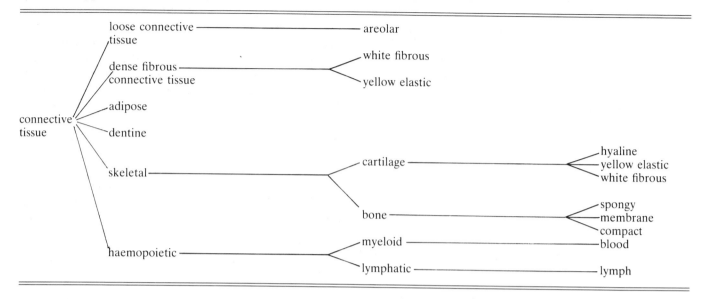

cartilage, and in addition it binds other tissues together, for example the skin with underlying structures, or the epithelia of mesenteries. This tissue also forms sheaths around the organs of the body, separating them so that they do not interfere with each other's activities, as well as embedding and protecting blood vessels and nerves where they enter or leave organs. Connective tissue is a composite structure made up of a variety of cells developed from mesenchyme originating in the embryonic mesoderm, several types of fibre which are non-living products of the cells, and a fluid or semi-fluid intercellular matrix consisting of hyaluronic acid, chondroitin, chondroitin sulphate and keratin sulphate.

The cells are usually widely separated from each other and their metabolic needs relatively small. An extensive vascular network is often present in various parts of the body (as in the dermis of the skin) but this is primarily concerned with supplying other structures, such as the epithelium, with oxygen and nutrients rather than the connective tissue itself. Connective tissue may be subdivided into a number of types as indicated in table 8.4.

This tissue fulfils many functions other than packing and binding other structures together, such as providing protection against wounding or bacterial invasion (areolar), insulation of the body against heat loss (adipose), providing a supportive framework for the body (cartilage and bone) and producing blood (haemopoietic tissue).

8.4.1 Loose connective tissue

This tissue contains cells widely dispersed in intercellular material and has fibres loosely woven in a random manner.

Areolar

Areolar tissue possesses a transparent semi-fluid matrix which contains a mixture of mucin, hyaluronic acid and chondroitin sulphate. Scattered throughout are numerous

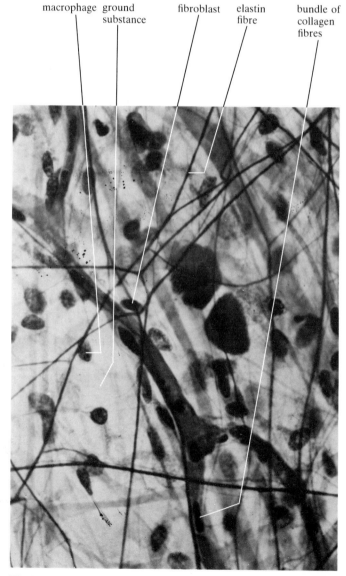

Fig 8.21 *Loose areolar tissue*

236

wavy bundles of **collagen fibres** and a loose anastomosing network of thin straight fibres of **elastin** (fig 8.21). Collagen fibres are flexible but inelastic whilst elastin fibres are flexible and elastic. Together the fibres endow the tissue with considerable tensile strength and resilience. Very fine and thread-like **reticular fibres** are present, located around blood vessels and nerves, and form the connective tissue covering around muscle fibres. It is thought that reticular fibres may be immature collagen fibres. Interspersed in the matrix are a variety of different cell types. They include fibroblasts, macrophages, mast cells, plasma cells, chromatophores, fat and mesenchyme cells. **Fibroblasts** are the cells which produce the fibres. They are flattened and spindle-shaped and contain an oval nucleus. Generally they lie closely applied to the fibres they synthesise, but can migrate towards wounded tissue and secrete more fibres in this region to effectively seal off the injured area. **Macrophages (histiocytes)** are polymorphic cells capable of amoeboid locomotion, which engulf bacteria or other foreign particles. Generally they are immobile, but at times cells wander to areas of bacterial invasion and therefore provide a means of defence for the body. Together with the reticular cells of the lymphatic system they comprise the **reticulo-endothelial system** of the body (section 14.11.2). **Mast cells** are oval-shaped, small and contain granular cytoplasm. They secrete the matrix as well as heparin and histamine, and are found in abundance close to blood vessels. **Heparin** is an anticoagulant present in all mammalian tissues. It neutralises the action of thrombin, preventing the conversion of prothrombin to thrombin. **Histamine** is released from tissues when they are injured or disrupted in any way. It causes vasodilation, contraction of smooth muscle and stimulates gastric secretion. **Plasma**

cells are rare and are the products of mitotic cell division by migratory lymphocytes. When present they produce antibodies which are important components of the body's immune system (section 14.14). **Chromatophores** are present in specialised areas, such as the skin and the eye. The cells are much branched and densely packed with melanin granules. Each **fat cell** contains a large lipid droplet which fills the bulk of the cell. The cytoplasm and nucleus are confined to the margins. **Mesenchyme cells** act as a reserve of undifferentiated cells for the tissue. They can be stimulated to transform into one of the above cell types as the need arises.

Areolar tissue is found around all the organs of the body, it connects the skin to the structures below, and binds sheets of epithelia to form mesenteries. It also ensheathes blood vessels and nerves where they enter or leave organs.

8.4.2 Dense (compact) fibrous connective tissue

This tissue has more fibres situated in the matrix than cells. The fibres may be irregularly arranged, or orientated such that the individual fibres lie more or less parallel to each other.

White fibrous

This is a tough, shiny tissue composed of numerous highly organised bundles of collagen fibres closely packed together and running parallel to each other (fig 8.22). Rows of fibroblasts are interspersed among the collagen and run alongside the bundles. Each bundle is bound to its neighbours by areolar tissue. The tissue is strong, flexible, yet inextensible and its tensile strength is achieved by the presence of collagen. Each strand of collagen possesses

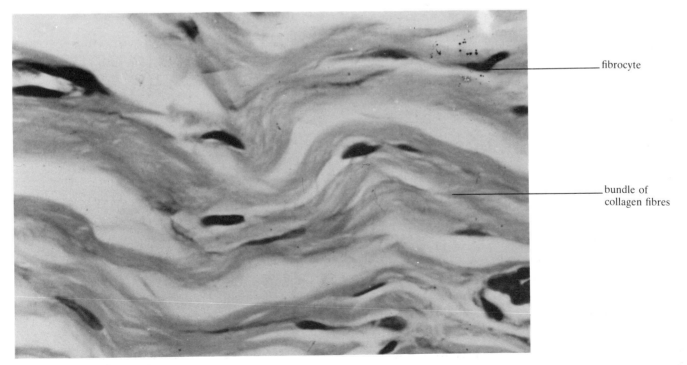

fibrocyte

bundle of
collagen fibres

Fig 8.22 *White fibrous tissue*

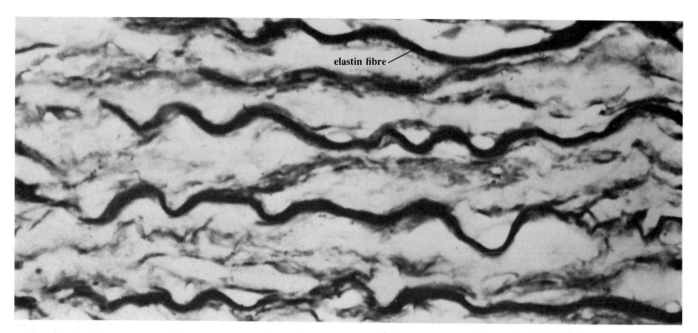

elastin fibre

three chains of tropocollagen plaited together as in a rope (section 5.5.3). The fibres are precisely organised so that they run parallel to the lines of stress which are encountered as a result of the functions carried out by the structures containing collagen.

White fibrous tissue is abundant in tendons, some ligaments, the sclerotic and cornea of the eye, the kidney capsule, and the perichondrium and periosteum of cartilage and bone respectively.

Yellow elastic

In contrast to white fibrous tissue, this possesses a loose network of irregularly arranged branched yellow elastic fibres (fig 8.23). The fibroblasts are randomly scattered throughout the matrix and some fine collagen fibres are also present. The elastic fibres endow the tissue with elasticity and flexibility and the collagen gives it strength. The tissue is located in ligaments, the walls of arteries, as a component of the lung and associated air passages, and in the great cords of the neck.

8.4.3 Adipose tissue

This tissue has no specific matrix of its own and is really areolar tissue containing large numbers of fat cells arranged into lobules. Each cell is filled almost entirely by a central fat droplet which squeezes the cytoplasm and nucleus to the periphery (fig 8.24).

In mammals, adipose tissue is found in the dermis of the skin, the mesenteries, and around the kidneys and heart. It provides a considerable energy reserve, acts as a shock absorber, and insulates against heat loss.

8.4.4 Skeletal tissues

Cartilage

Cartilage is a connective tissue consisting of cells embedded in a resilient matrix of chondrin. The matrix is

Fig 8.23 (above) *Yellow elastic tissue*
Fig 8.24 (below) *Adipose tissue*

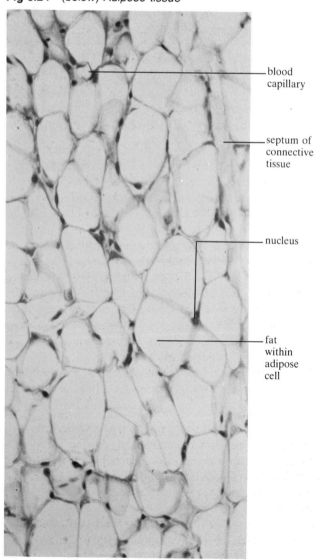

blood capillary

septum of connective tissue

nucleus

fat within adipose cell

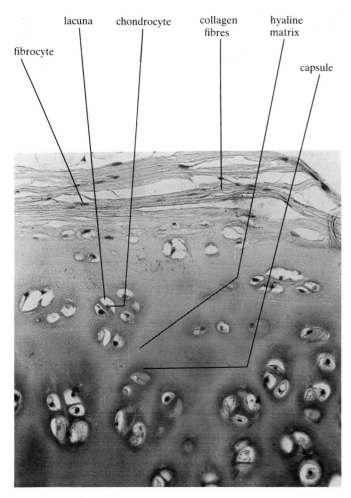

fibrocyte   lacuna   chondrocyte   collagen fibres   hyaline matrix   capsule

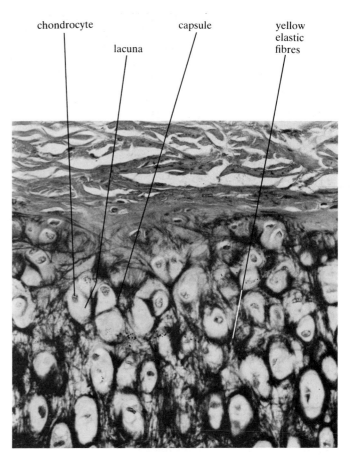

chondrocyte   lacuna   capsule   yellow elastic fibres

Fig 8.25 (left) *Hyaline cartilage*

Fig 8.26 (above) *Elastic cartilage*

deposited by cells called **chondroblasts** and possesses many fine fibrils mostly made up of collagen. Eventually the chondroblasts become enclosed in spaces called **lacunae**. In this condition they are termed **chondrocytes**. The margin of a piece of cartilage is enclosed by a dense layer of cells and fibrils called the **perichondrium**. From here new chondroblasts are produced, which are constantly added to the internal matrix of the cartilage.

Cartilage is a hard but flexible tissue. It is highly adapted to resist any strains that are placed upon it. The matrix is compressible and elastic and is able to absorb mechanical shocks such as frequently occur between the articular surfaces of bones. The collagen fibrils resist any tension which may be imposed on the tissue.

Three types of cartilage are recognisable. For each type the organic components of the matrix are quite distinct.

Hyaline cartilage (fig 8.25). The matrix is a semi-transparent material consisting of chondroitin sulphate and frequently fine collagen fibrils. The peripheral chondrocytes are flattened in shape whereas those situated internally are angular. Each chondrocyte is contained in a lacuna, and each lacuna may enclose one, two, four or eight chondrocytes.

Unlike bone, no processes extend from the lacunae into the matrix, neither are there blood vessels in this area. All exchange of materials between the chondrocytes and the matrix occurs by diffusion.

Hyaline cartilage is an elastic, compressible tissue located at the ends of bones, in the nose and air passages of the respiratory system and in parts of the ear. It is the only type of skeletal material found in elasmobranchs and forms the embryonic skeleton in bony vertebrates.

Yellow elastic cartilage (fig 8.26). The matrix is semi-opaque and contains a network of yellow elastic fibres. They confer greater elasticity and flexibility than is found in hyaline cartilage and permit the tissue to quickly recover its shape after distortion. It is located in the external ear, eustachian tube, the epiglottis and cartilages of the pharynx.

White fibrous cartilage (fig 8.27). This consists of large numbers of bundles of densely packed white collagen fibres embedded in the matrix. This provides greater tensile strength than hyaline cartilage, as well as a small degree of flexibility. White fibrous cartilage is located as discs between adjacent vertebrae (intervertebral discs) where it provides a cushioning effect. It is also found in the symphysis pubis (the region between the two pubic bones of the pelvis) and the ligamentous capsules of joints.

Bone

Bone is the most abundant of all animal skeletal materials

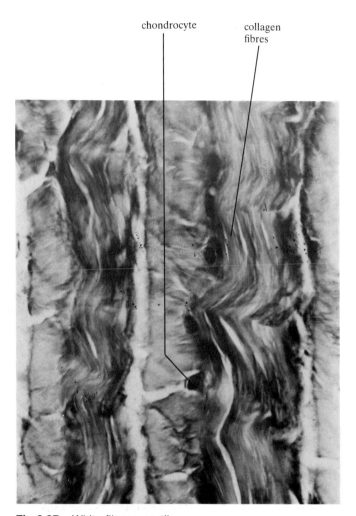

chondrocyte   collagen fibres

Fig 8.27 *White fibrous cartilage*

providing supportive, metabolic and protective functions for those possessing it. It is a calcified connective tissue made up of cells embedded in a firm matrix. About 30% of the matrix is composed of organic material consisting chiefly of collagen fibrils, whilst 70% is inorganic bone salts. The chief inorganic constituent of bone is hydroxy-apatite, $Ca_{10}(PO_4)_6(OH)_2$, but sodium, magnesium, potassium, chloride, fluoride, hydrogencarbonate and citrate ions are also present in variable amounts.

Bone cells, called **osteoblasts**, are contained in lacunae which are present throughout the matrix. They lay down the inorganic components of bone. Fine canals containing cytoplasm connect the lacunae to each other and blood vessels passing through them provide the means by which osteoblasts exchange materials.

The structure of bone is specially designed to withstand the compression strains falling upon it and to resist tension. When the bone fibrils are laid down they are impregnated by apatite crystals. This arrangement provides maximum strength for the bone.

Bone resorption and reconstruction processes enable a particular bone to adapt its structure to meet any change in the mechanical requirements of the animal during its development. Calcium and phosphate may be released into the blood as needed, under the control of two hormones, **parathormone** and **calcitonin** (sections 16.6.4 and 16.6.5).

Compact or dense bone (fig 8.28). A transverse section of compact bone shows it to consist of numerous cylinders of concentric bony **lamellae** each surrounding a central **Haversian canal**. One such cylinder plus its canal is termed an **Haversian system** or **osteon**.

Interspersed between the lamellae are numerous lacunae containing living bone cells called **osteoblasts**. Each cell is capable of bone deposition. Its cytoplasm possesses a well-defined rough endoplasmic reticulum and Golgi apparatus and is rich in RNA. When osteoblasts are not active they are termed **osteocytes**. In this condition they contain reduced quantities of cell organelles and often store glycogen. If structural changes in the bone are required they are activated and quickly differentiate into osteoblasts.

Radiating from each lacuna are many fine channels called **canaliculi** containing cytoplasm which may link up with the central Haversian canal, with other lacunae or pass from one lamella to another.

An artery and a vein run through each Haversian canal, and capillaries branch from here and pass via the canaliculi to the lacunae of that particular Haversian system. They facilitate the passage of nutrients, metabolic waste and respiratory gases towards and away from the cells. A Haversian canal also contains a lymph vessel and nerve fibres tightly packed with areolar tissue. Transverse Haversian canals communicate with the marrow cavity and also interconnect with the longitudinal Haversian canals. These contain larger blood vessels and are not encircled by concentric lamellae.

At the outer and inner surfaces of the bone the lamellae are not in the form of concentric cylinders but are orientated circumferentially over it. The **canals of Volkman** ramify these areas. The canals contain blood vessels which pass through them to link with those in the Haversian canals.

The matrix of compact bone is composed of bone collagen, manufactured by the osteoblasts, and hydroxy-apatite together with quantities of magnesium, sodium, carbonates and nitrates. The combination of organic with inorganic material produces a structure of great strength. The lamellae are laid down in a manner that is suited to the forces acting upon the bone, and the load that has to be carried.

Covering the bone is a layer of dense connective tissue called the **periosteum**. Bundles of collagen fibres called **Sharpey–Schafer** fibres from the periosteum pierce the bone and provide an intimate connection between the underlying bone and periosteum and act as a firm base for tendon insertions. The inner region of the periosteum is vascular and forms a layer which contains undifferentiated potential osteoblasts.

Spongy or trabecular bone (fig 8.29). Spongy bone consists of a meshwork of thin, interconnect-

(a)

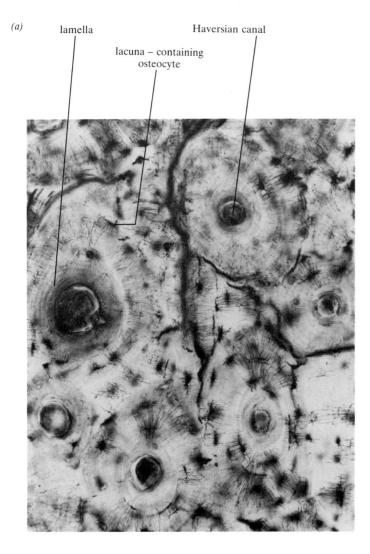

lamella

lacuna – containing
osteocyte

Haversian canal

(b)

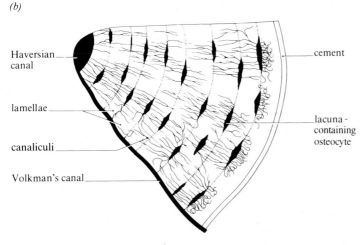

Haversian
canal

lamellae

canaliculi

Volkman's canal

cement

lacuna -
containing
osteocyte

Fig 8.28 *(a) Part of a transverse section of a long bone.
(b) TS Haversian system. The presence of large numbers of
lamellae provides the bone with great strength despite its light
weight*

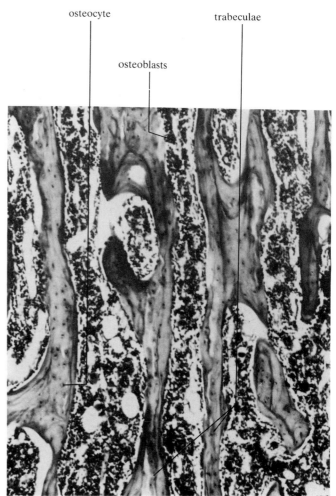

osteocyte

osteoblasts

trabeculae

Fig 8.29 *Spongy bone*

ing bony struts called **trabeculae**. Its matrix contains less
inorganic material (60–65%) than compact bone. The
organic material is primarily composed of collagen fibres.
The spaces between the trabeculae are filled with soft
marrow tissue. If the marrow is red, as at the epiphyses of
long bones such as the femur, the cells are predominantly
red blood cells, whereas if it is yellow, as in the diaphyses of
long bones, the cells are primarily fat cells. Three different
types of cell appear to be present in spongy bone, which
may be three different functional stages of the same cell
type. These are **osteoblasts** which synthesise the spongy
bone, **osteocytes** which are resting osteoblasts, and **osteoclasts** which can resorb the calcified matrix.

The trabeculae are orientated in the direction in which
the bone is stressed. This enables the bone to withstand
tension and compression forces effectively whilst at the
same time keeping the weight of the bone to a minimum.

Spongy bone occurs in the embryo, growing organisms,
and the epiphyses of long bones.

Membrane or dermal bone (fig 8.30). Such
bones have no cartilage forerunner but are formed directly
by intramembranous ossification in the dermis of the skin.
Aggregations of osteoblasts appear in this region and form

241

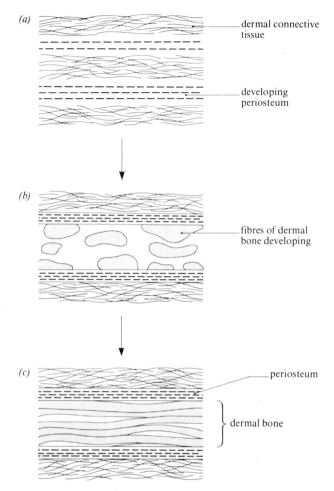

(a) dermal connective tissue

developing periosteum

(b) fibres of dermal bone developing

(c) periosteum

dermal bone

Fig 8.30 *Development of dermal bone*

rows of cells which manufacture bone trabeculae. In this way, flat bony plates are produced very close to the surface of the body. They increase in size when more bone is deposited on their inner and outer surfaces, and then may sink further into the body to become part of the skeleton. Membrane bones form components of the skull, jaws and pectoral girdle.

8.4.5 Dentine

The composition of dentine (commonly called ivory) is very much like that of bone. However it contains a higher inorganic content (75%) and is consequently harder. Dentine contains no lacunae or osteons, and the arrangement of osteoblasts is quite different from that in bone (fig 8.31). In dentine they are confined to the dentinal inner margins and perforate the matrix with many odontoblastic processes which contain **microtubules**, and frequently blood vessels and nerve endings sensitive to touch and low temperatures. Collagen fibres, manufactured and laid down at the apices of the processes ultimately become calcified by impregnation with apatite crystals to form new dentine. Dentine is located between the enamel and pulp cavity in teeth, above and below gum level.

8.4.6 Haemopoietic tissue

Haemopoietic tissue forms red and white blood cells and is located in the red bone marrow and lymphoid tissue of adult mammals. Bone marrow or myeloid tissue produces red blood cells and granulocytes,

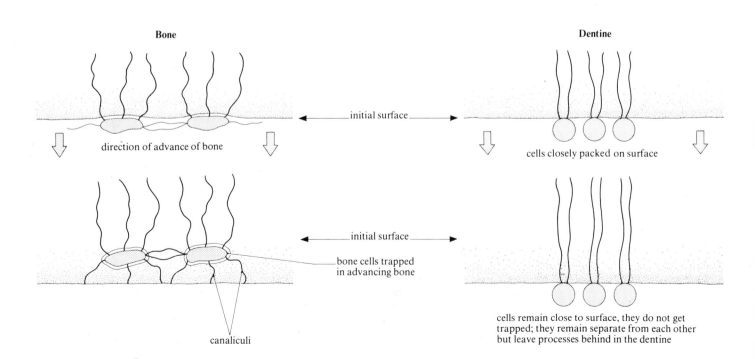

Bone

Dentine

initial surface

direction of advance of bone

cells closely packed on surface

initial surface

bone cells trapped in advancing bone

cells remain close to surface, they do not get trapped; they remain separate from each other but leave processes behind in the dentine

canaliculi

Fig 8.31 *Comparison of growth in bone and dentine (Modified after John Currey (1970) Animal Skeletons, Arnold.)*

whilst lymphocytes and monocytes are differentiated in lymphoid tissue. Haemopoietic tissue consists of free cells enmeshed in a stroma of loose scleroprotein fibres, often termed reticular connective tissue.

Bone marrow/myeloid tissue

The stroma here consists of very loose, reticular connective tissue permeated by wide intercellular spaces. It is traversed by numerous thin-walled, wide blood sinuses through which mature blood cells escape into the bloodstream. Lining the sinuses are phagocytic cells which form part of the body's reticulo-endothelial system.

It is thought that all blood cells are derived from primitive cells called **haemocytoblasts** which in turn differentiate into **erythroblasts**, the precursors of erythrocytes, **myelocytes**, the precursors of granulocytes, **lymphoblasts**, the precursors of lymphocytes, **monoblasts**, the precursors of monocytes, and **megakaryocytes** which produce platelets.

Further details of the structure and functions of these cells can be found in section 14.11.

Lymphoid tissue

This is responsible for the differentiation of lymphocytes. Three types of the tissue exist: **loose lymphoid tissue**, where the stroma of reticular connective tissue predominates over the free cells, **dense lymphoid tissue**, where there are many more free cells embedded in the stroma, and **nodular lymphoid tissue**, which possesses dense aggregates of free cells.

The free cells are composed primarily of lymphocytes of various sizes and functions. There are also plasma cells present which have developed from lymphocytes, and occasionally monocytes and eosinophils are evident. Some of the cells are phagocytic. Further details of the lymphatic system are given in section 14.12.1.

8.5 Muscle tissue

Muscle tissue makes up 40% of a mammal's body weight. It is derived from embryonic mesoderm and consists of highly specialised contractile cells or fibres held

Table 8.5 The similarities and differences between voluntary, involuntary and cardiac muscle.

Features	Voluntary	Involuntary	Cardiac
Other names	Striated, striped, skeletal	Unstriated, unstriped, smooth	Heart
Specialisation	Most highly specialised	Least specialised	More specialised than involuntary
Structure	Very long cells, usually called fibres, subdivided into units called sarcomeres. Fibres bound together by vascular connective tissue.	Consists of individual, spindle-shaped cells, associated in bundles or sheets	Cells terminally branched and connected to each other by special interdigitating surface processes, the **intercalated discs**. Arrangement of fibres is three-dimensional.
Nucleus	Several in variable positions near periphery of fibre	Single, elongated in shape and centrally placed	Several centrally placed
Cytoplasmic contents	Mitochondria in rows in periphery and between fibres, prominent SER forming network of tubules, T-system well developed, glycogen granules and some lipid droplets	Prominent mitochondria, individual tubules of the SER, glycogen granules	Numerous large mitochondria in columns between cells, poorly developed SER consisting of network of tubules, T-system well developed
Sarcolemma	Present	Absent	Present
Myofilaments/ myofibrils	Very conspicuous, length 1–40 mm, diameter 10–60 μm	Inconspicuous, length 0.02–0.5 mm, diameter 5–10 μm	Conspicuous, length 0.08 mm or less, diameter 12–15 μm
Innervation	Under control of the voluntary nervous system via motor nerves from the brain and spinal cord (neurogenic)	Under control of autonomic nervous system (neurogenic)	Myogenic, but rate of contraction can be influenced by the autonomic nervous system
Cross striations	Present	Absent	Present
Intercalated discs	Absent	Absent	Present
Activity	Powerful, rapid contractions, short refractory period, therefore fatigues quickly	Shows sustained rhythmical contraction and relaxation, as in peristalsis	Rapid rhythmical contraction and relaxation, long refractory period, therefore does not fatigue; contraction not sustained
Location	Attached to the skeleton in the trunk, limbs and head	In walls of intestinal, genital, urinary and respiratory tracts, and the walls of blood vessels	Found only in the walls of the heart chambers

SER – Smooth endoplasmic reticulum.

together by connective tissue. Three types of muscle are present in the body, classified according to their method of innervation and they are **voluntary** (striated), **involuntary** (unstriated) and **cardiac**. Table 8.5 shows the main points of similarity and difference between them. Further details can be found elsewhere in sections 14.12 and 17.4. (Also see figs 8.32–4.)

8.6 Nervous tissue

Nervous tissue is derived from embryonic mesoderm. It is composed of densely packed interconnected nerve cells called **neurones** (as many as 10^{10} in the human brain), specialised for conduction of nerve impulses, and accessory neuroglial cells. There is little intercellular space between them. Nervous tissue also contains receptor cells, and is frequently ensheathed by vascularised connective tissue.

8.6.1 Neurones

These are the functional units of the nervous system. Neurones are excitable cells, that is they are capable of transmitting electrical impulses, and this provides the means of communication between **receptors** (cells or organs which receive stimuli, such as the sensory cells in the skin) and **effectors** (tissues or organs which react to stimuli, such as muscles or glands). Neurones which conduct impulses towards the central nervous system (the brain and spinal cord) are called **afferent** or **sensory neurones**, whilst **efferent** or **motor neurones** conduct impulses away from the central nervous system. **Internuncial**, intermediate, relay or association neurones frequently interconnect afferent neurones with efferents. The structure of these neurones is shown in fig 8.35.

Each neurone possesses a cell body (perikaryon) 3–100 μm in diameter (fig 8.35), which contains a nucleus, the majority of the cell's other organelles embedded in a mass of cytoplasm, and a variable number of cytoplasmic processes extending from it. The arrangement of these processes forms the basis of one means of classifying neurones (fig 8.36), there being uni-, pseudouni-, bi-, and multipolar neurones. Processes which conduct impulses towards the cell body are called **dendrons**. They are small, relatively wide, and break up into fine terminal branches. Processes conducting impulses away from the cell body are termed **axons** or nerve fibres and may be several metres long. They are thinner than dendrons.

The terminal region of an axon is neurosecretory and breaks up into many fine branches with swollen endings. It communicates with adjacent neurones at sites called **synapses** which may be excitatory or inhibitory (section 16.1.2). The bulbous endings possess small vesicles containing transmitter substance (acetylcholine or noradrenaline) and many mitochondria, for these regions are extremely active metabolically. Nissl's granules, which are

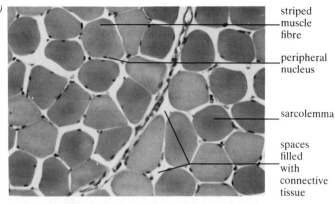

(a)

striped muscle fibre

peripheral nucleus

sarcolemma

spaces filled with connective tissue

(b)

striations

nucleus

striped muscle fibril

Fig 8.32 *(a) TS and (b) LS of voluntary (striped) muscle*

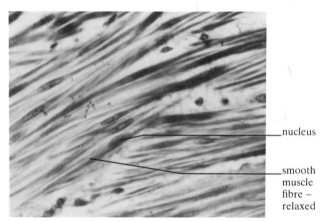

nucleus

smooth muscle fibre – relaxed

Fig 8.33 *LS involuntary (smooth) muscle*

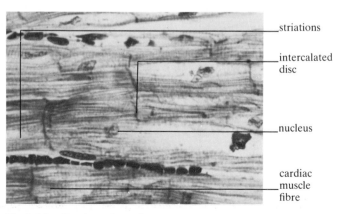

striations

intercalated disc

nucleus

cardiac muscle fibre

Fig 8.34 *Section of cardiac muscle*

groups of ribosomes associated with protein synthesis, and Golgi apparatus are present in the cell body (fig 8.37). Microtubules, neurofibrils, rough endoplasmic reticulum and mitochondria are present throughout the axoplasm of the neurone.

Nerve fibres may be **myelinated** (as, for example, in cranial and spinal nerves) or **non-myelinated** (as in autonomic nerves). In the former case, the fibres are completely surrounded by a fatty **myelin sheath** formed by many **Schwann cells**. The sheath is constricted at intervals along its length by **nodes of Ranvier** (fig 8.35). One Schwann cell nucleus is visible in the sheath between a pair of nodes. Surrounding the sheath is a tough inelastic membrane, the **neurilemma**.

Non-myelinated fibres do not possess nodes of Ranvier, and are incompletely enclosed by a Schwann cell (fig 8.38). Indeed there may be up to nine fibres partially shrouded by a single Schwann cell.

8.6.2 Nerves

These consist of bundles of nerve fibres ensheathed in connective tissue called the **epineurium**. Inward extensions of the epineurium, called the **perineurium**, divide the fibres into smaller bundles, whilst each fibre is itself surrounded by connective tissue called the **endoneurium** (fig 8.39). Nerves are classified according

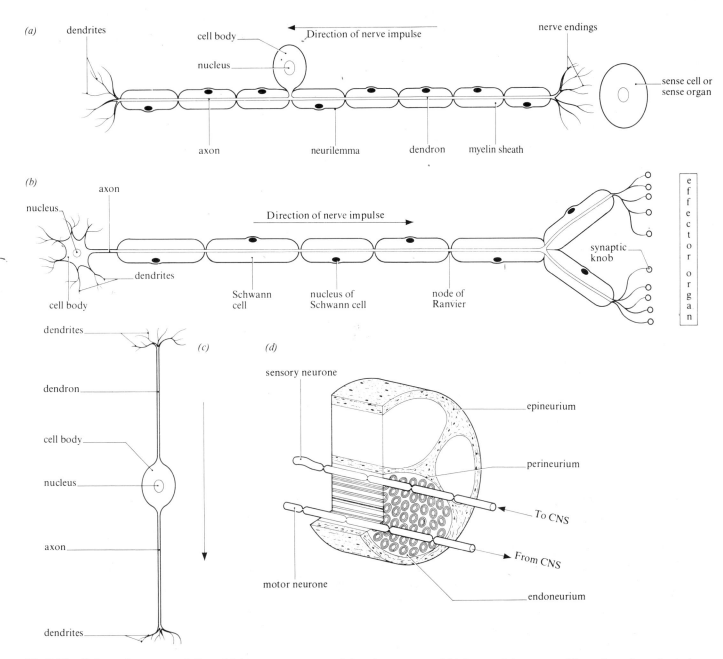

Fig 8.35 *Schematic representation of (a) sensory neurone, (b) motor neurone, (c) internuncial neurone, (d) section of myelinated nerve*

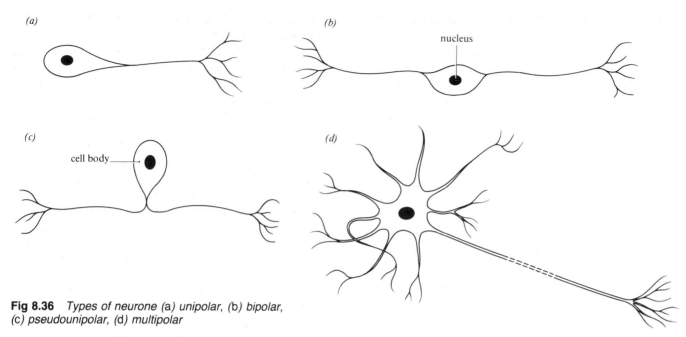

Fig 8.36 *Types of neurone* (a) *unipolar,* (b) *bipolar,*
(c) *pseudounipolar,* (d) *multipolar*

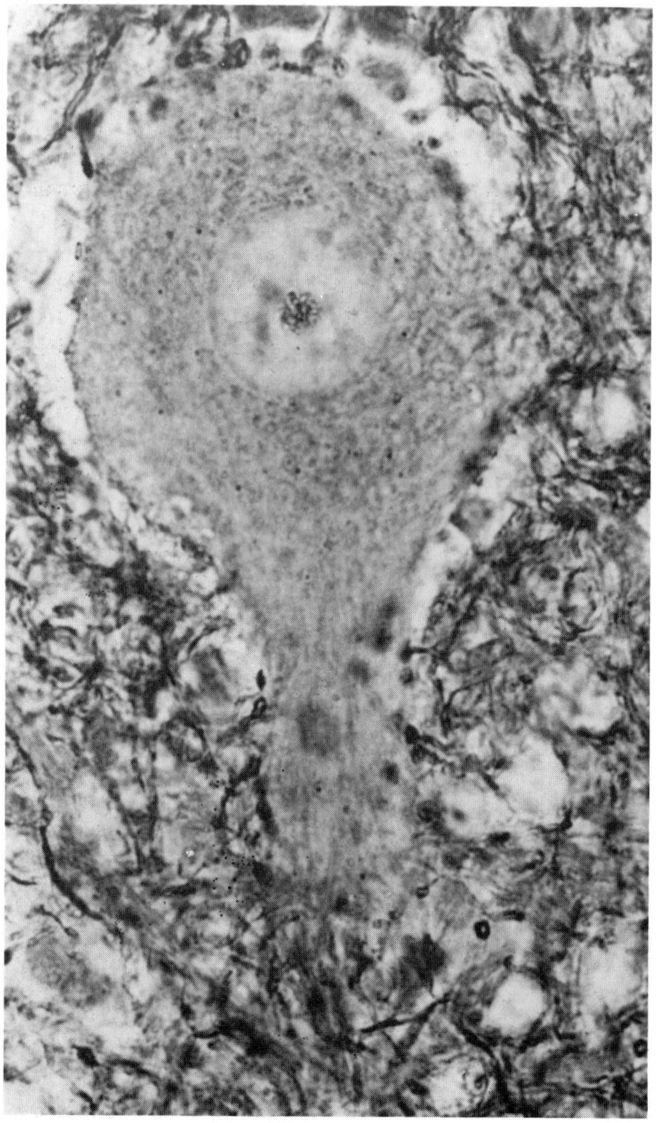

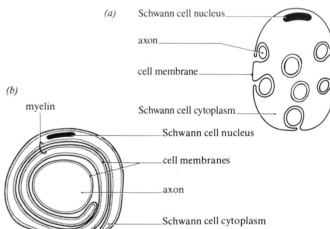

Fig 8.38 (a) *Non-myelinated nerve fibre.* (b) *Myelinated nerve fibre*

Fig 8.37 *Neurone with synapses*

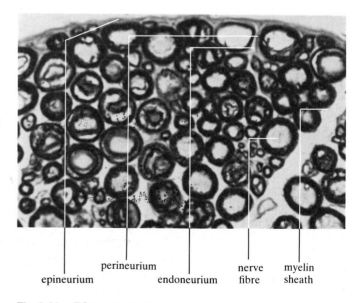

epineurium perineurium endoneurium nerve fibre myelin sheath

Fig 8.39 *TS myelinated nerve*

to the direction in which they convey impulses. Sensory or afferent nerves convey impulses into the central nervous system (such as the olfactory, optic and auditory nerves), whilst efferent or motor nerves conduct impulses away from the central nervous system (as in the oculomotor, pathetic and abducens). Mixed nerves convey impulses in both directions (for example the trigeminal, facial, glossopharyngeal, vagus and all spinal nerves).

8.6.3 Neuroglia

These cells are ten times more numerous than neurones and are found packed around the neurones throughout the central nervous system, thus supporting them mechanically by filling up the majority of interneurone space. It is thought that their metabolic activity is closely allied to that of the neurones they surround and that they might be involved in the memory processes by storing

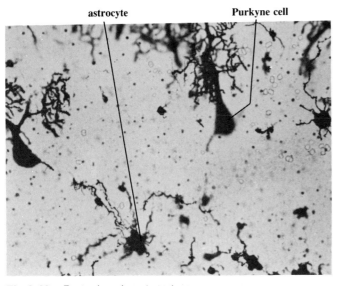

Fig 8.40 *Protoplasmic astrocytes*

information in the form of an RNA code. They may also nourish the cytoplasmic extensions of neurones. Satellite neuroglia, called **Schwann cells**, synthesise the myelin sheath of myelinated nerve fibres whilst others are phagocytic in function. The different kinds of neuroglial cells are classified as follows. **Ependymal cells** line the brain cavities and spinal canal and form an epithelial layer in the choroid plexus. They serve to connect the cavities with underlying tissues (section 16.2.4). **Macroglia** are divided into two categories, astrocytes and oligodendrocytes. **Protoplasmic astrocytes** are located in the grey matter (fig 8.40). Numerously branched, short thick processes radiate from the cell body which contains an ovoid nucleus and much glycogen. **Fibrous astrocytes** are located in the white matter. Fewer branched, long processes radiate from the cell body which itself possesses an ovoid nucleus and much glycogen. Some of the branches actually abut upon the walls of blood vessels. These cells convey nutrients from the bloodstream to the neurones. Both types of astrocytes are interconnected forming an extensive three-dimensional meshwork in which the neurones are embedded. They also divide frequently to form scar tissue if the central nervous system is injured.

Oligodendrocytes are located in grey and white matter. They are smaller than astrocytes and the single nucleus is spherical. Fewer, finer branches radiate from the cell body, which itself contains cytoplasm rich in ribosomes. Schwann cells are specialised oligodendrocytes which synthesise the myelin sheath of myelinated fibres.

Microglia are located in grey and white matter but are more numerous in the grey. A thick process arises from each end of the small, elongated cell body which itself contains lysosomes and a well-developed Golgi apparatus. All branches possess further lateral branches. When the brain is damaged these cells are stimulated to become phagocytic and move around in amoeboid fashion to combat invasion of foreign particles.

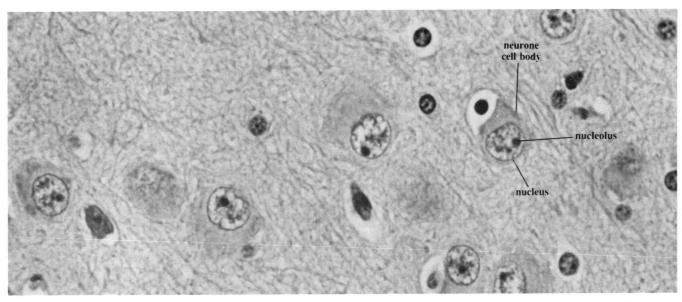

Fig 8.41 *Neurones and glial cells in human cortex*

WINDSOR & MAIDENHEAD COLLEGE

CLAREMONT ROAD

WINDSOR

Chapter Nine

Autotrophic nutrition

In chapters 9–11 we shall be concerned with living organisms as consumers of food, that is energy and materials. The process of *acquiring* energy and materials is called **nutrition** and this is the theme of chapters 9 and 10. In chapter 11 **respiration** is considered, the process whereby organisms *release* energy from the energy-rich compounds acquired by nutrition.

Energy can neither be created nor destroyed (**the law of conservation of energy**). It may occur in various forms, such as light, chemical, heat, electrical, mechanical and sound, and these can be converted from one form to another, that is they are **interconvertible**. A simple example would be striking a match, where, in the matchhead, chemical energy is converted to heat, light and sound energy.

Energy may be defined as the capacity to do work. All living organisms may be regarded as working machines which require a continuous supply of energy in order to keep working, and so to stay alive. This energy is required in order to carry out a variety of vital processes. The forms of work include:

chemical synthesis of substances for growth and repair;
active transport of substances into and out of cells;
electrical transmission of nerve impulses;
mechanical contraction of muscles (movement);
maintenance of a constant body temperature in birds and mammals;
bioluminescence (that is the production of light by living organisms, such as fireflies, glow-worms and some deep sea animals);
electrical discharge, as in the electric eel.

9.1 Grouping of organisms according to their principal sources of energy and carbon

Living organisms can be grouped on the basis of their source of energy or source of carbon. Carbon is the most fundamental material required by living organisms (section 5.1.1)

Energy source

Despite energy existing in several forms, only two are suitable as energy sources for living organisms, namely light and chemical energy. Organisms utilising light energy to synthesise their organic requirements are called **phototrophs** or **phototrophic** (*photos*, light; *trophos*, nourishment), while those utilising chemical energy are called **chemotrophs** or **chemotrophic**. Phototrophs are character-ised by the presence of pigments, including some form of chlorophyll, which absorb light energy and convert it to chemical energy. An alternative term for the process of phototrophism is **photosynthesis**.

Carbon source

In this alternative grouping organisms which have an inorganic source of carbon, namely carbon dioxide, are called **autotrophs** or **autotrophic** (*autos*, self) and those having an organic source of carbon are called **heterotrophs** or **heterotrophic** (*heteros*, other). Unlike heterotrophs, autotrophs synthesise their own organic requirements from simple inorganic materials.

Table 9.1 summarises the groupings and shows how they interact. An important principle to emerge is that chemotrophic organisms are totally dependent on phototrophic organisms for their energy, and heterotrophic organisms are totally dependent on autotrophic organisms for their carbon.

By far the most important groups are the photoautotrophic organisms, including all green plants, and the chemoheterotrophic organisms, including all animals and fungi. Ignoring a few bacteria the position is simplified by saying that heterotrophic organisms are ultimately dependent on green plants for their energy and carbon. Photoautotrophic organisms are sometimes described as **holophytic** (*holos*, whole; *phyton*, plant).

> **9.1** Define photoautotrophism and chemoheterotrophism.

In ignoring the two minor groups, it should be stressed that the activities of the chemosynthetic organisms are of great importance, as will be shown in sections 9.10 and 9.11.

A few organisms do not fit neatly into these groups. *Euglena*, for example, is normally autotrophic, but some species can survive heterotrophically in darkness if an organic carbon source is present. Fig 9.1 illustrates further the relationship between the two main nutritional categories. It also illustrates how energy flows and carbon is cycled through living organisms and the environment, themes which are important in ecology (chapter 12).

In the carbon cycle, the carbon is released as carbon dioxide by respiration and converted to organic compounds by photosynthesis. Further details of the carbon cycle, including the role of chemosynthetic organisms, are shown in fig 9.2.

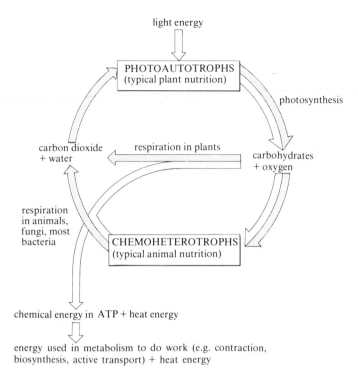

9.2 Examine fig 9.2. Which nutritional categories are indicated by (*a*) the darker background shading and (*b*) the white background?

9.3 What is the total natural annual productivity (turnover) of carbon in the carbon cycle?

9.2 Photosynthesis

Table 9.1 shows that there are two types of photosynthetic organisms, photoautotrophs and photoheterotrophs. The majority of these organisms are photoautotrophs and it is these that will be studied in detail in this chapter.

9.2.1 Importance of photosynthesis

All life on Earth depends on photosynthesis, either directly or, as in the case of animals, indirectly. Photosynthesis makes both carbon and energy available to living organisms and produces the oxygen in the atmosphere which is vital for all aerobic forms of life. Humans also depend on photosynthesis for the energy-containing fossil fuels which have developed over millions of years. One recent estimate of the annual fixation of carbon (not carbon dioxide) by photosynthesis is 75×10^{12} kg year^{-1} (fig 9.2). About 40% of this is contributed by phytoplankton living in the oceans. Of the total amount of solar radiation intercepted by our planet, about half reaches its surface after absorption, reflection and scattering in the atmosphere. Of this, only about 50% is of the right wavelength to stimulate photosynthesis and, although estimates vary, it is

Fig 9.1 (above) *Flow of energy (open arrows) and cycling of carbon (solid arrows) through photoautotrophs and chemoheterotrophs, and balance between photosynthesis and respiration. Light energy is converted to chemical energy in photosynthesis and used in the synthesis of organic materials from inorganic materials. Organic materials form the energy and carbon source for chemoheterotrophs and are released again in the process of respiration (also carried out by plants). Every energy conversion is accompanied by some loss of energy as heat*

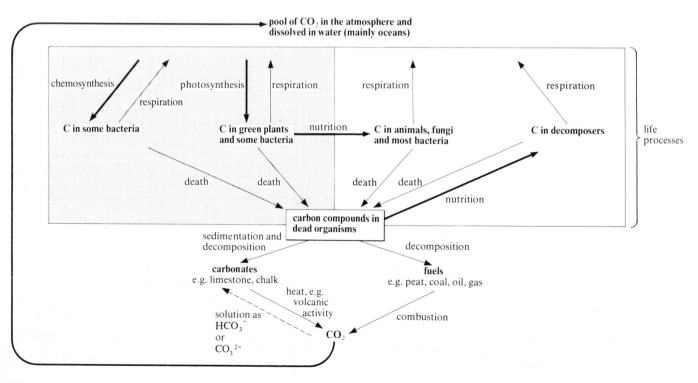

Table 9.1 Grouping of living organisms according to principal sources of carbon and energy.

		CARBON SOURCE	
		AUTOTROPHIC use carbon dioxide (inorganic)	HETEROTROPHIC use organic source of carbon
ENERGY SOURCE	PHOTOTROPHIC (PHOTOSYNTHETIC) use light energy	PHOTOAUTOTROPHIC all green plants, blue-green bacteria and green and purple sulphur bacteria	PHOTOHETEROTROPHIC few organisms, e.g. some purple non-sulphur bacteria
	CHEMOTROPHIC use chemical energy	CHEMOAUTOTROPHIC (CHEMOSYNTHETIC) a few bacteria, e.g. *Nitrosomonas* and some other nitrogen cycle bacteria	CHEMOHETEROTROPHIC all animals and fungi, most bacteria, some parasitic flowering plants, e.g. dodder (*Cuscuta*)

NB Majority of living organisms occur in photoautotrophic and chemoheterotrophic categories.

likely that only about 0.2% of this is used in net plant production (about 0.5% of the energy actually reaching plants). From this small fraction of the available energy virtually all life is sustained.

One potentially important use of photosynthesis is as an alternative source of energy to our depleting natural reserves of oil and gas. Attempts are currently being made to mimic the early stages of the photosynthetic process in plants whereby water is split into hydrogen and oxygen using light (solar) energy. Hydrogen could be burned as a fuel, and the waste material would be water. This system would therefore provide an attractive alternative or supplement to nuclear and other forms of energy.

Research into photosynthesis is also of great importance in agriculture because, as the figures above suggest, there is great scope for improving the efficiency of agriculture. New sources of food are being produced from micro-organisms such as algae and photosynthetic bacteria which are often more efficient as 'crops'. If grown on sewage or industrial waste, these substances could be purified or made use of at the same time as food is produced.

Fig 9.2 (left) *The carbon cycle. Heavy arrows indicate the dominant of the two pathways. Some rough estimates of actual quantities involved:*
Oceans (mainly phytoplankton): 40 × 10¹² kg carbon per year fixed as carbon dioxide by photosynthesis. Most of this is released in respiration
Land: 35 × 10¹² kg carbon per year fixed as carbon dioxide by photosynthesis
10 × 10¹² kg carbon per year released as carbon dioxide by respiration of plants and animals
25 × 10¹² kg carbon per year released as carbon dioxide by respiration of decomposers
5 × 10¹² kg carbon per year released as carbon dioxide by burning of fossil fuels, enough to be causing a gradual increase in carbon dioxide concentration in atmosphere and oceans NB 10³ kg = 1 tonne

9.4 What advantages would production of hydrogen fuel by the action of light on water have over nuclear power?

9.3 The structure of the leaf

In higher plants the major photosynthetic organ is the leaf. As with all living organs, structure and function are closely linked. From the equation for photosynthesis

$$CO_2 + H_2O \xrightarrow[\text{chlorophyll}]{\text{sunlight}} (CH_2O)_n + O_2$$

carbon dioxide — water — carbohydrate — oxygen

it can be deduced that first the leaf requires a source of carbon dioxide and water, secondly it must contain chlorophyll and be adapted to receive sunlight, thirdly oxygen will escape as a waste product and finally the useful product, carbohydrate, will have to be exported to other parts of the plant or stored. In its structure the leaf is highly adapted to satisfy these requirements. Figs 9.3 and 9.4 show labelled photomicrographs of leaf sections which will aid you in interpreting sections of monocotyledonous and dicotyledonous leaves. Fig 9.5 is a simplified drawing of a vertical section through a dicotyledonous leaf. The epidermis of different leaf types are shown in fig 8.2 and details of stomatal structure and function are dealt with in chapter 14.

The structure and function of different tissues in a dicotyledonous leaf are summarised in table 9.2.

9.5 Make a list of the ways in which the structure of the leaf contributes to its successful functioning.

251

Table 9.2 Structure and function of tissues in a dicotyledonous leaf.

Tissue	Structure	Function
Upper and lower epidermis	One cell thick. Colourless flattened cells. External walls covered with a cuticle of cutin (waxy substance). Contains stomata (pores) which are normally confined to, or more numerous in, the lower epidermis. Each stoma is surrounded by a pair of guard cells.	Protective. Cutin is waterproof and protects from desiccation and infection. Stomata are sites of gaseous exchange with the environment. Their size is regulated by guard cells, special epidermal cells containing chloroplasts.
Palisade mesophyll	Column-shaped ('palisade') cells with numerous chloroplasts in a thin layer of cytoplasm.	Main photosynthetic tissue. Chloroplasts may move towards light.
Spongy mesophyll	Irregularly shaped cells fitting together loosely to leave large air spaces.	Photosynthetic, but fewer chloroplasts than palisade cells. Gaseous exchange can occur through the large air spaces via stomata. Stores starch.
Vascular tissue	Extensive finely branching network through the leaf.	Conducts water and mineral salts to the leaf in xylem. Removes products of photosynthesis (mainly sucrose) in phloem. Provides a supporting skeleton to the lamina, aided by collenchyma of the midrib, turgidity of the mesophyll cells, and sometimes sclerenchyma.

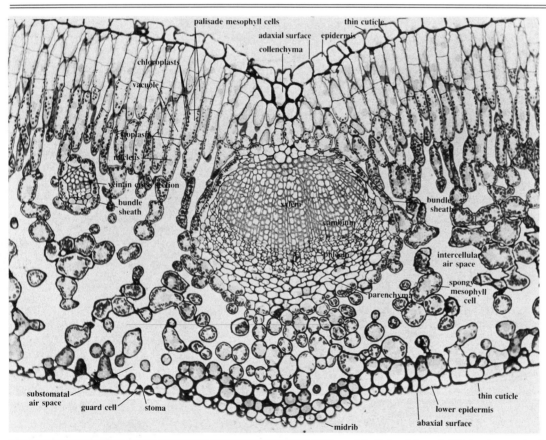

Fig 9.3 (above) *TS lamina and midrib of a privet leaf (Ligustrum), a typical dicotyledon*

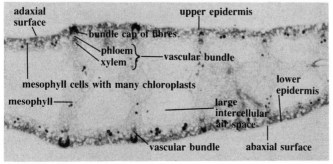

Fig 9.4 (left) *TS lamina of an Iris leaf, a typical monocotyledon*

252

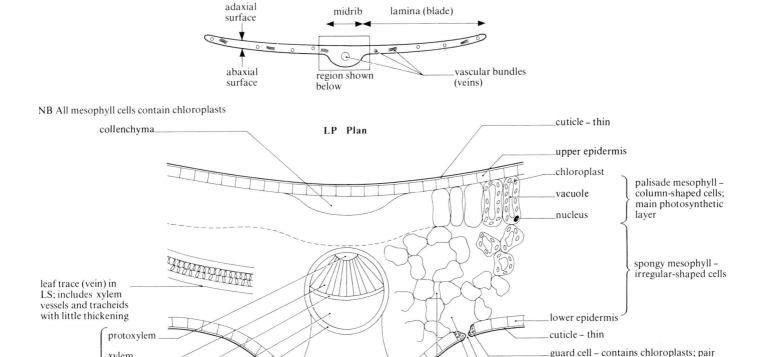

Plan (naked eye)

adaxial surface

midrib

lamina (blade)

abaxial surface

region shown below

vascular bundles (veins)

NB All mesophyll cells contain chloroplasts

LP Plan

collenchyma

cuticle – thin

upper epidermis

chloroplast

vacuole

nucleus

palisade mesophyll – column-shaped cells; main photosynthetic layer

spongy mesophyll – irregular-shaped cells

leaf trace (vein) in LS; includes xylem vessels and tracheids with little thickening

lower epidermis

cuticle – thin

guard cell – contains chloroplasts; pair control opening of stoma

stoma – more numerous in lower epidermis; allows gaseous exchange

substomatal air space for efficient gaseous exchange

vascular bundle

protoxylem

xylem

cambium

phloem

sheath – closely packed parenchyma or sclerenchyma

collenchyma of midrib

large intercellular air space

A final point to note is the arrangement of the leaves for minimal overlapping. Such leaf mosaics are particularly noticeable in some plants, such as ivy. Etiolation (rapid extension growth in the dark) and phototropism (growth towards light) are further phenomena which ensure that leaves reach the light.

9.3.1 Chloroplasts

In eukaryotes, photosynthesis takes place in organelles called chloroplasts, distributed in the cytoplasm in numbers varying from one (as in *Chlamydomonas* and *Chlorella*) to about 100 (palisade mesophyll cells). In higher plants chloroplasts are usually biconvex in section and circular in surface view. They are about 3–10 μm (average 5 μm) in diameter, and so are visible with a light microscope. They are more variable in form in the algae, for example they are spiral in *Spirogyra* and cup-shaped in *Chlamydomonas*, and usually possess pyrenoids, as in *Spirogyra* (section 3.2.4).

Chloroplasts arise from small, undifferentiated bodies called **proplastids** found in the growing regions of plants (meristems) and are surrounded by two membranes, which form the **chloroplast envelope**. They always contain chlorophyll and other photosynthetic pigments located on a system of membranes running through a ground substance, or stroma. Their detailed structure is revealed

Fig 9.5 *Diagrammatic transverse section of a typical dicotyledon leaf*

by electron microscopy. Fig 7.6 shows the typical appearance of chloroplasts in a leaf mesophyll cell as seen at low power in the electron microscope. Figs 9.6 and 9.8 show electron micrographs and fig 9.7 a diagram of a chloroplast, illustrating the membrane system. The membrane system is the site of the **light reactions** in photosynthesis (section 9.4.2). The membranes are covered with chlorophyll and other pigments, enzymes and electron carriers. The system consists of many flattened, fluid-filled sacs called **thylakoids** which form stacks called **grana** at intervals, with lamellae (layers) between the grana. Each granum resembles a pile of coins and the lamellae are often sheet-like (fig 9.8). Grana are just visible under the light microscope as grains.

The stroma is the site of the **dark reactions** of photosynthesis (section 9.4.3). The structure is gel-like, containing soluble enzymes, notably those of the Calvin cycle, and other chemicals such as sugars and organic acids. Excess carbohydrate from photosynthesis is also stored as grains of starch mainly in the light. Spherical lipid droplets are often associated with the membranes. They become more conspicuous as membranes break down during senescence, presumably accumulating lipids from the membranes. In chromoplasts they are often very large and accumulate carotenoid pigments.

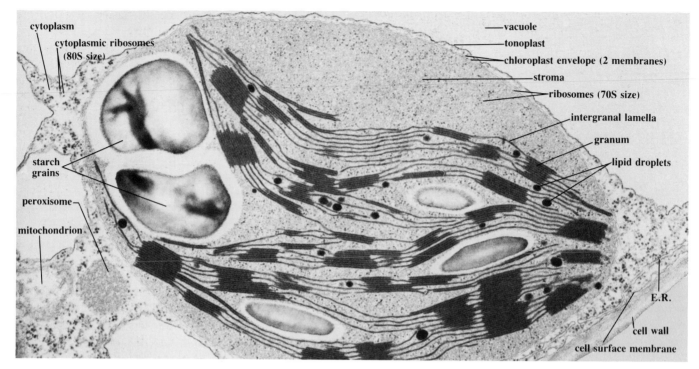

Fig 9.6 *Electron micrograph of a chloroplast (× 15 800)*

Labels (clockwise): cytoplasm; cytoplasmic ribosomes (80S size); starch grains; peroxisome; mitochondrion; vacuole; tonoplast; chloroplast envelope (2 membranes); stroma; ribosomes (70S size); intergranal lamella; granum; lipid droplets; E.R.; cell wall; cell surface membrane

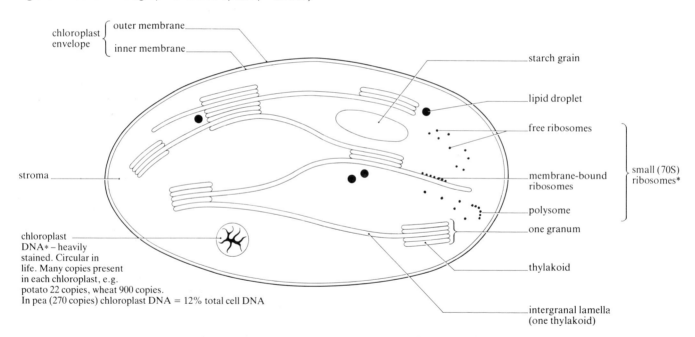

Labels: chloroplast envelope {outer membrane, inner membrane}; stroma; chloroplast DNA* – heavily stained. Circular in life. Many copies present in each chloroplast, e.g. potato 22 copies, wheat 900 copies. In pea (270 copies) chloroplast DNA = 12% total cell DNA; starch grain; lipid droplet; free ribosomes; membrane-bound ribosomes; polysome; small (70S) ribosomes*; one granum; thylakoid; intergranal lamella (one thylakoid)

Fig 9.7 *Chloroplast structure. The membrane system has been reduced for convenience (*prokaryote-like protein synthesising machinery)*

Protein-synthesising machinery and the endosymbiont theory

An interesting feature of chloroplasts, apart from photosynthesis, is their protein-synthesising machinery. During the 1960s it was shown that both chloroplasts and mitochondria contain DNA and ribosomes. This led to speculation that these organelles may be partially or completely independent of the control of the nucleus in the cells containing them. It was further suggested that they might represent prokaryotic organisms which invaded eukaryotic cells at an early stage in the history of life. Thus the organelles represent an extreme form of symbiosis, a

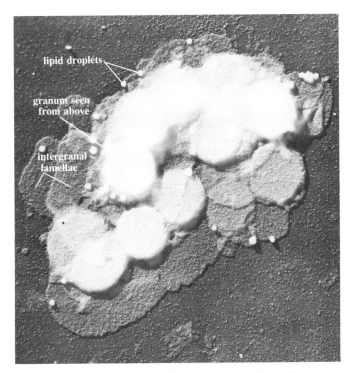

Fig 9.8 *Scanning electron micrograph of a 'stripped' chloroplast (a chloroplast whose outer envelope has been removed) looking down from above on the lamellae and grana which can be seen in three dimensions. Note that the lamellae are sheet-like and interconnect the grana. The preparation is a shadowed replica (see appendix 2)*

lipid droplets

granum seen from above

intergranal lamellae

theory known as the **endosymbiont theory**. Some of the evidence for this is presented in table 9.3.

Photosynthetic prokaryotes (blue-green bacteria and other photosynthetic bacteria) do not contain chloroplasts. Instead their photosynthetic pigments are located in membranes distributed throughout the cytoplasm. Thus the whole cell is similar to one chloroplast, and is approximately the same size. It is now believed that chloroplasts are the descendants of blue-green bacteria.

It has been shown that, while chloroplasts and mitochondria do code for and make some of their own proteins, they no longer contain enough DNA to code for all of them, and the task is shared with nuclear DNA.

9.3.2 Photosynthetic pigments

The photosynthetic pigments of higher plants fall into two classes, the chlorophylls and carotenoids. The role of the pigments is to absorb light energy, thereby converting it to chemical energy. They are located on the chloroplast membranes and the chloroplasts are usually arranged within the cells so that the membranes are at right-angles to the light source for maximum absorption. Table 9.4 shows the range of pigments found within each of the main groups of plants.

Chlorophylls

Chlorophylls absorb mainly red and blue-violet light, reflecting green light and therefore giving plants their characteristic green colour, unless masked by other pigments. Fig 9.9 shows the absorption spectra of chlorophylls *a* and *b* compared with carotenoids.

Chlorophylls are characterised by a porphyrin ring (fig 9.10) which is a structure found in several important biological compounds, such as the haem of haemoglobin, myoglobin and cytochromes. The porphyrin ring is a flat, square structure containing four smaller rings (I–IV), each possessing a nitrogen atom which can bond with a metal atom, such as magnesium in the chlorophylls and iron in haem. The 'head' is joined to a long hydrocarbon tail by an ester linkage formed between an alcohol group ($-OH$) at the end of phytol and a carboxyl group ($-COOH$) on the head. Different chlorophylls have different side-chains on the head and this modifies their absorption spectra.

Table 9.3 Comparison of prokaryotes, chloroplasts and mitochondria with eukaryotes.

	Prokaryotes, chloroplasts and mitochondria	*Eukaryotes*
DNA	Circular Not contained in chromosomes Not contained in nucleus	Linear Contained in chromosomes Contained in a nucleus
Ribosomes	Smaller (70S)	Larger (80S)
Sensitivity to antibiotics	Protein synthesis inhibited by chloramphenicol, not cycloheximide	Protein synthesis inhibited by cycloheximide, not chloramphenicol
Average diameter	Prokaryote cell: 0.5–3 μm Chloroplast: 3–5 μm Mitochondrion: 1 μm	Eukaryote cell: 20 μm

Table 9.4 The main photosynthetic pigments, their colours and distribution.

Class of pigment with examples	Colour	Distribution
Chlorophylls		
chlorophyll *a*	yellow-green	All photosynthetic organisms except some photosynthetic bacteria
chlorophyll *b*	blue-green	Higher plants and green algae
chlorophyll *c*	green	Brown algae, a few unicellular algae including diatoms
chlorophyll *d*	green	Some red algae
bacteriochlorophylls *a–d*	pale blue	Photosynthetic bacteria
Carotenoids (carotenes and xanthophylls)		
Carotenes		
β-carotene	orange	All photosynthetic organisms except photosynthetic bacteria
Xanthophylls (carotenols)		
Great variety	all yellow	Fucoxanthin helps give brown algae their colour. It has a very broad absorption spectrum

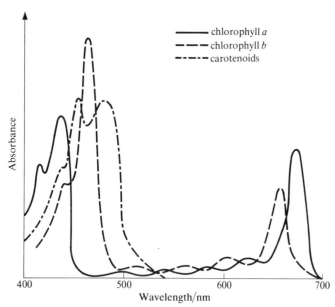

Fig 9.9 (above) *Absorption spectra of chlorophylls* a *and* b, *and carotenoids*

Fig 9.10 (right) *Structure of chlorophylls*

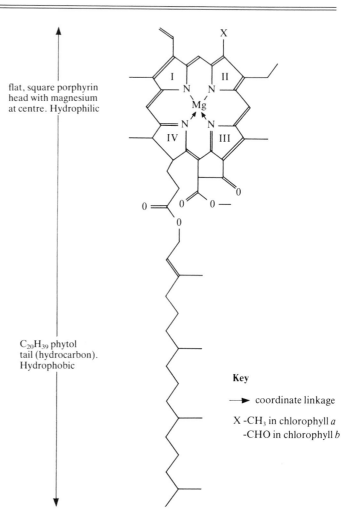

The structure is related to function in the following ways:
(*a*) the long tail is lipid soluble (hydrophobic) and so is anchored in the thylakoid membrane;
(*b*) the head is hydrophilic (water loving) and so generally lies in the surface of the membrane next to the aqueous solution of the stroma;
(*c*) the flat head is parallel to the membrane surface for light absorption;
(*d*) modifications of side-groups on the head cause changes in the absorption spectrum so that different energies of light are absorbed;
(*e*) absorption of light energy by the head causes changes in the energy levels of electrons within the head.

Chlorophyll *a* is the most abundant photosynthetic pigment and is the only one found in all photosynthetic plants due to its central role as a primary pigment. It exists in several forms, depending on its arrangement in the membrane. Each form differs slightly in its red absorption peak; for example, the peak may be at 670 nm, 680 nm, 690 nm or 700 nm.

9.6 How does the absorption spectrum of chlorophyll *a* differ from that of chlorophyll *b*?

Carotenoids

Carotenoids are yellow, orange, red or brown pigments that absorb strongly in the blue-violet range. They are usually masked by the green chlorophylls but can be seen in leaves prior to leaf-fall since chlorophylls break down first. They are also found in the chromoplasts of some flowers and fruits where the bright colours serve to attract insects, birds and other animals for pollination or dispersal; for example the red skin of the tomato is due to lycopene, a carotene.

Carotenoids have three absorption peaks in the blue-violet range of the spectrum (fig 9.9) and apart from acting as accessory pigments, they may also protect chlorophylls from excess light and from oxidation by oxygen produced in photosynthesis.

Carotenoids are of two types, carotenes and xanthophylls. **Carotenes** are hydrocarbons, the majority being

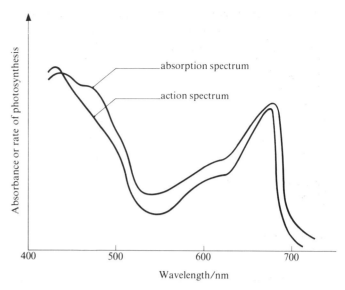

Fig 9.12 *Action spectrum for photosynthesis compared with absorption spectrum of photosynthetic pigments*

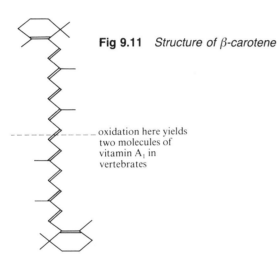

Fig 9.11 *Structure of β-carotene*

oxidation here yields two molecules of vitamin A₁ in vertebrates

C_{40} compounds (tetraterpenes). The most widespread and important is β-carotene (fig 9.11), which is familiar as the orange pigment of carrots. Vertebrates are able to break the molecule into two during digestion to form two molecules of vitamin A. **Xanthophylls** are chemically very similar to carotenes but contain oxygen.

Absorption and action spectra

When investigating a process such as photosynthesis that is activated by light, it is important to establish the action spectrum for the process and to use this to try to identify the pigments involved. An **action spectrum** is a graph showing the effectiveness of different wavelengths of light in stimulating the process being investigated, in this case photosynthesis, where the response could be measured for example in terms of oxygen production at different wavelengths. An **absorption spectrum** is a graph of the relative absorbance of different wavelengths of light by a pigment. An action spectrum for photosynthesis is shown in fig 9.12, together with an absorption spectrum for the

combined photosynthetic pigments. Note the close similarity, which indicates that the pigments, chlorophylls in particular, are those responsible for absorption of light in photosynthesis.

Excitation of pigments by light

Pigments are chemicals that absorb visible light and this causes the excitation of certain electrons to '**excited states**', that is the electrons absorb energy. The shorter the wavelength of light, the greater its energy and the greater its potential to promote electrons to these excited states. The excited state is usually unstable and the molecule returns to its '**ground state**' (original low energy state), losing its energy of excitation as it does so. This energy can be lost in several ways, including reversal of absorption either by **fluorescence** or **phosphorescence**. Here some of the energy is lost as heat and some as light. The emitted light has a longer wavelength (less energy) than the absorbed light. This can be observed when solutions of chlorophyll are irradiated and then observed in darkness.

In the light reactions of photosynthesis, the excited primary pigments lose electrons, leaving positive 'holes' in their molecules; for example

$$\text{chlorophyll} \xrightarrow{\text{light energy}} \text{chlorophyll}^+ + e^-$$
$$\text{(reduced form)} \qquad \text{(oxidised form)} \quad \text{electron}$$

Each electron lost is accepted by another molecule, the so-called **electron acceptor**, so this is an oxidation–reduction process (see section A1.2). The chlorophyll is oxidised and the electron acceptor is reduced. Chlorophyll is described as an **electron donor**.

257

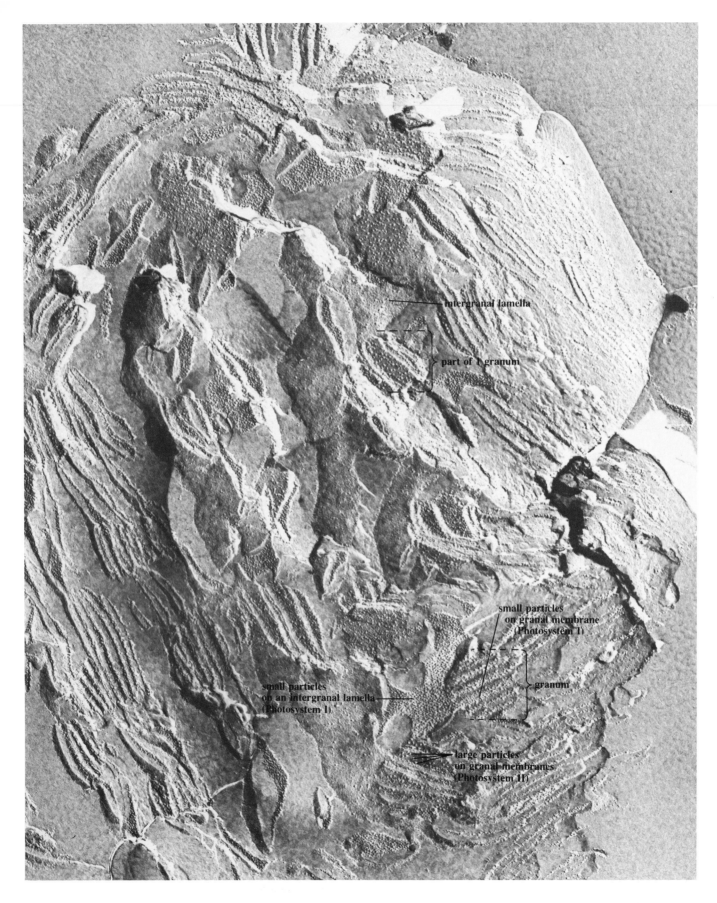

intergranal lamella

part of a granum

small particles
on granal membrane
(Photosystem I)

granum

small particles
on an intergranal lamella
(Photosystem I)

large particles
on granal membranes
(Photosystem II)

Fig 9.13 *Freeze-fractured isolated thylakoids of a chloroplast. The surfaces of fractured granal and intergranal membranes are visible. Note the aggregates of particles on the membranes*

Primary and accessory pigments

The photosynthetic pigments are of two types, **primary pigments** and **accessory pigments**. The latter pass the energy they emit to primary pigments. The electrons emitted by primary pigments are those that drive the reactions of photosynthesis.

There are two primary pigments, both forms of chlorophyll *a*; they are called P680 and P700 (see below). P stands for pigment. The accessory pigments are the other chlorophylls (including other forms of chlorophyll *a*) and the carotenoids.

> **9.7** Movement of energy from one pigment molecule to another must involve some loss of energy as heat, since energy cannot be transferred with 100% efficiency. Chlorophyll *b* passes energy to chlorophyll *a*. Can you predict whether chlorophyll *a* or chlorophyll *b* has the lower energy of excitation?

Photosystems and reaction centres

Over the last twenty years a great deal has been learned about the precise structural arrangement of the pigments and associated molecules in the thylakoid membranes. It is currently believed that there are two types of **photosystem**, called **photosystems I and II (PSI and PSII)** (photosystem I probably being the first to evolve). Evidence for this comes from both biochemical and electron microscopic observations. The latter come from the technique of freeze fracturing described in section A2.5, and are a good example of how this technique has contributed to biological research. Fig 9.13 shows the regular arrangement of two sizes of particle in the thylakoid membranes. It is thought that the small and large particles represent photosystems I and II respectively. Each has its own characteristic set of chlorophyll molecules, summarised in fig 9.14. PSII particles seem to be mostly associated with grana, and PSI particles with intergranal lamellae.

Each photosystem contains a collection of molecules involved in the light reactions of photosynthesis. Each contains about 300 chlorophyll molecules which act as a light-harvesting 'antenna'. A quantum of light energy absorbed by any of these molecules will pass downhill in energy terms to a **reaction centre** which consists of one specialised chlorophyll *a* molecule called **P680** or **P700**, depending on its absorption peak in nanometres. P680 and P700 are energy traps; they are the molecules which absorb light of the longest wavelength and hence the lowest energy. Other specialised forms of chlorophyll *a*, such as chlorophyll *a* 670, may be regarded as accessory pigments, like chlorophyll *b* 650. (Energy of excitation is transferred from one pigment molecule to another by resonance. The mechanism is known as **sensitised fluorescence** and is most efficient when the molecules are close together; hence their

arrangement in photosystems.) At the reaction centre energy from light causes electrons to leave P680 or P700, so driving a chemical reaction (see section 9.4.2). Thus it is here that light energy is converted to chemical energy and this is the energy conversion central to photosynthesis. The concept of a photosynthetic unit is sometimes used to include both photosystems and the electron transport chain that links them (section 9.4.2).

Computer program. LIGHT HARVEST, PHOTOSYNTHESIS: A double package. LIGHT HARVEST simulates the random walk of photons during light harvesting and demonstrates the concept of chlorophyll fluorescence. PHOTOSYNTHESIS is a simulation of the light phase of photosynthesis. The user controls the emission and wavelength of photons and can introduce blocking and uncoupling agents to explore some of the crucial experiments on which the light reaction is based.

9.4 Biochemistry of photosynthesis

A commonly used equation for photosynthesis is

$$6CO_2 + 6H_2O \xrightarrow[\text{chlorophyll}]{\text{light energy}} C_6H_{12}O_6 + 6O_2$$

carbon dioxide   water   sugar   oxygen
e.g. glucose

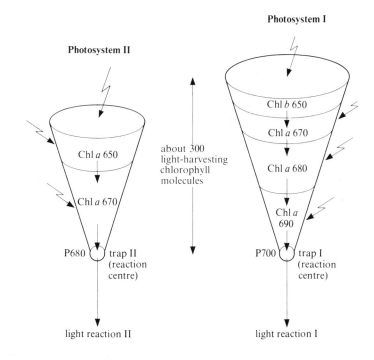

Fig 9.14 *Diagrammatic representation of energy traps in photosystems I and II*

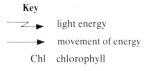

Key

⤳⟶ light energy

⟶ movement of energy

Chl chlorophyll

Chl *a* 670 chlorophyll *a* with absorption peak at 670 nm

P pigment, i.e. primary pigment molecule of chlorophyll *a*

This is useful for showing the formation of one molecule of sugar, but it should be realised that it is an overall summary of events. A better summary is

$$CO_2 + H_2O \xrightarrow[\text{chlorophyll}]{\text{light energy}} [CH_2O] + O_2$$

CH_2O does not exist as such, but represents a carbohydrate.

9.4.1 Source of oxygen

Looking at the equation a chemist would speculate about what type of reaction is involved, and a key question is whether the oxygen produced comes from carbon dioxide or water. The most obvious answer would seem to be carbon dioxide, so that the remaining carbon would be added to water to make carbohydrate. With the use of isotopes (section A1.3) in biology during the 1940s it became possible to answer the question directly.

The common isotope of oxygen has a mass number of 16 and is therefore represented as ^{16}O (8 protons, 8 neutrons). A rare isotope has a mass number of 18 (^{18}O). This is stable, but can be detected by virtue of its greater mass with a mass spectrometer, an important analytical instrument which distinguishes between different atoms and molecules according to their masses. In 1941 an experiment was carried out which produced results summarised in the following equation

$$CO_2 + H_2^{18}O \xrightarrow{\hspace{3cm}} [CH_2O] + {}^{18}O_2$$

The source of oxygen was thus shown to be water. The equation shows two atoms of oxygen coming from one molecule of water. So the balanced equation should be

$$CO_2 + 2H_2O \xrightarrow[\text{chlorophyll}]{\text{light energy}} [CH_2O] + O_2 + H_2O$$

This is the most accurate summary of photosynthesis and provides the extra information that water is produced, as well as used, in photosynthesis. This experiment confirmed indirect evidence put forward at about the same time by van Niel, who showed that bacteria do not produce oxygen during photosynthesis, although they use carbon dioxide. He concluded that all photosynthesising organisms need a source of hydrogen; for plants this is water, with oxygen being released; for sulphur bacteria, for example, it is hydrogen sulphide, sulphur being released instead of oxygen:

$$CO_2 + 2H_2S \xrightarrow[\text{chlorophyll}]{\text{light energy}} [CH_2O] + 2S + H_2O$$

This equation for sulphur bacteria is analogous to the plant equation.

These experiments provided a profound insight into the nature of photosynthesis because they showed that it takes place in two stages, the first of which involves acquiring hydrogen; in plants hydrogen is obtained by splitting water into hydrogen and oxygen. This requires energy which must be provided by light (hence the process used to be called **photolysis**: *photos*, light; *lysis*, splitting). Oxygen is released as a waste product. In the second stage, hydrogen combines with carbon dioxide to produce carbohydrate. Addition of hydrogen is an example of a type of chemical reaction called **reduction** (section A1.2).

The fact that photosynthesis is a two-stage process was first established in the 1920s and 1930s. The first stage was characterised by reactions requiring light and was called the **light reaction**. The second stage did not require light, and was called the **dark reaction**, although it takes place in light. It is now known that these are two sets of reactions which are also separated in space, the light reactions occurring on the chloroplast membranes and the dark reactions in the chloroplast stroma.

Having established that photosynthesis proceeds by light reactions followed by dark reactions, it remained in the 1950s to elucidate the nature of these reactions.

9.4.2 Light reactions

See reference to the computer program PHOTOSYNTHESIS, p. 259.

In 1958, Arnon and his co-workers showed that isolated chloroplasts, when exposed to light, could synthesis ATP from ADP and phosphate (**phophorlaytion**), reduce NADP to $NADPH_2$. and evolve oxygen.

He also showed that carbon dioxide could be *reduced* to carbohydrate *in the dark* if ATP and $NADPH_2$ (section 6.2.3) were provided. It therefore seemed that the role of the light reactions was to provide ATP and $NADPH_2$. Arnon noted the resemblance to respiration, where phosphorylation also occurs. This requires energy. In respiration it comes from oxidation of a food, usually glucose, and it is therefore called oxidative phosphorylation. In photosynthesis the energy comes from light and the process is therefore called photophosphorylation. Hence **oxidative phosphorylation** is the conversion of $ADP + P_i$ to ATP using chemical energy obtained from food by respiration, and **photophosphorylation** is the conversion of $ADP + P_i$ to ATP using light energy in photosynthesis. (P_i = inorganic phosphate.)

Arnon accurately predicted that photophosphorylation, like oxidative phosphorylation, would be coupled to the transfer of electrons in membranes. Electron transfer is fundamental to an understanding of both photosynthesis and respiration.

Cyclic and non-cyclic photophosphorylation

The role of the light reactions is to synthesise ATP and $NADPH_2$ using light energy. The process depends on a flow of electrons from primary pigments, and light provides the energy that causes this flow.

$$\text{chlorophyll } a \xrightarrow{\text{light energy}} \text{chlorophyll } a^+ + e^-$$
(reduced chlorophyll)   (oxidised chlorophyll)   excited electron

The fate of the electrons is summarised in fig 9.15. The pathway shown is sometimes known as the 'Z-scheme' from its shape. Remember that losing an electron is oxidation, gaining an electron is reduction (see section A1.2). In the Z-scheme two electrons are shown for convenience, though in practice they enter the scheme one at a time.

Fate of electrons. First, an electron from photosystem I or II is boosted to a higher energy level, that is it acquires excitation energy. Instead of falling back into the photosystem and losing its energy as, for example, fluorescence, it is captured by an electron acceptor (X or Y in fig 9.15). This represents the important conversion of light energy to chemical energy. The electron acceptor is thus reduced and a positively charged (oxidised) pigment is left in the photosystem. The electron then travels downhill, in energy terms, from one electron acceptor to another in a series of oxidation–reduction (redox) reactions. This electron flow is '**coupled**' to the formation of ATP in both cyclic and non-cyclic pathways; in addition, NADP is reduced in the non-cyclic pathway.

Non-cyclic photophosphorylation

Non-cyclic photophosphorylation is initiated by light shining on photosystems I and II. Excited electrons from P680 (PSII) and P700 (PSI) reduce electron acceptors X and Y respectively. P680 and P700 are now positively charged (oxidised). P680 is neutralised by electrons from water: electrons flow downhill from the latter to P680 via two electron carriers and oxygen is produced as a waste product of photosynthesis.

P700 is neutralised by electrons moving downhill from X via a chain of electron carriers, the energy from this flow being coupled to ATP production. Up to two ATP molecules may be made per pair of electrons, but this number is probably variable (two are shown in fig 9.15). Finally, electrons pass downhill from Y to NADP and combine with hydrogen ions to form $NADPH_2$. Note that the excess hydrogen ions are available from the 'splitting' of water.

Cyclic photophosphorylation

In cyclic photophosphorylation, electrons from Y are recycled back to P700 via the chain of electron carriers.

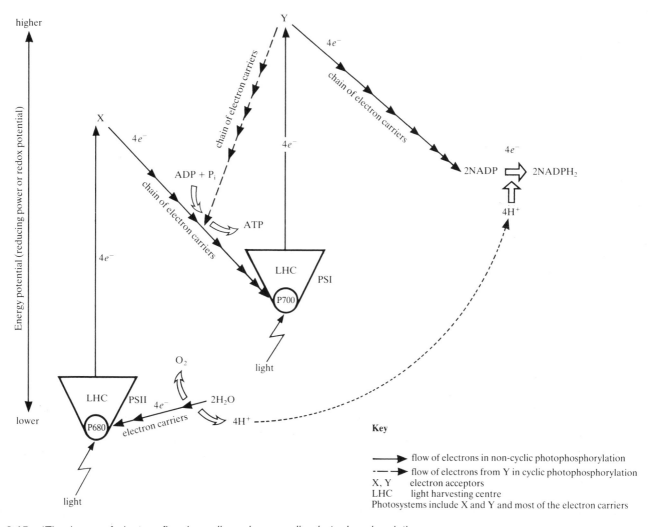

Fig 9.15 *'Z'-scheme of electron flow in cyclic and non-cyclic photophosphorylation*

Table 9.5 Comparison of cyclic and non-cyclic photophosphorylation.

	Non-cyclic	Cyclic
Pathway of electrons	Non-cyclic	Cyclic
First electron donor (source of electrons)	Water	Photosystem I (P700)
Last electron acceptor (destination of electrons)	NADP	Photosystem I (P700)
Products	Useful: ATP, NADPH$_2$ Waste: O$_2$	Useful: ATP only
Photosystems involved	I and II	I only

Their excitation energy is coupled to ATP production just as in non-cyclic photophosphorylation.

Table 9.5 shows the differences between cyclic and non-cyclic photophosphorylation.

The overall equation for non-cyclic photophosphorylation is

$$H_2O + NADP + 2ADP + 2P_i \xrightarrow[\text{chlorophyll}]{\text{light energy}} \tfrac{1}{2}O_2 + NADPH_2 + 2ATP$$

(maximum of 2ATP – may be less than 2)

Extra ATP can be made via cyclic photophosphorylation. The efficiency of energy conversion in the light reactions is high and estimated at about 39%.

The Hill reaction

In 1939 Robert Hill, working in Cambridge, discovered that isolated chloroplasts were capable of liberating oxygen in the presence of an oxidising agent (electron acceptor). This has since been called the Hill reaction. A number of so-called **Hill oxidants** substitute for the naturally occurring electron acceptor NADP, one of which is the blue dye DCPIP (2,6 dichlorophenolindophenol) that turns colourless when reduced:

$$\text{oxidised DCPIP} \xrightarrow{\text{light + chloroplasts}} \text{reduced DCPIP}$$

(blue)   H$_2$O   ½O$_2$ (colourless)

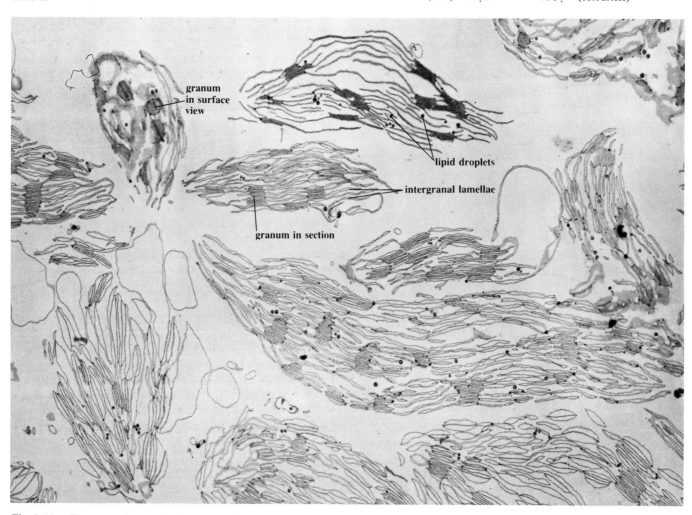

Fig 9.16 *Electron micrograph of chloroplasts after isolation in a dilute medium (× 13 485). Envelopes and stroma are lost*

Experiment 9.1: Investigating the Hill reaction

Isolation of chloroplasts

Materials

spinach, lettuce or cabbage leaves
scissors
cold pestle and mortar (or blender or food mixer)
muslin or nylon
filter funnel
centifuge and centrifuge tubes
ice–water–salt bath
glass rod

Solutions (see notes)

0.05 M phosphate buffer solution, pH 7.0
isolation medium
DCPIP solution (reaction medium)

Method

Chloroplasts can be isolated by grinding spinach, lettuce or cabbage leaves in a cold medium of suitable osmotic and ionic strength and pH, such as 0.4 M sucrose, 0.01 M KCl and 0.05 M phosphate buffer, pH 7.0. Solutions and apparatus must be kept cold during the isolation procedure if biochemical activity is to be preserved. The operation should also be performed as rapidly as possible, so study the method carefully and assemble the apparatus first.

Sufficient chloroplasts can be isolated using this method to supply several groups of students, if it is not practicable for all groups to prepare their own.

(1) Cut three small spinach, lettuce or cabbage leaves into small pieces with scissors, avoiding midribs and petioles. Place in a cold mortar or blender containing 20 cm³ of cold isolation medium (scale up quantities for blender if necessary).

(2) Grind vigorously and rapidly (or blend for about 10 s).

(3) Place four layers of muslin or nylon in a funnel and wet with cold isolation medium.

(4) Filter the homogenate through the funnel and collect in pre-cooled centrifuge tubes supported in an ice–water–salt bath. Gather the edges of the muslin and wring thoroughly into the tubes.

(5) Ensure that each centrifuge tube contains about the same volume of filtrate.

(6) If your bench centrifuge has a fixed speed, spin the filtrate for 2–5 min (a small pellet is required, but the time taken should be minimal).

   If a bench centrifuge with variable speed is available, spin the filtrate at 100–200 times gravity for 1–2 min. Respin the supernatant (the liquid above the sediment) at 1 000–2 000 times gravity for up to 5 min (sufficient time to get a small chloroplast pellet).

(7) Pour away the supernatant. Resuspend the pellet of one centrifuge tube in about 2 cm³ of isolation medium using a glass rod. Transfer the suspension from this tube to the second centrifuge tube and resuspend the pellet in that tube. (Alternatively, if more than one student group is to be supplied, use 2 cm³ in each tube and use one tube per group.)

(8) Store this chloroplast suspension in an ice–water–salt bath and use as soon as possible.

The Hill reaction

The chloroplast suspension can now be used to study the Hill reaction. The DCPIP solution should be used at room temperature.
Prepare the following tubes:

(1) 0.5 cm³ chloroplast suspension + 5 cm³ DCPIP solution. Leave in a bright light.

(2) 0.5 cm³ isolation medium + 5 cm³ DCPIP solution. Leave in a bright light.

(3) 0.5 cm³ chloroplast suspension + 5 cm³ DCPIP solution. Place immediately in darkness.

(4) It is useful to add 0.5 cm³ chloroplast suspension to 5 cm³ distilled water as a colour standard, showing what the final colour will be if the DCPIP is reduced.

Record your observations after 15–20 min.

If a colorimeter is available, the progress of the reaction can be followed by measuring the decrease in absorbance of the dye as it changes from the blue oxidised to the colourless reduced state. Prepare the mixtures given above for tubes (2) to (4) in colorimeter sample tubes. Insert a red (or yellow) filter and set the colorimeter at zero absorbance using tube (4) as a blank. Then set up tube (1) and immediately take a reading from this tube and return it to the light. Take further readings at 30 s intervals. Plot the rate of the reaction graphically. Once reduction is complete, take a reading from tube (3). Tube (2) can be checked for reduction of dye by first setting the colorimeter at zero with a blank of isolation medium. Ideally the time for complete reduction is about 10 min.

Notes

Prepare the solutions as follows.

0.05 M phosphate buffer solution, pH 7.0

Na₂HPO₄.12H₂O	4.48 g	(0.025 M)
KH₂PO₄	1.70 g	(0.025 M)

$Na_2HPO_4.12H_2O$ 4.48 g (0.025 M)
KH_2PO_4 1.70 g (0.025 M)

Make up to 500 cm³ with distilled water and store in a refrigerator at 0–4 °C.

Isolation medium

sucrose   34.23 g (0.4 M)
KCl   0.19 g (0.01 M)

Dissolve in phosphate buffer solution at room temperature and make up to 250 cm³ with the buffer solution. Store in a refrigerator at 0–4 °C.

DCPIP solution (reaction medium)

DCPIP   0.007–0.01 g (10^{-4} M approx.)
KCl   0.93 g (0.05 M)

Dissolve in phosphate buffer solution at room temperature and make up to 250 cm³. Store in a refrigerator at 0–4 °C. Use at room temperature.

(NB Potassium chloride is a cofactor for the Hill reaction.)

9.8 What change, if any, did you observe in tube (1)?

9.9 What was the purpose of tubes (2) and (3)?

9.10 What other organelles apart from chloroplasts might you expect in the chloroplast suspension?

9.11 What evidence have you that these were not involved in the reduction of the dye?

9.12 Why was the isolation medium kept cold?

9.13 Why was the isolation medium buffered?

9.14 What was (*a*) the electron donor, and (*b*) the electron acceptor, in the Hill reaction?

9.15 During the Hill reaction, DCPIP acts between X and PSI in the Z-scheme (fig 9.15) and oxygen is evolved. Does the Hill reaction involve cyclic or non-cyclic photophosphorylation, or both? Give your reasons.

9.16 Fig 9.16 shows the appearance of the chloroplasts after being used in the experiment. The photograph demonstrates the consequences of transferring chloroplasts from the hypertonic isolation medium containing sucrose to the hypotonic reaction medium.

(*a*) How do the chloroplasts in fig 9.16 differ in appearance from normal chloroplasts?

(*b*) Can you explain why transferring the chloroplasts to a medium lacking sucrose should bring about this change?

(*c*) Why was this change desirable before carrying out the Hill reaction?

9.17 What significance do you think the discovery of the Hill reaction might have had on the understanding of the photosynthetic process?

9.4.3 Dark reactions

The dark reactions which take place in the stroma do not require light and use the energy (ATP) and reducing power ($NADPH_2$) produced by the light reactions to reduce carbon dioxide. The reactions are controlled by enzymes and their sequence was determined by Calvin, Benson and Bassham of the USA during the period 1946–53, work for which Calvin was awarded the Nobel prize in 1961.

Calvin's experiments

Calvin's work was based on use of the radioactive isotope of carbon, ^{14}C (half-life 5 570 years, see section A1.3) which only became available in 1945. He also used paper chromatography, which was a relatively new but neglected technique. Cultures of the unicellular green alga *Chlorella* were grown in the now famous 'lollipop' apparatus (fig

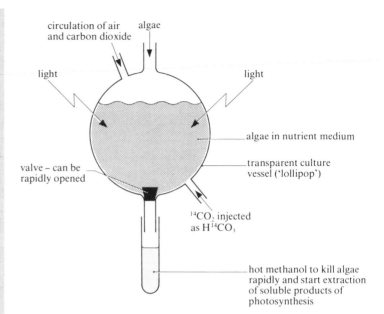

Fig 9.17 *Diagram illustrating the principle of Calvin's 'lollipop' apparatus. This comprises a thin, transparent vessel in which unicellular algae are cultured. Carbon dioxide containing radioactive carbon is bubbled through the algal suspension in experiments to determine the path taken by carbon in photosynthesis*

9.17). The *Chlorella* culture was exposed to $^{14}CO_2$ for varying lengths of time, rapidly killed by dropping into hot methanol, and the soluble products of photosynthesis extracted, concentrated and separated by **two-dimensional paper chromatography** (fig 9.18 and section A1.8.2). The aim was to follow the route taken by the labelled carbon through intermediate compounds into the final product of photosynthesis. Compounds were located on the chromatograms by **autoradiography**, whereby photographic film sensitive to radiation from ^{14}C was placed over the chromatograms and became darkened where radioactive compounds were located (fig 9.18). After only one minute of exposure to $^{14}CO_2$ many sugars and organic acids, including amino acids, had been made. However, using 5 s exposures or less, Calvin was able to identify the first product of photosynthesis as a 3C acid (an acid containing three carbon atoms), **glycerate-3-phosphate** (GP). He went on to establish the sequence of compounds through which the fixed carbon passed and the various stages involved are summarised below. They have since become known as the **Calvin cycle** (or Calvin–Benson–Bassham cycle).

9.18 What is the advantage of using a radioactive isotope with a long half-life in biological experiments?

9.19 What advantage might be gained by using *Chlorella* rather than a higher plant?

9.20 Why was the 'lollipop' vessel thin in section rather than spherical?

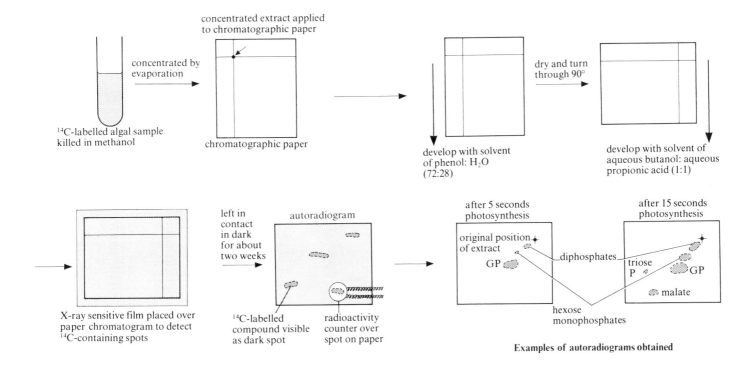

concentrated extract applied to chromatographic paper

concentrated by evaporation

^{14}C-labelled algal sample killed in methanol

chromatographic paper

dry and turn through 90°

develop with solvent of phenol: H_2O (72:28)

develop with solvent of aqueous butanol: aqueous propionic acid (1:1)

X-ray sensitive film placed over paper chromatogram to detect ^{14}C-containing spots

left in contact in dark for about two weeks

autoradiogram

^{14}C-labelled compound visible as dark spot

radioactivity counter over spot on paper

after 5 seconds photosynthesis

original position of extract

GP

diphosphates

after 15 seconds photosynthesis

triose P

GP

malate

hexose monophosphates

Examples of autoradiograms obtained

Fig 9.18 *(a) Detection of the products of $^{14}CO_2$ fixation in algae after brief periods of illumination by the use of paper chromatography and autoradiography. (b) Autoradiographs of the photosynthetic products from $^{14}CO_2$ added to algae illuminated for short periods of time*

Stages in carbon pathway

Acceptance of carbon dioxide (carbon dioxide fixation).

$$\text{RuBP} + CO_2 + H_2O \xrightarrow{\text{RuBP carboxylase}} \text{2GP}$$

(ribulose bisphosphate) 5C sugar

(glycerate-3-phosphate) 3C acid

first product of photosynthesis

The carbon dioxide acceptor is a 5C sugar (a pentose), **ribulose bisphosphate** (ribulose with two phosphate groups, formerly known as ribulose diphosphate, RuDP). Addition of carbon dioxide to a compound is called **carboxylation**; the enzyme involved is a **carboxylase**. The 6C product is unstable and breaks down immediately to two molecules of **glycerate-3-phosphate** (GP). The latter is the first product of photosynthesis. The enzyme ribulose bisphosphate carboxylase is present in large amounts in the chloroplast stroma, and is in fact the world's most common protein.

Reduction phase

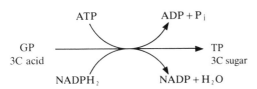

ATP

ADP + P$_i$

GP
3C acid

TP
3C sugar

NADPH$_2$

NADP + H$_2$O

GP is glycerate-3-phosphate, a 3C **acid**. It contains the acidic carboxyl group (–COOH). TP is triose phosphate or

glyceraldehyde-3-phosphate, a 3C **sugar**. It contains an aldehyde group (–CHO).

The reducing power of NADPH$_2$ and energy of ATP are used to remove oxygen from GP (reduction). The reaction takes place in two stages, the first using some of the ATP produced in the light reactions and the second using all the NADPH$_2$ produced in the light reactions. The overall effect is to reduce a carboxylic acid group (–COOH) to an aldehyde group (–CHO). The product is a 3C sugar phosphate (a triose phosphate), that is a sugar with a phosphate group attached. This contains more chemical energy than GP, and is the first carbohydrate made in photosynthesis.

Regeneration of the carbon dioxide acceptor, RuBP. Some of the triose phosphate (TP) has to be used to regenerate the ribulose bisphosphate consumed in the first reaction. This process involves a complex cycle, containing 3, 4, 5, 6 and 7C sugar phosphates. It is here that the remaining ATP is used. Fig 9.19 provides a summary of the dark reactions. In it the Calvin cycle is represented as a 'black box' into which carbon dioxide and water are fed and TP emerges. The diagram shows that the remaining ATP is used to phosphorylate ribulose phosphate to ribulose

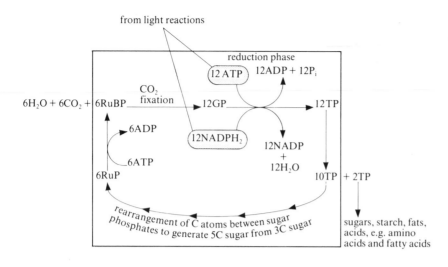

from light reactions

Fig 9.19 (left) *Summary of the dark reactions of photosynthesis (Calvin cycle). RuBP, ribulose bisphosphate; RuP, ribulose phosphate; GP, glycerate-3-phosphate; TP, triose phosphate*

Fig 9.20 (opposite) *Metabolism of GP and TP showing the relationship between photosynthesis and synthesis of food in plants. Main pathways only are shown. Some intermediate steps are omitted*

bisphosphate, but details of the complex series of reactions are not shown.

The overall equation which can be derived from fig 9.19 is

$$6H_2O + 6CO_2 \xrightarrow[\substack{12NADPH_2 \\ 18ATP}]{\substack{18ADP + 18P_i \\ 12NADP + 12H_2O}} 2TP$$

The important point to note is that six molecules of carbon dioxide have been used to make two molecules of a 3C sugar, triose phosphate. The equation can be simplified by dividing by six:

$$H_2O + CO_2 \xrightarrow[\substack{2NADPH_2 \\ 3ATP}]{\substack{3ADP + P_i \\ 2NADP}} [CH_2O] + 2H_2O$$

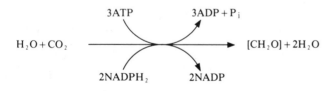

9.21 Redraw fig 9.19 showing only the numbers of C atoms involved; for example 6RuBP = 6 × 5C.
A summary of photosynthesis is given in table 9.6.

9.4.4 Metabolism of glycerate-3-phosphate and triose phosphate

Although triose phosphate (phosphoglyceraldehyde) is the end product of the Calvin cycle, it does not accumulate in large quantities since it is immediately converted to other products. The most familiar of these are glucose, sucrose and starch, but fats and organic acids (such as fatty acids and amino acids) are also rapidly made. Photosynthesis can strictly be regarded as complete once triose phosphate is made, because subsequent reactions can also occur in non-photosynthetic organisms, like animals and fungi. However, it is important to show here how glycerate-3-phosphate and triose phosphate can be used in the synthesis of all the basic food requirements of plants. Fig 9.20 summarises some of the main pathways involved and shows what a central position the reactions of glycolysis and Krebs cycle have in metabolism. The latter two pathways are discussed in chapter 11. Both glycerate-3-phosphate and triose phosphate are intermediates in glycolysis.

Synthesis of carbohydrates

Carbohydrates are synthesised in a process which is, in effect, a reversal of glycolysis. The two most common carbohydrate products are sucrose and starch. Sucrose is the form in which carbohydrate is exported from the leaf in the phloem (section 14.8). Starch is a storage product and is the most easily detected product of photosynthesis.

Synthesis of lipids

Glycerate-3-phosphate enters the glycolytic pathway and is converted to an acetyl group which is added to coenzyme A to form acetyl coenzyme A. This is converted to fatty acids in both cytoplasm and chloroplasts (not in mitochondria, where *breakdown* of fatty acids occurs). Glycerol on the other hand is made from triose phosphate.

Synthesis of proteins

Glycerate-3-phosphate and triose phosphate contain the

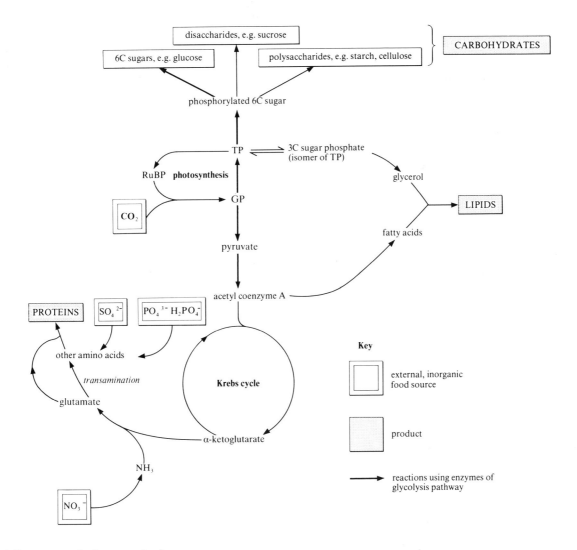

Table 9.6 Summary of photosynthesis.

	Light reactions	*Dark reactions*
Location in chloroplasts	Thylakoids	Stroma
Reactions	Photochemical, i.e. require light. Light energy causes the flow of electrons from electron 'donors' to electron 'acceptors', along a non-cyclic or a cyclic pathway. Two photosystems, I and II, are involved. These contain chlorophylls which emit electrons when they absorb light energy. Water acts as an electron donor to the non-cyclic pathway. Electron flow results in production of ATP (photophosphorylation) and $NADPH_2$. (See also table 9.5.)	Do not require light. Carbon dioxide is fixed when it is accepted by a 5C-compound ribulose bisphosphate (RuBP), to form two molecules of a 3C-compound glycerate-3-phosphate (GP), the first product of photosynthesis. A series of reactions occurs called the Calvin cycle in which the carbon dioxide acceptor RuBP is regenerated and GP is reduced to a sugar. (See also fig 9.19.)
Overall equation	$$2H_2O + 2NADP \xrightarrow[\text{chlorophyll}]{\text{light}} O_2 + 2NADPH_2$$ also $$ADP + P_i \longrightarrow ATP \text{ (variable amount)}$$	$$CO_2 + H_2O \quad \begin{matrix} 3ATP & 3ADP + 3P_i \\ \searrow & \nearrow \\ & \\ \nearrow & \searrow \\ 2NADPH_2 & 2NADP \end{matrix} \quad [CH_2O] + 2H_2O$$
Results	Light energy is converted to chemical energy in ATP and $NADPH_2$. Water is split into hydrogen and oxygen. Hydrogen is carried to $NADPH_2$ and oxygen is a waste product.	Carbon dioxide is reduced to carbon compounds such as carbohydrates, using the chemical energy in ATP and hydrogen in $NADPH_2$.
Combined equations		

$$\text{light} + \text{chlorophyll} \quad \begin{matrix} 2H_2O \\ \nearrow \quad \searrow \\ O_2 \quad 4H^+ + 4e^- \end{matrix} \quad \begin{matrix} [CH_2O] \\ \nwarrow \quad \nearrow \\ CO_2 + H_2O \end{matrix}$$

$$\text{Net equation: } CO_2 + H_2O \xrightarrow[\text{chlorophyll}]{\text{light}} [CH_2O] + O_2$$

elements carbon, hydrogen and oxygen. Nitrogen, sulphur and occasionally phosphorus are also needed if amino acids and hence proteins are to be made. Plants obtain these elements from the soil water (or surrounding water in aquatic plants) as inorganic salts (nitrates, sulphates and phosphates respectively).

Many plants are able to synthesise all their amino acids using ammonia or nitrate as the nitrogen source, and given a supply of glycerate-3-phosphate from photosynthesis. Mammals are unable to synthesise some of the common amino acids (the essential amino acids, see section 5.4) and have to rely on plants as the source. Glycerate-3-phosphate is first converted to one of the acids of the Krebs cycle via acetyl coenzyme A (fig 9.20). Subsequent synthesis of the amino acid is summarised below.

(1)

$$NO_3^- \xrightarrow[\text{nitrate reductase}]{\text{reduction}} NO_2^- \xrightarrow[\text{nitrite reductase}]{\text{reduction}} NH_3$$
(nitrate)   (nitrite)
from roots

(2)

$$NH_3 + \text{Krebs cycle acid} \xrightarrow[\text{+ reduction}]{\text{amination}} \text{amino acid}$$
(ammonia)

For example,

$$NH_3 + \alpha\text{-ketoglutarate} + NADPH_2 \underset{\text{transaminase}}{\rightleftharpoons} \text{glutamate} + NADP$$

Reaction (2) is the major route of entry of ammonia into amino acids. By a process called **transamination** other amino acids can be made by transferring the amino group ($-NH_2$) from one acid to another. For example,

$$\text{glutamate} + \text{oxaloacetate} \underset{\text{transaminase}}{\rightleftharpoons} \alpha\text{-ketoglutarate} + \text{aspartate}$$
(amino acid) (a Krebs cycle acid) (a Krebs cycle acid) (amino acid)

Other synthetic pathways for amino acids also occur. Some amino acids are made in the chloroplasts. About one-third of the carbon fixed and about two-thirds of the nitrogen taken up by plants are commonly used directly to make amino acids.

9.5   Factors affecting photosynthesis

The rate of photosynthesis is an important factor in crop production since it affects yields. An understanding of those factors affecting the rate is therefore likely to lead to an improvement in crop management.

Computer program. PHOTOPLOT generates graphs of oxygen production against a selected variable under different conditions for both C_3 and C_4 plants.

9.22 From the equation for photosynthesis what factors are likely to affect its rate?

9.5.1   The concept of limiting factors

The rate of a biochemical process which, like photosynthesis, involves a series of reactions, will theoretically be limited by the slowest reaction in the series. For example, in photosynthesis the dark reactions are dependent on the light reactions for $NADPH_2$ and ATP. At low light intensities the rate at which these are produced is too slow to allow the dark reactions to proceed at maximum rate, so light is a limiting factor. The principle of limiting factors can be stated thus:

when a chemical process is affected by more than one factor its rate is limited by that factor which is nearest its minimum value: it is that factor which directly affects a process if its quantity is changed.

The principle was first established by Blackman in 1905. Since then it has been shown that different factors, such as carbon dioxide concentration and light intensity, interact and can be limiting at the same time, although one is often the major factor. Consider one of these factors, light intensity, by studying fig 9.21 and trying to answer the following questions.

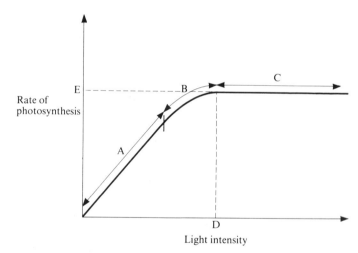

Fig 9.21 *Effect of light intensity on rate of photosynthesis*

9.23 In fig 9.21 (a) what is the limiting factor in region A?
(b) what is represented by the curve at B and C?
(c) what does point D represent on the curve?
(d) what does point E represent on the curve?

Fig 9.22 shows the results from four experiments in which the same experiment is repeated at different temperatures and carbon dioxide concentrations.

9.24 In fig 9.22 what do the points X, Y and Z represent on the three curves?

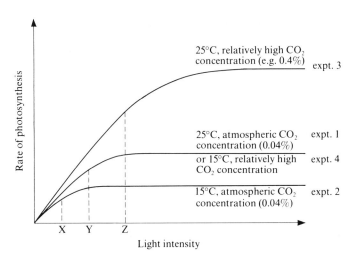

Fig 9.22 *Effect of various factors on rate of photosynthesis*

In fig 9.22, experiments 1–4 show that once light intensity is no longer limiting, both temperature and carbon dioxide concentration can become limiting. Enzyme-controlled reactions like the dark reactions of photosynthesis are sensitive to temperature; thus an increase in temperature from 15 °C to 25 °C results in an increased rate of photosynthesis (compare experiments 2 and 1, or 4 and 3) providing light is not a limiting factor. Carbon dioxide concentration can also be a limiting factor in the dark reactions (compare experiments 2 and 4, or 1 and 3). Thus in experiment 2, for example, both temperature and carbon dioxide concentration are limiting, and an increase in either results in increased photosynthetic rate.

9.5.2 Reaction rate graphs

The chief external factors affecting rate of photosynthesis are light intensity, carbon dioxide concentration and temperature. Graphs representing their effects all have the form of fig 9.21, with external factors plotted on the horizontal axis. All show an initial linear increase in photosynthetic rate where the factor being investigated is limiting, followed by a decrease in the rate of increase and stabilising of rate as another factor, or factors, becomes limiting.

In the following it is assumed that factors other than the one under discussion are optimal.

Light

When considering the effect of light on a process it is important to distinguish between the effects of light intensity, light quality and duration of exposure to light.

Light intensity. In low light intensities the rate of photosynthesis increases linearly with increasing light intensity (fig 9.21). Gradually the rate of increase falls off as the other factors become limiting. Illumination on a clear summer's day is about 100 000 lux (10 000 ft candles), whereas light saturation for photosynthesis is reached at about 10 000 lux. Therefore, except for shaded

plants, light is not normally a major limiting factor. Very high light intensities may bleach chlorophyll and retard photosynthesis, but plants naturally exposed to such conditions are usually protected by devices such as thick cuticles and hairy leaves.

Light duration (photoperiod). Photosynthesis only occurs during periods of light, but is otherwise unaffected by light duration.

Light quality (wavelength of colours). The effect of light quality is revealed by the action spectrum for photosynthesis (see fig 9.12).

Carbon dioxide concentration

Carbon dioxide is needed in the dark reactions where it is fixed into organic compounds. Under normal field conditions, carbon dioxide is the major limiting factor in photosynthesis. Its concentration in the atmosphere varies between 0.03% and 0.04%, but increases in photosynthetic rate can be achieved by increasing the percentage (see experiment 3, fig 9.22). The short-term optimum is about 0.5%, but this can be damaging over long periods; then the optimum is about 0.1%. This has led to some greenhouse crops, such as tomatoes, being grown in carbon-dioxide-enriched atmospheres. At the moment there is much interest in a group of plants which are capable of removing the available carbon dioxide from the atmosphere more efficiently, hence achieving greater yields. These 'C$_4$' plants are discussed in section 9.8.2, where the effects of high carbon dioxide concentrations on inhibiting photo-respiration, thus stimulating photosynthesis, are also discussed.

Temperature

The dark reactions and, to a certain extent, the light reactions are enzyme-controlled and therefore temperature-sensitive. For temperate plants the optimum temperature is usually about 25 °C. The rate doubles for every 10 °C rise up to about 35 °C, although other factors mean that the plant grows better at 25°C.

9.25 Why should the rate decrease at higher temperatures?

Water

Water is a reactant (raw material) in photosynthesis but so many cell processes are affected by lack of water that it is impossible to measure the direct effect of water on photosynthesis. Nevertheless, by studying the yields (amounts of organic matter synthesised) of water-deficient plants, it can be shown that periods of temporary wilting can lead to severe yield losses. Even slight water deficiency, with no visible effects, might significantly reduce crop yields. The reasons are complex and not fully understood. One obvious factor is that plants usually close their stomata in response to wilting and this would prevent access of carbon dioxide for photosynthesis. Abscisic acid, a growth inhibitor, has also been shown to accumulate in water-deficient leaves of some species.

Chlorophyll concentration

Chlorophyll concentration is not normally a limiting factor, but reduction in chlorophyll levels can be induced by several factors, including disease (such as mildews, rusts and virus diseases), mineral deficiency (section 9.12) and normal ageing processes (**senescence**). If the leaf becomes yellow it is said to be **chlorotic**, the yellowing process being called **chlorosis**. Chlorotic spots are thus often a symptom of disease or mineral deficiency. Iron, magnesium and nitrogen are required during chlorophyll synthesis (the latter two elements being part of its structure) and are therefore particularly important minerals. Potassium is also important. Lack of light can also cause chlorosis since light is needed for the final stage of chlorophyll synthesis.

Oxygen

Relatively high concentrations of oxygen, such as the 21% in the atmosphere to which plants are normally exposed, generally inhibit photosynthesis. In recent years it has been shown that oxygen competes with carbon dioxide for the active site in the carbon-dioxide-fixing enzyme RuBP carboxylase, thus reducing the overall rate of photosynthesis. In the subsequent reactions carbon dioxide is produced, again reducing net photosynthesis. These reactions comprise 'photorespiration' and are discussed in section 9.8.

Specific inhibitors

An obvious way of killing a plant is to inhibit photosynthesis, and various herbicides have been introduced with this intention. A notable example is DCMU (dichloro-phenyl dimethyl urea) which short-circuits non-cyclic electron flow in chloroplasts and thus inhibits the light reactions. DCMU has been useful in research on the light reactions.

Pollution

Low levels of certain gases of industrial origin, notably ozone and sulphur dioxide, are very damaging to the leaves of some plants, although the exact reasons are still being investigated. It is estimated, for example, that cereal crop losses as high as 15% may occur in badly polluted areas, particularly when compounded with dry conditions as in the British summer of 1976. Lichens are very sensitive to sulphur dioxide. Soot can block stomata and reduce the transparency of the leaf epidermis.

> **9.26** Suggest some habitats or natural circumstances in which (a) light intensity, (b) oxygen concentration or (c) temperature might be limiting factors in photosynthesis.

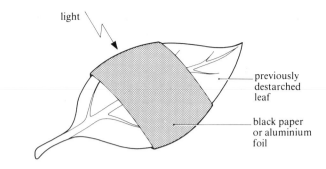

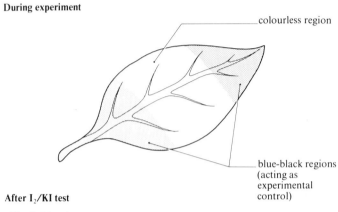

Fig 9.23 *Investigating the need for light in photosynthesis*

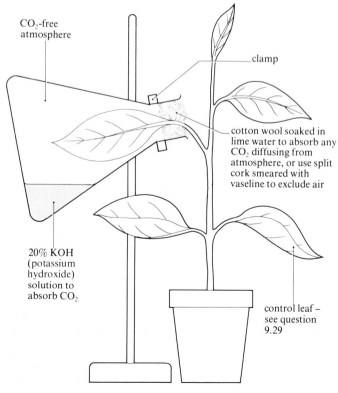

Fig 9.24 *Investigating the need for carbon dioxide in photosynthesis*

Experiments: To investigate conditions required for, and products of photosynthesis

As an indication that photosynthesis has occurred some product of the process can be identified. The first product is phosphoglyceric acid which is rapidly converted to a number of compounds, including sugars and thence starch. The latter can be tested for very easily and can be taken as an indication that photosynthesis has occurred, providing that the precaution is taken of starting the experiment with a destarched leaf or plant.

Destarching a plant

A plant can be destarched by leaving it in the dark for 24–48 h. It is advisable to check that destarching is complete before attempting the following experiment.

9.27 Why does this result in destarching?

Experiment 9.2: To test a leaf for starch

Materials

leaf to be tested	hot water bath
test tube	90% ethanol
forceps	iodine/potassium
white tile	iodide solution

Method

Starch can be detected using iodine/potassium iodide solution (I_2/KI) but the leaf must first be decolourised because the green colour of the chlorophyll masks the colour change. This is achieved by placing the leaf in a test-tube of boiling 90% ethanol in a water bath for as long as necessary (naked flames must be avoided because ethanol is highly inflammable).

The decolourised leaf is rinsed in hot water to remove ethanol and soften the tissues, spread on a white tile, and iodine solution poured on its surface. The red-brown solution stains any starch-containing parts of the leaf blue-black.

Experiment 9.3: To investigate the need for light

Materials

destarched leafy plant	black paper or metal foil
light source such as	starch test materials
a bench lamp	

Method

Although the destarching process itself demonstrates the need for light, the requirement can be investigated further by placing strips of black paper or metal foil over destarched leaves and exposing them to light for several hours. Procedure and expected results are shown in fig 9.23.

9.28 How would you criticise the experimental design and modify the experiment to take into account your criticisms?

Experiment 9.4: To investigate the need for carbon dioxide

Materials

destarched leafy plant such as potted geranium (*Pelargonium*)	starch test materials
	250 cm³ conical flask
	clamp and clamp stand
	limewater
light source such as a bench lamp	
cotton wool	
20% potassium hydroxide solution	

Method

Fig 9.24 illustrates a suitable procedure for investigating the need for carbon dioxide. The plant should be left for several hours in the light before testing the relevant leaves for starch.

9.29 Describe the conditions to which you would subject the control leaf.

A more satisfactory experiment showing the use of carbon dioxide is one involving the uptake of $^{14}CO_2$ (radioactively labelled carbon dioxide) into sugars and other compounds.

Experiment 9.5: To investigate the need for chlorophyll

Materials

plant with variegated leaves such as *Chlorophyton*, variegated ivy, geranium, maple or privet
starch test materials

Method

A number of plants have variegated leaves, that is leaves with green and non-green areas, the latter having no chlorophyll. Examples are given above. If the starch test is carried out on such a leaf, after careful mapping of the green and white (non-green) areas, it will be seen that only the green, chlorophyll-containing areas contain starch.

Experiment 9.6: To investigate the evolution of oxygen

Materials

Canadian pondweed
   (*Elodea*)
test-tube
glass funnel
light source such as a
   bench lamp

sodium hydrogen-
   carbonate
400 cm³ beaker
wooden splint
plasticine

Method

The simplest method to demonstrate that oxygen is a product of photosynthesis is to use a well-illuminated aquatic plant, such as *Elodea*, from which oxygen gas can be collected over water as shown in fig 9.25. A quantitative method is discussed in the next section.

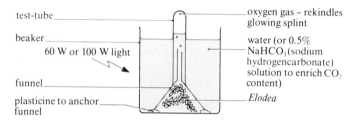

Fig 9.25 *Investigating the evolution of oxygen during photosynthesis*

9.6　Measuring rates of photosynthesis

9.30 From the equation for photosynthesis, what changes in the substances taken up and produced might be used to measure the rate of photosynthesis?

In section 9.5 certain external factors (such as light intensity, carbon dioxide concentration and temperature) were shown to affect the rate of photosynthesis. When a particular factor is being investigated, it is essential that other factors are kept constant and, if possible, at optimum levels so that no other factor is limiting.

9.6.1　The rate of oxygen evolution

Measuring the rate of oxygen evolution from a water plant is the simplest way to measure the rate of photosynthesis.

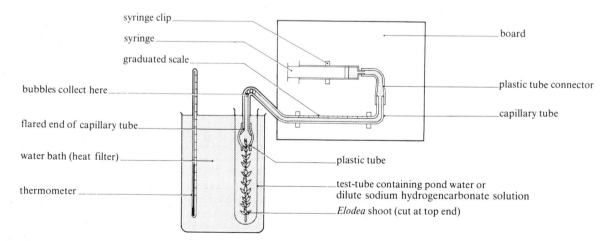

Fig 9.26 *Apparatus for measuring the rate of oxygen evolution by a water plant during photosynthesis*

Experiment 9.7: To investigate the effect of light intensity on the rate of photosynthesis

Materials

apparatus for collecting gas as shown in fig 9.26	metre rule
test-tube	stopclock
400 cm³ beaker	light source such as bench lamp
thermometer	Canadian pondweed (Elodea), previously well illuminated for several hours
mercury vapour lamp or projector lamp	
sodium hydrogen-carbonate	detergent (washing-up liquid)

Method

It is advisable to use *Elodea* that has been well illuminated and is known to be photosynthesising actively. The addition of 2–10 g of sodium hydrogencarbonate to each litre of pond water may stimulate photosynthesis if there are no obvious signs of bubbles being produced (this increases carbon dioxide availability). The water could also be aerated for an hour before the experiment.

(1) Cut the stem of a bubbling piece of *Elodea* to about 5 cm long with a sharp scalpel and place it, cut surface upwards, in a test-tube containing the same water that it has been kept in.

(2) Stand the test-tube in a beaker of water at room temperature. Record the temperature of the water, which acts as a heat shield, and check it at intervals throughout the experiment. It should remain constant and the water be renewed if necessary.

(3) Fill the apparatus with tap water, ensuring that no air bubbles are trapped in it and push the plunger well in to the end of the syringe (fig 9.26).

(4) Darken the laboratory. Place a bright light source 5 cm from the plant.

(5) Allow the plant to adjust to the light intensity (equilibrate) for 2–3 min. Ensure that the rate of bubbling is adequate (such as more than 10 bubbles per minute). A trace of detergent is sometimes sufficient to lower the surface tension to allow freer escape of bubbles.

(6) Position the *Elodea* so that its bubbles are collected in the capillary tube of the apparatus. Start timing.

(7) Collect a suitable volume of gas in a known period (for example 5–10 min). Measure the length of the bubble by drawing it slowly along the capillary tube by means of a syringe. The bubble can thus be positioned along the scale.

(8) Draw the bubble into the plastic tube connector where it will not interfere with subsequent measurements and repeat the procedure at increasing distances between the light source and *Elodea*, such as 10, 15, 20, 30, 40 and 80 cm. In each case allow time for the plant to equilibrate. The following three measurements are required under each condition: (*a*) the distance between plant and light source, (*b*) the time taken to collect the gas, and (*c*) the length of the gas bubbles collected (this measurement is directly proportional to volume and is used as a measurement of volume).

Results

The intensity of light falling on a given object is inversely proportional to the square of the distance from the source. In other words, doubling the distance between the weed and the lamp does not halve the light intensity received by the weed, but quarters it.

$$LI \propto \frac{1}{d^2}$$

where *LI* is the light intensity and *d* is the distance between object and light source. Plot a graph with rate of photosynthesis on the vertical axis (as length of gas bubble per unit time) and *LI* on the horizontal axis (as $1/d^2$ or, more conveniently, $1\ 000/d^2$).

> **9.31** (*a*) State the relationship between gas production and light intensity demonstrated by your results.
> (*b*) Why was the laboratory darkened and the temperature kept constant?
> **9.32** What are the main sources of inaccuracy in this experiment?
> **9.33** If the gas is collected and analysed it is found *not* to be pure oxygen. Can you account for this?
> **9.34** Why is it advisable to aerate the water before beginning the experiment?

If a simpler, quicker, though slightly less accurate method is required, the rate of oxygen evolution can be determined by counting the number of bubbles evolved from the cut end of a stem of *Elodea* in a given time period. This can be just as satisfactory, but errors may occur through variations in bubble size. This problem is less likely to arise if a trace of detergent is added to lower the surface tension (see (5) above). The *Elodea* can be anchored to the bottom of the tube with plasticine if necessary.

9.7　Compensation points

Photosynthesis results in uptake of carbon dioxide and evolution of oxygen. At the same time respiration uses oxygen and produces carbon dioxide. If light intensity is gradually increased from zero, the rate of photosynthesis gradually increases accordingly (fig 9.22). There will come a point, therefore, when photosynthesis and respiration exactly balance each other, with no net exchange of oxygen and carbon dioxide. This is called the **compensation point**, or more precisely the **light compensation point**, that is the light intensity at which net gaseous exchange is zero.

Since carbon dioxide concentration affects the rate of photosynthesis there also exists a **carbon dioxide compensation point**. This is the carbon dioxide concentration at which net gaseous exchange is zero for a given light intensity. The higher the carbon dioxide concentration, up to about 0.1% (1 000 ppm, parts per million), the faster the rate of photosynthesis. For most temperate plants the carbon dioxide compensation point, beyond which photosynthesis exceeds respiration, is 50–100 ppm, assuming light is not a limiting factor. Atmospheric carbon dioxide concentrations are normally in the range 300–400 ppm, and therefore under normal circumstances of light and atmospheric conditions this point is always exceeded.

Experiment 9.8: To investigate gaseous exchange in leaves

Materials

four test-tubes thoroughly cleaned and fitted with rubber bungs	unbleached cotton wool
	no. 12 cork borer
	water bath with
	test-tube clamps
forceps	bench lamp
test-tube rack	freshly picked leaves
2 cm³ syringe	hydrogencarbonate
aluminium foil	indicator

The hydrogencarbonate indicator (bicarbonate indicator) solution should be freshly equilibrated with the atmosphere by bubbling fresh air through it until cherry red. Hydrogencarbonate indicator is supplied as a concentrated solution and must be diluted by a factor of ten for experimental use. To equilibrate with atmospheric carbon dioxide, air from *outside* the laboratory should be pumped through the solution. A suitable method is to place the solution in a clear glass wash-bottle to which a tube is attached whose free end is hung from a window. A filter pump is then used to bubble air through the solution until there is no further colour change. The colour of the indicator at this stage is a deep red but will appear orange-red in the test-tubes. Time must be allowed for this procedure before the start of the experiment (100 cm³ of indicator will need to be aerated for at least 20 min).

Method

(1) Label four test-tubes A, B, C and D.
(2) Rinse the four tubes and a 2 cm³ syringe with a little of the indicator solution.
(3) Add 2 cm³ of the indicator solution to each tube by means of the syringe. Avoid putting fingers over the ends of the tubes since the acid in sweat will affect the indicator. Also avoid breathing over the open ends of the tubes.
(4) Cover the outside of the tubes A and C with aluminium foil.
(5) Set up the tubes as shown in fig 9.27, using two leaf discs per tube, cut from a fresh leaf with a number 12 cork borer.
(6) Arrange the tubes in such a way that they are equally illuminated by a bench lamp.
(7) Place a heat filter in the form of a glass tank of water between the tubes and the light source to prevent a rise in temperature during the experiment. Alternatively, the tubes can be clamped in a water bath.
(8) Note the colour of the indicator in each tube.
(9) At intervals shake the tubes gently and leave for at least 2 h, preferably overnight. Record the final colour of the indicator in each tube as seen against a white background.

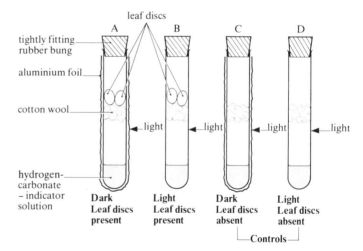

Fig 9.27　*Experiment to investigate gaseous exchange in leaf discs*

Results

Results can be interpreted using the following guide to colour changes.

yellow	orange	red	purple

← net carbon dioxide production ―| |― net carbon dioxide uptake →

←――――― increasing acidity ―――――→

―――――― increasing alkalinity ――――――→

If conditions become more acidic, this can be assumed to be the result of carbon dioxide being produced and dissolving in the indicator solution. If conditions become less acidic, this indicates a lowering of carbon dioxide concentration.

9.35 What can you conclude from your results and why were the controls necessary?

9.36 What is the name given to the equilibrium point where there is no further net uptake or production of carbon dioxide by the leaf discs in tube B?

Modifications of this experiment

(1) **Comparing rates of photosynthesis.** By using leaf discs as described, rather than whole leaves, comparative studies may be carried out using different light intensities, or, for example, old and young leaves on the same plant, yellow and green areas of variegated leaves, leaves of different species (such as a C_3 and a C_4 plant – see C_4 photosynthesis). To compare rates of photosynthesis, colours of the indicator solutions can be compared during or at the end of the experiment as appropriate. If light intensity is investigated, a mercury vapour lamp should be used. An interesting comparison can be made between shade-loving plants, such as enchanter's nightshade (*Circaea lutetiana*), and other species to determine whether the former are capable of photosynthesising at lower light intensities (that is they have lower light compensation points).

(2) **Using water plants instead of leaf discs.** Water plants such as *Elodea* may be used, providing they are washed well in distilled water to remove traces of dirt and pond water in order to minimise any contribution from micro-organisms. The plants should be placed directly in sufficient indicator solution to cover them. The solution has little effect on the plants during the course of the experiment.

9.8 Photorespiration and C_4 photosynthesis

9.8.1 Photorespiration

Photosynthesis is believed to have evolved in an atmosphere much richer in carbon dioxide than it is today, but one containing relatively little oxygen, probably about 0.02% oxygen compared with 21% today. Since 1920 it has been known that oxygen generally inhibits photosynthesis and the reason for this was discovered in 1971. It was shown that the carbon-dioxide-fixing enzyme, ribulose bisphosphate carboxylase (RuBP carboxylase) will accept not only carbon dioxide but also oxygen as a substrate. The two gases compete, in fact, for the same active site.

If oxygen is accepted the following reaction is catalysed:

$$(1) \quad O_2 + \underset{5C}{RuBP} \xrightarrow{\text{RuBP oxygenase}} \underset{2C}{\text{phosphoglycolate}} + \underset{3C}{GP}$$

Compare with the usual carbon dioxide-fixing process:

$$(2) \quad CO_2 + \underset{5C}{RuBP} \xrightarrow{\text{RuBP carboxylase}} \underset{2\times3C}{2GP}$$

Reaction (1) is called an **oxygenation**; the same enzyme is therefore called **RuBP oxygenase** in this reaction and RuBP carboxylase in reaction (2). The enzyme is therefore often called ribulose bisphosphate carboxylase-oxygenase, or RUBISCO. In reaction (1), one molecule each of glycerate-3-phosphate and phosphoglycolate are formed instead of the two GP molecules in reaction (2). Phosphoglycolate (phosphoglycolic acid) is converted immediately to glycolate (glycolic acid) by removal of the phosphate group.

Oxygen is therefore a **competitive inhibitor** (section 6.5) of carbon dioxide fixation and any increase in oxygen concentration will favour the uptake of oxygen rather than carbon dioxide, and so inhibit photosynthesis. Conversely an increase in carbon dioxide concentration will favour the carboxylation reaction.

The plant now has the problem of what to do with the glycolate and the pathway which deals with it is called **photorespiration**. Photorespiration is defined as a light-dependent uptake of oxygen and output of carbon dioxide. It is in no way related to normal respiration (now sometimes called **dark respiration** to avoid confusion) and only resembles it in that oxygen is used and carbon dioxide produced. It is light-dependent because a supply of RuBP is only available when photosynthesis is operating, RuBP being a product of the Calvin cycle. The function of photorespiration is to recover some of the carbon from the excess glycolate. The pathway involved is illustrated in fig 9.28. Details of the pathway can be ignored but the following four main points emerge.

(1) **Oxygen is used** (*a*) when glycolate is oxidised to glyoxylate in the peroxisome, and (*b*) when glycine is oxidised to serine in the mitochondrion.

(2) There is a **wasteful loss of carbon as carbon dioxide** when glycine is oxidised to serine.

(3) There is a **wasteful loss of energy** as $NADPH_2$ and ATP are used. Although ATP is produced when glycine is oxidised to serine, the overall process is energy-consuming.

(4) **Three different organelles are involved**, that is chloroplasts, peroxisomes and mitochondria. Peroxisomes are briefly described in chapter 7 (fig 7.6).

Overall, one molecule of PGA containing 3C atoms is produced from two molecules of glycolate ($2 \times 2C$ atoms), that is three carbon atoms out of four are recovered from the waste glycolate. Since intermediates such as glycine can be made more efficiently by other pathways, there seems to be no other function for the pathway.

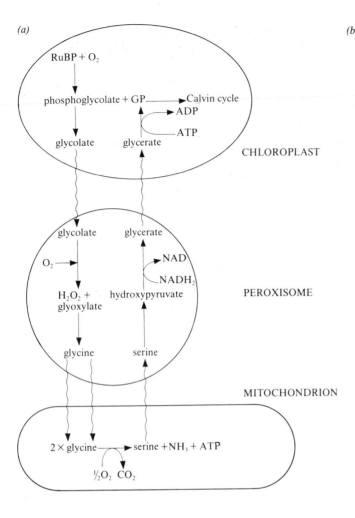

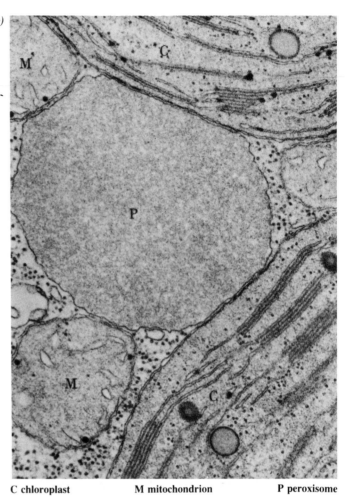

C chloroplast   **M mitochondrion**   **P peroxisome**

Fig 9.28 *(a) Pathway of photorespiration. Note oxygen is used and carbon dioxide produced. Also RuBP is used and this can only be produced by photosynthesis in the light. (b) Electron micrograph showing the intimate relationship between chloroplasts, peroxisomes and mitochondria typical of leaf mesophyll cells (× 38 700)*

The carbon lost represents carbon that had previously been fixed at the cost of energy. Also, the ammonia released when glycine is oxidised to serine must be reincorporated into amino acids at the expense of ATP.

9.37 How does a knowledge of photo-respiration help to explain the known effects of carbon dioxide and oxygen concentration on rates of photosynthesis?

9.38 What environmental conditions favour photorespiration?

Summary of photorespiration

(1) It is a light-dependent uptake of oxygen and output of carbon dioxide.

(2) It bears no relation to normal respiration ('dark respiration').

(3) It occurs as a result of RuBP carboxylase accepting oxygen as well as carbon dioxide; unwanted glycolate is produced as a result. The remaining reactions are a means of recovering some of the carbon from glycolate.

(4) Two molecules of glycolate (total four carbon atoms) are converted to one molecule of PGA (three carbon atoms) at the expense of energy. Oxygen is used and wasteful loss of the fourth carbon atom as carbon dioxide occurs.

(5) It reduces the potential yield of C_3 plants by 30–40%.

9.8.2   *C_4 photosynthesis*

In 1965 it was shown that the first products of photosynthesis in sugarcane, a tropical plant, appeared to be acids containing four carbon atoms (malic, oxaloacetic and aspartic) rather than the 3C-acid PGA of *Chlorella* and most temperate plants. Many plants, mostly tropical and some of great economic importance, have since been identified in which the same is true and these are called **C_4 plants**. Examples are the monocotyledons maize (*Zea*),

Sorghum, sugarcane (*Saccharum*) and millet (*Eleusine*); the dicotyledons include *Amaranthus* and some *Euphorbia* species. Plants in which the first product of photosynthesis is the C_3-acid GP are called C_3 plants. It is the biochemistry of the latter plants which has been described so far in this chapter.

In 1966, two Australian workers, Hatch and Slack showed that C_4 plants were far more efficient at taking up carbon dioxide than C_3 plants: they could remove carbon dioxide from an experimental atmosphere down to 0.1 ppm compared with the 50–100 ppm of temperate plants, that is they had **low carbon dioxide compensation points**. Such plants show no apparent photorespiration.

The new carbon pathway in C_4 plants is called the **Hatch–Slack pathway**. Subtle variations exist but the process in a typical C_4 plant, maize, will be described. C_4 plants possess a characteristic leaf anatomy in which two rings of cells are found around each of the vascular bundles. The inner ring, or **bundle sheath cells**, contains chloroplasts which differ in form from those in the **mesophyll cells** in the outer ring. The chloroplasts in the plants are therefore described as **dimorphic**. Figs 9.29 (*a*) and (*b*) illustrate this so-called 'Kranz' anatomy (Kranz means crown or halo, referring to the two distinct rings of cells). The biochemical pathway that takes place in these cells is summarised below and in fig 9.30.

Hatch–Slack pathway

The Hatch–Slack pathway is a pathway for transporting carbon dioxide and hydrogen from mesophyll cells to bundle sheath cells. Here carbon dioxide is fixed as in C_3 plants, as shown in fig 9.30, and reduced using the hydrogen.

Acceptance of carbon dioxide (carbon dioxide fixation) in mesophyll cells. Carbon dioxide is fixed in the **cytoplasm** of the mesophyll cells as shown below:

$$\text{PEP} + \text{CO}_2 \xrightarrow{\text{PEP carboxylase}} \text{oxaloacetate}$$
$$\text{(phosphoenolpyruvate)} \qquad\qquad 4C$$
$$3C$$

The carbon-dioxide-acceptor is phosphoenolpyruvate (PEP) instead of RuBP and the enzyme is PEP carboxylase instead of RuBP carboxylase. PEP carboxylase has two enormous advantages over RuBP carboxylase. First, it has a much higher affinity for carbon dioxide, and secondly it does not accept oxygen and hence does not contribute to photorespiration. Oxaloacetate is converted to malate or aspartate, both 4C-acids. They possess two carboxyl (—COOH) groups, that is they are **dicarboxylic acids**.

Malate shunt. Malate is shunted through plasmodesmata in the cell walls to the chloroplasts of the bundle sheath cells, where it is used to produce carbon dioxide (decarboxylation), hydrogen (oxidation) and pyruvate. The hydrogen reduces NADP to $NADPH_2$.

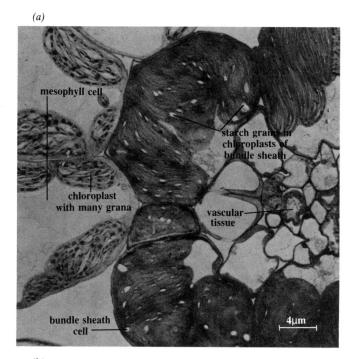

(a)

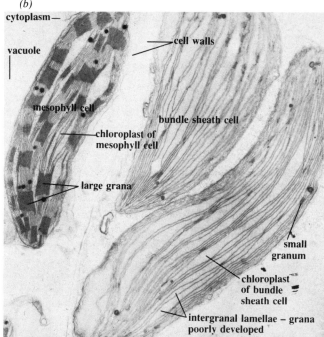

(b)

Fig 9.29 *(a) 'Kranz' anatomy, characteristic of C_4 plants. Micrograph of a crabgrass (Digitaria sargurnalis) leaf cross section to show the dimorphism between bundle sheath Chloroplasts and mesophyll chloroplasts. Grana in the bundle sheath are only rudimentary, whereas they are prominent in the mesophyll. Starch grains are present in both. (Magnification × 4 000). (b) Electron micrograph of maize leaf showing two types of chloroplasts found in bundle sheath cells and mesophyll cells (× 9 900)*

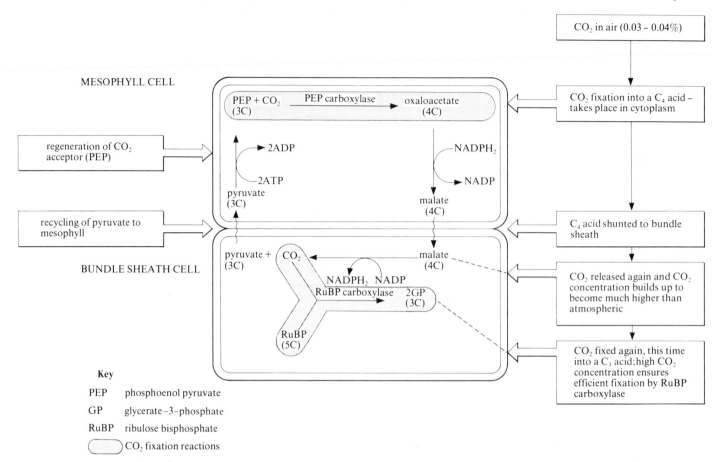

MESOPHYLL CELL

regeneration of CO₂ acceptor (PEP)

recycling of pyruvate to mesophyll

BUNDLE SHEATH CELL

CO₂ in air (0.03 – 0.04%)

CO₂ fixation into a C₄ acid – takes place in cytoplasm

C₄ acid shunted to bundle sheath

CO₂ released again and CO₂ concentration builds up to become much higher than atmospheric

CO₂ fixed again, this time into a C₃ acid; high CO₂ concentration ensures efficient fixation by RuBP carboxylase

Key

PEP   phosphoenol pyruvate

GP   glycerate–3–phosphate

RuBP   ribulose bisphosphate

CO₂ fixation reactions

Fig 9.30 *Simplified outline of C₄ pathway coupled with C₃ fixation of carbon dioxide. Transport of carbon dioxide from air to bundle sheath is shown, together with final fixation of carbon dioxide into the C₃ acid PGA*

Regeneration of the carbon dioxide acceptor. Pyruvate is returned to the mesophyll cells and is used to regenerate PEP by addition of phosphate from ATP. This requires the energy from two high-energy phosphate bonds.

Net result of C₄ pathway

The net result of the C_4 pathway is the use of two high energy phosphate bonds to transport carbon dioxide and hydrogen from the mesophyll cells to the chloroplasts of the bundle sheath cells.

Refixation of carbon dioxide in the bundle sheath cells

Carbon dioxide and $NADPH_2$ are produced, as well as pyruvate, in the bundle sheath chloroplasts (see malate shunt above). The carbon dioxide is refixed by RuBP carboxylase in the conventional C_3 pathway, and the latter also uses the $NADPH_2$

Since every carbon dioxide molecule has had to be fixed twice, the energy requirement is roughly double in C_4 photosynthesis compared with C_3 photosynthesis. At first sight then, the transport of carbon dioxide by the C_4 pathway seems pointless. However, fixation using PEP carboxylase in the mesophyll is so efficient that a high concentration of carbon dioxide accumulates in the bundle sheath. This means that the RuBP carboxylase works at an

advantage compared with the same enzyme in C_3 plants, where carbon dioxide is at atmospheric concentration. There are two reasons for this: firstly, like any enzyme it works much more efficiently at high substrate concentrations; and secondly, photorespiration is inhibited, because oxygen is competitively excluded from the enzyme by carbon dioxide.

The main advantage of C_4 photosynthesis therefore is that it improves the efficiency of carbon dioxide fixation and prevents wasteful loss of carbon by photorespiration. It is an addition rather than an alternative to the C_3 pathway. As a result, C_4 plants are photosynthetically more efficient because the rate of carbon dioxide fixation is normally the limiting factor in photosynthesis. C_4 plants consume more energy by using the C_4 pathway, but energy is not normally the limiting factor in photosynthesis, and C_4 plants grow in regions of high light intensity as well as having modified chloroplasts for making more efficient use of available energy (see below).

Table 9.7 Differences between mesophyll and bundle sheath chloroplasts in C₄ plants.

Mesophyll chloroplasts	Bundle sheath chloroplasts
Large grana	No grana (or very few and small)
Therefore photosystem II activity high, so plenty of ATP, NADPH₂ and O₂ generated.	Therefore photosystem II activity low, so little NADPH₂ or O₂ generated (some ATP from photosystem I)
Virtually no RuBP carboxylase so no CO₂ fixation (CO₂ fixation occurs in cytoplasm by PEP carboxylase)	High concentration of RuBP carboxylase so CO₂ fixation occurs
Little starch	Abundant starch grains

Mesophyll and bundle sheath chloroplasts

Table 9.7 summarises the important differences between mesophyll and bundle sheath chloroplasts, some of which are visible in fig 9.29.

9.39 Which type of chloroplast is specialised for light reactions and which for dark reactions?

9.40 Why is it an advantage that bundle sheath chloroplasts lack grana?

9.41 The malate shunt is, in effect, a carbon dioxide and hydrogen pump. What is the advantage of this?

9.42 What would be the effect of lowering oxygen concentrations on (a) C₃ photosynthesis, (b) C₄ photosynthesis? Explain your answers.

In addition to C₃ and C₄ plants, there is a small group of plants, mainly succulents, in which carbon dioxide is incorporated into organic compounds, such as citrate and malate, *during the hours of darkness*. In the light, these compounds undergo decarboxylation, releasing carbon dioxide within the leaf which is then taken up by the chloroplasts and built up into sugars via the C₃ pathway. These are referred to as **CAM** plants (**C**rassulacean, the family of plants into which many of the succulents are classified; **A**cid **M**etabolism which need not concern us further except to note that by having, in effect, an internal store of carbon dioxide, the stomata need to open for less time in the light, thereby reducing water loss).

Compared with C₃ plants, C₄ plants can cope with higher light intensities before light saturation occurs. They also have a higher affinity for carbon dioxide because the initial carboxylating enzyme is more efficient in C₄ plants. The overall optimum temperature for photosynthesis is higher for C₄ plants. As a result of the increased physiological efficiency, the stomata need to be open for shorter periods and hence in C₄ plants there is a higher rate of production of dry mass per unit of water taken up by the plants.

Table 9.8 Summary of the physiological differences between C₃ and C₄ plants.

	C_3	C_4
Representative species	Most crop plants cereals, tobacco, beans	Maize, sugar species
Rate of photo-respiration	High	Low
Light intensity for maximum rate of photosynthesis	10 000–30 000 foot candles	Not saturated at 10^5 lux
Effect of temperature on net rate (25 °C v. 35 °C)	No change or less at the warmer temperature	50% greater at the warmer temperature
'First product' produced in light	Glycerate-3-phosphate	Malate and oxaloacetate
Compensation point	40–60 ppm CO₂	Around zero
Water loss per g dry mass producer	450–950	250–350

Photorespiration is also reduced in C₄ plants and since

$$\begin{matrix} \text{net assimilation} \\ \text{rate} \end{matrix} = \begin{matrix} \text{total amount of} \\ \text{photosynthetic} \\ \text{products produced} \end{matrix} - \begin{matrix} \text{respiration losses} \end{matrix}$$

it follows that if photorespiration could be reduced in C₃ plants, higher yields would be produced overall. Attempts have been made, by the use of potential inhibitors sprayed on the leaf, to inhibit photorespiration. However, whilst these inhibitors have been shown to block certain enzyme steps in the photorespiration pathway, they have as yet been unsuccessful under field conditions; an example are the hydroxysulphonates which inhibit glycolic acid oxidase.

Attempts are being made to extend the environmental range of a number of crop and foliage plants having the C₄ type of metabolism by breeding new varieties and by selection within a species. For example, certain varieties of maize and soya bean have been shown to give consistently higher rates of photosynthesis than other varieties in different environmental conditions. C₄ plants do not appear to perform too well under temperate conditions.

9.8.3 Significance of photorespiration and the C₄ pathway

Photorespiration can be regarded as an unfortunate consequence of the increase in the oxygen concentration of the Earth's atmosphere (itself a result of photosynthesis) and of the ability of RuBP carboxylase to accept oxygen as well as carbon dioxide. It is wasteful of both carbon and energy and it has been estimated that it reduces the net rate of photosynthesis, and hence the potential yield, of C₃ plants by as much as 30–50%. Hence

Table 9.9 Comparison of C₃ and C₄ plants.

	C₃ plants	C₄ plants	
Carbon dioxide fixation	Occurs once	Occurs twice, first in mesophyll cells, then in bundle sheath cells	
Carbon dioxide acceptor	RuBP, a 5C-compound	**Mesophyll cells** PEP, a 3C-compound	**Bundle sheath cells** RuBP
Carbon dioxide – fixing enzyme	RuBP carboxylase, which is inefficient	PEP carboxylase which is very efficient	RuBP carboxylase, working efficiently because carbon dioxide concentration is high
First product of photosynthesis	A C₃ acid, GP	A C₄ acid, e.g. oxaloacetate	
Leaf anatomy	Only one type of chloroplast	'Kranz' anatomy, i.e. two types of cell, each with its own type of chloroplast	
Photorespiration	Occurs; therefore oxygen is an inhibitor of photosynthesis	Is inhibited by high carbon dioxide concentration. Therefore atmospheric oxygen is not an inhibitor of photosynthesis.	
Efficiency	Less efficient photosynthesis than C₄ plants. Yields usually much lower.	More efficient photosynthesis than C₃ plants. Yields usually much higher.	

it is of great economic significance, notably in crop plants. Various ways of inhibiting the process are being sought. One method would be to grow crops in atmospheres with artificially reduced oxygen concentrations, but this is difficult. Another is to artificially increase carbon dioxide concentrations to 0.1–1.5%, a five-fold increase over atmospheric, though this is commercially viable only for high-cash crops grown in greenhouses, such as tomatoes and flowers. Breeding C₄ genes into C₃ plants may prove possible and the techniques of genetic engineering may eventually prove useful.

The C₄ pathway is thought to be more recently evolved than the C₃ pathway and involves both a superior carbon-dioxide-fixing mechanism and a means of inhibiting photorespiration. Thus C₄ plants increase in dry mass more rapidly than C₃ plants and are more efficient crop plants.

They have evolved chiefly in the drier regions of the tropics, for which they are adapted in two major ways. First, their maximum rate of carbon dioxide fixation is greater; therefore the higher light intensities and temperatures of the tropics are more efficiently exploited. Secondly, C₄ plants are more tolerant of dry conditions. Plants usually reduce their stomatal apertures in order to reduce water loss by transpiration, and this also reduces the area for carbon dioxide entry. Carbon dioxide is fixed so rapidly in C₄ plants that a steep carbon dioxide diffusion gradient can still be maintained between external and internal atmospheres, thus allowing faster growth than C₃ plants. C₄ plants lose only about half the water that C₃ plants lose for each molecule of carbon dioxide fixed.

However, in cooler, moister, temperate regions with fewer hours of high light intensity, the extra energy (about 15% more) required by C₄ plants to fix carbon dioxide is more likely to be a limiting factor and C₃ plants may even have a competitive advantage in such situations.

9.9 Photosynthetic bacteria

Since photosynthesis probably first appeared in prokaryotes, details of the process in these organisms are of interest. In table 9.10 some of the more important comparisons between pro- and eukaryotes are made.

Four groups of photosynthetic bacteria occur as follows.

Green sulphur bacteria (for example *Chlorobium*). Anaerobic bacteria using hydrogen sulphide (H_2S) or other reduced sulphur compounds as hydrogen (electron) donors.

For example,

$$2H_2S + CO_2 \xrightarrow[\text{bacteriochlorophyll}]{\text{light}} [CH_2O] + 2S + H_2O$$

Sulphur is deposited.

Purple sulphur bacteria (for example *Chromatium*). Red and brown pigments (carotenoids) dominate bacteriochlorophyll making the cells appear purple. They are mostly anaerobic and details of photosynthesis are as above.

Purple non-sulphur bacteria (for example *Rhodospirillum*). Bacteria using organic compounds as a source of hydrogen to reduce either carbon dioxide (photoautotrophic) or an organic carbon source (photoheterotrophic).

Blue-green bacteria (see table 9.10). Resemble many plants in using water as a hydrogen donor.

Table 9.10 Comparison of photosynthesis in prokaryotes and eukaryotes.

	Prokaryotes	Eukaryotic plants
Bacteria	Blue-green bacteria	
No chloroplasts	No chloroplasts	Chloroplasts (each equivalent to a prokaryotic cell?)
Membranes present as extensions of the plasma membrane; called chromatophores	Membranes present throughout cytoplasm	Membranes in chloroplasts
Membranes not stacked	Membranes not stacked	Membranes usually stacked, forming grana in higher plants
Photosystem II absent; therefore no oxygen produced	Photosystem II present; therefore oxygen produced from water	Photosystem II present; therefore oxygen produced from water
Hydrogen donor variable e.g. H_2S, H_2, organic compounds, not water	Water acts as hydrogen donor	Water acts as hydrogen donor
Primary pigment is bacteriochlorophyll	Primary pigment is chlorophyll	Primary pigment is chlorophyll
No phycobilins	Also contain phycobilins (a third class of photosynthetic pigment)	Phycobilins only in red algae (a primitive feature)

9.10   Chemosynthesis

Chemosynthetic organisms (chemoautotrophs) are bacteria using carbon dioxide as a carbon source but obtaining their energy from chemical reactions rather than light. The energy is obtained by oxidising inorganic materials such as hydrogen, hydrogen sulphide, sulphur, iron(II), ammonia and nitrite.

Iron bacteria (for example *Leptothrix*).

$$Fe^{2+} \xrightarrow{\text{oxygen}} Fe^{3+} + \text{energy}$$

Full equation:

$$4FeCO_3 + O_2 + 6H_2O \longrightarrow 4Fe(OH)_3 + 4CO_2 + \text{energy}$$

Colourless sulphur bacteria (for example *Thiobacillus*).

$$\underset{\text{sulphur}}{S} \xrightarrow[\text{nitrate}]{\text{oxygen or}} \underset{\text{sulphate}}{SO_4^{2-}} + \text{energy}$$

Full equation: $2S + 3O_2 + 2H_2O \longrightarrow 2H_2SO_4 + \text{energy}$

Under anaerobic conditions some species use nitrate as a hydrogen acceptor, thus carrying out denitrification (see section 9.11.1).

Nitrifying bacteria (see section 9.11.1).

$$\underset{\text{ammonium}}{NH_4^+} \xrightarrow{\text{oxygen}} \underset{\text{nitrite}}{NO_2^-} + \text{energy (for example } \textit{Nitrosomonas})$$

Full equation: $2NH_3 + 3O_2 \longrightarrow 2HNO_2 + 2H_2O + \text{energy}$

$$NO_2^- \xrightarrow{\text{oxygen}} NO_3^- + \text{energy (for example } \textit{Nitrobacter})$$

Full equation: $2HNO_2 + O_2 \longrightarrow 2HNO_3 + \text{energy}$

In the above examples oxygen is an electron (hydrogen) acceptor and the bacteria are aerobic.

Chemosynthetic bacteria play important roles in the biosphere, principally in maintaining soil fertility through their activities in the nitrogen cycle.

9.11   Mineral cycles (biogeochemical cycles)

9.11.1   The nitrogen cycle

The atmosphere contains 79% by volume of nitrogen, yet nitrogen is relatively scarce in combined (fixed) form because it is rather inert chemically. Nitrogen is an essential constituent of amino acids, and hence proteins, and it limits the supply of food available in ecosystems more than any other plant nutrient. The only way in which it can be made available to living organisms is via **nitrogen fixation**, an ability confined to certain prokaryotes, although the techniques of genetic engineering may eventually lead to introduction of the relevant genes into green plants. The nitrogen cycle is summarised in fig 9.31.

Nitrogen fixation

Nitrogen fixation is energy-consuming because the two nitrogen atoms of the nitrogen molecule must first be separated. Nitrogen-fixers achieve this by an enzyme, nitrogenase, using energy from ATP. Non-enzymic separation requires the much greater energy of industrial processes or of ionising events in the atmosphere, such as lightning and cosmic radiation.

Nitrogen is so important for soil fertility, and the demand for food production so great, that colossal amounts of ammonia are produced industrially each year to be used mainly for nitrogenous fertilisers such as ammonium nitrate (NH_4NO_3) and urea ($CO(NH_2)_2$). The amounts of nitrogen fixed commercially are now roughly equal to the amounts fixed naturally. We are still relatively ignorant as to the effects which the gradual accumulation of fixed nitrogen, which is now occurring, will have in the biosphere. We have learned through experience of some of the problems, such as run-off of nitrate fertilisers into lakes

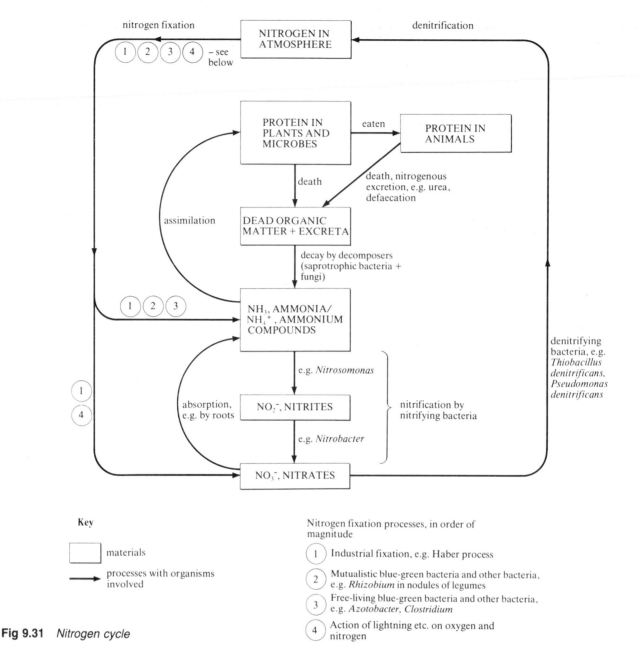

Fig 9.31 *Nitrogen cycle*

Key

☐ materials

→ processes with organisms involved

Nitrogen fixation processes, in order of magnitude

① Industrial fixation, e.g. Haber process

② Mutualistic blue-green bacteria and other bacteria, e.g. *Rhizobium* in nodules of legumes

③ Free-living blue-green bacteria and other bacteria, e.g. *Azotobacter*, *Clostridium*

④ Action of lightning etc. on oxygen and nitrogen

and rivers causing an imbalance of salts. This can result in the complete loss of life from the water.

A relatively small amount of fixed nitrogen (5–10%) is formed by ionising events in the atmosphere. The resulting nitrogen oxides dissolve in rain, forming nitrates.

The legumes, such as clover, soyabean, lucerne and pea, are probably the greatest natural source of fixed nitrogen. Their roots possess characteristic swellings called **nodules** which are caused by colonies of nitrogen-fixing bacilli (genus *Rhizobium*) living within the cells. The relationship is mutualistic because the plant gains fixed nitrogen in the form of ammonia from the bacteria and, in return, the bacteria gain energy and certain nutrients, such as carbohydrates, from the plants. In a given area legumes can contribute as much as 100 times more fixed nitrogen than free-living bacteria. It is not surprising, therefore, that they

are frequently used to add nitrogen to the soil, especially since they have the added benefit of making good fodder crops.

9.43 Farmers often say that legumes are 'hard on the soil', meaning that they place a large demand on soil minerals. Why should this be so?

All nitrogen-fixers incorporate nitrogen into ammonia, but this is immediately used to make organic compounds, mainly proteins (fig 9.31).

Decay and nitrification

Most plants depend on a supply of nitrate from the soil for their nitrate source. Animals in turn depend directly or indirectly on plants for their nitrogen supply. Fig 9.31 shows how nitrates are recycled from proteins in dead organisms by saprotrophic bacteria and fungi. The sequence from proteins to nitrate is a series of oxidations, requiring oxygen and involving aerobic bacteria. Proteins are decomposed via amino acids to ammonia when an organism dies. Animal wastes and excreta are similarly decomposed. Chemosynthetic bacteria (section 9.10) then oxidise ammonia to nitrate, a process call **nitrification**.

9.44 In which of the nutritional categories would you place bacteria and fungi which are decomposers?

Denitrification

Nitrification can be reversed by denitrifying bacteria (**denitrification**) whose activities can therefore reduce soil fertility. They only do this under anaerobic conditions, when nitrate is used instead of oxygen as an oxidising agent (electron acceptor) for the oxidation of organic compounds. Nitrate itself is reduced. The bacteria are therefore **facultative aerobes**. It should not be assumed that their activities on a global scale are detrimental to the biosphere because it has been estimated that most of the atmospheric nitrogen might now be in the oceans or locked up in sediments were it not for denitrification.

9.45 What natural areas or situations might favour denitrification?

9.46 Why should good drainage and ploughing increase soil fertility?

9.11.2 The sulphur cycle

Fig 9.32 shows the sulphur cycle. Sulphur is abundant in the Earth's crust and is available to plants principally as sulphate. It is an essential constituent of virtually all proteins.

As with nitrogen, animals depend ultimately on plants for their sulphur requirements. In addition to the natural sulphur cycle shown in fig 9.32, oxides of sulphur, such as sulphur dioxide (SO_2), are increasingly being added to the atmosphere as a result of burning fossil fuels and the smelting of sulphur ores. These are pollutants and when dissolved in rain make it acidic. A growing body of

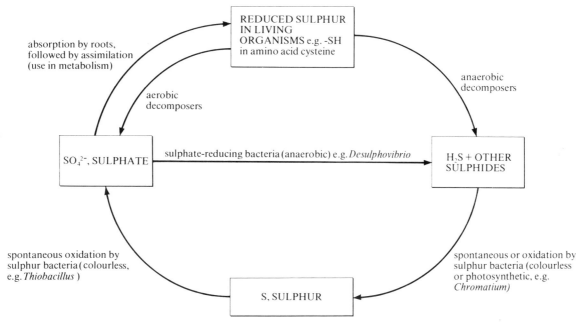

Fig 9.32 (above) *Sulphur cycle*

Fig 9.33 (below) *Phosphorus cycle*

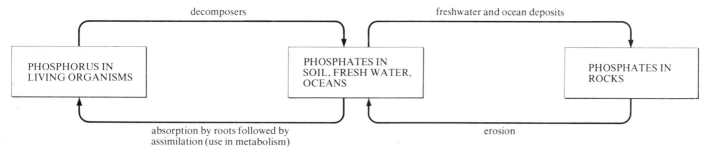

evidence is suggesting that acid rainfall can have widespread ecological repercussions.

9.11.3 The phosphorus cycle

Phosphorus is an essential constituent of nucleic acids, proteins, ATP and some other vital organic compounds. It is a relatively uncommon element and, like nitrogen and potassium, is often a limiting factor in the productivity of ecosystems. The cycle, shown in fig 9.33, is simple because phosphorus forms no natural gaseous compounds. Much of the phosphorus that finds its way to the oceans becomes locked up in sedimentary deposits.

9.11.4 The carbon and oxygen cycles

The carbon cycle is discussed in section 9.1 and shown in fig 9.2. The oxygen cycle is closely related.

9.11.5 Summary

The cycling of materials in the biosphere has been shown to involve complex nutritional relationships between living organisms. These form part of the study of ecology and certain aspects of them are discussed in more detail in chapter 12. Bacteria are an integral part of each cycle and their activities can therefore be seen to be essential in maintaining all life. The different modes of nutrition involve autotrophic, heterotrophic, photosynthetic and chemosynthetic activity. An understanding of these cycles is essential to humans if we are to make the best use of available materials and to understand the consequences of interference with them.

9.12 Mineral nutrition of plants and animals

Autotrophic nutrition involves not only the synthesis of carbohydrates from carbon dioxide and a hydrogen donor, such as water, but the subsequent use of minerals like nitrates, sulphates and phosphates to make other organic requirements, such as proteins, nucleic acids and so on. Heterotrophic organisms also require certain minerals to supplement their organic food. In many cases the same nutrients are required, and for the same reasons, so it is convenient to consider the whole area of mineral nutrition as a bridge between autotrophic nutrition (chapter 9) and heterotrophic nutrition (chapter 10).

A nutritional element essential for the successful growth and reproduction of an organism is called an **essential element**. The major essential elements for life are carbon, hydrogen, oxygen, nitrogen, sulphur, phosphorus, potassium, sodium, magnesium, calcium and chlorine. In addition, certain elements, the **trace elements**, are essential in trace amounts (a few parts per million). Of these, all organisms require manganese, iron, cobalt, copper and zinc; some also require combinations of molybdenum, vanadium, chromium and other heavy metals, as well as boron, silicon, fluorine and iodine (see table 5.1). All except carbon, hydrogen and oxygen are taken up as minerals from soil or water by green plants. The mechanism of uptake is discussed in chapter 14.

For heterotrophic organisms (animals and fungi) the trace elements (inorganic) are sometimes grouped with vitamins (organic) as **micronutrients**, since both are required in trace amounts and have similar fundamental roles in cell metabolism, often as enzyme cofactors. Vitamins are considered in chapter 10. Autotrophic organisms synthesise their own vitamins. The other essential elements are called **macronutrients**. Deficiency of any of the nutrients mentioned can lead to **deficiency diseases**.

Some examples of the functions of the major minerals are given in table 9.11. A study of the table will reveal that mineral elements are taken up by plants as separate ions, either anions (negatively charged) or cations (positively charged). This is also true of trace elements, though their ions are not shown in the table.

Cations fall into two broad categories, namely light metals, whose roles in cell metabolism are usually associated with their high mobility, and the heavy metals, such as iron and copper, which are generally fixed in the membranes of mitochondria and chloroplasts. Animals do not obtain all their essential elements in the form of minerals. Much of their nitrogen, for example, is ingested in the form of proteins.

The geographical distribution of the minerals, particularly trace elements, can vary enormously and is one of the factors in the environment determining the distribution of different plants, and hence, animal species. A balance of trace elements is essential for soil fertility. Extreme cases are known of plants thriving in areas of high metal contamination, such as on spoilage tips from mines or over natural mineral deposits, and such plants can prove toxic to grazing animals. Conversely, these plants can be useful to humans if they help to cover formerly unsightly areas.

9.12.1 Mineral element deficiencies

It is not always easy, or possible, to isolate the effects of individual minerals. In plants, for example, chlorosis (lack of chlorophyll) can be caused by lack of magnesium or iron, both having different roles in chlorophyll synthesis (table 9.11). A common deficiency disease of sheep and cattle called scour, which causes diarrhoea, is due to copper deficiency induced by high levels of molybdenum in the pastures. Different effects may occur in different organisms; lack of manganese, for example, causes grey speck in oats, marsh spot in beans and poor flavour in oats.

The close interaction and varied effects of mineral elements are due to their fundamental effects on cell metabolism. However, it is possible by various means, such

Table 9.11 Some essential mineral elements and examples of their uses in living organisms.

MACRONUTRIENTS

Element and symbol	Taken up by plants as	General importance	Common deficiency diseases or symptoms Plants	Humans	Common food source for humans
Nitrogen, N	Nitrate, NO_3^- Ammonium, NH_4^+	Synthesis of proteins, nucleic acids and many other organic compounds, e.g. coenzymes and chlorophyll.	Stunted growth and strong chlorosis, particularly of older leaves	Kwashiorkor due to lack of protein	Protein, e.g. lean meat, fish and milk. Milk is rich in phosphorus
Phosphorus, P	Phosphate, PO_4^{3-} Orthophosphate, $H_2PO_4^-$	Synthesis of nucleic acids, ATP and some proteins. Also, phosphate is a constituent of bone and enamel. Phospholipids in membranes.	Stunted growth, particularly of roots		
Potassium, K	K^+	Mainly associated with membrane function, e.g. conduction of nervous impulses, maintaining electrical potentials across membranes Na^+/K^+ pump in active transport across membranes, anion/cation and osmotic balance. Cofactor in photosynthesis and respiration (glycolysis). Common in cell sap of plant vacuoles.	Yellow and brown leaf margins and premature death	Rarely deficient	Vegetables, e.g. brussels sprouts (= buds), and meat.
Sulphur, S	Sulphate, SO_4^{2-}	Synthesis of proteins (e.g. keratin) and many other organic compounds, e.g. coenzyme A.	Chlorosis, e.g. 'tea-yellow' of tea		Protein, e.g. lean meat, fish and milk.
Sodium, Na	Na^+	Similar to potassium, but usually present in lower concentrations. Often exchanged for potassium.		Muscular cramps	Table salt (sodium chloride) and bacon.
Chlorine, Cl	Chloride, Cl^-	Similar to Na^+ and K^+, e.g. anion/cation and osmotic balance. Involved in 'chloride shift' during carbon dioxide transport in blood. Constituent of hydrochloric acid in gastric juice.		Muscular cramps	Table salt and bacon.
Magnesium, Mg	Mg^{2+}	Part of structure of chlorophyll. Bone and tooth structure. Cofactor for many enzymes, e.g. phosphatases (e.g. ATPase).	Chlorosis		Vegetables and most other foods.
Calcium, Ca	Ca^{2+}	Formation of middle lamella (calcium pectate) between plant cell walls and normal cell wall development. Constituent of bone, enamel and shells. Activates ATPase during muscular contraction. Blood clotting.	Stunted growth	Poor skeletal growth, possibly leading to rickets	Milk, hard water.

TRACE ELEMENTS – all cations except boron, fluorine and iodine

Element and symbol	Substance containing	Examples of functions	Common deficiency diseases or symptoms Plants	Humans	Common food source for humans
Manganese, Mn	Phosphatases (transfer PO_4 groups)	Bone development (a 'growth factor')	Leaf-flecking, e.g. 'grey-speck' in oats	Poor bone development	Vegetables and most other foods.
	Decarboxylases Dehydrogenases }	Oxidation of fatty acids, respiration, photosynthesis.			
Iron, Fe	Haem group in: haemoglobin and myoglobin }	Oxygen carriers.		Anaemia	Liver and red meat. Some vegetables, e.g. spinach.
	Cytochromes	Electron carriers, e.g. respiration, photosynthesis.			

TRACE ELEMENTS – all cations except boron, fluorine and iodine

Element and symbol	Substance containing	Examples of functions	Common deficiency diseases or symptoms Plants	Humans	Common food source for humans
Iron, Fe cont.	Catalase and peroxidases	Break down H_2O_2.	Strong chlorosis, particularly in young leaves		
	Other porphyrins	Chlorophyll synthesis.			
Cobalt, Co	Vitamin B_{12}	Red blood cell development.		Pernicious anaemia	Liver and red meat (as vitamin B_{12}).
Copper, Cu	Cytochrome oxidase	Terminal electron carrier in respiratory chain – oxygen converted to water.	Dieback of shoots		
	Haemocyanin	Oxygen carrier in certain invertebrates.			
	Plastocyanin	Electron carrier in photosynthesis.			
	Tyrosinase	Melanin production.		Albinism	
Zinc, Zn	Alcohol dehydrogenase	Anaerobic respiration in plants (alcohol fermentation).	'Mottle leaf' of *Citrus*		Most foods.
	Carbonic anhydrase	Carbon dioxide transport in vertebrate blood.	Malformed leaves, e.g. 'sickle leaf' of cocoa		
	Carboxypeptidase	Hydrolysis of peptide bonds in protein digestion.			
Molybdenum, Mo	Nitrate reductase	Reduction of nitrate to nitrite during amino acid synthesis in plants.	Slight retardation of growth; 'scald' disease of beans		
	Nitrogenase	Nitrogen fixation (prokaryotes).			
Boron, B	—	Plants only. Normal cell division in meristems. Mobilisation of nutrients?	Abnormal growth and death of shoot tips, 'heart-rot' of beet; 'stem-crack' of celery	Not needed	
Fluorine, F	Associated with calcium as calcium fluoride in animals	Component of tooth enamel and bone.		Dental decay more rapid	Milk, drinking water in some areas.
Iodine, I	Thyroxine (Probably not required by higher plants)	Hormone controlling basal metabolic rate.		Goitre; cretinism in children	Seafoods, salt.

as experimentally manipulating mineral uptake, to show that specific sets of symptoms are associated with deficiencies of certain elements.

Such knowledge is of importance in both medicine and agriculture because deficiency diseases are common worldwide, both in humans and in their crops and animals.

Experiments on plants were done in the late nineteenth and early twentieth century, particularly by German botanists, using the now classic water culture or sand culture techniques. In these experiments, plants are grown in prepared culture solutions of known composition. Many economically important plant deficiency diseases are now catalogued with the aid of colour photography, enabling rapid diagnosis. Although many of the essential roles of the mineral elements are established, further physiological and biochemical research remains to be done.

9.12.2 Special methods for obtaining essential elements

Insectivorous plants

Insectivorous or carnivorous plants are green plants which

are specially adapted for trapping and digesting small animals, particularly insects. In this way they supplement their normal autotrophic nutrition (photosynthesis) with a form of heterotrophic nutrition. Such plants typically live in nitrogen-poor habitats, and use the animals principally as a source of nitrogen. Having lured the insect with colour, scent or sweet secretions, the plant traps it in some way and then secretes enzymes and carries out extracellular digestion. The products, notably amino acids, are absorbed and assimilated.

Some of the plants are interesting for the elaborate nature of their trap mechanisms, notably the Venus fly trap (*Dionaea muscipula*), pitcher plants (*Nepenthes*) and sundews (*Drosera*). *Drosera* is one of the few British examples, most being tropical or subtropical. It is found on the wetter heaths and moors which are typically acid, mineral-deficient habitats. The details of the various mechanisms are outside the scope of this book.

Mycorrhizas

A mycorrhiza is a mutualistic association between a fungus and a plant root. It is likely that the great majority of land plants enter into this kind of relation with soil fungi. They are of great significance because they are probably the major route of entry of mineral nutrients into roots. The fungus receives organic nutrients, mainly carbohydrates and vitamins, from the plant and in return absorbs mineral salts (particularly phosphate, ammonium, potassium and nitrate) and water, which can pass to the plant root. Generally only young roots are infected. Root hair production either ceases or is greatly reduced on infection. A network of hyphae spreads through the surrounding soil, covering a much larger surface area than the root could, even with root hairs. It has been suggested that plants of the same species, or even different species may have common interconnections with mycorrhizas, a concept which could radically alter our view of natural ecosystems.

Two groups of mycorrhizas occur, the ectotrophic and endotrophic mycorrhizas. **Ectotrophic mycorrhizas** form a sheath around the root and penetrate the air spaces between the cells in the cortex, but do not enter cells. An extensive intercellular net is formed. They are found mainly in forest trees such as conifers, beech, oak and many others, and involve fungi of the Basidiomycota. Their 'fruiting bodies' (mushrooms) are commonly seen near the trees.

Endotrophic mycorrhizas occur in virtually all other plants. Like ectotrophic mycorrhizas, they also form an intercellular network and extend into the soil, but they appear to penetrate cells (although in fact they do not break through the cell surface membranes of the root cells).

As we learn more about mycorrhizas, it is likely that the knowledge will be applied with advantage to agriculture, forestry and land reclamation.

Root nodules

Nitrogen fixation in root nodules of leguminous plants has already been discussed in section 9.11.1 of this chapter. The bacteria which inhabit the nodules stimulate growth and division of the root parenchyma cells resulting in the swelling or nodule.

Chapter Ten

Heterotrophic nutrition

Heterotrophs are organisms that feed on complex ready-made organic food (their carbon source is organic, p. 249). They use it as a source of (i) energy for their vital activities, (ii) building materials, that is specific atoms and molecules for cell maintenance and repair and growth, and (iii) vitamins (coenzymes) that cannot be synthesised in the organism but which are vital for specific cellular processes.

The survival of heterotrophs is dependent either directly or indirectly on the synthetic activities of autotrophs. All animals and fungi and the majority of bacteria are heterotrophic. A few bacteria, such as purple non-sulphur bacteria, possess **bacteriochlorophyll** and are able to utilise energy to synthesise their organic requirements from other organic raw materials, and are called **photoheterotrophs** (table 9.1).

The manner in which heterotrophs procure and take in their food varies considerably; nevertheless the way in which it is processed into a utilisable form within the body is very similar in most of them. It involves two distinct processes: first a method of reducing large complex food molecules into simpler soluble ones (**digestion**), and secondly a means of **absorbing** the soluble molecules from the region of digestion into the tissues of the organism.

For convenience, the main forms of heterotrophic nutrition may be classified as holozoic, saprotrophic (or saprophytic), mutualistic and parasitic, although some overlap between groups may occur.

10.1 Modes of heterotrophic nutrition

10.1.1 Holozoic nutrition

All organisms feeding in this way take food into the body where it is then digested into smaller soluble molecules which can be absorbed and assimilated. The term holozoic is applied to mainly free-living animals which have a specialised digestive tract (alimentary canal) in which these processes occur. Most animals and insectivorous plants are holozoic.

The characteristic processes involved in holozoic nutrition are defined as follows.

Ingestion is the taking in of complex organic food.

Digestion is the breakdown of large complex insoluble organic molecules into small, simple soluble diffusible molecules. This is achieved by mechanical breakdown and enzymatic hydrolysis. Digestion may be either extra- or intracellular.

Absorption is the uptake of the soluble molecules from the

digestive region, across a membrane and into the body tissue proper. The food may pass directly into cells or initially pass into the bloodstream to be transported to appropriate regions within the body of the organism.

Assimilation is the utilisation of the absorbed molecules by the body to provide either energy or materials to be incorporated into the body.

Egestion is the elimination from the body of undigested waste food materials.

The stages involved in holozoic nutrition are summarised in fig 10.1.

Animals which feed on plants are called **herbivores**, those that feed on other animals **carnivores**, and those that eat a mixed diet of animal and vegetable matter are termed **omnivores**. If they take in food in the form of small particles the animals are **microphagous** feeders, for example

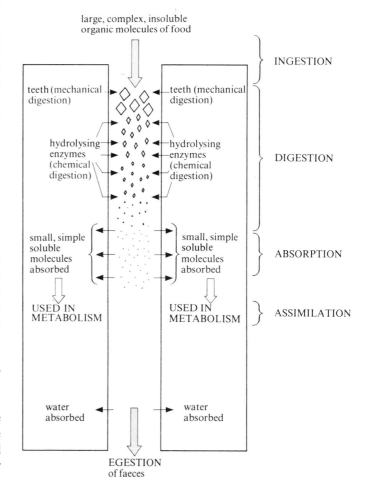

Fig 10.1 *Stages in the holozoic nutrition of a mammal*

earthworms, whereas if the food is ingested in liquid form they are classed as **fluid feeders**, such as aphids and mosquitoes. Animals which take in food in the form of large pieces are termed **macrophagous**. A summary of holozoic feeding methods is given below.

(1) Microphagous
  (*a*) pseudopodial
  (*b*) ciliary
  (*c*) filter-feeding – setose (setae are hair-like struc-
      tures)
     – ciliary
(2) Macrophagous
  (*a*) tentacular
  (*b*) scraping/boring
  (*c*) seizing prey
  (*d*) detritus/deposit-feeding
(3) Fluid-feeding
  (*a*) sucking
  (*b*) piercing and sucking

10.1.2 Saprophytic or saprotrophic nutrition
(*sapros*, rotten; *phyton* plant; *trophos*, feeder)

Organisms which feed on dead or decaying organic matter are called **saprophytes** or **saprotrophs**. Many fungi and bacteria are saprophytes, for example the fungus *Mucor hiemalis*. They were once regarded as plants, hence the use of 'phyte' in saprophyte. A more recent term, saprotroph, avoids this problem. Saprotrophs secrete enzymes onto potential food where it is digested. The soluble end-products of this extracellular chemical decomposition are then absorbed and assimilated by the saprotroph. Saprotrophs feed on the dead organic remains of plants and animals and contribute to the removal of such organic refuse by decomposing it. Many of the simple substances formed are not used by the saprotrophs themselves but are absorbed by plants. In this way the activity of the saprotrophs provides important links in nutrient cycles serving to return vital chemical elements from the dead bodies of organisms to living ones.

The saprotrophic nutrition of Mucor hiemalis

Nutritive fungal hyphae penetrate the substrate on which *Mucor* is growing and secrete hydrolysing enzymes from their tips which results in extracellular digestion as shown in fig 10.2. Carbohydrase and protease enzymes carry out the extracellular digestion of starch to glucose and protein to amino acids respectively. The thin, much-branched nature of the mycelium of *Mucor* ensures that there is a large surface area for absorption. Glucose is used during respiration to provide energy for the organism's metabolic activities whilst glucose and amino acids are used for growth and repair. Surplus glucose is converted to glycogen and fat, and excess amino acids to protein granules. These products are stored in the hyphal cytoplasm.

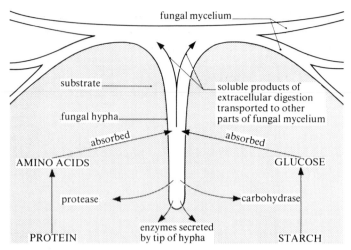

Fig 10.2 *Extracellular digestion and absorption in* Mucor hiemalis

10.1 Briefly describe the ways in which *Mucor* is economically important to humans.

10.1.3 Symbiosis: mutualism, parasitism and commensalism

The term symbiosis means literally 'living together' (*syn*, with; *bios*, life). It was introduced by the German scientist de Bary in 1879 and broadly defined as 'the living together of dissimilarly named organisms'; in other words, as an association between two or more organisms of different species. Since de Bary's time the term symbiosis has been restricted by many biologists to meaning a **close** relationship between two or more organisms of different species **in which all partners benefit**.

Since the 1970s symbiosis has assumed more importance as a topic in biology. For example, the endosymbiont theory of 1967 (section 9.3.1) launched the field of intracellular symbiosis; the ecological importance of symbiosis has received greater attention with the realisation that the great majority of plants obtain their minerals with the assistance of mycorrhizas, and that much nitrogen fixation is carried out by symbiotic bacteria; and rumen fermentation, involving symbiotic organisms, is of potential importance in increasing cattle productivity. At the same time biologists have become increasingly aware that degree of closeness and degree of benefit (or harm) of an association are two variables about which it is difficult to be precise and that there is a continuous spectrum of degrees of closeness and benefit or harm. Most modern biologists therefore prefer to use something like de Bary's original definition of symbiosis, a move approved by the Society for Experimental Biology in 1975*.

The following definitions will therefore be adopted in

* *References*: SEB Symposia XXIX, *Symbiosis*, CUP (1975) D.H. Jennings & D.L. Lee (eds.); G.H. Harper, (1985) 'Teaching symbiosis', *J.Biol.Ed.* **19** (3), 219–23; D.C. Smith & A.E. Douglas (1987) *The Biology of Symbiosis*, Arnold.

this book. For convenience (and to preserve existing terms) emphasis is placed on whether the relationship is beneficial or not to both partners, but **closeness** of association could equally well be a criterion for classification (see G.H. Harper*, footnote p. 290).

Symbiosis – the living together in close association of two (or more) organisms of different species. (Note: many associations involve three or more partners. Interactions commonly, but not necessarily, involve nutrition. 'Close' is difficult to define.)

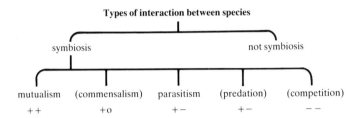

Key:
+, a partner benefits;
−, a partner receives harm;
o, a partner is unaffected;
( ) additional terms included by some biologists under symbiosis
Parasitism is regarded as a form of predation by some biologists.

Ectosymbiont – a symbiont external to its partner, e.g. leech, dodder (ectoparasites).
Endosymbiont – a symbiont within its partner –
    intracellular, e.g. chloroplasts, mitochondria
    extracellular, e.g. tapeworm (endoparasite).
Obligate symbiont – a symbiont which cannot survive without its partner.
Facultative symbiont – a symbiont which can survive without its partner.

Mutualism

Mutualism is a close association between two living organisms of different species which is beneficial to both partners. The larger organism may be referred to as the **host**. (This is the narrow definition of symbiosis, no longer recommended.) (Note, 'benefit' has been defined as increasing fitness for survival. Some biologists include brief interactions, such as pollination of flowers by insects, which is of benefit to flower and insect.) The association may be between two organisms of the same or different kingdoms. For example, the sea anemone *Calliactis parasitica* attaches itself to a shell used by the hermit crab *Eupagurus* (fig 10.3). The anemone obtains nourishment from the scraps of food left by the crab, and is transported from place to place when the crab moves. The crab is camouflaged by the anemone and may also be protected by nematocysts. It seems that the anemone is unable to survive unless attached

Fig 10.3 *Sea anemones,* Calliactis parasitica, *attached to a whelk shell inhabited by the hermit crab* Eupagurus bernhardus

to the crab's shell, and, if the anemone is removed, the crab will seek another anemone and actually place it on the shell it is inhabiting.

Herbivorous ruminants harbour a vast fauna of cellulose-digesting ciliates, such as *Entodinium* (see section 10.8.3). These can only survive in anaerobic conditions such as are found in a ruminant's alimentary canal. Here the ciliates feed on the cellulose contained in the host's diet, converting it into simple compounds which the ruminant is then able to further digest, absorb and assimilate itself. Some other examples of mutualism involving micro-organisms are given in table 10.1. See also mycorrhizas (section 9.12.2) and endosymbiosis (section 9.3.1).

Commensalism (com-, together; mensa, table)

Commensalism is a close association between two living organisms of different species which is beneficial to one (the **commensal**) and does not affect the other (the **host**). Commensalism means literally 'eating at the same table' and is used to describe symbiotic relationships which do not fit conveniently into the mutualism and parasitism categories. For example, the colonial hydrozoan *Hydractinia echinata* attaches itself to whelk shells inhabited by hermit crabs. It obtains nourishment from the scraps of food left by the crab after it has eaten. In this particular case the crab is totally unaffected by the association. An orchid or lichen (the commensal) growing on a tree (the host) would be another example.

Parasitism (para, beside; sitos, food)

Parasitism is a close association between two living organisms of different species which is beneficial to one (the **parasite**) and harmful to the other (the **host**). The parasite obtains food from the host and generally shelter. A successful parasite is able to live with the host without

Table 10.1 Some examples of mutualism involving micro-organisms (based on D.C. Smith & A.E. Douglas (1987) *The Biology of Symbiosis*, Arnold).

Host (larger partner)	Symbiont (smaller partner)	Benefit for host	Benefit for symbiont
Blue-green bacteria (cyanobacteria)	Aerobic bacteria	Nitrogen fixation enhanced (takes place in anaerobic conditions)	Oxygen from photosynthesis
Hydra	*Chlorella* (intracellular)	Maltose	All nutrients available
Legumes	*Rhizobium* spp. (intracellular)	Ammonia (fixed nitrogen)	Products of photosynthesis
Plants	Blue-green bacteria in soil	Ammonia (fixed nitrogen)	Products of photosynthesis
Plants	Blue-green bacteria in cells (= chloroplasts)	Photosynthesis	All nutrients available for protein formation
Plants _(Mycorrhiza formed)_	Fungi	Mineral nutrients taken up, particularly N, PO_4 and K; roots protected against pathogens; drought resistance enhanced	Products of photosynthesis
Lichen fungi _(Lichen formed)_	Blue-green bacteria	Glucose, ammonia (fixed nitrogen)	Shielded from high light intensity and water loss is reduced; some joint synthesis of metabolites
Lichen fungi _(Lichen formed)_	Algae	Polyhydric alcohols (closely related to carbohydrates)	
Vertebrates with a rumen (ruminants)	Various prokaryotes, protozoans and fungi living in rumen	Formation of food, e.g. cellulose converted to fatty acids; protection from some gut parasites	Food ingested by host
Some teleost fish, squids and tunicates	Luminescent bacteria	Luminous light organs, e.g. angler fish (tip of projection from head acts as bait)	Nutrients and oxygen

causing it any great harm. The degree of benefit or harm may be difficult to establish (see G.H. Harper*, footnote p. 290).

Parasites which live on the outer surface of a host are termed **ectoparasites** (for example ticks, fleas and leeches). Such organisms do not always live a fully parasitic existence. Those that live within a host are **endoparasites** (such as *Plasmodium* and *Taenia*). If the organism has to live parasitically at all times it is said to be an **obligate** parasite. **Facultative** parasites are fungi that feed parasitically initially, but having eventually killed their host continue to feed saprotrophically on the dead body. Some green plants are partial parasites; they photosynthesise but nevertheless obtain micronutrients from their host. Mistletoe is such an example; its haustoria penetrate the xylem of the host from where they absorb mineral salts and water.

The very nature of the parasitic niche means that parasites are highly specialised, possessing numerous adaptations, many of which are associated with their host and its mode of life. Table 10.2 shows some of the structural, physiological and reproductive modifications used by various parasites in order to cope with the rigours of their existence. Micro-organisms which cause disease may be regarded as parasites (tables 2.6, 2.7 and 3.4).

10.2 List the structural, physiological and reproductive features that make *Fasciola* (liver fluke) a successful parasite.

10.2 Feeding mechanisms in a range of animals

10.2.1 Microphagous feeders

Pseudopodial

Amoeba consumes rotifers, diatoms, desmids, bacteria, flagellates, ciliates and minute particles of debris. It ingests its food by means of phagocytosis. **Pseudopodia** envelop the material and enclose it, together with a variable amount of water, in a **food vacuole** (fig 10.4). The vacuole then becomes surrounded by many tiny lysosomes which ultimately fuse with its membrane and discharge their enzymatic contents into it. Hence digestion is intracellular. At this stage the vacuole becomes known as a **digestive vacuole**. Initially it decreases in size as water is withdrawn,

Table 10.2 Some structural, physiological and reproductive specialisations of parasites.

	Type of modification	Examples
Structural	Absence or degeneration of feeding and locomotory organs – characteristic of gut parasites.	*Fasciola* (liver fluke), *Taenia* (tapeworm)
	Highly specialised mouthparts as in fluid feeders.	*Pulex* (flea), *Aphis* (aphid)
	Development of haustoria in some parasitic green plants.	*Cuscuta* (dodder)
	Boring devices to effect entry into a host.	nematodes, fungi
	Attachment organs such as hooks or suckers.	*Taenia, Hirudo* (leech), *Fasciola*
	Resistant outer covering.	*Taenia, Fasciola*
	Degeneracy of sense organs associated with the constancy of the parasite's environment.	*Taenia*
Physiological	Exoenzyme production to digest host tissue external to parasite.	*fungi, Plasmodium* (a protozoan (Apicomplexa) which infects mammals and birds, and in the case of humans causes malaria)
	Anticoagulant production in blood feeders.	*Pulex, Hirudo*
	Chemosensitivity in order to reach the optimum location in the host's body.	*Plasmodium, Monocystis* (a protozoan (Apicomplexa) parasitic in the seminal vesicles of earthworms)
	Production of cytolytic substances to aid penetration into host.	*Cuscuta* (a flowering plant belonging to the family Convolvulaceae, which does not possess chlorophyll and parasitises a variety of green plants)
	Production of anti-enzymes.	gut parasites
	Ability to respire adequately in anaerobic conditions.	gut parasites
Reproductive	Hermaphrodite condition thus aiding possible self-fertilisation.	*Taenia, Fasciola*
	Enormous numbers of reproductive bodies, i.e. eggs, cysts and spores.	*Taenia, Fasciola*
	Resistance of reproductive bodies when external to the host.	*Monocystis, Phytophthora* (e.g. potato blight)
	Employment of specialised reproductive phases in life cycle.	*Fasciola*
	Use of secondary hosts as vectors.	*Taenia, Fasciola, Plasmodium*

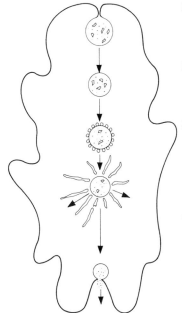

1 Food vacuole forms which contains food particles and water

2 Decreases a little in size as a result of water loss from vacuole. Increased acidity – pH 5.6

3 Enzymes discharged into food vacuole from lysosomes surrounding the food vacuole. This now becomes a digestive vacuole where intracellular digestion takes place. pH is 7.3

4 Fine canals radiate from digestive vacuole along which the soluble products of digestion pass into the surrounding cytoplasm

5 Exocytosis of any insoluble or indigestible material

Fig 10.4 *Ingestion, digestion and absorption in* Amoeba

and its contents become first acid (approximately pH 5.6) and then alkaline (about pH 7.3).

The enzymes poured into the digestive vacuole include carbohydrases, amino-, exo- and endopeptidases, esterase, collagenase and nuclease. They are secreted at different times so that their digestive effects are separated by time rather than spatially as in higher organisms. When digestion is complete the digestive vacuole membrane is drawn out into numerous fine canals. The soluble products of digestion are passed into the canals and finally into the surrounding cytoplasm of the animal by micropinocytosis. Undigested material is voided from the organism by **exocytosis** at any point on its surface.

Ciliary

The main diet of *Paramecium* consists of bacteria. Specialised tracts of cilia along the oral groove sweep the microorganisms in feeding currents towards the cytopharynx (fig 10.5). Any bacteria present are conveyed along the cytopharynx towards the cytostome or 'mouth' by the cilia of the undulating membrane. At the 'mouth' are a number of specialised cilia arranged in a criss-cross fashion which act as a filter to prevent large particles being taken in.

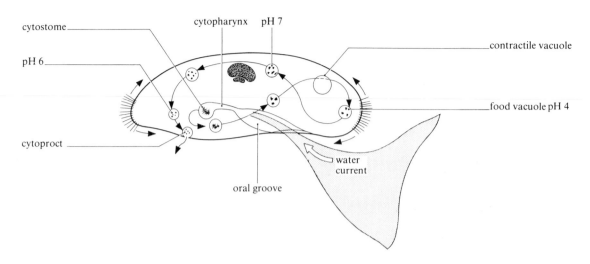

Fig 10.5 *Feeding currents and the pathway of food and digestive vacuoles in* Paramecium

Small particles together with a quantity of water pass into the plasmasol of the organism. Every so often the plasmasol actively segregates the collected food particles into a food vacuole which then follows a consistent pathway through the endoplasm. Once again the vacuole contents become acid initially (pH 2–4) and then alkaline (pH 7–8). The acid phase usually kills the prey, whilst most digestive activity takes place as the vacuole contents become less acid. For example, proteolytic enzymes work best at a pH of 5.7–5.8.

The soluble end-products of this intracellular digestion are finally absorbed into the cytoplasm of the organism and any undigested material is eliminated at the cytoproct by the process of **exocytosis**.

Experiment 10.1: To investigate ingestion of yeast cells and the formation of food vacuoles in *Paramecium*

Materials

Paramecium culture
cavity slides and cover-slips
dissecting needles
10% methyl cellulose
cotton wool
yeast culture stained in Congo red
monocular microscopes

Method

(1) Mount a drop of culture solution containing *Paramecium* on a cavity slide.
(2) Add one or two drops of 10% methyl cellulose and mix. This slows down the movements of *Paramecium*.
(3) Add a few cotton wool fibres. This supports the cover-slip when it is applied and also creates partitions which confine the movements of the animal and make it easier to observe.

(4) Add a drop of yeast suspension stained with Congo red dye. Congo red is an indicator in the pH range 3–5:

red/orange	pH 5.1
purple	pH 3–5
blue/violet	pH 3

(5) Cover the slide with a cover-slip and examine under the high power objective of the microscope.
(6) Observe the fate of the yeast cells as they enter the oral groove and are ingested at the cystosome. Food vacuoles containing yeast cells should be seen forming.
(7) Note if any colour changes take place in the food vacuoles. Comment on any changes that occur.

10.2.2 Filter feeding

Setose

Daphnia pulex, the common water flea, possesses a number of broad limbs with numerous stiff bristles (setae), all enclosed under a **carapace** (fig 10.6). When the limbs collectively move forward they draw water, containing suspended food particles, towards themselves. The bristles filter off the food from this feeding current and when the limbs move in a backward direction the food is propelled towards the mouth along a food groove by setae located at the base of each limb. At the mouth entrance the food particles become enmeshed by sticky mucus secretions prior to being swallowed.

Ciliary

Mytilus edulis, the common mussel, is found attached to rocks and stones in shallow coastal waters. It is a sedentary bivalve mollusc and possesses two 'gills', or ctenidia, covered with cilia on each side of its body. The movement of the cilia causes a current of water to enter the animal via an inhalant siphon and leave via an exhalant siphon. The

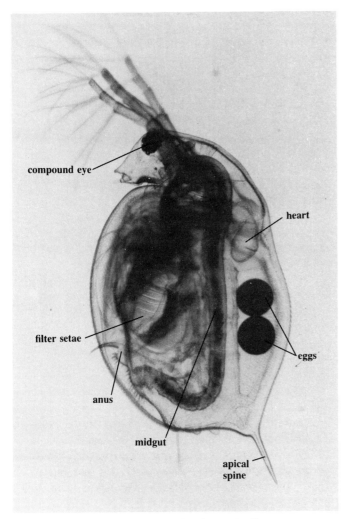

compound eye

heart

filter setae

anus

eggs

midgut

apical spine

Fig 10.6 *Lateral view of* Daphnia, *the water flea*

water which enters contains the food of the mussel such as microscopic protozoa and algae. Numerous secretory cells scattered among the cilia produce streams of sticky mucus which entangle the food particles. The trapped food is then swept by tracts of cilia towards the mouth which is located in a dorsal position near the anterior end of the 'gill'. Ciliated labial palps encircling the mouth sort out the food particles to some extent before they enter the mouth. The alimentary canal of the mussel consists of a stomach and short intestine which terminates via an anus located close to the exhalant siphon.

Experiment 10.2: To investigate feeding in *Daphnia pulex* (water flea)

Materials

Daphnia culture
cotton wool
cavity slides and cover-slips
dissecting needles
yeast culture coloured with neutral red
binocular and monocular microscopes

Method

(1) Place a culture of *Daphnia* on a cavity slide.
(2) Add cotton wool fibres to slow down the animal's movement.
(3) Add a drop of yeast solution coloured with neutral red. Neutral red is an indicator for the pH range 6–8:

red	pH 6.8
rose-red	pH 7.7
orange/yellow	pH 8.0

(4) Add a cover-slip and view a *Daphnia* in the lateral position under a microscope (fig 10.6).
(5) Note the nature and beating of the thoracic appendages and their setae under the carapace.
(6) Observe the movement of the yeast as it is swept towards the filtering region before ingestion.
(7) Note any colour change in the ingested yeast as it passes along the gut and comment on your observations.

10.2.3 Macrophagous feeders

Tentacular

The cnidarian *Hydra* feeds primarily on *Daphnia* and *Cyclops* and digestion is partly extracellular and partly intracellular. When these organisms brush against the projecting cnidocils of nematoblasts located on the tentacles of *Hydra* the nematocyst contents are automatically discharged. Penetrant nematocysts paralyse the prey whilst volvants and glutinants hold it tightly against the tentacle. The tentacle then bends over towards the 'mouth' which in turn opens widely enabling the prey to enter the enteron (fig 10.7).

Glandular **zymogen cells** in the endodermis secrete powerful proteolytic enzymes which initiate extracellular digestion. Endodermal flagellate cells and contractions of the body assist in the circulation of food and enzymes, and in breaking it up into fine particles. Extracellular digestion is completed in 4 h after which time the food particles are engulfed by the phagocytic action of endodermal amoeboid cells where digestion is completed intracellularly as in *Amoeba*.

The soluble products of digestion ultimately diffuse from the endodermis via the mesogloea to the ectodermis. Undigested material is egested via the single oral aperture.

Sepia officinalis, the cuttlefish, is a carnivore. The activity of pigment cells in its skin enables it to camouflage itself well. Suitably coloured, it lies in wait for its prey, which may be shrimps or crabs.

The cuttlefish has efficient eyesight and when it spots suitable prey it quickly extends two long prehensile tentacles which adhere tightly to the prey by means of their terminal suckers. The tentacles are then rapidly retracted towards the mouth carrying the prey with them. Sometimes a small quantity of toxic venom is injected into the prey from posterior salivary glands to assist in paralysing and killing it.

The other eight short tentacles of *Sepia* hold the prey against the mouth where a pair of beak-shaped horny jaws break up and bite off pieces of the prey (fig 10.8). Within the mouth is a radula which rasps the food into small pieces which are then swallowed. This mechanical breakdown of the food is assisted by proteases secreted from the salivary glands.

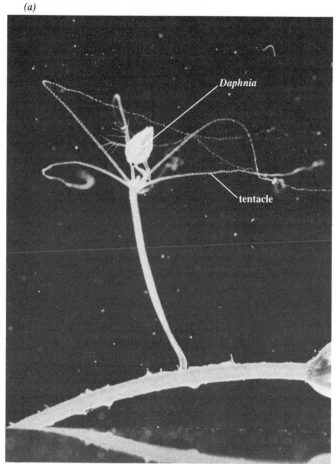

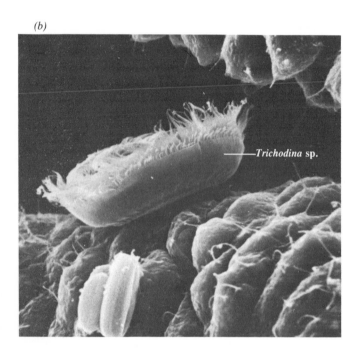

Fig 10.7 (above) *(a)* Hydra vulgaris *capturing a* Daphnia *(left) (b) Scanning electron micrograph of* Trichodina *lodged on the tentacles of a brown hydra*

Fig 10.8 (below) Sepia officinalis *(squid) showing tentacles covering mouth*

Scraping and boring

Helix aspersa, the common garden snail, feeds by using a rasping organ, the **radula**, in conjunction with a horny jaw plate (fig 10.9). The radula consists of about 150 rows of backwardly pointing 'teeth' with just over 100 teeth per row.

Leaves are held by the lips of the snail. The radula moves back and forth over the leaves with its teeth tearing the food whilst at the same time pressing it against the jaw plate. In this way minute fragments of vegetation are obtained which are gradually pushed backwards towards the pharynx. This type of activity wears down the front 'teeth' of the radula which become loose and eventually fall out to be swallowed with the food. They are rapidly and continuously replaced by new teeth. The rasping action of the radula ensures that the tough cellulose walls of the vegetation acted upon are broken down so that the cell contents are exposed to the hydrolytic action of enzymes, especially proteases, further along the digestive tract.

Biting and chewing mouthparts

Exoskeletal appendages in segments four, five and six form the feeding apparatus which surrounds a ventrally situated mouth in the grasshopper *Chorthippus*. The mouth is bordered anteriorly by the plate-like **labrum** or upper lip (fig 10.10). Beneath this lies a pair of stout, strong **mandibles** or jaws. Each mandible possesses an anterior ridged cutting surface and a posterior grinding surface which works against that of its partner and serves to cut, tear and crush food. A pair of **maxillae** is situated behind the mandibles. Each maxilla bears an **olfactory palp**. Hanging down behind the maxillae is an exoskeletal flap called the **labium** or lower lip. This assists in manoeuvring the food and also has a sensory function. The grasshopper is herbivorous and feeds mainly on leafy vegetation. It grips the leaf between its lips whilst the mandibles bite fragments from it. Activity of the maxillae and labium propels the food towards the mouth where it is swallowed. In the hypopharynx it is moistened with saliva secreted from

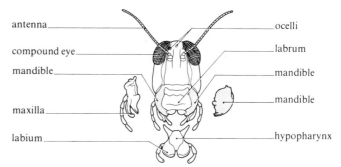

Fig 10.10 *Mouthparts of the common grasshopper* Chorthippus

salivary glands. The saliva contains amylase and sucrase, and so carbohydrate digestion begins immediately.

Seizing and swallowing

Scyliorhinus caniculus, the spotted dogfish, is a predatory carnivore and feeds on crustacea, shellfish, annelid worms, small fish and fragments of dead or dying animals. Its wide mouth is ventrally situated and enables the dogfish to swallow some animals whole. The buccal cavity is wide and flattened dorsoventrally. It possesses large, backwardly pointing **dermal denticles** which act as teeth and prevent prey from escaping once it has been seized in the mouth.

The buccal cavity leads into a wide pharynx which contains a tough muscular pad, the tongue. This assists in swallowing the food by moving it in an upward and backward direction into the oesophagus. The lining of the oesophagus is considerably folded. These folds extend around the food when it is swallowed and at the same time prevent much water from entering the gut. The shape of the stomach is asymmetrical, consisting of a dilated cardiac limb where the acid phase of digestion occurs, and a smaller, narrower pyloric limb. The alkaline phase of digestion takes place in the duodenum, which is relatively short and follows the stomach. The bile and pancreatic ducts open separately into the duodenum.

The duodenum leads into the ileum which contains the

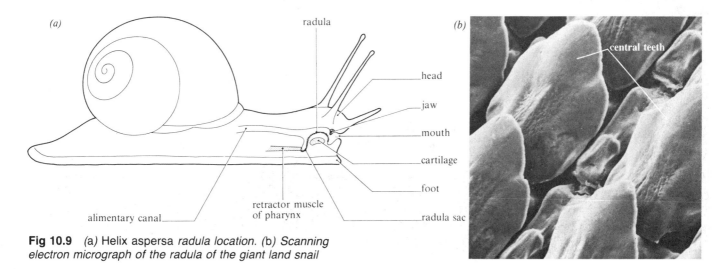

Fig 10.9 *(a)* Helix aspersa *radula location. (b) Scanning electron micrograph of the radula of the giant land snail*

297

spiral valve. This structure possesses many infoldings of the intestinal lining which slow down the passage of food and increase the surface area for its absorption. All absorbed food passes directly to the extremely large liver.

Throughout the length of the alimentary canal are numerous mucous glands whose secretions assist the smooth passage of the food. Undigested material is collected in the rectum and defaecated via the anus.

Detritus feeder

Lumbricus terrestris, the common earthworm, consumes fragments of fresh or decaying organic matter, especially vegetation, either at the soil surface, or after the food has been pulled into its burrow. Pieces of food are torn off by the mouth, moistened by alkaline secretions of the pharynx and drawn into the buccal cavity by the pumping action of the muscular pharynx. The food is then swallowed into the pharynx by **peristalsis**. Earthworms can also feed on organic material contained in the substrate which they swallow during burrowing activity.

The alimentary canal is straight and runs from mouth to anus. It is specialised at various points along its length for digestion and absorption of the ingested food. Table 10.3 indicates the sections of the alimentary canal involved in these activities, the segments which they occupy and their

structure and specific functions. Any undigested material is propelled to the posterior of the alimentary canal by peristaltic activity and voided via the anus as 'worm casts'.

Fluid feeding

Sucking. The housefly, *Musca domestica*, possesses a **proboscis tube** constructed from a highly modified labium. There are no mandibles, and the maxillae are reduced to a pair of palps. At the proximal end of the proboscis is a centrally placed mouth whilst at the distal end are two lobes called **labella**. Each of these contains numerous fine food channels termed **pseudotracheae** which ultimately converge into a central proboscis canal.

Generally the proboscis is held pressed against the underside of the insect's body, but when the insect feeds, it is extended by blood pressure so that the labella are placed on the food. If the food is solid, saliva from salivary glands is secreted onto it via an opening above the mouth. The saliva contains a number of enzymes which make the food soluble. When the food has been made soluble, or if the food is liquid in the first place, it passes into the pseudotracheae by capillary action. From here it is sucked up into the body by the activity of the muscles of the pharynx (fig 10.11*b*).

The feeding device of a butterfly such as *Pieris brassicae*

Table 10.3 Structure and functions of various regions of the earthworm gut.

Region of alimentary canal	Segments	Structure	Function
Mouth	–	–	Tears off pieces of food. Grips food as it is drawn into the worm's burrow.
Buccal cavity	1–3	Wide, thin-walled	Food moistened and softened by secretions of pharyngeal glands. Secretions include mucus and a proteolytic enzyme. Food eventually swallowed by peristalsis.
Pharynx	4–5	Dilatable, muscular and thick-walled	Possesses patches of glandular material, exudations of which help to soften the food.
Oesophagus	6–13	Narrow, tubular and thin-walled	Oesophageal pouches open into it from segment 10. Openings of two pairs of calciferous glands from segments 11 and 12; these glands secrete a fluid containing calcium carbonate particles. They are excretory in function rather than digestive and represent the manner in which excess calcium is removed from the body.
Crop	14–16	Wide and thin-walled	Acts as a storage chamber for the food. Some preliminary digestion occurs here.
Gizzard	17–19	Spherical, thick-walled, hard and muscular	Contains sharp fragments of stone. Mastication of food occurs here. Pieces of food are reduced in size by abrasion against the stones and the cuticularised lining of the gizzard.
Intestine	20	Surrounded by longitudinal and circular muscles. Its surface area is increased by the presence of a typhlosole.	A large surface area is presented for secretion of enzymes and absorption of digested food. Digestion is extracellular and food is absorbed into a network of capillaries lining the intestine. Three types of cell are present in the intestine: (1) glandular cells secrete proteolytic, amylolytic and lipolytic enzymes; (NB cellulase is secreted.) (2) ciliated cells help mix food with enzymes; (3) mucus cells lubricate food and protect gut lining from the digestive action of enzymes.
Anus	–	–	Faeces voided as worm casts.

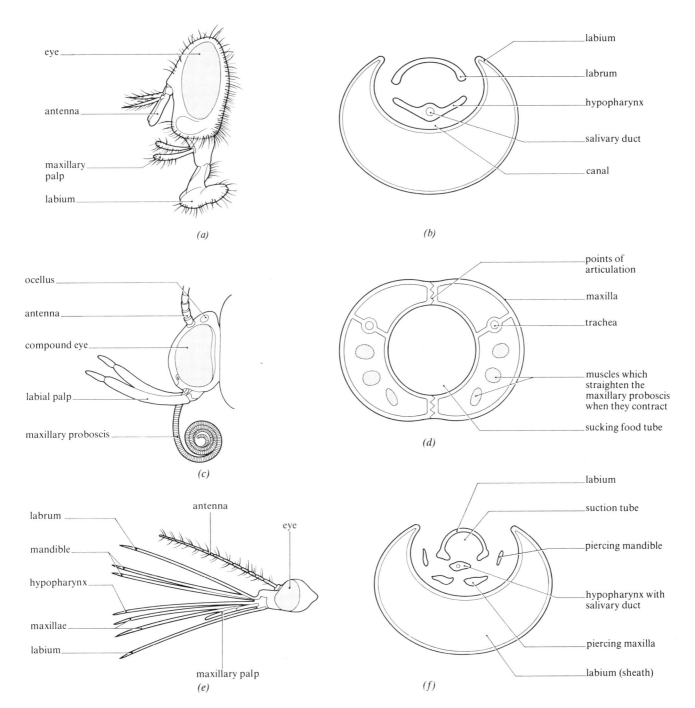

eye

antenna

maxillary palp

labium

(a)

labium

labrum

hypopharynx

salivary duct

canal

(b)

ocellus

antenna

compound eye

labial palp

maxillary proboscis

(c)

points of articulation

maxilla

trachea

muscles which straighten the maxillary proboscis when they contract

sucking food tube

(d)

labrum

antenna

eye

mandible

hypopharynx

maxillae

labium

maxillary palp

(e)

labium

suction tube

piercing mandible

hypopharynx with salivary duct

piercing maxilla

labium (sheath)

(f)

Fig 10.11 *(a) Mouthparts of the housefly* Musca domestica. *(b) TS mouthparts of* M. domestica. *(c) Mouthparts of the large white butterfly* Pieris brassicae. *(d) TS mouthparts of* P. brassicae. *(e) Mouthparts of the female mosquito* Anopheles sp. *(f) TS mouthparts of* Anopheles *sp.*

is its proboscis. In contrast to that of the housefly the proboscis is formed from the two maxillae. The part of each maxilla which together form the long tube through which the food is sucked (**galea**) is greatly elongated and grooved on its inner surface. These two structures fit together to form the proboscis tube. Mandibles are absent and the maxillary palps are either absent or poorly developed.

At rest, the proboscis is in the form of a coiled tube held under the head. When *Pieris* feeds, reflex contraction of the oblique galea muscles uncoils the proboscis. The proboscis is extended into the corolla of a flower and its tip placed directly on the food which is nectar, a dilute solution of sugar. It is frequently the case that the depth of the corolla tube corresponds to the length of the butterfly's proboscis. Muscles in the pharynx then begin to contract causing the nectar to be sucked into the mouth of the insect (fig 10.11*d*).

Piercing and sucking mouthparts. The **female mosquito** such as *Anopheles* sp., feeds on the blood of mammals. In order to obtain its meal it has to pierce the mammal's skin. This it does by using its mandibles or maxillae which have been highly modified into four sharp **stylets**. The stylets are contained in a grooved sheath, the proboscis sheath, formed by a greatly elongated labium.

Also present in the proboscis are the deeply grooved labrum and the hypopharynx. When the hypopharynx presses against the labrum it forms a food channel along which the fluid food is pumped. The hypopharynx also contains a salivary duct. During feeding, saliva containing an **anticoagulant** is secreted into the blood to prevent it clotting as it is sucked up into the pharynx via the narrow food channel (fig 10.11*f*).

10.3 Nutrition in mammals

10.3.1 Dietary requirements

Every mammal requires a daily supply of energy-providing foods (carbohydrates and fats), growth-promoting foods (proteins) and sufficient amounts of mineral salts, water, roughage and vitamins. For a diet to be adequate and balanced, these foodstuffs must be ingested in the correct proportions. Such a diet does not necessarily prevent illness but it certainly reduces the chances of the individual contracting a nutritionally based disease. The optimum nutritional value of any particular meal intake will vary markedly in different individuals depending on their sex, age, activity, body size and the temperature of their external environment (less food is eaten per individual in warm climates).

The Netherhall Education Software program *Balance your Diet* enables the user to enter their own diet and to compare it with a standard recommended diet. Protein, carbohydrate, fat and certain mineral and vitamin contents of diet can be examined separately.

> **10.3** Why does a mouse require a larger number of joules per unit weight than a man?

10.3.2 Measurement of the energy value of foodstuffs

An adequate diet must contain sufficient energy for the body's daily metabolic needs. This energy is measured as heat energy and expressed as joules. The energy value of a foodstuff can be calculated by burning a known mass of it in oxygen in a **bomb calorimeter**. The heat generated by this oxidation is transmitted to a known mass of water whose corresponding temperature rise is measured. Using the knowledge that 4.18 J of heat energy raises the temperature of 1 g of water by 1 °C, the number of joules generated by the burning of the food can be calculated (fig 10.12).

Table 10.4 indicates the recommended daily intake of energy and nutrients for humans in the UK for a variety of ages and activities and for both sexes. Table 10.5 shows the composition of a selection of foods per 100 g edible portion and is based on values published in the *Manual of Nutrition*, HMSO, 1976.

> **10.4** How many kilojoules are produced if 1 g of sugar burned in oxygen raises the temperature of 500 g of water by 7.5 °C?

10.3.3 Measurement of energy expenditure by humans

In order to calculate energy expenditure in humans a method of 'indirect calorimetry' is used. Accurate measurements of oxygen consumption, carbon dioxide excretion, and sometimes nitrogen excretion in the urine are used in the calculation of energy expenditure. The theory behind this method is that the same quantity of heat is released, oxygen consumed, and carbon dioxide and water produced when one gram of foodstuff is burned in the air as when it is burned in the body. However, this is only an approximate value since complete oxidation of food materials does not occur in the body.

> **10.5** It has been calculated that 1 g glucose combines with 774 cm³ oxygen releasing 15.8 kJ heat, and 1 g long-chain fatty acid combines with 2012 cm³ oxygen releasing 39.4 kJ heat. Why does 1 g of fatty acid give rise to more than twice as much heat as 1 g glucose?
>
> **10.6** Why is it that protein and fats subjected to combustion in a bomb calorimeter liberate more heat than when exactly the same weight of each foodstuff is burned in the body?

10.3.4 Malnutrition

This situation arises when an organism is deficient in (**undernutrition**) or receives excess (**overnutrition**) of one or more nutrients or energy-providing foods over a long period of time. In many underdeveloped parts of the world undernutrition is the most common form of malnutrition, whereas in the industrialised West obesity, coronary heart disease and reduced life expectancy are all symptoms of overnutrition.

10.3.5 Carbohydrates, proteins and fats

Detailed information concerning the structure and functions of these foodstuffs is provided in chapter 5. However, a note about the quality of dietary protein is necessary here. The nutritional value of a protein depends upon the composition of its amino acids and whether or not it can be digested by the animal concerned. Vegetable

Table 10.4 Recommended daily intakes of energy and nutrients for humans in the UK (Department of Health and Social Security (1969)) for a variety of ages and activities, and for both sexes.

Age range and occupational category	Body wt kg	Energy kcal	MJ	Protein[1] g	Thiamin mg	Riboflavin mg	Nicotinic acid mg equivalents	Ascorbic acid mg	Vitamins A μg retinol equivalents[2]	Vitamins D μg cholecalciferol	Calcium mg	Iron mg
BOYS and GIRLS												
0 up to 1 year	7.3	800	3.3	20	0.3	0.4	5	15	450	10	600	6
2 up to 4 years	13.5	1400	5.9	35	0.6	0.7	8	20	300	10	500	7
4 up to 7 years	20.5	1800	7.5	45	0.7	0.9	10	20	300	2.5	500	8
BOYS												
9 up to 12 years	31.9	2500	10.5	63	1.0	1.2	14	25	575	2.5	700	13
12 up to 15 years	45.5	2800	11.7	70	1.1	1.4	16	25	725	2.5	700	14
15 up to 18 years	61.0	3000	12.6	75	1.1	1.7	19	30	750	2.5	600	15
GIRLS												
9 up to 12 years	33.0	2300	9.6	58	0.9	1.2	13	25	575	2.5	700	13
12 up to 15 years	48.6	2300	9.6	58	0.9	1.4	16	25	725	2.5	700	14
15 up to 18 years	56.1	2300	9.6	58	0.9	1.4	16	30	750	2.5	600	15
MEN												
18 up to 35 years												
sedentary	65	2700	11.3	68	1.1	1.7	18	30	750	2.5	500	10
moderately active		3000	12.6	75	1.2	1.7	18	30	750	2.5	500	10
very active		3600	15.1	90	1.4	1.7	18	30	750	2.5	500	10
35 up to 65 years												
sedentary	65	2600	10.9	65	1.0	1.7	18	30	750	2.5	500	10
moderately active		2900	12.1	73	1.2	1.7	18	30	750	2.5	500	10
very active		3600	15.1	90	1.4	1.7	18	30	750	2.5	500	10
65 up to 75 years (assuming a sedentary life)	63	2350	9.8	59	0.9	1.7	18	30	750	2.5	500	10
75 and over (assuming a sedentary life)	63	2100	8.8	53	0.8	1.7	18	30	750	2.5	500	10
WOMEN												
18 up to 55 years												
most occupations	55	2200	9.2	55	0.9	1.3	15	30	750	2.5	500	12
very active		2500	10.5	63	1.0	1.3	15	30	750	2.5	500	12
55 up to 75 years (assuming a sedentary life)	53	2050	8.6	51	0.8	1.3	15	30	750	2.5	500	10
75 and over (assuming a sedentary life)	53	1900	8.0	48	0.7	1.3	15	30	750	2.5	500	10
pregnancy, 2nd and 3rd trimester		2400	10.0	60	1.0	1.6	18	60	750	10	1200	15
lactation		2700	11.3	68	1.1	1.8	21	60	1200	10	1200	15

[1] Recommended intakes calculated as providing 10% of energy.

[2] 1 retinol equivalent = 1 μg retinol or 6 μg β-carotene or 12 μg other biologically active carotenoids.

foods generally contain small quantities of protein and the amino acids present are rarely in the proportions required by animal tissues. Therefore there is a danger of malnutrition if only one vegetable food forms the major component of the diet. However, a good vegetarian diet can be worked out which provides the complete range of protein requirements by using a wide variety of protein-containing vegetable foods. These include cereals, legumes, nuts, fruit and other vegetables. An exception to the low protein content of many vegetable foods is soya bean protein which is as good as most animal proteins. Many, though not all animal proteins contain a high proportion of essential amino acids in balanced amounts and are termed '**first-class**' proteins.

10.3.6 Mineral salts

A wide variety of inorganic elements is present in the body, all of which must be obtained from food or drink consumed by the mammal concerned. They are required for many metabolic activities and in the structure of a number of tissues. The different functions of the elements are considered in table 9.11.

10.3.7 Water

Water is essential to mammals as all bodily metabolic reactions take place in solution. Since water makes up 65–70% of the total body weight, and this weight remains relatively constant each day, it follows that the 2–3 dm³ of water lost daily from the body must be replaced by fluids or food consumed by the mammal each day. The importance of water to life becomes clear when one

Table 10.5 Composition of selected foods per 100g edible portion, based on values published in *Manual of Nutrition* (1976).

	Energy kcal	kJ	Protein g	Fat g	Carbo-hydrate g	Minerals Ca mg	Fe mg	Vitamins A µg	D µg	B₁ mg	B₂ mg	Nicotinic acid equivalents mg	C mg
Almonds	580	2397	20.5	53.5	4.3	247	4.2	0	0	0.32	0.25	4.9	0
Apples	46	197	0.3	0	12.0	4	0.3	5	0	0.04	0.02	0.1	5
Apricots, canned	106	452	0.5	0	27.7	12	0.7	166	0	0.02	0.01	0.3	5
Bacon, rashers, cooked	447	1852	24.5	38.8	0	12	1.4	0	0	0.40	0.19	9.2	0
Bananas	76	326	1.1	0	19.2	7	0.4	33	0	0.04	0.07	0.8	10
Beans, canned in tomato sauce	63	266	5.1	0.4	10.3	45	1.4	50	0	0.07	0.05	1.4	3
Beans, runner	23	100	2.2	0	3.9	27	0.8	50	0	0.05	0.10	1.4	20
Beef, average	226	940	18.1	17.1	0	7	1.9	0	0	0.06	0.19	8.1	0
Beef, corned	216	905	26.9	12.1	0	14	2.9	0	0	0.01	0.23	9.0	0
Beer, bitter, draught	30	127	0	0	2.3	11	0	0	0	0	0.05	0.7	0
Beetroot, boiled	44	189	1.8	0	9.9	30	0.7	0	0	0.02	0.04	0.4	5
Biscuits, plain, semi-sweet	431	1819	7.4	13.2	75.3	126	1.8	0	0	0.17	0.06	2.0	0
Bread, brown	230	981	9.2	1.4	48.3	88	2.5	0	0	0.28	0.07	2.7	0
Bread, white	251	1068	8.0	1.7	54.3	100	1.7	0	0	0.18	0.03	2.6	0
Bread, wholemeal	241	1025	9.6	3.1	46.7	28	3.0	0	0	0.24	0.09	1.9	0
Brussels sprouts, boiled	17	75	2.8	0	1.7	25	0.5	67	0	0.06	0.10	1.0	41
Butter	731	3006	0.5	81.0	0	15	0.2	995	1.25	0	0	0.1	0
Cabbage, boiled	15	66	1.7	0	2.3	38	0.4	50	0	0.03	0.03	0.5	23
Carrots	23	98	0.7	0	5.4	48	0.6	2000	0	0.06	0.05	0.7	6
Cauliflower	13	56	1.9	0	1.5	21	0.5	5	0	0.10	0.10	1.0	64
Cheese, Cheddar	412	1708	25.4	34.5	0	810	0.6	420	0.35	0.04	0.05	5.2	0
Chicken, roast	148	621	24.8	5.4	0	9	0.8	0	0	0.08	0.19	12.8	0
Chocolate, milk	578	2411	8.7	37.6	54.5	246	1.7	6.6	0	0.03	0.35	2.5	0
Coconut, desiccated	608	2509	6.6	62.0	6.4	22	3.6	0	0	0.06	0.04	1.8	0
Cod, fried in batter	199	834	19.6	10.3	7.5	80	0.5	0	0	0.04	0.10	6.7	0
Cod, haddock, white fish	76	321	17.4	0.7	0	16	0.3	0	0	0.08	0.07	4.8	0
Coffee, instant	155	662	4.0	0.7	35.5	140	4.0	0	0	0	0.10	45.7	0
Cornflakes	354	1507	7.4	0.4	85.4	5	0.3	0	0	1.13[a]	1.41[a]	10.6[a]	0
										0.04[b]	0.10[b]	0.8[b]	
Cream, double	449	1848	1.8	48.0	2.6	65	0	420	0.28	0.02	0.08	0.4	0
Cream, single	189	781	2.8	18.0	4.2	100	0.1	155	0.10	0.03	0.13	0.8	0
Eggs	147	612	12.3	10.9	0	54	2.1	140	1.50	0.09	0.47	3.7	0
Fish fingers	178	749	12.6	7.5	16.1	43	0.7	0	0	0.09	0.06	3.1	0
Flour, white	348	1483	10.0	0.9	80.0	138	2.1	0	0	0.30	0.03	2.7	0
Fruit cake, rich	368	1546	4.6	15.9	55.0	71	1.6	57	0.80	0.07	0.07	1.2	0
Ham, cooked	269	1119	24.7	18.9	0	9	1.3	0	0	0.44	0.15	8.0	0
Honey	288	1229	0.4	0	76.4	5	0.4	0	0	0	0.05	0.2	0
Ice-cream, vanilla	192	805	4.1	11.3	19.8	137	0.3	1	0	0.05	0.20	1.1	1
Jam	262	1116	0.5	0	69.2	18	1.2	2	0	0	0	0	10
Kipper	184	770	19.8	11.7	0	60	1.2	45	22.20	0.02	0.30	6.9	0
Lamb, roast	291	1209	23.0	22.1	0	9	2.1	0	0	0.10	0.25	9.2	0
Lettuce	8	36	1.0	0	1.2	23	0.9	167	0	0.07	0.08	0.4	15
Liver, fried	244	1020	24.9	13.7	5.6	14	8.8	6000	0.75	0.27	4.30	20.7	20
Luncheon meat	313	1298	12.6	26.9	5.5	15	1.0	0	0	0.07	0.12	4.5	0
Margarine	734	3019	0.2	81.5	0	4	0.3	900[c]	8.00	0	0	0.1	0
Marmalade	261	1114	0.1	0	69.5	35	0.6	8	0	0	0	0	10
Milk, liquid, whole	65	274	3.3	3.8	4.8	120	0.1	44[d] 37[e]	0.05[d] 0.01[e]	0.04	0.15	0.9	1
Oils, cooking and salad	899	3696	0	99.9	0	0	0	0	0	0	0	0	0
Onions	23	98	0.9	0	5.2	31	0.3	0	0	0.03	0.05	0.4	10
Oranges	35	150	0.8	0	8.5	41	0.3	8	0	0.10	0.03	0.3	50
Parsnips	49	210	1.7	0	11.3	55	0.6	0	0	0.10	0.09	1.3	15
Peaches, canned	88	373	0.4	0	22.9	4	1.9	41	0	0.01	0.02	0.6	4
Peanuts, roasted	586	2428	28.1	49.0	8.6	61	2.0	0	0	0.23	0.10	20.8	0
Pears, fresh	41	175	0.3	0	10.6	8	0.2	2	0	0.03	0.03	0.3	3
Peas, fresh or quick frozen, boiled	49	208	5.0	0	7.7	13	1.2	50	0	0.25	0.11	2.3	15
Pineapple, canned	76	325	0.3	0	20.0	13	1.7	7	0	0.05	0.02	0.3	8
Plain cake, Madeira	426	1785	6.0	24.0	49.7	67	1.4	82	1.20	0.08	0.11	1.7	0
Plums	32	137	0.6	0	7.9	12	0.3	37	0	0.05	0.03	0.6	3
Pork, average	330	1364	15.8	29.6	0	8	0.8	0	0	0.58	0.16	6.9	0
Potatoes, boiled	80	339	1.4	0	19.7	4	0.5	0	0	0.08	0.03	1.2	4–15[f]
Potato chips, fried	236	1028	3.8	9.0	37.3	14	1.4	0	0	0.10	0.04	2.2	6–20[f]

Table 10.5 (*cont.*)

	Energy kcal	kJ	Protein g	Fat g	Carbo-hydrate g	Ca mg	Fe mg	A μg	D μg	B₁ mg	B₂ mg	Nicotinic acid equivalents mg	C mg
Potatoes, roast	111	474	2.8	1.0	27.3	10	1.0	0	0	0.10	0.04	2.0	6–23(f)
Prunes	161	686	2.4	0	40.3	38	2.9	160	0	0.10	0.20	1.7	0
Raspberries	25	105	0.9	0	5.6	41	1.2	13	0	0.02	0.03	0.5	25
Rhubarb	6	26	0.6	0	1.0	103	0.4	10	0	0.01	0.07	0.3	10
Rice	359	1531	6.2	1.0	86.8	4	0.4	0	0	0.08	0.03	1.5	0
Rice pudding	142	594	3.6	7.6	15.7	116	0.1	96	0.08	0.05	0.14	1.0	1
Sausage, pork	367	1520	10.6	32.1	9.5	41	1.1	0	0	0.04	0.12	5.7	0
Soup, tomato, canned	55	230	0.8	3.3	5.9	17	0.4	35	0	0.03	0.02	0.2	6
Spaghetti	364	1549	9.9	1.0	84.0	23	1.2	0	0	0.09	0.06	1.8	0
Spinach	21	91	2.7	0	2.8	70	3.2	1000	0	0.12	0.20	1.3	60
Spirits, 70% proof	221	914	0	0	0	0	0	0	0	0	0	0	0
Steak and kidney pie, cooked	304	1266	13.3	21.1	14.6	37	5.1	126	0.55	0.11	0.47	6.0	0
Strawberries	26	109	0.6	0	6.2	22	0.7	5	0	0.02	0.03	0.5	60
Sugar, white	394	1680	0	0	105.0	1	0	0	0	0	0	0	0
Sweet corn, canned	79	336	2.9	0.8	16.1	3	0.1	35	0	0.05	0.08	0.3	4
Syrup	298	1269	0.3	0	79.0	26	1.4	0	0	0	0	0	0
Tomatoes, fresh	12	52	0.8	0	2.4	13	0.4	117	0	0.06	0.04	0.7	21
Watercress	14	60	2.9	0	0.7	222	1.6	500	0	0.10	0.16	2.0	60
Wine, red	67	277	0	0	0.3	6	0.8	0	0	0.01	0.02	0.2	0
Yoghurt, fruit	96	410	4.8	1.0	18.2	160	0.2	10	0.02	0.05	0.23	1.2	1
Yoghurt, natural	53	224	5.0	1.0	6.4	180	0.1	10	0.02	0.05	0.26	1.3	0

(a)

Notes:
Carbohydrates are given as monosaccharides.
Vitamin A is given in the form of μg retinol equivalents, i.e. values include estimated contributions from carotene precursors. 1 μg retinol is equivalent to 3.33 i.u. (international units).
Vitamin D is given as μg: 1 μg vitamin D is equivalent to 40 i.u.
Vitamin B₁ is *thiamin.*
Vitamin B₂ is *riboflavin.*
Nicotinic acid is given in two forms; *total* being the nicotinic acid present as the vitamin itself. Nicotinic acid *equivalents* include the contribution estimated to be supplied by tryptophan, assuming 60 mg tryptophan gives rise to 1 mg nicotinic acid.
(*a*) fortified; (*b*) unfortified; (*c*) some margarines contain carotene; (*d*) summer value; (*e*) winter value; (*f*) high in new potatoes falling during storage.

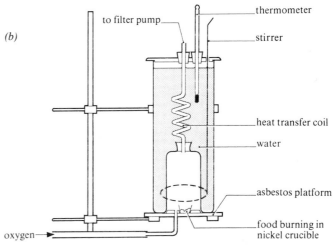

(b)

Fig 10.12 (a) *Calorimeter designed to investigate the energy content of food materials. The small electric heating coil is used to ignite the food.* (b) *Diagram to show the labelled parts*

considers that a man deprived of food may live for more than 60 days, but for only a few days if deprived of water. A full account of the functions of water in both animals and plants can be found in chapters 5 and 14.

10.3.8 Roughage

In humans, roughage or dietary fibre consists of indigestible cellulose of the cell walls of plants. It has a water-holding capacity and provides bulk to the intestinal contents, especially in the large intestine. Here, stretching of the colon wall stimulates reflex peristaltic activity and hence aids movement of the colon contents towards the rectum, and promotes defaecation. Absence of dietary fibre can lead to constipation and other disorders of the large intestine.

10.3.9 Milk

The only dietary item that most mammals receive during the first weeks of their lives is milk. It provides an almost complete diet during this stage of their development, containing carbohydrate, protein, fat, minerals (especially calcium, magnesium, phosphorus and potassium) and a variety of vitamins. The one major element that milk lacks is iron, a constituent of haemoglobin in blood. However, this problem is overcome by the embryo accumulating iron from its mother and storing a sufficient quantity of it in its body prior to birth. This sustains embryonic development, and development after birth until the offspring begins to ingest solid food.

> **10.7** Early this century, Frederick Gowland Hopkins in Cambridge performed a famous experiment where he took two sets of eight young rats and fed both on a diet of pure casein, starch, sucrose, lard, inorganic salts and water. The first set received additionally 3 cm³ of milk per day for the first 18 days. On day 18, the extra milk was denied the first set, but given to the second set of rats instead. The result of the experiment is shown in fig 10.13.
> (a) What hypothesis can you deduce from the graph?
> (b) Support your answer with comments.
> (c) Why is a diet of milk inadequate for an adult?

10.3.10 Vitamins

Vitamins are complex organic compounds present in very small quantities in natural food and absorbed into the body from the small intestine. They possess no energy value but are essential for the good health of the body and in maintaining its normal metabolic activities. If the diet is deficient in a particular vitamin, metabolic activity is impaired. This produces a disorder symptomatic with that particular vitamin deficiency, which is termed a **deficiency disease**. When the dietary intake of a

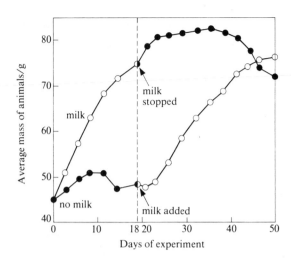

Fig 10.13 *Gowland Hopkins's experiment on feeding milk to rats*

particular vitamin is inadequate, the deficiency disease it might cause can be avoided by supplementing the diet with the necessary vitamin. Table 10.6 indicates some of the sources and functions of the principal vitamins required in the human diet and the deficiency diseases caused by a lack of them.

10.4 The alimentary canal in humans

Digestion and absorption occurs in the alimentary canal, digestive or gastrointestinal tract, or more plainly the gut, which runs from the mouth to the anus. As the gut wall is continuous with the outside surface of the body, the food it contains is considered to be external to the body in a positional and physiological sense. Food can only be absorbed into the body when its highly complex molecules are broken down physically by the teeth and muscles of the gut wall, and chemically by its enzymes into molecules of a suitably small size to be absorbed into the blood capillaries surrounding the small intestine. From here they are delivered to the cells of the body tissues where they undergo assimilation. The pattern of enzymatic digestion can be seen in fig 10.14.

The gut is a locally differentiated structure, that is it is specialised at various points along its length, with each region designed to carry out a different role in the overall processes of digestion and absorption. It begins with the mouth and buccal cavity which are followed by the pharynx, oesophagus, stomach, the small intestine comprising the duodenum and ileum, the large intestine consisting of the caecum, bearing the appendix, colon and rectum and terminating at the anus (fig 10.15 and table 10.7).

Whilst each different portion of the digestive tract possesses its own special characteristics, all conform to a basic common structure as shown in fig 10.16. This consists

Table 10.6 Sources and functions of principal vitamins required in the human diet.

Name of vitamin and its designated letter	Principal sources	Function	Deficiency diseases and symptoms
Fat-soluble vitamins			
A Retinol	Halibut and cod-liver oil; ox liver; milk and derivatives; carrots; spinach; watercress	Controls normal epithelial structure and growth. The aldehyde form of vitamin A, retinal, is essential for the formation of the visual pigment rhodopsin, which aids 'night vision'.	Skin becomes dry, cornea becomes dry and mucous membranes degenerate. Poor 'night vision'. Serious deficiency results in complete night blindness. **Xerophthalmia** – permanent blindness may occur if the vitamin is not added to the diet.
D Calciferol	Halibut and cod-liver oil; egg yolk; margarine; made by the action of sunlight on lipids in the skin; milk	Controls calcium absorption from the digestive tract, and concerned with calcium metabolism. Important in bone and tooth formation. Aids absorption of phosphorus.	**Rickets** – this is the failure of growing bones to calcify. Bow legs are a common feature in young children and knock knees in older ones. Deformation of the pelvic bones in adolescent girls can occur which may lead to complications when those girls give birth. **Osteomalacia** – an adult condition where the bones are painful and spontaneous fractures may occur.
E Tocopherol	Wheat germ; brown flour; liver; green vegetables	In rats, it affects muscles and the reproductive system and prevents haemolysis of red blood corpuscles. Function in humans is unknown.	Can cause sterility in rats. Muscular dystrophy. **Anaemia** – increased haemolysis of red blood corpuscles.
K Phylloquinone	Spinach; cabbage; brussels sprouts; synthesised by bacteria in the intestine	Essential for final stage of prothrombin synthesis in the liver. Therefore it is a necessary factor for the blood-clotting mechanism.	Mild deficiency leads to a prolonged blood-clotting time. Serious deficiency means blood fails to clot at all.
Water-soluble vitamins			
B_1 Thiamin	Wheat or rice germ; yeast extract; wholemeal flour; liver; kidney; heart	Acts as a coenzyme for decarboxylation. Aids chemical changes in respiration, especially in Krebs cycle.	**Beriberi** – nervous system affected. Muscles become weak and painful. Paralysis can occur. Heart failure. Oedema. Children's growth is impaired. Keto acids, e.g. pyruvic acid, accumulate in the blood.
B_2 Riboflavin	Yeast extract; liver; eggs; milk; cheese	Forms part of the prosthetic group of flavoproteins which are used in electron transport.	Tongue sore. Sores at the corners of the mouth.
B_6 Pyridoxine	Eggs; liver; kidney; whole grains; vegetables; fish	Phosphorylated pyridoxine acts as a coenzyme for amino acid and fatty acid metabolism.	Depression and irritability. Anaemia. Diarrhoea. Dermatitis.
B_5 Pantothenic acid	In most foods	Forms part of coenzyme A molecule which is involved in activation of carboxylic acids in cellular metabolism.	Poor neuromotor coordination. Fatigue. Muscle cramp.
B_3(pp) Nicotinic acid (niacin)	Meat; wholemeal bread; yeast extract; liver	Essential component of the coenzymes NAD, NADP which operate as hydrogen acceptors for a range of dehydrogenases. Also a part of coenzyme A.	**Pellagra** – skin lesions, rashes. Diarrhoea.
M or Bc Folic acid	Liver; white fish; green vegetables	Formation of red blood corpuscles. Synthesis of nucleoproteins.	**Anaemia** – particularly in women during pregnancy.
B_{12} Cyanocobalamin	Meat; milk; eggs; fish; cheese	RNA nucleoprotein synthesis. Prevents pernicious anaemia.	Pernicious anaemia.
H Biotin	Yeast; liver; kidney; egg white; synthesis by intestinal bacteria	Used as a coenzyme for a number of carboxylation reactions. Involved in protein synthesis and transamination.	Dermatitis. Muscle pains.
C Ascorbic acid	Citrus fruits; green vegetables; potatoes; tomatoes	Concerned with the metabolism of connective tissue and the production of strong skin. Essential for collagen fibre synthesis.	**Scurvy** – skin of gums becomes weak and bleeds. Wounds fail to heal. Connective tissue fibres fail to form. Anaemia. Heart failure.

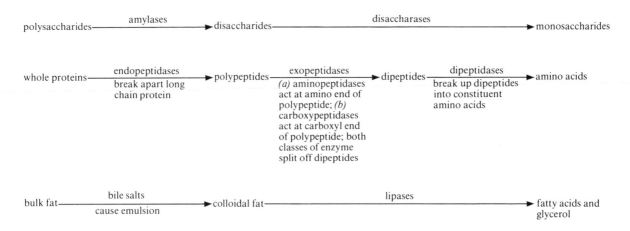

Fig 10.14 *General pattern of enzyme digestion in the human alimentary canal*

Table 10.7 Summary of the functions of the different parts of the human digestive system.

Specialised part	Function
Buccal cavity	Ingestion, mastication
Pharynx	Swallowing
Oesophagus	Links pharynx to stomach
Stomach	Food storage and digestion of protein
Duodenum	Digestion and absorption
liver (bile)	Emulsification of fats
pancreas	
(pancreatic juice)	Digestion of starch, protein and fat
Ileum	Completion of digestion and absorption of food
Colon	Absorption of water
Rectum	Formation and storage of faeces
Anus	Egestion

of four distinct layers: the mucosa, submucosa, muscularis externa and serosa.

Mucosa. This is the innermost layer of the gut and is composed of glandular epithelium which secretes copious quantities of mucus and possess enzymes embedded in the brush border. The mucus lubricates the food and facilitates its easy passage along the digestive tract. It also prevents digestion of the gut wall by its own enzymes. The epithelial cells rest on a basement membrane beneath which is the **lamina propria**, containing connective tissue, blood and lymph vessels. Outside this is a thin layer of smooth muscle, the **muscularis mucosa**.

Submucosa. This is a layer of connective tissue containing nerves, blood and lymph vessels, collagen and elastic fibres. It may contain some mucus-secreting glands which deposit their contents onto the surface via ducts, such as Brunner's glands in the duodenum.

Muscularis externa. This layer is composed of an inner circular and an outer longitudinal layer of

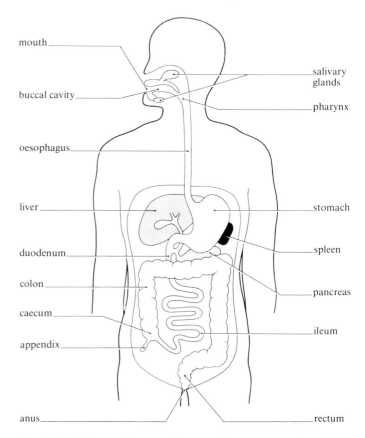

Fig 10.15 *General layout of human alimentary canal*

smooth muscle. Coordinated movements of the two layers provide the wave-like peristaltic activity of the gut wall which propels food along. At a number of points along the gut the circular muscle thickens into structures called **sphincters**. When these relax or contract they control the movement of food from one part of the alimentary canal to another. They are found at the junctions of the oesophagus and stomach (cardiac sphincter), stomach and duodenum (pyloric sphincter), ileum and caecum, and at the anus.

Between the circular and longitudinal muscle layers is

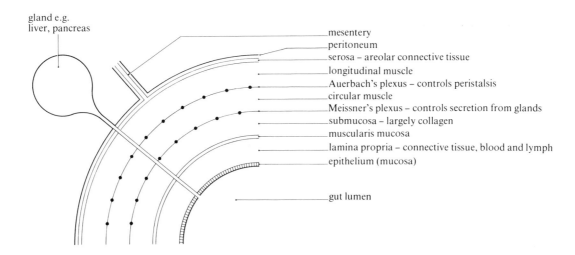

Fig 10.16 *General plan of gut structure as seen in transverse section*

Auerbach's plexus. This consists of nerves from the autonomic nervous system which control peristalsis. Impulses travelling along sympathetic nerves cause the gut muscles to relax and the sphincters to close, whilst impulses travelling via the parasympathetic nerves stimulate the gut wall to contract and the sphincters to open (section 16.2). Between the circular muscle and submucosa is another nerve plexus, **Meissner's plexus.** This controls secretion from glands in the gut wall.

Serosa. This is the outermost coat of the gut wall. It is composed of loose fibrous connective tissue.

The whole of the outer surface of the gut is covered by **peritoneum.** This tissue also lines the abdominal cavity, where most of the gut is located, and constitutes the **mesenteries** which suspend and support the stomach and intestines from the dorsal body wall. Mesenteries consist of double layers of peritoneum containing nerves, blood vessels and lymph vessels that pass to and from the gut. The peritoneum cells are moist and help to reduce friction when the gut wall slides over other portions of itself or other organs.

10.4.1 Human dentition

In humans there are two jaws, the fixed upper jaw and the movable lower jaw, with the tongue between each half of the lower jaw. Both jaws bear teeth which are used to chew or **masticate** food into smaller pieces. This is mechanical digestion and increases the surface area of food for efficient enzyme attack. The teeth are very hard structures and ideally suited to their task. Humans have two successive sets of teeth, a condition called **diphyodont.** The **deciduous** or milk teeth appear first, only to be progressively replaced by the **permanent** teeth. Human teeth have different shapes and sizes and possess uneven biting surfaces, which is known as **heterodont.** This is in contrast to the **homodont** condition of fish and reptiles where all teeth are similar and usually cone-shaped. Humans possess 32 permanent teeth consisting of eight incisors, four canines, eight premolars and twelve molars. These replace the eight incisors, four canines and eight premolars of the deciduous dentition. The arrangement of the teeth can be conveniently expressed in the form of a **dental formula.** Human permanent dentition is:

$$2 \left[i\,\frac{2}{2} \quad c\,\frac{1}{1} \quad pm\,\frac{2}{2} \quad m\,\frac{3}{3} \right]$$

where the letters indicate the type of tooth, the numerator represents the number of each type of tooth in the upper jaw on one side of the head and the denominator represents the teeth in the lower part of the jaw on the same side (fig 10.17).

The number, size and shape of the teeth differ within the buccal cavity of humans and also between different mammals. This can be correlated with their different functions and different diets respectively. The basic structure and function of each type of tooth is as follows. **Incisors** are situated at the front of the buccal cavity. They have flat, sharp edges which are used for cutting and biting food (fig 10.18a). **Canines** are prominently pointed teeth (fig 10.18b). They are poorly developed in humans, but highly developed in carnivores where they are designed for piercing and killing prey, and tearing flesh. **Premolars** possess one or two roots and two cusps (projections on the surface of a tooth) (fig 10.18c). They are specialised for crushing and grinding food, although in humans they may also be used to tear food. **Molars** have more than one root; upper molars have three roots, lower molars two (fig 10.18d). Each has four or five cusps. They are used to crush and grind food. They are not present in the deciduous dentition of humans.

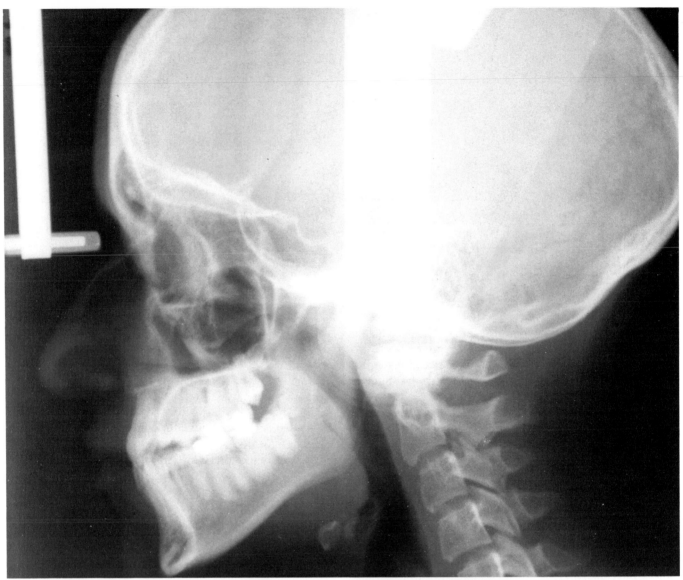

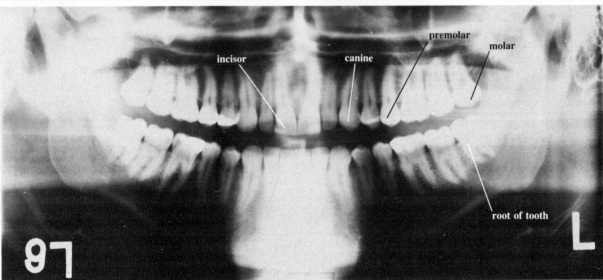

Fig 10.17 (a) X-ray of side of human head to show permanent dentition on one side.

(b) X-ray from the front to show a complete permanent dentition. Dental formula $2\left[\text{i}\,\dfrac{2}{2}\quad \text{c}\,\dfrac{1}{1}\quad \text{pm}\,\dfrac{2}{2}\quad \text{m}\,\dfrac{3}{3}\right]$

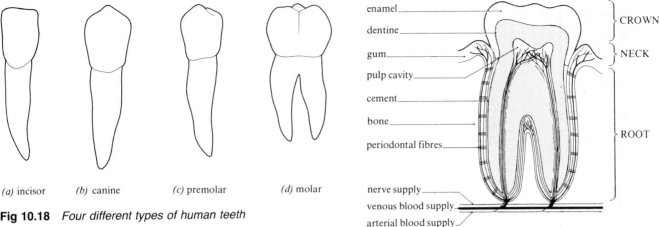

(a) incisor   (b) canine   (c) premolar   (d) molar

Fig 10.18 *Four different types of human teeth*

Fig 10.19 *Vertical section of a premolar tooth*

10.4.2   Generalised structure of a tooth

The visible part of the tooth, termed the **crown**, is covered with enamel (fig 10.19), the hardest substance in the body. It is relatively resistant to decay. The neck of the tooth is surrounded by the **gum** whilst the root is embedded in the jawbone. Beneath the enamel is **dentine** which forms the bulk of the tooth. Though tough, it is not as hard as enamel or as resistant to decay. It is ramified by numerous canaliculi containing cytoplasmic extensions of **odontoblasts** (section 8.4.5), the dentine-producing cells. The **pulp cavity** contains odontoblasts, sensory endings of nerves and blood vessels which deliver nutrients to the living tissues of the tooth and remove their waste products.

The root of the tooth is covered with **cement**, a substance similar to bone. Numerous **periodontal fibres**, connected to the cement at one end and the jawbone at the other, anchor the tooth firmly in place. However it is still able to move slightly and this reduces the chances of it being sheared off during chewing.

10.4.3   The development of teeth in humans

Teeth begin to form in the embryo after about the sixth week of its development. Cells in the buccal epithelium divide to form tooth buds. Each bud extends into the mesoderm and is termed an **enamel organ**. Gradually this organ becomes concave and develops the characteristic shape of a tooth. The enamel organ differentiates into outer and inner epithelia. The inner cells enclose a mesodermal core, the **dental papilla**, from which the dentine-producing cells and pulp develop. When dentine begins to form, the inner enamel layer differentiates into **ameloblasts** which produce enamel. Finally the jawbone becomes fashioned into a tooth socket to house the growing tooth. Fig 10.20 shows the stages in tooth development. The first tooth appears through the gum of a baby at about six months after birth.

Deciduous dentition is usually complete by the age of

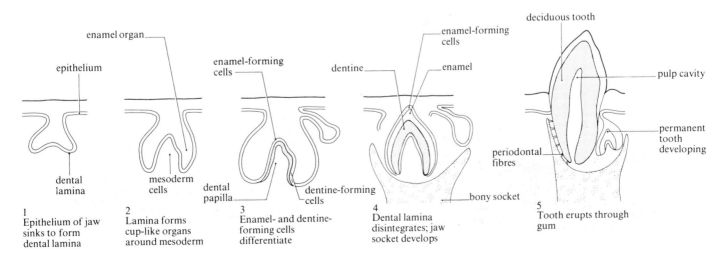

Fig 10.20 *Stages in the development of a tooth*

309

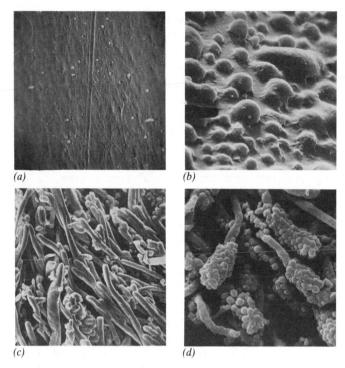

(a)

(b)

(c)

(d)

Fig 10.21 (above) *Development of dental plaque. (a) Coccal bacteria deposit as pioneer species and then multiply to form a film. (b) Organisms embedded in a matrix of extracellular polymers of bacterial and salivary origin. (c) The complexity of the community increases and rod- and filament-shaped populations appear. (d) In the climax community many unusual associations between different populations can be seen, including 'corn cob' arrangements*

three years but the jaws continue growing and these teeth soon become too small. At the age of six years the deciduous teeth begin to loosen and are replaced by the first of the larger permanent teeth. By the age of 13, 28 of the permanent teeth should have appeared. The four

(a)

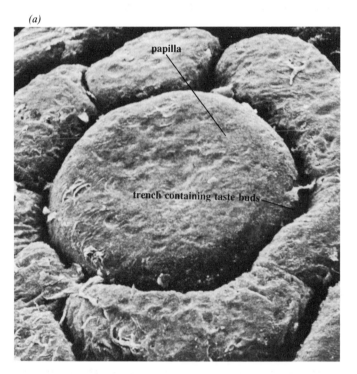

papilla

trench containing taste buds

Fig 10.22 (above) *(a) Scanning electron micrograph of circumvallate papilla of a three-week-old puppy. The taste buds are in the trenches surrounding the surface papillae* (below) *(b) VS taste buds in the tongue*

(b)

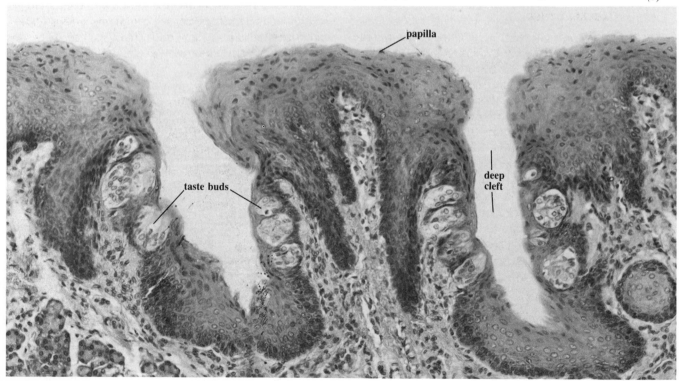

papilla

taste buds

deep cleft

remaining molars, or wisdom teeth, usually appear after the age of 17.

In humans and carnivores the hole at the base of the pulp cavity closes when each tooth reaches a particular size and prevents it from growing any further. However, in herbivores, this does not occur to the same extent and nutrients continue to pass to the tooth which grows continuously.

10.4.4 Dental disease

Two major dental diseases exist, periodontal disease and dental caries. Both are caused by **plaque** which is a mixture of bacteria and salivary materials. If allowed to accumulate, the bacteria cause inflammation of the gums (**periodontal disease**). Plaque also combines with certain chemicals in the saliva which make it harden and calcify to form deposits of **calculus** which cannot be removed by brushing. Some of the bacteria in plaque convert sugar into acid which initiates the process of **dental caries** (fig 10.21).

Periodontal disease

This is a disease of the gums caused by micro-organisms that are normally present in the mouth in dental plaque, especially in the areas between the gums and the teeth. Neglect of oral hygiene creates favourable conditions for the spread of this disease. Initially periodontal disease causes inflammation of the gums. If this condition, which is generally painless, is allowed to continue, the inflammation may spread to the root of the tooth and destroy the periodontal fibres which anchor it in place. Eventually the tooth becomes loose and may have to be extracted.

Dental caries

The micro-organisms in dental plaque convert sugar in the buccal cavity to acid. Initially the enamel is slowly and painlessly dissolved by the acid. However when the dentine and pulp of the tooth are attacked this is accompanied by severe pain or 'toothache', and the possible loss of teeth.

Several factors contribute to the spread of dental caries. They include prolonged exposure to sugary foodstuffs, disturbance of saliva composition, lack of oral hygiene and low levels of fluoride in drinking water. Prevention of dental caries may be aided by adding fluoride to drinking water, fluoridation of some foods such as milk, children taking fluoride tablets, brushing teeth with fluoridated toothpaste, good oral hygiene and regular visits to the dentist and oral hygienist and care with the composition of the diet.

10.4.5 Buccal cavity

The buccal cavity is the region enclosing the jaws and tongue, and is lined by stratified squamous epithelium. During mastication the muscular tongue moves food around the mouth and mixes and moistens it with saliva. The tongue possesses **taste buds** (fig 10.22) that are sensitive to sweet, salty, sour and bitter substances and this aids in food discrimination. In humans the tongue is also important in speech.

About $1.5\,dm^3$ of saliva are produced by humans each day by three pairs of **salivary glands** (fig 10.23) and numerous **buccal glands** in the mucosa of the buccal cavity. Saliva is a watery secretion containing the enzymes **salivary amylase** and **lysozyme**, sodium chloride and sodium hydrogencarbonate, phosphates, carbonates, Ca^{2+}, K^+, Mg^{2+}, sulphocyanide and mucus. The mucus moistens and lubricates the food and makes it easier to swallow. Salivary amylase begins the digestion of starch first to dextrins, shorter polysaccharides, and then to the disaccharide maltose. Carnivorous mammals which rapidly gulp large chunks of food into their stomachs possess no digestive enzymes in their saliva. Lysozyme helps in keeping the buccal cavity clear of pathogenic micro-organisms by catalysing the breakdown of their cell walls. Ultimately the semi-solid, partially digested food particles are stuck together by **mucin** and moulded into a **bolus** (or pellet) by the tongue, which then pushes it towards the pharynx. From here as a result of a reflex action it is swallowed into the oesophagus via the pharynx.

10.4.6 Swallowing

This is initially a voluntary action, but once begun it continues involuntarily to its completion. Fig 10.24 shows the stages involved in swallowing in humans.

10.4.7 Oesophagus

This is a narrow muscular tube lined by stratified squamous epithelium containing mucus glands. In humans it is about 25 cm long and quickly conveys food and fluids by peristalsis from the pharynx to the stomach.

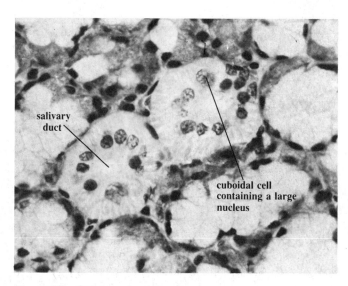

salivary duct

cuboidal cell containing a large nucleus

Fig 10.23 *Salivary gland tissue*

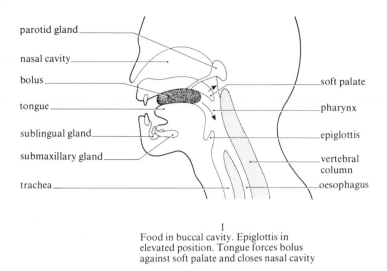

parotid gland

nasal cavity

bolus

tongue

sublingual gland

submaxillary gland

trachea

soft palate

pharynx

epiglottis

vertebral column

oesophagus

1
Food in buccal cavity. Epiglottis in elevated position. Tongue forces bolus against soft palate and closes nasal cavity

tongue elevated

bolus

glottis

epiglottis depressed

2
The glottis (opening into the larynx) is closed by the epiglottis, so bolus enters the oesophagus not the trachea

Fig 10.24 *Actions involved in the swallowing of food in humans*

The upper part of the oesophagus contains striated muscle, the middle section a mixture of striated and smooth muscle and the lower region purely smooth muscle. Carnivores which gulp their food in large pieces possess striated muscle along the whole length of the oesophagus.

10.4.8 Stomach

The stomach in humans is situated below the diaphragm and on the left side of the abdominal cavity. It is a distensible muscular bag whose function is to store and partially digest food. Unlike the other regions of the gut it consists of three smooth muscle layers, the outer longitudinal, middle circular and inner oblique layers which serve to churn and mix the food with the gastric secretions. When undistended the stomach lies in folds, but when fully distended it can hold nearly $5\,dm^3$ of food. The thick mucosa is liberally supplied with mucus-secreting epithelial cells and possesses numerous gastric pits (figs 10.26 and 10.27). These possess zymogen cells, oxyntic cells and enzymes and hydrochloric acid. Collectively the secretions of the stomach are called **gastric juice**. The mucus provides a barrier between the stomach mucosa and gastric juice and prevents the stomach self-digesting. The cardiac sphincter, at the junction between the oesophagus and upper cardiac region of the stomach, and pyloric sphincter, at the junction of the stomach and the duodenum, prevent the uncontrolled exit of food from the stomach. Both act as valves and serve to retain food in the stomach for periods of up to 4 h. Periodic relaxation of the pyloric sphincter releases small quantities of the food into the duodenum. The mucosa of the cardiac region of the stomach contains only mucus glands, whilst the main body of the stomach, the fundus, contains many long, tubular **zymogen** or **chief glands** (fig 10.27). These possess zymogen cells, oxyntic cells and argentaffine cells.

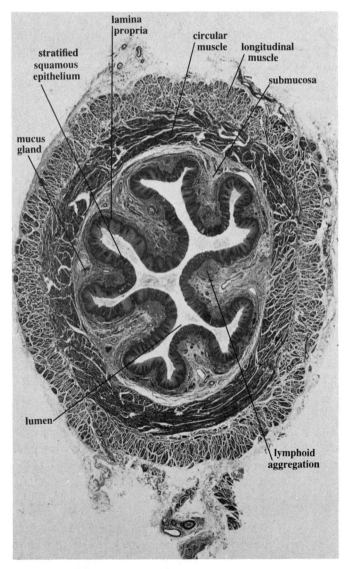

stratified squamous epithelium

lamina propria

circular muscle

longitudinal muscle

submucosa

mucus gland

lumen

lymphoid aggregation

Fig 10.25 *TS human oesophagus*

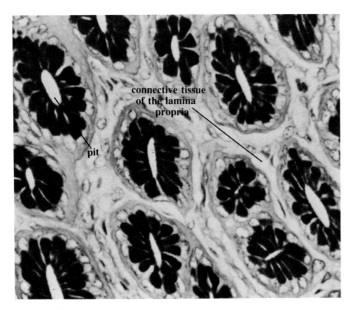

Fig 10.26 *TS gastric pits of a mammal*

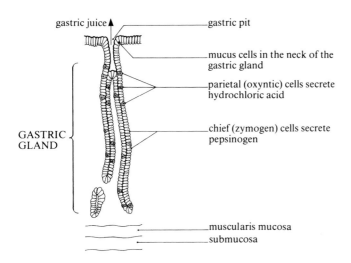

Fig 10.27 *VS stomach wall showing gastric gland*

Zymogen or chief cells. These secrete the inactive enzymes pepsinogen and prorennin.

10.8 Why is it necessary for pepsin to be secreted in an inactive state?

Oxyntic or parietal cells. These secrete a 0.04–0.05% solution of hydrochloric acid which makes the pH of the stomach contents 1–2.5, ideal for the optimum activity of the stomach enzymes. The acid kills many bacteria, loosens fibrous and cellular components of tissue and promotes the conversion of pepsinogen to its active form pepsin. Pepsin hydrolyses protein into smaller polypeptides and converts more molecules of pepsinogen to pepsin, a process known as **autocatalysis**. Hydrochloric acid converts prorennin to rennin which, in turn, coagulates caseinogen, the soluble protein of milk, into the

insoluble calcium salt of casein, in the presence of calcium ions. This calcium salt is then digested by pepsin. Hydrochloric acid renders calcium and iron salts suitable for absorption in the intestine, begins hydrolysis of sucrose to glucose and fructose and splits nucleoproteins into nucleic acid and protein.

Argentaffine cells. These produce the **intrinsic gastric factor** which aids absorption of molecules of the vitamin B_{12} complex.

The muscles of the stomach wall thoroughly mix up the food with gastric juice and eventually convert it into a semi-liquid mass called **chyme**. Gradually the stomach empties the chyme into the duodenum via the relaxed pyloric sphincter.

10.4.9 Small intestine

This consists in humans of an upper tube, 20 cm long, called the **duodenum**, into which open the pancreatic and bile ducts. The duodenum leads on to the **ileum** which is about 5 m long (fig 10.28a and b). The submucosa of the small intestine is thrown into many folds. The mucosa possesses numerous finger-like projections called **villi** whose walls are richly supplied with blood capillaries and lymph vessels and contain smooth muscle (fig 10.30). They are able to constantly contract and relax, thus bringing themselves into close contact with the food in the small intestine. The individual cells on the surface of the villi possess tiny microvilli on their free surfaces (fig 10.30 and section 7.2.11).

10.9 (*a*) List the features of the small intestine which increase its surface area.
(*b*) Why is this an advantage to the animal concerned?

Throughout the small intestine, certain mucosal cells secrete mucus. Submucosal Brunner's glands also secrete mucus and alkaline fluid in the first part of the duodenum, so protecting the intestinal mucosa against the acid pH of the stomach and providing an optimum pH of 7–8 at which the intestinal enzymes are active.

10.10 What would happen to the activity of the intestinal enzymes if the pH in the duodenum remained at 2?

Until the mid-1960s it was believed that digestion of foodstuffs was completed by a group of enzymes collectively called the succus entericus secreted into the lumen of the small intestine by the epithelial cells lining this region. However, this is now known not to be the case.

The disaccharase and peptidase enzymes involved in the final digestive process are, in fact, bound to membranes of the microvilli of the epithelial mucosa (fig 10.30b). Other peptidases are located within these cells. It is at these sites

313

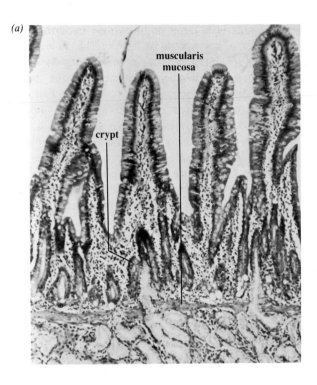

(a)

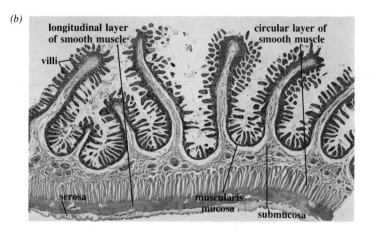

(b)

Fig 10.28 *(a) VS duodenum. (b) VS ileum*

Table 10.8 Summary of digestive secretions and their action.

Secretion	Enzymes	Site of action	Optimum pH	Substrate	Products
Saliva (from salivary glands)	Salivary amylase	Buccal cavity	6.5–7.5	Amylose in starch	Maltose
Gastric juice (from stomach mucosa)	(Pro)rennin (in young)	Stomach	2.00	Caseinogen in milk	Casein
	Pepsin(ogen)	Stomach	2.00	Proteins	Peptides
	Hydrochloric acid (not an enzyme)	Stomach	—	Pepsinogen	Pepsin
				Nucleoproteins	Nucleic acid and protein
Membrane-bound enzymes in small intestine	Amylase	Microvilli of	8.5	Amylose	Maltose
	Maltase	brush border	8.5	Maltose	Glucose
	Lactase	of	8.5	Lactose	Glucose + galactose
	Sucrase	epithelial	8.5	Sucrose	Glucose + fructose
(exopeptidases)	{ Aminopeptidase	mucosa of	8.5	Peptides and	Amino acids
	{ Dipeptidase	small intestine	8.5	dipeptides	Amino acids
Intestinal juice	Nucleotidase	Small intestine	8.5	Nucleotides	Nucleosides
	Enterokinase	Small intestine	8.5	Trypsinogen	Trypsin
Pancreatic juice (from pancreas)	Amylase	Small intestine	7.00	Amylose	Maltose
	Trypsin(ogen)	Small intestine	7.00	{ Proteins	Peptides
				{ Chymotrypsinogen	Chymotrypsin
(endopeptidases)*	{ Elastase	Small intestine	7.00	Proteins	Peptides
	{ Chymotrypsin(ogen)	Small intestine	7.00	Proteins	Amino acids
(exopeptidase)*	Carboxypeptidase	Small intestine	7.00	Peptides	Amino acids
	Lipase	Small intestine	7.00	Fats	Fatty acids + glycerol
	Nuclease	Small intestine	7.00	Nucleic acid	Nucleotides
	Bile salts (not enzymes)	Small intestine	7.00	Fats	Fat droplets

* Exopeptidases split off terminal amino acids from proteins (polypeptides)

Endopeptidases break bonds between amino acids within proteins thus producing smaller peptides

Collectively these enzymes break up polypeptides into their constituent amino acids so that they can be absorbed by the villi of the ileum

that the final hydrolysis of disaccharides, dipeptides and some tripeptides occurs (fig 10.29). The end-products are monosaccharides and amino acids respectively, which are liberated into the lumen of the small intestine. A full list of the enzymes involved can be found in table 10.8.

Also present in the small intestine is nucleotidase, which converts nucleotides to nucleosides, and enterokinase, a non-digestive enzyme which converts inactive trypsinogen of the pancreatic juice into active trypsin. In addition to its own set of enzymes the small intestine receives alkaline pancreatic juice and bile from the pancreas and liver respectively. Bile, produced by liver cells (section 18.4), is stored in the gall bladder and contains a mixture of salts, notably sodium glycocholate and taurocholate, which reduce the surface tension of fat globules and emulsify them into droplets, so increasing their total surface area. In this form they are acted upon more effectively by lipase. Further information about the composition of bile is given in section 18.4.

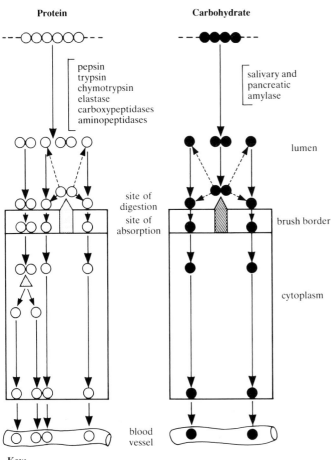

Key:
○ a single amino acid
⌂ a brush border peptidase
△ an intracellular peptidase
● a monosaccharide (mainly glucose)
▧ sucrase, maltase or lactase

Fig 10.29 *Schematic diagram of two epithelial cells, the one on the left bringing about the final phase of protein digestion, with the subsequent absorption of amino acids; and the one on the right the corresponding processes for carbohydrates*

The pancreas is a large gland whose exocrine tissue resembles that of salivary glands (fig 10.31). This tissue is composed of groups of cells called **acini** (singular acinus) which produce a variety of digestive enzymes that are poured into the duodenum via the pancreatic duct. They include amylase to convert amylose to maltose, **lipase** to convert fats to fatty acids and glycerol, **trypsinogen**, which when converted to trypsin by enterokinase digests proteins into smaller polypeptides and more trypsinogen into trypsin, **chymotrypsinogen** which is converted to chymotrypsin to digest proteins to amino acids, **carboxypeptidases** to convert peptides to amino acids and **nucleases** to convert nucleic acids to nucleotides.

A summary of the enzymes secreted by the human gut and their action is given in table 10.8. Table 10.9 indicates the differences in structure between the major regions of the alimentary canal in humans.

10.4.10 Absorption of food in the small intestine

Absorption of the end-products of digestion occurs through the microvilli (section 7.2.11) of the epithelial cells of villi lining the ileum. The structure of the villus is ideally suited for this function as can be seen in fig 10.30. Monosaccharides, dipeptides and amino acids are absorbed either by diffusion or active transport into the blood capillaries (section 7.2.2).

> **10.11** Why is it important that active transport is employed in the absorption of the foodstuffs monosaccharides, dipeptides and amino acids?

From the villi the blood capillaries converge to form the hepatic portal vein which delivers the absorbed food to the liver. Fatty acids and glycerol enter the columnar epithelial cells of the villi. Here they are reconverted into fats. These fats then enter the lacteals. Proteins present in these lymph vessels coat the fat molecules to form lipoprotein droplets called **chylomicrons**. These pass into the bloodstream via the thoracic lymphatic duct. The lipoproteins are subsequently hydrolysed by a blood plasma enzyme and enter cells as fatty acids and glycerol where they may be used in respiration or stored as fat in the liver, muscles, mesenteries or subcutaneous tissue.

Inorganic salts, vitamins and water are also absorbed in the small intestine.

10.4.11 Peristalsis in the alimentary canal of humans

Whilst the food is in the alimentary canal it is subjected to a number of peristaltic movements. Alternate rhythmic contractions and relaxation of its wall produce **segmenting movements** which constrict parts of the small intestine and bring chyme and the absorptive mucosal lining close together. **Pendular movements** are produced when loops of the intestine suddenly shorten vigorously,

(a)

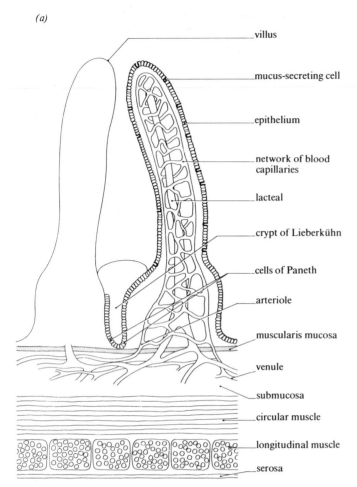

villus

mucus-secreting cell

epithelium

network of blood capillaries

lacteal

crypt of Lieberkühn

cells of Paneth

arteriole

muscularis mucosa

venule

submucosa

circular muscle

longitudinal muscle

serosa

(b)

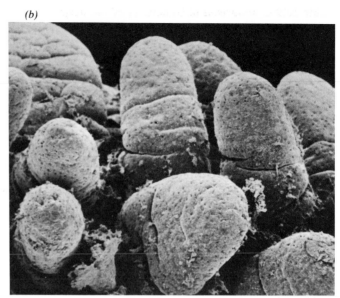

(c)

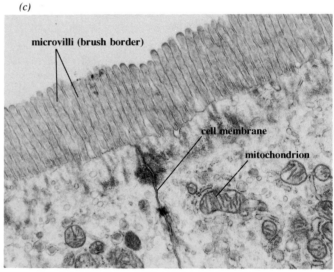

microvilli (brush border)

cell membrane

mitochondrion

Fig 10.30 *(a) TS wall of human small intestine showing a villus. (b) Scanning electron micrograph showing villi on surface of small intestine. (c) Electron micrograph of mucosal cell showing microvilli*

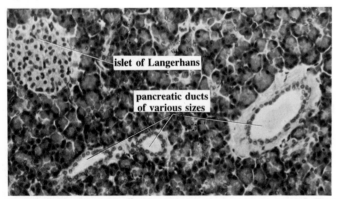

islet of Langerhans

pancreatic ducts of various sizes

Fig 10.31 *TS pancreas tissue showing pancreatic ducts and one islet of Langerhans*

throwing food from one end to the other thus thoroughly mixing it. As well as this, normal peristaltic activity occurs which propels the intestinal contents further along the alimentary canal. The ileocaecal sphincter opens and closes from time to time, to allow small amounts of residue from the ileum to enter the large intestine.

10.4.12 Large intestine

In the human large intestine the bulk of the water and any remaining inorganic nutrients are absorbed, whilst some metabolic waste and inorganic substances, notably calcium and iron, in excess in the body are excreted as salts. Mucosal epithelial cells secrete mucus which

Table 10.9 Comparison of structures of the major regions of the alimentary canal in humans.

Layer	Oesophagus	Stomach	Small intestine	Large intestine
Mucosa (a) epithelium lining lumen	Stratified, squamous	Simple columnar	Simple columnar, absorptive and mucus cells	Simple columnar, absorptive and mucus cells
	Specialisation – a few mucus glands located in the lamina propria and sub-mucosa	Specialisation – gastric glands located in lamina propria, four cell types: (i) mucus (ii) parietal (iii) peptic (iv) endocrine	Specialisation – (i) intestinal glands in crypts of Lieberkühn (ii) Paneth cells (iii) endocrine cells (iv) duodenal mucus glands	Specialisation – intestinal glands in lamina propria
(b) lamina propria	Some mucus glands	Many gastric glands	Intestinal glands and prominent lacteals	Tubular glands
(c) muscularis mucosa	Present	Present	Present	Present
Submucosa	Some deep mucus glands	Present	Duodenal glands	Intestinal glands
Muscularis inner circular, outer longitudinal	Transitional from striated muscle in upper region to smooth muscle in lower region in Man	With additional innermost layer of oblique muscle. Circular muscle forms cardiac and pyloric sphincters	Present	Present
Serosa	Present	Present	Present	Incomplete serosa

lubricates the solidifying food residue or faeces. Many symbiotic bacteria present in the large intestine synthesise amino acids and some vitamins, especially vitamin K, which are absorbed into the bloodstream.

In humans the appendix is a blindly ending pouch leading from the caecum and possesses no known function. It is, however, of great significance in herbivores (section 10.8.3). The bulk of the faeces consists of dead bacteria, cellulose and other plant fibres, dead mucosal cells, mucus, cholesterol, bile pigment derivatives and water. Faeces can remain in the colon for 36 h before being passed on to the rectum where it is stored briefly before egestion via the anus. Two sphincters surround the anus, an internal one of smooth muscle and under the control of the autonomic nervous system, and an outer one of striated muscle controlled by the voluntary nervous system.

In a young baby reflex defaecation occurs when distension of the rectum causes relaxation of the internal sphincter. Gradually the child learns to bring this reflex under control of its higher nervous centres and defaecation usually only occurs when the external sphincter is relaxed.

Experiment 10.3: To investigate the anatomy of the stomach, small and large intestines and to analyse the amino acid content of each region

Materials

freshly killed adult rat	dissection kit
mammal Ringer's solution	distilled water
specimen tubes	thin-layer chromatogram plate
binocular microscope	wire loop
chromatogram jar and lid	fume cupboard
n-butanol	ninhydrin spray
glacial ethanoic acid	oven at 100 °C

Method

(1) Dissect the freshly killed rat under Ringer's solution. Ligature and remove separately the stomach, small intestine and large intestine, open them and wash out the contents of each region with mammal Ringer's solution into separate labelled specimen tubes.

(2) Cut small sections of the wall of each region and examine under a binocular microscope. Observe and make notes on the nature of their internal surfaces.

(3) Add a small quantity of distilled water to the contents of the specimen tubes, shake well and allow to settle for 15 min.

(4) Take a drop of the clear liquid above the sediment from each specimen tube and place it on separate, prepared thin-layer chromatogram plates and dry. Repeat the procedure to increase the amount of specimen present. When the spots are dry, place them in a chromatogram jar containing a solvent composed of 40 parts n-butanol, 10 parts glacial acetic acid and 15 parts distilled water to about 1 cm depth. Place a lid on the chromatogram jar (fig 10.32).

(5) Allow to run for 90 min, then dry the plates in a fume cupboard and spray with ninhydrin.

(6) Heat the plates to 100 °C in an oven.

(7) Examine each plate for purple and other coloured spots. The spots indicate the presence of amino acids in the solutions being examined.

(NB This practical is based on Nuffield A-level Biology, *Maintenance of the organism*.)

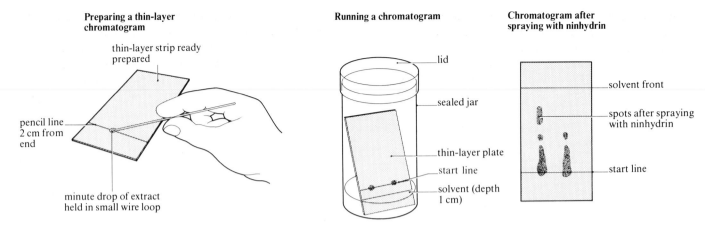

Preparing a thin-layer chromatogram

thin-layer strip ready prepared

pencil line 2 cm from end

minute drop of extract held in small wire loop

Running a chromatogram

lid

sealed jar

thin-layer plate

start line

solvent (depth 1 cm)

Chromatogram after spraying with ninhydrin

solvent front

spots after spraying with ninhydrin

start line

Fig 10.32 *Preparing a thin-layer chromatogram*

Experiment 10.4: To investigate digestion and absorption using a model gut

Materials

freshly killed adult rat
watchglasses
1% starch solution
three 15 cm lengths of visking tubing
cotton
boiling tubes
paper clips
water bath at 30 °C
stopclock

iodine/potassium iodide solution
Benedict's solution
bunsen burner
asbestos mat
test-tubes and rack
teat pipette

Method

(1) Dissect a freshly killed rat and ligature and extract the stomach, small intestine and large intestine separately.

(2) Put each in a watchglass and cover with 1% starch suspension. Mix and break up the tissue whilst in the watchglass.

(3) Take three 15 cm lengths of visking tubing (1 cm diameter) and tie a knot at one end.

(4) Pour the macerated tissue of each region into the three lengths of visking tubing, one for each region, and top up with more starch solution until the total volume is about 10 cm³. Mix each solution thoroughly.

(5) Lower the three tubes into separate boiling tubes containing 10 cm³ of distilled water and prevent each open end falling into the water by attaching it to the side of the boiling tube with a paper clip.

(6) Set up a control with the visking tubing containing 10 cm³ of 1% starch solution only.

(7) Incubate all tubes in a water bath at 30 °C for 30 min.

(8) After 30 min perform the following tests on each tube and its contents:

(a) take a drop of solution from each of the visking tubing contents and test each separately for the presence of starch with iodine/potassium iodide solution;

(b) take a further drop of solution from each of the visking tubing contents and test for reducing sugar using Benedict's solution. Record your results;

(c) perform separate starch and reducing sugar tests on the water surrounding the visking tubing in each tube.

(9) Construct a table of your results.

(10) Is there any evidence that gut tissue changes starch to reducing sugar? If so which region produces most reducing sugar?

(11) What do your observations indicate about the nature of visking tubing?

10.5 The control of digestive secretions

Secretion of digestive enzymes is an energy-consuming process, and it would be extremely wasteful if the body was constantly producing them, especially in the absence of food. Instead the bulk of digestive juice is produced only when there is digestive work to be done. In this way the overall control of digestive activity is coordinated and regulated as an orderly sequence during the digestion of food.

In the buccal cavity salivary secretion is released by two reflex reactions. First an unconditional cranial reflex occurs when food is present in the buccal cavity. Contact with the taste buds of the tongue elicits impulses which travel to the brain and from there to the salivary glands which are stimulated to secrete saliva. Secondly, there are the conditioned reflexes of seeing, smelling or thinking of food.

Secretion of gastric juice occurs in three phases. The first is the **nervous** (or vagus) phase. The presence of food in the buccal cavity and its swallowing initiate impulses which pass via the vagus nerve to the stomach whose mucosa is stimulated to secrete gastric juice. This takes place before the food has reached the stomach and therefore prepares it to receive food. The nervous phase of gastric secretion lasts for approximately 1 h. The second phase is the distension phase in which distension of the stomach by the food it contains also stimulates the flow of gastric juice. Thirdly, there is the gastric (or humoral) phase, in which the presence of food in the stomach stimulates the pyloric mucosa to produce a hormone, gastrin, which reaches the rest of the stomach mucosa via the bloodstream and stimulates it to produce gastric juice rich in hydrochloric acid for about 4 h.

The phases of gastric secretion can be seen in fig 10.33.

Enterogastrone is another hormone that is released by the stomach mucosa in response to the presence of fatty acids in the food. This generally inhibits hydrochloric acid secretion, slows down stomach peristalsis and delays its emptying. The intrinsic gastric factor which aids absorption of vitamin B_{12} is also released by the gastric mucosa.

When acidified chyme enters and makes contact with the walls of the duodenum it triggers the duodenal mucosa to secrete intestinal juice and also to produce two hormones **cholecystokinin–pancreozymin** and secretin. The pancreozymin component of the former hormone is conveyed by the bloodstream to the pancreas as is secretin which also reaches the liver along with cholecystokinin. Pancreozymin induces the formation of pancreatic juice rich in enzymes. Secretin stimulates the flow of pancreatic juice rich in hydrogencarbonate, and bile synthesis by the liver. Cholecystokinin causes contraction of the gall bladder and subsequent release of bile into the duodenum.

Table 10.10 summarises the endocrine control of the various secretions of the alimentary canal and its associated organs.

10.6 The fate of the absorbed food materials

Carbohydrates and amino acids are both absorbed into the bloodstream surrounding the small intestine and passed to the liver via the hepatic portal vein. Most of the glucose is stored here or in muscle as glycogen

and fats, though some leaves via the hepatic vein to be distributed round the body where it is oxidised during respiration. Between meals, if the body requires more energy, glycogen can be reconverted to glucose and transported by the blood to those tissues in need.

Amino acids are used for the synthesis of new protoplasm, the repair of damaged parts of the body and the formation of enzymes and hormones. Surplus amino acids cannot be stored and are deaminated in the liver. Their amino (NH_2) groups are removed and converted to urea which is delivered via the bloodstream to the kidneys and excreted in the urine. The remainder of the amino acid molecule is converted to glycogen and stored.

Absorbed fats bypass the liver and enter the venous bloodstream via the thoracic lymphatic duct. Fats represent the major energy store of the body. Normally, however, glucose is in adequate supply and the fats are not required for energy production. In this case they are stored in subcutaneous adipose tissue, around the heart and kidneys and in the mesenteries. Some fat is incorporated into cell and nuclear membranes.

10.7 Regulation of food intake in humans

The regulation of food intake in humans is under the general control of two centres in the **hypothalamus** of the brain, the **hunger** and **satiety** centres. Stimulation of the hunger centre causes the individual to seek and eat food, whilst stimulation of the satiety centre inhibits food intake. The most important factor influencing both these centres is the level of glucose in the blood. This is monitored by the hypothalamus and gives a good indication of the nutritional condition of the body.

Shortly after a meal, blood glucose level is high and this stimulates the satiety centre to inhibit the body from eating any further food. A long time after a meal the blood glucose level will be low, and this condition triggers the hunger centre into action (fig 10.34).

This seemingly quite simple explanation is not by any means the complete story. Whether or not food is

Table 10.10 Summary of endocrine control of the secretions of the alimentary canal and its associated organs in humans.

Hormone	Site of production	Stimulus for secretion	Target organ	Response
Gastrin	Stomach mucosa	Distension of stomach by food	Stomach	Increased secretion of HCl
Enterogastrone	Stomach mucosa and small intestine	Fatty acids in food	Stomach	Inhibits HCl secretion, slows peristaltic activity, delays emptying
Pancreozymin	Duodenal mucosa	Food in the duodenum	Pancreas	Increased secretion of pancreatic enzymes
Cholecystokinin	Duodenal mucosa	Fatty food in the duodenum	Gall bladder	Contraction of gall bladder to release bile
Secretin	Duodenal mucosa	Acid and food in the duodenum	Pancreas	Increased flow of hydrogencarbonate in pancreatic juice
			Liver	Synthesis of bile

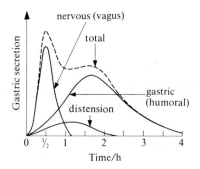

Fig 10.33 (left) *Phases of gastric secretion (From J. H. Green (1968) An introduction to human physiology, Oxford Medical Publications.)*

Fig 10.34 (below) *Hypothalamic control of food intake*

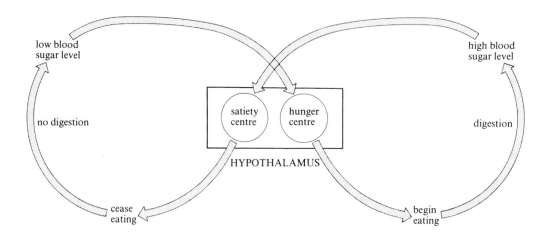

consumed is also affected by other subsidiary factors which include stretch reflexes in the alimentary canal, the psychological state of the individual, conscious habits and damage to or impairment of the brain. Any of these may upset the activity of the hypothalamic centres and cause abnormal responses towards food intake.

10.8 Variations in the mammalian alimentary canal

10.8.1 A carnivore – the cat

The cat is a carnivore and has teeth adapted for catching and breaking down animal food. The dental formula is:

$$2 \left[i \frac{3}{3} \quad c \frac{1}{1} \quad pm \frac{3}{2} \quad m \frac{1}{1} \right]$$

The closely fitting incisors are small and chisel-shaped and used to tear away flesh near the bone surface. The enlarged canines are curved and fang-like and used to seize and kill prey, and tear off flesh. Two cheek teeth on each side of the jaws are considerably enlarged – with prominent ridges running parallel with the line of the jaw. These are the **carnassial** teeth (third upper pm and first lower m). They act rather like the two blades of a pair of scissors with the

inner surfaces of the teeth of the upper jaw moving closely against the outer surfaces of the teeth in the lower jaw as they shear flesh from the prey. The other cheek teeth are flattened and possess sharp edges used for cutting flesh and cracking bones.

The jaw joint operates as a closely fitting hinge and permits only up-and-down movement. The cheek teeth, which require the greater force for their operation, are placed nearest the joint.

Contraction of the **temporal** muscle closes the lower jaw. It is attached to a prominent bony extension from the lower jaw which projects upwards towards the ears. This particular arrangement provides efficient leverage on the food as it is being sheared by the teeth, or when the cat's mouth is snapping shut whilst killing its prey. Another muscle, the **masseter**, pulls the base of the lower jaw upwards and reduces the strain on the jaw joint (fig 10.35).

10.8.2 A herbivore – the sheep

A sheep eats grass and its dentition is closely correlated with its feeding habits and diet. Its dental formula is:

$$2 \left[i \frac{0}{3} \quad c \frac{0}{1} \quad pm \frac{3}{2} \quad m \frac{3}{3} \right]$$

Upper incisors and canines are absent. In their place is a horny pad against which the chisel-shaped lower incisors

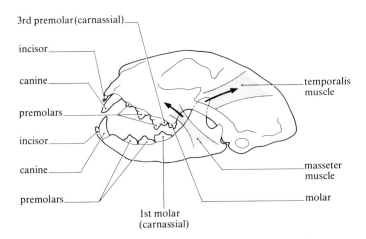

Fig 10.35 *Jaws, dentition and musculature of the cat*

and canines bite when the sheep is cropping grass. Between the front and cheek teeth is a large gap, the **diastema**, which provides space for the tongue to manipulate the cropped grass in such a way that grass being chewed is kept apart from that freshly gathered.

The cheek teeth possess broad grinding surfaces whose area is further increased by the surfaces of the upper teeth being folded into a W-shape and those of the lower teeth being folded into an M-shape. The ridges of the teeth are composed of hard enamel whilst the troughs are of dentine.

The jaw joint is very loose and allows forward, backward and sideways movement. During chewing the lower jaw moves from side to side, with the W-shaped ridges of the upper cheek teeth fitting closely into the grooves of the M-shaped lower teeth as they grind the grass. The masseter muscle is large, and the temporal small, the reverse arrangement of that of the cat (fig 10.36).

10.8.3 Cellulose digestion in ruminants (such as sheep and cattle)

Ruminants possess highly complex alimentary canals. A number of compartments precede the true stomach, the first of which is the **rumen**. This acts as a fermentation chamber where food, mixed with saliva,

undergoes fermentation by mutualistic micro-organisms. Many of these produce cellulases which digest cellulose. Their presence is absolutely essential to the ruminant as it is unable to manufacture cellulase itself. The end-products of fermentation are ethanoic, propanoic and butanoic acids, carbon dioxide and methane. The acids are absorbed by the host, who uses them as a major source of energy during oxidative metabolism. In return the micro-organisms obtain their energy requirements through the chemical reactions of fermentation and an ideal temperature in which to live.

A ruminant is able to regurgitate and rechew partially digested material from the rumen. This is called rumination or 'chewing the cud'. The food is then reswallowed and undergoes further fermentation. Eventually the partially digested food is passed through the initial compartments of the alimentary canal until it reaches the **abomasum** which corresponds to the stomach in humans. From here onwards food undergoes digestion by the usual mammalian digestive enzymes (fig 10.37).

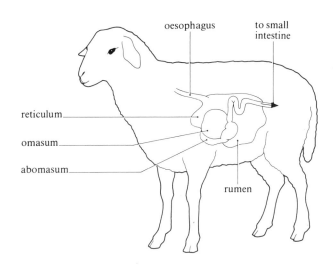

Fig 10.37 *Complex arrangement of compartments preceding the small intestine in a ruminant*

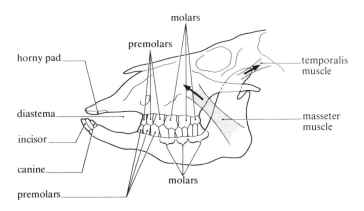

Fig 10.36 *Jaws, dentition and musculature of the sheep*

Chapter Eleven

Energy utilisation

Each living cell is a complex entity characterised by a high degree of order within its system. Experiments have revealed that the cell interior as a whole is continuously active, with materials constantly entering and leaving it. All reactions proceeding in the cell can be grouped into two categories. **Anabolic** reactions are the synthesis of large molecules from smaller, simpler molecules. Energy is used in this process (that is the process is **endergonic**).

$$A + B \rightarrow AB \quad [+\Delta G]$$

(Where ΔG = free energy change of a reaction.)

Catabolic reactions are the breakdown of large molecules to smaller, simpler molecules usually accompanied by release of energy (so the process is **exergonic**). Sometimes the simpler molecules can be used for biosynthesis again.

$$AB \rightarrow A + B \quad [-\Delta G]$$

[NB It is important to note that not every catabolic reaction liberates energy. Some degradations that the cell performs in order to eliminate unwanted substances are actually endergonic.]

The sum total of catabolic and anabolic reactions occurring at any time in a cell represents its **metabolism**:

catabolism + anabolism = metabolism

Organic compounds that enter a cell provide it with two essentials. These are small 'building' molecules, which are used for biosynthesis of new cellular components or the replacement of components past their useful life, and chemical energy. Generally when nutrients are degraded within a cell, energy is liberated. Much of it is utilised by the cell to maintain its own life processes. It is transferred to various sites in the cell and converted into different forms. Each form of energy may then be used for a particular job of work within the cell. This may be biosynthesis, mechanical work, cell division, active transport, osmotic activity, and, in some specialised cells, muscular contraction, bioluminescence or electrical discharge (fig 11.1). Chemical energy is most appropriate for use by living cells as it can be transferred within and between cells quickly, and released in economically regulated amounts as, and when, required. All energy is derived from the sun. Energy is readily convertible into its different forms by organisms, but to enter the food chains solar energy must be absorbed by green plants (**autotrophs**), and converted by their chlorophyll-containing cells into chemical energy (contained in the simple sugar glucose or polysaccharide starch) by the process of photosynthesis. A portion of this energy is liberated and used by plants for their own requirements. Animals have to

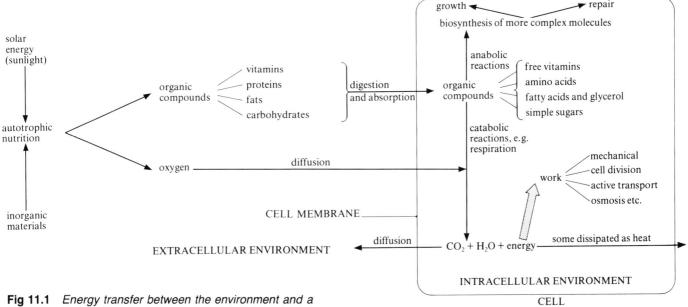

Fig 11.1 *Energy transfer between the environment and a heterotrophic cell. It is important to remember that whenever one form of energy is converted into another, a certain proportion of it is dissipated as heat*

find a ready-made source of energy (that is foodstuffs). Some of them obtain this by feeding on plants (**herbivores**), whilst others (**carnivores**) devour the tissues of herbivores for their energy supply (see chapter 12).

11.1   Role of respiration

Respiration may be defined as generally any process that liberates chemical energy when organic molecules are oxidised. Where the process occurs within cells it is called **internal**, **tissue** or **cell respiration**. If it requires oxygen, it is **aerobic** respiration; whereas if the reaction takes place in the absence of oxygen, it is **anaerobic** respiration.

Organic molecules (usually carbohydrate or fat) are broken down bond by bond, by a series of enzyme-controlled reactions. Each releases a small amount of energy, much of which is channelled into molecules of a chemical nucleotide called **adenosine triphosphate** (ATP).

Tissue respiration must not be confused with the processes of acquiring and extracting oxygen from, and discharging carbon dioxide into, the environment. These are collectively termed **external respiration**, or preferably **gas exchange**. They may involve organs or structures with specialised surfaces for the efficient exchange of gases, over which air or water is pumped by various respiratory movements (section 11.6).

11.2   ATP (adenosine triphosphate)

ATP is composed of the purine adenine linked to the 5C sugar ribose and three phosphate groups (fig 11.2). When the bonds of the two end phosphate groups of ATP are hydrolysed, the free energy yield for each is of the order of 30.6 kJ, whereas if the third phosphate group is hydrolysed the energy yield is only 13.8 kJ (table 11.1). It is because of this that ATP and ADP (adenosine diphosphate) are popularly, though erroneously, believed to possess 'energy-rich' bonds (often signified ($\sim$)). The reason why ATP releases more energy on hydrolysis than many other compounds is not clear. However it is thought to involve the distribution of charges within the molecule.

Table 11.1 Free energy of hydrolysis of phosphate compounds.

Compound	ΔG (free energy change) (kJ mol^{-1})
Phosphoenolpyruvate	-62.1
1,3 diphosphoglycerate	-49.5
Creatine phosphate	-43.3
ATP (to ADP and phosphate)	-30.6
ADP (to AMP and phosphate)	-30.6
AMP (to adenosine and phosphate)	-13.8
Glucose-6-phosphate	-13.8

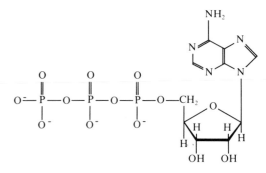

Fig 11.2 *Structure of ATP. The two end phosphate groups are attached by pyrophosphate bonds, which when hydrolysed yield a large quantity of free energy*

More chemical energy is said to be required to maintain its integrity. This is not contained in any one bond, but is a property of the whole molecule.

> **11.1**   Table 11.1 shows that ATP is not by any means the most 'energy-rich' compound in cells. What is the significance of ATP lying in an intermediate position in the table?

11.2.1   Importance of ATP

ATP is the standard unit in which the energy released during respiration is stored. To make one ATP molecule from ADP and phosphate 30.6 kJ of energy are required. Therefore it can only be formed from reactions that yield more than 30.6 kJ mol^{-1}. Any energy liberated in excess of 30.6 kJ mol^{-1}, and all that from reactions that yield less than 30.6 kJ mol^{-1}, cannot be stored in ATP and is lost as heat.

Because all the chemical energy is in one form (ATP), the energy-consuming processes need only one system that can accept chemical energy from ATP. Thus a great economy of mechanism is achieved.

ATP is an instant source of energy within the cell. It is mobile and transports chemical energy to energy-consuming processes anywhere within the cell. When the cell requires energy, hydrolysis of ATP is all that has to occur for the energy to be made available. ATP is found in all living cells, and hence is often known as the **universal energy carrier**.

ADP may be rephosphorylated to ATP by respiratory activity (fig 11.3), or by another 'high-energy' compound, such as creatine phosphate which is present in muscle cells. If all available ADP of a muscle cell has been converted to ATP, phosphate is transferred from ATP to creatine to form creatine phosphate. This releases a quantity of ADP which can combine with more phosphate to make extra ATP. The reverse occurs when ATP levels decrease: phosphate is transferred from creatine phosphate to ADP thus restoring ATP stocks (fig 11.4).

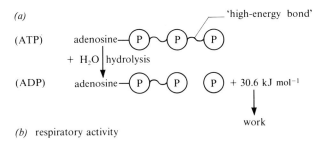

(a)

(ATP)

'high-energy bond'

+ H_2O | hydrolysis

(ADP)

+ 30.6 kJ mol^{-1}

work

(b) respiratory activity

energy yield
of 30.6. kJ
or more from
respiration

(ADP)

(ATP)

Fig 11.3 (left) *(a) Hydrolysis, and (b) rephosphorylation, of ATP by respiratory activity*

Fig 11.4 (below) *Energy-phosphate transfer between ATP and creatine*

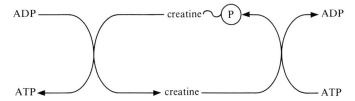

ADP

creatine

ADP

ATP

creatine

ATP

A third way to rephosphorylate ADP is by photophosphorylation by chlorophyll-containing cells of green plants (section 9.4).

The metabolic role of ATP is significant in that it lies at the centre of cellular activity, acting as a common intermediate between respiration and energy-requiring processes, with phosphate being consistently removed and replaced.

> **11.2** 'The role of ATP can be compared with that of a battery.' Explain this statement.

11.3 Biological oxidation

In general, cell oxidations are of three types:

(1) $$A + O_2 \rightarrow AO_2$$

direct oxidation by molecular oxygen;

(2) $$AH_2 + B \rightarrow A + BH_2$$

where A is oxidised at the expense of B;

> **11.3** What is this type of oxidation called?
>
> **11.4** What are enzymes which carry out this type of oxidation called?

(3) $$Fe^{2+} \rightarrow Fe^{3+} + e^-$$

where an electron transfer occurs, such as the oxidation of one ionic form of iron (Fe^{2+}) to another (Fe^{3+}).

Each type of oxidation is to be found in the series of reactions which are collectively termed **aerobic respiration**.

11.3.1 Cell respiration in outline

Cell respiration involves oxidation of a substrate to yield chemical energy (ATP). Organic compounds which are used as substrates in respiration are carbohydrates, fats and proteins.

Carbohydrates. These are usually the first choice of most cells. In fact brain cells of mammals cannot use anything but glucose.

Polysaccharides are hydrolysed to monosaccharides before they enter the respiratory pathway:

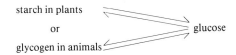

starch in plants
or
glycogen in animals

glucose

Fats. They form the 'first reserve' and are mainly used when carbohydrate reserves have been exhausted. However in skeletal muscle cells, if glucose and fatty acids are available, these cells respire the acids in preference to glucose.

Proteins. Since proteins have other essential functions, they are only used when all carbohydrate and fat reserves have been used up, as during prolonged starvation.

When glucose is the substrate, its oxidation can be divided into three distinct phases: glycolysis (the Embden–Meyerhof pathway); oxidative decarboxylation (Krebs, or citric acid cycle or TCA, tricarboxylic acid cycle); oxidative phosphorylation (respiratory chain incorporating hydrogen and electron transfer). Glycolysis is common to anaerobic and aerobic respiration, but the other two phases only occur when aerobic conditions prevail. Details of each of the processes are to be found later in this chapter, but an outline is given below.

11.3.2 Glycolysis and the Krebs cycle

During aerobic respiration glucose is oxidised by a series of dehydrogenations. At each dehydrogenation, hydrogen is removed and used to reduce a coenzyme:

$$AH_2 + B \xrightarrow{\text{dehydrogenase}} A + BH_2$$

| AH_2 reduced respiratory substrate | B coenzyme (hydrogen acceptor) | A oxidised respiratory substrate | BH_2 reduced coenzyme |

Most of these oxidations occur in the mitochondrion, where the usual coenzyme hydrogen acceptor is NAD (nicotinamide adenine dinucleotide):

$$NAD + 2H \rightarrow NADH_2$$

or, more accurately,

$$NAD^+ + 2H \rightarrow NADH + H^+$$

$NADH_2$ then enters the respiratory chain to be reoxidised.

11.3.3 The respiratory chain and oxidative phosphorylation

$NADH_2$ is oxidised back to NAD and the hydrogen released is passed along a chain of at least five carrier substances to the end of the chain where the hydrogen combines with molecular oxygen to form water. The passage of hydrogen along this 'respiratory chain' of carriers involves a series of redox reactions. The energy released from some of these is sufficient to make ATP, a process called **oxidative phosphorylation**. The net yield per molecule of glucose completely oxidised to water and carbon dioxide is 38 molecules of ATP, synthesised from ADP and inorganic phosphate. Glycolysis yields two ATP, Krebs cycle two ATP and the respiratory chain 34 ATP (fig 11.5).

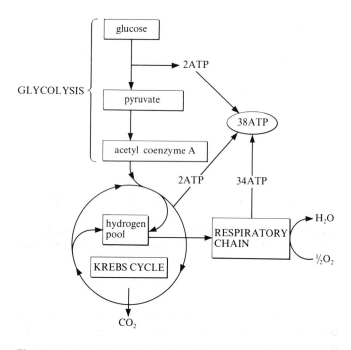

Fig 11.5 *Aerobic respiration in outline*

11.3.4 Glycolysis in detail

Glycolysis represents a series of reactions in which a glucose molecule is broken down into two molecules of pyruvate (fig 11.6). It occurs in the cytoplasm of cells, not in the mitochondria, and does not require the presence of oxygen. The process may be sub-divided into

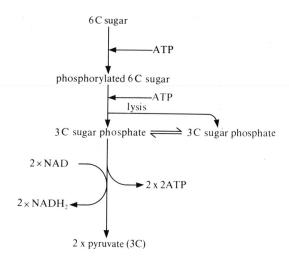

Fig 11.6 *Glycolysis in outline*

two steps, first the conversion of glucose into fructose 1,6-diphosphate, and secondly the splitting of fructose-1,6-diphosphate into 3C sugars which are later converted into pyruvate. Two ATP molecules are used up for phosphorylation reactions in the first step, whilst four ATP molecules are produced in the second step. Therefore there is a net gain of two ATP molecules. Four hydrogen atoms are also released. Their fate will be discussed later. The equation of the overall reaction is:

$$C_6H_{12}O_6 \rightarrow 2C_3H_4O_3 + 4H + 2ATP$$
glucose   pyruvic acid   (net gain)

The input and output of materials during glycolysis is shown in table 11.2.

Table 11.2 Input and output of materials during glycolysis.

Total input	Total output
1 molecule of glucose (6C)	2 molecules of pyruvate (2×3C)
2 ATP	4 ATP
4 ADP	2 ADP
$2\times NAD$	$2\times NADH_2$
$2\times P_i$	$2\times H_2O$

The ultimate fate of pyruvate depends on the availability of oxygen in the cell. If it is present, pyruvate will enter a mitochondrion and be completely oxidised into carbon dioxide and water (**aerobic respiration**). If oxygen is unavailable, pyruvate will be converted into ethanol or lactate (**anaerobic respiration**).

11.5 Study fig 11.7 carefully and answer the following questions:
(*a*) What is the process occurring at B and D?
(*b*) What class of enzyme controls reaction C?
(*c*) Name the processes occurring at E.
(*d*) Which vitamin contributes to the molecule of NAD?

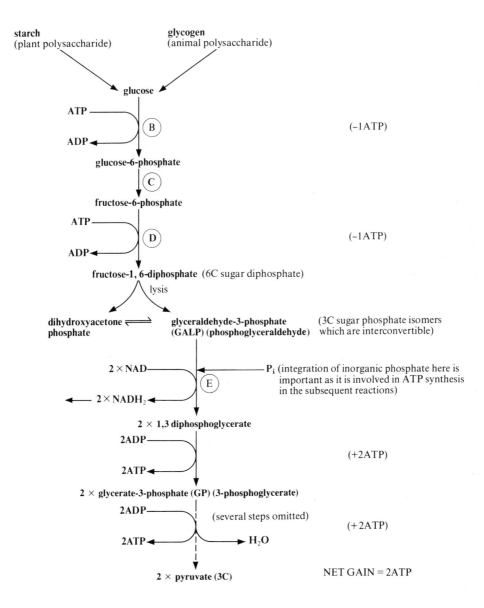

starch
(plant polysaccharide)

glycogen
(animal polysaccharide)

glucose

ATP
ADP
(B)
(–1 ATP)

glucose-6-phosphate

(C)

fructose-6-phosphate

ATP
ADP
(D)
(–1 ATP)

fructose-1, 6-diphosphate (6C sugar diphosphate)

lysis

dihydroxyacetone ⇌ glyceraldehyde-3-phosphate
phosphate   (GALP) (phosphoglyceraldehyde)

(3C sugar phosphate isomers
which are interconvertible)

$2 \times$ NAD

P_i (integration of inorganic phosphate here is
important as it is involved in ATP synthesis
in the subsequent reactions)

(E)

← $2 \times$ NADH$_2$

$2 \times$ 1,3 diphosphoglycerate

2ADP
2ATP
(+2 ATP)

$2 \times$ glycerate-3-phosphate (GP) (3-phosphoglycerate)

2ADP
(several steps omitted)
2ATP
→ H_2O
(+2 ATP)

$2 \times$ pyruvate (3C)

NET GAIN = 2ATP

11.3.5   Aerobic respiration

There are two phases involved in aerobic respiration. First, if sufficient oxygen is available, each pyruvate molecule enters a mitochondrion where its oxidation is completed by aerobic means. This involves oxidative decarboxylation of pyruvate, that is the removal of carbon dioxide together with oxidation by dehydrogenation. During these reactions pyruvate combines with a substance called coenzyme A (often written CoAS—H) to form acetyl coenzyme A. Sufficient energy is released to form an 'energy-rich' bond in the acetyl CoA molecule. In reality the complete reaction is much more complex than this description suggests and involves five different coenzymes and three different enzymes.
The overall reaction is:

$$CH_3COCOOH + CoAS—H + NAD \rightarrow$$
$$CH_3CO{\sim}S—CoA + CO_2 + NADH_2$$
acetyl CoA

Fig 11.7 *Glycolysis in detail. General biochemistry of glycolysis. It is important to note that because two 3C compounds are formed when fructose-1,6-diphosphate is cleaved, four ATP are produced during subsequent reactions, two for each 3C compound converted to pyruvate*

The NADH$_2$ formed as a result of acetyl CoA formation is collected and channelled into the respiratory chain in the mitochondrion.

The second phase is the Krebs cycle (named after its discoverer, Sir Hans Krebs). The acetyl component of acetyl CoA possesses two carbons and is passed into the Krebs cycle when acetyl CoA is hydrolysed. The acetyl component combines with oxaloacetate, a 4C compound, to form citrate (6C). This reaction requires energy which is provided at the expense of the energy-rich bond of acetyl CoA. A cycle of reactions follows during which the acetyl groups fed in by acetyl CoA are dehydrogenated to release

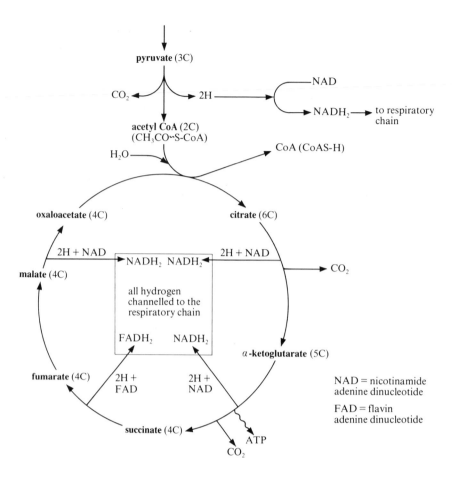

pyruvate (3C)

CO_2 ← → 2H → NAD

NADH$_2$ → to respiratory chain

acetyl CoA (2C)
($CH_3CO\sim S\text{-}CoA$)

H_2O → CoA (CoAS-H)

oxaloacetate (4C)   citrate (6C)

2H + NAD   2H + NAD → CO_2

malate (4C)

NADH$_2$ NADH$_2$

all hydrogen channelled to the respiratory chain

FADH$_2$ NADH$_2$

a-ketoglutarate (5C)

fumarate (4C)

2H + FAD   2H + NAD

NAD = nicotinamide adenine dinucleotide

FAD = flavin adenine dinucleotide

succinate (4C)

ATP

CO_2

Fig 11.8 *Simplified diagram of the Krebs cycle*

four pairs of hydrogen atoms and decarboxylated to form two molecules of carbon dioxide. During the latter process oxygen is taken from two molecules of water and used to oxidise two carbon atoms to carbon dioxide. This is termed **oxidative decarboxylation**. At the end of the cycle oxaloacetate is regenerated and able to link up once again with another molecule of acetyl CoA, and so the cycle continues. One molecule of ATP, four pairs of hydrogen atoms and two molecules of carbon dioxide are released per molecule of acetyl CoA oxidised. The hydrogen atoms are accepted by NAD or FAD (section 6.2.3) and are eventually passed into the respiratory chain. As two molecules of acetyl CoA are formed from one oxidised glucose molecule, Krebs cycle must rotate twice for each molecule respired. Therefore the net result is two ATP synthesised, four carbon dioxide liberated and eight pairs of hydrogen atoms released for entry into the respiratory chain (fig 11.8).

The overall reaction for glycolysis, acetyl CoA formation and Krebs cycle is:

$$C_6H_{12}O_6 + 6H_2O \rightarrow 6CO_2 + 4ATP + 12 \square H_2$$

where $\square$ = hydrogen acceptor.

11.3.6 Oxidative phosphorylation and the respiratory chain

The pairs of hydrogen atoms removed from respiratory intermediates by dehydrogenation reactions during glycolysis and the Krebs cycle are ultimately oxidised to water by molecular oxygen with accompanying phosphorylation of ADP to form ATP molecules. This is accomplished when hydrogen, released from NADH$_2$ or FADH$_2$, is passed along a chain of at least five intermediate substances, which include flavoprotein, coenzyme Q and a number of different cytochromes, until at the end the hydrogen combines with molecular oxygen to form water. As a result of the passage of hydrogen the intermediate carriers undergo a series of **redox** reactions, and they are arranged in such a way that at three points in the chain, each time the hydrogen atoms are passed from one intermediate to another, a small amount of energy is liberated and incorporated into a molecule of ATP. Fig 11.9 represents the respiratory chain (or electron transport chain). In fact the initial part of the chain effects mainly hydrogen transfer whilst the latter portion operates purely electron transfer. Carriers X, Y and Z are **cytochromes**. X and Y contain a protein pigment with an iron-containing prosthetic group called **haem**, as occurs also in haemoglobin. During each redox reaction the iron ion is alternately in its oxidised (Fe^{3+}) and reduced (Fe^{2+}) forms. Finally, at the terminal stage, carrier Z, which contains copper and is commonly called **cytochrome oxidase** (cytochrome a/a_3), promotes the reduction of molecular oxygen to water. This stage of cell oxidation can be inhibited by potassium cyanide or carbon monoxide.

11.3.7 Hydrogen and electron carriers

NAD and NADP (nicotinamide adenine dinucleotide (phosphate))

These are closely related coenzymes, both derived from nicotinic acid (vitamin B complex). Each molecule is electropositive (lacks one electron) and can carry an electron as well as a hydrogen atom. When a hydrogen pair is accepted, one hydrogen atom dissociates with its electron and proton:

$$\begin{array}{ccc} \text{H} & \longrightarrow & \text{H}^+ + e^- \\ \text{hydrogen} & & \text{proton electron} \\ \text{atom} & & \text{or} \\ & & \text{hydrogen ion} \end{array}$$

The other hydrogen atom remains whole and becomes attached to NAD(P).

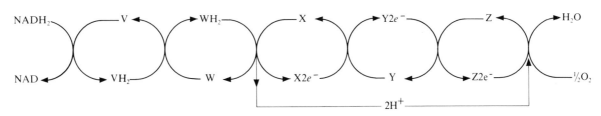

Fig 11.9 (above) *Diagrammatic representation of the respiratory chain which effects hydrogen-electron transfer. Points at which enough energy is liberated to form ATP are not shown. Not all carriers are shown*

Fig 11.10 (below) *Respiratory chain. Each cytochrome can only carry one electron, and it is thought that there are two ranks of cytochromes in each respiratory pathway. Only one is shown in the figure, but values have been doubled in order to obtain the correct end-product of the reaction. The electrons flow downhill in energy terms*

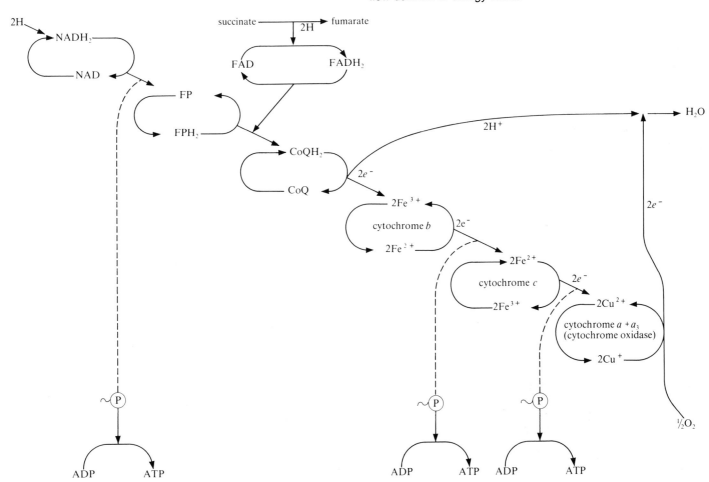

329

The overall reaction is:

$$NAD(P)^+ + H \quad [H^+ + e^-] \longrightarrow NAD(P)H + H^+$$

| whole atom | dissociated hydrogen atom | reduced coenzyme | free proton in medium |

or, more simply,

$$NAD(P) + H_2 \longrightarrow NAD(P)H_2$$

The free proton is used later to reoxidise the coenzyme when the hydrogen is released.

Flavoproteins

These are coenzymes derived from vitamin B_2. FAD (flavin adenine dinucleotide) is the prosthetic group, whilst the protein part of the molecule acts as an enzyme. In the respiratory chain the protein part acts as **NAD dehydrogenase** and catalyses the oxidation of reduced NAD. The hydrogen is carried by the flavoprotein in the form of whole atoms.

In the Krebs cycle the protein part of FAD acts as **succinic dehydrogenase**. It catalyses the oxidation of succinate to fumarate. Reduced FAD enters the respiratory chain at a point after the first ATP synthesis site. Therefore only two ATP will be formed from its reoxidation (fig 11.10).

Coenzyme Q

This has a 6C ring structure. It accepts hydrogen from flavoprotein and passes it on to cytochrome b.

Cytochromes

All are proteins of relatively low molecular mass. They possess tightly bound haems as prosthetic groups, and carry electrons rather than hydrogen atoms. The electron-carrying component of cytochromes is the iron of the haem group. It normally exists in its oxidised state (Fe^{3+}), but when it accepts an electron it is reduced to its ferrous state (Fe^{2+}). What happens is that each hydrogen atom passing from coenzyme Q dissociates into a hydrogen ion and an electron:

$$H \rightarrow H^+ + e^-$$

The electron is then accepted by an ion of iron:

$$Fe^{3+} + e^- \rightleftharpoons Fe^{2+}$$
oxidised   reduced

The hydrogen ions are temporarily deposited in the surrounding medium until they are required at the end of the respiratory chain.

The electron is passed from cytochrome b to c and finally to cytochrome $a + a_3$, a tight-knit complex of two cytochromes commonly known as cytochrome oxidase. This complex contains copper as well as iron, and undergoes a redox reaction when cytochrome a_3 finally passes on electrons to oxygen (fig 11.10). Only one electron at a time can be carried by a cytochrome and it is thought

that there are two ranks of cytochromes in each respiratory chain handling pairs of electrons:

$$H_2 \longrightarrow 2H^+ + 2e^- \begin{array}{c} \nearrow e^- + Fe^{3+} \longrightarrow Fe^{2+} \\ \searrow e^- + Fe^{3+} \longrightarrow Fe^{2+} \end{array}$$

With this information about hydrogen and electron carriers you should now be able to appreciate a much more detailed study of the respiratory chain which is given in fig 11.10.

Final analysis of aerobic respiration

(1) $C_6H_{12}O_6 + 6H_2O \xrightarrow[\text{Krebs cycle}]{\text{glycolysis}} 6CO_2 + 12H_2 + 4ATP$

(2) $12H_2 + 6O_2 \xrightarrow[\text{chain}]{\text{respiratory}} 12H_2O + 34ATP$

Add (1) and (2):

$$C_6H_{12}O_6 + 6O_2 \longrightarrow 6CO_2 + 6H_2O + 38ATP$$

11.3.8   Anaerobic respiration

A variety of microorganisms (anaerobes) employ anaerobic respiration as their major ATP-yielding process. Indeed, some bacteria are actually killed by substantial amounts of oxygen and of necessity have to live where there is no oxygen. They are termed **obligate anaerobes** (for example *Clostridium botulinum* and *C. tetani*).

Other organisms such as yeasts and alimentary canal parasites (such as tapeworms), can exist whether oxygen is available or not. These are called **facultative anaerobes**. Also some cells that are temporarily deprived of oxygen (such as muscle cells) are able to respire anaerobically. (See reference to computer program in section 16.4)

With no oxygen available to accept the hydrogen atoms released during glycolysis, an alternative acceptor must be used instead of $NADH_2$. Pyruvate becomes that acceptor, and depending on the metabolic pathways within the organisms or cells themselves the end-products of anaerobic respiration will either be ethanol and carbon dioxide (as in yeasts, for example):

$$CH_3COCOOH \longrightarrow CH_3CHO + CO_2$$
pyruvic acid   ethanal

$$CH_3CHO + NADH_2 \longrightarrow CH_3CH_2OH + NAD$$
ethanal   ethanol

(this process is termed **alcoholic fermentation**) or lactate, as in animals when cells are temporarily deprived of oxygen and in some bacteria, for example **lactate fermentation** in muscle cells:

$$CH_3COCOOH + NADH_2 \longrightarrow CH_3CHOHCOOH + NAD$$
pyruvic acid   lactic acid

A summary of the pathways of anaerobic respiration is given in fig 11.11.

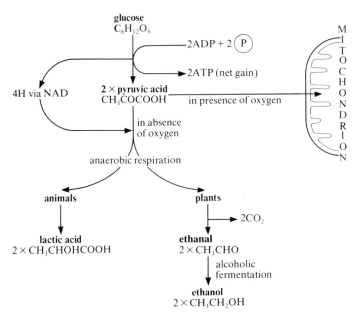

glucose
$C_6H_{12}O_6$

2ADP + 2 (P)

2ATP (net gain)

4H via NAD

$2 \times$ pyruvic acid
$CH_3COCOOH$ — in presence of oxygen

in absence of oxygen

anaerobic respiration

animals

plants

→ $2CO_2$

lactic acid
$2 \times CH_3CHOHCOOH$

ethanal
$2 \times CH_3CHO$

alcoholic fermentation

ethanol
$2 \times CH_3CH_2OH$

M I T O C H O N D R I O N

Fig 11.11 *Summary of pathways of anaerobic respiration*

No further ATP is produced by either process and so the energy yield per glucose molecule respired in this way is two ATP. In fact a considerable amount of energy remains trapped in ethanol and lactate. Hence when later compared with aerobic respiration (section 11.3.9) it must be regarded as an inefficient process (fig 11.11). The lactate has to be removed from muscle cells by the blood to prevent fatigue. It is reconverted to glucose and then to glycogen in the liver aerobically. Further details of this process are discussed in section 16.4.

11.3.9 Energy conversion efficiency of aerobic and anaerobic respiratory processes

Aerobic respiration

$C_6H_{12}O_6 + 6O_2 \rightarrow 6CO_2 + 6H_2O + 38ATP \quad \Delta G = -2880 \text{ kJ mol}^{-1}$

Therefore efficiency $= \dfrac{38 \times -30.6}{-2880} = 40.37\%$

(where -30.6 kJ represents the free energy liberated on hydrolysis of ATP to ADP).

Anaerobic respiration

(1) Yeast (alcoholic) fermentation

$C_6H_{12}O_6 \rightarrow 2C_2H_5OH + 2CO_2 + 2ATP \quad \Delta G = -210 \text{ kJ mol}^{-1}$

Therefore efficiency $= \dfrac{2 \times -30.6}{-210} = 29.14\%$

(2) Muscle glycolysis (lactate fermentation)

$C_6H_{12}O_6 \rightarrow 2CH_3CHOHCOOH + 2ATP \quad \Delta G = -150 \text{ kJ mol}^{-1}$
$\phantom{C_6H_{12}O_6 \rightarrow 2}$ lactic acid

Therefore efficiency $= \dfrac{2 \times -30.6}{-150} = 40.80\%$

Study of the above figures indicates that the efficiency of each system is relatively high when compared with petrol engines (25–30%) and steam engines (8–12%). The amount of energy captured as ATP during aerobic respiration is 19 times as much as for anaerobic respiration. This is because a great deal of energy remains locked within lactate and ethanol. The energy in ethanol is permanently unavailable to yeast, which clearly indicates that alcoholic fermentation is an inefficient energy-producing process. However, much of the energy locked in lactate may be liberated at a later stage if oxygen is made available. In the presence of oxygen, lactate is converted to pyruvate in the liver. Pyruvate then enters Krebs cycle and is fully oxidised to carbon dioxide and water, releasing many more ATP molecules in the process (section 17.4.8).

11.3.10 Fermentation in industry

Fermentation processes are commercial or experimental processes in which micro-organisms are cultured in containers, called **fermenters** or **bioreactors**, in a liquid or solid medium. The term was originally applied only to anaerobic cultures, as in brewing, but it is important to realise that it is now more loosely applied to both anaerobic and aerobic processes, such as the growth of *Penicillium* in the penicillin industry, involving the culture of micro-organisms in artificial vessels.

Fig 11.12 shows a typical fermenter (one that is sealed from the atmosphere during operation) and gives some information on its use. The contents of most fermenters are stirred during operation, but this is not always the case, as with the production of the single cell protein 'Pruteen' by ICI, where air introduced at high velocity at the bottom of the vessel is used to achieve mixing. The **product** is either the cells themselves (biomass) or some useful cell product. *All* operations must be carried out under *sterile conditions* to avoid contamination of the culture. In addition, all inlets and outlets of the fermenter must be capable of being kept sterile. The fermenter and the medium used are sterilised before use, either together or separately. Stock cultures of the organism to be used in the fermentation are kept in an inactive form (for example stored frozen). A sample is re-activated, grown up to sufficient bulk using aseptic techniques (**scale-up**), and then added to the fermenter, a process known as **inoculation**. Once inside the fermenter, the organism grows and multiplies, using the nutrient medium.

Two basic types of fermentation are possible, **batch fermentation** (or **closed system**) and **continuous culture** (or **open system**). In the more common batch fermentation the process is stopped once sufficient product has been formed. The contents of the fermenter are removed, the product isolated, the micro-organism discarded and the fermenter is then cleaned and set up for a fresh batch. Continuous culture involves continuous long-term operation over many weeks, during which nutrient medium is added as fast as it is used, and the overflow is harvested. Continuous

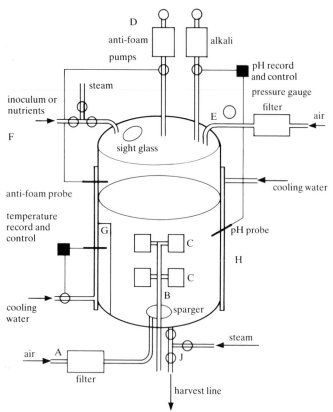

Fig 11.12 *Typical enclosed, aerated, agitated, cylindrical fermentation vessel (bioreactor) (adapted from S.B. Primrose (1987) Modern Biochemistry, Blackwell.)*

Size of the vessel is very variable, ranging from 1 dm³ (experimental) to 500 000 dm³ for commercial production. Shape and material used in construction are also variable, although cylindrical, stainless steel vessels are common

Key:

A *Air main – most fermentations are aerobic, requiring large volumes of sterile air. The 'sparger' is a specially designed part that releases air. Air bubbles may help the mixing process, provide oxygen for aerobic respiration and aid release of volatile waste products.*

B *Stirrer shaft – present in most fermenters. Agitation needed to*

   (i) increase rate at which O₂ dissolves;
   (ii) maintain diffusion gradients of O₂ and nutrients into cells and products out of cells;
   (iii) prevent clumping of cells or mycelia of fungi;
   (iv) promote heat exchange between medium and cooling surfaces.
   Shaft bearings must be strong and sterile.

C *Stirrer paddles – usually flat and vertical.*

D *Alkali and anti-foam inlets – alkali added if, as is usual, acidity increases during fermentation (to maintain a constant pH). Aeration and agitation generate foam, particularly from proteins, and prevent escape of contents through the exhaust, so anti-foaming agents are added.*

E *Exhaust – contents of fermenter are under pressure; therefore pressure gauge and safety valve attached.*

F *Top of fermenter has a part for addition of medium, inoculum (micro-organism), access for cleaning etc.*

G *Baffle – vertical fin on inside wall; helps to prevent vortex formation as culture is rotated.*

H *Cooling jacket – reduces temperature; needed because culture generates heat.*

J *Harvest – samples may be taken during fermentation in order to monitor process.*

culture has found only limited application, but is used for the production of single cell protein where a large biomass of cells is required. ICI, for example, produce a single cell protein (SCP), that is a microbial protein, called 'Pruteen' from the bacterium *Methylophilus methylotrophus* (see also section 2.5.4). The cell is provided with methanol, oxygen (in air), nitrogen in the form of ammonia and inorganic nutrients such as calcium and phosphorus. The plant at Billingham has the largest continuous culture fermenter in the world, with a $1\frac{1}{2}$ million litre capacity. The temperature must be carefully monitored within the range 30–40 °C and the pH kept at 6.7. The final product, Pruteen, is 72% protein and also has a high vitamin content. It is ideal for use in animal feeds. The fermenter has been run continuously for as long as 100 days and can produce 150 tonnes per day. Unfortunately, it is not economic to produce SCP for animal feeds, other sources of protein being cheaper at present in developed countries. Developing countries cannot afford to run such large-scale technology, and do not have the necessary expertise.

Downstream processing

Downstream processing is the name given to the phase following fermentation when the desired product is recovered and purified. Many techniques are used, including precipitation, chromatography (for example with streptomycin and interferon), electrophoresis, centrifugation, distillation (as with propanone, ethanoic acid, and spirits such as whisky), concentration, drying, solvent extraction (as with penicillin) and filtration (as with beer and soft drinks). As an indication of the importance of downstream processing, it involves over 90% of the 200 staff employed by Eli Lilly in their human insulin producing plant.

An enormous range of products is now produced by fermentation processes, some of which are shown in table 11.3. Living cells have the advantage over traditional chemical technology in that they can operate at lower temperatures, neutral pH, produce higher yields, show greater specificity, include production of particular isomers, and very often produce chemicals such as antibiotics and hormones that cannot be produced easily (if at all) by any other means. However, there are specialist techniques, such as aseptic techniques and complex methods of separation, which can make the process more technically demanding.

11.4   Shuttle systems

Although the sequence of events for aerobic respiration shown in this text indicates that 38 ATP are produced for each glucose molecule oxidised, it must be stated that the total number of ATP molecules produced in aerobic respiration can vary according to the tissue involved. Two NADH₂ complexes are produced during glycolysis in the cytoplasm. Cytoplasmic NADH₂ cannot pass through the mitochondrial membrane and therefore

Table 11.3 Fermentation products according to industrial sectors. Adapted from table 3.1, John E. Smith (1988), *Biotechnology*, New Studies in Biology, 2nd ed. Arnold.

Sector	Activities
Chemicals	
Organic (bulk)	Ethanol, acetone, butanol
Organic (fine)	Enzymes
	Perfumeries
	Polymers (mainly polysaccharides)
Inorganic	Metal beneficiation, bioaccumulation and leaching (Cu, U)
Pharmaceuticals	Diagnostic agents (enzymes, monoclonal antibodies)
	Enzyme inhibitors
	Steroids
	Vaccines
	Antibiotics, e.g. penicillin, streptomycin
Energy	Ethanol (gasohol)
	Methane (biogas)
Food	Dairy products (cheeses, yoghurts, fish and meat products)
	Beverages (alcoholic, tea and coffee)
	Baker's yeast
	Food additives (antioxidants, colours, flavours, stabilisers)
	Novel foods (soy sauce, tempeh, miso)
	Mushroom products
	Amino acids, vitamins
	Starch products
	Glucose and high-fructose syrups
	Functional modifications of proteins, pectins
Agriculture	Animal feedstuffs (SCP)
	Veterinary vaccines
	Composting processes: silage for cattle fodder
	Microbial pesticides
	Rhizobium and other N-fixing bacterial inoculants
	Mycorrhizal inoculants
	Plant cell and tissue culture (vegetative propagation, embryo production, genetic improvement)

the electrons derived from glycolysis have to enter via indirect routes or **shuttles**. According to which shuttle operates, the number of ATP molecules produced from cytoplasmic $NADH_2$ can be four or six, thus making the total either 36 or 38 ATP. The principle of this shuttle is illustrated in fig 11.13. X acts as a carrier molecule, carrying hydrogen from the cytoplasm into the mitochondrion. It can pass through the mitochondrial membranes, whereas $NADH_2$ cannot.

$NAD \rightarrow NADH_2$ in heart and liver cells, giving three ATP on reoxidation.
$FP \rightarrow FPH_2$ in muscle and nerve cells, giving two ATP on reoxidation.
Therefore 38 ATP are formed in heart and liver cells, whilst 36 ATP are formed in muscle and nerve cells.

The shuttles constantly transfer electrons from the cytoplasm to mitochondria and at the same time reoxidise cytoplasmic $NADH_2$. This effectively prevents a build-up of hydrogen atoms in the cytoplasm and explains why no lactate accumulates during aerobic respiration.

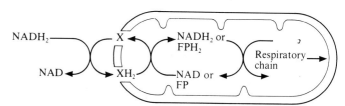

Fig 11.13 *Model of shuttle system (see text for full explanation). NAD, nicotinamide adenine dinucleotide; FP, flavoprotein*

Experiment 11.1: To investigate the oxidation of a Krebs cycle intermediate

The most efficient way of releasing energy from a substrate and storing this for future use is by a series of smaller reactions, each one reversible and enzyme-mediated. One of the intermediate reactions involved is the oxidation, by removal of hydrogen, of succinic acid to fumarate.

There are substances which accept such hydrogen atoms and, in doing so, change colour. One example is 2:6 dichlorophenolindophenol (DCPIP). It is blue in its oxidised form but loses its colour when reduced.

If the coloured form of DCPIP is decolorised by a tissue extract, one explanation could be that it has accepted hydrogen atoms from succinic acid. If the rate of decolorisation increased when succinic acid was added this would tend to confirm the hypothesis that DCPIP was a hydrogen acceptor of atoms from succinic acid.

As most living processes are governed by enzymes, these must be present before the oxidation will occur. The enzyme succinic dehydrogenase reduces succinic acid and further experiments could reveal the presence of the enzyme. In this experiment mitochondria are isolated from germinating mung bean seedlings and a suspension of these used as a source of enzyme. It is essential to carry out the extraction as quickly as possible. Once cells are disrupted, further metabolism is short-lived.

The experiment is divided into two parts. The first part consists of the extraction of the enzyme required and the second uses the extracted enzyme to oxidise succinic acid. DCPIP is used to indicate that a reaction has or has not occurred.

Ideally all the apparatus concerned with the first part of this experiment (the preparation of the enzyme extract) should be placed in a refrigerator for at least one hour before it is required for the experiment.

Materials

4 centrifuge tubes (capacity 15 cm³)
2 glass rods
2×10 cm³ graduated pipettes
2×1000 cm³ beakers (polythene preferably)
ice
salt
mung beans

test-tubes and rack
1 × 1 cm³ graduated pipette
stopclock

Solutions (see notes)
buffer/sucrose solution
buffer/sucrose + succinic acid solution (succinate solution)
0.1% DCPIP (solution made up in buffer/sucrose solution)
distilled water

Method

(1) Germinate some mung beans by placing the dry beans on damp cotton wool in the dark for 3–4 days (24 beans are needed for the whole experiment per student or group).
(2) Prepare an ice bath by placing ice in a 1000 cm³ polythene beaker and adding a little salt to lower the temperature further.
(3) Place the flask containing buffer/sucrose solution and two centrifuge tubes in the ice bucket.
(4) Take 12 mung beans and remove their testas and radicles.
(5) Place six beans in each centrifuge tube.
(6) Add 1 cm³ of buffer/sucrose solution which does not contain succinic acid to each tube.
(7) Crush the beans thoroughly using a cold glass rod, keeping the tubes in the ice bucket.
(8) Add a further 10 cm³ of buffer/sucrose solution to each centrifuge tube.
(9) Place the centrifuge tubes on opposite sides of the centrifuge head and spin the tubes at maximum speed for 3 min.
(10) Place the centrifuge tubes back in the ice bucket.
(11) Pipette 15 cm³ of distilled water into a test-tube and mark the position of the meniscus.
(12) Pour off the distilled water and carefully fill the tube to the mark with supernatant from the centrifuge tubes.
(13) The next step must be carried out very quickly: add 0.5 cm³ of DCPIP solution to the reaction tube and mix the contents by placing a thumb over the end of the tube and inverting the tube.
(14) Start the stopclock as the solutions are mixing.
(15) Note the colour of the solution after 20 min.
(16) Repeat the entire experiment using the buffer/sucrose solution containing succinic acid.

The experiment can be monitored colorimetrically. This is done as follows.

(1) Using a red filter, switch on the colorimeter and allow it to warm up for 5 min.
(2) Add 0.5 cm³ of DCPIP solution to 15 cm³ of supernatant as before.
(3) Mix the solutions and start the stopclock.
(4) Place the tube in the colorimeter and adjust the needle to 0% transmission.
(5) Take readings after 1, 2, 5, 10 and 20 min.

(6) Repeat the experiment using buffer/sucrose containing succinic acid.
(7) Plot a graph of percentage transmission (vertical axis) against time.
(8) Draw your own conclusions from the results that you obtain.

Notes on how to make solutions

Buffer/sucrose solution (100 cm³)

disodium hydrogen phosphate (Na_2HPO_4)	0.76 g
potassium dihydrogen phosphate (KH_2PO_4)	0.18 g
sucrose	13.60 g
magnesium sulphate	0.10 g

Buffer/sucrose + succinic acid (100 cm³)

As for buffer/sucrose, plus

succinic acid	1.36 g
sodium hydrogencarbonate	1.68 g

The best method for making up these solutions is to make up enough buffer/sucrose solution for both halves of the experiment (solutions are made up in distilled water). Divide the solution into two and add succinic acid and sodium hydrogencarbonate (in the correct concentration) to one half.

There will be effervescence when succinic acid and sodium hydrogencarbonate are added to the buffer/sucrose solution. The solution should be shaken well to get rid of as much carbon dioxide as possible as this could affect the experiment.

DCPIP solution

Use 0.1 g of dichlorophenolindophenol in 10 cm³ of buffer/sucrose solution (without succinic acid for both experiments). The solid does not dissolve very well and so after thorough mixing the suspension should be filtered.

11.5 Mitochondria

Mitochondria are present in all eukaryotic cells and are the major sites of aerobic respiratory activity within cells. They were first seen as granules in muscle cells by Kolliker in 1850. Later, in 1898, Michaelis demonstrated that they played a significant role in respiration by showing experimentally that they produced a colour in redox dyes.

The number of mitochondria per cell varies considerably and depends on the type of organism and nature of the cell. Cells with high energy requirements possess large numbers of mitochondria (for example, liver cells contain upwards of 1 000 mitochondria) whilst less active cells possess far fewer. Mitochondrial shape and size are also tremendously variable. They may be spiral, spherical, elongate, cup-shaped and even branched, and are usually larger in active cells than in less active ones. Their length ranges from 1.5–10 μm, and width 0.25–1.00 μm, but their diameter does not exceed 1 μm.

11.6 Why should the diameter of mitochondria remain fairly constant when the length is so variable?

Mitochondria are able to change shape, and some are able to move to areas in the cell where a lot of activity is taking place. This is facilitated by **cytoplasmic streaming** and provides the cell with a large concentration of mitochondria in areas where ATP need is greater. Other mitochondria assume a more fixed position (as in insect flight muscle, fig 11.14).

11.5.1   Structure of mitochondria

Mitochondria can be extracted from cells in pure fractions by cell homogenisation and ultracentrifugation techniques. Once isolated they may be examined with an electron microscope using various techniques such as sectioning or negative staining. Each mitochondrion is bounded by two membranes (an envelope), the outer one being separated from the inner by a space some 6–10 nm wide. A semi-rigid matrix is enclosed by the inner membrane which itself is folded inwards into a number of shelf-like **cristae** (fig 11.15). Techniques using ultrasonic vibration and detergent action can be used to separate the two membranes, making it easier to study their individual structure and activity. Even so, knowledge of the outer membrane is still scarce. It is said to be permeable to substances with molecular weights below 21 000, and that such molecules are able to diffuse across it. The cristae of the inner membrane effectively increase its surface area, serving to provide abundant space for **multi-enzyme systems**, and greater access to enzymes present in the matrix. The inner membrane exhibits selectivity over what materials are allowed through it, and it is known that active transport mechanisms involving **translocase** enzymes are responsible for the movement of ADP and ATP across it. Negative staining techniques which stain the space around structures rather than the structures themselves (fig 11.15b) indicate the presence of **elementary particles** on the matrix side of the inner membrane. Each particle consists of a head piece, stalk and base. Whilst the photograph (fig 11.15e) suggests that the particles stick out from the membrane into the matrix, it is generally recognised that this is an artefact produced by the method of preparation, and that probably the particles are tucked into the membrane. The head piece is associated with ATP synthesis and is a coupling enzyme, ATPase (formerly termed F_1), which acts to link the phosphorylation of ADP to the respiratory chain. At the base of the particle, and extending through the inner membrane, are the components of the respiratory chain itself. They are arranged in precise positions relative to each other. The mitochondrial matrix contains most of the enzymes controlling the Krebs cycle and fatty acid oxidation. In addition, mitochondrial

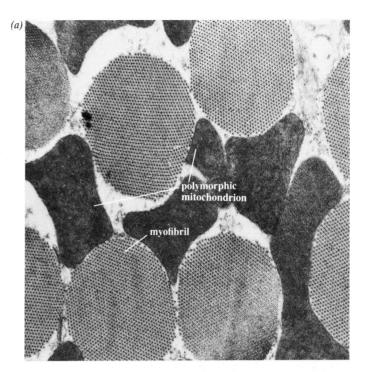

(a)

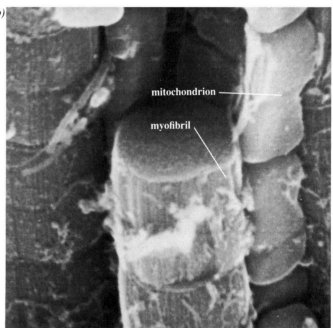

(b)

Fig 11.14   *(a) Transmission electron micrograph, and (b) scanning electron micrograph of the flight muscle from the house fly* Musca *to show that each myofibril is surrounded by polymorphic mitochondria.*

DNA, RNA and ribosomes are present as well as a variety of small proteins (section 7.2.12).

11.7 What chemical substances would be exchanged between the cytoplasm and the mitochondria? Indicate whether they are entering or leaving the mitochondria.

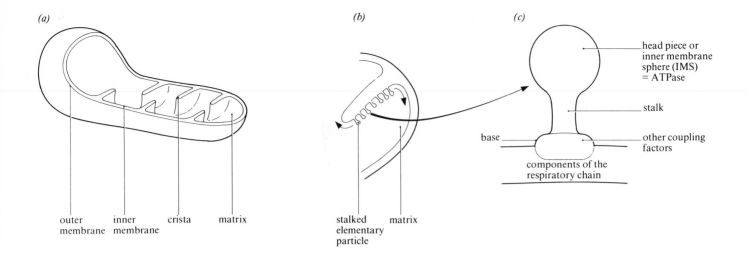

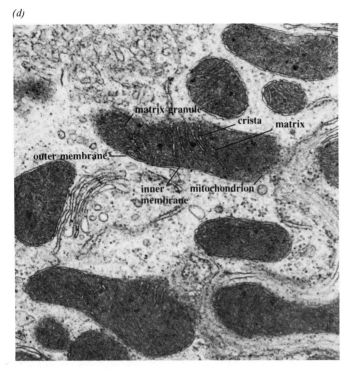

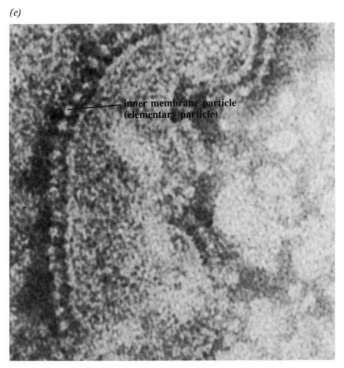

Fig 11.15 *Structure of mitochondrion: (a) Diagram of mitochondrion. (b) Diagram of crista showing inner membrane particles. (c) Structure of inner membrane particle. (d) Low power electron micrograph of mitochondrion.*

(e) Transmission electron micrograph of inner membrane particles (F_1-F_0 ATPase) from osmotically disrupted mitochondria of the house-fly (Musca)

11.5.2 Mitochondrial assembly

Whilst mitochondrial DNA carries enough information for the synthesis of about 30 proteins, this is not enough for it to be able to build all the proteins that are required to make a new mitochondrion. Therefore reproduction of a mitochondrion must rely to some extent on nuclear DNA, cytoplasmic enzymes and other molecules supplied by cells. Fig 11.16 summarises current information concerning the interaction between the mitochondrion and the rest of the cell during mitochondrial assembly.

11.5.3 Evolution of mitochondria – the endosymbiont theory

It is suggested that mitochondria were originally independent prokaryotic, bacteria-like organisms which gained access, by accident, to a host cell and entered into a succesful mutualistic (**symbiotic**) union with it. Presumably conditions in the host cell were favourable for the prokaryote, whilst in return the prokaryote provided a greatly increased capacity to manufacture ATP, and conferred upon the host an ability to respire aerobically. A number of observations support this theory. First,

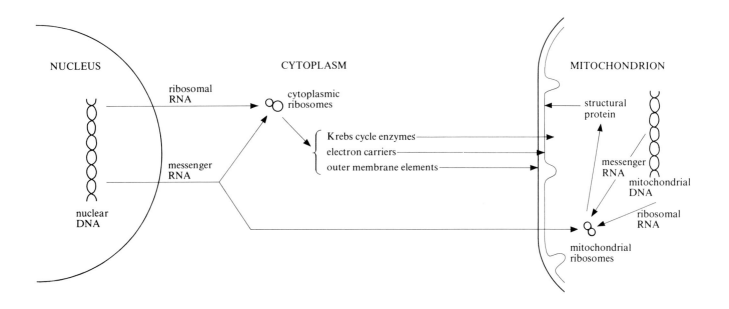

Fig 11.16 *Control of mitochondrial assembly (From Tribe & Whittaker,* Chloroplasts and mitochondria, *Series in Biology, No. 31, Arnold.)*

mitochondrial DNA is generally circular (fig 11.17). This is very much like that found in present-day bacteria. Secondly, mitochondrial ribosomes are smaller than those of the cytoplasm but equivalent in size to bacterial ribosomes. Thirdly, mitochondrial and bacterial synthesis mechanisms are sensitive to different antibiotics when compared with the cytoplasmic mechanism, for example chloramphenicol and streptomycin inhibit mitochondrial and bacterial protein synthesis, whereas cycloheximide inhibits cytoplasmic protein synthesis (see section 9.3.1 for endosymbiotic theory and chloroplasts).

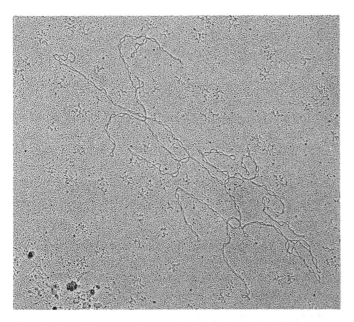

Fig 11.17 *Electron micrograph of mitochondrial DNA from the brewer's yeast* Saccharomyces carlsbergensis. *The molecule is a 'supercoiled' circle of double-strand DNA with a circumference of 26 micrometres. It is made up of some 75 000 nucleotides*

11.5.4 ATP synthesis

The mechanism for the coupling of ATP synthesis to electron transport has been the subject of intensive research for over 30 years. Experimental evidence is now overwhelmingly in favour of the **chemiosmotic** hypothesis (now given the status of a theory) which was unpopular when first postulated by the British biochemist Mitchell in 1961. Mitchell was awarded the Nobel Prize for his work in 1978. The theory applies to both mitochondria and chloroplasts (and to ATP generation in bacteria). Mitchell argued that ATP synthesis was intricately associated with the ways in which electrons and protons are passed along the respiratory chain. Certain conditions are necessary if the theory is correct. They are summarised as follows.

(1) The inner mitochondrial membrane must be intact and impermeable to movement of protons (hydrogen ions) from the outside to the inside.

(2) Respiratory chain activity results in protons being drawn into the electron transport chain from the internal matrix and then removed to the space between the inner and outer membranes of the mitochondrion. (From here protons pass to the outside of the mitochondrion since the outer membrane is freely permeable to small molecules.)

(3) Movement of the protons to the outside of the mitochondrion causes an accumulation of hydrogen ions and creates a pH gradient across the mitochondrial inner membrane. It is thought that this occurs because the electron carriers involved are located in the inner

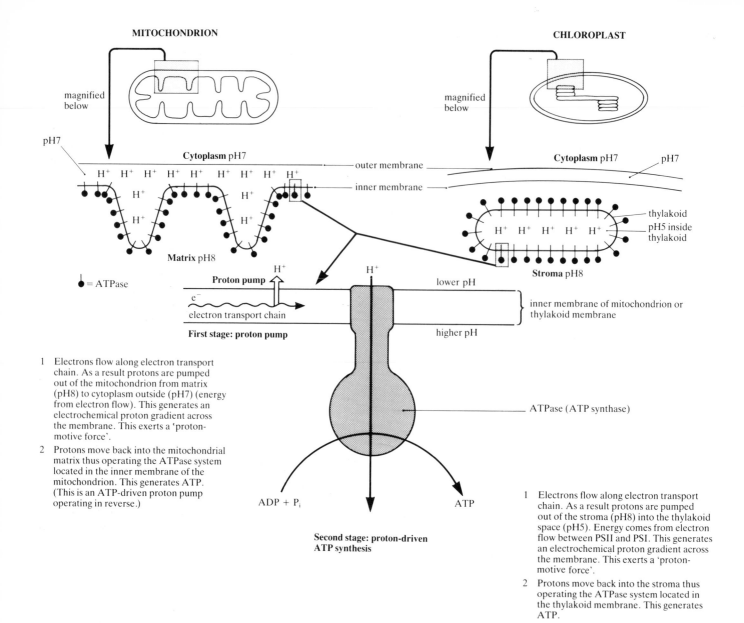

MITOCHONDRION

CHLOROPLAST

magnified below

magnified below

pH7

Cytoplasm pH7

outer membrane

inner membrane

H⁺ H⁺ H⁺ H⁺ H⁺ H⁺ H⁺ H⁺ H⁺

H⁺

H⁺

H⁺

H⁺

Matrix pH8

= ATPase

Cytoplasm pH7

pH7

thylakoid
pH5 inside
thylakoid

H⁺ H⁺ H⁺ H⁺ H⁺

Stroma pH8

Proton pump H⁺

e⁻
electron transport chain

First stage: proton pump

1 Electrons flow along electron transport chain. As a result protons are pumped out of the mitochondrion from matrix (pH8) to cytoplasm outside (pH7) (energy from electron flow). This generates an electrochemical proton gradient across the membrane. This exerts a 'proton-motive force'.

2 Protons move back into the mitochondrial matrix thus operating the ATPase system located in the inner membrane of the mitochondrion. This generates ATP. (This is an ATP-driven proton pump operating in reverse.)

H⁺

lower pH

inner membrane of mitochondrion or thylakoid membrane

higher pH

ATPase (ATP synthase)

ADP + P$_i$

ATP

Second stage: proton-driven ATP synthesis

1 Electrons flow along electron transport chain. As a result protons are pumped out of the stroma (pH8) into the thylakoid space (pH5). Energy comes from electron flow between PSII and PSI. This generates an electrochemical proton gradient across the membrane. This exerts a 'proton-motive force'.

2 Protons move back into the stroma thus operating the ATPase system located in the thylakoid membrane. This generates ATP.

Fig 11.18 *ATP generation by chemiosmosis in mitochondria and chloroplasts (Mitchell's hypothesis)*

membrane in such a position as to allow only uptake from the inside and loss to the outside.

(4) Normally a pH gradient could not be maintained, as hydrogen ions would pass back into the mitochondrion by diffusion. Therefore the maintenance of this gradient is energy-requiring. The energy obtained from the transfer of electrons down the electron (respiratory) transport chain is thought to provide the energy needed.

(5) The combined pH gradient and electrical potential across the membrane is referred to as the **electrochemical proton gradient**. This gradient can be used to do work; this is usually referred to as **proton-motive force (PMF)**. In respiration (and photosynthesis)

protons move back across the membrane, down their concentration gradients, through specific sites where the enzyme ATPase, otherwise known as ATP synthase, is located (see section 11.5.1). The energy released by the protons is used to drive ATP synthesis (see fig 11.18).

Proton-motive force can power certain processes other than ATP synthesis, for example transport of small molecules across the mitochondrial inner membrane and rotation of bacterial flagella.

Mitchell's hypothesis explains why the membrane must be intact (because the proteins required are located in the membrane and alteration of the structure of the membrane would also alter the protein positions and structure). It also explains why the membrane needs to be impermeable to hydrogen ions (from outside to inside), because if the membrane was completely permeable a pH gradient would not form, and a pH gradient is vital to this theory.

11.5.5 Other respiratory pathways

Pentose phosphate shunt (hexose monophosphate shunt)

This particular pathway requires oxygen and is a major source of 5C sugars which are components of important nucleotides (ATP, NAD, FAD) and nucleic acids. The shunt may operate simultaneously alongside the normal glycolytic pathway and in different cells can contribute between 10 and 90% of energy supplied by carbohydrate respiration.

In essence, six molecules of glucose-6-phosphate are initially dehydrogenated and then decarboxylated. NADP acts as the hydrogen acceptor molecule. Six molecules of ribulose-5-phosphate (5C sugar phosphate) are the result of these processes, and six carbon dioxide molecules are formed as by-products. The ribulose-5-phosphate molecules then undergo an intricate series of reactions which finally result in the resynthesis of five molecules of glucose-6-phosphate. Glyceraldehyde-3-phosphate is also formed and may be redirected into the glycolytic pathway, converted into pyruvate and finally passed into the Krebs cycle. The net result of the shunt is a yield of 36 ATP. This compares favourably with the 38 ATP formed by the glycolytic and Krebs pathway. Fewer reactions are involved in the shunt and consequently fewer enzymes required. The overall reaction is:

$$6 \times \text{glucose-6-P} + 12\text{NADP}^+ + 6\text{H}_2\text{O}$$
$$\downarrow$$
$$5 \times \text{glucose-6-P} + 12(\text{NADPH} + \text{H}^+) + 6\text{CO}_2 + \text{P}$$
$$6\text{O}_2 \longrightarrow \downarrow$$
$$12\text{NADP}^+ + 12\text{H}_2\text{O}$$

The shunt also functions to generate NADPH_2 which serves as a hydrogen and electron donor (a reducing agent) in the synthesis of a number of biochemicals. For instance, in adipose tissue the shunt operates to generate large amounts of NADPH_2 which are in turn consumed in the reduction of acetyl CoA to fatty acids during lipid synthesis.

Glyoxylate cycle

This occurs in seeds that possess tissues rich in fat, and enables stored fat to be converted into carbohydrate. The enzymes responsible for the cycle are mostly contained in organelles called **glyoxysomes** (a type of peroxisome because they contain catalase), which are active when seeds are germinating, although the cycle does occur in other organelles. However, once all the fat reserves have been consumed, the cycle ceases to operate.

During germination fats are hydrolysed to fatty acids and glycerol. Subsequently fatty acids are broken down by the process of β-oxidation (section 11.5.6) producing substantial amounts of acetyl CoA. Acetyl CoA enters the Krebs cycle as usual, but the glyoxylate cycle is added to the reactions of the Krebs cycle. For each turn of the glyoxylate cycle the overall reaction shown below occurs:

$$2 \text{ acetyl CoA} \longrightarrow \text{succinate} + 2\text{H} + 2\text{CoA}$$
$$2 \times 2\text{C} \qquad\qquad\qquad 4\text{C}$$

The pair of hydrogen atoms released passes to oxygen in the respiratory chain causing production of ATP. Succinate may be used to supply the carbon skeleton in the manufacture of a range of compounds. Thus the glyoxylate cycle uses a 2C compound in the form of acetyl CoA as a fuel and provides energy and 4C intermediates for biological synthesis.

11.5.6 Fat as a respiratory substrate

Some animal tissues such as liver, and seeds possessing large deposits of fat, are able to use fat as a respiratory substrate without first converting it to carbohydrate. Initially fat is hydrolysed by enzymes called lipases into fatty acids and glycerol.

Glycerol

Glycerol is first phosphorylated by ATP into glycerol phosphate and then dehydrogenated by NAD to the sugar dihydroxyacetone phosphate. This is next converted into its isomer glyceraldehyde-3-phosphate (fig 11.19). As can be seen the process consumes one ATP, but yields three ATP when hydrogen is transferred to oxygen along the respiratory chain. Glyceraldehyde-3-phosphate is subsequently incorporated into the glycolysis pathway and Krebs cycle, liberating a further 17 ATP. Therefore the yield per one molecule of glycerol aerobically respired is $20 - 1 = 19$ ATP.

Fatty acid

Each fatty acid molecule is oxidised by a process called β-**oxidation** which, in essence, involves 2C fragments of acetyl coA being split off from the acid so that the long fatty acid molecule is shortened 2C atoms at a time. Each acetyl CoA formed can enter Krebs cycle as usual to be oxidised to carbon dioxide and water (fig 11.22). The process takes place in the matrix of the mitochondrion. A great deal of energy is released from each fatty acid molecule thus oxidised, for instance 147 ATP per molecule of stearate. Not surprisingly, therefore, fatty acids are important energy sources, contributing, for example, at least half the normal energy requirements of heart muscle, resting skeletal muscle, liver and kidneys.

11.5.7 Protein as a respiratory substrate

Very occasionally, when carbohydrate and fat reserves have been exhausted, proteins are utilised as respiratory material. They are first hydrolysed into their constituent amino acids and then deaminated (their amino groups are removed). This may occur in two ways, oxidative deamination or transamination.

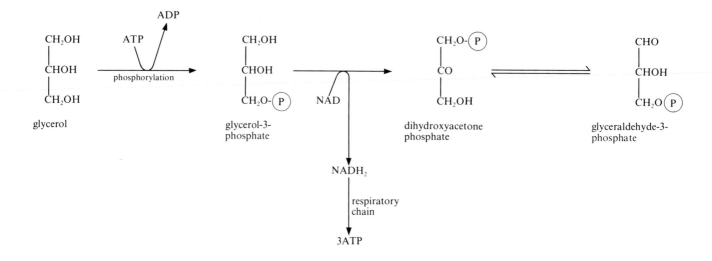

Fig 11.19 *Conversion of glycerol into glyceraldehyde-3-phosphate*

Oxidative deamination

This takes place in vertebrate liver cells. An ammonia molecule is removed from the amino acid by dehydrogenation and hydrolysis. Later it is excreted as ammonia, uric acid or urea depending on the animal concerned (section 18.5). The deaminated amino acid is an α-keto acid. According to the nature of its R group, it may be respired like a carbohydrate (**glucogenic**) or via the fatty acid pathway (**ketogenic**) (fig 11.20).

Transamination

This occurs in all cells and is controlled by **transaminase** enzymes. It is the transfer of an amino group from an amino acid to a keto acid. In this way one amino acid can be converted into another. The process also produces α-**keto acids** which are able to enter the normal respiratory pathways. Some examples are shown in fig 11.21. A summary of the major metabolic pathways in respiration is shown in fig 11.22.

11.6 Gaseous exchange

Whether aerobic or anaerobic respiration is occurring, the constant passage of gases between organisms and their environments has to be maintained. Aerobes require oxygen for oxidation of foodstuffs and energy release, whilst aerobes and most anaerobes must expel carbon dioxide, a waste product of respiration. Exchange of carbon dioxide and oxygen between environment and organism is termed **gaseous exchange**, and the area where gaseous exchange actually takes place is called the **respiratory surface**. Gaseous exchange takes place in all organisms by the physical process of **diffusion**. In order for this to occur effectively the respiratory surface must satisfy the following criteria:

it must be **permeable**, so that gases can pass through;

it must be **thin**, because diffusion is only efficient over distances of 1 mm or less;

it should possess a **large surface area** so that sufficient amounts of gases are able to be exchanged according to the organism's need.

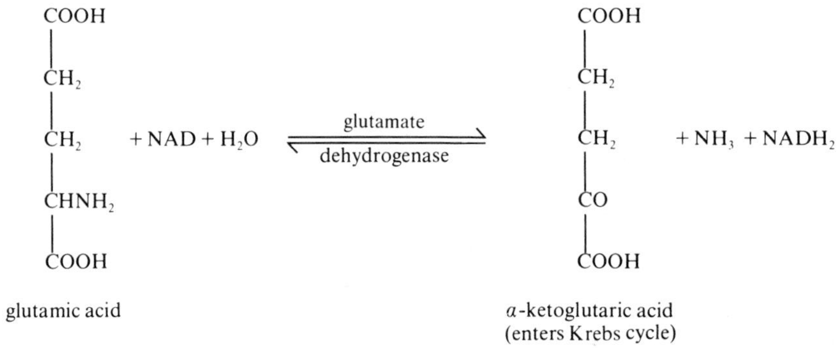

$$\begin{array}{c}\text{COOH}\\|\\\text{CH}_2\\|\\\text{CH}_2\\|\\\text{CH}_2\\|\\\text{CHNH}_2\\|\\\text{COOH}\end{array} + NAD + H_2O \underset{\text{dehydrogenase}}{\overset{\text{glutamate}}{\rightleftharpoons}} \begin{array}{c}\text{COOH}\\|\\\text{CH}_2\\|\\\text{CH}_2\\|\\\text{CO}\\|\\\text{COOH}\end{array} + NH_3 + NADH_2$$

glutamic acid

α-ketoglutaric acid
(enters Krebs cycle)

Fig 11.20 *Oxidative deamination of glutamic acid*

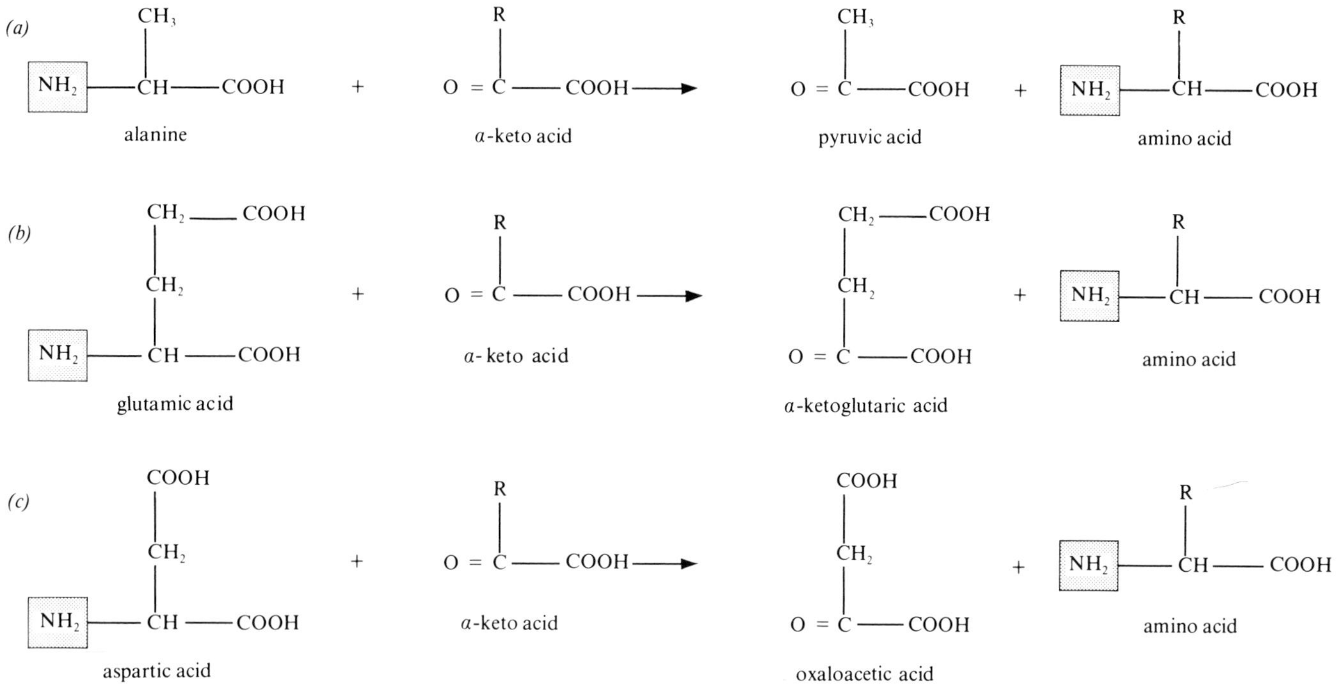

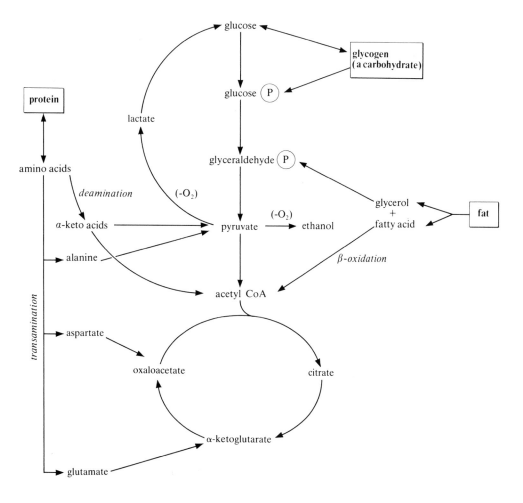

Fig 11.21 *Examples of transaminations*

Fig 11.22 *Summary of major metabolic pathways in respiration*

Organisms acquire their oxygen either direct from the atmosphere or from oxygen dissolved in water. There are marked differences in the oxygen content of air and water. A unit volume of air contains far more oxygen in it than an equal volume of water. Therefore it follows that an aquatic organism such as a fish must pass a correspondingly much greater volume of water over its gaseous exchange surface than a terrestrial vertebrate passes air in order to obtain sufficient oxygen for its metabolic needs.

11.6.1 Protoctista

The rhizopod *Amoeba proteus* measures less than 1 mm in diameter and possesses a large surface area to volume ratio. Diffusion of gases occurs over the whole surface of the animal via the cell membrane, and is enough to satisfy its metabolic needs.

11.6.2 Cnidaria

In the multicellular, diploblastic *Hydra* and *Obelia*, all cells are in contact with the surrounding aquatic medium, and each is able to exchange gases sufficient for its own needs through the cell membrane adjacent to the surrounding water.

11.6.3 Platyhelminthes

A free-living platyhelminth, such as *Planaria*, acquires the oxygen it needs by means of diffusion over its body surface. This is facilitated by the worm's extremely flattened body (no more than 0.06 cm thick) which increases the surface area to volume ratio, and by the fact that it generally lives in well-aerated streams or ponds.

Many platyhelminths such as the tapeworm *Taenia* are internal parasites, surviving in regions where there is little available oxygen. In this case they operate as **anaerobes**. Their size or shape is not limited by any need for oxygen diffusion, although they still retain a large surface area to volume ratio.

> **11.8** What other advantage does a large surface area to volume ratio confer on the platyhelminths?

11.6.4 The need for special respiratory structures and pigments

As animals increase in size, so their surface area to volume ratio decreases, rendering simple diffusion over the body surface inadequate to supply oxygen to cells of the organism not in direct contact with the surrounding medium. Also the increased metabolic activity of many of these larger animals increases their rate of oxygen consumption.

In order to cope with the increased demand, certain regions of the body have developed into specialised respiratory surfaces. Different organisms possess different types of gaseous exchange surface. Each is designed to work efficiently in a specific environment. They can be classified as shown in fig 11.23. Generally their surface area is greatly increased and often associated with a transport system, such as a blood vascular system. The possession of a transport system puts the respiratory surface in contact with all other tissues of the organism and enables oxygen and carbon dioxide to be continuously exchanged between the respiratory surface and cells. The presence of a respiratory pigment in the blood further increases the efficiency of the blood's oxygen-carrying capacity. In addition there may be special ventilation movements which assist in ensuring a rapid exchange of gases between the animal and the surrounding environment by maintaining steep diffusion gradients.

Respiratory pigments

Blood that contains any form of respiratory pigment is a more efficient oxygen carrier than blood without one. This is because the pigment permits far greater amounts of oxygen to be taken up and transported. The pigment may be in the blood plasma or enclosed in specific cells. It is interesting to note that the relative molecular masses of pigments confined to cells are relatively low when compared with those of pigments in plasma. In fact, the plasma-based pigments are aggregates of many small molecules behaving as one large molecule. This arrangement permits an increase in the amount of pigment in the blood without increasing the number of dissolved molecules in solution.

> **11.9** Why is this important?
>
> **11.10** What is the advantage of confining the pigment to cells?

All known respiratory pigments are linked to protein molecules. They are able to bind reversibly to oxygen. At high oxygen concentrations, the pigment unites easily with oxygen, whereas at low oxygen concentrations the oxygen is quickly released. A more detailed account of the transport of oxygen by haemoglobin can be found in chapter 14. The common respiratory pigments of animals are shown in table 11.4.

11.6.5 Annelida

There are no organ systems especially designed for gaseous exchange, and consequently respiratory exchange takes place by diffusion over the whole body surface. Any such systems would appear to be unnecessary, as the general cylindrical shape of the worms maintains a high surface area to volume ratio, and their relative inactivity necessitates only a small oxygen consumption rate per unit body mass.

Table 11.4 Table of common respiratory pigments.

Pigment	Metal	Colour $+O_2 \rightleftharpoons -O_2$	Animal groups	Location in blood	cm³ of O₂ carried per 100 ml of blood
Haemocyanin	copper	blue ⇌ colourless	some snails	plasma	2
			crustacea	plasma	3
			cephalopods	plasma	8
Haemerythrin	iron	red ⇌ colourless	some annelids	always in cells	2
Chlorocruorin	iron	red ⇌ green	some annelids	plasma	9
Haemoglobin	iron	orange ⇌ purple red red	some molluscs	plasma	2
			some annelids	plasma or cells	7
			fishes	cells	9
			amphibia	cells	11
			reptiles	cells	10
			birds	cells	18
			mammals	cells	25

NB Compare the above figures with the fact that the amount of physically dissolved oxygen is 0.2 cm³ of O₂ per 100 ml of blood.

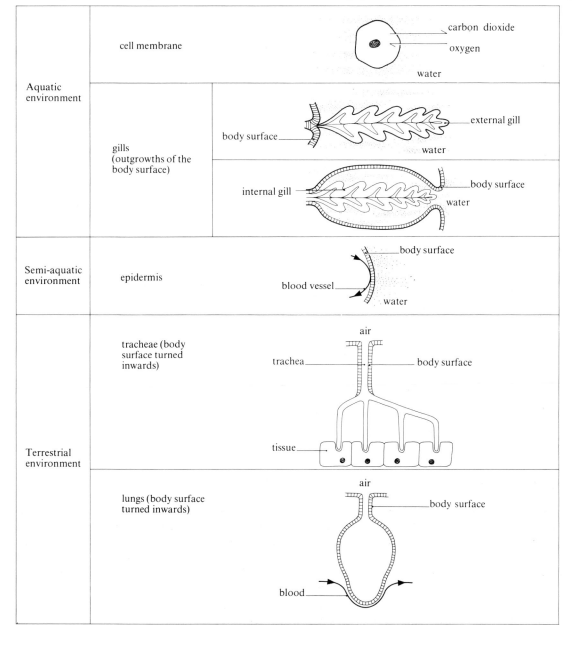

Fig 11.23 *Types of respiratory surface*

The worms do possess a blood vascular system which contains the respiratory pigment **haemoglobin** in solution. Contractile pumping activity by the blood vessels facilitates the passage of blood and dissolved gases round the body and maintains steep diffusion gradients.

Terrestrial oligochaetes (such as *Lumbricus*) keep their thin cuticles moist by glandular secretions from the epidermis and fluid exuded from the dorsal pores. **Looped blood capillaries** are present in the epidermis immediately below the cuticle. The distance between body surface and blood vessels is small enough to enable rapid diffusion of oxygen into the blood. Earthworms have little protection against desiccation and consequently their behavioural responses are designed to confine them to moist conditions.

Aquatic polychaetes (such as *Nereis* sp.) possess pairs of segmental **parapodia** along the lengths of their bodies (fig 4.19). They are mobile extensions of the body wall which are heavily vascularised, and serve to increase the respiratory surface of the animal. Once again, the blood in the parapodia is close enough to the body surface for rapid diffusion of gases to occur.

11.6.6 Arthropoda

A basic form of the arthropodan gaseous exchange system can be seen in the insects. Here gaseous exchange occurs via a system of pipes called the **tracheal system**. This allows gaseous oxygen to diffuse from the outside air directly to the tissues without the need for transportation by blood. This is much faster than diffusion of dissolved oxygen through the tissues and permits high metabolic rates.

Pairs of holes called **spiracles**, found on the second and third thoracic, and first eight abdominal, segments lead into air-filled cavities. Extending from these are branched tubes called **tracheae** (fig 11.24). Each trachea is bounded by squamous epithelium which secretes a thin layer of

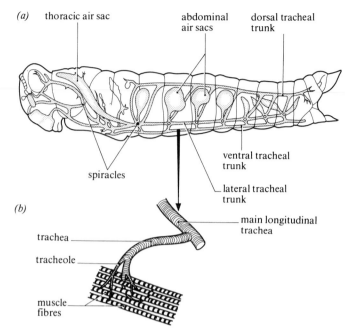

Fig 11.24   (a) VLS tracheal system of grasshopper. (b) Structure of insect trachea

chitinous material over itself. This is usually further strengthened by spiral or annular patterns of thickening which function to maintain an open pipeline even when the lumen of the trachea is subjected to reduced pressure (compare the cartilage hoops in trachea and bronchi of humans). In each segment the tracheae branch into numerous smaller tubes called **tracheoles** which ramify among the insect tissues, and in the more active ones, such as flight muscle, end blindly within cells. Tracheoles lack a chitinous lining; moreover the degree of branching may be adjusted according to the metabolic needs of individual tissues.

At rest the tracheoles are filled with watery fluid (fig 11.25) and diffusion of oxygen through them, and carbon

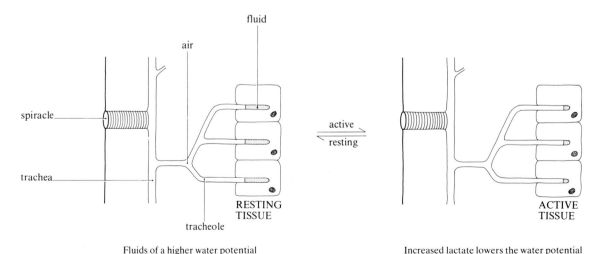

Fluids of a higher water potential surround the the tracheole; fluids diffuse into the tracheole

Increased lactate lowers the water potential of the surrounding fluid; fluid withdrawn from the tracheoles; air moves in to replace it

Fig 11.25   *Conditions in resting and active insect tissues – the functioning of tracheoles*

dioxide in the reverse direction, satisfies the insect's requirements. However, during exercise, increased metabolic activity by the muscles leads to accumulation of metabolites, especially lactate, so increasing the tissue's solute potential. When this occurs the fluid in the tracheoles is drawn osmotically into the tissues, causing more air and therefore more oxygen to enter the tracheoles and come into close contact with the tissues just at the time when it is required.

The overall flow of air in and out of the insect is regulated by a spiracular closing mechanism. Each spiracular opening is controlled by a system of valves operated by tiny muscles. It also has hairs around its edges which prevent foreign bodies entering and undue loss of water vapour. The size of the aperture is adjusted according to the level of carbon dioxide in the body.

Increased activity leads to increased carbon dioxide production. This is detected by chemoreceptors and the spiracles are opened accordingly. Ventilation movements by the body may also be initiated by the same stimulus, notably in larger insects. Dorso-ventral muscles contract and flatten the insect body, decreasing the volume of the tracheal system, thus forcing air out (expiration). Inspiration (intake of air) is achieved passively, when the elastic nature of the body segments returns them to their original shape.

There is evidence to suggest that the thoracic and abdominal spiracles open and close alternately, and that this, in conjunction with ventilation movements, provides a unidirectional flow of air through the animal, with air entering in through the thorax and out via the abdomen.

However, even though the tracheal system is a highly effective means of gaseous exchange it must be realised that it relies solely on diffusion of gaseous oxygen through the body. Since this can only occur efficiently across small distances, it imposes severe limitations on the size that insects can attain. Diffusion is only effective over distances of up to 1 cm; therefore, even though some insects may be up to 16 cm in length, they cannot be more than 2 cm broad!

11.6.7 Chondrichthyes

Cartilaginous fish (for example, the dogfish)

Situated on either side of the pharynx of the dogfish are five pairs of gill pouches, each of which contains a gill. Typically each gill is supported by a vertical rod of cartilage called the **branchial arch**. The septum overlying the arch is extended into a series of horizontal folds called **lamellae**. Each lamella in turn possesses further vertical folds on its upper and lower surface; these are termed secondary lamellae or **gill plates** (fig 11.26). The free edge of each gill septum is considerably elongated and forms an effective flap valve. It is used to close periodically the gill slit immediately posterior to it during ventilation activity. The spaces between the flap valves and gill lamellae represent the parabranchial cavities.

Deoxygenated blood from the ventral aorta is passed into each gill via an afferent branchial artery. In the gill plate region the artery branches repeatedly into many fine capillaries. It is here that gas exchange takes place. The capillaries finally reunite into efferent branchial arteries and leave the gill at its base.

Ventilation movements, taking the form of a buccal pressure pump operating in front of the gills and a suction pump behind them, serve to draw an almost continuous stream of water across the gills. Water enters the fish through the mouth and spiracles when the floor of the buccal cavity and pharynx is lowered. This is because the increased volume of the pharyngeal region reduces its pressure and consequently water rushes in (fig 11.27a). At the same time, the reduced pressure developed in the pharynx pulls the flap valves of the gills securely over the gill slits so preventing entry of water from this quarter. Whilst this is happening the suction pump mechanism is also working. Lateral movement of the flap valves causes the parabranchial cavities to expand and develop a lower hydrostatic pressure than that in the pharynx. Hence because of this differential pressure gradient, water not

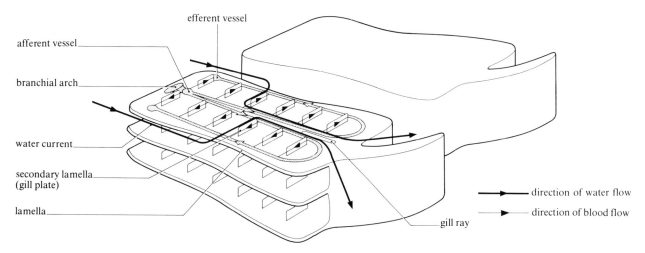

Fig 11.26 *Water flow over the lamellae in the dogfish*

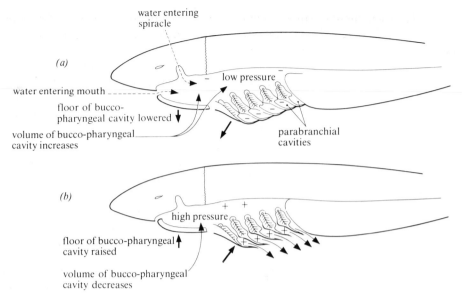

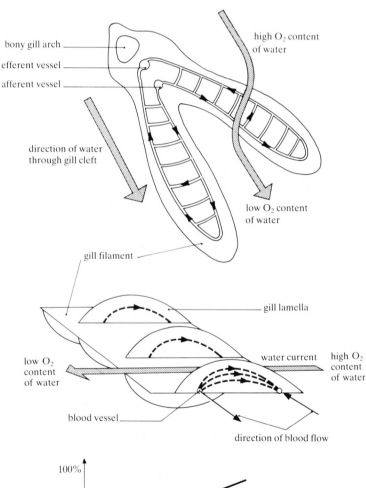

Fig 11.27 (above) *Side view of dogfish to show path of respiratory current. Pressures in the cavities are indicated with respect to zero pressure outside the fish. (a) Intake of water into bucco-pharyngeal cavity. (b) Expulsion of water via gills*

only moves into the pharynx but also simultaneously moves between the gill filaments, creating an almost continuous flow through the fish.

When the pharyngeal cavity is full of water, the mouth and spiracles close, the flap valves open, and the floor of the buccal cavity and pharynx is raised. This action forces water through the gill pouches over the respiratory epithelium of the lamellae and finally out of the fish through the gill slits (fig 11.27b). Whilst the water is being passed through the pharyngeal region and oesophageal sphincter muscle contracts, closing the oesophagus, thus preventing water from passing into the alimentary canal. Even though oxygen is absorbed by haemoglobin found in red blood cells, less than 50% of it is actually extracted by the dogfish, whereas 80% is absorbed by bony fish. This is because dogfish have relatively small gill surfaces when compared with bony fish and also because much of the water that flows over the gills travels in a direction parallel to the blood flow.

> **11.11** Why should blood flowing in the same direction as the water current be a relatively inefficient mechanism for exchange of gases?

11.6.8 Osteichthyes
Bony fish (for example, the herring)

A bony fish possesses four branchial arches on either side of the pharynx separating five pairs of gill clefts (or gill slits). Each gill on the arches is composed of two rows of fragile gill filaments arranged in the shape of a V (fig 11.28). The

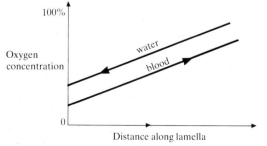

Fig 11.28 *Gill filaments of a bony fish*

filaments possess lamellae of a design similar to those found in the dogfish, and have a rich supply of blood capillaries. A movable gill cover, the operculum, which is reinforced with thin layers of bone, encloses and protects the gills in an opercular cavity. It also plays a part in the fish's ventilation mechanism.

During inspiration the buccal cavity expands, and this decreases the pressure within, causing water to be drawn in. Simultaneously the outside water pressure presses the valve at the posterior end of the operculum shut preventing entry of water from this region. However, also active at this time are opercular muscles which contract causing the opercular cavity to be enlarged. The pressure in the opercular cavity is less than that in the buccal cavity and hence water is drawn from the buccal cavity over the gills into the opercular cavity. Therefore gaseous exchange is able to continue even when the fish is engaged in taking in a fresh volume of water.

When expiration takes place the mouth closes, as does the oesophageal entrance, and the floor of the buccal cavity is raised. This forces water over the gills, through the gill slits and ultimately to the outside via the now open posterior end of the operculum. The coordinated activity of the buccal cavity and the opercular muscles ensures that a continuous flow of water passes over the gills for the majority of the time.

Adjacent gill filaments overlap at their tips, providing resistance to water flow. This slows down the passage of water over the gill lamellae thus increasing the time available for gaseous exchange to take place. The blood in the lamellae flows in the opposite direction to that of the water. Such a countercurrent system ensures that blood will constantly meet water with a relatively higher concentration of dissolved oxygen in it, and that a concentration gradient wll be maintained between blood and water throughout the entire length of the filament and across each lamella. In this way bony fish are able to extract 80% of the oxygen in water.

11.6.9   Amphibia

A frog is able to exchange gases in three different ways: through its skin by **cutaneous respiration**, via the epithelium of the buccal cavity in **buccal respiration**, and in a lung in **pulmonary respiration**.

Cutaneous respiration   The skin of the frog is richly supplied with blood capillaries and maintained in a moist condition by secretions of mucus from mucus glands. Thus atmospheric oxygen is able to dissolve in the mucus and subsequently diffuse into the blood. This is called cutaneous respiration. Oxygen uptake through the skin is almost constant throughout the year, due to the constant oxygen concentration in the atmosphere maintaining a constant diffusion gradient. In winter it supplies almost all the oxygen required by the animal, whereas in the spring when the frogs are most active it may represent only a

quarter of the frog's needs. The extra oxygen required is taken in via the buccal cavity and lungs.

Buccal respiration   Visible bucco-pharyngeal movements of the throat maintain a constant exchange of gases between the buccal cavity and the atmosphere. This is called buccal respiration. For inhalation to occur the mouth and glottis must be closed, the nostrils open, and the floor of the buccal cavity lowered, this being achieved by contraction of **sternohyoid muscles** attached to the hyoid cartilage (fig 11.29a). The buccal cavity is lined with moist, heavily vascularised epithelium, and it is here that gaseous exchange occurs. Exhalation follows when **petrohyoid muscles**, also attached to the hyoid cartilage, contract, raising the floor of the buccal cavity.

Pulmonary respiration.   The lungs are a pair of hollow sacs which hang down in the abdominal cavity. Their surface is extremely folded, but even so they present a relatively small surface area when compared with the lungs of a mammal. The epithelial lining of the lungs is moistened with mucus and profusely supplied with blood. Leading from each lung is a short tube, the **bronchus**. The two bronchi join forming the **trachea**. This is connected to a

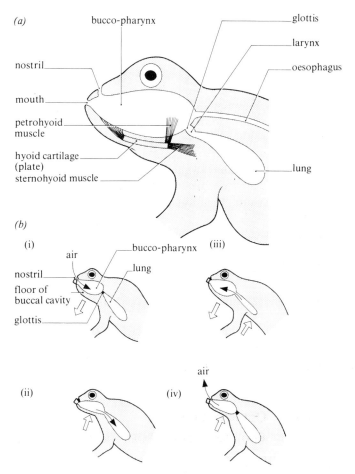

Fig 11.29 (a) VS through head of frog to show hyoid cartilage (plate) and associated musculature. (b) Ventilation of the lungs in a frog

small chamber, the **larynx**, which runs into the buccal cavity via a structure called the **glottis** (fig 11.29*a*).

At infrequent intervals violent swallowing movements are observed to occur in the frog. This is pulmonary ventilation working. The sequence of events is as follows:

(1) (fig 11.29*b* (i)) With the mouth closed, the nostrils open, the glottis closed, the floor of the buccal cavity is lowered. Air enters the buccopharynx.

(2) (fig 11.29*b* (ii)) The nostrils close, the glottis opens. Air from the lungs is forced into the buccopharynx by muscle action and elastic recoil of the lungs, thus mixing it with freshly inhaled air.

(3) (fig 11.29*b* (iii)) The floor of the buccal cavity is raised, and accompanied by vigorous gulping movements. This forces the mixed air into the lungs. When the lungs are full of air, the glottis closes and air is trapped there for some time. It is here that gas exchange between the blood capillaries in the epithelium of the lungs and the inspired air occurs.

(4) After a short interval, exhalation occurs. The nostrils close, the glottis opens and the floor of the buccal cavity is once again lowered. Air is sucked into the buccal cavity from the lungs (fig 11.29*b* (iv)). The nostrils open, the glottis closes and the floor of the buccal cavity is raised. This forces air out of the body via the nostrils.

11.6.10   Reptilia

Reptiles possess a horny body covering which is generally impermeable to gases and therefore not able to be used in a respiratory capacity. Gaseous exchange occurs exclusively via the lungs. These are sac-like in construction and exhibit much more complex folding than the amphibian lungs. Reptiles possess ribs, but no true diaphragm separates the thorax from the abdomen. Ventilation occurs when the ribs are moved by intercostal muscle contractions. The mechanism is very similar to that found in mammals.

11.6.11   Aves

Birds are homeothermic organisms and require a high rate of metabolism in order to maintain their body temperature. An efficient respiratory mechanism has been developed in order to make this possible.

Bird lungs are small, compact structures composed of numerous branching air tubes called **bronchi**. The smallest of these, the **parabronchi**, are heavily vascularised and it is here that gaseous exchange occurs. Extending from the lungs are large, thin-walled **air sacs**. These are poorly vascularised and do not take part in gaseous exchange. Functionally the air sacs can be divided into an anterior group and a posterior group, and they serve to move air in and out of the respiratory system.

Ventilation movements are complex, but basically they occur in the following sequence:

first inhalation – air flows directly to the posterior sacs (fig 11.30);

first exhalation – air in the posterior sac is passed to the lungs;

interclavicular sac

anterior group of air sacs

cervical sac

lung

abdominal sac

posterior group of air sacs

Fig 11.30 *The lung and air sac system of a bird*

second inhalation – air passes from the lungs to the anterior sacs;

second exhalation – air from the anterior sacs is forced to the outside.

This method of ventilation ensures that there is a unidirectional flow of air from the posterior sacs, through the lungs to the anterior sacs, before it leaves the bird's body.

Quiet respiratory movements are carried out in a similar manner to mammals, with intercostal and abdominal muscles producing inspiratory and expiratory movements respectively. When a bird is in flight, ventilation of the lungs is augmented by vigorous contraction of the pectoral muscles acting on a large keeled sternum. In these circumstances greater volumes of air are exchanged between bird and atmosphere in order to satisfy its metabolic needs.

11.7   Gaseous exchange in a mammal

The respiratory system of a mammal consists of a series of air tubes connecting a pair of lungs, which are situated in the thoracic cavity, to the atmosphere. Air is passed into the lungs through these tubes in the following sequence: nasal passages, pharynx, larynx, trachea, bronchi, bronchioles, alveoli of lungs. Twelve pairs of bony ribs surround and protect the lungs and heart in the thoracic cavity. Dorsally, each rib articulates with a thoracic vertebra and is able to be moved up or down. The anterior ten pairs of the ribs join to a bony plate, the **sternum**. The remaining pairs of ribs are called free or floating ribs. **Intercostal muscles** are attached to the ribs, and a large **diaphragm**, which effectively separates the thorax from the abdomen, are also vital parts of the system.

Air enters the body through two external nostrils, each of which possesses a border of large hairs which trap particles in the air and filter them out of the system. The walls of the passage are lined with **ciliated epithelial** and **mucus**-secreting goblet cells. Mucus serves two functions. First it traps any particles that have managed to pass through the hairs of the external nostrils. Epithelial cilia then beat in such a fashion as to carry the trapped particles to the back of the buccal cavity where its expulsion from the respiratory tract is completed by swallowing. Secondly, it moistens the incoming air, which incidentally is also warmed by the temperature of the superficial blood vessels in the nasal channels of the organism. In the roof of the posterior part of the nasal cavity is a mass of olfactory epithelium consisting of neurosensory and supporting cells, richly supplied with blood. Here odours in the air are detected. At the end of the nasal passages the air enters the pharynx via two internal openings. By the time this occurs the air has been generally freed from particles, warmed, moistened and its odour detected.

Next the air must traverse the pharynx in order to enter the larynx. As both food and air pass through the **pharynx**, the slit-like opening to the larynx, the **glottis**, has to be protected against the entry of food which could block the air breathing channels. This is achieved by a triangular flap of cartilaginous tissue, the **epiglottis**.

The **larynx** consists of a collection of nine cartilages forming a box-like structure at the entrance of the trachea. Muscles attached to the cartilages move them relative to each other. Within the box are two sets of horizontally aligned fibro-elastic ligaments, the **vocal cords**. When air is expelled over the vocal cords and through the glottis, sound waves are formed. By varying the tension of the cords, the pitch of the sound can be changed.

Air from the larynx enters the **trachea**. This is a tube which lies directly in front of the oesophagus and extends into the thoracic cavity. The wall of the tube is strengthened and held open by horizontally arranged C-shaped cartilages. The open section of the C is applied against the oesophagus. The cartilage prevents collapse of the tube during inspiration (fig 11.31). Lining the trachea is a carpet of pseudostratified, ciliated columnar epithelium. Mucus from goblet cells interspersed in the epithelium traps dust and germs, and the rhythmic beating of the cilia in a direction towards the back of the buccal cavity removes unwanted materials from the trachea.

At its lower end the trachea splits into two **bronchi**. The right bronchus further divides into three bronchi. They extend separately into the three lobes of the right lung. The left bronchus divides into two bronchi, and similarly these penetrate the two lobes of the left lung (fig 11.32). Within the lungs each bronchus subdivides many times into much smaller tubes called **bronchioles**. The C-shaped cartilages of the bronchi are initially replaced in the smaller tubes by irregularly shaped plates of cartilage, but when the internal diameter of the bronchioles is less than 1 mm, cartilaginous support ceases altogether. At this point, the thin bronchiole wall consists merely of smooth muscle, connective tissue possessing elastic fibres which promotes inflation and recoil of the bronchiole, and an inner lining of ciliated epithelium interspersed by mucus-secreting cells. The smallest tubes, called **respiratory bronchioles**, are about 0.5 mm in diameter. They, in turn, divide repeatedly into many **alveolar tubes** lined with cuboidal epithelium, which terminate in hollow, lobed air sacs called **alveoli** (figs 11.33 and 11.34). Collectively the alveoli form the gaseous exchange surface of the mammal.

There are over 700 million alveoli present in the lungs of a mammal, representing a total surface area of 80–90 m². The wall of each alveolus is only 0.0001 mm thick. On its outside is a dense network of blood capillaries, all of which have originated from the pulmonary artery and will ultimately rejoin to form the pulmonary vein. Lining each alveolus is moist squamous epithelium. Collagen and elastic fibres are also present and they provide flexibility for the alveoli, enabling them to expand and recoil easily during breathing.

Giant alveolar cells in the alveolus wall provide a secretion that is released inside the alveolus and lines its

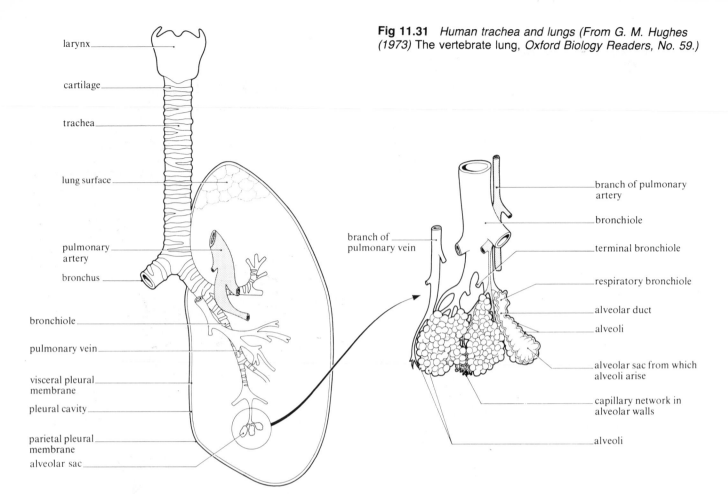

larynx

cartilage

trachea

lung surface

pulmonary artery

bronchus

bronchiole

pulmonary vein

visceral pleural membrane

pleural cavity

parietal pleural membrane

alveolar sac

branch of pulmonary vein

branch of pulmonary artery

bronchiole

terminal bronchiole

respiratory bronchiole

alveolar duct

alveoli

alveolar sac from which alveoli arise

capillary network in alveolar walls

alveoli

Fig 11.31 *Human trachea and lungs (From G. M. Hughes (1973)* The vertebrate lung, *Oxford Biology Readers, No. 59.)*

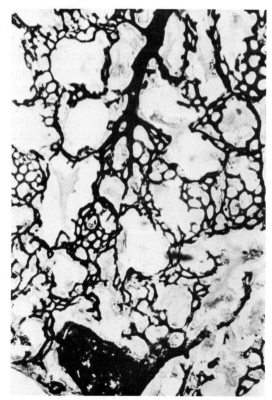

Fig 11.32 *Human lung injected to show airways and pulmonary circulation*

walls. This secretion contains a detergent-like lipoprotein called a **surfactant**. It lowers the surface tension of the alveoli, and hence the amount of effort needed to breathe in and inflate the lungs.

Surfactant also speeds up the transport of oxygen and carbon dioxide between the air and liquid phases and has a bactericidal effect, helping to remove any potentially harmful bacteria which reach the alveoli. Surfactant is constantly being secreted and reabsorbed in a healthy lung.

Surfactant is lacking in premature babies, resulting in the condition known as **respiratory distress syndrome**, in which breathing is very difficult and which may result in death.

The oxygen diffuses across the thin membranous barrier represented by the alveolar epithelium and capillary endothelium, and passes initially into the blood plasma. It then combines with haemoglobin in the red blood corpuscles to form oxyhaemoglobin. Carbon dioxide diffuses in the reverse direction from the blood to the alveolar cavity.

The diameter of the alveolar capillaries is smaller than the diameter of the red blood corpuscles passing through them. This means that the corpuscles have to be squeezed through the capillaries by blood pressure. During this process more of their surface area is exposed to the gaseous exchange surface of the alveolus permitting greater uptake of oxygen. Progress of the corpuscles is also relatively slow, thus increasing the time available for gaseous exchange to take place. When blood leaves the alveolus it possesses the

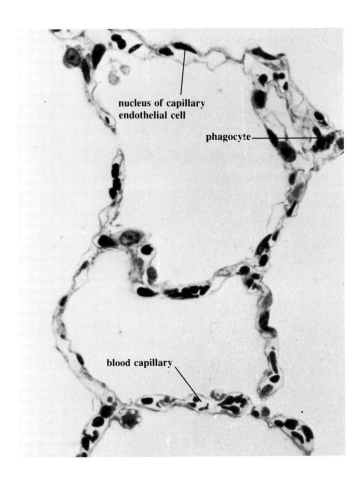

nucleus of capillary endothelial cell

phagocyte

blood capillary

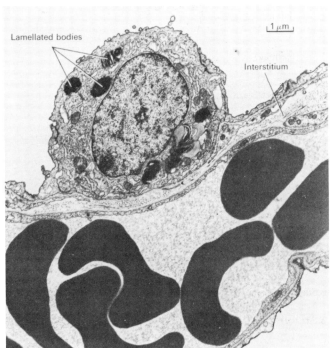

Lamellated bodies

Interstitium

1 μm

Fig 11.34 (above) *Electron micrograph of a longitudinal section through an alveolar capillary in dog lung (From G. M. Hughes (1973) The vertebrate lung, Oxford Biology Readers, No. 59).*

Fig 11.33 (left) *Histology of lung alveolus*

same partial pressure of oxygen and carbon dioxide as alveolar air.

11.7.1 Structure of the thorax

Each lung is surrounded by a **pleural cavity**. This is a space lined by two flexible, transparent pleural membranes (**pleura**). The inner visceral membrane is in contact with the lungs, whilst the outer parietal membrane lines the walls of the thorax and diaphragm. The pleural cavity contains a fluid, secreted by the membranes. This lubricates the pleura, thus reducing friction as the membranes rub against each other during breathing movements. The cavity is air-tight and its pressure stays at 3–4 mmHg lower than that in the lungs. This is important for it causes the lungs almost to fill the thorax. The negative pressure of the pleural cavity is maintained during inspiration and this allows the alveoli to inflate and fill any extra available space provided by the expanding thorax.

11.7.2 The mechanism of ventilation

Air is passed in and out of the lungs by movements of the intercostal and diaphragm muscles which alter the volume of the thoracic cavity. There are two types of intercostal muscle between each rib. The **external intercostals** slant forwards and downwards, whilst the **internal intercostals** slant backwards and downwards

(fig 11.35). The diaphragm consists of circular and radial muscle fibres arranged around the edge of a circular inelastic sheet of white fibres.

Inspiration is an active process. The external intercostal muscles contract and the internal intercostals relax. This produces a forward and outward movement of the rib cage away from the vertebral column. Simultaneously, the diaphragm contracts and flattens. Both actions increase the volume of the thorax. As a result the pressure in the thorax,

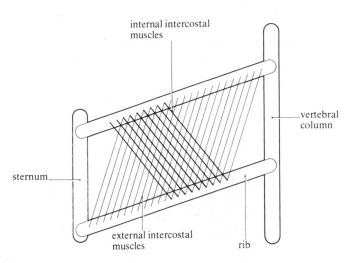

internal intercostal muscles

vertebral column

sternum

external intercostal muscles

rib

Fig 11.35 *Diagrammatic representation of the position of the intercostal muscles*

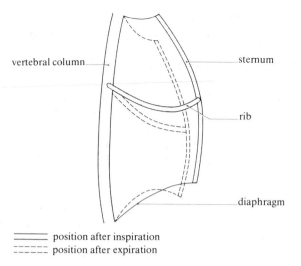

vertebral column — sternum

rib

diaphragm

— position after inspiration
----- position after expiration

Fig 11.36 *Side view of thorax to show movements during breathing (only one rib shown)*

and hence the lungs, is reduced to less than atmospheric pressure, so permitting air to rush in and inflate the alveoli, until the air pressure in the lungs is equal to that of the atmosphere (fig 11.36).

Expiration is largely a passive process under resting conditions and is brought about by the elastic recoil of the lung tissue and respiratory muscles. The external intercostal and diaphragm muscles relax and return to their former size and position, whilst the internal intercostals contract. This reduces the volume of the thorax and raises its pressure above that of the atmosphere. Consequently air is forced out of the lungs and expiration completed. Under conditions of exercise, forced breathing occurs. When this happens additional muscles are brought into action and expiration becomes a much more active, energy-consuming process. The internal intercostals contract more strongly and move the ribs vigorously downwards. The abdominal muscles also contract strongly, causing more active upward movement of the diaphragm.

11.7.3 Control of ventilation

Involuntary control of breathing is carried out by a **breathing centre** located in the medulla of the brain (fig 11.37). The ventral portion of the breathing centre acts to increase inspiratory rate and is called the **inspiratory centre**, whilst its dorsal and lateral portions cut off inspiratory activity and promote expiration. These regions are collectively termed the **expiratory centre**. The brea-

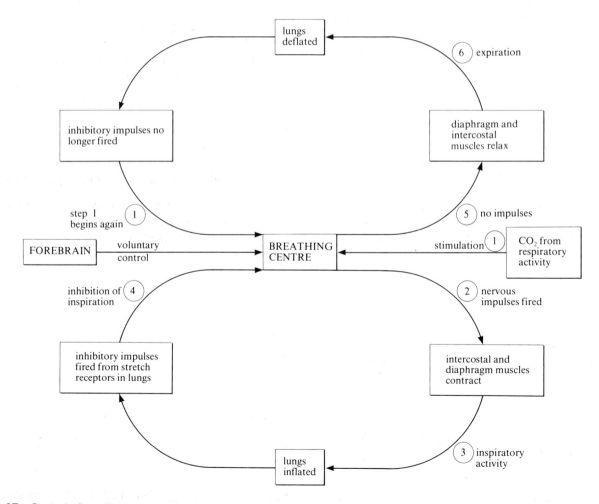

Fig 11.37 *Control of ventilation*

352

thing centre communicates with the diaphragm via phrenic and thoracic nerves. The bronchial tubes and alveoli are innervated by branches of a cranial nerve, the vagus.

The main stimulus that controls the breathing rate is the concentration of carbon dioxide in the blood. When carbon dioxide levels increase, **chemoreceptors** in the carotid and aortic bodies of the blood system are stimulated to discharge nerve impulses which pass to the inspiratory centre. The inspiratory centre then sends out impulses via the phrenic and thoracic nerves to the diaphragm and intercostal muscles causing them to increase the rate at which they contract. This automatically increases the rate at which inspiration takes place. Inspiratory activity inflates the alveoli, and stretch receptors located here and in the bronchial tree are stimulated to discharge impulses to the expiratory centre which automatically cuts off inspiratory activity. The respiratory muscles therefore relax and expiration takes place. After this has occurred, the alveoli are no longer stretched and the stretch receptors no longer stimulated. Therefore the expiratory centre becomes inactive and inspiration can begin again. The whole cycle is repeated rhythmically throughout the life of the organism.

Within limits, the rate and depth of breathing are under voluntary control. When such control is being exerted, impulses originating in the cerebrum pass to the breathing centre which then carries out the appropriate action. Oxygen concentration also has an effect on the breathing rate (section 11.8). However, under normal circumstances there is an abundance of oxygen available, and its influence is relatively minor.

11.7.4 Lung volumes and capacities

The average man has a lung capacity of approximately 5 dm³ (fig 11.38). During quiet breathing he will breathe in and out about 450 cm³ of air. This is called the **tidal volume**. If after a normal tidal inspiration he continues to inhale, he can take in a further 1 500 cm³ of air. This is called his **inspiratory reserve volume**. If after a tidal expiration the man continues to exhale he can force out a further 1 500 cm³ of air. This is termed his **expiratory**

reserve volume. The amount of air exchanged after a forced inspiration followed immediately by a forced expiration is termed the **vital capacity**. Even after forced expiration 1 500 cm³ of air remain in the lungs. This cannot be expelled and is called **residual air**.

During inspiration about 300 cm³ of the tidal volume reaches the lungs, whilst the remaining 150 cm³ remains in the respiratory tubes, where gaseous exchange does not occur. When expiration follows, this air is expelled from the body as unchanged room air and is termed **dead space air**. The air that reaches the lungs mixes with the 1 500 cm³ of air already present in the alveoli. Its volume is small compared to that of the alveolar air and complete renewal of air in the lungs is therefore a necessarily slow process. The intermittent slow exchange between fresh air and alveolar air affects the composition of gases in the alveoli to such a small extent that they remain relatively constant at 13.8% oxygen, 5.5% carbon dioxide and 80.7% nitrogen. Comparison of the composition of gases of inspired, expired and alveolar air is interesting (table 11.5). It is clear that one-fifth of the oxygen inspired has been retained for use by the body, and 100 times the amount of carbon dioxide expelled. The air that comes into close contact with the blood is alveolar air. It contains less oxygen than inspired air, but more carbon dioxide.

Table 11.5 Percentage composition by volume of gases in inspired, alveolar and expired air.

Gas	Inspired air	Alveolar air	Expired air
Oxygen	20.95	13.8	16.4
Carbon dioxide	0.04	5.5	4.0
Nitrogen	79.01	80.7	79.6

11.7.5 Measurement of respiratory activity

An instrument commonly used in schools, laboratories and hospitals for measuring the volume of air which enters and leaves the lungs is the spirometer. Essentially it consists of an air-filled box with a capacity of

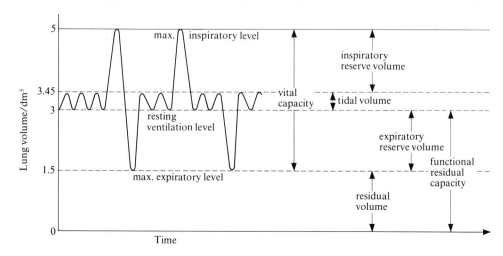

Fig 11.38 *Lung volumes and capacities*

six or more litres, suspended freely over water. The air in the box is connected by a series of pipes to the subject so that the air in the subject's lungs and the air in the box is a sealed system.

The box is counterbalanced so that when gas is passed in or out, the box rises or falls accordingly. When the subject breathes out, the box is raised; when he breathes in, the box is lowered. A pen attached to the box writes on a slowly rotating drum (kymograph) recording all the movements of the box.

Detailed instructions for the operation of the spirometer are supplied by the manufacturer and will not be dealt with here. However, it is important to be able to analyse the tracings recorded by the spirometer, and to understand what information can be derived from them.

From spirometer tracings the metabolic rate, respiratory quotient, tidal volume, rate of breathing and consumption of oxygen can be measured.

The **respiratory rate** is calculated as the number of breaths taken per minute. **Pulmonary ventilation (PV)** is expressed in terms of the respiratory rate multiplied by the tidal volume:

$$PV = \text{respiratory rate} \times \text{tidal volume}$$

For example, if respiratory rate is 15 breaths per minute and tidal volume is 400 cm³, then PV = 15 × 400 cm³ = 6 000 cm³ per minute (that is 6 000 cm³ of air will be exchanged between subject and outside environment each minute).

Alveolar ventilation (AV) is the volume of air that actually reaches the lungs. It is less than that of the pulmonary ventilation.

$$AV = \text{respiratory rate} \times (\text{tidal volume} - \text{dead space air})$$

For example if TV = 400 cm³, dead space air = 150 cm³ and respiratory rate = 15 breaths per minute,

then AV = 15 × (400 − 150) cm³
       = 15 × 250 cm³
       = 3 750 cm³ per minute (that is 3 750 cm³ of air will be exchanged between the lungs of the subject and outside environment each minute).

> **11.12** Why is the volume of alveolar air less than that of the pulmonary ventilation?

Measuring the metabolic rate of an organism

As respiration is directly involved with most metabolic activities within the body, its measurement gives a relatively accurate indication of metabolic activity. The metabolic rate can be calculated by measuring the rate of oxygen consumption.

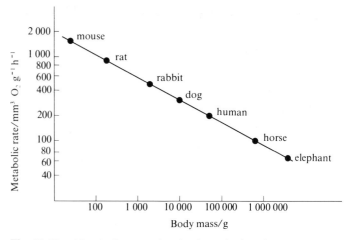

Fig 11.39 *Metabolic rate of animals, calculated per gram body mass, plotted on logarithmic coordinates*

> **11.13** Consider fig 11.39. It can be seen that the smaller the mammal the higher its metabolic rate. Why is this so?
>
> **11.14** How can you compare the metabolic rates of mammas of different size?

11.7.6 The basal metabolic rate (BMR)

The BMR of an organism is the minimum rate of energy conversion required just to stay alive during complete rest or sleep. Before the BMR of human subjects is measured they undergo a standardised rest period of 12–18 h physical and mental relaxation. No meal is eaten during this time. This ensures that the alimentary canal is empty before measurements are taken. The BMR varies with age, sex, size and state of health of the individual and is clearly correlated with body surface area to volume ratio.

11.7.7 Respiratory quotient (RQ)

Consider the equation:

$$C_6H_{12}O_6 + 6O_2 \longrightarrow 6CO_2 + 6H_2O + \text{energy}$$

From this it is quite clear that in a given time the volume of carbon dioxide produced during respiration of carbohydrate is equal to the volume of oxygen consumed (remember, one mole of any gas occupies the same volume under the same conditions of temperature and pressure). The ratio of $CO_2:O_2$ is called the respiratory quotient, and for metabolism of carbohydrate its value is 1,

that is RQ = $\dfrac{\text{volume of } CO_2 \text{ evolved}}{\text{volume of } O_2 \text{ absorbed}}$
(from direct observations)

Or $\dfrac{\text{moles or molecules of } CO_2 \text{ evolved}}{\text{moles or molecules of } O_2 \text{ evolved}}$
(from equations)

Therefore, from the above equation, $RQ = \dfrac{CO_2}{O_2} = \dfrac{6}{6} = 1$

11.15 The equation for respiration of the fat tripalmitin is:

$$2C_{51}H_{98}O_6 + 145O_2 \longrightarrow 102CO_2 + 98H_2O$$

What is the RQ for tripalmitin?

11.16 What is the RQ when glucose is respired anaerobically to ethanol and carbon dioxide?

Analysis of respiratory quotients can yield valuable information about the nature of the substrate being used for respiration and the type of metabolism that is taking place (table 11.6).

11.17 Why is the usual RQ for humans between 0.7 and 1.0?

11.18 From the spirometer trace given in fig 11.40 calculate
(a) respiratory rate;
(b) tidal volume;
(c) pulmonary ventilation;
(d) oxygen consumption.

Table 11.6 Respiratory quotients of a variety of substrates.

RQ	Substrate	
>1.0	Carbohydrate plus some anaerobic respiration	
1.0	Carbohydrates	
0.9	Protein	
0.7	Fat, such as tripalmitin	
0.5	Fat associated with carbohydrate synthesis	The carbon dioxide released during respiration is being put to other uses and therefore not released from the body
0.3	Carbohydrate with associated organic acid synthesis	

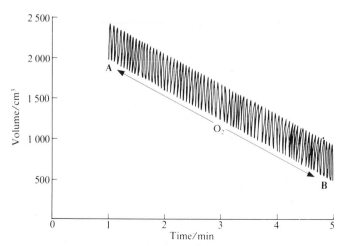

Fig 11.40 *Spirometer tracing of subject at rest*

Experiment 11.2: To measure the oxygen uptake in small terrestrial non-vertebrates such as woodlice

The oxygen uptake by the non-vertebrates in this experiment is measured using a manometer. Fig 11.41 shows the apparatus which is used.

In respiration, oxygen is taken up and carbon dioxide is given off, and so to ensure that the manometer is recording oxygen consumption alone, soda-lime is incorporated in the apparatus in order to absorb the carbon dioxide evolved.

The apparatus is used to investigate the effect of temperature on oxygen uptake. A water bath is used in order to keep the temperature of the atmosphere surrounding the organisms constant whilst readings are being taken.

Materials

manometer	clamps and stands
manometer fluid	water bath
1 cm³ syringe	thermometer
2 boiling tubes	stopclock
2 pieces of zinc gauze (to fit the diameter of a boiling tube)	graph paper
	small non-vertebrates such as woodlice or blow-fly larvae
glass beads (or any equivalent non-absorbent material of equal volume to the invertebrates)	soda-lime

Method

(1) Half fill a manometer with fluid and connect a 1 cm³ syringe to the three-way tap attached to one arm of the manometer.
(2) Place equal volumes of soda-lime in the bottom of each of two boiling tubes and then place a zinc-gauze platform 1 cm above the soda-lime.
(3) Place some invertebrates in one boiling tube (experimental) and an equal volume of glass beads in the other tube (control). The animals must not come into contact with the soda-lime and so the platform must be an absolute barrier between the animals and the soda-lime.
(4) Connect the manometer to the two boiling tubes as shown in fig 11.41, and adjust the three-way tap and screw-clip so that the apparatus is open to the atmosphere.
(5) Clamp the apparatus so that the boiling tubes are in a water bath at 20 °C and leave the apparatus at this temperature with the taps open for at least 15 min.
(6) Close the tap and screw-clip, note the position of the manometer fluid and start the stopclock.
(7) At regular intervals, read off the position of the manometer fluid against the scale.
(8) At the end of the experiment open the tap and screw-clip again.

Fig 11.41 *Apparatus used in the investigation of oxygen uptake in small terrestrial non-vertebrates*

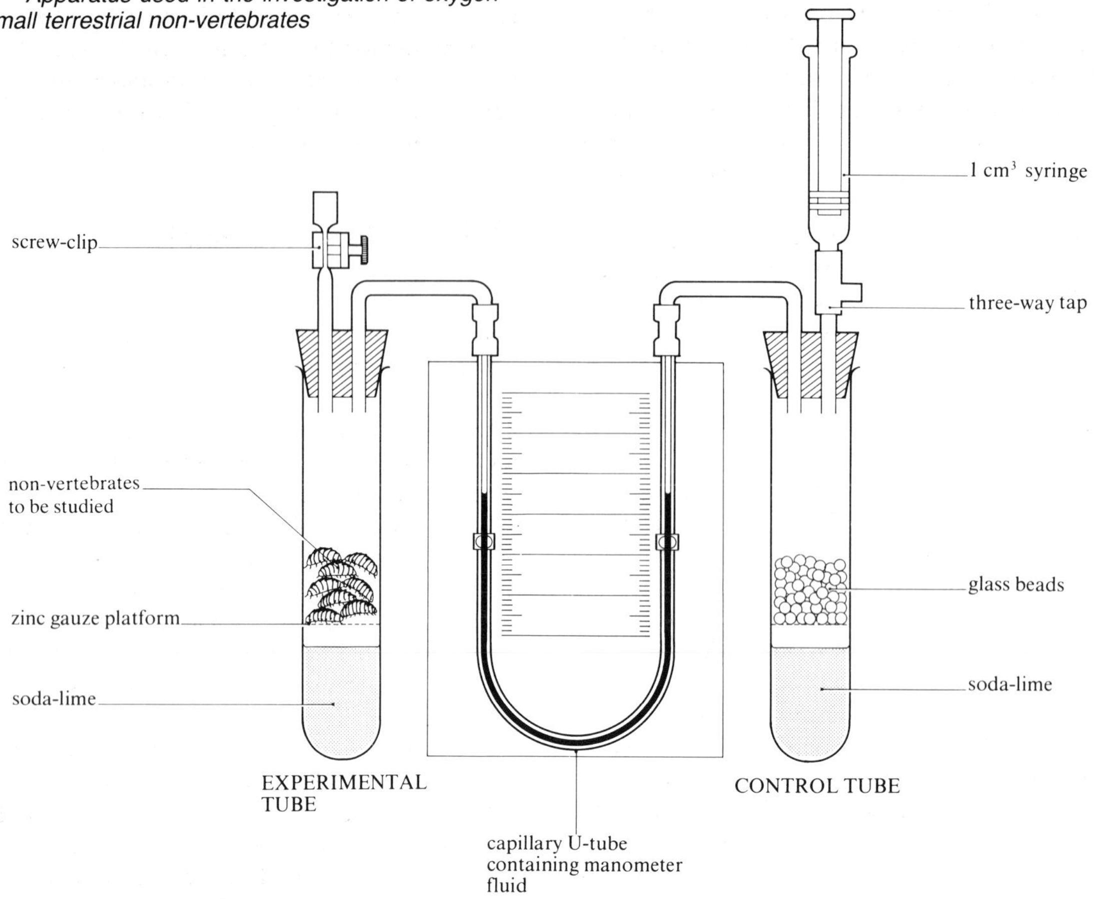

(9) Plot a graph of the change in fluid level against time.

(10) Calculate the rate of oxygen uptake.

(11) Repeat the experiment several times over a range of temperatures, such as 20, 25, 30, 35 and 40 °C.

(12) Plot a graph of rate of oxygen consumption against temperature.

Notes

(1) The fluid that is used in the manometer can be dyed water, oil or mercury. The less dense the fluid, the greater the displacement in the manometer.

(2) In order to measure the change in the manometer fluid levels, a scale must be attached to the manometer. This can be done by attaching the manometer U-tube to a piece of hardboard on which a scale or graph paper has been glued. Alternatively an adhesive metric scale can be attached to the arm of the manometer itself. The tape is available from Philip Harris Ltd.

(3) Before any readings are taken in the experiment, the apparatus must be checked to ensure that it is air-tight. This can be done by pushing air into the apparatus using the syringe, causing the manometer fluid to be displaced. The tap should then be used to close off the apparatus to the atmosphere and if the apparatus is air-tight, the difference in levels of fluid should not decrease.

11.8 Unusual conditions in ventilation

11.8.1 The effects of altitude and acclimatisation

As mountaineers ascend high mountains they suffer from inadequate oxygenation of the blood. This condition is known as **anoxia** or hypoxia. It occurs because the partial pressure of oxygen, along with the other gases of the atmosphere falls with increasing altitude, for example at 5 450 m the barometric pressure of the atmosphere is 0.5 bar which is exactly half that at sea level. Although the percentage of oxygen by volume is the same at this altitude when compared with the other gases of the atmosphere, the actual amount is half that found at sea level.

Respiratory activity is stimulated by chemoreceptors, and at high altitudes, in the quest for more oxygen, the increased ventilation expels large quantities of carbon dioxide from the lungs causing the acidity of the blood to decrease. The increase in alkalinity (greater pH) causes a condition known as **alkalaemia**. This greater pH inhibits the activity of the chemoreceptors. Consequently pulmonary ventilation is hampered and becomes relatively inadequate, causing great discomfort and fatigue.

Given time, the respiratory and circulatory systems are able to adjust somewhat to the lower partial pressure of oxygen at altitude. After several days alkaline urine is expelled from the body thus reducing the alkalaemia. With the chemoreceptors no longer inhibited, pulmonary ventilation will increase, and carbon dioxide concentration will once again become the main chemical stimulus in the regulation of breathing. At the same time the bone marrow is stimulated to produce more red blood cells (erythrocytes). This raises the oxygen-carrying capacity of the blood, offsetting to some extent the incomplete oxygen saturation of the blood caused by the low partial pressure of oxygen. When these adjustments are complete the body is said to be **acclimatised** to its new conditions.

11.8.2 Diving mammals

Seals frequently remain submerged in the water for periods of up to 15 min. Their blood has a much higher oxygen-carrying capacity than human blood, being able to carry between 30 and 40 cm^3 of oxygen per 100 cm^3 of blood.

During the dive, considerable modifications to the circulatory and respiratory system occur which ensure the effective distribution of oxygen within the animal and its efficient use. Generally this is what happens. At the onset of a dive, a nervous reflex decreases the rate at which the heart contracts, thus slowing blood flow. Blood pressure in the arteries is maintained because the blood vessels constrict. Certain blood vessels constrict completely and bring about a redistribution of the blood supply so that only the vital organs, such as the heart, brain and other parts of the nervous system, receive blood. Such modification means that oxygen in the blood will be used up slowly but is always available to those organs which are very sensitive to anoxia. With little oxygen available to them, the muscles of the seal respire anaerobically, accumulating lactate. Because there is little blood passing from the muscles to the general circulation, no great quantity of lactate will be distributed around the body and the body will not suffer any harmful effects. On returning to the surface, the first

breath taken is the signal for the heartbeat to increase, and blood circulation to return to normal. When this happens lactate is passed into circulation and metabolised in the liver. The air in the lungs is rapidly replaced as each breath can exchange as much as 80% of the air in the lungs.

11.9 Flowering plants

Plants require less energy per unit mass than animals as they possess lower metabolic rates. Whereas some small plants can carry out gaseous exchange by diffusion over their whole surfaces, large flowering plants exchange gases through stomata in their leaves and on their green stems (herbaceous stems), or if the stems are woody. through cracks in the bark or via lenticels (section 21.6.6).

Once inside the plant, movement of oxygen is determined by the diffusion gradients that exist in the intercellular air spaces. In this way oxygen travels towards the cells and dissolves in the surface moisture of their walls. From here it passes by diffusion into the cells themselves. Carbon dioxide leaves the plant by the same pathway but in the reverse direction.

The whole situation becomes more complex in chlorophyll-containing cells if the plant is also photosynthesising at the same time. Here oxygen produced by the chloroplasts may be immediately used up by mitochondria contained in the same cell, and carbon dioxide issuing from the mitochondria consumed by the chloroplasts.

Further information concerning gaseous exchange in flowering plants can be found in chapter 9.

11.19 (*a*) Construct a table showing the major differences between photosynthesis and aerobic respiration. (*b*) Make a list of similarities (including biochemical similarities), between photosynthesis and aerobic respiration.

Chapter Twelve

Organisms and their environment

Ecology is the study of the relationships of living organisms to each other and their surroundings. As stated by Cousins (*New Scientist*, 4 July 1985), 'Ecology is a science, from which much is expected. It provides the foundations of our understanding of agriculture, forestry and fisheries and is called upon to predict the effects of everything from pollutants to the construction of dams.'

The term 'ecology' was first used by the German biologist Ernst Haeckel in 1869 and is derived from the Greek roots *oikos* meaning a 'house' or 'living-place' and *logos* meaning the 'study' or 'science' of. Thus, literally, ecology means the study of the Earth's households. It adopts a **holistic** approach, that is one in which a whole picture is built up which is more important than the parts. Although the various parts often have to be analysed separately, as described in chapter 13, it is the synthesis of all the available information into an overall picture of the living systems and their surroundings that is important.

Ecology has its roots in natural history and has ranked alongside physiology, genetics and other disciplines as a branch of biology from about 1900. Since the mid-1950s the scope and importance of ecology have increased greatly, and modern ecology is best viewed as an important interdisciplinary science linking physical, biological and social sciences. Its relationship with other branches of biology is summarised in fig 12.1, which shows that living organisms can be studied at different levels of organisation. Ecology spans the right-hand portion of the diagram, which includes organisms, populations and communities. Ecologists regard these as the living part (**biotic component**) of a system called the **ecosystem**. This also includes a non-living part (the **abiotic component**) which contains matter and energy. Populations, communities and ecosystems are terms which have precise meanings in ecology, and they are defined in fig 12.1. The different ecosystems are united to form the **biosphere**, or **ecosphere**, which includes all the living organisms and the physical environment with which they interact. Thus, the oceans, land surface and lower parts of the atmosphere all form part of the ecosphere.

Within ecology, studies have traditionally been of two types, namely autecology and synecology. **Autecology** focusses on the relationships between an organism or population and the environment, whereas **synecology** looks at communities and the environment. Thus, while in an autecological study one might examine the ecology of a single oak tree, or the species *Quercus robur* (pedunculate oak), or the genus *Quercus* (oak), in a synecological study it is the whole oakwood community which is examined (sections 13.4 and 13.5).

As Southwood noted in 1981* in a useful review summarising the changing nature of ecology, the first ecological article in the journal *New Scientist* (in 1956) dealt with the reintroduction of reindeer into Scotland in an essentially autecological study. More recently, emphasis in ecological work has shifted to ecosystem (that is community and environment) and even ecosphere (whole planet) studies. It is at these levels of organisation that the important contribution to ecology of sciences other than biology, notably chemistry, physics, pedology (the study of soils) and hydrology (the study of water), as well as the various social sciences, can most clearly be understood.

* Southwood, T. R. E. *New Scientist*, **92**, 512–4 (19 Nov 1981).

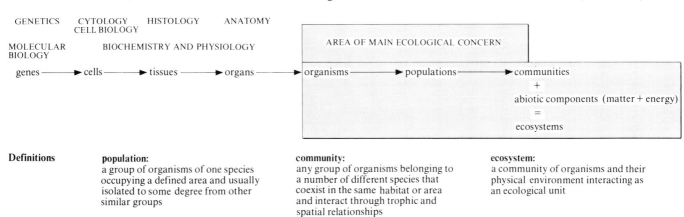

Fig 12.1 *Levels of organisation from genes to ecosystems*

The growing public awareness of ecology, particularly of the interaction between human societies and the environment, is seen in many ways. In the USA developers are required by law to prepare Environmental Impact Statements (EIS) before any new project may begin. In Britain, though this is not yet a legal requirement, major industries and environmental groups who may oppose developments prepare similar assessments of new proposals. No company can afford to alienate public opinion, and planning officials are unlikely to allow development without a proper environmental appraisal. An example is BP's extensive environmental appraisal of the Wytch Farm on-shore oil development in Dorset. Many local planning authorities now employ ecologists or seek advice from County Naturalists Trusts and ecologists have increasingly found a place in government committees dealing with the environment, such as the permanent Royal Commission on Environmental Pollution (RCEP).

Ecology has also formed the basis for a new political perspective, called 'environmentalism'. In Europe, this is encapsulated by the various national Green parties. Under the West German electoral system *die Grünen* (the Greens) have achieved formal political representation. In Britain, the Green party has adopted a lobbyist approach and seeks, with other major environmental organisations such as Greenpeace and Friends of the Earth, to inform and persuade politicians and industrialists of the importance of environmental concerns. Recent well-publicised industrial accidents as at Seveso (Italy), Bhophal (India) and Chernobyl (USSR) have heightened public and political awareness of the relevance of ecology to industry. The current controversy in Britain over nuclear waste disposal, the growing international concern over toxic chemical wastes (as highlighted by the case of *Karin B*, summer 1988), and the failure, in European eyes, of the British government to act effectively on acid rain or respond wholeheartedly to initiatives on North Sea pollution, further emphasise the political relevance and international dimension of ecology.

One positive response has been the emergence over the last two decades of environmental consultancies and agencies supplying advice to industry and other interested parties. Designing in an environmentally benign manner and thus avoiding the need for subsequent costly modification may make the difference between profit and loss, success and failure, in an ever more environmentally aware market. The success of enterprises such as Environmental Data Services, established in 1978, which has trebled the circulation of its journal during the 1980s testifies to this process.

E.P. Odum's assertion in 1971 that 'ecology is the branch of science most relevant to the everyday life of all people' thus seems vindicated by events of the last 20 years.

Student action
1 Look at the job advertisements in *New Scientist* or the quality daily newspapers on 'Public appoint-ments' day (for example *Guardian, Times, Independent*). How many jobs require ecological knowledge?
2 Watch the main evening television news programme every day for one week and list the environmentally related issues reported.

12.1 Approaches to ecology

The holistic approach (see introduction above) is the distinctive characteristic of ecological science. A proper understanding of ecology requires simultaneous consideration of all factors interacting in a particular place. The sheer scope of this task presents problems, and in practice most ecologists adopt one of several main approaches when undertaking a new investigation. These approaches may be summarised as the ecosystem approach, community approach (synecology), population approach (autecology), habitat approach and the evolutionary–historical approach.

These five approaches to ecology interact and overlap to some extent. However, they provide a useful framework for study. In this text it is not possible to consider all in equal depth. Instead, attention will be focussed on the ecosystem, community and population approaches which between them include the essence of the subject. A brief summary of the chief characteristics and contributions to ecological thought, and the problem-solving possibilities of the five main approaches is given below.

Ecosystem approach

The ecosystem was first defined by Tansley in 1935 as the living world and its habitat (see also definitions in fig 12.1 and section 12.2.1). The ecosystem approach focusses on the **flow of energy** and **cycling of matter** between living and non-living components of the ecosphere. A systems ecologist, therefore, is usually concerned with the functional relationships (such as feeding) between organisms and between them and their environment, rather than with species composition of communities and identification of rarities or special variations. The ecosystem approach also highlights the similarity in organisation of all living systems irrespective of differing taxonomy or habitat. A simple comparison of aquatic and terrestrial ecosystems, as shown in fig 12.2, serves to stress this point. Note the similarities in structure and functional units between the two systems despite the widely differing habitats and species contained within them.

At the same time, the ecosystem approach introduces the concept of **homeostasis** (self-regulation) in living systems, shows how this operates by means of feedback mechanisms (see section 12.4, biogeochemical cycles), and makes clear that breakdown of the regulatory mechanisms, for example by pollution, may lead to biological imbalance, such as excess of an organism as with a plankton bloom. The ecosystem approach is also relevant to the future development of scientifically sound agricultural practices.

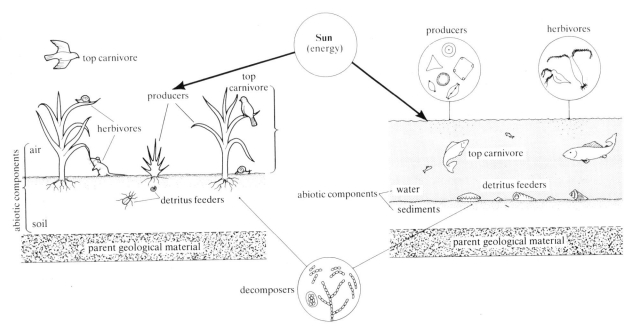

Fig 12.2 *A simple comparison of the gross structure of an aquatic (freshwater or marine) and a terrestrial ecosystem. Necessary units for function are: abiotic components (basic inorganic and organic compounds); producers (vegetation on land, phytoplankton in water); animals – direct or 'grazing herbivores' (grasshoppers, meadow mice etc. on land,* *zooplankton etc. in water); indirect or detritus-feeding consumers (soil non-vertebrates on land, bottom non-vertebrates in water); 'top' carnivores (hawks and large fish); decomposers (bacteria and fungi of decay). Animals and decomposers comprise the consumers. (Modified from E.P. Odum (1975)* Ecology, *2nd ed., Holt, Rinehart & Wilson.)*

Community approach

The term community is defined in fig 12.1.

Community ecology focusses in particular on the biotic components of ecosystems. It is synonymous with a synecological approach (sections 12.6 and 13.4). Thus in community studies one examines the plants, animals and microbiology of a recognisable biotic unit such as a woodland, grassland or heathland. Limiting factors may be identified, but functional aspects of the physical environment, such as weathering, are not usually studied in any detail. Emphasis is placed instead on the identification and description of the species present, as well as an examination of factors which control their presence, such as dispersal, competition and resource partitioning.

One important aspect of community studies is the concept of **succession** and **climax communities** (section 12.6) which is an important consideration for rational conservation management.

Population approach

The term population is defined in fig 12.1.

Population ecology covers the area studied in autecology (sections 12.7 and 13.5). Modern population studies are concerned with the characteristic mathematical forms of the growth, maintenance and decline of species populations. They embrace a number of important concepts, such as natality (birth rate), survivorship and mortality. Population ecology provides the theoretical basis for understand-

ing the 'outbreaks' of pests in agriculture and medicine, the possibilities for biological control methods (the control of pest organisms by biological means, such as the introduction of their predators or parasites), and the critical numbers of individuals needed for continued survival of a species. This latter point is important for the design and management of nature reserves and game parks, and also links strongly with evolutionary and historical ecology.

Recent application of ecological ideas to palaeontology has given new insights into how species interacted in fossil communities. Population ecology provides an important theoretical basis for examining species expansion and extinction since the very beginnings of life on this planet.

Habitat approach

Habitat is a spatial concept. It describes the typical environment of a particular organism, population, community or ecosystem. On the grand scale, Earth has four major habitats, that is marine, estuarine, freshwater and terrestrial. Each is characterised by physical conditions or limiting factors, such as salinity or temperature, that influence the presence and survival of organisms in that habitat and their distribution within it. The organism is adapted to the physical conditions of the habitat. On the local scale, habitat studies give unity to, and a readily understood basis for, field studies. Local habitats could include a hedgerow, freshwater pond, oakwood or rocky shore. Details of the application of the habitat approach to ecological investigation are given in chapter 13. Some communities are so intimately linked with particular habitats that they cannot meaningfully be studied in any other context, such as sand dune or saltmarsh communities. However, an adequate study of, for example, sand dune ecology will incorporate the four main approaches. Particular locations within the same overall habitat may have their own special conditions and are sometimes referred to as **microhabitats**, such as the bark of a rotting log in an oakwood.

An organism's habitat is not solely a matter of *physical conditions*. Its environment may be modified, or even primarily determined, by *other living organisms*. A related, useful and important concept is the **ecological niche** which combines the ideas of spatial habitat with the functional relationships of an organism, especially its feeding activities and other interactions with other species. An organism's niche thus describes both its location and function ('address' and 'profession') within a particular community or ecosystem. To understand more completely why an organism not only exists but flourishes in a particular place one must study its absolute physical and biotic constraints, which will determine its **potential niche**, and its preferences and behaviour, which will determine the more restricted range of its **realised niche** (actual niche). (See reference to computer program ECOSPACE, p. 373.)

If two species occupy the same niche they will generally compete with one another until one is displaced. Similar habitats have similar ecological niches and in different parts of the world may contain morphologically similar, but taxonomically different, animal and plant species. Open grassland and scrub, for example, will typically provide a niche for fast running herbivores, but these may be horses, antelope, bison, kangaroos, and so on.

The habitat approach is also convenient for studying those characteristics of the physical environment which are intimately linked with plants and animals such as soils, moisture and light. Here, links with the ecosystem and community approaches are particularly strong. The continued development of related sciences such as hydrology, pedology, meteorology, climatology and oceanography has opened important new interdisciplinary areas of study. Unfortunately it also makes the challenge too wide for reasonable study by any one individual, so teamwork is important and individual ecologists generally focus on one aspect of plant or animal and environment interactions, such as forest hydrology, crop climatology and derelict land reclamation. Again it is expected that functional approaches (ecosystem, community and population approaches) will be incorporated into the studies.

Evolutionary and historical approach

By studying how ecosystems, communities and populations have changed over time we gain important insights into why changes occurred and a good basis for predicting the likely nature of future changes. **Evolutionary ecology** views the changes over time since life evolved. Most importantly it gives us an understanding of the nature of the ecosphere before humans became a major environmental influence. It may be studied using, for example, fossil evidence or sedimentary sequences. Typical studies include attempts to reconstruct the ecosystems of the past, comparisons of ancient and modern distributions of taxa, and analysis of the interplay of genetic and environmental change. Evolutionary ecology thus has strong links with geological science. The application of ecosystem ideas to the past is a relatively new development in traditional palaeontology. Evolutionary ecology in general is an expanding and fruitful field of study.

Historical ecology is concerned with change since the developing technology and culture of the human species made human activity a major influence on ecological systems. In Britain and the rest of Europe the major technological and cultural advance that led to the beginning of widespread forest clearance dates from the New Stone Age or Neolithic times, which began about 5000 BP (before present). Pollen analysis and archaeological artefacts form the main evidence for pre-documentary times (in cultural terms 'pre-historic'). Written records, early maps and tree-ring analysis are important additional tools for more recent historical times.

The long-term perspective is a valuable aid to ecological understanding. Trends and strategies may be identified which contemporary studies alone cannot reveal, for example the periodicity of drought in the Sahel and the possibility of cyclical phases of abundance and decline in North Sea herring stocks.

Another example is provided by the checkerspot butterfly (*Euphydryas ethida*). In parts of its range it lays eggs on just one plant, a species of Indian paint brush *Castilleja linariifolia*. Other plants around, and used elsewhere in its range, are ignored. The area concerned experiences occasional drought phases which return periodically between 50–100 years. Chance opportunity to observe a drought phase showed that unlike the alternative host species, *C. linariifolia* was drought-resistant. The survival benefits of the checkerspot's highly selective egg-laying strategy in this part of its range thus became clear.

Historical ecology is a vital aspect of effective conservation management of plagioclimax communities, such as the lowland heaths of Britain and the rest of Europe (see section 12.6.1). These heaths are a result of forest clearance followed by centuries of grazing and burning management. Their present species and landscape value cannot be conserved without maintaining or simulating traditional land-use practice. Without this, succession to some form of scrub or woodland is inevitable.

Evolutionary studies of species dispersal, adaptive radiation and extinction, like historical studies, form the basis of models of nature reserve design and species conservation. The study of past communities and populations has traditionally been part of palaeontology, but the application of ecosystem ideas to the past is relatively recent.

12.2 The ecosystem

12.2.1 Definitions and key concepts

Ecosystems are made up of living and non-living components, known as **biotic** and **abiotic** components respectively. The organisms which comprise the biotic component are collectively known as the **community**. The term 'ecosystem' was first used by Tansley in 1935 to describe 'the whole complex of living organisms living together as a sociological unit and its habitat,' or, to put it more simply, 'the living world and its habitat'. The terms 'microcosm' in North America and 'biogeocenose' in Soviet and Central European literature embrace similar ideas. Numerous restatements of the concept have tried to improve or elaborate on Tansley's original definition. Two useful examples are given below.

Lindeman (1942): 'a system composed of physical–chemical–biological processes active within a space–time unit of any magnitude'.

Odum (1963): 'the basic functional unit of nature including both organisms and their non-living environment, each interacting with the other and influencing each other's properties and both necessary for the maintenance and development of the system'.

Several key points about ecosystems emerge from these definitions.

(a) Living (biotic) and non-living (abiotic) components are equally important and therefore demand equivalent study and understanding.

(b) The close association of living and non-living components; both affect each other.

(c) Ecosystems can be studied at any scale. This makes individual systems difficult to define since boundaries are not clear-cut. The 'fuzzy' boundaries, however, make us constantly aware of the potential importance of factors beyond the immediate focus of interest. For example, damming the River Nile 700 miles upstream at Aswan reduced freshwater discharge and sediment load downstream and led to increased salinity and decreased nutrient status in the eastern Mediterranean. An associated decline in the eastern Mediterranean fisheries and rapid recession of the delta coastline was seen.

(d) All organisms and all features of the physical environment are necessary for the maintenance and flourishing of the system. Any change will bring a reaction (homeostasis). This may not necessarily be detrimental, but human manipulation of ecosystems has often brought unwished-for 'side-effects', such as pests in simplified agricultural ecosystems. Changes may not be immediately obvious. Some scientists consider that we are only now beginning to see the consequences of continuing human ignorance of this point, for example global warming due to increased atmospheric carbon dioxide from accelerating use of fossil fuels, chlorofluorocarbons (CFCs) and the ozone hole, pollutants and acid rain. These three examples in particular reflect failure to understand the importance of the abiotic environment for the well-being of living systems. These issues will be discussed more fully in sections 12.4 and 12.10.

12.2.2 Overall structure of ecosystems

Ecosystems are complex, but a convenient starting point is fig 12.2 which shows in a simplified way the overall structures of a terrestrial and an aquatic ecosystem.

The biotic component can usefully be subdivided into autotrophic and heterotrophic organisms. All living organisms fit into one of these two categories, as has been described in table 9.1. Autotrophic organisms synthesise their own organic requirements from simple inorganic molecules and, with the exception of chemosynthetic bacteria, do this by photosynthesis, using light as an energy source. Heterotrophic organisms require a source of organic food and, with the exception of a few bacteria, rely on a chemical source of energy derived usually from the organic food they consume. It will be shown that heterotrophs are dependent on autotrophs for their existence, and that an understanding of their relationships is essential to an understanding of ecosystems. It is also fundamental to human manipulation of ecosystems as, for example, in agriculture.

The non-living or abiotic component of an ecosystem is principally divided into soil or water, and climate. Soil and water contain a mixture of inorganic and organic nutrients. The underlying bedrock from which soil is partly derived and on which it is based contributes to the properties of the soil. Climate includes such environmental variables as light, temperature and water, which are important in determining the types of living organisms that can flourish in the ecosystem. In aquatic ecosystems salinity is another major variable.

The essence of ecosystem studies, however, lies in understanding how connections between the different organisms and their abiotic environment work. Two important ways in which these connections are seen are energy flow and nutrient cycling. These aspects are discussed more fully in the following sections.

With the ecosystem approach, the ultimate aim is a synthesis of our knowledge to give an understanding of a complex system. This should always be kept in mind as different parts of the system are analysed separately.

12.3 Energy and nutrient relationships

12.3.1 Energy flow and nutrient cycling

The organisms of an ecosystem are linked by their energy and nutrient relationships and the distinction between energy and nutrients must be fully appreciated. Chapters 9–11 have already dealt with living organisms as consumers of energy and nutrients and it is strongly recommended that the opening sections of chapter 9 are read so that this is fully understood.

In section 9.1 energy is defined as the capacity to do work and living organisms are likened to machines in that they require energy to keep working, that is to stay alive. The ecologist can regard the whole ecosystem as one machine which is kept working by an input of energy and nutrients (materials or matter). Nutrients are derived originally from the abiotic components of the ecosystem, to which they eventually return by way of the decomposition of waste products or dead bodies or organisms. Thus a constant recycling of nutrients occurs within an ecosystem. Both living and non-living components are involved, so the cycles are called **biogeochemical cycles**. They are described further in section 12.4.

The energy to drive these cycles is supplied ultimately by the Sun. In the biotic component, photosynthetic organisms utilise the Sun's energy directly and pass it on to other organisms. The net result is a flow of energy and a cycling of nutrients through the ecosystem as illustrated in fig 12.3. It should also be pointed out that in the abiotic component climatic factors such as temperature, movement of the atmosphere, evaporation and rainfall are also regulated by the input of solar energy.

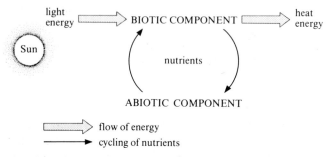

Fig 12.3 *Flow of energy and cycling of nutrients through an ecosystem*

To understand why energy flows through the ecosystem in a linear fashion, rather than being recycled and re-used like the nutrients, a brief consideration of thermodynamics is necessary.

Energy can be thought of as existing in various forms, such as mechanical, chemical, heat and electrical energy, all of which are interconvertible. The change of energy from one form to another, known as the **transformation of energy**, is governed by the laws of thermodynamics. The first law, **the law of conservation of energy**, states that

energy may be transformed from one form into another, but can be neither created nor destroyed. The second law states that, in performing work, no energy conversion can be 100% efficient and that some energy must escape as heat. Heat is the result of the random movement of molecules, whereas work always involves a non-random (ordered) use of energy.* The concept of 'work' can be applied to any energy-consuming process carried out by living organisms, from processes at the cellular level, such as the maintenance of electrical gradients across membranes and protein synthesis, to processes at the level of the whole organism, such as growth, development, repair and reproduction.

Thus living organisms are energy transformers and each time an energy transformation is carried out some energy is lost as heat. Ultimately *all* the energy that enters the biotic component of the ecosystem is lost as heat. It may be imagined that since heat can be used to do work, as in a steam locomotive, there is no reason why heat should not be recycled. However, the process that generates the heat requires more energy than can be reclaimed by recycling the heat, so that an overall rundown of useable energy would still occur. In practice, living organisms do not use heat as a source of energy to do work, but use light or chemical energy.

The study of energy flow through ecosystems is called **energetics** and since this, together with nutrient cycling, is a dominant theme in the study of ecosystems it is important to become familiar with the energy units used.

12.3.2 Energy units

The SI unit of energy is the joule, though the traditional unit of the calorie is still in common use. Both units are defined in table 12.1, which also includes references to the energy content of representative foods and organisms, and to daily food requirements of representative organisms.

> **12.1** Why are the figures for energy content in table 12.1 quoted for dry mass rather than fresh (wet) mass?
>
> **12.2** Account for the large difference in daily energy requirements of humans and small birds or mammals on a weight for weight basis.

12.3.3 The Sun as a source of energy

The ultimate source of energy in ecosystems is the Sun. The Sun is a star which releases vast amounts of solar energy into space. The energy travels through space as electromagnetic waves and a small fraction of it, about 1/2000 millionth, amounting to $10.8 \times 10^6 \, \text{kJ m}^{-2} \text{yr}^{-1}$,

---

* The increase in disorder that is involved in energy transformations is measured as a term called **entropy**. In the universe as a whole entropy is continuously increasing.

Table 12.1 Units of energy and energy content of some living organisms and biological molecules.

Energy units

calorie (cal) or gram calorie	– the amount of heat (or energy) needed to raise the temperature of one gram of water through 1 °C (14.5 °C to 15.5 °C)
kilocalorie (kcal or Cal)	– 1 000 cal
joule (J)	– 10^7 ergs: 1 erg is the amount of work done when 1 newton moves through 1 metre (1 newton (N) is a unit of force) Alternatively 981 ergs is the work done in raising a gram weight against the force of gravity to a height of 1 cm
kilojoule (kJ)	– 1 000 J

$$1 \text{ J} = 0.239 \text{ cal} \qquad 1 \text{ cal} = 4.186 \text{ J}$$

Energy content (averages or approximations)

	joules per gram dry mass (energy value)
carbohydrate	16.7
protein	20.9
lipid	38.5
terrestrial plants	18.8
algae	20.5
non-vertebrates (excl. insects)	12.6
insects	22.6
vertebrates	23.4

(differences between these groups of organisms are due partly to different mineral contents)

Daily food requirements	*kJ per kg live body mass*
humans	167 (about 12 500 kJ day^{-1} for a 70 kg adult)
small bird or mammal	4 186
insect	2 093

Based on data from table 3.1, Odum, E. P. (1971) *Fundamentals of Ecology*, 3rd ed. Saunders.

is intercepted by the Earth. Of this, about 40% is reflected immediately from the clouds, dust in the atmosphere and the Earth's surface without having any heating effect. This is termed the planetary **albedo**. A further 15% is absorbed and converted to heat energy in the atmosphere, particularly by ozone in the stratosphere, and by water vapour. The ozone layer absorbs almost all short-wave ultraviolet radiation which is important because such radiation is lethal to exposed living material. The remaining 45% of incoming energy penetrates to the Earth's surface. This represents an average of about $5 \times 10^6 \text{ kJ m}^{-2} \text{ yr}^{-1}$, though the actual amount for a given locality varies with latitude and local features such as aspect. Just under half the radiation striking the Earth's surface is in the photosynthetically active range (PAR), the visible wavelengths. However, under optimum conditions only a very small proportion, about 5% of incoming radiation (or 10% PAR) is converted in photosynthesis into gross primary produc-

tivity (GPP). A more typical figure for good conditions is 1% of total radiation (2% PAR) while the biosphere average is about 0.2% of total incident radiation. Net primary productivity (NPP) (the net gain of organic material in photosynthesis after allowing for losses due to respiration) varies between 50% and 80% of gross primary productivity (see section 12.3.7).

As a global average the energy fix by Earth's green plants is only 0.1% of total sunlight. Terrestrial systems, which cover one-third of the planetary surface, fix half the total sunlight captured. Cultivated crops achieve higher rates of GPP and NPP during their short cultivation periods, but so far it has proved impossible to achieve higher rates of photosynthetic fixation on a sustained basis under normal field conditions.

12.3.4 Energy transfers: food chains and trophic levels

Within the ecosystem, the energy-containing organic molecules produced by autotrophic organisms are the source of food (materials and energy) for heterotrophic organisms; a typical example is a plant being eaten by an animal. This animal may in turn be eaten by another animal, and in this way energy is transferred through a series of organisms, each feeding on the preceding organism and providing raw materials and energy for the next organism. Such a sequence is called a **food chain**. Each stage of the food chain is known as a **trophic level** (*trophos*, feeding), the first trophic level being occupied by the autotrophic organisms, the so-called **primary producers**. The organisms of the second trophic level are usually called **primary consumers**, those of the third level are **secondary consumers**, and so on. There are usually four or five trophic levels, and seldom more than six for reasons stated in section 12.3.7 and obvious from fig 12.12. Further characteristics of each link in the chain are as follows, and the sequence is summarised in fig 12.4.

Primary producers

The primary producers are autotrophic organisms, and are mainly green plants and algae. Some prokaryotic organisms, namely a few bacteria (including blue-green bacteria), are also photosynthetic but their contribution is relatively small. Photosynthetic organisms transform solar energy (light energy) to chemical energy which is contained within the organic molecules that make up their tissues. A small contribution is also made by chemosynthetic bacteria which obtain their energy from inorganic compounds.

The major primary producers of aquatic ecosystems are algae, often minute unicellular algae that make up the phytoplankton of the surface layers of oceans and lakes. On land the major primary producers are the larger plants, namely angiosperms and conifers, which form forests and grasslands.

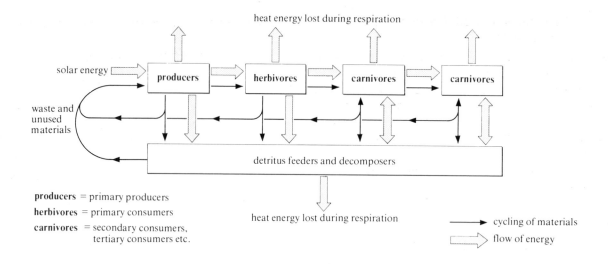

Fig 12.4 *Flow of energy and cycling of materials through a typical food chain. Note that a two-way exchange is possible between carnivores and detritus feeders/decomposers. The latter feed on dead carnivores; carnivores may eat living detritus feeders/decomposers*

Primary consumers

Primary consumers feed on primary producers and are therefore **herbivores**. On land, typical herbivores include insects as well as reptiles, birds and mammals. Two important groups of herbivorous mammals are rodents and ungulates. The latter are hoofed grazing animals such as horses, cattle and sheep that are adapted for running on the tips of their digits.

In aquatic ecosystems (freshwater and marine) the herbivores are typically small crustaceans and molluscs. Most of these organisms, such as water fleas, copepods, crab larvae, barnacles and bivalves (for example mussels and clams) are filter-feeders and extract the minute primary producers from the water as described in section 10.2.2. Together with protozoa, they make a large contribution to the zooplankton which feed on the phytoplankton. Life in the oceans and lakes is almost totally dependent on plankton since it is found at the beginning of virtually all food chains.

Primary consumers also include parasites (fungi, plants or animals) of plants.

Secondary and tertiary consumers

Secondary consumers feed on herbivores and are therefore carnivores. Tertiary consumers feed on secondary consumers and are also carnivores.

Secondary and tertiary consumers may be **predators**, which hunt, capture and kill their prey; **carrion feeders**, which feed on corpses; or **parasites**, in which case they are smaller than their hosts. Parasite food chains are excep-

tional and are included with pyramids of numbers in section 12.3.6 (see questions 12.4 and 12.5).

In a typical predator food chain the carnivores get larger at successive trophic levels:

plant (such as nectar)→fly→**spider→shrew→owl**

rosebush sap→aphid→**ladybird→spider→insectivorous bird→hawk**

In a typical parasite food chain the parasites get smaller at successive levels (questions 12.4 and 12.5). Some further examples of food chains are given below.

Marine:	diatom (unicellular algae forming important constituents of phytoplankton, e.g. *Chaetoceros*) →	copepod (small crustaceans forming important constituent of zooplankton, e.g. *Calanus*) →	herring *Clupea harengus*
Seashore:	seaweed e.g. *Fucus* →	flat periwinkle *Littorina littoralis* →	oystercatcher *Haematopus ostralegus*
Freshwater:	diatom e.g. *Navicula* →	mayfly larva *Baetis rhodani* →	caddis fly larva *Rhyacophila* sp.
Woodland:	woodland plant e.g. *Rubus* sp. (blackberry) →	bank vole *Clethrionomys glareolus* →	tawny owl *Strix aluco*

Decomposers and detritivores (detritus food chains)

Two basic types of food chain exist, namely grazing food chains and detritus food chains. The examples quoted so far are of **grazing food chains**, in which the first trophic level is occupied by a green plant (or alga), the second by a 'grazing' animal (herbivore) (the term grazing being used in a broad sense to include the eating of a plant by any animal), and subsequent levels by carnivores. When plants and animals die their bodies still contain energy and raw materials, as do the waste products such as urine and faeces which they deposit during lives. These organic materials are decomposed by micro-organisms, namely fungi and bacteria, which live saprotrophically on the remains. They are called **decomposers** and secrete digestive enzymes onto the dead or waste material, subsequently absorbing the products of digestion. The rate of decomposition varies with substrate and climate. The organic matter of animal urine, faeces and corpses is consumed within a matter of weeks, whereas fallen trees and branches may take many years to decompose. Essential to the breakdown of wood (and other plant material) is the action of fungi which produce cellulase, softening the wood and allowing small animals to penetrate and ingest material.

Decomposition is rapid in warm and moist environments, such as tropical rain forest, but takes place slowly in cool and/or dry conditions. The virtual absence of litter from the rain forest floor and the low content of humus in rain forest soils by comparison with the conspicuous litter layer and significant humus content of soils in temperate oakwoods or beechwoods illustrates this point. This has important implications for human use of these systems.

Fragments of decomposing material are called **detritus**, and many small animals feed on these, contributing to the process of breakdown. They are called **detritivores**. Because the combined activities of the true decomposers (fungi and bacteria) and detritivores (animals) lead to breakdown (decomposition) of materials, they are sometimes all referred to collectively as decomposers, although strictly the term decomposer relates to saprotrophic organisms.

Detritivores may in turn be fed upon by larger organisms, building up another type of food chain which can be regarded as starting with detritus. Thus:

detritus→detritivore→carnivore.

Some detritivores of woodland and seashore communities are shown in fig 12.5.

Two typical detritus food chains of woodlands are:

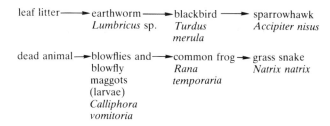

Fig 12.5 *Some woodland and seashore detritivores*

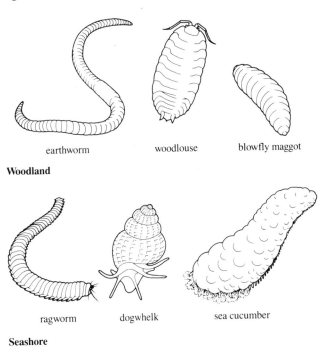

Woodland — earthworm, woodlouse, blowfly maggot

Seashore — ragworm, dogwhelk, sea cucumber

Some typical terrestrial detritivores are earthworms, woodlice, millipedes and the smaller (<0.5 mm) animals such as mites, springtails, nematode worms and enchytraeid worms. Methods for isolating and examining these are given in section 13.2.

12.3.5   Food webs

In food chains each organism is depicted as feeding on only one other type of organism. However, the feeding relationships within an ecosystem are more complex than this because each animal may feed on more than one organism in the same food chain, or may feed in different food chains. This is particularly true of carnivores at the higher trophic levels. Some animals, most notably humans, feed on plants, animals and fungi and are called **omnivores**. In reality then, the food chains interconnect in such a way as to produce a **food web**. Fig 12.6 illustrates woodland and freshwater food webs. Only some of the many possible interrelationships can be shown on such diagrams and it is usual to include only one or two carnivores at the highest level. Such diagrams illustrate the feeding relationships among organisms in an ecosystem and provide a basis for more quantitative studies of energy flow and exchange of material through the biotic component of ecosystems. This is discussed more fully in the next two sections.

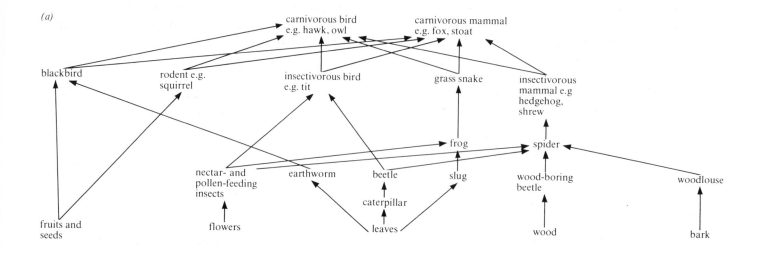

(a)

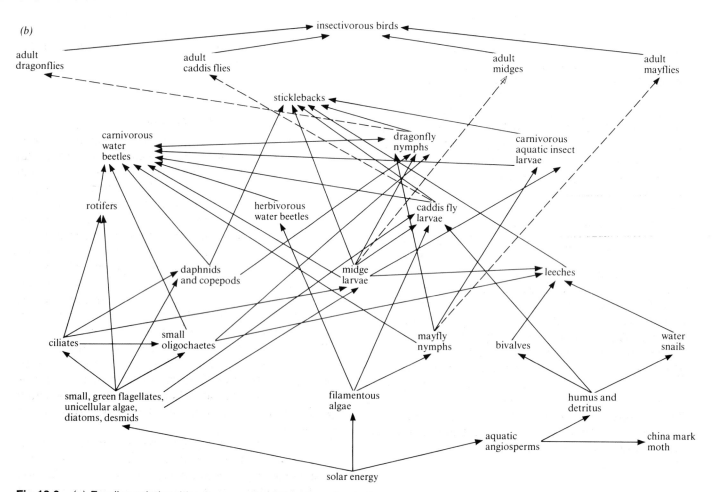

(b)

Fig 12.6 *(a) Feeding relationships in a woodland, forming a food web. (b) Food web of a freshwater habitat (Based on Popham (1955)* Some aspects of life in fresh water, *Heinemann.)*

12.3.6   Ecological pyramids

Feeding relationships and energy transfer through the biotic component of ecosystems may be quantified and shown diagrammatically as ecological pyramids. These give an apparently simple and fundamental basis for comparing different ecosystems, or even seasonal variation of pollution-induced change within a single system. However, ecologists working on the International Biological Programme (IBP), which aimed to gather data for quantitative comparison of all major world ecosystems during the decade 1964–74, have questioned the usefulness of pyramids as summaries of energy flow. One difficulty lies with deciding the trophic level of an organism. This and other problems will be considered later (see 'Criticism of ecological pyramids').

Pyramid of numbers

In a given area of ecosystem, small organisms usually outnumber large organisms. A pyramid of numbers of animals in different size classes can readily be constructed (fig 12.7a) (autotrophic organisms are omitted). This is the original basis of the pyramid of numbers as first proposed by Charles Elton in the 1920s. Elton also noted that typically predators are larger than their prey, and his original model based on *size* classes was therefore adapted by Lindeman to a trophic model, that is one based on the *feeding* levels of the organisms concerned irrespective of their size (figs 12.7b, c, d). Though theoretically this may seem a reasonable extension of Elton's ideas, in practice it is much more difficult to collect data for trophic levels than for size classes (see 'Criticism of ecological pyramids' later). Fig 12.7e shows how Elton's size-based and Lindeman's trophic models are related.

In a trophic-level based pyramid of numbers the organisms of a given area are first counted and then grouped into their trophic levels (as best one can). When this is done it is usually found that there is a progressive decrease in the number of animals at each successive level. Plants in the first trophic level also often outnumber animals at the second trophic level, though this will depend on the relative sizes of organisms, such as a tree compared with phytoplankton. A *pyramid* of numbers typically results.

For diagrammatic purposes the number of organisms in a given trophic level can be represented as a rectangle whose length (or area) is proportional to the number of organisms in a given area (or volume if aquatic). Figs 12.7b, c and d illustrate three types of naturally occurring pyramids of numbers. The carnivores in the highest trophic level are known as the **top carnivores**.

12.3   In pyramid (b) (fig 12.7) the primary producers (plants) are small in size and outnumber the herbivores. Describe and explain the difference in pyramid (c) compared with (b).

12.4   *Leptomonas* is a parasitic flagellate protozoan, thousands of which may be found in a single flea. Construct a pyramid of numbers based on the following food chain:

grass→herbivorous mammal→flea→*Leptomonas*

12.5   Give a possible explanation of the difference between pyramids (b) and (d) in fig 12.7.

Although the data needed to construct pyramids of numbers may be relatively easy to collect by straightforward sampling techniques, there are a number of complications associated with their use. Three important problems are as follows.

(1) Deciding to which trophic level an organism belongs.
(2) The producers vary greatly in size, but a single grass plant or alga, for example, is given the same status as a single tree. This explains why a true pyramid shape is often not obtained. Also parasitic food chains may give inverted pyramids (see questions 12.3–12.5).
(3) The range of numbers is so great that it is often difficult to draw the pyramids to scale, although logarithmic scales may be used.

Pyramid of biomass

The disadvantages of using pyramids of numbers can be overcome by using a **pyramid of biomass** in which the total mass of the organisms (**biomass**) is estimated for each trophic level. Such estimates involve weighing representative individuals, as well as recording numbers, and so are more laborious and expensive in terms of time and equipment. Ideally, dry masses should be compared. These can either be estimated from wet masses or can be determined by destructive methods (experiment 13.1). The rectangles used in constructing the pyramid then represent the masses of organisms at each trophic level per unit area or volume. Fig 12.8a shows a typical pyramid of biomass in which biomass decreases at each trophic level.

The biomass at the time of sampling, in other words at a given moment in time, is known as the **standing biomass** or **standing crop biomass**. It is important to realise that this figure gives no indication of the *rate* of production (**productivity**) or consumption of biomass. This can be misleading in two ways.

(1) If the rate of consumption (loss through being used as food) more or less equals the rate of production, the standing crop does not necessarily give any indication of productivity, that is the amounts of material and energy passing from one trophic level to the next in a given time period such as one year. For example, a fertile, intensively grazed pasture may have a smaller standing crop of grass, but a higher productivity, than a less fertile and ungrazed pasture.

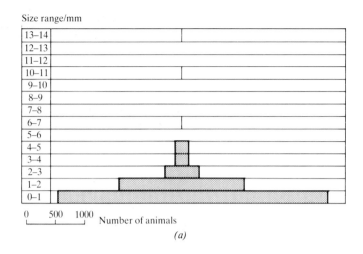

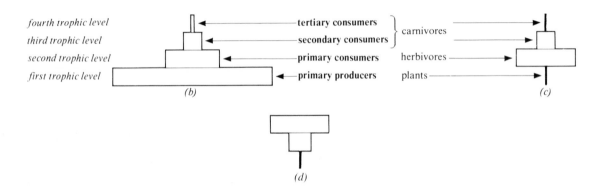

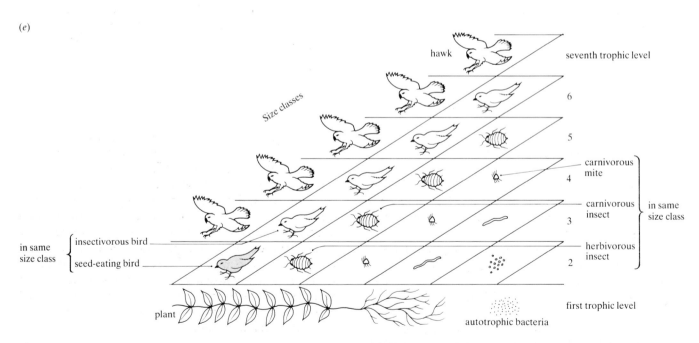

Fig 12.7 *(a) An Eltonian pyramid of numbers based on size class, in this case for animals on the floor of a forest in Panama. (From S. Cousins,* New Scientist, *4.7.85, p. 51). Elton noted that predators are typically larger than their prey, but the size difference is never so great that prey are too small to see and handle easily or too big to catch and eat without difficulty. In short, the size of prey is 'about right' for the predator. Two common types of pyramids of number, (b) and (c), and an inverted pyramid (d). These are explained in the text. (e) The Lindeman and Elton models, rearranged so that they relate to one another: a hawk feeds at five trophic levels. Organisms in the same trophic level are in the same horizontal row. Size classes are arranged diagonally and omit autotrophic organisms. Note that trophic levels usually contain organisms of a variety of sizes, and that the hawk, for example, should be divided between trophic levels 3 to 7.*

(a) old field, Georgia USA

(b) English Channel

Fig 12.8 (above) *Pyramids of biomass. Type (a) is the more common. Type (b) is an inverted pyramid (see text). Figures represent g dry mass m⁻². (From E. P. Odum (1971)* Fundamentals of ecology, *3rd ed., W. B. Saunders.)*

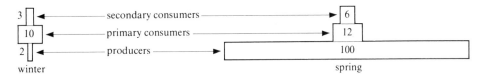

Fig 12.9 (above) *Seasonal changes in pyramids of biomass for an Italian lake. Figures represent mg dry mass m⁻³. (From E. P. Odum (1971)* Fundamentals of ecology, *3rd ed., W. B. Saunders.)*

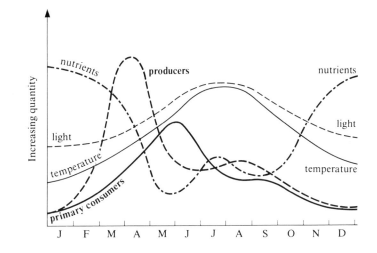

Fig 12.10 (right) *Changes in standing crop biomass of producers and primary consumers and in certain environmental variables in a lake during one year. (From M. A. Tribe, M. R. Erant & R. K. Snook (1974)* Ecological principles, *Basic Biology Course 4, Cambridge University Press.)*

(2) If the producers are small, such as algae, they have a high turnover rate, that is a high rate of growth and reproduction balanced by a high rate of consumption or death. Thus, although the standing crop may be small compared with large producers such as trees, the productivity may be the same, trees accumulating their biomass over a long time period. Put another way, an amount of phytoplankton with the same productivity as a tree would have a much smaller biomass than a tree, even though it could support the same amount of animal life. In general, the larger, longer-lived plants and animals have lower 'turnover rates' than the smaller, shorter-lived plants or algae and animals, and accumulate materials and energy over a longer time period. One possible consequence of this is shown in fig 12.8b, where an inverted pyramid of biomass is illustrated for an English Channel community. The zooplankton is shown to have a higher biomass than the phytoplankton on which it feeds. This is characteristic of ocean and lake planktonic communities at certain times of year;

phytoplankton biomass exceeds zooplankton biomass during the spring 'bloom', but at other times the reverse might be true. Such apparent anomalies are avoided by using pyramids of energy as described below.

Equally, these differences may highlight useful information. For example, persistence of an algal bloom and a broad-based biomass pyramid in an aquatic ecosystem may indicate the onset of eutrophication (a form of water pollution discussed more fully in section 12.4.1). Similarly, the frequent inversion of marine pyramids suggests that for these systems harvesting plant or algal rather than animal biomass is not the sensible strategy it appears to be for many terrestrial ecosystems.

12.6 Consider the two pyramids of biomass shown in fig 12.9. They represent plankton in an Italian lake at two different times of year, spring and winter. Account for the inversion of the pyramid during the year.

371

Pyramid of energy

The most fundamental and ideal way of representing relationships between organisms in different trophic levels is by means of a pyramid of energy. This has a number of advantages.

(1) It takes into account the *rate* of production, in contrast to pyramids of numbers and biomass which depict the standing states of organisms at a particular moment in time. Each bar of a pyramid of energy represents the amount of energy per unit area or volume that flows through that trophic level in a given time period. Fig 12.11 shows a pyramid of energy for an aquatic ecosystem. Note that the units used are for energy flow.

carnivores to top carnivores — 88
herbivores to carnivores — 1 603
producers to herbivores — 14 098
gross production — 87 110

Fig 12.11 *Pyramid of energy for Silver Springs, Florida. Figures represent energy flow in kJ m⁻² yr⁻¹. (From E. P. Odum (1971) Fundamentals of ecology, 3rd ed., W. B. Saunders.)*

(2) Weight for weight, two species do not necessarily have the same energy content, as table 12.1 indicates. Comparisons based on biomass may therefore be misleading.

(3) Apart from allowing different ecosystems to be compared, the relative importance of populations within one ecosystem can be compared and inverted pyramids are not obtained. This is illustrated by table 12.2, where energy flow (energy output) through primary consumers of different biomass is compared. Note, for example, that the great importance of soil bacteria in terms of energy flow is not obvious from their small biomass.

(4) Input of solar energy can be added as an extra rectangle at the base of a pyramid of energy.

Although pyramids of energy are sometimes considered the most useful of the three types of ecological pyramid, they are the most difficult to obtain data for because they require even more measurements than pyramids of

Table 12.2 Density, biomass and energy flow of five primary consumer populations.

	Approximate density (no. m⁻²)	Biomass (g m⁻²)	Energy flow (kJ m⁻² day⁻¹)
Soil bacteria	10^{12}	0.001	4.2
Marine copepods (*Acartia*)	10^{5}	2.0	10.5
Intertidal snails (*Littorina*)	200	10.0	4.2
Salt marsh grasshoppers (*Orchelimum*)	10	1.0	1.7
Meadow mice (*Microtus*)	10^{-2}	0.6	2.9
Deer (*Odocoileus*)	10^{-5}	1.1	2.1

From Odum, E. P. (1971).

biomass. One extra piece of information needed is the energy values for given masses of organisms. This requires combustion of representative samples. In practice, pyramids of biomass can sometimes be converted to pyramids of energy with reasonable accuracy, based on previous experiments.

Criticisms of ecological pyramids

The pyramids of numbers, biomass and energy described depend on assigning living organisms to trophic levels. While the correct level is obvious for plants and obligate herbivores, many carnivores and omnivores eat a varied diet and thus their trophic level varies according to the food selected, which may in turn be an animal with a range of possible trophic levels. This point is clear from the discussion of food webs and from figs 12.6a and b.

From question 12.9 the problem of assigning a hawk to a particular trophic level is self-evident. Even if its typical diet was studied, one would still need to know the feeding histories of the prey organisms. Thus, even for the hawk alone, a great deal of fieldwork would be needed; on an ecosystem basis one can only resort to very approximate generalisations.

Another major problem is where to place dead and waste material. Such material is ecologically important as a food source. For example, it is estimated that up to 80% of the energy fixed by terrestrial plants enters decay pathways rather than being passed on to herbivores, yet it is hard to fit dead material (detritus) and its consumers into conventional pyramids. One solution is to regard dead material as a new trophic level 1, and to split the ecosystem into a herbivore pathway and a detritus pathway.

These and other problems mean that our ideas about ecological pyramids are currently subject to radical reappraisal, and this should be borne in mind when studying pyramids of numbers, biomass and energy.

Computer program. PYRAMID and ECO-SPACE: A double package. PYRAMID allows population numbers, biomass or energy to be plotted for different trophic levels in the traditional bar format, but the bars expand and contract as the user steps through the changing seasons for selected ecosystems. ECOSPACE allows the concept of the ecological niche to be visualised.

12.3.7 Efficiency of energy transfer: production ecology

The study of productivity is known as production ecology, and involves the study of energy flow through ecosystems.

Energy enters the biotic component of the ecosystem through the primary producers, and the rate at which this energy is stored by them in the form of organic substances which can be used as food materials is known as **primary productivity**. This is an important parameter to measure as it determines the total energy flow through the biotic component of the ecosystem, and hence the amount (biomass) of life which the ecosystem can support.

12.10 In considering the primary productivity of an ecosystem, which groups of organisms other than plants make a contribution?

As mentioned in section 12.3.3, the amount of the Sun's radiation intercepted by Earth's surface varies with latitude and with details of location such as aspect and altitude. The amount intercepted by plants also varies with light quality and the organisation and amount of vegetation cover. In Britain, incident radiation on plants averages about 1×10^6 kJ m^{-2} yr^{-1}. Of this, as much as 95–99% is immediately lost from the plant by reflection, radiation or heat of evaporation. The remaining 1–5% of incoming radiation is absorbed by the chlorophyll and used in the production of organic molecules. The rate at which this chemical energy is stored by plants is known as **gross primary productivity (GPP)**. Between 20–50% of the GPP is used by the plant in simultaneous respiration and photorespiration, leaving a net gain known as the **net primary productivity (NPP)** which is stored in the plant. It is this energy which is potentially available to the next trophic level.

In the example given in fig 12.12 NPP is given as 8 000 kJ m^{-2} yr^{-1}, or 0.8% of received radiation (1% of absorbed radiation and 80% of the GPP). All these figures are on an annual basis, productivity being higher in the summer and lower during winter.

When herbivores and carnivores consume other organisms, food (materials and energy) is thereby transferred from one trophic level to the next. Some of the food remains undigested and is lost immediately in the process of egestion. For animals with alimentary canals this is usually in the form of faeces. Collectively, egested substances are known as **egesta**. They contain energy, as do

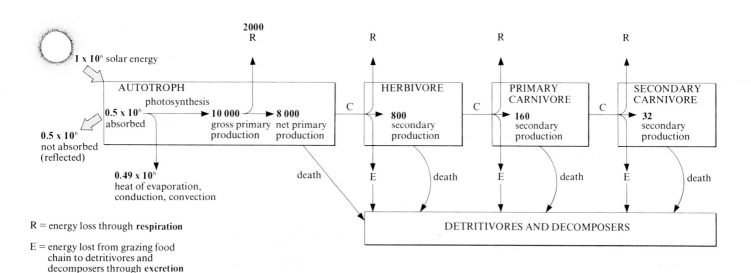

R = energy loss through **respiration**

E = energy lost from grazing food chain to detritivores and decomposers through **excretion** (e.g. urine) and **egestion** (e.g. faeces)

C = **consumption** by organisms at the higher trophic level   all energy values given in kilojoules (kJ)

Fig 12.12 *Energy flow through a grazing food chain, such as a grazed pasture. Figures represent kJ m^{-2} yr^{-1}*

373

organic excretory products (**excreta**) such as urea. Excreta are products of the animal's own metabolism, unlike egesta. Animals, like plants, also lose energy as a result of respiration. The energy remaining in heterotrophs after losses through egestion, excretion and respiration is available for production, that is growth, repair and reproduction.

Production by heterotrophs is called **secondary production** (whatever the trophic level). The following word equation summarises the fate of energy consumed by an animal:

food consumed = growth + respiration + egesta + excreta

Some of these terms can be measured easily in domestic animals or in laboratory studies of wild animals. Growth is measured as increase in biomass, or better as increase in energy value of the body, with time. Faeces and excreta can be collected, weighed and subtracted from the mass of food consumed to determine food retained and used for growth and respiration.

Measurement of productivity

As a result of photosynthesis, there is an increase in dry mass. The **relative growth rate** (R) is defined as the gain in mass per unit of plant mass in unit time.

$$R = \frac{\text{increase in dry mass in unit time}}{\text{dry mass of plant}}$$

The increase in dry mass in unit time is equal to

$$\frac{W_t - W_0}{t}$$

where W_t is the dry mass after time t and W_0 is the dry mass at the start of the time period.

The **net assimilation rate** (NAR, also called unit leaf rate) relates increase in dry mass to leaf area.

$$\text{NAR} = \frac{\text{increase in dry mass in unit time}}{\text{leaf area}}$$

Note that the increase in dry mass is equal to the difference between the real rate of photosynthesis minus the losses in dry mass due to respiration.

Measured rate (or apparent = true rate − respiratory losses rate of photosynthesis)

Biomass is the total dry mass of all organisms in an ecosystem.

Total biomass	=	biomass of primary producers	+	biomass of consumers	+	biomass of decomposers	+	biomass of dead organisms

Even though it is difficult to measure biomass accurately, biomass provides a useful comparison between different land areas or ecosystems.

Not all the organic material produced by a crop is suitable for commercial sale. For example, in cereals the aim is to produce maximum yield of grain and the remainder of the plant is not of general economic importance (although some straw will be used for the bedding of animals). The **harvestable dry matter**, as the name suggests, is the dry mass of the crop useful for commercial exploitation. It may therefore be advantageous to alter the distribution of the newly synthesised carbohydrate in the plant, more being directed to storage organs. The **harvest index** represents the economic yield compared with total yields. Indolyl acetic acid (IAA) has been shown to increase the movement of synthesised materials into storage tissue. Further, a reduced oxygen level retards reproductive growth and deflects carbohydrates into storage regions, that is economically important regions.

Solar radiation is intercepted by the leaves. In a tropical rain forest the tree canopy will be very dense, but in other situations the available leaf area for absorption of light may be quite small and most light falls on the base soil.

The **leaf area index** is a measure of the leaf area per unit of ground area, that is plants with a large index will absorb most of the light. The index for clover plants has been cited as around 33%, that for *Holium regidum* (a grass) around 16%.

Fig 12.12 shows clearly that energy is lost at every stage in the food chain and the length of the food chain is obviously limited by the extent of these losses. The proportion of energy lost in the first transfer of energy from solar energy received to net primary production is high. Subsequent transfers are at least ten times more efficient than this initial transfer. The average efficiency of transfer from plants to herbivores is about 10% and from animal to animal is about 20%. In general, herbivores make less efficient use of their food than do carnivores because plants contain a high proportion of cellulose and sometimes wood (which contains cellulose and lignin) which are relatively indigestible and therefore unavailable as energy sources for most herbivores.

Energy lost in respiration cannot be transferred to other living organisms. However, the energy lost from a food chain in the form of excreta and egesta is not lost to the ecosystem because it is transferred to detritivores and decomposers. Similarly, any dead organisms, fallen leaves, twigs and branches and so on will start detritus and decomposer food chains. The proportion of net primary production flowing directly into detritus and decomposer food chains varies from one system to another. In a forest ecosystem most of the primary production enters the detrital rather than the grazing pathway with the result that the litter and humus on the forest floor is the centre of much of the consumer activity, even though the organisms involved are mostly inconspicuous. However, in an ocean ecosystem or an intensively grazed pasture more than half the net primary production may enter the grazing food chain.

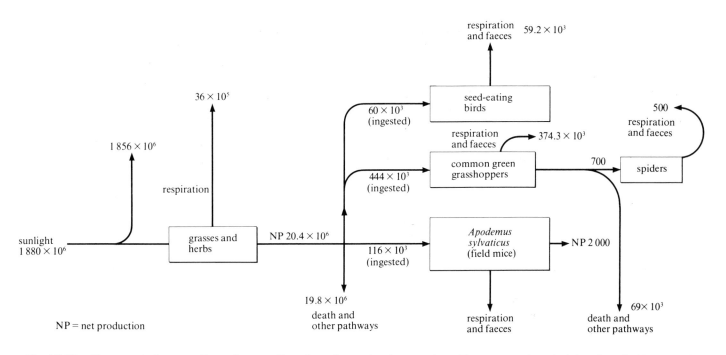

Fig 12.13 *Movement of energy through a small portion of grassland ecosystem. Figures are given in kJ m^{-2} yr^{-1}. (From M. R. Tribe et al. (1974) Ecological principles, Basic Biology Course 4, Cambridge University Press.)*

12.11 Fig. 12.13 shows the energy flow through a small portion of a grassland ecosystem. The figures are given in kJm^{-2} yr^{-1}.
(a) What is the gross primary production of grasses and herbs?
(b) What is the photosynthetic efficiency (that is the efficiency of conversion of incident solar energy to gross primary production)?
(c) What is the net production of the seed-eating birds, spiders and common green grasshoppers respectively?
(d) How much energy is lost via respiration and faeces by field mice?
(e) Which of the organisms are producers, primary consumers and secondary consumers?
(f) Which of the organisms are heterotrophic?
(g) What are the 'other pathways' likely to be? (Name three.)
(Modified from Tribe, M. A. *et al.* (1975) *Ecological Principles*, CUP).

Detrital pathways are often complex and are less well understood than the conventionally described grazing pathways. Nevertheless they are just as important and, in terms of energy flow, frequently more important than grazing pathways. Globally, 80% of the NPP enters the detrital pathway. Most intensive agricultural systems ignore the potential value of detritus-based food production.

The figures quoted in this section have been on an annual basis. If the ecosystem is stable and not increasing overall in biomass, the total biomass at the end of the year will be the same as at the beginning. All the energy that went into primary production will then have passed through the various trophic levels and none retained as net production. Normally, however, a system will be in a process of change. A young forest, for example, would retain some of the energy input in the form of increased biomass at the end of the year. A year is a useful period over which to express productivity because it takes into account seasonal variations where these exist. For example, primary productivity is usually greater in the part of the year when new plant or algal growth commences and secondary production increases later.

One of the reasons for studying energy flow through ecosystems is that it has important implications for the way in which humans obtain their own food and energy requirements. It opens the way to analysing traditional systems of agriculture for their efficiency, and suggests where improvements can be made. Since energy is lost at each trophic level, it is clear that, for omnivores like humans, eating plants is a more efficient way of extracting energy from a system (table 12.3). However, in suggesting improved methods for providing food, other factors must be considered. For example, animal protein is generally a better source of the essential amino acids, though some pulse crops, such as soya bean, are richer sources than most plants. Also animal protein is more easily digested, since the tough plant cell walls must first be broken down before

Table 12.3 Outputs of agricultural food chains in UK.

Food chain		Example	Energy yield of food to humans (kJ $\times$ 10^3 ha^{-1})	Protein yield of food to humans (kg ha^{-1} yr^{-1})
(a) Cultivated plant crop→humans		Monocultures of wheat and barley	7 800–11 000	42
(b) Cultivated plant crop→ livestock→humans		Barley-fed beef and bacon pigs	745–1 423	10–15
(c) Intensive grassland→ livestock→humans		Intensive beef herd on carefully managed pasture		
	Meat		339	4
	Milk		3 813	46
(d) Grassland and crops→ livestock→humans		Mixed dairy farm		
	Milk		1 356	17

Data from Duckham, A. N. & Mansfield, G. B. (1970) *Farming Systems of The World*, Chatto and Windus.

the plant protein is released. Finally, there are many ecosystems where animals can concentrate food from large areas where it would be difficult to grow or harvest plant crops. Examples are grazing on poor quality pasture land, such as by sheep in Britain, reindeer in Scotland and Scandinavia and eland in East Africa, or taking fish from aquatic ecosystems.

Rational cropping of ecosystems

Cropping is the removal of any organism from an ecosystem for food, whether plant or animal. Rational cropping is using the ecosystem to produce food in the most efficient way. This may mean increasing the productivity of the crop and decreasing the effects of disease and predation by other animals, or by using a crop which is better adapted to conditions in the ecosystem.

Increasing productivity of plant crops may be achieved by adding fertilisers to the soil, and by adding water to the soil by irrigation or removing excess water by drainage, as necessary. Disadvantages of carrying out these processes are that, over a long period of time, the use of man-made fertilisers can lead to deterioration of the soil structure (section 12.5.1), which will eventually lead to a decrease in productivity, and addition of fertilisers or water is often energy-expensive.

Decreasing the effects of predation, generally insects and birds in the case of plant crops, and disease is usually carried out by the selection of resistant genetic strains of the crop or the use of chemicals to kill the pest or disease-producing organisms. Using such chemicals must be done with care; persistent chemicals, those not rapidly broken down in the ecosystem, may have disastrous effects on other trophic levels (sections 12.3.8 and 12.8.9), and effective pesticides used too frequently may bring about outbreaks of resistant strains of the pest which are more difficult to control, as in the attempts to control aphids on chrysanthemums and early season pests on cotton.

The use of C_4 plants, such as sugarcane and maize, rather than C_3 plants in conditions of relatively high light intensity and temperature is an example of using crops better suited to the environment. In these conditions C_4 plants photosynthesise more efficiently, and therefore have a higher productivity, than C_3 plants (section 9.8.2). A further example is the possible use of wild ungulates rather than domestic livestock as the 'crop' in East Africa. The wild ungulates, such as eland, have a greater year-long biomass (82.2–117.5 $\times$ 10^5 kg ha^{-1} compared with 13.2–37.6 $\times$ 10^5 kg ha^{-1} for domestic livestock and cattle) as they are more efficient at digesting and assimilating nutrients from poor quality herbage, which is particularly important during the dry season in the savanna, and possibly also because they have a higher resistance to diseases which occur in the area.

An important concept in the management of animal 'crops' which take several years to reach sexual maturity is that of **maximum sustainable yield**. This is discussed more fully in section 12.7.7.

12.3.8 Concentration effects in food chains

Since the Second World War there has been a dramatic increase in the number of man-made chemicals released into the environment. These include herbicides and pesticides designed to kill those organisms, particularly weeds and insects, that are harmful to crops, livestock and humans themselves. Among the first of the successful pesticides was a group of chlorinated hydrocarbons (organochlorines) which included DDT (dichlorodiphenyltrichloroethane), dieldrin and aldrin. These chemicals are toxic to a broad spectrum of animal species, including humans, although birds, fish and non-vertebrates are worst affected. To the surprise of many scientists, it was reported in the mid-1960s that DDT had been detected in the livers of penguins in the Antarctic, a habitat very remote from areas where DDT might have been used.

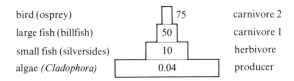

bird (osprey)	75	carnivore 2
large fish (billfish)	50	carnivore 1
small fish (silversides)	10	herbivore
algae (Cladophora)	0.04	producer

Fig 12.14 *Biomass and amounts of DDT at different trophic levels in a food chain. Figures represent amounts of DDT in parts per million (ppm)*

12.12 Fig 12.14 shows the amount of DDT at different levels in a food chain, the data for which were collected in the USA.

(a) If the concentration of DDT in the water surrounding the algae was 0.02 ppm, what was the final concentration factor for DDT in passing **from water** into (i) primary producers, (ii) small fish, (iii) large fish, (iv) the top carnivore.

(b) What conclusions can you draw from your answer to (a)?

(c) At which trophic level (i) is DDT likely to have the most marked effect, (ii) would DDT be most easily detected, (iii) are insect pests of crops found (a typical target of DDT)?

(d) Suggest ways in which the penguins might have come to contain DDT.

(e) Clear Lake, California, is a large lake used for recreational activities such as fishing. Disturbance of the natural ecosystem by eutrophication (nutrient enrichment, see sections 3.2.7 and 12.4) led to increased populations of midges during the 1940s and these were treated by spraying with DDD, a close relative of DDT, in 1949, 1954 and 1957. The first and second applications killed about 99% of the midges but they recovered quickly and the third application had little effect on the population.

Analysis of small fish from the lake showed levels of 1–200 ppm of DDD in the flesh eaten by humans, and 40–2 500 ppm in fatty tissues. A population of 1 000 western grebes that bred at the lake died out and levels of 1 600 ppm of DDD were found in their fatty tissues.

(i) Suggest a reason why the DDD did not succeed in eradicating the midges and why they recovered so quickly after the third application.

(ii) It has been observed that many animals die from DDT poisoning in times of food shortage. Suggest a reason for this based on the data given so far.

(f) In Great Britain, the winters of 1946–7 and 1962–3 were particularly severe. The death toll of birds was high in both winters, but much higher in 1962–3. Suggest a possible reason for this in the light of the data given about DDT.

Pesticide poisoning has had devastating effects on some top carnivores, most notably birds. The peregrine falcon, for example, has disappeared completely from the eastern USA as a result of DDT poisoning. Birds are especially vulnerable because DDT induces hormonal changes that effect calcium metabolism and result in the production of thinner egg shells with a consequently high loss of eggs through breakage. Levels of DDT in human body fat are 12–16 ppm in the USA, where the upper legal limit of DDT content for sale of food is 7 ppm.

In more recent years, some powerful but non-persistent pesticides have been developed, such as organophosphates (for example malathion), and the use of DDT has been severely reduced. However, DDT is relatively cheap to produce and continues to be more suitable for certain tasks, such as malaria control. When considering whether to use pesticides it is often a case of choosing the lesser of two evils. DDT has completely eradicated malaria in many parts of the world. In Mauritius, for example, although the birth rate has not changed significantly since 1900, a population explosion has occurred because far fewer babies are dying from malaria. Infant mortality fell from 150 per 1 000 to 50 per 1 000 in 10 years as a result of post-war spraying with DDT.

12.13 In trying to develop a new pesticide what properties would you ideally want it to have?

12.4 Biogeochemical cycles – the cycling of matter

Biogeochemical cycling is the second major function of ecosystems (along with energy flow). Each cycle summarises the movement of chemical elements through the living component of the ecosystem, namely the build-up in food chains of complex organic molecules incorporating the element and the breakdown in decomposition to simpler organic and subsequently inorganic forms which can be used again to make the living material of living organisms. As well as this actively cycling pool of an element, all cycles have a larger reservoir pool. Exchanges between the reservoir and active cycling pools are typically limited and often slow processes, for example the chemical weathering of phosphate rock, and fixation by lightning of nitrogen into nitrates during thunderstorms.

Cycles can be recognised for all chemical elements that occur in living systems. The biogeochemical cycles for carbon (C), nitrogen (N), sulphur (S) and phosphorus (P), four important macronutrients, were summarised diagrammatically in figs 9.2 and 9.31–33. These cycle diagrams, section 9.11 and table 9.10, which deal with mineral nutrition, should be referred to when reading this section (see also fig 12.15).

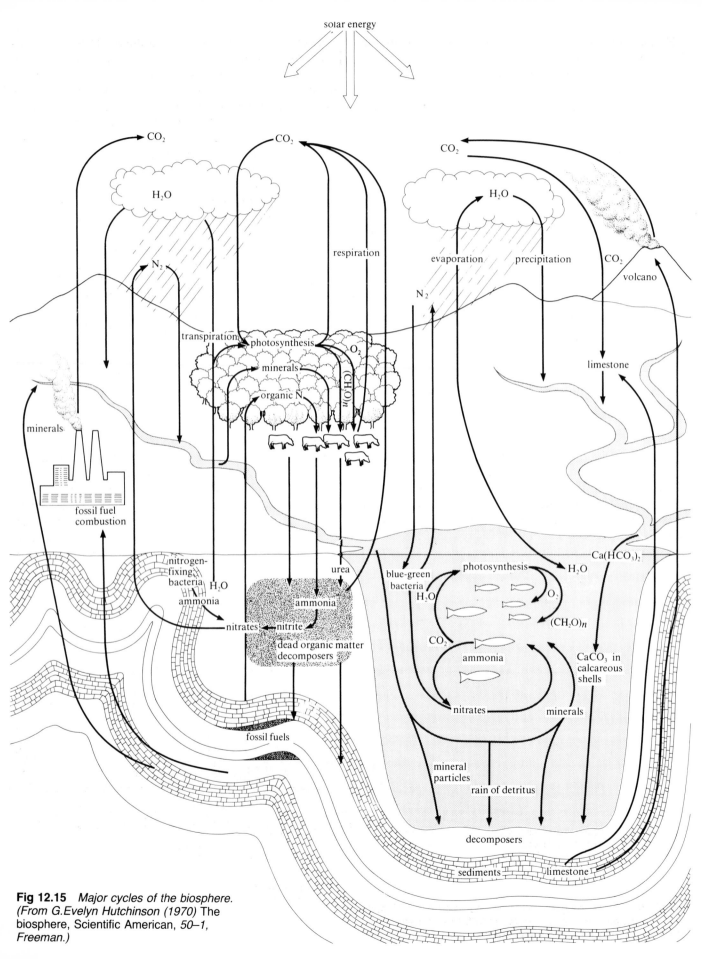

Fig 12.15 *Major cycles of the biosphere.
(From G.Evelyn Hutchinson (1970) The
biosphere, Scientific American, 50–1,
Freeman.)*

Humans require as many as 40 different elements for their biological well-being. Furthermore, the highly technological culture of modern human societies makes demands on virtually all other elements and many new synthetic substances. Some of these have become incorporated into living organisms and their potential for, and rate of, breakdown and recycling is an important contemporary concern. An understanding of biogeochemical cycling and maintenance of effective cycling is thus important for human society. Human activity generally speeds movement of material through the cycles and may fundamentally upset the balance of cycles. This may lead to build-up of material at one point in the cycle, in other words, **pollution**. Typically build-up occurs when more of an element (or new synthetic substance) is injected into one stage of a cycle than can be removed by the natural counterbalancing removal mechanisms. An obvious and topical example is the build-up of carbon as carbon dioxide in the atmosphere. This is discussed more fully in section 12.4.1.

As pointed out, all biogeochemical cycles have an active or cycling pool and a reservoir pool of an element. Reservoirs may be broadly divided into **sedimentary reservoirs**, as for phosphorus, and **fluid reservoirs**, the latter including both the oceans and the atmosphere, which displays some of the properties of fluids. Examples of elements with fluid reservoirs are nitrogen (atmosphere), and carbon with both atmospheric and oceanic reservoirs. Hydrogen, an important macronutrient not previously discussed, cycles in the water cycle (**hydrological cycle**) (fig 12.16). This cycle interacts in an important and complex way with the global energy budget because it plays a role in weather systems and is a gas which is important in the greenhouse effect (section 12.4.1).

Transfers in cycles are termed **fluxes**. These are measured as the quantity of nutrient exchanged per unit time per unit area or volume. On a global scale such measurement is difficult and our knowledge of the quantities of nutrients in different parts of cycles is, at best, only very approximate. Some of the more reliable estimates come from designed and accidental experiments using radioactive tracers (for example from nuclear weapons testing and from leaks and accidents at nuclear power plants such as Sellafield, Three Mile Island and Chernobyl). A useful and related idea is **turnover rate**. Turnover rate measures the flux to and from a particular pool in relation to the quantity of nutrient held in that pool. For example, if fixation of carbon as a result of photosynthesis took place in a field at the rate of 2 g of carbon m^{-2} day^{-1}, this would be termed the flux of carbon. If there were 200 g of carbon m^{-2} of field, the turnover rate of carbon would be 2/200 = 0.01 (the **turnover time** would be 100 days).

Human-induced turnover rate measures the additional rate of movement to or from a major reservoir due to human activity. In some cases human-induced turnover exceeds the natural turnover rate. The human-induced turnover rates for copper, zinc, lead and phosphorus are all more than ten times the natural rate. The chief concern with high human-induced turnover rates lies with the biologically active elements, such as lead, which is readily ingested and stored in the body and which is thought to affect behaviour and learning as well as being directly toxic in larger quantities. In the case of phosphorus the rate of loss of this important macronutrient from the exchange pool to the deep sea sedimentary reservoir (from which there is no practicable prospect of recovery on a human time scale) exceeds gains from natural weathering and new

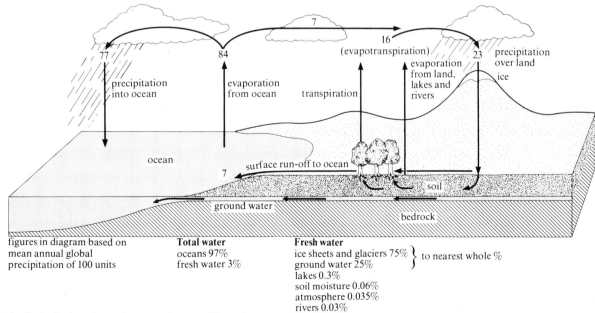

Fig 12.16 *The hydrological cycle and water storage. (Based on R.J. Chorley & P. Haggett (eds.) (1967) Physical and information models in geography, Methuen.)*

deposition in guano deposits. Furthermore, rock phosphate in a form readily recoverable in current technological and economic frameworks is limited. Future supplies depend substantially on recovery and recycling from dead material and sewage. This would simultaneously help to prevent phosphate enrichment of waters, which is a major cause of accelerated eutrophication in aquatic ecosystems (see section 12.4.1) and reduce permanent losses to deep sediments.

In general, cycles such as the nitrogen and carbon cycles with large numbers of negative feedback mechanisms, that is actions which tend to favour maintenance of the status quo, are considered more stable than those with fewer negative feedback mechanisms such as the phosphorus cycle. Unfortunately, it is often easier to understand the simpler cycles and to predict the likely effects of human activity. The cycles of the individual elements are not isolated but tend to interact, another factor that makes accurate prediction very difficult. In a classic experiment at Hubbard Brook,* the effects of land-use change on biogeochemical cycling in small water-shed ecosystems were monitored. After several years measuring the undisturbed systems the woodland cover of one small catchment was felled and the area treated with herbicide to prevent 'weed' growth before planting with commercial timber species. Since renewed plant growth was suppressed, exchange of soil nitrogen in the ammonium form with new plants (refer to fig 9.31) was prevented and breakdown to nitrite and nitrates took place more rapidly than in the undisturbed system. The process released hydrogen ions to the soil (H^+) which in turn displaced important cations (positively charged ions) from the soil. High levels of sodium, potassium and subsequently calcium and magnesium occurred in waters draining from the disturbed catchment, resulting in less fertile soil. This important study highlights the value of a continued vegetation cover, even 'weeds', in maintaining soil fertility.

12.4.1   Disrupted cycles and pollution

The carbon cycle – carbon dioxide and the planetary greenhouse effect

The main exchange pathways in the carbon cycle were summarised in fig 9.2. The additional human-induced turnover is estimated at $5 \times 10^{12}\ kg\ yr^{-1}$ released by human use of fossil fuels. Some scientists think that forest clearance is an equally important source of additional atmospheric carbon dioxide and that the total annual rate of human-induced release to the atmosphere may be nearer $10 \times 10^{12}\ kg\ yr^{-1}$.

---

* F.H. Bormann & G.E. Likens (1967) 'Nutrient cycling', *Science*, **155**, 424–9.

Carbon dioxide is normally present in the lower atmosphere, the troposphere, in very small amounts, about 300 ppm or 0.03% by volume (see fig 12.17). Its importance lies in its contribution to the planetary greenhouse effect. Carbon dioxide is transparent to incoming short-wave radiation from the Sun, but absorbs strongly the long-wave radiation which the Earth typically re-radiates into Space. It therefore 'traps' outgoing radiation, warming the lower atmosphere which in turn radiates energy back to the surface (as well as losing some in upward and lateral radiation). Ultimately, of course, any given 'package' of incoming energy will eventually be dissipated and lost to Space, but the atmosphere–surface exchanges induced by

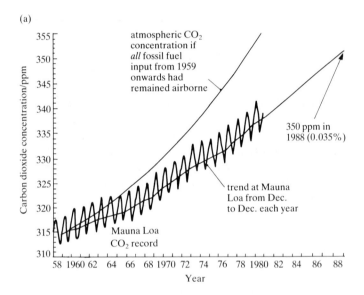

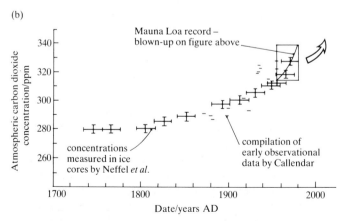

Fig 12.17 *Changes in atmospheric carbon dioxide concentrations since the year 1700: (a) 1958 to 1982 with projection to known level of 1988; (b) 1700 to 2000 (prediction). (From A. Crane & P. Liss (1985)* New Scientist *21.11.85.)*

the presence of carbon dioxide (and other near-surface greenhouse gases) are sufficient to raise planetary surface temperatures about 40 °C above those that would otherwise occur. It is important to realise that without this basic greenhouse effect, which has varied little for millions of years, living systems as we know them would not exist. The contemporary concern lies with the clear evidence that carbon dioxide levels (and those of other greenhouse gases, notably carbon monoxide, methane and chlorofluorocarbons (CFCs)), are rising at a rate unprecedented in recent Earth history and their increased presence will logically favour an increasingly warmer surface environment (see fig 12.18). This may in turn lead to increased evaporation and a greater atmospheric water vapour content. Since water vapour also acts as a powerful long-wave absorber, this may further increase surface temperatures. The resulting rise in surface temperatures will cause changes in the distribution pattern and intensity of the major planetary weather systems which may profoundly affect human activities.

Some meteorologists and governments have discounted the greenhouse warming threat, suggesting that increased dustiness of the atmosphere (linked with accelerating soil erosion associated with forest clearance, agricultural intensification, hedgerow removal, overgrazing and so on) will have a compensating cooling effect*, giving negligible change overall. However, unusual and socially catastrophic weather effects on a planetary scale during the summer of 1988, such as severe drought in the mid-western USA, floods in China and Sudan, more rain-bearing depressions than usual over Britain (the track is usually further north in summer), together with the now undisputed effects of CFCs on the ozone layer, have led to wider recognition of the possibility of pollution-induced global climatic changes and concern for the potential severity of its consequences. Whether or not these current events are directly caused by greenhouse warming, government action to control pollutant greenhouse gases is a welcome trend since the natural removal mechanisms for these gases are either too slow to alleviate the problem, as in the case of carbon dioxide, or induce further problems, as with the destruction of ozone in the stratosphere by CFCs (see section 12.10.2).

Carbon dioxide is removed from the atmosphere by photosynthesis, by exchange with the oceans, and by deposition as carbonates. The chief process is exchange with the oceans. The surface 75 m of the oceans is a well-mixed layer, heated by the Sun and agitated by wind. These properties combined with its carbonate chemistry make it a relatively rapid absorber of carbon dioxide. Ultimately the rate of carbon dioxide uptake is determined by exchange between this layer and the deep ocean waters renewing the capacity of the surface waters to absorb further carbon dioxide. Exchange with the deep waters is in most areas a slow process. The vast cold (<5 °C) oceanic deeps are isolated from the surface mixed waters by a

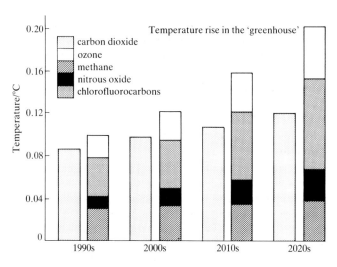

Fig 12.18 *Contributions of different greenhouse gases to predicted rise in global temperatures between now and the 2020s. (From M. McElroy (1988)* The challenge of global change, New Scientist, ***119***, 1623, 34–6.)

stagnant zone of decreasing temperature and increasing density, the **oceanic thermocline** (see fig 12.19), which extends to about 1000 m from the surface. It is estimated that over hundreds of years surface waters see only about 10% of total ocean water. Thus, although the oceans have the potential for absorbing all the human-induced excess atmospheric carbon dioxide, in practice the slow transfer between surface and deep waters prevents realisation of its full potential as a carbon dioxide sink, at least within a human time-scale. It is also possible that as the surface oceanic waters become saturated with carbon dioxide, the proportion of atmospheric carbon dioxide that the oceans can absorb may be reduced, making the carbon dioxide problem even worse.

Predictions of future levels of atmospheric carbon dioxide are difficult. This is partly because of our poor understanding of the detailed chemistry of oceanic absorption (more seems to be absorbed than the chemical models predict) and also because increasing environmental awareness may both curtail deforestation and accelerate development of alternative energy sources, reducing rates of increase in fossil fuel consumption. Nevertheless it is widely assumed that atmospheric carbon dioxide will rise to 600 ppm by the mid-twenty-first century, more than double the level for the late nineteenth century (see fig 12.17). Climatic consequences are also hard to predict due to the large number of possible feedback effects. Most models predict an overall global warming by the mid-twenty-first century of 3±1.5 °C. In high latitudes the problem may be made worse due to positive feedback mechanisms. As snow and ice melt, surface albedo (reflectivity) decreases, leading to increased absorption of radiation, a consequent increase in evaporation and finally increased water vapour in the atmosphere. Since water vapour is itself a greenhouse gas, this will mean that the temperature rise in high

---

* Dusts in the atmosphere reduce incoming short-wave radiation more than outgoing long-wave.

latitudes for a doubling of atmospheric carbon dioxide concentration should be two or three times greater than that at the tropics. Possible environmental consequences range from extensive flooding of coastal lowlands (serious, for example, for S.E. England, the Netherlands, Bangladesh and tropical islands on coral atolls such as the Maldives) to major shifts in world climatic zones. This could have profound implications for food production and the balance of trade and political power. Some small compensation may be gained from increased terrestrial photosynthesis (carbon dioxide is usually a limiting factor in photosynthesis).

Marine phytoplankton production may increase, especially in coastal waters where pollutants generate additional nitrates and phosphates, the key limiting factors for phytoplankton productivity. While this may aid carbon dioxide removal from the atmosphere, its desirability is questionable. Such increased productivity favours rapid growing species, reducing phytoplankton variety, and new dominant species may be less palatable to consumers. This does not favour secondary production in the grazing pathway and resulting excess decomposition may lead to oxygen depletion and death among consumer organisms.

A review of the greenhouse effect is presented in *New Scientist* of 22nd October 1988.

Eutrophication

Eutrophication means nutrient enrichment. Over a long time period, typically several thousand years, lake ecosystems classically show a natural progression from an **oligotrophic** (few nutrients) to a **eutrophic** or even **dystrophic** (rich in nutrients) state (see table 12.4). In the twentieth century, however, rapid eutrophication has occurred in many lakes, such as Lake Erie (N. America), Lake Zurich (Switzerland) and Lough Neagh (Northern Ireland); in semi-enclosed seas such as the Baltic and Black Seas; and in river systems worldwide. This is due to human activity. The main factors are heavy use of nitrogen fertilisers on agricultural land and the increased discharge of phosphates from sewage works. The phosphate problem reflects not only a larger human population but also the modern tendency for greater concentration of settlement into urban areas and the development of mains sewage systems. Thus, not only has the absolute amount of phosphate-containing waste increased but also its distribution has been locally concentrated with direct and more immediate discharge to rivers, lakes and seas compared with traditional systems (such as septic tanks). Sewage discharge into lake ecosystems generates particularly acute problems as, for example, seen in Lake Washington, USA. Here dense phytoplankton blooms occurred in the mid-1950s, accompanied by severe oxygen depletion in deep waters (hypolimnion – see fig 12.19). These changes were directly attributed to an increased sewage discharge from the developing city of Seattle. The situation was remedied by diverting all sewage to the Pacific via Puget Sound and at the same time improving the quality of the discharge to

Table 12.4 The general characteristics of oligotrophic and eutrophic lakes.

	Oligotrophic	*Eutrophic*
Depth	Deeper	Shallower
Summer oxygen in hypolimnion	Present	Absent
Algae and blue-green bacteria	High species diversity, with low density and productivity, often dominated by green algae	Low species diversity with high density and productivity, often dominated by blue-green bacteria
Blooms	Rare	Frequent
Plant nutrient flux	Low	High
Animal production	Low	High
Fish	Salmonids (e.g. trout, char) and coregonids (whitefish) often dominant	Coarse fish (e.g. perch, roach, carp) often dominant

(Source: C.F. Mason (1981) *Biology of freshwater pollution*, Longman.)

Note
In the classic mode of natural lake eutrophication a newly formed deep lake (a classic situation would be following retreat of an ice sheet) contains few nutrients since there has been no opportunity for weathering and sediment removal from the surrounding catchment. Primary and secondary productivity are hence low, the waters are clear, and oxygen status is good throughout.

With time, as weathering proceeds, nutrient status increases, primary and secondary productivity rise, organic and inorganic sediments accumulate and the lake becomes shallower. The more productive waters are less clear and the hypolimnion (see fig 12.19) may become seasonally oxygen-depleted.

A **dystrophic** lake is one which receives large quantities of organic matter from terrestrial plants giving the water a brown colouration. Such lakes typically have peat-filled margins and may develop into peat bogs.

prevent acute pollution in Puget Sound itself. Problems of eutrophication have been reported for lake systems throughout the world. Wherever possible, sewage discharge to lakes or virtually enclosed waters, such as coastal bays, should be avoided.

This human-induced accelerated eutrophication process is properly termed 'cultural eutrophication' (though it is commonly referred to as eutrophication). It reflects disruption of the normal nitrogen and phosphorus cycles. Human activity has increased the active or cycling part of these cycles beyond the self-regulatory (homeostatic) capacity of the systems, at least in the immediate time-scale. Cultural eutrophication generates acute economic as well as ecological problems. Good quality water resources are important for many industrial processes, vital for human and livestock drinking water supplies, essential for commercial and recreational fisheries and necessary for the maintenance of recreational amenities and navigation routes on major waterways (see table 12.5).

Nitrates and phosphates are the nutrients most commonly limiting primary productivity in aquatic ecosystems. Additional nitrate and phosphate, therefore, favours an increase in the more rapidly growing competitive planktonic species, such as *Oscillatoria rubescens*, and an overall

Table 12.5 The main effects of eutrophication on the receiving ecosystem and the problems for human societies associated with these effects.

Effects
(1) Species diversity decreases and the dominant biota change
(2) Plant, algal and animal biomass increases
(3) Turbidity increases
(4) Rate of sedimentation increases, shortening the life span of the lake
(5) Anoxic conditions may develop

Problems
(1) Treatment of drinking water may be difficult and the supply may have an unacceptable taste or odour
(2) The water may be injurious to health
(3) The amenity value of the water may decrease
(4) Increased vegetation may impede water flow and navigation
(5) Commercially important species (such as salmonids and coregonids) may disappear

(Source: C.F. Mason (1981) *Biology of freshwater pollution*, Longman.)

reduction in phytoplankton diversity. Such changes often also result in a disproportionate increase in less palatable species (species less suitable as food for consumers). This factor and the typically longer life histories and hence larger response times to environmental change of secondary producers or consumer organisms means that not all the increased primary production is eaten by the consumer organisms. Instead the excess material enters the decomposition pathway. Breakdown to simple inorganic nutrients is an oxygen-demanding process. Dissolved oxygen levels may be reduced below those necessary for successful growth and reproduction of sensitive species such as salmon and trout (*Salmo* spp). In extreme cases fish death and the subsequent decomposition in turn impose a further oxygen demand making the situation increasingly worse (positive feedback). This may not just be a problem for the immediately affected area. Zones of oxygen depletion on an otherwise unaffected river system may be sufficient to disrupt breeding in migratory species such as salmon and eels.

The problem of eutrophic oxygen depletion may be made worse by seasonal thermally induced water stratification in deep lakes. In mid-latitudes thermal stratification typically establishes in early summer (fig 12.19). This prevents diffusion of new oxygen supplies to the deep waters of the hypolimnion until the stratification pattern breaks down with seasonal cooling and higher wind speeds in autumn. The waters of streams and rivers entering the lake, since they are relatively shallow, are warm and therefore the dissolved nutrients and oxygen they contain mix only with the waters of the epilimnion (fig 12.19). Fauna requiring stable temperatures are mainly found in the hypolimnion. Excess phytoplankton production, encouraged by the warmth and increased nutrient status of the epilimnion, on death falls to the hypolimnion where its decomposition makes an additional oxygen demand on a restricted oxygen resource. Providing sufficient oxygen is present to meet this extra demand, and the needs of existing fauna, no major problem arises. However, if the situation is not carefully monitored, sudden and catastrophic fish kills may result in late summer when oxygen supplies approach exhaustion.

Case study

In the Norfolk Broads nitrate, and especially phosphate, levels have risen dramatically since the nineteenth century, especially during the last 40 years (table 12.6). At the same time clear waters with low phytoplankton productivity and many bottom-growing plants, such as stonewort (*Chara* spp.), have changed to phytoplankton-rich, turbid (cloudy) waters with few plants. At Barton Broad, transparency as measured by a Secchi disc*, was only 11 cm in 1973. Although most of the pre-1800 fauna of broadland,

* Disc with four segments, alternately black and white. The disc is lowered on a graduated string until the contrasting segments can no longer be distinguished. This gives an index of water turbidity.

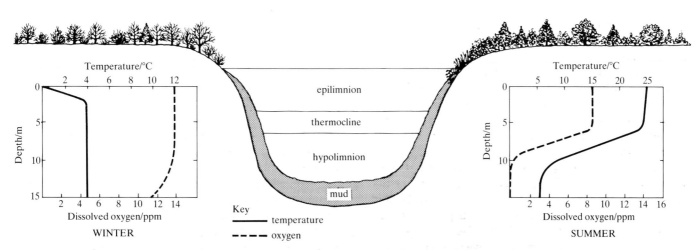

Fig 12.19 *Thermal stratification in a mid-latitude lake (data from Linsley ponds, Conn.). In summer a warm oxygen-rich circulating layer of water, the epilimnion, is separated from cold oxygen-poor hypolimnion waters by a broad zone of rapid temperature change called the thermocline. A similar gradient of oxygen status is also evident – see text for explanation. (Modified from E.P. Odum (1971)* Fundamentals of ecology, *Saunders p. 310.)*

Table 12.6 Phosphate levels in lakes and the Norfolk Broads since 1800.

General data	$\mu g\,dm^{-3}$ phosphate
Clear water upland lake	5
Naturally fertile lowland lake	10–30
Norfolk Broads pre-1800	10–20
Norfolk Broads 19th century following land enclosure	up to 80
Norfolk Broads 20th century following conversion to modern sewage disposal methods and population increase	max. 2 000 typically 150–300

(Source: B. Moss (1979) *Alarm call for the Broads*, Geographical Magazine.)

Changing phosphate levels at Barton Broad, Norfolk and the principal sources of phosphate

Date	$\mu g\,dm^{-3}$ phosphate	Sewage	Land drainage
1800	13.3	zero	13.3
1900	52.0	(separate data not available)	
1920	72.0	(separate data not available)	
1940	119.0	72	47
1975	361.0	287	74

(Source: B. Moss, 1980, *Ecology of fresh waters*, Blackwell Scientific Publications.)

including game fish (such as trout), have been lost, by the mid-twentieth century the area was a world-renowned centre for coarse fish and angling. However, the continued steep rise in nutrients, especially phosphorus, together with other complexly interacting factors have meant that these coarse fisheries are now also threatened. Additional threats include periodic marine flooding, now made worse by river bank erosion, and the physical disturbance of sediments and destruction of fringing reed banks. The latter protect river banks and are important fish spawning sites. Their destruction is due to the expanding and changing holiday and boating industry, where yachting is being replaced by activities with diesel-powered launches.

Recent experiments suggest that eutrophication can be reversed. Access points to two small broads, Alderfen and Cockshoot, were dammed, blocking new supplies of nutrient-rich river water*. At Alderfen no further action was taken. Initial improvement and return to a clear-water system was lost when a complex interaction occurred between existing sediments and new organic decomposition associated with a new spring phytoplankton bloom. This triggered renewed phosphate release from the nutrient-rich sediments. At first algal growth was restricted by nitrate availability but eventually nitrogen-fixing planktonic species became established and by summer 1985, six years after the experimental closure, the waters were again turbid. At Cockshoot damming was accompanied by dredging and removal of the phosphate-rich sediment. Aquatic plants are steadily recolonising this broad and fish stocks are increasing. Dredging Cockshoot, in 1982, cost

* The Broads are former peat-cuttings abandoned due to flooding during the late thirteenth and fourteenth centuries. Many typically have narrow and shallow connections to the main river system, hence damming is a practicable proposition.

£75 000. A more typical-sized broad would cost £250 000. However, in many cases navigation rights would preclude damming and isolation of a broad, especially on a permanent basis.

General effects and remedies

Physical isolation of water bodies will often be impracticable, but the experiments described above show that recovery is possible, if complex. They also increase our scientific understanding of the eutrophication process. Phosphate removal at sewage works is now widely used on sewage outfalls entering broadland rivers and at many other seriously affected locations. This involves chemical precipitation and removal of phosphorus using lime or iron salts. This can greatly aid recovery of a receiving waterway, especially when accompanied by initial dredging. Sediment removed may be beneficially used as fertiliser, at least partly offsetting costs. Dredging also benefits the system by deepening lakes and river channels. This favours increased flow which should improve oxygenation. In lake systems it may lead to increased water volume, thus having a further diluting effect on nutrients. In deeper stratified lakes an associated increase in hypolimnion volume and hence potential oxygen supply may result. Deepening may even induce stratification and thus limit nutrient supply to the epilimnion, in turn reducing phytoplankton productivity. Nutrient levels may also be reduced by bottom sealing using polythene sheeting, or by chemical removal by precipitation, such as with aluminium sulphate.

Algal blooms may be directly controlled chemically using algicides but this poses problems of unwanted harmful effects on non-target species which may be difficult to predict.

Eutrophication in water reservoirs is a serious economic problem. Algae may block filters at purification plants. This seriously reduces the throughput of water and cleaning may necessitate temporary shut-down of a plant. Small algae may persist into consumer supplies where subsequent decomposition in feeder pipes can impart an unpleasant (though harmless) taste and smell to drinking water supplies.

Most efforts in eutrophication control are concentrated on phosphate reduction. This is mainly because phosphates are typically added at clearly identifiable 'point-sources' whereas the main sources of additional nitrate are overland flow linked with agricultural practice. Nitrate entry to water systems is thus more diffuse, and control or remedy, aside from a radical change in farming practice, is less practicable. Nitrate levels in drinking water supplies are, however, a cause for concern. The European health standards set thresholds of $50 \, \text{mg dm}^{-3}$ nitrate for drinking water supplies with levels up to $100 \, \text{mg dm}^{-3}$ acceptable. In lowland England, river water intakes frequently exceed $100 \, \text{mg dm}^{-3}$ nitrate and must be mixed with waters with low nitrate values before entering public supply.

Monitoring eutrophication

Changes associated with eutrophication can be monitored biologically and chemically. This gives the opportunity for remedial action before catastrophic ecosystem damage occurs. Changes in phytoplankton species present may be indicative of eutrophication. Blue-green bacterial blooms are common, for example *Oscillatoria rubescens* at Lake Washington and *Anabaena flos-aquae* at Lough Neagh. Table 12.7 summarises characteristic plankton groups for oligotrophic and eutrophic lakes. Eutrophic waters characteristically show high abundance and low species diversity of phytoplankton. Reference slides to aid basic plankton identification are available from Philip Harris Ltd. Another approach monitors chlorophyll *a* abundance as an index of algal biomass. Mean summer values for oligotrophic lakes lie between $0.3–2.5 \, \text{mg m}^{-3}$, whereas for eutrophic lakes values range from $5–140 \, \text{mg m}^{-3}$.

Field records of large non-vertebrate fauna are useful indicators of river water quality. A simple five-point scheme is shown in table 12.8. Care must be taken to standardise sample procedure and ensure adequate replication for representative samples of each location. In Britain the two schemes most widely used in the water industry and by other professionals are the Trent Biotic Index (TBI) (table 12.9) and the Chandler Biotic Score (CBS). The simpler five-part scheme is based on these. The TBI monitors the presence or absence of key species together with species richness but does not include abundance estimates for different species. This makes calculation rapid but has the disadvantage that a single individual of a sensitive species may have a disproportionate influence on the index. In estimating the Chandler score, the value for each individual species is weighted according to five levels of abundance. A criticism of the method is, however, that assignment of

Table 12.7 Characteristic algal associations of oligotrophic and eutrophic lakes.

	Algal group	Examples
Oligotrophic lakes	Desmid plankton	*Staurodesmus, Staurastrum*
	Chrysophycean plankton	*Dinobryon*
	Diatom plankton	*Cyclotella, Tabellaria*
	Dinoflagellate plankton	*Peridinium, Ceratium*
	Chlorococcal plankton	*Oocystis*
Eutrophic lakes	Diatom plankton	*Asterionella, Fragillaria crotonensis, Stephanodiscus astraea, Melosira granulata*
	Dinoflagellate plankton	*Peridinium bipes, Ceratium, Glenodinium*
	Chlorococcal plankton	*Pediastrum, Scenedesmus*
	Myxophycean plankton	*Anacystis, Aphanizomenon, Anabaena*

(Source: Mason, C. F. (1981) *Biology of freshwater pollution*, Longman.)

Table 12.8 Five-point scale for water pollution studies using presence and absence indicator species.

Level of pollution	Oxygen concentration	Indicator organisms
(A) Clean water or very low pollution levels	High	stonefly nymph mayfly nymph
(B) Low pollution levels		caddis fly larva freshwater shrimp
(C) High pollution levels		water louse bloodworm
(D) Very high pollution levels	Low	sludgeworm rat tailed maggot
(E) Extreme pollution levels	No oxygen	No apparent life

This point scheme is used in the Philip Harris and Griffin water pollution study packs. The Philip Harris scheme includes colour photographs of key indicator species. The Griffin package has an excellent series of black and white drawings of a wider range of indicator organisms as well as procedures for calculating the Trent Biotic Index and other simple pollution indicator tests. The drawings are reproduced from *The biology of polluted waters* by H. B. N. Hynes, Liverpool Press 1971.

abundance scores is somewhat arbitrary. Further, since taxa must be counted as well as identified the Chandler Score takes much longer to determine than the Trent Biotic Index. (Appropriate microcomputer packages can be used for this work.) Full details of these methods, including worked examples, are given in C.F. Mason (1981) *Biology of Freshwater Pollution*, Longman.

A useful chemical indicator of eutrophication is the **biochemical oxygen demand (BOD)**. The BOD measures the rate of oxygen depletion by organisms. It is assumed this primarily reflects micro-organism activity in decomposing organic matter present in waters. (As discussed, organic matter typically increases as waters become nutrient-enriched.) Oxygen consumption by algal respiration will inevitably also be included in the test. In practice this is normally not important, but in some cases it may account for up to 50% of the total BOD. BOD is thus an approximate rather than precise guide to water quality. It is

Table 12.9 The Trent Biotic Index.

Summary table

			Total number of groups present				
			0–1	2–5	6–10	11–15	16+
	Indicator species		*Trent*	*Biotic*	*Index*		
Clean	Plecoptera nymph present	More than one species	–	7	8	9	10
Organisms in order of tendency to disappear as degree of pollution increases		One species only	–	6	7	8	9
	Ephemeroptera nymph present	More than one species*	–	6	7	8	9
		One species only*	–	5	6	7	8
	Trichoptera larvae present	More than one species†	–	5	6	7	8
		One species only†	4	4	5	6	7
	Gammarus present	All above species absent	3	4	5	6	7
	Asellus present	All above species absent	2	3	4	5	6
	Tubificid worms and/or red chironomid larvae present	All above species absent	1	2	3	4	–
	All above types absent	Some organisms such as *Eristalis tenax* not requiring dissolved oxygen may be present	0	1	2	–	–
Heavily polluted							

Groups list
The term 'group' here denotes the limit of identification which can be reached without restoring to lengthy techniques.
Thus the groups are as follows:
Each known species of Platyhelminthes (flatworms).
Annelida (worms) excluding genus *Nais*.
Genus *Nais* (worms).
Each known species of Hirudinae (leeches).
Each known species of Mollusca (snails).
Each known species of Crustacea (*Asellus*, shrimps).
Each known species of Plecoptera (stone-fly).
Each known genus of Ephemeroptera (may-fly) excluding *Baetis rhodani*.
Baetis rhodani (may-fly).
Each family of Trichoptera (caddis-fly).
Each species of Neuroptera (alder-fly).
Family Chironomidae (midge larvae) except *Chironomus thummi*.
Chironomus thummi (blood worms).
Family Simulidae (black-fly larvae).
Each known species of other fly larvae.
Each known species of Coleoptera (beetles and beetle larvae).
Each known species of Hydracarina (water-mites).

* *Baetis rhodani* excluded
† *Baetis rhodani* (Ephemeroptera) is counted in this section for the purpose of classification.
The maximum value is 10. Biotic indices are effectively marks out of ten with zero representing virtually lifeless heavily polluted waters.

Procedure
(1) Sort each sample, separating the animals according to group (see above groups list). Count the total number of groups present.
(2) Note which indicator species are present, starting from the top of the list.
(3) To find the Trent Biotic Index, take the highest indicator species, e.g. caddis fly, *Trichoptera*, and work along the line. Note from the top the group number and read off the Trent Biotic Index.

e.g.		
	Highest indicator animal	Trichoptera
	Number of indicator species	more than one
	Total numbers of groups	7
	Trent Biotic Index	6

Source: The Griffin Pollution test kit: handbook for users

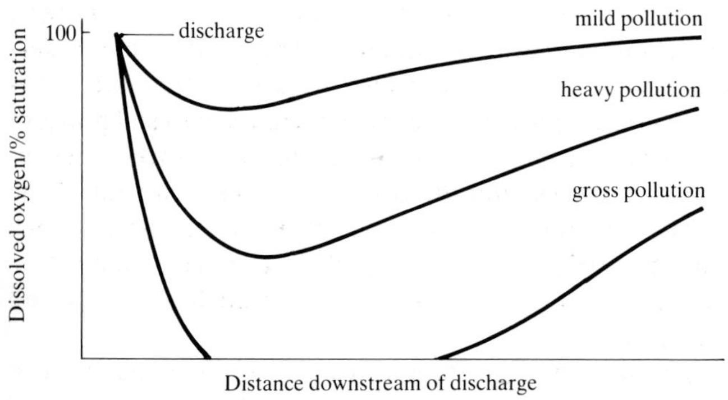

Fig 12.20 *The effect of an organic discharge on the oxygen content of river water. (From C.S. Mason (1981)* Biology of fresh water pollution, *Longman.)*

most useful when used in conjunction with other water quality indicators. The standard measure of BOD is the weight of oxygen in milligrams consumed from a decimetre cubed of sample when stored in darkness for five days at 20 °C. Clean river water normally has a BOD of 3 mg dm^{-3} compared with 10 mg dm^{-3} for a badly polluted stream. Typical BOD values for domestic sewage are between 250–350 mg dm^{-3} (see table 12.10).

Deoxygenation of a river caused by organic wastes is a slow process so that the point of maximum deoxygenation may occur considerably downstream of a discharge (see fig 12.20).

> **12.15** List the factors that will determine the degree of deoxygenation.

In the River Thames in 1967, at low flow in autumn, minimum oxygen conditions prevailed for 40 km downstream of London Bridge, whereas in spring with high flow only 12 km had minimum oxygen. Furthermore, depletion was immediately associated with this major conurbation at low flow but did not occur until 22 km downstream at high flow.

Table 12.10 (a) A guide to water quality as measured by BOD. (b) Comparative BOD strengths of some typical industrial liquid wastes.

(a)

Rivers		Sewage	
BOD (mg dm^{-3})	Quality	BOD (mg dm^{-3})	Quality
1	very good	600	strong sewage
2	good	350	moderate sewage
3	fairly good	200	weak sewage
5	doubtful	20	maximum Royal Commission
10	poor		standard effluent assuming ×
20	very poor		10 dilution in receiving stream

(b)

Waste type	5 day BOD (mg dm^{-3})
cotton	200–1 000
tannery	1 000–2 000
laundry	1 600
brewery	850
distillery	7 000
dairy	600–1 000
cannery: peas	570
fruit (citrus)	2000
farm waste	1 000–2 000
silage	50 000
paper board	100–450
coke oven	780
oil refinery	100–500

(Adapted from Open University, *Clean and dirty water*.)

12.5 The abiotic component of ecosystems: the physical habitat factors

The abiotic, or non-living, component of an ecosystem may be conveniently divided into edaphic factors (concerning soil), climatic factors, topographic factors and other physical factors that might be operating, such as wave action, ocean currents and fire.

As we have seen in the discussion of biogeochemical cycling, the abiotic components are often intimately linked with the living components of an ecosystem. On the grand scale the physical factors of the environment determine the extent of the biosphere (that part of the planet which is suitable for life). At a regional and local level they influence the distribution of species and, in close association with biotic controls such as competition and predation, influence the structure of communities and the nature of ecosystems. Biotic factors are more fully discussed in the sections on communities and population ecology.

12.5.1 Edaphic factors

The scientific study of soil is called **pedology**. Early work on soils stressed their importance as sources of nutrients for plants, or examined them from a geological viewpoint. Work was therefore concentrated on the physics and chemistry of soils and they were seen as inert inorganic substances. The first scientist to describe soil as a dynamic rather than an inert medium was the Russian Dokuchaev in his classic work on Russian soils dating from 1870. He saw soil as a constantly changing and developing material, a dynamic zone in which physical, chemical and biological activities occur. He recognised five major soil-forming factors, namely climate, parent material (geology), topography (relief), organisms and time.

Although soil is included here with abiotic factors, it is better to regard it as a vital link between the biotic and abiotic components of terrestrial ecosystems. The term 'soil' is applied to the layer of material overlying the rocks of the Earth's crust. A suitable nutrient content and structure are essential for successful crop production. Though modern technology can, with some success, overcome poor soil conditions, a proper scientific understanding is vital to avoid unwanted environmental side-effects. Examples of these are the depletion of soil nutrient reserves so often associated with forest clearance, especially in the tropics, and the constant over-enrichment of the soil by fertilisers in many intensively farmed areas which leads to nutrient rich run-off and associated eutrophication of waters, such as in the East Anglian broadlands and fenlands.

Soil has four important structural components, namely the mineral skeleton (typically 50–60% of the total soil composition), organic matter (up to 10%), air (15–25%) and water (25–35%). Methods of analysing these components are given in section 13.1.1. In addition, there is a biotic component which has been considered in the previous section, and is also dealt with further in chapter 13.

Mineral skeleton (inorganic content)

The mineral skeleton of soil is the inorganic component and is derived from the parent rock by weathering.

Soil texture. The mineral fragments comprising the soil skeletal material vary in size from boulders and stones down to sand grains and minute clay particles. The skeletal material is usually divided arbitrarily into **fine earth** (particles <2 mm) and larger fragments. Particles <1 μm in diameter are termed **colloidal**. The mechanical and chemical properties of soil are largely determined by the fine earth material. The distribution of soil particle sizes within the fine earth is examined by mechanical analysis in the laboratory (section 13.1.1) or, with experience, by 'feel' methods in the field. Fig 12.21 summarises two of the most widely used thresholds for sand, silt and clay. In all cases clay consists of particles <0.002 mm (2 μm) in diameter.

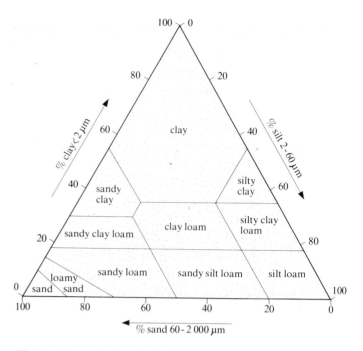

Fig 12.22 *Triangular diagram of soil textural classes as used by the Soil Survey of England and Wales, 1974*

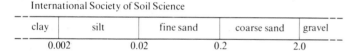

International Society of Soil Science

clay	silt	fine sand	coarse sand	gravel
0.002	0.02	0.2	2.0	

British Standards System (used by the Soil Survey of England and Wales)

clay	silt	fine sand	medium sand	coarse sand	stones
0.002		0.06	0.2	0.6	2.0

Diameter of soil particle/mm
(log scale)

Fig 12.21 *Particle-size classes of soil*

The relative proportions of sand, silt and clay determine the soil texture. Fig 12.22 shows a standard soil texture triangle and indicates the limits of 11 major soil textural classes in routine use by the Soil Survey of England and Wales.

Soil texture is important agriculturally. Medium- and fine-textured soils such as clays, clay loams and silt loams are generally more suitable for plant growth because they have the most satisfactory nutrient and water retention. Sandy soils are faster draining and lose nutrients through leaching, but may be advantageous in obtaining early crops as the surface dries more rapidly than that of a clay soil in early spring, resulting in a warmer soil. Stone content of the soil (particles >2 mm) may also have importance agriculturally since it will affect wear and tear on agricultural implements, and will modify the drainage characteristics of fine earth. Generally, as the stone content of a soil increases, its water holding capacity decreases.

The implications of soil texture can be partly understood by comparing the properties of pure sand and clay, as shown in table 12.11. An ideal soil would contain roughly equal quantities of clay and sand, combined with a range of intermediate particle sizes. In these circumstances a porous crumb structure is formed and such soils are called **loams**. They usually combine the advantages and eliminate the disadvantages of the extreme soil types. Thus mechanical analysis of a given soil, which is easily carried out, is a useful guide to the soil's properties.

Agriculturally, soils are sometimes referred to as heavy (clays) or light (sands) reflecting the power needed to work the soil with agricultural implements. Heavy soils are poorly drained and are usually wet and sticky. Clodding and compacting make it difficult to obtain a fine tilth. Light soils are well drained and a fine tilth is readily obtained as the particles separate easily.

Chemistry of the mineral skeleton. Soil chemistry is partly determined by the mineral skeleton and partly by the organic matter, which is described later. A high proportion of the minerals in soil are present as crystalline structures, left as the resistant products of weathering of the original parent rock. Sand and silt consist mainly of the mineral quartz, SiO_2, which is extremely inert and is otherwise known as silica. Silica is also the basis of silicate ions, SiO_4^{4-}, which typically combine with cations, particularly aluminium (Al^{3+}) and iron (Fe^{3+}, Fe^{2+}), to form electrically neutral crystals. Silicates are the predominant soil minerals.

Table 12.11 Comparison of the properties of sand and clay.

Property	Sand	Clay
Texture	Coarse, particle size >0.06 mm	Fine, particle size <0.002 mm
Structure	Structureless	Forms large sticky masses (clods) when wet, becomes hard and cracks on drying
Porosity	Pore spaces relatively large, good aeration, rapid drainage	Pore spaces relatively small, poor aeration, slow drainage
Water-holding capacity	Poor water retention, little water held by capillarity (surface tension) or adsorption, does not become waterlogged	Good water retention, relatively large amounts of water held by capillarity (surface tension) and adsorption, easily waterlogged
Temperature	Warm due to low moisture content*	Cold due to high moisture content*
Nutrient retention	Low, rapidly leached (sandy soils tend to become acidic as bases are leached out and humic acids accumulate)	High, not leached, clay particles attract cations and some anions

* Water has a high specific heat capacity and a high latent heat of evaporation.

One particularly abundant and important group of minerals affecting nutrient and water retention are the clay minerals. Most of these occur as minute flat crystals, often hexagonal in shape, which form a colloidal suspension in water. Each crystal contains layers of silicate sheets combined with sheets of aluminium hydroxide, with spaces between the layers. The combined surface area of the layers with their spaces is very large compared with the volume of the crystals ($5–800 \text{ m}^2 \text{ g}^{-1}$ clay). The important feature of these minerals is that they have a permanent negative charge which is neutralised by cations adsorbed from the soil solution. The cations are thereby prevented from being leached out of the soil and remain available for exchange with other cations in the soil solution and in plants. The extent to which cations are freely exchangeable is referred to as the **cation exchange capacity** and is an important indicator of soil fertility. Water is also attracted into the spaces, causing hydration and swelling of the clay.

Organic matter

The organic content of the soil is derived from the decay of dead organisms, parts of organisms (such as shed leaves), excreta and egesta. The dead organic matter is utilised as food by a combination of detritivores, which ingest and help to break down the material, and decomposers (fungi and bacteria) which complete the process of decomposition. Undecomposed material is called **litter** and the final, fully decomposed amorphous material in which the original material is no longer recognisable is called **humus**. Humus is a dark brown to black colour and chemically very complex and variable in composition, consisting of many types of organic molecule. These consist mainly of phenolic acids, carboxylic acids and esters of fatty acids. Humus, like clay, is in a colloidal state. Some of it adheres strongly to the clay to form a **clay–humus complex**. Like the clay,

the humus has a large surface area and high cation exchange capacity, the anions in humus being carboxyl and phenolic groups. This capacity is particularly important in soils with a low clay content. Humus is very important for soil structure, generally improving aeration and water and nutrient retention through its chemical and physical properties.

At the same time as **humification** (formation of humus) occurs, essential elements pass from organic compounds into inorganic compounds, such as nitrogen into ammonium ions (NH_4^+), phosphorus into orthophosphate ions ($H_2PO_4^-$) and sulphur into sulphate ions (SO_4^{2-}). This process is called **mineralisation**. Carbon is released as carbon dioxide as a result of respiration (see carbon cycle, fig 9.2).

In order for any type of humus to develop the soil must be reasonably well drained, because decay is extremely slow under waterlogged conditions where lack of oxygen restricts growth of aerobic decomposers. Under such conditions the structure of the animal and plant remains is thus preserved for long periods of time and they gradually become compressed to form peat which may accumulate to great depths. The method of determining the humus content of a soil sample is described in experiment 13.2.

Air content

The soil atmosphere, together with the soil water, occupies the pores between soil particles. Porosity (pore space) varies with different soils, increasing in the series from clays through loams to sands. There is free exchange of gases between the soil and external atmosphere with the result that the air in both has a similar composition. Generally, the soil air has a slightly lower level of oxygen and a higher level of carbon dioxide than outside owing to respiration by soil organisms. Oxygen is required by the roots of plants, by soil animals and by decomposers. Some of the soil gases are in solution and it is in this form that they are exchanged with living organisms. If the soil becomes waterlogged, the air spaces fill with water and the soil becomes anaerobic. Minerals like those of iron, sulphur and nitrogen will tend to exist in their reduced states (Fe^{2+}, sulphide, sulphite, nitrite) and may trap any oxygen that becomes available by becoming oxidised. The soil will become acidic because anaerobic organisms continue to produce carbon dioxide. Humus turnover is reduced, so humic acids also accumulate. Unless the soil is rich in bases it may become extremely acidic and this, like the oxygen depletion, adversely affects the soil micro-organisms. Changes in the oxidation state of iron affect the colour of the soil. The oxidised form of iron, Fe^{3+}, imparts yellow, red and brown colours while the reduced form, Fe^{2+}, gives the grey colour characterstic of waterlogging. A prolonged period of anaerobic conditions will result in plant death. The method of determining the air content of a soil sample is described in experiment 13.3.

Water content

Some water is retained around the soil particles, while the remaining water, sometimes called **gravitational water** (fig 12.23), is free to drain downwards through the soil. The latter is important in causing the phenomenon known as **leaching**, which is the washing-out of minerals, including nutrients. The level to which gravitational water drains is called the **water table** and this may fluctuate in depth depending on rainfall.

Water may be retained as a thin, tightly bound film around individual colloidal particles. This is sometimes called **hygroscopic water** (fig 12.23). It is adsorbed by hydrogen bonding, for example, to the surfaces of silica and clay minerals or to the cations that are associated with clay minerals and humus. It is the water least available to plant roots and is the last water to remain in very dry soils. The amount present depends on the colloidal content of the soil and is therefore much greater in clay soils (about 15% by weight) than in sandy soils (about 0.5% by weight).

As water layers build up around soil particles, water begins to fill the finer pores between soil particles and spreads to larger and larger pores. The hygroscopic water grades into **capillary water** which is the water held round soil particles by surface tension (fig 12.23). This is the water that can move upwards through fine pores and channels from the water table by capillarity, a phenomenon caused by the high surface tension of water. Capillary water is easily utilised by plants and is their most important regular supply of water. It is easily lost by evaporation, unlike hygroscopic water. Fine-textured soils like clays hold more capillary water than coarse-textured soils like sands.

The total amount of water that can be retained by a soil (determined by adding water until it drains out and then stops dripping) is called the **field capacity** (section 13.1.1).

Water is required by all living organisms in the soil, and enters living cells by osmosis (section 14.1). It is also

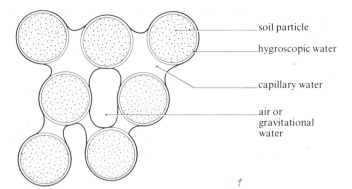

Fig 12.23 *Three types of soil water*

important as a solvent for nutrients and respiratory gases which are taken up from an aqueous solution by plant roots. It contributes to the weathering process of parent rocks in ways described in the next section.

12.5.2 Formation of soil

Soil is formed as a result of the interaction of many variables, the five most important of which are climate, parent material (geology), topography (relief or surface morphology), organisms, and time. These are considered separately below.

Climate and weathering

Weathering is the action of climate and, to a lesser extent, living organisms in bringing about the physical and chemical breakdown of the parent rock or material. The two most important factors in weathering are water and temperature and so it is most conveniently dealt with in the context of climate. The more general influence of climate on soil formation will be described after weathering.

Physical weathering can be caused by temperature changes, as when rocks alternately expand and contract in response to diurnal (daily) changes in temperature. This process causes rocks to shatter. If water is present in the cracks and alternate freeze and thaw cycles occur, the expansion of ice can cause tremendous pressures to build up, also shattering the rock. Other causes of physical weathering are particles of various sizes like sand, which, when carried by wind, water or glaciers, scour exposed surfaces.

Chemical weathering may be influenced by biological and climatic factors. It is generally accelerated by higher temperatures. Water acts as a solvent and also as a reagent in the hydrolysis of certain rock minerals. Rocks which are slightly water soluble, such as the calcium carbonate of chalk and limestone, are particularly liable to erosion by solution. The effectiveness of water may be increased when carbon dioxide from the atmosphere, or from the respiration of soil organisms, dissolves to form the dilute acid carbonic acid. Lichens are among the few living organisms that can grow on bare rock surfaces, and they contribute to chemical erosion by extracting certain nutrients from weathering rocks.

As with weathering, temperature and precipitation (such as rainfall and snowfall) are the two key climatic factors operating. The amount of precipitation is particularly important. If it exceeds evaporation, the soil may be subject to leaching, or if drainage is poor to waterlogging. If evaporation exceeds precipitation, capillarity will occur from the water table and movement of soluble materials will be in a predominantly upward direction, again affecting the nature of the soil that develops. A dry climate will result in sparse vegetation cover which, in turn, reduces humus production. Soils may be less acidic as a result. Lack of water also inhibits chemical weathering and leaching.

Some of the effects of climate are shown in the summary diagram of fig 12.24.

Parent material

All rocks can be traced back in origin to the solidification of molten magma which was released from below the Earth's crust. The rocks formed directly by the cooling of this magma are called **igneous rocks**, and may be acidic, such as granite, or basic, such as basalt, according to their mineral composition. When acid rocks weather, a high proportion of silicon remains as SiO_2, or quartz, which is seen as sand grains. When basic rocks weather, however, much of the silicon is present in a rich variety of silicates which pass through various sequences of weathering, giving rise to the different minerals of silt and clay.

Sedimentary rocks are formed by the deposition of material derived from the weathering of other rocks, or from the remains of living organisms. The material accumulates and becomes compacted, possibly over millions of years. Typical sedimentary rocks are sandstones, chalk and limestone and their effect on soil development is described in section 12.5.3.

Metamorphic rocks are rocks which have been changed after their formation by periods of heating and recrystallisation, such as slate and marble. They are generally more resistant to weathering than other rocks.

Finally, some soils develop on transported materials such as wind-blown sand, alluvial deposits and glacial moraines. The influence of parental material here is obviously related to the original parent rock from which these materials were derived.

Topography

Topography, or relief, exerts its influence chiefly through altitude, steepness of slope and aspect. These affect local climate and drainage. Drainage is generally better on slopes, getting progressively poorer towards lowland and valley bottoms where peats may develop if soils are permanently waterlogged. On slopes, a certain proportion of water will be lost as run-off and this causes losses of weathered rock and soil. Soils on slopes tend to be thinner as a result. Soil is also lost as it creeps slowly downhill in response to gravity. This process is called **solifluction** and

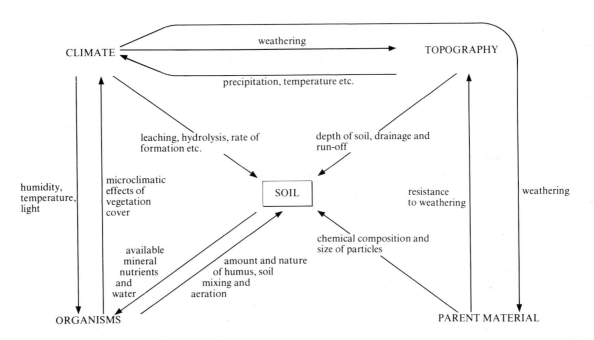

Fig 12.24 *Interactions of the four direct agents of soil formation. Note the major influence of climate. Time is a fifth, indirect factor which is not included*

leads to the accumulation of soil on lower slopes. Vegetation cover may limit these processes.

Climatic factors such as precipitation and temperature are affected by topography, which thereby indirectly influences the soil-forming process. Aspect can be an important factor as well as altitude in influencing the climate. Temperatures are higher on the sunnier aspects, resulting in faster soil development, drier soils and often influencing the vegetation. Windward slopes are generally wetter than leeward slopes.

Organisms

Organisms contribute the organic part of the soil (litter and humus), the physical and chemical properties of which have been described already.

Vegetation may influence the development of soils as well as climate and parent material. There is some evidence, for example, that podzols (section 12.5.3) are more likely to form under heath and coniferous forest communities than under deciduous forest communities. This is probably related to the nature of the chemicals leached from the plant litter and their inhibiting effects on decomposition by soil micro-organisms. Also, the acid litter of heaths and pine forests is unfavourable for earthworms, unlike that of deciduous forests.

The activity of soil detritivores, particularly earthworms in temperate regions, is important in a number of ways, and these in turn are influenced by vegetation. In general, detritivores speed up the process of decomposition by breaking up and increasing the surface area of litter. In addition, their faeces and excretory products contribute to the volume of processed organic and inorganic material. Soil, in passing through the gut of an earthworm, is made finer in texture and the mineral and organic components are mixed in a way that improves the structure of the soil. Earthworm burrows improve aeration and provide passages for root growth. Similarly, burrowing vertebrates such as moles and rabbits improve soil mixing and aeration.

Human influence.
The delicate and dynamic balance of a mature soil is potentially at risk when humans use soil for agricultural purposes, and in many cases destruction of soil structure, depletion of nutrients and soil erosion have been the consequences.

Soil structure is particularly at risk when heavy agricultural machinery is used for working the land, soil compaction having a number of consequences such as reducing pore size and hence reducing aeration and drainage. Soil compaction may also be caused by large numbers of livestock. Repeated use of the soil, particularly for the same crops, may reduce the levels of important nutrients. Ploughing soil may increase its erosion by wind and water, particularly on slopes. Similarly, removal of a protective layer of vegetation through overgrazing by livestock, or by forest clearance, can expose soil to greater erosion. Clearance of tropical rain forest can have very serious consequences on the thin soils typical of such

regions. On bare soil there is increased surface run-off which may not only increase soil erosion but also wash out soil nutrients. The consequences for humans are a decreased soil fertility and, in some extreme cases, a change towards barren conditions is triggered resulting in a large drop in productivity of the ecosystem. For example, in some areas of Malaya and parts of the Brazilian Amazon there has been massive erosion accompanied by silting of rivers. Soil management is therefore one of the key areas of interest to agriculturalists and soil scientists.

Time

Time has an indirect but important influence on soil formation. The rate of formation is very variable, from a few decades on volcanic ash to several thousand years starting from a bare rock surface in a temperate climate. Over geological periods of time, climatic and topographic changes occur which in turn influence soil development.

To summarise, it should be stressed again that the soil-forming factors mentioned do not act in isolation. Each, apart from time, has a direct effect on soil formation, but may also act indirectly by influencing the other factors. Fig 12.24 summarises the interactions of these factors.

12.5.3 Types of soil

Soils have been shown to be the products of the interaction between the biotic and abiotic components of the ecosystem. As a result they show great complexity and variety, and they are very difficult to classify. Of the many soil types that have been recognised and described, three very different types which are found in moist temperate zones like Britain are chosen here as examples, namely podzols, brown earths and rendzinas. Soils can be examined by digging soil pits or using a soil auger as described in section 13.1.1. There is a standard nomenclature for the characteristic layers or **horizons** seen when a soil is viewed in profile. The nomenclature for freely drained soils, such as the three examples chosen, is shown and explained in fig 12.25.

Podzols

The name originates from the Russian words *pod*, meaning under, and *zola*, meaning ash, and a typical profile is shown in fig 12.26.

Podzols are characteristic of sandstone, gravel deposits (from glacial drift) and alluvial deposits, and are associated mainly with heathland and coniferous forests in cool, humid climates. The litter layer is deep as it contains materials with a high concentration of phenols, such as conifer 'needles', which are resistant to microbial decay.

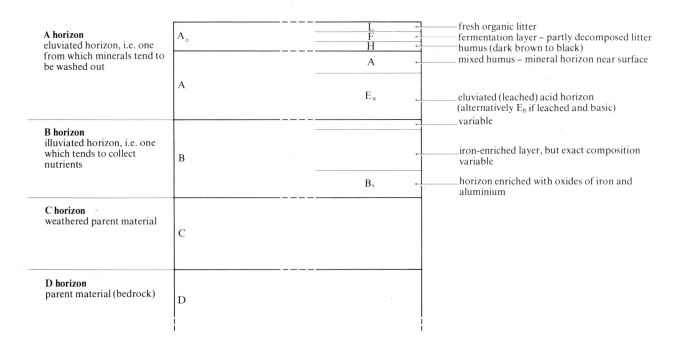

Fig 12.25 (above) *Generalised profile for a typical, freely drained soil showing the nomenclature used for soil horizons*

Fig 12.26 (below) *Profile of a podzol*

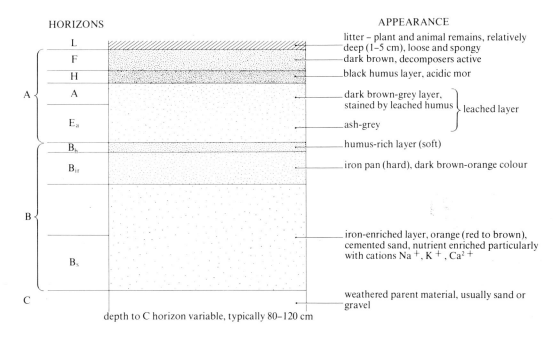

The calcium content of the soil is low so conditions are too acidic (pH 3–6.5) to support significant numbers of earthworms and other animals which ingest soil. Hence decay of the litter is mainly carried out by fungi, which is a relatively slow process, and results in a deep acidic humus layer, known as **mor**, which remains quite separate from the lower layers.

The E_a horizon is characteristically ash grey in colour and stands out clearly from the layers above and below. It consists mainly of separate grains of sand and is heavily leached, and hence nutrient deficient, as a result of either (or both) very high precipitation and very pervious bedrock. Organic acids from the humus contribute to the leaching process by carrying cations down through the soil.

Beneath the leached layer collect the iron, humus and other materials that have been washed down through the soil. Humus and iron oxide may cement into a '**hard pan**' or '**iron pan**' which may be so hard and thick that it prevents the penetration of roots to the layers below. Between the pan and the parent rock is a nutrient-rich layer where the leached materials accumulate.

The leached layer, in which most of the plant roots grow, is acidic (pH 3–4) and low in nitrates and other plant nutrients. Podzols are therefore relatively infertile soils. Much of the nutrient in a coniferous forest is present in the biomass of trees and litter rather than in the soil. Cycling of nutrients and decomposition are slow. Such soils tend to be a dominant factor in determining the vegetation of an area because they favour acid-loving plants (**calcifuges**). There are relatively few calcifuge species and these include many typical heathland species such as the heathers (*Calluna vulgaris*, *Erica tetralix*, *Erica cinerea*), heath bedstraw (*Galium saxatile*), sheep's sorrel (*Rumex acetosella*) and *Rhododendron*.

Brown earths (brown forest soils)

Brown earths have a relatively simple appearance in profile, consisting of uniform brown or dark brown A and B horizons which grade in conspicuously into each other (fig 12.27). They are associated with temperate deciduous forests, often in warmer areas of lower altitude than the coniferous forests and podzols. Decomposition rates are greater than on podzols owing to the higher temperature and lower resistance of deciduous leaves to decomposition, so there is a faster turnover of litter. This layer is usually shallow as the soil is rich enough in calcium (giving a pH of 4.5–8) to support large numbers of organisms, such as earthworms, which mix the top layers of soil. The humus formed is of the type known as **mull**. Nutrient richness increases towards the B horizon (as in podzols). The absence of a pan means that plant roots have unrestricted

access to the deeper, more nutrient-rich layers and the soils are generally relatively fertile. They weather to a loamy texture.

The mineral nutrient content of deciduous forest biomass and of the soil is higher than in coniferous forest and there is a more rapid cycling of nutrients.

Rendzinas

The name originates from the Polish word *rzedzic*, to tremble, as the soil is shallow and implements used in farming strike the rock below.

Rendzinas are very shallow soils consisting of a thin dark brown or black calcium-rich layer covering a calcareous bedrock (fig 12.28). They are characteristic of limestone and chalk.

Climate has little influence on their development and they are therefore found in a wide range of climatic conditions. Chalk and limestone generally form hills (such as the Downs and Cotswolds) since, being very permeable, water tends to penetrate the rock rapidly rather than form streams and cause erosion.

Rendzinas generally form steep slopes where the soil remains, in effect, permanently underdeveloped due mainly to soil creep. The rapid water penetration of the soil means that over a period of time the calcareous material, which is slightly soluble, is leached out, leaving behind any siliceous material in the surface layer of the soil. There is no B horizon and no definite layers can be distinguished. The important features of the soil from a biological viewpoint are its high calcium carbonate content (up to 80%), which gives it a basic character with a pH value greater than 8.0, and its rapid drainage which causes drying out. Plants growing on rendzinas are, therefore, usually specialised for

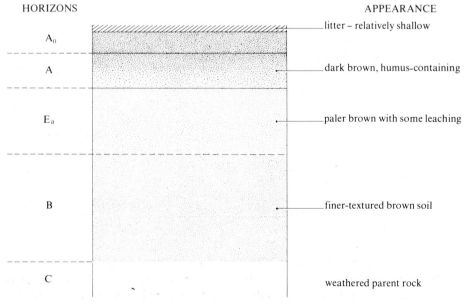

HORIZONS

A_0 — litter – relatively shallow

A — dark brown, humus-containing

E_a — paler brown with some leaching

B — finer-textured brown soil

C — weathered parent rock

APPEARANCE

Fig 12.27 *Profile of a brown earth*

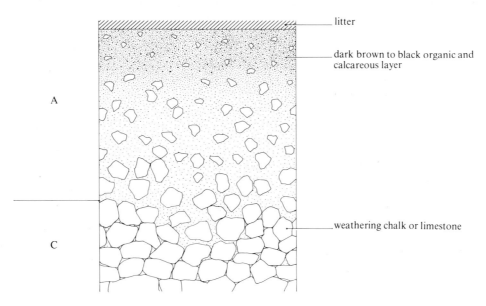

litter

dark brown to black organic and calcareous layer

weathering chalk or limestone

Fig 12.28 *Profile of a rendzina*

reducing water loss and for making maximum use of available water, often possessing laterally extensive roots or deep roots which penetrate the bedrock. Some plants thrive particularly well in conditions of high pH and are called **calcicoles**. Examples are *Clematis* (traveller's joy or old man's beard), milkwort (*Polygala calcarea*), wild marjoram (*Origanum vulgare*) and basil-thyme (*Acinos arvensis*). Rendzinas generally have a relatively rich flora. Like podzols, they are an example of a soil which has a particularly strong influence on the biotic component of the ecosystem.

12.5.4 Climatic factors

The chief climatic variables in an ecosystem are light, temperature, water availability and wind. Their effects on soil formation have already been described in section 12.5.2 and their direct effects on the biotic component of the ecosystem are described below.

Light

As the source of energy for photosynthesis light is essential for life, but it influences living organisms in many other ways. In considering its effects it is useful to remember that the intensity, quality (wavelength or colour), and duration (photoperiod) of light can all have different effects.

Light intensity is affected by the angle of incidence of the Sun's rays to the surface of the Earth. This varies with latitude, season, time of day and aspect of slope.

Photoperiod or daylength is a more or less constant 12 h at the equator but at higher latitudes it varies seasonally. Plants and animals of higher latitudes typically show photoperiodic responses that synchronise their activities with the seasons, such as flowering and germination of plants (section 15.4), migration, hibernation and reproduction of animals (section 16.8.5).

The need for light by plants has an important effect on the structure of communities. Aquatic plants are confined to surface layers of water and in terrestrial ecosystems competition for light favours certain strategies such as gaining height through growing tall or climbing, and increasing leaf surface area. In woodland this results in stratification as explained in section 12.6.1.

Some of the major processes in which light is involved, and which are discussed in other chapters, are summarised in table 12.12.

Temperature

The main source of heat is the Sun's radiation, with geothermal sources being important only in a minority of habitats, such as the growth of bacteria and blue-green algae in hot springs.

A given organism will survive only within a certain temperature range for which it is metabolically and structurally adapted. If the temperature of a living cell falls below freezing, the cell is usually physically damaged and killed by the formation of ice crystals. At the other extreme, if temperatures are too high, enzymes become denatured. Between the extremes enzyme-controlled reactions, and hence metabolic activity, double in rate with every 10°C rise. Most organisms are able to exert some degree of control over their temperatures by a variety of responses and adaptations so that extremes and sudden changes of environmental temperature can be 'smoothed

Table 12.12 Major processes of plants and animals in which light is involved.

Photosynthesis (see chapter 9 and section 12.3.4)
  On average 1–5% of the radiation incident on plants is used in photosynthesis
  Source of energy for rest of food chain
  Light is also needed for chlorophyll synthesis

Transpiration (see section 14.3)
  About 75% of the radiation incident on plants is wasted in causing water to evaporate thereby causing transpiration
  Important implications for water conservation

Photoperiodism (see sections 15.4 and 16.8.5)
  Important for synchrony of plant and animal behaviour (particularly reproduction) with seasons

Movement (see section 15.1 and chapter 17)
  Phototropism and photonasty in plants; important for reaching light
  Phototactic movements of animals and unicellular plants; important for locating suitable habitat

Vision in animals (see section 16.5)
  One of the major senses

Other roles
  Synthesis of vitamin D in humans
  Prolonged exposure to ultra-violet damaging, particularly to animals, therefore pigmentation, avoidance behaviour, etc.

out' (sections 18.3 and 18.4). Aquatic environments undergo less extreme temperature changes, and therefore provide more stable habitats, than terrestrial environments owing to the high heat capacity of water.

As with light intensity, temperature is broadly dependent on latitude, season, time of day and aspect of slope. However, local variations are common, particularly in microhabitats, which have their own microclimates. Vegetation usually has some microclimatic effect on temperature, as in forests (section 12.6.3) or on a smaller scale within individual clumps of plants or the shelter of leaves and buds of individual plants.

Moisture and salinity

Water is essential for life and is a major limiting factor in terrestrial ecosystems. It is precipitated from the atmosphere as rain, snow, sleet, hail or dew. There is a continuous cycling of water, the **hydrological cycle** (section 12.4), which basically governs water availability over land surfaces. For terrestrial plants water is absorbed mainly from the soil. Rapid drainage, low rainfall and high evaporation, or a combination of these factors, can result in dry soils, whereas the opposite extremes can lead to permanent waterlogging. The amount of water in the soil therefore depends on the water-retaining capacity of the soil itself, and the balance between precipitation and the combined effects of evaporation and transpiration (evapotranspiration), evaporation taking place from the surface of wet vegetation as well as from the soil surface.

Plants can be classified according to their ability to tolerate water shortage as xerophytes (high tolerance), mesophytes (medium tolerance) and hydrophytes (low tolerance/water-adapted). Some of the xeromorphic adaptations will be discussed in section 14.3.7 with transpiration and in section 19.3.2, and are also summarised in table 12.13. Similarly, terrestrial animals show adaptations for gaining and conserving water, particularly in dry habitats (see section 19.3.4 and table 12.13).

Aquatic organisms also have problems of water regulation (section 19.4.6). Salinity of water is relevant, as a comparison between freshwater and marine species will reveal. Relatively few plants and animals can withstand large fluctuations in salinity. Those that can are usually associated with estuaries or salt marshes, such as the snail *Hydrobia ulvae* which can survive a range of salinities from 50–1 600 mmol dm^{-3} of sodium chloride. Salinity may also be relevant in terrestrial habitats; if evaporation exceeds precipitation soils may become saline and this is a problem in some irrigated areas.

Atmosphere

The atmosphere is a major part of the ecosphere, to which it is linked by a number of biogeochemical cycles which have gaseous components, principally the carbon, nitrogen, oxygen and hydrological cycles. Its physical properties are also of importance, for its low resistance to movement and lack of physical support for terrestrial organisms has had a direct influence on their structure and, in addition, several animal groups have exploited flight as a means of locomotion. The atmosphere, like the oceans, is constantly circulating. This is a mass flow phenomenon, the energy for which comes from the Sun.

On a large scale, atmospheric circulation is particularly relevant to the distribution of water vapour because this can be picked up locally (by evaporation), carried by mass flow in moving air masses and deposited locally (by precipitation). If the release of other gases into the atmosphere is local, for example pollutant gases such as sulphur dioxide from industrial areas, then the pattern of atmospheric circulation will affect their distribution and eventual precipitation as solutions in rainfall.

Wind can interact with other environmental variables to affect growth of vegetation, particularly trees in exposed places, where they may become stunted and distorted on their windward sides. Wind is also important in increasing evapotranspiration under conditions of low humidity.

Dispersal of spores, seeds, and so on through the atmosphere, aided by wind, increases the spread of non-motile organisms like plants, fungi and some bacteria. Winds may also influence the dispersal or migration of flying animals.

Another atmospheric variable is atmospheric pressure, which decreases with altitude. This reduces oxygen availability and affects animals as described in section 11.8.1, and increases transpiration from plants, the latter resulting in adaptations to conserve water, as, for example, those shown by many alpine plants.

Table 12.13 Adaptations of plants and animals to dry conditions.

	Examples
Reducing water loss	
Leaves reduced to needles or spines	Cactaceae, Euphorbiaceae (spurges), conifers
Sunken stomata	*Pinus, Ammophila*
Leaf rolls into cylinder	*Ammophila*
Thick waxy cuticle	Leaves of most xerophytes; insects
Swollen stem with large volume to surface area ratio	Cactaceae and Euphorbiaceae ('succulents')
Hairy leaves	Many alpine plants
Leaf-shedding in drought	*Fouquieria splendens* (ocotillo or candle plant)
Stomata open at night, close during day	Crassulaceae (stonecrops)
Efficient carbon dioxide fixation at night with partial opening of stomata	C_4 plants, e.g. *Zea mais*
Uric acid as nitrogenous waste	Insects, birds and some reptiles
Long loop of Henlé in kidneys	Desert mammals, e.g. camel, desert rat
Tissues tolerant of high temperatures, reducing sweating or transpiration	Many desert plants, camel
Burrowing behaviour	Many small desert mammals, e.g. desert rat
Spiracles covered with flaps	Many insects
Increasing water uptake	
Extensive shallow root system and deep roots	Some Cactaceae, e.g. *Opuntia*, and Euphorbiaceae
Long roots	Many alpine plants, e.g. *Leontopodium alpinum* (edelweiss)
Burrow for water	Termites
Water storage	
In mucilaginous cells and cell walls	Cactaceae and Euphorbiaceae
In specialised bladder	Desert frog
As fat (water is a product of oxidation)	Desert rat
Physiological tolerance of water loss	
Apparent dehydration can occur without death	Some epiphytic ferns and clubmosses, many bryophytes and lichens, *Carex physiodes* (sedge)
Loss of high proportion of body mass with rapid recovery when water is available	*Lumbricus terrestris* (70% loss of body mass), camel (30% loss)
Evasion	
Pass unfavourable season as seed	Californian poppy
Pass unfavourable season as bulbs or tubers	Some lilies
Seed dispersal, with some reaching more favourable conditions	Soil organisms, e.g., mites, earthworm
Escape behaviour	Earthworm, lungfish
Aestivation in mucus-containing sheath	

Microclimates

The particular climatic conditions associated with a habitat or microhabitat may be different from those of surrounding areas and are known as **microclimates**. An understanding of microclimates and microhabitats is important to an understanding of the complexity of ecosystems. One example will serve to illustrate the general principle. A group of crustaceans known as isopods (woodlice and related forms) contains species such as *Porcellio* which are confined to the moist air of leaf litter and other species such as *Armadillidium* which are able to withstand drier conditions and wander more freely.

12.5.5 Topography

The influence of topography is intimately connected with the other abiotic factors since it can strongly influence local climate and soil development, as already discussed.

The main topographic factor is altitude. Higher altitudes are associated with lower average temperatures and a greater diurnal temperature range, higher precipitation (including snow), increased wind speeds, more intense radiation, lower atmospheric pressures and reduced gas concentrations, all factors which have an influence on plant and animal life. As a result, vertical zonations are common, as shown in fig 12.29.

Mountain chains can act as climatic barriers. As air rises over mountains it cools and precipitation tends to occur. Thus a rain 'shadow' occurs on the leeward side of the mountains where air is drier and precipitation is less. This affects the ecosystems. Mountains also act as barriers to dispersal and migration and may play important roles as isolating mechanisms in the process of speciation as described in section 25.7.

Another important topographic factor is aspect. In the northern hemisphere south-facing slopes receive more sunlight, and therefore higher light intensities and temperatures, than valley bottoms and north-facing slopes (the reverse being true in the southern hemisphere). This has

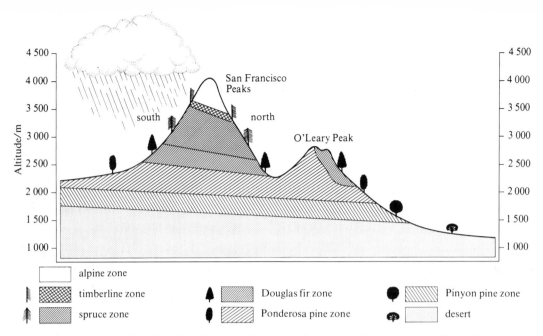

Fig 12.29 *Zonation of vegetation on San Francisco Peaks, Arizona, as viewed from the south-east. (From Merriam (1890). Redrawn and modified from W. D. Billings (1972)* Plants, man and the ecosystem, *2nd ed., Macmillan: London and Basingstoke.)*

striking effects on the natural vegetation and on land use by humans.

Steepness of slope (**inclination**) is a third topographic factor. Steep slopes generally suffer from faster drainage and run-off and the soils are therefore thinner and drier, with more xeromorphic vegetation. At slope angles in excess of 35° soil and vegetation are typically unable to develop, and screes of loose material form.

12.6 Community ecology and ecological succession

A **community** is a group of interacting populations living in a given area, and represents the living part of an ecosystem. It functions as a dynamic unit with trophic levels, a flow of energy and a cycling of nutrients through it as described in section 12.4.

Some of the interactions have also been mentioned in section 12.3, such as predator–prey relationships (including grazing) and parasitism. Others may exist, such as mutualism (section 2.5.3) and competition (section 12.7.6).

The structure of a community is always built up over a period of time. An example which can be used as a model for the development of a community is the invasion and colonisation of bare rock, as on a recently created volcanic island. Trees and shrubs cannot grow on bare rock since there is insufficient soil. Algae and lichens, however, can invade and colonise such areas, coming in by various methods of dispersal and forming the **pioneer community**. The accumulation of dead and decomposing organisms, and the erosion of rock by weathering, leads to the accumulation of sufficient soil for invasion and colonisation by larger plant species such as mosses or ferns. Ultimately these plants will be succeeded by even larger and more nutrient-demanding plants such as seed-bearing plants,

including grasses, shrubs and trees. Fig 12.30 shows a typical succession.

Such replacement of some species by others through time is called an **ecological succession**. The final stable and self-perpetuating community, which is in equilibrium with its environment, is called the **climax community** and is the most productive that environment **can sustain** (see also section 12.6.2). The animals of such a community will also have shown a succession, to a large extent dictated by the plant succession, but also being influenced by what animals are available to migrate from surrounding communities.

The type of succession described above, in which there is an initial colonisation of bare rock or other surface lacking organic soil, such as sand dunes or glaciated surfaces, is called **primary succession. Secondary succession** is said to occur when the surface is completely or largely denuded of vegetation but has already been influenced by living organisms and has an organic component, for example a cleared forest or a previously burned or farmed area. Seeds and spores and organs of vegetative reproduction, such as rhizomes, may be present in the ground and thus influence the succession. In both primary and secondary succession the flora and fauna of surrounding areas are major factors influencing the types of plants and animals entering the successions through chance dispersal and migration.

> **12.16** What factors are likely to affect the number and diversity of species reaching an area?

The complete succession is sometimes called a **sere**, the sere being made up of a series of **seral communities** (**seral stages**). Seres of particular environments tend to follow similar successions and may therefore be classified according to environment; for example a **hydrosere** develops in an aquatic environment as a result of the colonisation of open water and a **halosere** develops in a saltmarsh.

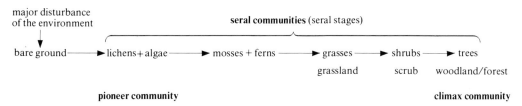

major disturbance
of the environment

seral communities (seral stages)

bare ground ——→ lichens + algae ——→ mosses + ferns ——→ grasses ——→ shrubs ——→ trees

grassland   scrub   woodland/forest

pioneer community   climax community

Fig 12.30 *A typical terrestrial succession*

The climax community is often described as having one **dominant** or several **co-dominant** species. The term **dominance** is rather subjective but normally refers to those species with the greatest collective biomass or productivity, although physical size of individuals is also usually considered important. In practice, however, the concept of dominance is often of little value, as will become apparent in section 12.6.2.

The idea of succession was first discussed in detail in 1916 by Clements, who studied communities in North America and came to the conclusion that climate is the dominant factor in determining the composition of the climax community. His view, known as the **monoclimax hypothesis**, was that only one climax community was ultimately possible for a given climate, and this was called the **climatic climax community**. Any interruption in the progression towards this climax caused by local conditions of topography, microclimate, land use, and so on might still lead to a stable community, but this could not be regarded as a climax because, in the long term, climatically determined factors should lead to the theoretical climax. Mountains and hills, for example, would gradually be eroded away with consequent changes in community structure. The more modern concept is that of the **polyclimax**, which is a climax influenced by all physical factors, one or several of which may be dominant, such as drainage, soil, topography or incidence of fire. The community is regarded as a true climax community if it is 'stable' with respect to time. Any change occurring is relatively slow compared with the time taken to establish the climax through succession.

Zonation

Within a community at any one time species may be spatially distributed according to variations of the physical environment. This is called **zonation**. A good example is the zonation of seaweeds and marine animals that occurs on rocky shores from low-tide to high-tide level and into the splash zone. Physical conditions vary through these zones, notably length of exposure to air between successive tides, and each zone is occupied by species adapted for its particular conditions. Zonation on a seashore is described and illustrated in section 13.4. Another good example of zonation is the vertical zonation that occurs on mountains with increasing altitude (fig 12.29). Superficially zonations may resemble succession, but it is important to recognise the basic difference, namely that with zonation the species vary in space (spatially), whereas with succession the species vary in time (temporally).

12.6.1 Examples of natural and human-made climax communities in Britain

As a result of pollen analysis it has been possible to build up a picture of vegetational changes in Britain since the end of the last Ice Age about 10 000 years ago. At the beginning of the Neolithic period, about 5 000 years ago, much of lowland Britain was covered with a climax community that was forest, predominantly oak. Neolithic Man introduced agriculture and large-scale clearing of forest for the first time. He cultivated mainly the drier upland areas, such as the chalk of the South Downs and the limestone of the Cotswolds. The areas were reduced to grassland for grazing and were sometimes tilled for crops. By the Iron Age there appears to have been fairly extensive cultivation and grazing of these areas. A typical climax community of, for example, Roman times (when Britain was in the late Iron Age) would have been a lowland forest with oak as the dominant species, or a mixed oak–beech climax, and herbivores such as deer, omnivores such as bears and carnivores such as wolves and lynxes. The Weald (the area between the North and South Downs) and Midland Plain are examples of areas that were covered with such forests. Pine and birchwoods would have occupied poorer, sandier soils and higher altitudes. Ash was common on limestone and chalk hills, alder and willow at wetter sites.

The modern distribution of beech shows a link with chalk and limestone areas. This reflects past human use of beech and oak and the greater climatic sensitivity of beech in Britain. Beech can grow on a wide variety of soils and will dominate oak on well-drained loams. Its patchy modern distribution on loamy soils in part reflects human preference for oak throughout history. Oak timber, for structural purposes, and bark, for tanning, were highly valued. Adequate supplies of beech, used in furniture making, could be obtained from beechwoods in areas where oak could not compete (the shallow soils of calcareous areas – see below). Furthermore, oak produces acorns every three or four years whereas heavy beech 'mast' (seed) years are less frequent in Britain. Frost susceptibility and cool summers are key factors affecting seed production, germination and establishment of beech in Britain. Oak regeneration thus occurred more regularly, reinforcing the human-induced tendency for more widespread and abundant oakwoods.

Cultivation spread from the hills to the more fertile and easily worked lowland soils. There has been a progressive clearing of natural forest and increasing interference in natural communities until the present day, when there are only a few 'relic' woodlands left that can be regarded as near natural. As a result, some of the apparently natural vegetation that occurs today is not a natural climax vegetation but a man-made climax or a 'sub-climax' (sometimes called a **plagioclimax**). Two good examples are chalk grassland and lowland heaths.

Much of the vegetation of chalk hills, such as the North and South Downs, is described as grassland and is notable for the rich variety of its flowers, many of which, like certain orchids, are relatively rare. The vegetation could be regarded as a **biotic climax** because the dominant influence in maintaining its stability has been the biological effects of grazing. Sheep and cattle were the dominant herbivores and, since its introduction by the Normans, the rabbit. Rabbits became a serious pest on parts of the chalk downs by cropping the grass too short for use by other herbivores, but in 1953 the viral disease myxomatosis was 'accidentally' introduced into England. The disease spread so rapidly through the rabbit populations of Britain that by the end of 1955 the rabbit population had decreased by 90%. The disease still recurs at intervals and the rabbit population is still much smaller than formerly. As a result of the reduction of both rabbit and domestic grazing, young saplings of shrubs and trees survived. This, and possibly other factors, mean that **regeneration** of woodland is now a

'problem' in some areas where conservationists are trying to preserve the chalk grassland. The typical regeneration sequence (really a secondary succession) is through scrub in which shrubs dominate (such as bramble, hawthorn, dogwood, juniper and blackthorn) to a climax consisting of beech woodland. Beech is able to dominate oak in the drier chalk soils and is particularly characteristic of the steeper slopes. It is a good competitor partly because it shades out other species. Furthermore, on the drier shallow chalk soils the characteristically dense surface root mat of beech successfully absorbs water and nutrients and gives stability, whereas oak with its deep tap-root system cannot survive. On gentler slopes with deeper, moister soils, ash may precede beech in the succession and on heavier, wetter clay soils, oak is a better competitor than beech and becomes dominant. Here, then, the edaphic factor is important in the succession. Fig 12.31a summarises a succession from chalk grassland to woodland in south-east England.

Fig. 12.31b shows a succession from heath to woodland. Lowland heaths also tend to be associated with a particular soil type, being typical of acid, sandy soils. Like chalk grassland, they are also a man-made climax. In the past they were used for grazing by horses, cattle, pigs and rabbits, the dominant plants being heathers, particularly ling (*Calluna vulgaris*). The heath was managed by burning off older heaths to allow regeneration of young succulent shoots from underground parts that survive the fire. Fire is

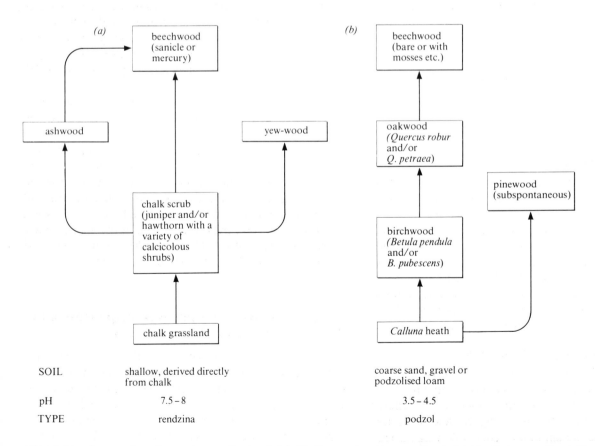

SOIL	shallow, derived directly from chalk	coarse sand, gravel or podzolised loam
pH	7.5 – 8	3.5 – 4.5
TYPE	rendzina	podzol

Fig 12.31 *Successions in south-east England. (a) Chalk grassland to woodland. (b) Heath to woodland (From A. G. Tansley (1968)* Britain's green mantle, *2nd ed., Allen & Unwin.)*

thus an important abiotic factor in development of this community. If unmanaged, the heath gradually reverts to woodland, passing through a scrub stage in which gorse, broom, hawthorn and other species may dominate. Birch is often the first tree to become established, followed by oak, the whole succession taking about 200 years. Pine, and less commonly beech, may also compete with oak. If lowland heaths are to be maintained, then active measures to remove shrubs and trees are necessary. Such management is desirable in the context of conservation because they contribute to the diversity of habitats found in Britain and certain rare species, such as the natterjack toad and the Dartford warbler (in Dorset), are found on heaths or heath scrubs.

A climax woodland

A typical terrestrial climax community is a deciduous woodland. Such communities are very rich in species. In a wood near Oxford, which has been the subject of a long and detailed study, about 4 000 animal species have been found. Part of the reason for species richness is the complex structure of the woodland which contains many niches and microhabitats.

The food web of an oak woodland has already been shown in fig 12.6a. Fig 12.32 illustrates the structure of a typical deciduous woodland. An important feature is the layering, or **stratification**, that is present. Most of the primary production occurs in the tree canopy and most of the decomposition at ground level, with animals occupying niches in all layers.

12.6.2 Underlying features of succession

A number of attempts have been made to explain succession in terms of some underlying principle which determines the progress towards the climax.

Ecological dominance

The observations of ecological dominance suggest that successions might always progress towards a climax in which one species is dominant, such as oak woodland, or a few species are codominant, such as beech–oak woodland. These impose restrictions, such as shading, on the species that can grow in association with them. However, the concept of dominance is difficult to apply in some situations, as in planktonic communities, among animals, and in the tropical rain forests where several hundred tree species may be found in roughly equal numbers.

Productivity and biomass

Lindeman in 1942 proposed that succession involved increasing productivity until a climax community was reached in which the maximum efficiency of energy conversion occurred. This is a logical and attractive hypothesis because the amount of energy flowing through an ecosystem is a major limiting factor in determining the number and biomass of the organisms it can support.

Evidence shows that the later stages of successions do become more productive, but that there is usually a decline in gross productivity associated with the climax community. Thus older forests have lower productivities than younger forests, which in turn may have lower productivities than the more species-rich herb layers that precede them. A similar decline has been observed in some aquatic systems. Maximum productivity is now believed to be attained very quickly in many successions, although data are still few. The reasons for the decrease in productivity can only be speculated upon, but using a forest as an example, older trees might be expected to be less productive than younger trees for several reasons. One is that the accumulation of nutrients in the increasing standing crop biomass may lead to a reduction in nutrient recycling. However, a simple reduction in vigour as the average age of the individuals in the community increases to a constant point would presumably cause a reduction in productivity.

Although it does not seem possible to argue that successions lead to maximum efficiency of energy conversion (maximum gross productivity) it *is* possible to argue that they lead to maximum accumulation of biomass. This is most obvious in the case of forest communities, where the plants become larger and larger during the succession, but the accumulated biomass of other climax communities is also normally greater than in the successional stages. Changes in gross productivity, respiration and biomass during a typical succession are summarised in fig 12.33. This shows that in the climax community these terms become more or less constant. It also shows that an upper limit of biomass is reached when total respiratory losses (R) from the system almost equal gross primary productivity (P), that is the P/R ratio is almost equal to 1. These and other trends that can be expected to occur in successions are summarised in table 12.14.

During a succession more and more of the available nutrients become locked up in the biomass of the community with a consequent decrease in nutrients in the abiotic component of the ecosystem (such as soil and water). The amount of detritus produced also increases and detritus feeders take over from grazers as the main primary consumers. Appropriate changes in food webs occur and detritus becomes the main source of nutrients.

12.6.3 Interactions between organisms and their abiotic environment

Early in a succession the most important interactions are those between living organisms and the abiotic environment. The latter is changed as a result of the activity of the organisms. A striking example of this is the formation of sand dunes as a result of the accumulation of wind-blown sand around shoots of *Ammophila* (marram grass or *Psamma*). This is described in fig 12.34. The stabilisation of sand dunes takes place over a period of a

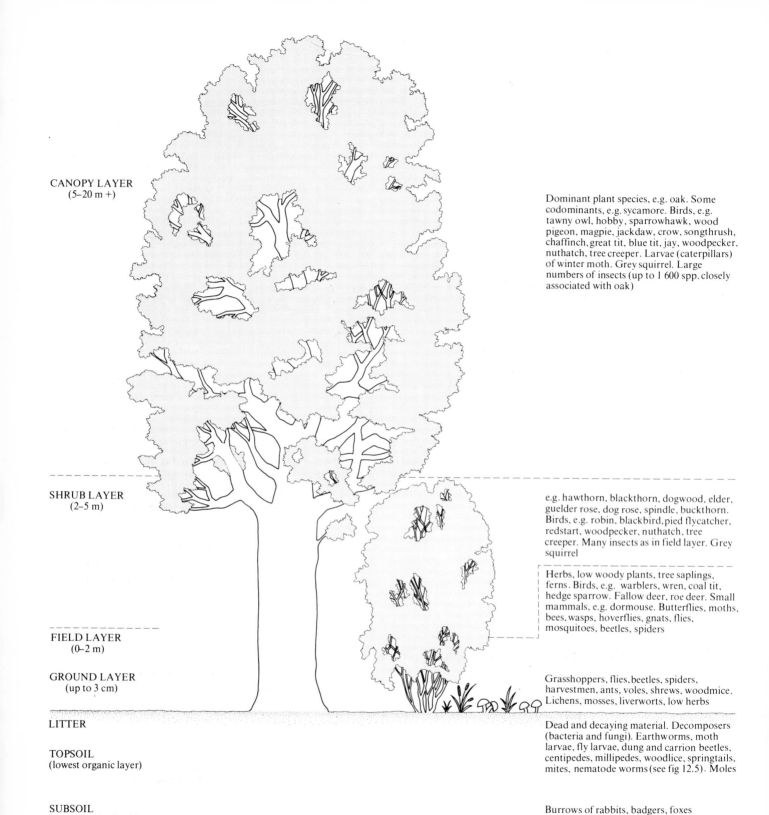

CANOPY LAYER
(5–20 m +)

Dominant plant species, e.g. oak. Some codominants, e.g. sycamore. Birds, e.g. tawny owl, hobby, sparrowhawk, wood pigeon, magpie, jackdaw, crow, songthrush, chaffinch, great tit, blue tit, jay, woodpecker, nuthatch, tree creeper. Larvae (caterpillars) of winter moth. Grey squirrel. Large numbers of insects (up to 1 600 spp. closely associated with oak)

SHRUB LAYER
(2–5 m)

e.g. hawthorn, blackthorn, dogwood, elder, guelder rose, dog rose, spindle, buckthorn. Birds, e.g. robin, blackbird, pied flycatcher, redstart, woodpecker, nuthatch, tree creeper. Many insects as in field layer. Grey squirrel

Herbs, low woody plants, tree saplings, ferns. Birds, e.g. warblers, wren, coal tit, hedge sparrow. Fallow deer, roe deer. Small mammals, e.g. dormouse. Butterflies, moths, bees, wasps, hoverflies, gnats, flies, mosquitoes, beetles, spiders

FIELD LAYER
(0–2 m)

GROUND LAYER
(up to 3 cm)

Grasshoppers, flies, beetles, spiders, harvestmen, ants, voles, shrews, woodmice. Lichens, mosses, liverworts, low herbs

LITTER

TOPSOIL
(lowest organic layer)

Dead and decaying material. Decomposers (bacteria and fungi). Earthworms, moth larvae, fly larvae, dung and carrion beetles, centipedes, millipedes, woodlice, springtails, mites, nematode worms (see fig 12.5). Moles

SUBSOIL
(weathered bedrock)

Burrows of rabbits, badgers, foxes

BEDROCK

Fig 12.32 *Layered structure of a typical deciduous woodland community. NB Some animals move between layers. For example grey squirrels forage on ground and sleep, breed and move among trees; birds may rest in one layer and feed in another, such as the tawny owl which takes mammals from the field and ground layers and nests in the canopy*

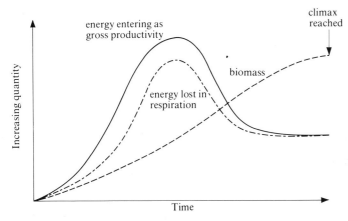

Fig 12.33 *Changes in gross productivity, respiration and biomass during a typical succession (Modified from M. R. Tribe et al. (1974)* Ecological principles, *Basic Biology Course 4, Cambridge University Press).*

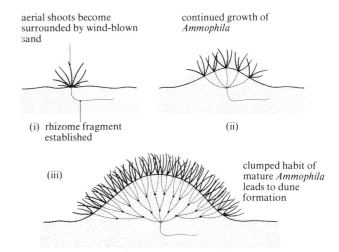

Fig 12.34 *Dune formation by gradual deposition of wind-blown sand around aerial shoots of* Ammophila

few years. A sand dune succession around Lake Michigan in North America has been well documented and here dune stabilisation allows the growth of larger plants with low nutritional requirements such as cottonwood (a tree) and pine trees. Over a long period of time a soil develops, typically accumulating organic carbon from the litter layer. Also an increase in nitrogen-fixing organisms leads to an increase in nitrogen in the soil. In parts this has allowed growth of oak which has become the dominant species. Once oak has become fully established soil conditions apparently remain constant over an indefinite period with nutrient demand being balanced by nutrient input from recycling of leaf litter and the like. Here edaphic factors have been a dominant influence in the succession and have themselves been modified during the succession. It may take a thousand years, however, for the soil to reach a stable chemical composition and for the succession to be regarded as complete.

The relatively short period for early stages of succession compared with later stages is an important general principle illustrated by this example. Further development of a climax community will obviously occur in response to changes in environmental conditions, either sudden or long term. The succession of Ice Ages over the last 1.8 million years is a good example of climatic change that profoundly changed the communities of affected areas.

There are other ways in which the biotic community can influence the physical environment, and hence the succession. Local modifications of climate may be induced, forming microclimates within the ecosystem.

12.17 In what ways might a group of trees influence the environment under the canopy in terms of (*a*) light, (*b*) temperature, (*c*) wind and (*d*) moisture?

Table 12.14 Summary of changes in an ecosystem during a typical secondary succession.

	Stage of ecosystem development	
Characteristic	Immature (early)	Mature (late)
Gross production/community respiration (P/R ratio)	high (>1)	approaches 1
Net community production	high	low
Food chains	linear, mainly grazing	web-like, mainly detritus-feeding
Total organic matter (or biomass)	small	large
Species diversity	low	high*
Structure of community	simple	complex (stratification, many microhabitats)
Niche specialisation	broad	narrow
Size of organism	small	large
Strategies adopted by species (section 12.7.4)	'r-strategy'	'K-strategy

* Most plant and some animal examples of succession show a peak of species diversity before climax.

Similarly, microhabitats may be generated, such as dead wood (which in a given area may support about 200 species of animals), dung (inhabited by more than 300 animal species, chiefly beetles and flies) and carrion (also colonised by beetles and flies).

12.6.4 Interactions between organisms within the community

In the latter stages of succession, biotic interactions become more important in forming the detailed community structure. The variety of living organisms tends to increase, so inevitably their interactions become more complex. The tropical rain forest communities, which are among the longest established climax communities, are renowned for their species richness and the extreme complexity of their biotic interactions. What we are concerned with in community ecology are the dynamic interactions between species rather than between members of the same species. A number of different kinds of interaction occur. Predator–prey interactions, including grazing, have already been mentioned with food chains and food webs in section 12.3, where the dependence of some organisms on others was made clear. Parasitic, symbiotic and other mutualistic relationships are also important. Studies of these interactions are bound to overlap with population ecology as some of the examples given in section 12.7.6 show.

What gradually occurs is the filling of the available ecological niches and there may be competition for a given niche (section 12.1). The chances of an animal becoming established will depend on many factors, including the likelihood of its chance migration into the area, the availability of suitable food, its ability to find a suitable niche and if necessary the ability to compete effectively for this niche. The more specialised it becomes for a particular niche, the less chance there is of direct competition.

A useful concept related to that of the ecological niche is **resource partitioning**, which is the sharing of the available resources among the different species of the community. Specialisation by different species to make use of different resources leads to less competition and a more stable community structure. Resource partitioning may take several forms, for example:

(1) specialisation of morphology and behaviour for different foods, such as the beaks of birds which may be modified for picking up insects, drilling holes, cracking nuts, tearing flesh, and so on;
(2) vertical separation, such as canopy dwellers and forest floor dwellers;
(3) horizontal separation, such as the occupation of different microhabitats.

One or a combination of these three factors would serve to separate the organisms into groups of species with reduced competition between the groups because each occupies its own niche. For example, ecological groupings have been devised for birds based on feeding location (air,

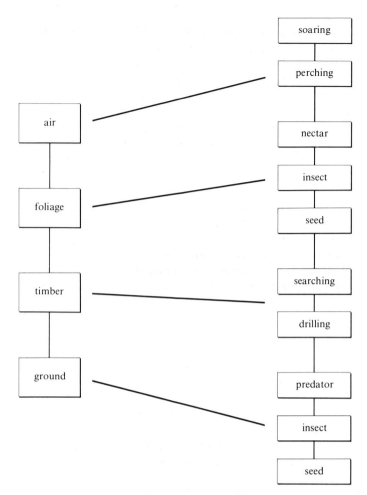

Fig 12.35 *Ecological groupings used to analyse three California bird communities. (From B. D. Collier* et al. *(1974) Dynamic ecology, Prentice Hall.)*

foliage, timber or ground) and subdivided according to major food type as shown in fig 12.35.

When plants are considered, each species will have its own set of optimum conditions of moisture, altitude, soil pH, and so on, in which it will be favoured, once again restricting direct competition.

Despite a tendency for each species to evolve its own particular niche, some direct competition between species for available resources is inevitable. This is discussed in section 12.7.6.

12.7 Population ecology

In the introduction to this chapter, a **population** was defined as a group of organisms of one species occupying a defined area and usually isolated to some degree from other similar groups by geographical or topographical boundaries, or even arbitrarily delimited by an investigator. Thus one might study, for example, the roe deer of a particular woodland, the frogs of a particular valley or trout of a particular lake. Many biogeographers widen this definition and use similar principles to study major plant and animal taxa, rather than single species, on a global or regional scale. In studying populations we are concerned not just with the numbers of a given species living in a given area at a given moment in time, but rather with an understanding of how populations grow, are maintained and decline in response to their environments. This aspect of population ecology is called **population dynamics**. It focusses on properties of a group rather than the individuals within the group. Thus one examines characteristics such as density, natality, mortality, survivorship, age structure, biotic potential, dispersion and growth form. Populations also have properties related to their ecology such as adaptiveness and reproductive fitness. Population interactions such as competition, predation and parasitism not only regulate growth of a given population but influence the structure of communities. Successions (section 12.6) may also be examined from the viewpoint of population interactions. Knowledge of how numbers of individuals of different species change through time can also be used in biomass and energy flow studies and therefore in building up knowledge of ecosystems. As discussed in section 12.1, the study of populations has many important applications, as in the control of pests and in species conservation.

Apart from some long-term studies on phytoplankton, most work on population dynamics has been done on animals and micro-organisms. With plants, especially crop plants, it is more useful, and easier in some cases, to collect data on biomass rather than numbers. The dynamics of plant populations exert a strong influence on those of animal populations so, for example, the study of human populations (**demography**) is usually studied in geography alongside the availability of food resources based on plants.

Some information on how to sample animal and plant populations with the aims of collecting data on numbers and distribution is given in section 13.2.

12.7.1 Birth rate (natality) and death rate (mortality)

Population size may increase as a result of **immigration** from neighbouring populations, or by reproduction of individuals within the population. One measure of reproduction is known as **fecundity** and concerns the numbers of offspring produced by individual females of the species. Fecundity is expressed in different ways, depending on convenience and the species involved. It could be defined, for example, as the average number of fertilised eggs produced in an average breeding cycle or in a lifetime. For mammals, fecundity is expressed as **birth rate** or **natality**, the number of young produced per female per unit time (usually per year). In the case of humans, birth rate is usually expressed as the number of births per thousand head of population per year (table 12.15) and generally less-developed countries, where birth control is not practised extensively, have birth rates about twice those of

Table 12.15 Birth rate statistics from selected countries and regions. Figures are the number of live births per year per thousand head of population.

Countries		Regions	
Kenya	53	Africa	46
Congo	45	Latin America	32
Iran	44	Asia	29
Egypt	41	North America	16
India	36	Australia	16
South Africa	36	Europe	14
Mexico	33		
Malaysia	31		
Jamaica	27		
Israel	25		
Chile	22		
China	18		
United States	16		
Cuba	15		
Japan	14		
East Germany	14		
United Kingdom	13		
Italy	12		
Sweden	12		
West Germany	10		

Data from the 1981 World Population data sheet.

Table 12.16 Death rate statistics from selected countries and regions. Figures are the number of deaths per year per thousand head of population.

Countries		Regions	
Ethiopia	25	Africa	17
Congo	19	Asia	11
India	15	Europe	10
East Germany	14	Latin America	9
Iran	14	North America	9
Kenya	14	Australia	7
South Africa	12		
United Kingdom	12		
West Germany	12		
Egypt	11		
Sweden	11		
Italy	9		
United States	9		
Mexico	8		
Malaysia	8		
Chile	7		
Israel	7		
China	6		
Cuba	6		
Jamaica	6		
Japan	6		
Costa Rica	4		

Data from the 1981 World Population data sheet.

better developed countries. Sociologists refer to the cultural change by a society from the higher to the lower rate as the **demographic transition**.

Population size may decrease as a result of **emigration** or death (**mortality**). In population biology, mortality strictly means *rate* of death and may be expressed in terms of per cent, or numbers per thousand, dying per year. Table 12.16 shows that generally speaking the well-developed countries have a lower mortality rate than the less-developed countries, owing to their better medical care and nutrition. Where medical and food aid programmes have been introduced, as in India since the Second World War, the statistics reveal a decreasing mortality rate. The very low mortality rates of places like Japan and Costa Rica are due to a relatively high proportion of young people.

12.7.2 Survivorship curves

The percentage of individuals that die before reaching reproductive age (**pre-reproductive mortality**) is one of the chief factors affecting population size, and for a given species is much more variable than fecundity. If population size is to remain constant, on average only two offspring from each male–female pair must survive to reproductive age.

If we start with a population of newborn individuals and the decrease in numbers of survivors is plotted against time, a **survivorship curve** is obtained. On the vertical axis, actual numbers of survivors may be plotted, or percentage survival:

$$\frac{\text{number of survivors}}{\text{number in original population}} \times 100\%$$

Different species have characteristic survivorship curves, depending partly on their pre-reproductive mortality. Some representative examples are shown in fig 12.36.

Most animals and plants exhibit a phenomenon called senescence or ageing, manifested as a declining vigour with increasing age beyond maturity. Once senescence begins, there is increasing likelihood of death occurring within a given time period. The immediate cause of death can vary, but the underlying cause is a reduced resistance to external factors such as disease. Curve (*a*) in fig 12.36 shows an almost ideal curve for a population in which senescence is the major factor affecting mortality. An example would be a human population in a modern industrialised country in which high standards of medicine and nutrition are maintained. Most people live to old age, but little can yet be done to prolong life expectancy beyond about 75 years. The main deviation of curve (*a*) from the ideal is due to infant mortality, shown by the dip at the start of the curve. Although infant mortality is much lower in industrialised countries, there are still above-average risks to life in early infancy. Another factor which will combine with senescence to affect the curve to some extent is accidental death, the cause of which may vary with age. In England, for example, deaths through car accidents reach a peak among people in their early twenties. A curve like (*a*) would also be obtained for an annual crop plant such as wheat, where all the plants in a given field senesce simultaneously.

Curve (*b*) is for a population with a high mortality rate early in life, such as might occur for mountain sheep or for humans in a country in which starvation and disease are prevalent. Curve (*c*) shows the kind of smooth curve that would be obtained if there was a constant mortality rate throughout life (50% per unit time). Such a curve is obtained if chance is the major factor influencing mortality and the organisms die out before senescence becomes evident. A curve similar to this was once obtained for a population of glass tumblers in a cafeteria. Some animal populations show survivorship curves which approximate closely to this model curve, for example *Hydra*, where there is no special risk attached to being young. Most non-vertebrates and plants show a curve similar to (*c*) but with high juvenile mortality superimposed so that the initial part of the curve descends even more steeply.

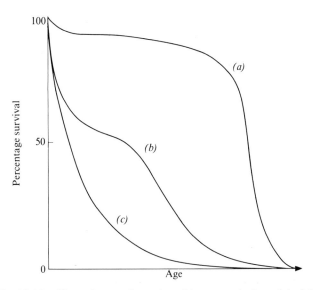

Fig 12.36 *Three types of survivorship curve. Letters (a), (b) and (c) are explained in the text*

Minor variations in survivorship curves may occur within species for various reasons, a common one being sexual differences. In humans, for example, female life expectancy is slightly greater than for males, although the precise reasons for this are unknown.

By plotting survivorship curves of species it is possible to determine the mortality rates of individuals of different ages and hence to determine at which ages they are most vulnerable. By identifying the factors causing death at these ages, an understanding can be gained of how population size is regulated.

12.21 The following figures apply to sockeye salmon in a Canadian river system. Each female salmon lays 3 200 eggs in a gravelly shallow in the river in autumn. 640 fry (young fish derived from these eggs) enter a lake near the shallow in the following spring. 64 smolts (older fish survivors from the fry) leave the lake one year later and migrate to the sea. Two adult fish (survivors of these smolts) return to the spawning grounds $2\frac{1}{2}$ years later; they spawn and then die. Calculate the percentage mortalities for sockeye salmon for each of the following periods:

(a) from laying eggs to movement of fry into the lake six months later;

(b) from entering the lake as fry to leaving the lake as smolts 12 months later;

(c) from leaving the lake as smolts to returning to the spawning grounds as adult salmon 30 months later;

Draw a survivorship curve for the sockeye salmon in this river system (plot percentage survival against age). What is the pre-reproductive mortality for these salmon?

Source of question: Open University, S100, Unit 20, *Species and Populations*, p. 71.

12.7.3 Population growth and growth curves

Populations grow and decline in characteristic ways. The size of population increase will be determined by the reproductive potential of the organisms concerned and by environmental resistance. The maximum reproductive potential is the rate of reproduction given unlimited environmental resources. This is termed the intrinsic rate of natural increase or the **biotic potential** (and is symbolised r in the growth curve equation). Biotic potential will vary according to the age structure of a population and obviously may also be influenced by male:female ratios. Thus, theoretically, one can identify the population structure that maximises biotic potential (r-max). This point is put to practical use in fisheries management. By regulating catches the population age structure can be manipulated to human advantage to maximise recruitment and growth rates.

In practice the full biotic potential of an organism is not realised. The difference between the actual occurring rate of increase and the biotic potential reflects environmental resistance. **Environmental resistance** means the sum total of limiting factors, both biotic and abiotic, which act together to prevent the biotic potential from being realised. It encompasses external factors such as predation, food supply, heat, light and space, and internal regulatory mechanisms such as intraspecific competition and behavioural adaptations. Strong feedback links exist between all these factors, for example intraspecific competition arises in response to some resource (such as space) which is in limiting supply.

Growth curves

Two basic forms of growth curve can be identified, the **J-shaped** growth curve and the **S-shaped** or sigmoidal growth curve. These contrasting forms may be combined or modified in various ways according to the particular circumstances of an organism's environment and life history. Human intervention may deliberately, or by chance, modify a population's growth form. An example is that of the Kaibab deer, more fully discussed in 12.7.6. In this case the population growth form changed from an S-shaped curve (density-dependent – see later) to a J-shaped form (density-independent – see later) primarily due to the deliberate removal by humans of predatory control. Eventually a new equilibrium was established marking a return to density-dependent control. Similarly many 'pest' problems arise as a result of human removal of a predator or competitor species, triggering a shift from sigmoidal to J-shaped population growth.

The S-shaped or sigmoidal growth curve describes a situation in which, in a new environment, the population density of an organism increases slowly initially, as it adapts to new conditions and establishes itself, then increases rapidly, approaching an exponential growth rate. It then shows a declining rate of increase until a zero population growth rate is achieved where rate of reproduction (natality) equals rate of death (mortality) (fig 12.37*a*). The declining rate of increase reflects increasing environmental resistance, which becomes proportionately more important at higher population densities. In other words, as numbers increase so competition for essential resources, such as food or nesting materials, increases until eventually feedback in terms of increased mortality and reproduction failures (fewer matings, stress-induced abortion) reduces population growth to zero with natality and mortality in approximate equilibrium.

This type of population growth is said to be **density-dependent** or density-conditioned since, for a given set of resources, growth rate depends on the numbers present in the population. The point of stabilisation or zero growth rate is the maximum **carrying capacity** of the given environment for the organism concerned. It represents the point where the sigmoidal curve levels off (the upper **asymptote**) and is symbolised K in population growth equations (see fig 12.37*a* and table 12.17 equation (*a*)).

The growth of a great variety of populations representing micro-organisms, plants and animals, under both laboratory and field conditions has been shown broadly to follow this basic sigmoidal pattern. A useful example is the model used in chapter 2 to describe bacterial population growth, in which a fresh culture medium is inoculated with bacteria (fig 2.7; see also section 21.1.2 and fig 21.1*d* for yeast). Phytoplankton in lakes and oceans may show sigmoidal growth in spring as could insects such as flour beetles or mites introduced into a new habitat with abundant food and no predators. In plants and animals which have complicated life histories and long periods of development there are often delays in build-up of numbers and limiting factors, and hence delayed density-conditioning. Such time lags may, for example, derive from the time of approximately one generation which elapses before the depression of birth rates at high densities shows up as a decrease in the adult population. An approximate way of incorporating such time delays into the equation for S-shaped growth is shown in table 12.17, equation (*b*). If the time-delay factor is long compared with the natural response time there will be a tendency for the population to overshoot carrying capacity and then overcompensate, giving an oscillatory rather than smooth return to the equilibrium point K, the carrying capacity.

The J-shaped growth curve describes a situation in which, after the initial establishment phase (lag phase), population growth continues in an exponential form until stopped *abruptly*, as environmental resistance becomes *suddenly* effective (fig 12.37*b*, table 12.17, equation (*c*)). Growth is said to be **density-independent** since regulation of growth rate is not tied to the population density until the final crash. The crash may be triggered by factors such as seasonality or the end of a breeding phase, either of the organism itself or of an important prey species. It may be associated with a particular stage in the life cycle, such as seed production, or it may be induced by human intervention as when an insecticide is used to control an insect pest

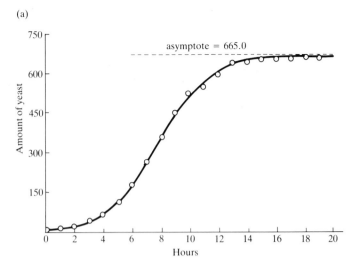

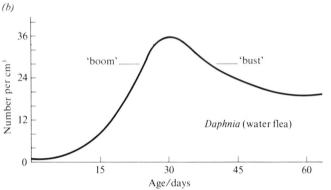

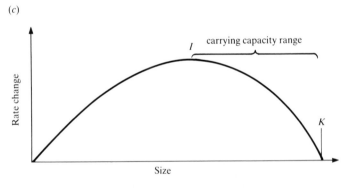

Fig 12.37 *Types of population growth curve. (a) The sigmoidal growth curve (S-shaped curve). The growth of yeast in a culture. A simple case of the sigmoid growth form in which environmental resistance (in this case detrimental factors produced by the organisms themselves) is linearly proportional to the density. (From E.P. Odum (1983) Basic ecology, Holt-Saunders International.) (b) 'Boom-and-bust' (J-shaped) curve of water fleas grown in culture. (Based on A.S. Boughey (1971) Fundamental ecology, International Textbook Co.) (c) S-shaped curve redrawn to show changing rate of increase in relation to changing population size. I, the inflexion point, represents the maximum level (highest growth rate) and is thus the theoretical point for maximum sustainable yield for a game or fish population. The range between I and K represents a secure or desirable density. It maximises returns but minimises risk of sudden environmental change triggering either underproduction or overshooting carrying capacity and damaging the environment as might occur for an organism with a long generation time and hence delayed response to environmental change. (Source as in a.)*

Table 12.17 Equations for the sigmoid (S-shaped) and J-shaped ('boom-and-bust') growth curves.

Symbols: r is the intrinsic rate of increase or biotic potential
N is the number of individuals in the populations
t represents time
d is the conventional mathematical symbol for instantaneous change
K represents carrying capacity
T represents the lag effect as in generation time (explained in text)

(a) Sigmoid growth curve $\dfrac{dN}{dt} = rN\dfrac{(K-N)}{K}$ or $rN(1-N/K)$

(b) Time-delayed regulation of sigmoidal growth $\dfrac{dN}{dt} = rN\,[1-N(t-T)/K]$

(c) J-shaped growth curve $\dfrac{dN}{dt} = rN$ with a definite limit on N

Note the clear relationship between the J-shaped and S-shaped curves.

The expressions $\dfrac{(K-N)}{K}$ or $(1-N/K)$ are alternative ways of indicating environmental resistance created by the growing population itself, which brings about an increasing reduction in the potential reproduction rate as the populations size approaches carrying capacity (K). In word form, equation (a) above may be summarised as:

rate of population increase	equals	maximum possible rate of increase times the numbers present in the population	times	degree of realisation of maximum rate of increase

population. Following the crash, such populations typically show a fluctuating recovery pattern giving the 'boom-and-bust' cycles characteristic of some insect species and

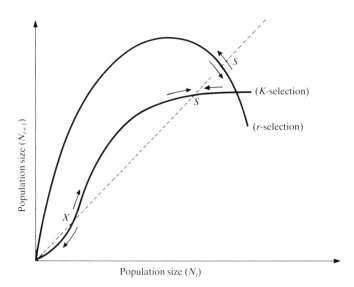

Fig 12.38 *The population growth curves of r- and K-strategists showing stable (S) equilibrium points and an unstable point, the extinction point (x) for K-strategists. Note that though an equilibrium point exists for r-selected species it is not monotonic as for K-species but subject to continuous marked oscillations (the boom-and-bust cycles). (Taken from T.R.E. Southwood (1976) Bionomic strategies and population parameters in Theoretical ecology, principles and applications, ed. R.M. May, Blackwell.)*

associated with algal blooms. In very general terms the J-shaped growth form may be considered an incomplete sigmoid curve where a sudden limiting effect comes in to play before the self-limiting effects within the population assume importance. Studies of thrips on roses illustrate this point. A J-shaped growth form is evident in favourable years when seasonality triggers a halt in population increase. In climatically less favourable years population growth is more sigmoidal in form as limiting effects within the population, such as competition for food resources, become operative before seasonality intervenes.

As discussed, the maximum population of an organism that a particular environment can sustain is termed the carrying capacity. This is identified theoretically as the K value (or upper asymptote) of the sigmoidal growth curve. Populations showing J-shaped growth do not achieve stability in this way, instead showing characteristic boom-and-bust cycles (see also fig 12.38, section 12.7.4). In practical terms carrying capacity implies continuing yield at maximum population densities without environmental damage. However, as fig 12.37c shows, greater productivity, or rates of population increase, will be achieved at lower densities than carrying capacity. Thus in fisheries or livestock management, where a regular annual crop of good size and quality is sought, maximum sustained yield will be achieved at population densities below carrying capacity. A good strategy aims to maintain the population between I (maximum population growth) and K (carrying capacity). If carrying capacity is exceeded then environ-

409

mental resources will be depleted. Carrying capacity, at least temporarily, is reduced. The resulting lower population density then allows recovery of environmental resources. If human intervention or some other agencies cause carrying capacity to be greatly exceeded, then a permanent environmental change may result leading to an environment which is no longer suitable for that organism or which is capable of permanently sustaining only a much smaller population. Grazing-induced or grazing-exacerbated desertification of large areas of tropical rangelands reflects this process.

12.7.4   Population strategies

In the last section the terms *r* and *K* were used in equations for population growth. Species that reproduce rapidly and have a high value of *r* are termed *r*-species or *r*-strategists. They are generally opportunist species and represent the typical pioneer species of new and disturbed habitats. Migration and dispersal are an important part of their strategy. Selection pressure in such species favours high reproduction rate and short generation time. Since they may even be the sole colonisers of an area, competitive ability has little importance and *r*-strategists are typically small in size. Mortality rates are high. Although extinction is regularly the fate of individual populations (as habitats are modified and *K*-strategists arrive and out-compete), the species as a whole is very resilient. Mobility and low generation time ensure species survive as individuals, seeds or propagules, move on to new situations, such as newly dug ground, and rapidly colonise.

K strategists reproduce relatively slowly, for example trees and humans. The dominant thrusts of their survival strategy are mortality avoidance and competitive ability. Thus they are characteristic of stable, relatively undisturbed habitats where competitive ability rather than reproductive speed is a major survival attribute. They tend to be more typical of the later stages of succession. Such species are not very adapted to recover from population densities significantly below their equilibrium level (the *K* value or carrying capacity). If population numbers are greatly depressed these species tend towards extinction (fig 12.38). In contemporary times, when human activity often triggers rapid habitat change, such as deforestation, extreme *K*-strategists require active conservation measures rather more often than predominant *r*-strategists.

These two strategies, *r* and *K*, represent two different solutions to the same problem, which is long-term survival. Both are effective and the precise strategy of any species, or species population, will reflect the interplay of factors such as habitat variability in space and time and the longevity, size and fecundity of the organism concerned. In consequence a broad continuum of strategies from small, opportunist extreme *r*-selected species to large, dominant extreme *K*-selected species occurs (sometimes termed the **r-K continuum**). Some of the characteristics typical of extreme *r*- and *K*- strategists are summarised in table 12.18.

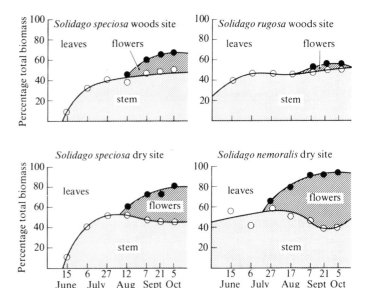

Fig 12.39 *Percentage biomass allocated to production of leaves, stems and flowers in four goldenrod populations (From R. M. May (ed.) (1976) Theoretical ecology, Blackwell.)*

12.22   Study fig 12.39 which shows the allocation of standing crop biomass to reproduction (flowers) in three species of goldenrod (*Solidago*) in the eastern USA. They grow either on woodland sites or on open, dry, disturbed sites of earlier successional stages, as indicated in the figure.
(*a*) Which species devotes most biomass to reproduction?
(*b*) Which species is most heavily r-selected?
(*c*) Which species is most heavily K-selected?
(*d*) What general conclusion can be made by comparing the reproductive behaviour of *S. speciosa* on woodland and dry sites?

Computer program.   POPULATION MODELLER: A program concerning population dynamics. Population size vs. time for one species (e.g. *r*-strategy species, *K*-strategy species), or two species (e.g. competitors, predator–prey).

Table 12.18 Characteristics of r- and K-species.

r-species (opportunist species)	K-species (equilibrium species)
Reproduce rapidly (high fecundity, short generation time); therefore high value of r (the intrinsic rate of increase)	Reproduce slowly (low fecundity, long generation time); therefore low value of r
Reproduction rate not sensitive to population density	Reproduction rate sensitive to population density, rising rapidly if density falls
Investment of energy and materials spread over many offspring	Investment of energy and materials concentrated on a few offspring, with parental care in animals
Population size fluctuates greatly showing marked oscillations about a theoretical balance point which is not maintained over a long period (fig 12.38)	Population size stays close to equilibrium level determined by K
Species not very persistent in a given area	Species persistent in a given area
Disperse widely and in large numbers; with animals, migration may occur every generation	Disperse slowly
Reproduction is relatively expensive in terms of energy and materials	Reproduction is relatively inexpensive in terms of energy and materials; more energy and materials devoted to non-reproductive (vegetative) growth
Small size	Large size; woody stems and large roots if plants
Individuals short-lived	Individuals long-lived
Can occupy open ground	Not well adapted to growing in open sites
Habitats short-lived (e.g. ripe fruit for *Drosophila* larva)	Habitats stable and long-lived (e.g. forest for monkeys)
Poor competitors (competitive ability not required)	Good competitors
Relatively lacking in defensive strategies	Good defence mechanisms
Do not become dominant	May become dominant
More adaptable to changes in environment (less specialised)	Less resistant to changes in environmental conditions (highly specialised for stable habitat)
Examples	*Examples*
bacteria	large tropical butterflies
Paramecium	condor (large bird of prey)
aphids	albatross
flour-beetles	humans
annual plants	trees

12.7.5   Factors affecting population size

In section 12.7.3 population growth was examined. Once a population has finished its initial growth phase there usually continues to be fluctuations in population size from generation to generation. An example is shown in fig 12.40 for caterpillars of the winter moth (*Operophthera brumata*) in an oakwood near Oxford.

Important influences are likely to be variations in climatic conditions (such as temperature), food supply and predation. Sometimes fluctuations are regular and may be called **cycles**. Their study is laborious and time-consuming since data for field organisms often must be collected over several years. In some cases data have been obtained by using laboratory organisms with short life cycles, such as fruit flies, rats and mites. These are used as model populations where conditions can be controlled more rigorously than in the field.

Basically, population size may change as a result of changes in fecundity or mortality, or possibly both. When

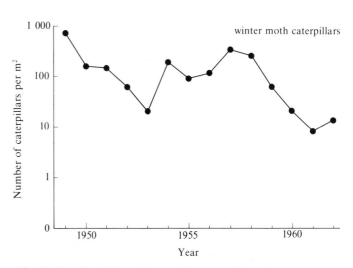

Fig 12.40 *Population changes for caterpillars of the winter moth* (Operophthera brumata) *in Wytham Wood near Oxford. Note that the numbers are plotted on a logarithmic scale. (From Open University Science Foundation Course (S100) Unit 20 (1971) Open University Press.)*

411

studying change it is usual to search for the '**key factor**', that is the factor responsible for the greatest proportion of the observed change from one generation to the next. In most cases studied this is a factor affecting mortality.

Whereas fluctuations in population size might be expected to be purely random, in practice some factors operate to regulate population size within certain limits. These are factors which reduce numbers by increasing mortality or reducing fecundity, and which become more effective as population density increases. Hence they may be described as **density-dependent factors**. Food shortage and increased predation are two factors which sometimes operate in this way. They have direct effects on mortality for obvious reasons. Two regulatory mechanisms which have been well studied and which affect fecundity are territorial behaviour and the physical effects of overcrowding.

Territorial behaviour

Territorial behaviour or **territoriality** occurs in a wide range of animals, including certain fish, reptiles, birds, mammals and social insects. It has been particularly well studied in bird populations. Either the male bird, or both the male and the female, of a pair may establish a breeding territory which they will defend against intruders of the same species. The song of the bird and sometimes a visual display, such as that of the robin's red breast, are means of asserting territorial claims, and intruders usually retreat, sometimes after a brief 'ritual fight' in which neither competitor is seriously damaged (section 16.8.6). This has obvious advantages over a series of 'real' fights. There is little or no overlap between neighbouring territories of the same species and, in areas where the territory includes the food of the species, it will contain sufficient food to support the birds and their young. As population sizes grow, territories usually become smaller and able to support fewer new birds. In extreme cases, some birds may be unable to establish territories and therefore fail to breed. Regulation is therefore due to spatial interactions.

Overcrowding

Another form of regulation in which space is an important factor is that due to overcrowding. Laboratory experiments with rats show that when a certain high population density is reached, fecundity is greatly reduced even if there is no food shortage. Various hormonal changes occur which affect reproductive behaviour in a number of ways; for example, failure to copulate, infertility, number of abortions and eating of young by the parents all increase, and parental care decreases. The young abandon the nest at an earlier age, with consequent reduction in chances of survival. There is also an increase in aggressive behaviour. Changes like these have been demonstrated for a number of mammals and could operate under natural circumstances outside the laboratory. Natural populations of voles, for example, show a similar kind of regulation.

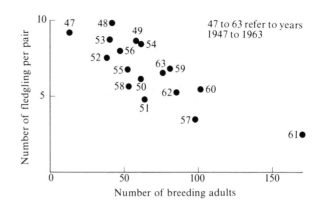

Fig 12.41 *Breeding success of the great tit in Marley Wood, near Oxford, in relation to the number of breeding pairs present from 1947–63. (From M. E. Solomon (1976)* Population dynamics, *2nd ed., Studies in Biology No. 18, Arnold.)*

There are many examples where breeding success and population density have been shown to be related. Cyclic variations in population size and density-dependent breeding success was shown during an investigation of great tits in Marley Wood near Oxford, as shown in fig 12.41.

In the laboratory it has been shown that the number of eggs laid per day by fruit flies (*Drosophila melanogaster*) decreases as the population density of flies increases. With plants, the number of seeds produced by each plant may similarly be reduced at higher plant densities. In these examples it is not always obvious what the regulatory factors are, but food availability, competition for territory, mutual physical disturbance with possible hormonal changes are all factors which may be involved.

The regulatory factors help to smooth out and compensate for other factors which might have a random, non-regulatory effect on population size, such as climate. For example, the particularly low number of breeding great tits in Marley Wood in the year 1947 (fig 12.41) might have been due to high mortality during the very cold winter of 1946–7. This may have been compensated for by the relatively high breeding successes of 1947 and 1948, but the reasons for these successes are probably complex.

It was noted above that the key factor affecting population size is often one that affects mortality. This is clearly shown in the case of the winter moth, whose life cycle is illustrated in fig 12.42. Its numbers have been studied over many years in an oakwood near Oxford. Table 12.19 shows the average number of individuals killed by each of six factors affecting mortality.

The table shows that on average each pair of adults replaces itself each year with two surviving offspring, but fig 12.43 shows that in reality the population varies from year to year. Assuming that fecundity is constant, at least one of the six mortality factors must fluctuate and be density-dependent since the average size of the population remains constant. Fig 12.43 shows how the mortality factors fluctuated between 1950 and 1961. The key factor

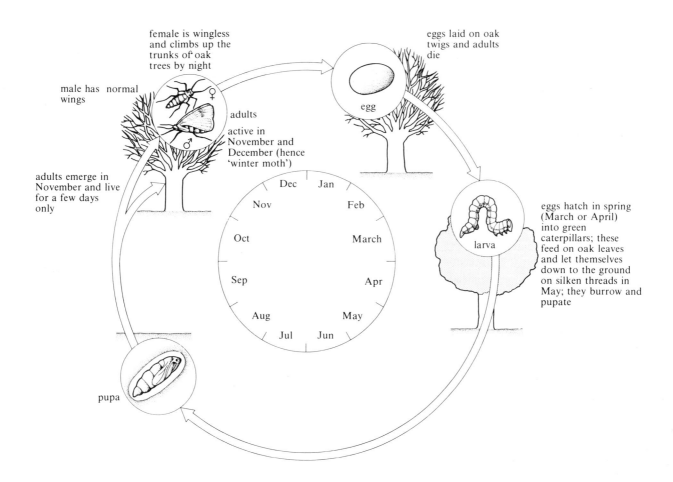

Fig 12.42 *Annual life cycle of the winter moth. (Based upon Open University Science Foundation Course Unit 20 (1971) Open University Press.)*

Table 12.19 Average number of individuals of the winter moth killed by six mortality factors acting in succession.

Number of eggs laid by the female moth	200

Mortality factor	Number killed
Winter disappearance (death of a few eggs and very high mortality of newly hatched caterpillars)	184
Parasitic fly living on caterpillars	1
Other parasites living on caterpillars	1.5
Disease of caterpillars	2.5
Predators killing pupae in soil (beetles, shrews)	8.5
Parasitic wasp living on pupae	0.5
Total	198

Therefore number of adults surviving to breed:	2

Based on Open University S100 Unit 20 (see fig 12.42).

affecting total mortality (pre-reproductive mortality) is winter disappearance because this accounts for the greatest proportion of deaths and is most closely correlated with total mortality. The high mortality at this stage is explained by the fact that the hatching of the young caterpillars is closely, but not always exactly, synchronised with bud burst and young leaf growth of the oak trees. If the caterpillars hatch slightly early many starve, and if too late many will not have completed development before the leaves become too tough and full of tannin to be eaten. If, however, exact synchrony is achieved, the population of caterpillars may reach pest proportions and trees may be completely defoliated.

Although winter disappearance is the key factor affecting mortality, some other factor appears to be acting to regulate population size because fluctuations in mortality are less dramatic than those for winter disappearance in most years (the years 1960 and 1961 being exceptions). The regulating factor therefore causes relatively more deaths in years when, as a result of reduced winter disappearance, the population becomes larger. In other words, the regulatory factor is density-dependent. Both pupal predation and pupal parasitism show a relationship with total mortality which suggests that they could be regulatory

413

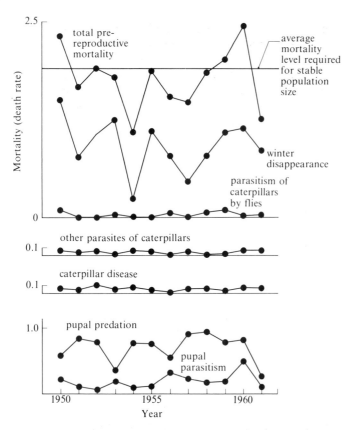

Fig 12.43 *Total pre-reproductive mortality and contributions to total mortality made by six different factors for winter moth in Wytham Wood between 1950 and 1961. (From Open University Science Foundation Course Unit 20 (1971) Open University Press.)*

Some factors regulating population sizes such as climate have been regarded as density-independent, but it is inevitable that they interact with other factors that are density-dependent. Thus, although the terms are useful, it is better to avoid stressing the difference between density-dependent and density-independent factors and to study each case on its merits with the awareness that complex interactions may occur. The study of population fluctuations and regulation is a complex area of ecology and the examples quoted are chosen merely to illustrate some of the more obvious ways in which these factors operate.

12.7.6 Factors affecting population size: interspecific factors

It is seldom possible to confine studies of population dynamics to single species. It has already been shown, for example, that an understanding of the fluctuations in winter moth populations depends on a knowledge of parasites of the moth. A number of well-recognised types of interaction may occur between populations of different species (interspecific interactions). At a given trophic level there may be **interspecific competition**, that is competition between members of different species, for available resources such as food and space. It is here that the study of niches is important in community ecology. Populations from different trophic levels may also interact, as, for example, in the cases of predator–prey relationships and host–parasite relationships. Other relationships exist, some of which are subtle and complex, including some symbiotic relationships where both partners benefit.

Interactions between members of the same species (**intraspecific interactions**) also occur, such as territoriality and other forms of agonistic behaviour (section 16.8.8).

Predator–prey relationships

(See reference to computer program, p. 410.)

In trying to understand interspecific interactions, simple and well-controlled situations can be set up in the laboratory. It is hoped that they represent 'models' of real situations.

A commonly used and simple model of predator–prey relationships is one that has been well illustrated by laboratory experiments with two mites, one predatory (*Typhlodromus*) and one herbivorous (*Eotetranychus*). Fig 12.44 shows the cyclic fluctuations that occur in their numbers, the cycles for the two species being slightly out of phase with each other.

The explanation for these cycles is that an increase in numbers of the prey supports a subsequent increase in numbers of the predator. The predators then cause a crash in numbers of the prey, followed by an inevitable decline in numbers of predators. The cycles are completed when the decline in predators allows an increase in numbers of the prey. Each cycle occurs over a number of generations. These laboratory models must be applied with caution to

factors. Pupal predation is highest in years when numbers of pupae are high (low winter disappearance). The relationship with parasitism is more complex because parasite density tends to be out of phase with host density. The parasite of winter moth pupae is a wasp whose larvae feed on the caterpillars. Parasite–host relationships are complex, but briefly a year of high host density will result in successful parasite breeding followed a year later by higher parasite density and hence higher pupal parasitism.

Very accurate models are now available for predicting winter moth numbers. Such models have practical application when the species is a pest, as is the winter moth in eastern Canada. The models suggest ways in which the pest might be controlled and, in the case of winter moth, enabled the consequences of introducing parasites as a means of biological control to be predicted with success.

Another density-dependent factor that may affect the population size of a species is migration (or dispersal). For example, at high aphid densities not only does the rate of reproduction of the aphids decrease but a higher proportion develop wings and leave the plant on which they are feeding.

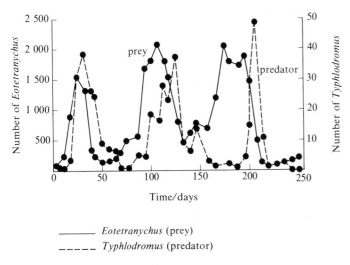

Eotetranychus (prey)

————— Typhlodromus (predator)

Fig 12.44 *Oscillations in the populations of the predatory mite* Typhlodromus *and its prey, the plant mite* Eotetranychus, *in a laboratory habitat. (From M. K. Sands (1978) Problems in ecology, Mills & Boon Ltd).*

natural predator–prey relationships, which will be subject to other influences. Some herbivorous mammals show regular cycles of numbers in the populations, an example being the lemmings of North America and Scandinavia, which have a four-year cycle. Their predators, the arctic fox and the snowy owl, consequently show a similar cycle, but the factor causing the lemming cycle could be overgrazing rather than predation. Occasionally, when populations of Scandinavian lemmings reach exceptionally high densities, they emigrate *en masse*, many plunging to their deaths in fjords or drowning in rivers. Other cyclic changes in herbivores may also be influenced more by availability of food plants than by predators. Death through disease is another factor which could operate in a density-dependent manner since epidemics are more likely to spread through populations of high density. There is some evidence that this might contribute to the population cycles of the snowshoe hare of northern Canada. Here again one of its main predators, the Canadian lynx, follows a similar, but out-of-phase, cycle. Hares form 80–90% of the diet of the lynx.

Although it may not be the only factor, there is no doubt that predation plays an important part in regulating natural populations. Some indication of the importance of predator–prey relationships in this respect, and the long-term advantage it has for the prey, can be gained from the fate of a population of deer on the Kaibab Plateau in Arizona. In 1906 the area was declared a wild-life refuge, and in order to protect the deer their predators, such as pumas, wolves and coyotes, were systematically exterminated over the next 30 years. Up to 1906 the deer population had remained stable at about 4000, but subsequently a 'population explosion' occurred, as shown in fig 12.45 with the result that the carrying capacity of the environment, estimated to be about 30 000 deer, was exceeded. The

population reached an estimated 100 000 by 1924, following a 'boom-and-bust' curve, but the overgrazing resulted in starvation and, together with disease, a subsequent population crash. In addition, the vegetation had been seriously damaged and did not recover to its 1906 level, with the result that the carrying capacity of the environment dropped to 10 000 deer.

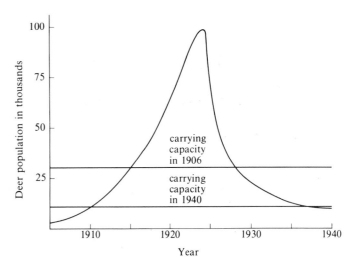

Fig 12.45 *Deer population on the Kaibab Plateau following eradication of predators*

Host–parasite relationships

In some of the cases studied host and parasite populations show similar out-of-phase cycles to those described above, notably when insects parasitise other insects (see also the case of the winter moth, section 12.7.5).

Interspecific competition

(See reference to computer program, p. 410.)

Competition may occur between populations within an ecosystem for any of the available resources, such as food, space, light or shelter. If two species occur at the same trophic level then they are likely to compete with each other for food. Adaptive radiation by one or both species may then occur over a period of time with the result that they come to occupy separate niches within the trophic level, thus minimising the extent of competition. Alternatively, if the competitors occupy the same niche, or strongly overlapping niches, an equilibrium situation may be reached in which neither succeeds as well as it would in the absence of the competitor, or one of the competitors declines in numbers to the point of extinction. The latter phenomenon is known as **competitive exclusion**. It is difficult to study in wild populations but some classic work on laboratory populations has been done, originally by the Russian biologist Gause in 1934 who worked on competition between several species of *Paramecium*. Some of his results are shown in fig 12.46.

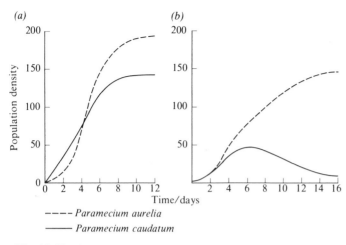

Fig 12.46 *Population growth of two species of* Paramecium. *(a) Cultured separately. (b) Cultured together*

> **12.23** With reference to fig 12.46,
> (a) what type of population growth curve is shown by the two species when grown in isolation?
> (b) what resources might the two species be competing for in the mixed culture?
> (c) what factors give *P. aurelia* a competitive advantage over *P. caudatum*?

When the two species are cultured together *P. aurelia* has a competitive advantage over *P. caudatum* for gaining food and after five days the numbers of *P. caudatum* start to decrease until, after about 20 days, the species has become 'extinct', that is it has been competitively excluded. *P. aurelia* takes longer to reach the stationary phase of growth than when grown in isolation, so is also affected adversely by the competition, even though it is more successful than its competitor. This helps to explain the selection pressure for competitors to adapt to separate niches. Under natural circumstances, the less successful competitor rarely becomes extinct, but merely becomes rare and may even increase in numbers again before achieving an equilibrium position.

The **competitive exclusion principle** (or **Gausian exclusion principle**) has since been confirmed in further animal experiments carried out by other workers. Competitive exclusions have been shown to occur in plant populations; in mixed cultures of duckweed (*Lemna*) species; *L. gibba* was capable of excluding *L. polyrrhiza*.

The study of natural populations is made more complex by the larger number of interacting populations and by the fact that the environmental variables such as temperature, moisture and food supply cannot be controlled.

Among plant populations one form of competition which has attracted a lot of interest is allelopathy. It is part of the more general phenomenon of allelochemistry whereby plants and micro-organisms produce as secondary products of metabolism a variety of complex organic molecules which affect the growth of other living organisms. They include antibiotics and growth inhibitors, such as penicillin which is produced by the fungus *Penicillium* and has antibiotic properties against Gram positive bacteria (section 2.2.2). Chemical competition among micro-organisms is very intense and the relationships between them very complex. When chemicals affect other populations at the same trophic level the interspecific competitive effect that results is termed **allelopathy**. The odours of aromatic plants are volatile secondary products which are sometimes involved in interactions of this kind. For example, Muller in 1966 showed that the volatile terpenes released from aromatic plants of the Californian chaparral (a type of scrubland) are adsorbed on to surrounding soil particles and inhibit the germination or growth of surrounding plants. Phenolic compounds that are leached into the soil from the litter of some plants have a similar effect. Many plants of chalk grassland are aromatic and this may be an additional factor to grazing in maintaining the rich variety of herbaceous plants present. Recent studies of allelochemical interactions suggest that they are probably widespread in occurrence, both within and between trophic levels. Among the important plant chemicals involved are phenolics, terpenoids and alkaloids. They are extremely interesting because they are often physiologically active in animals. One example of allelopathy involving animals is provided by the monarch butterfly (*Danaus plexippus*). Its caterpillar stage feeds on the milkweed plant which produces chemicals that act as strong cardiac (heart) poisons in vertebrates. This, or its unpleasant taste, presumably discourages grazing herbivores from eating the plant. However, the caterpillars can tolerate the poison, and can store it and carry it into the adult butterfly stage. The butterfly, in turn, gains protection from predatory birds. The bright colouration and striking markings of the butterfly therefore serve as a warning to potential predators and act as a protective device. An incidental consequence of this is that some butterflies mimic the markings of the monarch, thus also gaining protection. This case study serves to show just how complex species interactions can become. Generally speaking, many plants subjected to grazing contain toxic chemicals to which only a few herbivore species are tolerant.

Two other important processes, in which interaction occurs between flowering plants and animals, are discussed in section 20.2, namely seed dispersal and pollination. The latter, in particular, is a good example of another general principle, namely that of **co-evolution**, where species become adjusted to each other's presence and develop over a period of time varying degrees of mutual dependence and benefit. In the case of toxic plants, the most beneficial and stable arrangement would be for the toxin-tolerant herbivores to provide a service to the plant, such as the pollination by butterflies whose caterpillars feed on the plant.

12.7.7   Applications: fisheries management

The practical relevance of population ecology has already been mentioned, as in species extinction, pest control, conservation and game and fisheries management. A further discussion of fisheries management, with particular reference to North Sea fisheries, is given below. The use of population models in pest control is discussed in section 12.8.

Fisheries, especially marine fisheries, are an important source of human food. Unlike most other food species, fish have not been domesticated nor are they owned by individuals. Modern fishing still depends on hunter-gathering techniques and fishermen compete for stock. Biological, economic, social and political factors complexly interact in the story of modern fishing and quota systems.

Over the last century technological advances have greatly improved human hunting ability. Furthermore, developments in refrigeration and freezing methods have meant that vessels can fish to capacity without the pressure to return quickly to port to ensure landings arrive in good condition. A summary of the main changes in fishing practice is given in table 12.20. Fifty years ago it was already evident that problems existed in the North Sea haddock fisheries. The proportion of small fish in the catch was rising markedly (fig 12.47). Furthermore, twentieth-century records for haddock catches show a downward trend in peacetime with peaks immediately following the two World Wars during which fishing activities had largely ceased (fig 12.48). The trend for world landings of sea fish since the Second World War is similar, the graph flattening out in the late 1970s despite increasing fishing effort (fig 12.49).

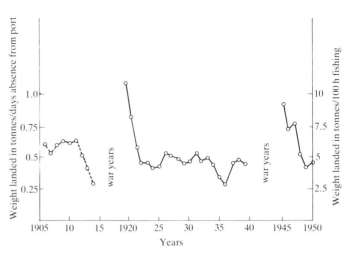

Fig 12.47 *Haddock catch per unit of fishing effort by Scottish trawlers in the North Sea. Dashed line, landings/days absence; full line, landings/100h. (From M. Graham (1956) Sea fisheries – their investigation in the UK, Arnold.)*

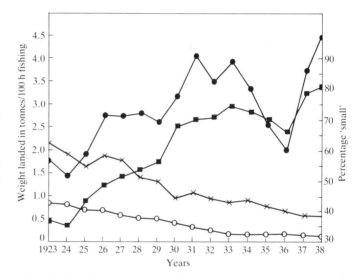

Fig 12.48 *Catch, per unit of fishing effort of North Sea haddock, and percentage of the 'small' category, 1923–38. Open circles, large; crosses, medium; full circles, small; open squares, percentage 'small'. (From M. Graham (1956) Sea fisheries – their investigation in the UK, Arnold.)*

Table 12.20 The main changes in fishing practice since the late-nineteenth century.

Major innovation	Consequent benefits
(1) Steam ships replace sailing ships	(*a*) Larger nets can be used (*b*) Greater independence of wind and tide results
(2) Diesel replaces steam	(*a*) Even faster, more powerful ships can search larger areas (*b*) Even less weather-dependent
(3) Development of plastic nets which are stronger, lighter and longer-lasting	(*a*) The use of even larger nets is possible (*b*) By using different densities of synthetic fibres in different parts of the net, improved control and manipulation is possible
(4) Development of radar and sonic detection techniques	(*a*) Improved hunting and detection of fish shoals (*b*) Better net handling
(5) Development of large-scale on-board refrigeration and freezing facilities	(*a*) Fishing boats can travel further from port and continue fishing without deterioration of the catch

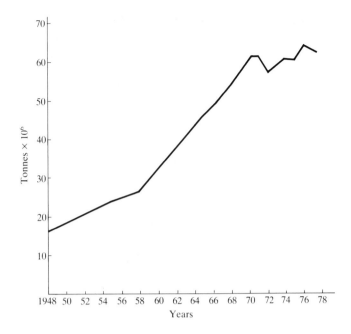

Fig 12.49 *World landings of sea fish 1948–77. (From R.V. Tait (1981)* Elements of marine ecology, *3rd ed., Butterworths, p. 305)*

In the 1940s Russell* proposed a model for fisheries management based on the ideas of population dynamics. His summary equation is:

$$S_2 = S_1 + (A + G) - (C + M)$$

S_1 = weight of stock at the beginning of a year,
S_2 = weight of stock at the end of that year,
A = annual increment by recruitment of young fish to the stock,
G = annual increment due to growth of all fish in the stock,
C = total weight of fish removed during the year by fishing, 'the catch',
M = weight of fish lost during the year by death from all the other causes, that is the natural mortality.

The amount by which stock weight increases if there is no fishing is the natural yield, $A + G - M$ in Russell's formula. If the stock remains unchanged then annual recruitment and growth $(A + G)$ must be balanced by mortality and catch $(M + C)$. In short, catch must equal natural yield. Relatively simple data on fish landings and fishing techniques can enable reasonable prediction (plus or minus 20%) of the balanced catch. Quotas, permissible mesh sizes and other key criteria can be adjusted accordingly. Furthermore, equilibrium between catch and natural yield can be established for any population size. The maximum possible

* E.S. Russell (1942) *The Overfishing Problem*, CUP.

sustained yield will be achieved when fishing effort is adjusted so that the population is maintained at or just above inflexion point, I, the maximum growth rate described for S-shaped population growth in section 12.7.3 (see fig 12.37c). In other words the fishery must be neither overfished nor underfished if a maximum yield is to be achieved and sustained.

Underfishing

Under conditions of light fishing a fish population grows to the limits of food supply, the carrying capacity (K) in the absence of significant human predation. Competition for food restricts the numbers surviving and the size to which individuals grow. Slow-growing older fish compete successfully with new recruits for food. Fish catches show a predominance of older fish but, due to age, persistent undernourishment and associated disease vulnerability, they are typically of poor quality. Such a fishery is underfished and landings may command a poor price. The population could sustain a larger more profitable fishery of better quality fish if more fish were caught (that is, if the balance were shifted back from K towards I in fig 12.37c). Reduction in population size by removing the excess older fish would achieve a better growth rate through the remaining stock and improve the condition of the fish.

Overfishing

Heavy fishing, by contrast, generates a population of mainly young small specimens, since fish are caught as soon as they reach a catchable size. Again such fish may fetch poor prices since they have little edible flesh in relation to bone. (Scarcity value may to some extent offset this tendency for price reduction.) The stock is overfished. These young fish would make rapid growth if left longer in the sea and would soon reach a more valuable size, giving heavier landings of better quality fish per unit fishing effort, giving an improved profit margin. Productivity and prices would thus improve if fishing were reduced. However, individual fishermen are tempted to achieve weight quotas and improve immediate income by catching more rather than fewer fish. This triggers positive feedback favouring further reductions in fish stock and size, requiring more effort (cost) to achieve reasonable landings. If this reaches the point where fish are caught before spawning so that the reproductive capacity of the stock is seriously impaired then a catastrophic reduction in numbers occurs and local extinction may result. Overfishing which involves fish being caught before reproductive age is called **recruitment overfishing**; overfishing of fish before optimum age is called **growth overfishing**, and typically occurs before recruitment overfishing.

These basic fish management models may be further refined by studying the fate of particular age cohorts rather than the population overall. This requires more detailed measurement of fish landings but gives valuable insights into population structure and a more-sensitive basis for prediction. Current events in the North Sea cod fisheries illustrate this point. Catches during 1987 were good, prompting fishermen to ask for increased quotas. On the other hand, scientists advising the Council of European Fisheries Ministers decided that cod catches should be reduced by about one-third. According to the scientists, during the past 15 years fishermen have been taking an increasingly higher proportion of cod stocks from the North Sea. Up to 60% of the cod aged over 18 months were being caught each year and this has lead to a reduction in spawning stock to its lowest post-war level (figs 12.50 *a* and *b*). How far the spawning stock can decline before the

reproduction rate crashes is unknown for North Sea cod (that is the point of no return or X on fig 12.38). This worrying picture is complicated by extreme variability in the rate at which 18-month-old fish enter the fisheries, the recruitment rate (fig 12.50*c*). Scientists suspect that these vast natural variations are masking a general downward trend. Hence advice on quotas tends to be cautious. Low recruitment in 1986 and 1987 suggests future problems unless quotas are reduced. Paradoxically, this advice has coincided with a temporary, it is assumed, abundance of cod in the North Sea reflecting good recruitment in 1985. Age cohort data are clearly fundamental to prediction in this instance.

If the biological basis of fisheries management seems complex, the intricacies of fisheries economics and social factors equally cannot be ignored. A biologically optimum fishing rate may glut the market, reducing prices; high catches per unit fishing effort leave expensive equipment and skilled personnel idle once quotas have been attained. The issues of fishing rights and quotas are matters of international politics. Scientific studies can usefully supply the basis for good decision-making; however, a viable policy must be workable in social and economic terms as well as scientifically ensuring the maintenance of valuable food resources.

(a)

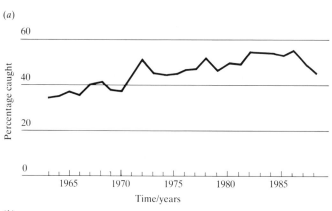

(b)

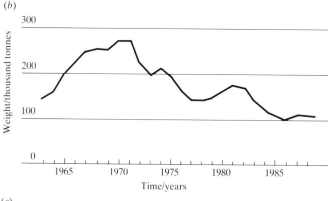

(c)

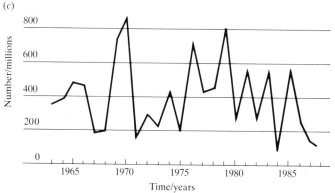

Fig 12.50 *(a) Percentage of North Sea cod stock caught annually. (b) Spawning stock of cod in the North Sea. (c) Recruitment of cod to North Sea stock as monitored by the number of one-year-old fish. (From Richard North,* The Independent, *Feb. 1988)*

12.8　Agriculture and horticulture

Tremendous advances in crop production have taken place throughout the world in the last 20 years as a result of the widespread implementation of the results of technological and scientific research. A number of examples are given below.

12.8.1　Ploughing

Used for the breaking down of the soil to produce seed beds, and converting grassland to arable crops by turning over or burying the grass sward. In the UK this is carried out in the autumn, the winter frosts breaking down the soil clods.

Curved steel blades are pulled by a tractor through the soil. Modern hydraulic devices on the tractor automatically control the working depth of the ploughing. Disc ploughs are then used to break up the hard soil to produce a good tilth and final preparation of the seed bed may involve the use of steel-framed or disc harrows.

12.8.2　Direct drilling

After a cereal harvest, the stubble is treated with a herbicide, such as Paraquat, to destroy any weeds. Using a seed drill which is able to break through the undisturbed soil, the seeds are sown directly thus avoiding the cost of ploughing (thereby reducing fuel costs).

12.8.3 Crop rotation

Historically, cultivation in the UK was based on a three-year cycle; winter wheat was followed by spring wheat and then the field left unplanted for one year to preserve and restore soil fertility. With the development of root crops and the recognition of the importance of legumes (nitrogen fixation and a source of green manure), a four-course rotation was developed of roots, wheat, grassland or legumes, and barley; this rotation was modified as a wider variety of crops was introduced, such as kale and sugar beet. The rotation not only maintained soil fertility but also prevented the build-up of harmful pests. With the development of pesticides however, monoculture has become possible. Here only one type of crop is grown successfully on the same land from one year to another; an intensive system of agriculture.

In developing countries, or in opening-up new areas for cultivation, a system described as **slash-and-burn** may be carried out. The existing vegetation is cut down and then burnt, a method returning important mineral nutrients to the soil. The clearing formed may be cultivated continuously for several years until soil fertility falls, and then the site is abandoned for a number of years and other areas brought into cultivation. Cropping is usually mixed, several different kinds of crop being grown on the same plot.

12.8.4 Harvesting

This depends on the type of crop and ranges from the modern use of combine harvesters to the hand-picking of fruits. The crop is harvested when, by experience, it is judged to be at a suitable stage of development (percentage moisture content for grain crops; ripeness in fruit, allowing for further changes during transit to the markets). However, in modern intensive cultivation the grain may be dried artificially to the correct level. The crop may be delivered directly to the commercial market or factories (such as sugar beet) or undergo a period of storage.

12.8.5 Storage

The main aims are to prevent or reduce deterioration of the quality of the crop whilst awaiting favourable market conditions. Losses are due mainly to insect and fungal attack. Details of storage again depends on the crop, such as cereals in silos and grain stores, potatoes and root crops in clamps, fruit in special fruit stores.

It was found that the storage life of fruits is inversely related to their rate of respiration. This rate can be reduced by refrigeration and in some instances by circulating an air stream enriched with 10% carbon dioxide. However, ethene (which is produced by ripening fruits) hastens the onset of ripening and reduces storage life. Various volatile substances can be added to the air stream circulating through potato stores to prevent premature sprouting, and similarly fungicides may be used in fruit stores.

12.8.6 Weed control

A weed may be defined as a plant growing in the 'wrong' place. Weeds are characterised by rapid seed germination and growth and the production of a large number of fertile seeds. The seeds may remain dormant for several years. The adult plant, by virtue of tap roots or rhizomes and rapid regeneration of injured tissue, may withstand hoeing, grazing and contact herbicides thereby making eradication difficult.

Weeds, by virtue of these characteristics, compete with planted crops, particularly in the seedling stage, for water, nutrients and light so successfully that crop yields may be reduced. At the later stage of harvesting, the weed seeds may contaminate the harvest of cereal crops, for example false oat plants in cereal crops.

12.8.7 Herbicides

One of the earliest goals of agricultural research was to increase the yield of cereal crops using plant hormones. Application of auxins led to the discovery of herbicide action and to the use of the phenoxyacetic acids such as MCPA. Many different herbicides have now been developed, superceding the old 'contact' herbicides such as sodium chlorate. These more recent compounds are called systemic herbicides (systemic because they are translocated about the plant) and act as selective weedkillers. For example, paraquat and diquat interfere with the photosynthetic processes and can be used to kill weeds amongst well-established woody species of plants. Atrazine and simazine remain in the soil and prevent seed germination.

12.8.8 Insecticides

Insecticides, like other pesticides, are effective by inhibiting some metabolic process. They differ widely in composition, effectiveness, mode and speed of action and the dosage required. Ideally they should have a very limited range of action so that other organisms are not directly affected (table 12.21).

12.8.9 Pesticides and environment

Pesticides are chemical substances used by humans to control pests, those living organisms thought to be harmful to human interests. The term pesticide is an all-embracing word for **herbicides** (kill plants), **insecticides** (kill insects), **fungicides** (act on fungi), and so on. Most pesticides are poisons and aim to kill the target species, but the term also includes chemosterilants (chemicals causing sterility) and growth inhibitors. In Britain pesticide use is mainly associated with agriculture and horticulture, though pesticides are also widely used in food storage and to protect wood, wool and other natural products. In many countries pesticides are used in forestry and they are also used extensively to control human and animal disease

Table 12.21 Examples of the three main insecticides and their major uses.

Type	Example	Comments
(1) *Contact* – penetrate through the cuticle of the insects	(a) *Natural* nicotine (developed from tobacco) pyrethrum (extracted from the flowers of *Chrysanthemum* sp.) rotenone (from the roots of *Derris*)	Useful for destroying insects such as aphids which pierce the epidermal layers to suck out the plant juices
	(b) *Synthetic* compounds are now widely developed and are constantly being modified as resistant strains are built up as a result of selection, e.g. DDT, gammexane, parathion, malathion	Earlier examples used were chlorinated hydrocarbons (DDT) but these were persistent in soil and very toxic. Use is banned in some countries and replaced by organophosphates (e.g. parathion)
(2) *Systemic* – absorbed through the alimentary canal. Often stomach poisons. Newer insecticides interfere with the transmission of nervous impulses	Some of the contact poisons have been used as systemic insecticides, e.g. parathion, DDT	Used against insects with biting mouthparts. May be used in foliage against leaf-eating insects or as poison-bait ingredients against locusts
(3) *Fumigants* – compounds which, when volatilised, destroy insects	Formaldehyde, ethene oxide	Used in greenhouses

vectors (see section 12.3.8, malaria). Pesticides are used worldwide and in a wide variety of habitats. The more persistent disperse into all environments including those which are never sprayed, for example the open oceans and subpolar regions. Most life on Earth is thus in contact with pesticides. Natural organic pesticides, such as pyrethrum, and inorganic substances, such as $CuSO_4$ in Bordeaux mixture (as applied to vines), have been used for centuries, but widespread use of new synthetic organic substances (about 90% of all present pesticide applications) is a development of the post-Second World War era. The world's species have no previous evolutionary experience of these substances. Pesticides are thus a new environmental factor. The ecological effects of pesticides can be hard to predict, making pesticide ecology a fascinating and important new area of study. Our understanding of unwanted 'side-effects', such as loss of non-target species, pest resurgence and chronic effects on human populations, is still rudimentary.

The important ecological characteristics of pesticides are toxicity, persistence and their non-specific and density-independent mode of action. Toxicity and persistence are linked in that a lethal but non-persistent chemical may in the long run do less damage than a sublethal persistent chemical. This is because the latter has more opportunity for incorporation into food chains where it may be metabolised to a more toxic form or more typically accumulate to toxic concentrations in predators at the top of the chain (see section 12.3.8). Pesticide applications as dusts and sprays lead to their wide dispersal in air and water and incorporation into food chains in non-target areas. The organochlorines (for example DDT) which are both persistent and fat-soluble are probably the most widely distributed pesticides. Organochlorine residues have been found in all British bird species and, more noteworthy perhaps, are present in the bodies of Antarctic penguins.

Though persistence is generally an undesirable quality, particularly on food crops, in some instances, for example in the control of animal parasites and soil-borne diseases, some degree of persistence is an important practical and economic requirement. Toxicity for a particular species is commonly defined by the **lethal dose 50** (LD_{50}). This is the single dose administered orally which kills half an experimental laboratory population. In the field when the organism is subjected to additional environmental stresses a higher proportion may die. Nevertheless, by definition, some survive. (In short the aim in agriculture is to reduce crop injury to an acceptable level, the balance being largely a question of economics.) Unfortunately the survivors form the basis of a resistant pest population and, in organisms such as insects with a rapid life cycle, resistance and pest resurgence are common problems. The typical response is to develop an alternative pesticide, but this is both expensive and, many ecologists would argue, inevitably destined for a similar fate.

12.24 Why are pesticide resistance and pest resurgence associated with the use of pesticides?

Most pesticides are discovered empirically, that is by trial and error or chance observation. It is not always

421

known how they cause death in the target species. Toxicity may vary greatly between taxa and even between closely related species, for example in mammals the LD$_{50}$ for DDT is four times greater per unit body weight for sheep than for rats. In some organisms a sexual difference is seen. The LD$_{50}$ for male mice for DDT is 500 mg kg^{-1} compared with 550 mg kg^{-1} for females. These differential responses and especially those between different taxa and major organism groups can be put to practical advantage so that, for example, dicotyledonous weeds in cereal crops can be selectively eliminated without apparently harming the crop plant or its human consumers. Fleas and similar animal parasites may be controlled without harming their animal host. However, there is need for caution with these assumptions. Recent reports of illness among farm workers regularly exposed over many years to organophosphates* as, for example, used in sheep dips, suggest a need for a radical reappraisal of safety standards. Long-term consequences of pesticide exposure, even at low doses, and possible synergistic links with other contaminants or disease vectors are little known due to the relative newness of most pesticides. There is mounting concern that the 'harmless' traces of pesticide metabolites left as residues on food, though not directly toxic and certainly not lethal, may, nevertheless, lower disease resistance or be biologically accumulated to significant levels. Pesticide residues in the North Sea (especially PCBs, polychlorinated biphenyls) are thought by many scientists to be linked with the rapid spread of viral disease in the common seal population during summer 1988.

Agriculturally, when non-target organisms affected by pesticides are also predators of target species serious economic problems may result due to pest resurgence. A good example of this was seen in a study of the use of DDT to control cabbage white butterfly, *Pieris rapae*, on brussels sprouts. Initial pesticide application gave good control but subsequently numbers of butterfly larvae *exceeded* those in an unsprayed control area. This effect was even more pronounced following repeat applications of DDT to 'control' the new infestation. Examination of the crop ecosystem showed that pesticide concentrations on the leaves were rapidly reduced due to subsequent growth of existing and new leaves. However, levels in the soil remained high, especially if crop residues were ploughed in. Thus eggs deposited by adults from surrounding areas after spraying were little affected by the pesticide, but the main larval predators, the soil-dwelling ground beetle *Harpalus rufipes* and harvestman *Phalangium opilio*, showed reduced numbers and survivors fed less frequently. Larval predation was thus significantly reduced and larval numbers exceeded pre-spraying levels. Further applications of DDT merely made this situation worse (fig 12.51).

Predatory species are generally most disadvantaged by pesticide use. This is because they occur in lower numbers than their prey and therefore the population is more vulnerable and recovers more slowly. This is made worse when the predator also feeds on resistant prey with residues in the body fat. In addition, poisoned, dead, dying or ailing and behaviourally conspicuous prey are more readily

* Organophosphates – nerve poisons now widely used in place of the more persistent organochlorines.

(a) 1964 (after one application of DDT, 6 July 1964)

(b) 1965 (after three applications of DDT, 6 July, 20 August 1964 and 28 June 1965)

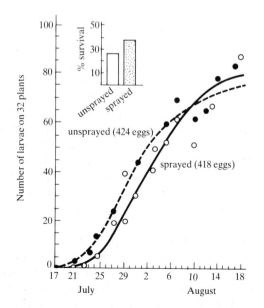

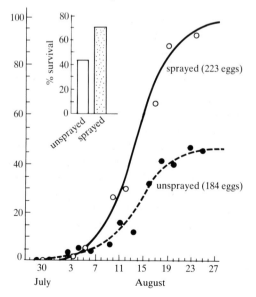

Fig 12.51 *An illustration of the differential effects of DDT on crop and soil fauna.* Pieris *lives on crops; spraying with DDT to control it is effective for only a very short period in the first year (a). Because the soil-living predators of* Pieris *are affected by residuals in the soil,* Pieris *in fact increases markedly after repeated spraying (b). (From J.P. Dempster (1968)* The control of Pieris rapae *with DDT: II Survival of the young stages of* Pieris *after spraying. J. Appl. Ecol.,* 5, *451–62.)*

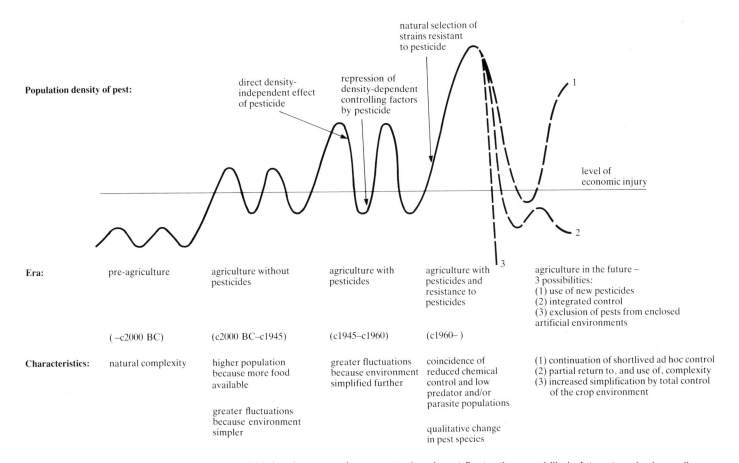

Population density of pest:

- natural selection of strains resistant to pesticide
- direct density-independent effect of pesticide
- repression of density-dependent controlling factors by pesticide
- level of economic injury

1
2
3

Era:	pre-agriculture	agriculture without pesticides	agriculture with pesticides	agriculture with pesticides and resistance to pesticides	agriculture in the future – 3 possibilities: (1) use of new pesticides (2) integrated control (3) exclusion of pests from enclosed artificial environments
	(–c2000 BC)	(c2000 BC–c1945)	(c1945–c1960)	(c1960–)	
Characteristics:	natural complexity	higher population because more food available greater fluctuations because environment simpler	greater fluctuations because environment simplified further	coincidence of reduced chemical control and low predator and/or parasite populations qualitative change in pest species	(1) continuation of shortlived ad hoc control (2) partial return to, and use of, complexity (3) increased simplification by total control of the crop environment

Fig 12.52 *Diagram illustrating the historical development of pest control and pest fluctuations and likely future trends depending on the control methods. (From N.W. Moore (1987)* A synopsis of the pesticide problem, Advances in Ecological Research, *Blackwell.)*

caught than healthy unaffected organisms. The general effect of pesticide use is to reduce species diversity. It also tends to increase productivity at lower levels of ecosystems and lessen productivity at higher levels. The effects on decomposer organisms are poorly understood and the implications of all these for nutrient cycling and soil fertility need further study. Fig 12.53 summarises the main ways in which pesticides affect ecosystems. You should consider the implications of these changes (see also fig 12.52).

Problems such as those outlined, especially resistance, resurgence and health risk to the human population, have led to more extensive consideration of alternative control techniques. The main alternatives, **biological control** and **integrated control** (more carefully targeted use of pesticides linked with biological control), are briefly considered below. Two useful sources of further information are the Studies in Biology booklets numbers 50 and 132.*

* H.F. van Emden (1974) *Pest control and its ecology*, Studies in Biology 50, Edward Arnold.
M.J. Samways (1981) *Biological control of pests and weeds*, Studies in Biology 132, Edward Arnold.

Biological control

Biological control of pests has traditionally meant regulation by natural enemies: predators, parasites and pathogens. As such it represents a form of population management preventing unchecked exponential growth of pest species (see section 12.7.3, growth curves). Some scientists take a wider view and include other techniques, such as genetic manipulation, within the scope of biological control. **Cultural control** methods such as crop rotation, tillage, mixed cropping, removal of crop residues and adjustment of harvest or sowing times to favour crop or natural enemies rather than pest may also be considered biological control. Classic biological control has been most successfully applied to introduced species which may lack natural controls, either biological or physical, as in the form of seasonal checks, in a new environment. The control of cottony cushion scale, *Icerya purchasi*, on newly established citrus plantations in California in the late-nineteenth century is the first truly successful example of scientifically planned biological control. This pest was introduced with nursery stock from its native Australia. Field searches in Australia identified two natural enemies, a parasitic fly *Cryptochetum iceryae* and a predatory lady beetle *Rodolia cardinalis*, commonly known as Vedalia. These were taken

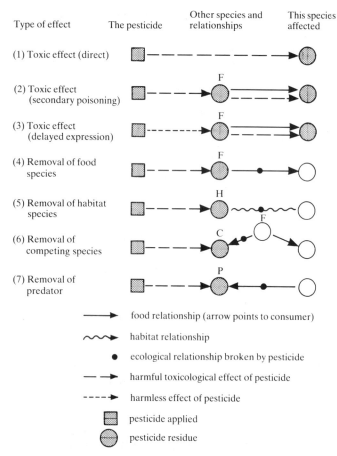

Type of effect	The pesticide	Other species and relationships	This species affected

(1) Toxic effect (direct)

(2) Toxic effect (secondary poisoning)

(3) Toxic effect (delayed expression)

(4) Removal of food species

(5) Removal of habitat species

(6) Removal of competing species

(7) Removal of predator

→ food relationship (arrow points to consumer)

∿→ habitat relationship

• ecological relationship broken by pesticide

— → harmful toxicological effect of pesticide

---→ harmless effect of pesticide

▨ pesticide applied

⊚ pesticide residue

Fig 12.53 *The main ways in which pesticides affect ecosystems. Note C, competing species; F, food species; H, habitat species; P, predator species. (From N.W. Moore (1967) A synopsis of the pesticide problem, Advances in Ecological Research. J.B. Cragg (ed.), pp. 75–126, Blackwell.)*

caterpillar-moth *Cactoblastis cactorum*. This New World cactus became a serious pest of warm grazing lands in the Old World and Australia. Entomologists searching for a suitable natural enemy found 150 possible species. Of these 51 were introduced into quarantine in Australia, 19 were released and 12 became established, but long-term control was achieved by just one species, *C. cactorum*. With plant-feeding insects, a high degree of specificity is needed and it is preferable if the insect feeds on only one type of plant. Once released into a new environment the predator cannot be re-called. Exhaustive testing and monitoring in the native environment is necessary to ensure that there is no chance that the introduced predator will develop an affinity for a plant of economic value.

Biological control is also widely used in commercial glasshouses. Tomatoes and cucumbers are grown in this way in most temperate countries. Two major glasshouse pests are the two-spotted spider mite, *Tetranychus urticae*, and the glasshouse white fly, *Trialeurodes vaporariorum*. Soil sterilisation in winter kills all beneficial predatory controls. New plants introduced the following spring typically bring in some white fly, and new spider mites emerge from hibernation in the glasshouse structure. With no natural enemies present, populations grow rapidly unless pesticides are used or natural enemy control is re-established. Fig 12.54 compares the resultant population fluctuations of spider mite using these two alternatives. Growers opting for biological control must act rapidly for the method to be effective. The natural white fly enemy, the parasite *Encarsia formosia*, and the spider mite

to California and, after careful study and successful controlled release in canvas tents, were released into the wild. Both parasite and predator spread rapidly and effective and persistent control was achieved within months.

Success is not always so immediate. Careful matching of climatic conditions and monitoring of interactions with native species is essential. In attempts to control the walnut aphid, *Chromaphis juglandicola*, in California a parasitic wasp, *Trioxys pallidus*, was introduced from Cannes in France. It was moderately successful in coastal areas but in the hot interior, the main walnut-growing area, the wasp died out after one season. Ten years later after extensive searching in similar hot and dry environments a new strain of *Trioxys* was introduced from Iran. It successfully overwintered and within a year over 50 000 square miles were cleared of the pest and parasitisation exceeded 90%.

Biological control can be particularly effective against exotic (foreign) weeds that threaten to smother indigenous vegetation or pastureland. A classic example is the control of prickly pear, *Opuntia*, in Australia by the introduced

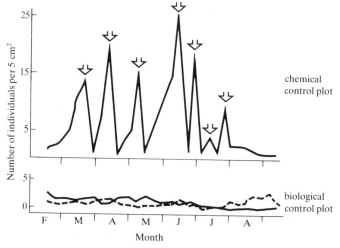

Fig 12.54 *Control of the two-spotted spider mite, Tetranychus urticae (full line), on glasshouse cucumbers in Finland by chemical control using dicofol (arrowed) and biological control employing the predatory mite Phytoseiulus persimilis (dashed line). With chemical control the spider mite population fluctuates wildly, whereas with biocontrol the pest population is maintained at a fairly low and steady level. (After M. Markkula, K. Tiittanen, & M. Nieminen, (1972) Ann. Agr. Fenn., 11, 74–8.)*

predator *Phytoseiulus persimilis* are bred commercially to ensure abundant supplies. In the 'pest-in-first' method of control, the white fly and spider mite pests are first deliberately introduced followed, after about seven days, by the parasite and predator once sufficient food (that is pest) has established for their survival. There is some understandable grower resistance to this 'unnecessary' introduction of pest species. An alternative approach is to make repeated releases of the natural enemy at about 10-day intervals around the time that infestation is typically expected. For successful control it is vital to have natural enemies present in the first stages of infestation. If that critical period is missed natural enemies never seem to catch up and overtake the pests.

Integrated control

Integrated control is pest population management which combines and integrates biological and chemical controls in a sensitive way. Pesticides are used as necessary in a manner least disruptive to complementary biological control. Integrated control programmes have the twin objectives of economic and ecologically acceptable management of pest populations.

They aim to keep pest populations below the level of economic injury, or even prevent their development, while causing minimum harm and disruption to a crop (agroecosystem) or 'natural' ecosystems and especially the beneficial natural enemies of the crop or host species. Most programmes seek to limit the use of expensive and environmentally damaging chemicals and target their use more carefully. One way of achieving this is to design more-selective chemicals. For example, the development of the aphicide pirimor, a carbamate insecticide which is highly selective for aphids but causes minimal harm to their natural enemies, has enabled development of an integrated control programme for peach–potato aphid, *Myzus persicae*, and red spider mite, *Tetranychus*, on glasshouse chrysanthemums in Britain. Both were formerly controlled by organophosphate insecticides, which are relatively unselective and to which the aphid was developing resistance. The spider mite is now controlled biologically with the predatory mite *Phytoseiulus*, and the aphid controlled by pirimor.

However, for economic reasons (research and development costs versus the smaller market for selective chemicals) a more practicable universal approach is better targeted use of broad-spectrum chemicals. Targeting may be improved spatially and by better timing of pesticide applications. The development of pheromones* has greatly aided spatial targeting. Sex attractants can be used to lure insects to some other control measure, a pesticide or chemosterilant. Alternatively, a pheromone might be used to aggregate a population in an area to be treated with a

* Synthetic production of chemical substances used by insects to signal behavioural activity such as mating and aggregation.

pesticide. In this way much less pesticide is used with minimal effects on non-target species and the wider environment. Another promising use of pheromones alone is to inhibit behavioural responses such as mating by saturating the atmosphere with the appropriate pheromone (very low concentrations will do this). Insects become habituated to the constant stimulus and the appropriate reaction is suppressed. Frequency of pesticide application can be reduced by careful timing to cause maximum damage to the pest (such as at mating) while minimising effects on associated species. This requires close study of the life histories of all the species concerned and a clear understanding of the ecosystems involved.

Successful integrated control schemes have been used on many crop ecosystems where previous blanket spraying of broad-spectrum chemicals has destroyed the natural balance of pest and natural enemy controls. For example the combined use of pheromones and pesticide has enabled recovery of the natural enemies of the cabbage white butterfly and development of new control measures for this pest on *Brassica* crops.

Integrated control measures are now widely used in commercial cotton growing where on a world-wide scale the use of broad-spectrum chemicals to control early season pests such as boll weevil, *Anthonomus grandis*, led to depletion of natural enemies and emergence of new pests, often pesticide-resistant. Economic costs were high. In the Canete valley, Peru, spraying of cotton monocultures resulted in the collapse of associated faunal control giving economic and ecological disaster. Legislative controls in the mid-1950s restricted use of broad-spectrum chemicals and instigated control based on cultural practices including mixed cultivations, reintroduction of beneficial insects and return to older, more-sensitive insecticides. By the late 1950s cotton yields rose dramatically. Similar schemes have been used in other badly affected cotton-growing areas. Worldwide integrated control measures have been developed for many crops of major economic and social importance such as alfalfa, apple, citrus, oil palm, rubber and cocoa, and furthermore such techniques are used extensively on glasshouse crops.

12.9   The Green Revolution

Originally this term was applied to the successful introduction of high-yielding rice and wheat varieties as a means of increasing production in developing countries.

In the late 1940s short-stem varieties of wheat from Japan were crossed with semi-dwarf American varieties. With dwarf varieties the problem of lodging (plants collapsing with a heavy ear) is greatly reduced so that increased application of nitrogenous fertilisers (to boost crop yield) and improved irrigation methods could be exploited. This resulted in yields increasing by a factor of 2.

The success of the wheat programme led to the setting up of the International Rice Research Institute in the Philippines. Successful breeding of tall, vigorous varieties from Indonesia with dwarf varieties from Taiwan produced the very successful strain 1R8. This responded to high nitrogen fertiliser application by early maturing and enabled two crops to be taken in one growing season.

However, in some respects these innovations proved disappointing since success depended on a high input of expensive fertilisers. They were therefore of no particular advantage to the peasant farmers. Furthermore, due to unfamiliar taste and texture, the rice produced by the new varieties was often unpopular with the indigenous population.

The choice of food crop may be greatly influenced by factors unconnected with primary food production. The commonly grown cereal crops may be deficient in certain of the essential amino acids, for example maize is low in the amino acids lysine and tryptophan. Affluent societies prefer to obtain their amino acids from animal meat proteins rather than plant proteins (such as beans and cereals). The grain is fed to animals which adds an extra consumer to the food chain and is therefore a relatively inefficient system of converting plant protein into high class animal proteins.

A solution to these problems is being sought by:

(1) intensive breeding programmes to select mutants and to develop varieties better adapted to climatic conditions or able to withstand temporary conditions such as drought;

(2) genetic engineering by which favourable genes can be transferred to the host plants, for example to confer disease resistance. Attempts are being made to transfer the NIF genes (the nitrogen-fixation genes) involved in the legume/*Rhizobium* interaction to non-leguminous crops. If successful, this will reduce the requirement for expensive nitrogenous fertilisers.

Plants which have considerable salt tolerance are being developed for semi-arid regions; for example some species have genes which allow, as a detoxication mechanism, the accumulation of ions in the vacuole. Other plants appear to tolerate higher levels of sodium ions in the leaf without their adversely affecting plant growth as in other plants.

Toleration of desiccation in some plants may be related to inherent properties of the cellular contents, such as cytoplasmic factors which are in some way able to limit the damage caused by desiccation and so can be rapidly reactivated on rehydration. However, other investigators have claimed that successful tolerance depends on the mechanical structural properties of the cell wall, or on the size and shape of the cells and on the size of the vacuole.

The term 'Green Revolution' is now used more broadly for the successful application of scientific agricultural principles to tropical crop plants.

Agricultural techniques resulting in increased production of organic materials have been responsible for sustaining large increases in world population. Neverthe-

less in Europe, the potential for food production by conventional and alternative agricultural methods is undergoing extensive review (EEC Agricultural Policy) and reorganisation as a consequence of over-production resulting in the 'food mountains'. However, worldwide there remains the question of underproduction with a marked dependence of final crop yield on climatic factors, particularly rainfall.

Claims have been made that the world can be divided into three main groups according to their increase in population.

Group 1: Where the population growth is less than 1.5% per year, for example North America, Western Europe and the USSR, which overall represents about 26% of the total world population.

Group 2: Where the growth rate lies between 1.5 and 2.5% per year, such as China, Japan, South Africa, representing about 28% of the total world population.

Group 3: Where the growth rate is greater than 2.5% per year, for example India, Pakistan, Indonesia and most of Africa, representing about 46% of the world population. It has been estimated that India requires an increase of about 3 million tonnes of grain per annum merely to cope with the increase in population. In addition, it is these areas which are most subject to sudden climatic disasters (such as flooding, lack of rain) and therefore where famine conditions are most likely to occur.

Agricultural systems attempt to improve the efficiency of the photosynthetic system in order to increase production of dry matter for human purposes (the harvest index). This efficiency is based on the conversion of solar radiation into production. The choice of crop species, however, is heavily influenced by factors unconnected with this production, such as food preferences, ease of cultivation, religious and social pressures and secondary constraints such as climate, particularly temperature ranges, and prevalence of pests. As a result of the need to provide food to satisfy population increases, particularly in the developing countries, there is a pressure to push the growth of desirable crops towards the geographical limits of their climatic adaptations, that is to exploit new growing regions, and to develop crops capable of growing successfully in these marginal regions.

Western agriculture depends on a high energy input (mechanisation) and on high rates of application of nitrogenous fertilisers. The production of these fertilisers, involving the reduction of atmospheric nitrogen, also requires a high energy input. Intensive cultivation is therefore possible only in a small percentage of agricultural lands. Worldwide, therefore, there is an emphasis on breeding new varieties of plants capable of producing more food of higher quality, but requiring lower energy input and less dependence on the use of nitrogenous fertilisers. This emphasis is coupled with the need to exploit new habitats for the growth of crops, and to encourage diversification by sowing a wider range of crops, for example in the UK, rape seed is now a common crop and the possibility of growing sunflowers (for oil) is being promoted.

Intensive agricultural programmes based upon artificially optimising environmental conditions have enabled high crop yields to be obtained. The use of greenhouses and hydroponics highlight many of the principles associated with such intensive programmes.

12.9.1 Greenhouses

In addition to the physiological and genetic factors already mentioned, the rate of photosynthesis under field conditions has been shown to be influenced by:

(1) mineral nutrition of the plant;
(2) water management, such as availability;
(3) presence of plant pathogens;
(4) heat stress;
(5) wind;
(6) age of the plant and leaves: factors which delay senescence lead to higher rates. For example, cytokinins delay the onset of senescence in leaves. Leaves sprayed with cytokinins (benzyl adenine and kinetin) have higher rates of photosynthesis compared with control plants.

To obtain higher crop yields, it is important that these factors should be maximised and controlled. By growing crops under glass or inexpensive plastics, such as polyethylene films, greater control of the conditions becomes possible, enabling crops to be grown out of season and thereby exploiting lucrative markets. Greenhouses are also useful for the propagation of seedlings.

The management of greenhouse crops is influenced by the type of plants being grown but general principles include the following.

(1) *Temperature control.* Most plants have an optimum temperature range. Control can be achieved by additional heating in winter, shading in summer and by controlled ventilation. In large commercial greenhouses, temperature is carefully controlled by automatically opening and closing ventilators.
(2) *Ventilation.* This ensures an adequate supply of fresh air, replenishing carbon dioxide levels, and also controls atmospheric moisture levels.
(3) *Water levels and humidity.* Plants vary considerably in their response to moisture; for example cacti thrive best in dry air conditions, foliage plants require higher humidities. Control of humidity is also needed to control the spread of fungal diseases. Humidity levels can be monitored and increased by sprays and mist propagators. Many of these operations are now con-

trolled by computers. Attempts have been made to increase artificially the levels of carbon dioxide in closed greenhouses. This has led to increased yields but the technique has been expensive to operate.

Sheets or cones of plastic film, often called hot caps, are sometimes placed over the tops of plants growing in the field and this has been used especially in Israel for winter vegetable production using plastic tunnels, the higher temperature in the tunnel favouring early crop production.

12.9.2 Hydroponics

This technique was based on earlier experiments investigating the mineral nutrition of plants by growing plants in a range of nutrient solutions. The plants are now usually grown in greenhouses with a continuous liquid culture which is circulated around the plants by pumps. These also give adequate aeration to the nutrient solutions. The nutrient content can be monitored and continuously adjusted (by computers) to maintain optimum growth conditions. The root system is supported in a sand and gravel base. Hydroponics is useful in areas where (*a*) the agriculture soil is poor, (*b*) water is limited, (*c*) there is a demand for off-season speciality crops of high market value.

Table 12.22 Reported increases in crops grown hydroponically.

	Usual agriculture methods	By hydroponics
tomatoes kg/plant	5.4	7.3–10.2
potatoes tonnes/hectare	12.1	26.3
rice kg/hectare	551.0	1 652

12.10 Conservation

Conservation may be defined as the human management of Earth's physical and biological resources in such a way as to give all forms of life, including humans, the best chances in a shared future. Conservation implies both taking action and expressing concern to protect the environment and keep it healthy. It is grounded in the belief, developed through the study of ecology, that well-managed resources of water, land and wildlife will continue to support human requirements for food, shelter, industrial products, recreation, knowledge and health well into the future.

The term conservation is often used to mean preservation of such things as historic buildings and other artefacts (human-made structures). However, conservation in its biological and environmental sense is a more dynamic and interactive process that implies a concern for the natural world which is itself always changing. We need to understand what is happening in our environment to be able to know how to look after it. This knowledge comes from ecology and the other environmental sciences.

Implicit in conservation practice and thinking is an attitude to the world we live in. This is called the **conservation ethic**. Here are just three ways in which it has been expressed.

'We have not so much inherited the world from our parents as borrowed it from our children' (a 'Green' slogan)

'Cheat the Earth and the Earth will cheat you' (an old Chinese proverb)

'Extinction is forever!' (a 'T' Shirt to 'Save the Whale')

Conservation, therefore, often has a moral and political dimension.

Effective conservation management requires integration of scientific and social objectives. In the long term the well-being of human societies and the health of the biosphere are inseparable. In the immediate time scale, however, the objectives of conservation can often seem in conflict with human aims for development and technological progress. Conservation issues are thus often highly contentious; they are also very wide-ranging. Some examples of what conservation may mean in practice are given below.

Conservation of genetic variety. This is one argument for tropical rain forest conservation since these forests have high species diversity. It is also important for maintaining domesticated plants and animals. Worldwide establishment of seed and sperm banks ensures survival of potential material from which to develop new varieties of food crops and domestic livestock in the event of major physical or biotic environment change, such as changing patterns of global climate associated with 'greenhouse' warming, or emergence of a new pest or disease due to natural or human-induced mutation.

Restoration of land, such as mine-waste or gravel pits for wildlife, recreation, agriculture and landscape enhancement, or even new building land thus relieving pressure on high-quality, relatively undisturbed countryside.

Waste recycling to avoid excessive use of resources. Recycling paper, often needlessly burned or thrown away, reduces timber needs for the paper industry. Domestic refuse can be used to generate power, particularly for local district heating schemes, though care must be taken when incinerating wastes to avoid air pollution problems.

Maintenance of forest cover on watersheds thus preventing erosion and rapid run-off and flooding and at the same time maintaining timber stocks for future human use.

Avoidance of overcropping both in the seas (see section 12.7.7) and on the land.

12.10.1 Spaceship Earth

For thousands of years humans were in equilibrium with their environment. Whenever a resource ran out locally the human population moved on. Another clearing could be made in the forest or a new search could bring ample supplies of fresh food. This balance between environmental requirements and human needs was found in hunter–gatherer societies; most recently it has been seen amongst the Kalahari bushmen, Australian aboriginies, or Inuit Indians (eskimos).

A series of revolutions in human history have broken down this balance between humans and nature. The origins of agriculture and of animal husbandry 10 000 years ago caused a huge expansion of farming people which led eventually to the first large settlements. Since then further revolutions in food production, disease control and industrialisation have changed a tiny global human population into a huge expanding and technological society that places ever greater demands on Earth's resources. In the 1960s, when humans first ventured off the planet, Earth was seen from space for the first time. This had a profound effect on people emphasising the finite limits to population growth and the exploitation of resources.

Ecologists have come to view Earth as a 'spaceship' that is completely equipped for long-term travel, but without any support other than its own original resources and the radiant energy of its nearest star, the Sun. Life has been sustained on this planet for an estimated 3 500 million years; there is no reason to suppose that it will not go on for at least as long into the future, if we look after it.

As discussed in section 12.3, the energy for life comes principally from the Sun, as light, feeding useful energy into food chains by photosynthesis of green plants and driving the biogeochemical cycles. Solar energy also powers the planetary weather systems and ocean currents that more immediately generate environments amenable to life.

In 1980 three worldwide bodies (UNEP, IUCN and WWF) drew up a **World Conservation Strategy** in which they presented an ecological analysis of how to run the spaceship, Earth. This document identified the conservation of Earth's physical environments and plant communities as the first essential of the life support system. The atmosphere, oceans and soils sustain plant life and plants sustain all animals. Furthermore on the terrestrial (land) surface it is largely plant cover that in turn conserves the soil from erosion.

Secondly the World Conservation Strategy recognised the need to conserve the Earth's genetic diversity. Evolution has produced a vast variety of life forms, both naturally and artificially selected. The extinction of species represents a permanent loss of biological resources and future opportunities. The World Conservation Strategy set out, for the first time, a programme for the sustained use of Earth's ecosystems. It argued that it must be possible to give every human being a satisfactory quality of life, whilst not abusing the life support system or threatening the other inhabitants of the spaceship with extinction. Many of the wild plants and animals of Earth may well prove more useful in the future than we suspect at present.

12.10.2 Abuse of the life support system – pollution

Until very recent Earth history, living systems evolved in rough balance with the atmosphere, hydrosphere and lithosphere, unaffected by human activity. Since the development of agriculture and technology, an increasing human impact on environment has occurred. In the last two centuries especially, widespread industrialisation has led to potentially damaging environmental pollution.

Pollution may be defined as the release into the environment of substances or energy in such quantities and for such duration that they cause harm to people or their environment. Pollution can affect all aspects of environment, human-made and natural, physical and biotic, and is readily transferred between components of the life support system. For example, pollutant gases in the atmosphere generate 'acid rain' (hydrosphere) which may in turn lead to soil acidification (lithosphere) and which also has profound effects on forests and aquatic ecosystems (biosphere), as well as damaging buildings and other artefacts (human-made structures).

Atmospheric pollution

Until the 1960s air pollutants were generally considered a local problem associated with urban and industrial centres. Subsequently it has become apparent that pollutants may be transported long distances in the air, causing adverse effects in environments far removed from the source of emission. Air pollution and its control is thus a global issue demanding international cooperation. Important atmospheric pollutants include gases such as chlorofluorocarbons (CFCs), sulphur dioxide (SO_2), hydrocarbons (HCs) and the oxides of nitrogen (NO_x). Gases occurring naturally in the atmosphere may be seriously depleted by pollution, such as ozone (O_3) in the stratosphere. Paradoxically in some locations ozone is occurring with increasing frequency as a ground-level air pollutant with mean monthly concentrations up to 200 ppm compared with normal monthly values peaking at 0.04 ppm. At these higher levels ozone damages many crop plants, for example tomatoes in California (total crop losses due to O_3 damage in California amount to about $1 billion per annum). In association with other hydrocarbons and NO_x pollutants, ozone may generate a direct human health hazard and is an important constituent of photochemical smog. Dusts, noise, waste heat, radioactivity and electromagnetic pulses may also pollute the atmosphere. Detailed consideration of the full range of pollutants, their effects and controls, is beyond the scope of this text*. The rising levels of atmospheric carbon dioxide and its role, with other gases, in the planetary 'greenhouse' effect was discussed in section 12.4.1. Two further topical global issues, ozone depletion and 'acid rain' are considered now.

Ozone depletion. The atmosphere provides a thermal blanket and radiation shield to the Earth. In the upper atmosphere, 15–50 km above the Earth, oxygen and ozone absorb much of the incoming short-wave radiation. These are ultraviolet (UV), X- and gamma-rays which are in the main very harmful to living organisms, damaging their genetic material. In the USA it has been estimated that a 5% reduction in stratospheric ozone would lead to a 7.5–15% increase in ground-level UV radiation† which might increase the incidence of skin cancer by tens or hundreds of thousands of cases a year. Furthermore radiation absorption by stratospheric ozone warms the stratosphere creating a deep temperature inversion layer (temperature usually *decreases* with height in the troposphere, the lower atmosphere). This effectively limits convective motion in the atmosphere. Any change or weakening of this temperature inversion layer, by extending the depth through which convection may operate, would profoundly alter global weather patterns and hence Earth surface climates.

High in the atmosphere oxygen molecules (O_2) are dissociated by radiation into oxygen atoms (O) which combine with oxygen molecules to make ozone (O_3). The reaction is reversible by sunlight ($O_3 + O \rightleftharpoons 2O_2$). Ozone exists at an equilibrium level in the 'ozone layer' at a concentration of 1 ppm.

Chlorofluorocarbons are a group of chemicals including carbon tetrachloride and chloroform. These are commonly used as solvents, aerosol propellants and refrigerator coolants. They are not readily broken down in the troposphere where they may contribute to increased 'greenhouse' warming as discussed in section 12.4.1. They rise eventually into the stratosphere. Above 25 km CFCs are broken down by sunlight, releasing chlorine and fluorine. These react with ozone, one atom of chlorine or fluorine destroying 10^5 molecules of ozone, and break it down into oxygen faster than it can be reformed from oxygen into ozone.

CFC pollution shifts the oxygen–ozone equilibrium. At the present levels of CFC pollution a 10% depletion of ozone may occur over the next 20 years, and it has been suggested that as much as two-thirds of the ozone could be destroyed in half a century. In 1987 a seasonal but complete depletion of the ozone layer occurred above Antarctica for the first time.

Acid rain. Acid rain is neither a simple nor a single phenomenon. The acid gases sulphur dioxide (SO_2) and oxides of nitrogen (NO_x) are produced by burning

---

* See D. Elsom. (1987) *Atmospheric Pollution*, Blackwell, for a fuller discussion of climatic and health effects.
† UV radiation of certain wavelengths also has an important *beneficial* effect converting skin steroids to vitamin D.

fossil fuels. Incomplete combustion of these fuels releases hydrocarbons. These may have effects as dry gases or they may be washed out of the atmosphere to produce acid precipitation in rain and snow (see figure 12.55). The most industrialised areas of the world, such as the eastern USA, western Europe, north-east China and Japan, have all experienced rainfall with a pH well below 4.0. (pH 5 is the conventionally accepted lower limit for natural rainfall acidity*.)

Acid rainfall (pH<5) is often accompanied by major changes in ecosystems and damage to buildings. This often happens in countries bordering those which are major sources of pollutants. Norway and Sweden, for example, receive acid rain as a result of air pollutants emitted in the UK and industrial centres of Europe which are transported by prevailing high-level winds (see figure 12.56). Acid rainfall in central Sweden and southern Norway has affected salmon and trout fisheries (see figure 12.57) and damaged forests. Tree injury associated with acid pollution is now widespread in Europe and evidence for damage to beech and yew has been recorded in Britain.

It is commonly found that where the soil on which precipitation falls does not neutralise the acid (being derived from carbonate-poor bedrock such as granite) the fauna of lakes and rivers suffers. Young fish fry and spawn are particularly susceptible. In Scandinavia, winter accumulation of acid pollutants in snow cover which is released as an acid 'pulse' in spring melt water, coinciding with the appearance of spawn and fry, makes this problem worse. Magnesium and calcium are leached from soils and from

damaged leaves, eventually aluminium, manganese and heavy metals come into solution and may reach toxic concentrations, causing damage to tree roots and the breakdown of mycorrhizas. This decreases the capacity of the tree to take up water and nutrients. Disease induced by mineral deficiencies becomes common, a situation made worse by dry conditions. 'Acid rain' also includes the phenomenon of ozone-acid mists, thought to be an important cause of die-back in the Black Forest in Germany, and dry deposition of acid pollutants. Ozone is generated as a ground-level pollutant as hydrocarbons react with nitrogen oxides in sunny conditions. Even at levels where they are individually harmless, ozone, sulphur dioxide, nitrogen oxides, heavy metals and other photo-oxidants may together produce severe and damaging reductions of plant growth. Combined with climatic stress, especially drought, such pollution 'cocktails' can lead to tree death.

The expression 'acid rain' therefore describes more than one phenomenon. Cures, such as adding lime to lakes (Sweden) and forests (W. Germany) can only be viewed as temporary stop-gaps. The remedy lies in reducing the release of pollutant gases. Attention has been focussed on reducing sulphur dioxide emissions since these have significant and clearly identifiable industrial sources, most notably coal-fired electricity generators. Furthermore the desulphurisation technology is available and effective, though costly*. In the long term, however, it may be equally important to reduce hydrocarbon and nitrogen oxide emissions.

* Carbon dioxide naturally present in the atmosphere can reduce rainfall pH to 5.6. Sea- or volcanic-derived sulphur may further reduce natural rainfall pH to as low as pH 4.7.

* High cost particularly applies when desulphurisation units are fitted retrospectively rather than included in initial plant design.

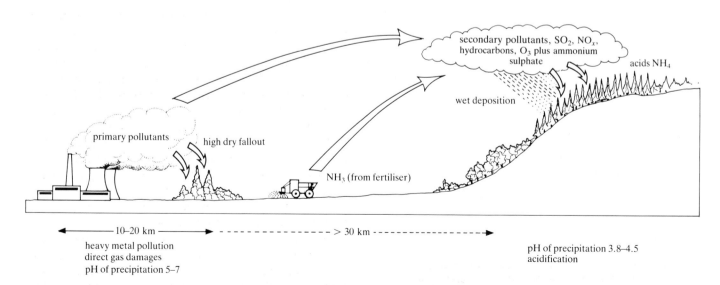

Fig 12.55 *The complexity of 'acid rain', a schematic representation of how air pollutants interact in complex ways to produce different effects in different areas. (From C. Rose (1985)* Acid rain falls on British woodland, New Scientist, *108, 1482, 52–7.)*

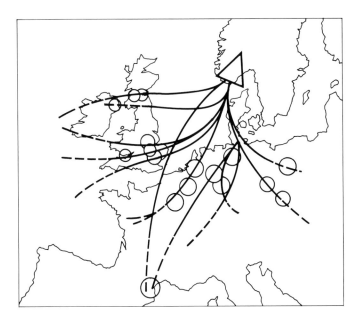

Fig 12.56 *Accumulation of acid pollutants over southern Norway. The arrows show the prevailing winds on 12 successive days of January 1974. The open circles represent centres of population or industry in western Europe. (From J.H. Ottaway (1980)* The biochemistry of pollution, *Studies in Biology, 123, Edward Arnold.)*

Fig 12.57 *Trends in the salmon catch from Norwegian rivers. (a) Southern rivers, the area most affected by acid pollution. (b) 68 other rivers of Norways. (From F. Pearce (1986)* Unravelling a century of acid pollution, New Scientist, *11, 1527, p. 23.)*

Water pollution

Until recently water pollution has been a relatively local problem of the developed world. The problem of eutrophication associated with excessive use of fertilisers in intensive agriculture and phosphate-rich sewage effluents was discussed in section 12.4.1. Such problems are increasingly occurring on a world-wide basis and affect marine as well as freshwater ecosystems. For example an algal bloom, many miles in extent, covered parts of the North Sea approaches to the Baltic in the summer of 1988. Sewage from coastal settlements discharges, sometimes untreated, into coastal waters where it generates a direct health hazard for recreational bathers as well as marine organisms. Land drainage from urban areas, industrial and waste disposal sites is often contaminated with heavy metals or hydrocarbons. Biological concentration by heavy metals in marine food chains may give lethal doses as occurred following the industrial discharge of mercury into coastal waters at Minimata in Japan. Concentrations in fish led to deaths of many humans and other animal predators. At sub-lethal levels heavy metals and contaminants such as pesticide and oil derivatives may weaken disease resistance. In the North Sea polychlorinated biphenyls (PCBs), which are also carcinogens, are present at concentrations of 0.000002 ppm. Dolphins at the end of the marine food chain have levels of 16 ppm. Measures to control and eventually stop toxic waste dumping and incineration at sea have recently been introduced by countries bordering the North Sea. Legislative controls of oil pollution associated with oil terminals and the discharge of ballast water from oil tankers were introduced much earlier after the problems were highlighted by major oil spills from oil tankers such as the Torrey Canyon (1967). Studies have shown that the less spectacular, small but persistent spills and leakages have a more damaging influence on marine ecosystems than the well-published major incidents. Use of detergents to disperse oil slicks is often more damaging to the environment than the oil itself. Further research and better marine pollution monitoring is urgently needed. Legislation to control marine pollution is difficult to police, particularly in open oceans. Easy profits can be made from criminal dumping, especially since waste disposal has become a major problem for human societies. Noxious chemicals and radioactive wastes are of course eventually dispersed by ocean currents and wave action, but if discharged into the shallow seas of the continental shelf, and especially the intertidal zone, they may damage ecosystems before effective dispersal can occur. Small seas are particularly vulnerable. The Irish Sea is a more radioactive marine environment than other coastal waters due to leaks from the Sellafield nuclear power station.

Another major problem relates to soil erosion on the continental surfaces. This increases the silt load of rivers and coastal waters. The silt may beneficially enrich fisheries, though even this is debatable (see section 12.4.1). Silt deposition is certainly leading to coral reef destruction. For example the Australian Great Barrier reef is adversely influenced by deforestation on the continental mainland.

Global deforestation

Forests are the natural climax vegetation of many parts of the world covering, until recent years, a third of the land surface. Although temperate forests are not significantly decreasing, tropical forests are declining at a rate that will decrease their 1950 extent (15% of global land surface) to 300 million hectares (7% of global land surface) by the year 2000 AD. Twelve million hectares of forest, an area the size of England, are disappearing annually and a further ten million hectares are being degraded by removal of good timber species, inappropriate management and inattention to conservation needs. Tropical forests are in urgent need of conservation. The responsibility for their care is international for their destruction is linked with our global economic system and the problem of Third World poverty.

Traditional practices of forest clearance for farming, by 'slash and burn,' did no long-term damage to forests at low human population densities. It is important to realise that, given time, the period of 'bush fallow' between clearances provided a natural rotation system that allowed the forest and its soils time to recover and did not threaten the forest's wildlife or endanger its soil in the long term. Human population expansion and competing land uses, such as plantation agriculture and hydroelectric power development, have forced a reduction in this fallow period and, because there is not enough land for people to farm, this is leading to the decline of forests. Not only are forests under much pressure from peasant farming systems but also importantly from fuel-wood gathering, from commercial logging for tropical hardwoods (such as teak and mahogany) and from wholesale burning and clearance for cattle ranching. The Third World countries alone are not to blame for this situation; global deforestation is a product of both population expansion (which increases the demands of poor farmers for more land and fuel) and the exploitative demands of rich countries for the very best timber and beef at cheap prices.

Loss of forests is serious for many reasons.

(a) There is a loss of traditionally harvested products such as timber, poles, twine, fuel-wood, honey, fruit, game animals and herbs, that at one time supplied local people with their needs.

(b) The demand for softwood timber (for building), pulp wood (for making paper) and tropical hardwood (for furniture) is rising globally. Long-term supplies are very much threatened.

(c) Forests are often on uplands and on watersheds, catching large amounts of rain and releasing the water slowly into streams and rivers. Deforestation of uplands is a major cause of floods in the plains below. India spends more than $1 000 million on flood damage repairs each year. In Bangladesh, in the summer of 1988, flooding occurred on an unprecedented scale affecting most of the country, largely due to the

deforestation in the mountains to the north in India and Nepal. International aid and cooperation is essential for an effective remedy.

(d) Deforestation results in soil erosion which causes economic losses and hunger. Clean water supplies may be fouled with resulting disease, as occurred in Bangladesh. The silting of reservoirs reduces their useful life, whilst harbours and estuaries must be continually dredged to keep them open.

(e) Deforestation increases global carbon dioxide (see section 12.4.1) and increases Earth surface albedo. Both of these changes may have long-term effects on the global climate.

(f) Removal of forest may alter the amount and frequency of rainfall and as a result sustained spring water supplies to communities may be lost.

(g) Forests have the most species-rich and diverse wildlife communities. Their destruction will lead to innumerable extinctions of little-known forms of life with the consequent loss of genetic variety and potential resources.

Soil erosion and the loss of agricultural land

Deforestation is not the only cause of soil erosion. Mismanagement of farmland and grassland may lead to a rapid loss of soil. Soil formation is a slow process (see section 12.5) and, globally, soil destruction currently far outreaches new soil formation. High rainfall, hilly areas with steep slopes and regularly cultivated soils are especially susceptible to soil erosion. Parts of south-east Asia have long-established systems of terracing that have proved very effective in holding soil. These are in areas where forests have been cleared. Principal soil conservation measures are terrace cultivation, contour ploughing and tying ridges to stop run-off of rain. Examples of such good conservation practices are found in Bali, whereas in neighbouring Java rapid population expansion and poor farming practice have led to considerable devastation. Annually, five million hectares of farmed land are coming out of crop production, worldwide, because of erosion losses.

Grasslands that are overgrazed by livestock frequently lose the plant cover that holds the topsoil. Plants are eaten to their roots and die, with the result that water running freely across the land surface causes sheet erosion, carrying off the topsoil. Channelled rainwater forms gullies which cut deep into the land surface. About seven million hectares of grazing land are lost this way each year and much of this will become desert.

Deserts may form naturally, but their creation may be accelerated by human activity in a process called **desertification**. One third of the global land surface is arid or semi-arid, and in these regions some 700 million people live a precarious and marginal existence. Much soil erosion is due to the activities of relatively poor people living on marginal land because they have few other choices. The Sahel, a semi-arid belt across the continent of Africa south of the Sahara, has increased in extent as a result of overpopulation by people and overgrazing by their cattle. Desertification may be checked and deserts made 'green' again by conservation measures: excluding people and their cattle, and planting trees. However, this is administratively, socially, economically and politically difficult. If deserts are 'made green' again they may generate higher and more regular rainfall through changed albedo effects on climate.

> **Thought question**
> 1(a) For any one of the topics discussed in this section or section 12.4.1 (the greenhouse effect or eutrophication), prepare a summary table showing who pays for the pollution and who benefits.
> (b) When polluters are forced to pay for pollution control the costs previously borne by society (as pollution) are included in the production process as pollution control. This is termed the **polluter pays principle**. What are the main likely benefits and difficulties associated with this idea?

12.10.3 Conservation of genetic diversity
Rare and endangered species

Throughout evolution species have been extinguished in the struggle for existence. Extinction is forever: the dodo and the dinosaurs cannot be brought back. Today there is an accelerating rate of extinction largely due to the fast rate of habitat destruction, the powerful forces exploiting wild populations and the shrinking of available spaces for wildlife in the face of human expansion. It is estimated that a species a day is lost at the present time. Many, most notably tropical rain forest species, are being lost before they can be described and named by taxonomists. Most losses are of insects. For every ten described species of higher plant or animal there is one in imminent danger of extinction. The plight of endangered species is well illustrated by the giant panda.

The giant panda, *Ailuropoda melanoleuca*, symbol of the Worldwide Fund for Nature (WWF), is found in eastern Tibet and southwest China. Its doleful face and cuddly-looking nature have endeared it to millions of sympathetic humans! At one time it was very much endangered since its habitat, of bamboo forest, was being encroached upon increasingly by the human population. Since interest in its survival has developed and forest reserves for it have been made, the panda population has increased to nearly a thousand members. The giant panda feeds almost exclusively on arrow bamboo, a plant that goes through cycles of abundance and scarcity. In Sichuan province in 1983 large areas of forest suddenly died back and at least 59 pandas died of starvation. Without human help this endangered species might at such a time have become extinct altogether. It has only been in the last few years that sufficient studies have been made of this species to begin to understand its physiology and ecology in the wild. It is now easy to see how ill-equipped giant pandas are to compete with people in the modern world. Pandas live solitary lives and only come together for breeding. Adults spend 95% of their waking hours feeding on bamboo, which is a poor diet, and will rarely eat anything else in preference. Although breeding captive pandas in zoos in this country has been a failure, the Chinese have been more successful at the 'Panda Farm' at Wolong in Sichuan province. Use has been made of the latest techniques in reproduction technology, such as artificial fertilisation and sperm banking, and several young have been born in captivity. It is now more possible to predict the female oestrus and ensure mating at the right moment. With environmental protection and a little human help the giant panda will survive.

There are many well-documented examples of endangered species and many programmes have been developed to save them. Some species, like the Californian condor, became endangered because of pesticide poisoning; others, such as the Philippines monkey-eating eagle, are declining because their habitat is disappearing. Each of these eagles requires more than 50 km² of forest and there are less than 300 of them left in the whole of southeast Asia. For some species decline is due to incompatibility with humans. The tiger in India, traditionally the enemy of humans, was close to extinction until several large tiger reserves were created. Poaching for ivory and rhinoceros horn have reduced elephant and rhino numbers to very low levels, a trend increased by the availability of automatic weapons. Other rare animals, such as mountain gorillas, are particularly sensitive to human disturbance. All endangered species of both plants and animals are documented in the 'Red Data Books' produced by the International Union for the Conservation of Nature and Natural Resources (IUCN). Successful species conservation requires public education, publicity, ecological research and financial support for game parks, nature reserves, zoos and botanic gardens.

Genetic resources for human use

Recently it has been realised that there is a considerable extinction of valuable alleles if an existing species' gene pool becomes too reduced. This is certainly happening with crop plants and domesticated animals. Of the estimated 80 000 plants that are edible, only 20 species have been cultivated on a large scale for human food. It is very common in developed countries for a very few varieties of these species to be grown. Selected, attested high-yielding varieties are often all that is available in bulk for sowing (half of all the wheat on the Canadian wheatlands is of one variety, 'Neepawa'). Plant breeders have for some time been aware of the genetic erosion that crop monoculture produces. They are therefore collecting many old crop varieties and conserving them, either in cultivation or in seed banks. The largest plant breeders 'gene bank' is at the International Rice Research Institute in the Philippines where thousands of traditional rice varieties are being kept (see section 12.9). Many of these are unproductive and do not respond well to fertilisers, but on the credit side they possess alleles for pest and disease resistance.

In Britain the Rare Breeds Survival Trust maintains a considerable variety of native forms of farm animal in approved collections. The rare Cornish chicken was recently used to provide an input of fast growth genes to modern poultry varieties. The Tamworth pig has proved highly successful as a free-range pig in Australia.

Undomesticated species are also a potential resource for humankind, and their conservation is only prudent. The rosy periwinkle, for example, came originally from the forests of Madagascar and produces a drug used in cancer therapy. Recently a variant of this plant was found in the West Indies that had the capacity to produce ten times more of the valuable cancer drug.

It has become increasingly clear that the best guarantee for the survival of wild species of plants and animals lies in conserving the environments in which the species are found. This requires an ecosystem approach. Each nation is at present seeking to set aside representative environments and will give them a high conservation status so that the genetic diversity of the world's plants and animals may be the better looked after (see section 12.10.6).

Thought question
2 How could sperm banks be used to enlarge the gene pool of a species like the African rhinoceros, populations of which must become increasingly isolated and unable to meet each other?

12.10.4   Thinking 'Green'

Ecological knowledge and the perception of the Earth's life support systems and its evolved genetic diversity has forced many people to look critically at our management of the Earth's resources. We need to 'think green' if we are to make sustained use of plant and animal communities and maintain environmental quality.

Sustainable use of plant and animal resources

Rational cropping of ecosystems was discussed in section 12.3.7. The relevance of population dynamics to sustained and profitable use of fish resources was considered in section 12.7. In the long-term economic prosperity and ecological well-being are inseparable.

Forest management illustrates the need to take a long-term view, for the timber needed by one generation has to be planted at least a generation in advance. Furthermore where land use is changed to accommodate human needs, as when forest is cleared for new agricultural land, it is important that the new system should be sustainable in the long term. In parts of Amazonia agricultural success following rain forest clearance has been very transient. Even schemes with substantial external financial support and agricultural expertise have been abandoned after 10 or 15 years. Weed invasion, pest and disease problems, and, above all, declining soil fertility make projects no longer economically viable. For a variety of reasons mature rain forest rarely re-establishes at such sites. In the intervening period soil nutrients are lost and structure is impaired; no forest seed store remains and, even when dispersal brings seeds of mature rain forest species, germination is poor in the exposed conditions. (Rain forest species require the high humidity found beneath a mature canopy for successful germination.) At best a secondary forest of lower productivity than the original rain forest is established in which only very low intensity shifting cultivation can be practised. Intensive agriculture is clearly not a sustainable use of these areas. It wastes resources since the replacing forest is less productive than the original. At the same time many mature forest species are permanently lost. Sustainable use of tropical moist forests which also allows economic and social development of the region concerned is a major contemporary challenge for the world community.

Recycling of materials

An important aspect of conservation is the need to imitate the cycling processes of nature. Materials are used over and over again in nature; why not with human materials? If commodities are uncommon, like rare metals, this may be done, but generally ours is a 'throw-away society'. During the Second World War Britain recycled a great deal of paper effectively. Poor Third World countries do the same today. Since we are rich enough to pay the world price for wood pulp relatively little is recycled in Britain and many other developed countries. Paper recycling technology is improving to such an extent that the wasteful use of paper could at least be halved.

Sewage processing is a technological use of the decay micro-organisms, normally present in the soil and fresh water, to break down human wastes. Sewage works employ this natural process and produce, as a by-product, nutrient-rich dried sludge. Providing it has not become contaminated with heavy metals (such as by lead from petroleum wastes) this may be returned to the land as fertiliser. Sewage sludge is also often used in reclamation schemes since it improves 'soil' structure as well as providing vital missing nutrients. Energy from the organic sewage may be used to generate methane (biogas) to provide the energy to run the sewage processing works.

New sources of energy

An awareness of the rate of depletion of fossil fuels and an unease about nuclear power has prompted many people to think creatively about alternative sources of energy.

The Sun may be considered to be a nuclear fusion reactor. It has the advantage of being 150 million kilometres away. The heat of the Sun indirectly causes the wind and waves of the sea. It also runs the hydrological cycle and provides all the energy for photosynthesis in the biosphere. Harnessing wind and wave power, hydroelectric power and direct solar power are all options open to us for the future. Even the use of plant materials to provide energy for fuel is feasible. Brazil has large industrial plants for converting sugar, from cane, into ethanol (gasohol) to run its cars. Such uses of 'biomass' to produce energy are potentially cleaner and less exploitive of the fossil fuel reserves of the planet. These should not be burned for they may be very valuable to us in the long-term and should be regarded as finite resources. They cannot last forever.

Reclamation from industrial and urban dereliction

Conservation may imply acting to restore the use of some exploited part of the environment for some useful biological production or human amenity or industrial use. Abandoned mining wastes and industrial sites often pose a considerable hazard, as demonstrated by the land-slip disaster at Aberfan, S. Wales in 1963. Establishing a vegetation cover on such areas will ameliorate the extreme drainage, flood, water quality and slope stability problems they pose.

The main problems for revegetating abandoned industrial and urban sites lie with establishing a suitable stable 'soil' medium for germination and sustained growth. Temperature and moisture extremes commonly impair germination. Deficiencies of nitrogen and phosphorus are major problems for plant growth. Mine spoil and industrial wastes may also be contaminated with heavy metals. Returning these wastes to agricultural or forestry use is difficult due to the risk of bioaccumulation of metal toxicity in food chains. Extremes of pH and recurring extreme acidity, as found in colliery spoil containing iron pyrites, also require amendment before a successful and sustained vegetation sward can be established. However, these areas can be reclaimed for new industrial sites, thus relieving pressure on remaining countryside resources.

Spoil heaps from mines in the Midlands and north of England have been landscaped and sown with mineral-tolerant grasses. To many people they now look like ordinary hills. Old gravel pits are very commonly landscaped and managed for wildlife, or for fishing, boating or waterskiing. In Northamptonshire 5 700 hectares of open cast mining have been reclaimed by landscaping, by clearing the boulders and by planting grass and clover mixtures. After 20 years the build-up of soil has been sufficient to return the land to arable farming once again, providing wild areas of countryside in the mean time.

12.10.5 United Kingdom conservation concerns

Britain has almost no 'natural' communities. Since Neolithic times people have been an important influence shaping the countryside (see section 12.6.1). The upland moors and lowland heaths (considered in 12.6.1) have been formed by centuries of deforestation and subsequent grazing and burning management. The structure and species composition of British woodlands has been modified by grazing and silvicultural practice. Thus, concern for maintaining the 'natural landscape' is in fact a demand for the retention of a pattern of rural appearances only a few hundred years old. Within this framework there are many seminatural communities that are worthy of conservation which will die out if they are not managed well. These are described below.

Broad-leaved woodland

As discussed in section 12.6.1, oak was the dominant species in the primeval forests of Britain. Beech was also important in the south and like ash sometimes dominated on limestone and chalk hills. Lime was particularly important during the warmest phase since the last Ice Age. Birch and, in the north, pine were characteristic on nutrient-poor sandy soils. Alder, another warmth-loving species, favoured wetter fertile sites with willow more prevalent on wet nutrient-poor sites.

A very few contemporary woodlands are primary forest remnants and even these have been modified by human activities. Most broad-leaved woodlands are secondary, having developed from formerly cleared sites. 'Ancient woodlands' are those that date from before 1600 AD; those with a later origin are termed 'recent'. Ancient woodlands have the highest conservation value. Although they may not be original primary forest, they often contain great species diversity, particularly of slow-growing long-lived plants such as lichens, and they bear the marks of human management back to very early times. Most woodlands were managed in one of two traditional systems as wood pastures or coppice. **Wood pastures** were common on poorer soils and livestock grazing as well as timber produce was important in these woodlands. Grazing pressure restricted regeneration, a problem overcome by **pollarding** (cutting the main stem above the browse line (c. 2 m) to encourage new growth of lateral branches which could be periodically harvested). The Ancient and Oriental woodlands of the New Forest, Hampshire are good examples of wood pastures. These woodlands are typically less species-rich than former coppice woodlands. Over centuries the more palatable species have been lost and browse-tolerant species such as holly and hawthorn have become dominant in the understorey. The future of many of these woodlands depends on regeneration success now largely dependent on the control of grazing pressure.

On richer soils the **coppice system** predominated. Understorey species such as ash and hazel were cut close to ground level on a 5–20 year cycle depending on the rate of regrowth and the products required. The larger coppice poles were used to provide materials for house building, furniture and fencing; lighter materials were used for basket weaving, sheep hurdles or wattle (to hold wall plaster) or were used in thatching. Woods provided much fuel and of all the products very little was wasted. Occasional standard trees were allowed to grow to near full maturity. These were felled on a longer time scale and sawn into planks or fashioned into beams. This management pattern, of **coppice with standards**, resulted in periodic expanses of flowers on the woodland floor. Such woods have a species-rich ground flora, with many rare and historically interesting plants. In the late nineteenth and early twentieth centuries economic and social changes led to the decline of coppice management and many woods were untouched for over 50 years. The re-establishment of coppice cycles, mainly by voluntary conservation groups, has restored the interest and beauty of many of these woodlands. Conserving ancient woodland helps to retain the rich plant and animal communities of the primary forests, many members of which are rare. Such places provide an experimental 'control' or baseline from which to judge other environmental change as well as being a living museum more ancient than our built historical monuments.

Hedgerows

Hedgerows may have high value to wildlife conservation. Hedges were used by the Saxons to divide and demarcate land plots and to contain livestock. Hedgerows are traditionally managed by periodic laying, in which upright stems are selected in the line of the hedge, half cut through and layed down, weaving them between stakes. The resulting structure, a layed hedge, is both alive and stockproof. Over the centuries such hedgerows have increased their species richness. One new species of woody shrub is added to every 27 m length of hedgerow, on average, every hundred years. Ancient hedgerows have therefore the greatest diversity of wildlife and some are even the only places in an arable area where plants typical of ancient woodland are found. Many thousands of kilometres of hedgerow were lost in lowland Britain between 1950 and 1980; between 1957 and 1969 the rate of removal was fastest. The enlarged field sizes released many hectares of land to arable farming, freed farmers from having to maintain them by trimming or proper laying, facilitated ploughing and harvesting by large machines, and removed from the countryside what farmers perceived as a reservoir of weeds, pests and diseases. Hedgerows, however, undoubtedly conserved many species of animals which act as predators of many plant pests. They reduce windspeeds providing important areas of livestock shelter in their lee and helping to reduce soil erosion. Hedgerows also provide a refuge and home for song birds, game birds and other woodland species.

Grasslands and meadows

Today most of the grazing used by farmers are sown leys of productive species, typically rye grass (*Lolium*) mixtures. Ancient grassland is much richer in plant species, and hence retains many more species of other wildlife. The richest of these ancient grass communities are the chalk grasslands (see also section 12.6.1) found on calcareous soils, and the ill-drained lush grass pastures of the lowland wetlands. Traditionally grass may be grazed (pasture) or mown for hay (meadow) or left for winter grazing (foggage). Management systems vary, but to conserve the diversity of grassland a traditional system should be maintained, avoiding the use of herbicides and nutrient enrichment with fertilisers. The addition of nitrogen will quickly deplete pastures of orchids, for example. Where a ley may have barely ten species of plant in it, an ancient meadow may have over a hundred. Many old meadows and pastures, characterised by such plants as cowslips, ox-eye daisies and ladies bedstraw, have disappeared under the plough. Wetlands have been drained and turned to profitable arable farming. The Halvergate marshes in East Anglia, Somerset levels and the flood meadows of the Test and Avon valleys (Hampshire), some of the best remaining wet grazing lands, have all been threatened. Once lost, such plant and animal communities are almost impossible to regain.

Upland moors (see section 12.6.1 for lowland heaths)

The northern uplands of Britain are characterised by a virtually treeless region of short vegetation, some of which is dominated by heather (*Calluna vulgaris*). It covers over a million hectares of Scotland and northern England on acid free-draining (podzolic) soils; it is rare on the continent of Europe where similar communities have been afforested. A distinction is sometimes made between upland heaths and moors based on the damper thicker peat of the latter. However, there is little difference in vegetation apart from an increased frequency of rare liverworts and lichens on the wetter moors. Moorland is partly man-made, being a product of the deforestation of primary upland forest followed by intensive sheep grazing and occasional burning. Moorlands have been conserved for their scenic beauty, recreational value to walkers and because game shooting (grouse and red deer) are still major recreations for the minority. Moorlands are rich in non-vertebrate species and many nationally rare birds, such as the curlew, the golden eagle and the merlin falcon, favour such open habitats. A trend in recent years has been towards planting forests on these uplands. There has also been much drainage of water-logged bogs that formerly defied any alternative land use. This has been seen in the Flow country of Scotland. The planting of coniferous monocultures of lodgepole pine (*Pinus contorta*) and sitka spruce (*Picea sitchensis*) is welcomed by those who see them as a means of meeting national timber needs, but opposed by those who see the plantations as an eyesore to the landscape, of doubtful real economic benefit, and an impoverishment of native wildlife. Satisfying a diversity of recreational needs together with economic and wildlife conservation imperatives requires careful long-term planning. The voice of conservation dictates that the existing diversity should be maintained without loss of species.

12.10.6   Conservation agencies

There are many individuals, groups and organisations concerned with conservation from local level up to international level. The groups to which individuals belong are diverse and specialised in their particular interest. All contribute in different ways to changing the way we think about and act towards the environment.

Local and national non-governmental bodies

Every county and major urban area in Britain has a Trust for Nature Conservation which owns and manages small nature reserves. Trusts, at the local level, provide a considerable protection for species-rich sites and involve many of their members in wildlife recording and fund raising. All such Trusts, which collectively have about 1 400 nature reserves, are affiliated to the Royal Society for Nature Conservation. Their junior club is called 'WATCH'.

The Royal Society for the Protection of Birds, has a larger membership, reflecting the popularity of birds in Britain. They too own a small number (126) of generally larger 'bird reserves'. These not only provide great publicity for the conservation movement, but importantly also conserve a wide variety of other wildlife as well. The RSBP junior membership is in the Young Ornithologist Club.

The British Trust for Conservation Volunteers unites a large number of groups of predominantly young adults working at physical conservation tasks as unpaid volunteers. This Trust provides training in woodland management practices (such as coppicing), hedge-laying, dry-stone walling and the use of power tools, and so on.

Other conservation charities, such as the National Trust (NT), are not solely concerned with wildlife and landscape conservation, although the NT has 340 properties of SSSI status (see below). Others, such as the Wildfowl Trust (ducks, geese and swans), the Rare Breeds Trust (conserving domesticated animal varieties) and the Woodland Trust, are more specific in their concerns. Groups such as Friends of the Earth and Greenpeace (which is also an important international agency) are concerned with political lobbying and direct action at a wider level.

Statutory conservation bodies

Although the voluntary movement is powerful in Britain there are also many statutory (government and state-funded) bodies concerned with conservation and the environment. The Nature Conservancy Council has powers to safeguard Sites of Special Scientific Interest (SSSIs) through notification procedures to local authorities and is responsible for running National Nature Reserves (NNRs) which tend to be larger and of higher conservation status than most wildlife sanctuaries. Many are on private land and have limited public access. There is just one Marine Nature Reserve (MNR) established around Lundy in 1986. The Countryside Commission (CC) is responsible for the wider countryside. It is an advisory and promotional body but does not own land or facilities. Formed in 1968, it replaced the National Parks Commission and still maintains an important advisory role for management and countryside issues within the National Parks. The Countryside Commission's chief function today is the designation of Areas of Outstanding Natural Beauty (AONBs), the definition of heritage coasts and the establishment of long distance footpaths (see table 12.23). It advises local and regional planning authorities on countryside matters, taking a particular interest in urban fringe areas. It has encouraged the setting up of Country Parks whose recreational and educational emphasis has relieved pressure on the more-sensitive and scientifically important sites managed by the NCC.

Each National Park is administered by a National Park Authority (NPA) with funding (about 75%) from the Department of the Environment. Since 1987 government funds (from MAFF, Ministry of Agriculture, Fisheries and Food) have also been available for Environmentally Sensitive Areas (ESAs) giving support to farmers to maintain traditional agricultural practices in countryside that might otherwise be spoilt by modern farming methods. The government research body most concerned with wildlife conservation is the Natural Environment Research Council. This body does outstanding work on all aspects of nature conversation and environmental protection, and coordinates with those also doing research in universities and polytechnics. The Forestry Commission is a statutory body whose responsibilities relate to timber production. It is required to take account of conservation interests on its estates and cooperates with the NCC and Country Trusts for Nature Conservation.

International conservation concerns

Within Britain there are voluntary organisations that concern themselves with conservation issues internationally. The Fauna and Flora Preservation Society funds much specific species conservation research and protection, whilst the Worldwide Fund for Nature (formerly the World Wildlife Fund, WWF) raises very large sums of money for both research into and purchases of endangered environments. The International Union for the Conservation of

Table 12.23 Protected landscapes in England and Wales

A. National Parks (see footnote (iii) re. Scotland)

Park	Area designated (km²)	Date of confirmation
Peak District	1 404	1951
Lake District	2 243	1951
Snowdonia	2 171	1951
Dartmoor	945	1951
Pembrokeshire Coast	583	1952
North York Moors	1 432	1952
Yorkshire Dales	1 761	1954
Exmoor	686	1954
Northumberland (including Roman Wall)	1 031	1956
Brecon Beacons	1 344	1957
Total	13 600	

(i) Since 1.1.88 the East Anglian Broads have been granted equivalent status to the National Parks. The area is administered by the Broads Authority which operates similarly to the individual National Parks authorities.

(ii) The Countryside Commission, Nature Conservancy Council and Forestry Commission all regard the New Forest, Hampshire as an area of equivalent status to a National Park and top grade NNR.

(iii) Scotland has no National Parks. There are however 40 designated (1980) National Scenic Areas. These were identified by the Countryside Commission for Scotland (CCS) on a subjective assessment of their beauty. They give some limited protection from development but many ecological and recreational agencies think a better defined and more rigorous system is needed. For a fuller discussion see MacEwen, A. & MacEwen, M. (1983). *National Parks: conservation or cosmetics?*, George Allen & Unwin.

B. Other Countryside Commission designated areas in England and Wales (1988)

38	AONBs
39	heritage coasts
13	long distance footpaths
227	country parks
251	picnic sites

About 22% of the land surface lies within a National Park or AONB.

Nature and Natural Resources (IUCN) is a wider network and forum for those concerned with conservation, allowing them to coordinate action and monitor the survival of endangered species all over the world. The latter are described in the 'Red Data Books', which list in detail animals and plants on the verge of extinction. At intergovernmental level the United Nations Environment Programme (UNEP) has done much, particularly for the marine environment, that is not within any one country's national sovereignty. UNEP, IUCN and WWF published an important charter in 1980 called *The World Conservation Strategy* in which many conservation ideas, of importance to the world's future, were discussed.

Chapter Thirteen

Quantitative ecology

The principles of ecology, as outlined in the previous chapter, are based on qualitative and quantitative data obtained from studies carried out on animals, plants, micro-organisms and the abiotic environment. This chapter deals with both qualitative and quantitative aspects of ecological investigation and presents a general introduction to some of the methods and techniques of obtaining, presenting and analysing data relating to the abiotic and biotic environments.

Before attempting any ecological investigation it is essential to identify the exact aims and objectives of the study and the degree of accuracy required. These, in turn will clarify the methods and techniques to be employed and will ensure that the data collected is relevant to the study and is adequate to form a basis for valid conclusions. In many cases, it simplifies the methods and techniques and reduces the time, money, resources and effort needed for the study. However it must be stressed that investigations frequently have to be modified in the light of problems encountered during the investigation.

13.1 Methods of measuring environmental factors

The main environmental factors which must be studied in order to complement biotic analyses, are edaphic, topographic and climatic factors such as water, humidity, temperature, light and wind. Many of the methods used to measure environmental factors are included below in experiments. Other methods of quantitative study are described in outline only.

13.1.1 Edaphic factors

Soils vary considerably in structure and chemical composition as described in section 12.4. In order to obtain a basic idea of the structure or profile of the soil, a pit is dug so that a clean-cut vertical section of the soil can be seen. The various thicknesses of clearly differentiated bands (horizons), shown in terms of colour and texture, can be measured directly, and samples removed from these horizons and used for the various analyses described below.

Alternatively a soil auger, which is an elongated cork-screw implement, is screwed into the ground to the desired depth and then removed. Soil trapped in the threads of the screw at various levels is removed into separate polythene bags for subsequent analysis. When using this method of obtaining a soil sample, it is important to keep a record of the level each part of the sample occupied in the ground. This information should be recorded on the relevant bag.

Experiment 13.1: To investigate the water content of a soil sample

Materials

about 80 g soil
aluminium foil pie dish
balance accurate to 0.1 g
thermostatically controlled oven
thermometer reading up to 150 °C
desiccator
tongs

Method

(1) Weigh aluminium foil pie dish while still empty. Record the mass (a).
(2) Add a broken-up soil sample to the pie dish and weigh. Record the mass (b).
(3) Place the pie dish containing the soil sample in the oven at 110 °C for 24 h.
(4) Remove the sample from the oven and cool in a desiccator.
(5) Weigh the sample when cool, and record the mass.
(6) Return the sample to the oven at 110 °C for a further 24 h.
(7) Repeat stages (4) and (5) until consistent weighings are recorded (constant mass). Record the mass (c).
(8) Calculate the percentage water content as follows:

$$\frac{b - c}{b - a} \times 100$$

(9) Retain the soil sample in the desiccator for experiment 13.2.

Notes

The value obtained in the experiment is the percentage total water present. This amount will depend upon recent rainfall. Alternative estimates of water content include field capacity and available water. The **field capacity** is the amount of water retained in the soil after excess water has drained off under the influence of gravity. To obtain this value the soil in the field should be flooded until surface water persists for several minutes, 48 h before the sample is removed for investigation. The **available water** is the water which is available to be taken up by plants and may be estimated by drying the weighed sample to constant mass at room temperature. The difference between wet mass and dry mass is the amount of available water present.

Experiment 13.2: To investigate the organic (humus) content of a soil sample

Materials

dried soil sample from experiment 13.1 in desiccator
crucible and lid
tripod, Bunsen burner, asbestos mat, fireclay triangle
desiccator
tongs

Method

(1) Heat the crucible and lid strongly in the Bunsen flame to remove all traces of moisture. Place in the desiccator to cool. Weigh and record the mass (*a*).
(2) Add the dried soil sample (kept from the previous experiment) from the desiccator and weigh. Record the mass (*b*).
(3) Heat the soil sample in the crucible, covered with the lid, to red-heat for 1 h to burn off all the organic matter. Allow to cool for 10 min and remove to the desiccator.
(4) Weigh the crucible and sample when cool.
(5) Repeat (3) and (4) until constant mass is recorded.
(6) Calculate the percentage organic content as follows:

$$\frac{b - c}{b - a} \times 100$$

(7) Repeat the experiment on soil samples taken from different areas to demonstrate variations in organic content.

Note

The percentage organic content obtained in this experiment is relative to dried soil and not to fresh (wet) soil. The organic content of a soil may be quoted as a percentage of fresh (wet) soil using the data obtained in experiment 13.1.

> **13.1** 60 g of a fresh sample of soil produced the following data on analysis. After repeatedly heating at 110 °C and cooling in a desiccator, the consistent readings of dry mass of 45 g were obtained. The dry soil was heated repeatedly to red-heat in a crucible, cooled in a desiccator and weighed. The mass was now found to be 30 g. Calculate the water content and organic content of the fresh soil sample.

Experiment 13.3: To investigate the air content of a soil sample

Materials

tin can of volume about 200 cm³
500 cm³ beaker
water
chinagraph pencil
metal seeker

Method

(1) Place the empty can open end uppermost into the 500 cm³ beaker and fill the beaker with water above the level of the can. Mark the water level in the beaker.
(2) Carefully remove the can containing the water and measure this volume of water in a measuring cylinder. Record the volume (*a*). The water level in the beaker will fall by an amount corresponding to the volume of water in the can.
(3) Perforate the base of the can using a drill, making about eight small holes.
(4) Push the open end of the can into soil from which the surface vegetation has been removed until soil begins to come through the perforations. Gently dig out the can, turn it over and remove soil from the surface until it is level with the top of can.
(5) Place the can of soil, with open end uppermost, gently back into the beaker of water and loosen soil in the can with seeker to allow air to escape.
(6) The water level in the beaker will be lower than the original level because water will be used to replace the air which was present in the soil.
(7) Add water to the beaker from a full 100 cm³ measuring cylinder until the original level is restored. Record volume of water added (*b*).
(8) The percentage air content of the soil sample can be determined as follows:

$$\frac{b}{a} \times 100$$

(9) Repeat the experiment on soil samples from different areas.

Experiment 13.4: To investigate the approximate relative proportions of solid particles (soil texture) in a soil sample

Materials

500 cm³ measuring cylinder
100 cm³ soil sample
300 cm³ water

Method

(1) Add the soil sample to the measuring cylinder and cover with water.
(2) Shake the contents vigorously.
(3) Allow the mixture to settle out, according to density and surface area of particles, for 48 h.
(4) Measure the volume of the various fractions of soil sample.

Results

A gradation of soil components is seen. Organic matter floats at the surface of the water, some clay particles remain in suspension, larger clay particles settle out as a layer on top of sand and stones which are layered according to their sizes.

Experiment 13.5: To investigate the pH of a soil sample

Materials

long test-tube (145 mm) and bung
test-tube rack
barium sulphate
BDH universal indicator solution and colour chart
soil sample
spatula
distilled water
10 cm³ pipette

Method

(1) Add about 1 cm of soil to the test-tube and 1 cm of barium sulphate, which ensures flocculation of colloidal clay.
(2) Add 10 cm³ of distilled water and 5 cm³ of BDH universal indicator solution. Seal the test-tube with the bung. Shake vigorously and allow contents to settle for 5 min.
(3) Compare the colour of liquid in the test-tube with the colours on the BDH reference colour chart and read off the corresponding pH.
(4) Repeat the experiment on soil samples from different areas.

Note

pH is one of the most useful measurements which can be made on a soil. Although a simple measurement, it is a product of many interacting factors and is likely to be a good guide to nutrient status and to types of plants (and therefore animals) that flourish. Acid soils tend to be less nutrient rich (poorer cation-holding capacity).

13.1.2 Climatic factors

Water, air and light are the most important climatic parameters to measure and this section outlines some of the basic practical methods used in their measurement.

Experiment 13.6: To investigate the pH of a water sample

Materials

universal indicator test paper or pH meter
water sample

Method

(1) Dip a piece of universal indicator test paper into the water sample and compare the colour produced with the colour chart. Read off the pH value.

OR

(2) Rinse the probe of the pH meter with distilled water, dip it into the water sample and read off the pH value. (This method is more accurate, but the meter must be accurately calibrated using prepared solutions of known pH before the experiment begins.) Rinse the probe with distilled water before returning it to buffer solution for storage.
(3) Repeat the experiment on water samples from different sources.

Experiment 13.7: To investigate the chloride content of a water sample (giving a rough estimate of salinity)

Materials

water sample
10 cm³ pipette
burette
distilled water in a wash bottle
3 conical flasks
white tile
potassium chromate indicator
50 cm³ silver nitrate solution (2.73 g 100 cm⁻³)

Method

(1) Place 10 cm³ of the water sample into a conical flask and add two drops of potassium chromate indicator solution.
(2) Titrate silver nitrate solution from the burette, shaking the conical flask constantly.
(3) The end-point of the titration is given by a reddening of the silver chloride precipitate.

(4) Repeat the titration on a further two 10 cm³ water samples. Calculate the mean volume of silver nitrate used.

(5) The volume of silver nitrate solutions used is approximately equal to the chloride content of the water sample (in g dm⁻³).

Experiment 13.8: To investigate the oxygen content of a water sample

Dissolved oxygen

The technique described here is the Winkler method which gives an accurate measure of oxygen content but requires many reagents. A simpler but less accurate method is described in Nuffield Advanced Science, Biological Science.

Materials

10 cm³ of alkaline iodide solution (3.3 g NaOH, 2.0 g KI in 10 cm³ distilled water) (CARE)
10 cm³ of manganese chloride solution (4.0 g $MnCl_2$ in 10 cm³ distilled water)
5 cm³ of concentrated hydrochloric acid (CARE)
starch solution (as indicator)
distilled water in a wash bottle
0.01 M sodium thiosulphate solution (see point (8) in method)
3 × 5 cm³ graduated pipettes
burette
white tile
3 conical flasks
250 cm³ water sample in glass bottle with ground glass stopper

Method

(1) Collect the water sample carefully without splashing and stopper the sample bottle under water to prevent entry of air bubbles.

(2) Add 2 cm³ of manganese chloride solution and 2 cm³ of alkaline iodide solution to the sample using pipettes whose tips are placed at the bottom of sample bottle. The heavier salt solutions will displace an equal volume of water from the top of the sample bottle.

(3) Add 2 cm³ of concentrated hydrochloric acid and stopper the bottle so that no air bubbles are trapped. Shake the bottle thoroughly to dissolve the precipitate. This leaves a solution of iodine in an excess of potassium iodide. The dissolved oxygen is now fixed and exposure to air will not affect the result.

(4) Remove a 50 cm³ sample of this solution and place it in a conical flask. Titrate with 0.01 M sodium thiosulphate solution from the burette as follows:
  (a) add thiosulphate solution whilst shaking conical flask until the yellow colour becomes pale:

(b) add three drops of starch solution and continue to titrate and shake until the blue-black colouration of the starch disappears.
Record the volume of thiosulphate used.

(5) Repeat stage (4) with two further 50 cm³ samples of water and obtain the mean volume used ($\bar{x}$).

(6) Using these solutions, 1 cm³ of 0.01 M thiosulphate solution corresponds to 0.056 cm³ of oxygen at STP (standard temperature and pressure).

(7) Calculate the concentration of oxygen per litre of water using the following formula:

$$\text{oxygen in cm}^3 \text{ dm}^{-3} = \frac{0.056 \times \bar{x} \times 1000}{50} \text{ at STP}$$

Where $\bar{x}$ = volume of thiosulphate solution required for the titration of 50 cm³ of samples.

(8) In comparative studies for water pollution work and estimating BOD, dissolved oxygen levels are commonly expressed in mg dm⁻³. Calculation of the final result is simpler if a working solution of 0.0125 M sodium thiosulphate is used. Then 1 cm³ sodium thiosulphate solution is equivalent to 0.1 mg oxygen.
  (a) Prepare a stock solution of 0.1 M sodium thiosulphate. To do this dissolve 24.82 g $Na_2S_2O_3.5H_2O$ in distilled water. Add a pellet of NaOH and dilute to 1 dm³ (litre). Store in a brown bottle. This solution may be kept for two or three weeks.
  (b) Prepare, as needed, a working solution of 0.0125 M sodium thiosulphate. To do this take 125 cm³ of stock solution and dilute to 1 dm³ (×8 dilution).
  (c) Carry out the method following the procedure outlined above but using 0.0125 M sodium thiosulphate in step (4).

$$\text{mg O}_2 \text{ in dm}^3 \text{ (1 litre) sample} = \frac{\bar{x} \times 0.1 \times 1000}{50}$$

$$\text{or } \bar{x} \times 2$$

$\bar{x}$ = mean volume of 0.0125 M thiosulphate solution required for the titration of 50 cm³ of sample.

NB It is quite common to use 25 cm³ water samples, thus saving on reagent with appropriate adjustment of the final calculation (mg O_2 dm⁻³ = $\bar{x} \times 4$).

Biochemical oxygen demand (BOD)

Materials

Either
(1) Reagents and glassware as described for the Winkler method in experiment above (dissolved oxygen),
or
an appropriately calibrated oxygen electrode.
(2) 500 cm³–1 dm³ water sample.

Method

A. Pre-checks

(1) If necessary adjust the sample pH to the range of 6.5–8.5 (to optimise micro-organism activity).

(2) If the oxygen content of the sample is known to be very low (e.g. already measured dissolved oxygen) the sample should be oxygenated for 5–10 min. This is important since the test measures the rate of oxygen consumption and organism activity. The results will be misleading if there is an insufficient initial oxygen supply.

(3) If high organic contamination is suspected, prepare sample dilutions* (see footnote at end of method) before incubating. Remember to check that the BOD of the dilution water itself is negligible. To do this incubate dilution water in the same way as samples. If necessary (that is if there is a significant decrease in dissolved oxygen) adjust results for oxygen loss in dilution water controls as well as for dilution factor itself.

Test procedure

(1) Place portions of the sample (dilute if necessary) into three glass stoppered bottles of 125 cm^3 or 250 cm^3 capacity. Pour carefully to avoid trapping air bubbles. Ensure bottles are completely full.

(2) Immediately determine the oxygen content of one bottle (express as mg dm^{-3}).

(3) Incubate the remaining two bottles *in the dark* (no photosynthesis) at a standard temperature (20 °C) or the temperature of the original sample for 1–5 days. The standard procedure is to incubate in darkness at 20 °C for five days.

(4) Determine the oxygen content of the incubated bottles (mg dm^{-3}).

(5) Subtract the mean value for the incubated samples from the original sample. This gives the sample BOD in mg dm^{-3} unless the sample was diluted before incubation. In this case use the following formula:

$$BOD = (x-y)(a+1)\,mg\,dm^{-3}$$

where x is the initial dissolved oxygen in mg dm^{-3}.

y is the mean final dissolved oxygen in mg dm^{-3}.

a is the volume(s) of dilution water to 1 volume of sample.

Water current

The simplest method of measuring water current is to record the time taken for a floating object to cover a known distance. In order to eliminate the effects of wind it is preferable to use an object which is mainly submerged. Alternatively an L-shaped tube 50 cm high, 10 cm long and 2 cm in diameter can be placed in a stream with the short end facing upstream. By measuring the height to which water rises in the long limb the velocity of the current can be measured using the formula:

$$v = \sqrt{(2hg)}$$

where v is the speed of the current (cm s^{-1}), g is the acceleration due to gravity (981 cm s^{-2}) and h is the height of the column (cm).

Humidity

The relative humidity of air is a measure of the moisture content of air relative to air fully saturated with water vapour. Relative humidity varies with temperature, since air expands on heating and can hold more water vapour. This is measured by a **whirling hygrometer** consisting of a wet and a dry thermometer mounted on a wooden frame resembling a football rattle (fig 13.1). It is whirled around until both thermometers give constant temperature readings. These temperatures are then examined in hygrometer tables and the corresponding relative humidity read off.

Temperature

Air, water and soil temperatures can be measured using a mercury thermometer, but measurements of temperature at a point in time provide little real information of ecological significance. It is the range of temperatures over a period of time which have more significance in ecological studies. Hence sophisticated time-based recordings of temperature are normally used or the maximum and minimum temperatures recorded using a maximum–minimum thermometer. Temperatures in microhabitats and inaccessible habitats, such as the centre of a tree, are measured using a **thermistor** (fig 13.2). This is an electrical device which can be miniaturised to fit into the tip of a ballpoint pen and whose resistance varies with temperature. By measuring the resistance of the thermistor and comparing this with previous temperature-calibrated resis-

---

* River water does not usually require dilution. A badly polluted stream or pond might require up to four parts dilution water to one part sample. Such contaminated water is a health risk and requires great care in handling and is best avoided for student class work. Tapwater was formerly commonly used for dilution but high chlorination now often makes this unsuitable. Synthetic dilution water is preferable (distilled or deionised water with appropriate chemicals added). Advice on the preparation of synthetic dilution waters is given in H.L. Golterman, R.S. Clymo & M.A.M. Ohnstad (1978) *Methods for physical and chemical analysis of fresh waters*, IBP Handbook No. 8, Blackwell Scientific Publications 2nd edition.

Any samples absorbing more than 6 mg dm^{-3} oxygen or having a final dissolved oxygen content less than 40% saturation should be diluted.

In some cases a considerable part of the BOD may be due to oxidation of ammonia. If wished this nitrification can be inhibited by adding 1 cm of 0.5 g dm^{-3} solution of allylthiourea to each sample. For a fuller discussion see Golterman *et al.* as cited above.

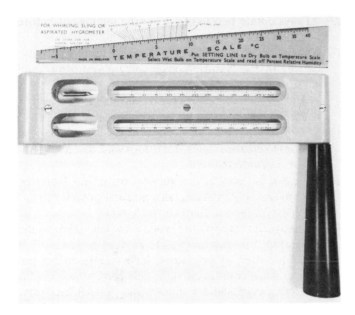

Fig 13.1 *Whirling hygrometer*

Fig 13.2 *A thermistor in use*

Light

Light varies in intensity, duration and quality (wavelength). Measurements of all three aspects are required to provide the information relevant to ecological study and specialised techniques are required to record them. For practical purposes some indication of intensity related to particular areas is generally required, so that the incident light in different areas can be compared. For this purpose an ordinary photographic exposure meter is adequate. Light intensities over a given period of time are recorded using Ozalid papers which have a cumulative sensitivity to light.

Wind speed and direction

The wind speed in a habitat at a given point in time is not as ecologically significant as the degree of exposure to wind experienced by the habitat. In this respect wind frequency, intensity and direction are all important. However, for most practical purposes a simple wind-gauge indicating the direction of the wind and a simple anemometer (fig 13.3) indicating wind speed are adequate for comparing features of wind in different habitats.

13.2 Biotic analysis

In analysing the organisms living in a given habitat (the biotic component of the ecosystem) the community structure must be determined in terms of species present in the habitat and numbers within each population. It is obviously impractical to attempt to find and count all the members of a given species, and so sampling techniques have to be devised which will give

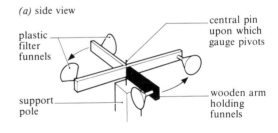

(a) side view

plastic filter funnels

central pin upon which gauge pivots

support pole

wooden arm holding funnels

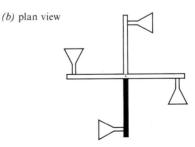

(b) plan view

Fig 13.3 *Simple anemometer which may be used to determine wind speed in terms of the rate of turning of the wooden arm painted black*

tances the environmental temperature can be obtained. The temperature extremes of microhabitats (microclimates) are also useful in ecological studies since they can often explain the disappearance of a particular species from an area, such as frost-sensitive plants.

indications of species present and their numbers. Generally speaking, the more accurate the results required the more time-consuming the method, so it is necessary to be clear about objectives. Also, if possible, non-destructive techniques should be used.

In all cases reliable methods of sampling (recording and/or collecting) organisms are required and it is safe to say that 'no stone should be left unturned' (providing it is replaced!) since organisms will occupy almost all available microhabitats. For example, at first sight a square metre of grass, soil, sand, rocky shore or stream bed may not appear to support many species, but closer examination, involving hand-sorting the soil, grass and weeds, turning over stones and examining roots, stems, flowers and fruits of plants and holdfasts of seaweeds, will reveal many more species.

In recording data, as many species as possible (plant and animal) should be identified in the field, using keys if necessary. Only if the species are obviously common locally and not known to be rarities should they be collected. Over-collection can have serious effects on local communities. In the case of collected animals, attempts should be made to keep them alive and to release them in a similar microhabitat to that in which they were collected. It is necessary to identify the organisms as accurately as possible, that is to the level of species. This cannot always be done but it should be possible to identify them at least as far as class, order or family. Identification of specimens depends upon familiarity with keys. The principles of classification, key construction and details of how to use a key are described in appendix 3.

A list of all the species in the habitat gives some indication of the diversity of structure of the community, the **species richness** or **diversity**.

(There are various numerical ways of expressing species richness using mathematical formulae. The numerical value is called the **diversity index** but details of this will not be considered here.)

These data provide information enabling possible food chains and food webs to be constructed, but are inadequate in providing information related to quantitative aspects of the community. The extent of the diversity is only fully revealed when the numbers of organisms within each species, that is the population sizes, are determined. This information enables a more detailed picture of the community to be constructed, such as a pyramid of numbers (section 12.3.6).

Obtaining the qualitative and quantitative data of a habitat depends on specific methods of collecting, sampling and estimating organisms within the habitat, and the method chosen is related to the mode of life, behaviour and size of the organism.

13.2.1   Methods of collecting organisms

There are several points to consider when collecting organisms and these are summarised below.
(1) Observe the Countryside Code at all times.

(2) Always obtain permission from the landowner before beginning an ecological study in an area.
(3) Consult the local Natural History Society, university, college or the Nature Conservancy about where and what you are to collect.
(4) Never remove organisms from their habitat or destroy them unnecessarily.
(5) Leave the habitat as undisturbed as possible, for example replace stones, turf, logs and so on to their original positions.
(6) Where it is necessary to remove organisms from the habitat for identification, take as few as possible and, if practicable, return them to the habitat.
(7) Keep specimens separate when removing them to the laboratory for identification to prevent contamination or being eaten by predators, for example do not put ragworm and crabs in the same collecting vessel. Useful collecting equipment includes jam jars, Kilner jars, polythene bottles, specimen tubes and polythene bags.
(8) Always record as much information as possible concerning the topography of the habitat and climate at the time of collection as the information may have a bearing on what is collected:
  (a) nature of rock or substratum (grass, mud, soil etc.);
  (b) nature of aspect (for example flat, south-facing, angle of slope etc.);
  (c) drainage;
  (d) soil, mud or sand profile;
  (e) temperature of substrate, water and air;

Table 13.1 Field booking sheet for recording edaphic, physiographic and climatic features.

Area .... Grid reference .... Date ....
(1) **Underlying rock**
(2) **Substratum/soil**
  (a) surface feature ....
  (b) depth of horizon A ....
  (c)   ,,   ,,   ,,   B ....
  (d)   ,,   ,,   ,,   C ....
  (e) pH ....
  (f) temperature ....
(3) **Topography**
  (a) aspect, direction .... angle ....
  (b) height above sea level ....
  (c) relief ....
  (d) drainage ....
  (e) land use ....
  (f) high or low water, time .... height ....
(4) **Climate**
  (a) air temperature, range ....
  (b) rainfall ....
  (c) cloud cover/sunlight ....
  (d) relative humidity ....
  (e) wind direction ....
  (f) wind speed ....
  (g) light intensity (horizontal), N ..., S ..., E ..., W ....
  (h) time of day ....

Table 13.2 A summary of various methods used to collect organisms.

Collecting method	Structure and function	Organisms collected
beating tray	A fabric sheet of known area is attached to a collapsible frame and held under a branch which is beaten with a stick or shaken. Organisms fall onto the sheet and are removed using a pooter (see later notes).	non-flying insects, larval stages, spiders
kite net	A muslin net is attached to a handle and swept through the air. Organisms become trapped in the net. All netting techniques must be standardised to ensure uniformity of sampling, e.g. eight, figure-of-eight sweeps per examination of the net.	flying insects
sweep net	A nylon net is attached to a steel handle and swept through grass, bushes, ponds or streams.	insects, crustaceans
plankton net	A bolting silk net is attached to a metal hoop and rope harness and towed through the water. A small jar is attached to the rear of the net to collect specimens.	plankton
sticky trap	Black treacle and sugar are boiled together and smeared onto a sheet of thick polythene which is then attached to a piece of chipboard with drawing pins. This can be hung in various situations and at various heights. Jam and beer can be added to the sticky substances to act as attractants.	flying insects
pitfall trap	A jam jar or tin is buried in the soil with the rim level with ground level. This is best placed where the ground falls away from rim level to prevent water entering the jar. A piece of slate supported on three stones acts as a lid preventing rainwater from entering. The trap can be baited with either sweet foods such as jam or decaying meat. Traps should be regularly cleared (fig 13.4).	walking/crawling insects, myriapods, spiders, crustacea
light trap	A mercury vapour light trap attracts flying organisms which hit baffles and fall down into the base and become trapped in cardboard egg boxes or crumpled-up paper. Cotton wool soaked in chloroform is added before examining the contents to anaesthetise or kill the organisms (fig 13.5).	night-flying insects, particularly moths and caddis flies
mammal trap	A Longworth mammal trap (fig 13.6) is left in a runway and filled with bedding material. Bait, e.g. grain or dried fruit, can be left outside and inside the trap. The trap can be left unset for some time until organisms become accustomed to it and then set. Animals are captured alive so the trap must be visited regularly. Some animals may remain 'trap shy' and never enter it, whereas others become 'trap happy' and visit it regularly. These two patterns can present problems when using the technique to estimate population sizes.	shrews, voles and mice
kick sampling	This is used for collecting in running fresh water. An open sweep or plankton net is held vertically downstream of the area being sampled by turning over stones and scraping off organisms which are then swept into the net. Alternatively the area being sampled is agitated by kicking and stamping so that organisms are displaced vertically and swept into the net by the current.	aquatic insects and crustaceans
pooter	This is used to collect small insects from beating trees or directly off vegetation for closer examination and/or counting (fig 13.10).	aphids, small insects and spiders
hand-sorting	Samples of soil or vegetation, e.g. grass, leaf litter, pond and seaweed, are placed at one end of a tray and small amounts of material are systematically examined between the fingers, specimens are removed to a collecting jar and sorted material passed to the other end of the tray. The sample is then examined as it is moved back to the original end of the tray.	mites, enchytraeid worms, insect larvae and small insects
extractions	5 cm^3 of 4% formaldehyde are added to 50 cm^3 of water and used to water a square metre of lawn or grassland. Earthworms are driven out from their burrows and collected and immediately washed in water to remove the formaldehyde.	earthworms
flotation	Add a known mass of soil to a beaker of saturated salt solution, stir vigorously for several minutes and allow soil to settle. Organisms float to surface in dense salt solution. Pour off surface layer of fluid into a Petri dish and examine under binocular microscope. Remove all specimens into another Petri dish containing 70% alcohol to kill and fix the specimens. Mount each specimen separately in glycerine on a microscope slide, cover with a cover-slip and identify under binocular microscope or low power of compound microscope.	mites, insects, eggs, cocoon, larval and pupal stages
Tullgren funnel (dry extraction)	Many soil and leaf litter-dwelling organisms move away from a source of heat and towards moister conditions. A soil or leaf litter sample is placed in the sieve about 25 cm below a 100 W bulb in a metal reflector (fig 13.7). Every two hours the bulb is moved 5 cm nearer to the sample until the bulb is 5 cm from the soil sample. The apparatus is left for a total of 24 h. All small arthropods move downwards and drop through the metal gauze into the alcohol beneath.	small arthropods e.g. millipedes, centipedes, mites, springtails and collembola
Baermann funnel (wet extraction)	A soil sample is placed in a muslin bag, submerged in a funnel containing water and suspended 25 cm from a 100 W in a metal reflector (fig 13.8). The apparatus is left for 24 h. The water and the gentle heating encourage organisms to leave the sample, move out into the water and sink to the base of the funnel. They are removed at intervals by opening the clip in the apparatus and allowing them to fall into the alcohol.	small arthropods, enchytraeid worms and nematodes

(f) substrate or water pH;

(g) cloud cover and rainfall;

(h) relative humidity of air;

(i) light intensity (such as shaded or open, possibly a meter reading);

(j) wind speed and direction (such as still, gentle breeze, gale, south-west);

(k) time of day and date.

An example of how some of these features may be recorded is shown in table 13.1.

There are a variety of methods of collecting specimens. A summary of methods and their applications is shown in table 13.2 and in figs 13.4–9.

Specimens should be collected from traps at regular intervals, identified, counted and, where possible, released. In the case of pitfall traps it should be realised that if natural predators and prey are collected it is probable that the prey will not be present when the trap is emptied. Where this is believed to be happening, 70% alcohol should be placed in the trap to kill the organisms as they fall in. Imagination and ingenuity are required in collecting specimens.

Generally, sites where specimens are collected are not randomly chosen and consequently the results obtained from the collections must be interpreted in the light of biased selection of collecting site. Whilst this may not affect the species of organism collected, so that community structure will be accurately represented, it is likely to give biased indications of numbers present. For example, the use of baits and lures to attract organisms to sticky traps, pitfall traps and mammal traps will influence the results, and conclusions based on quantitative data will reflect this

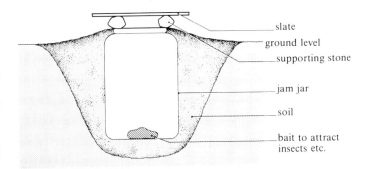

slate
ground level
supporting stone
jam jar
soil
bait to attract insects etc.

Fig 13.4 *Simple pitfall trap made by sinking a jam jar into the soil*

Fig 13.5 (below) *Mercury vapour lamp in use attracting insects*

Fig 13.6 (at bottom of page) *Longworth mammal trap*

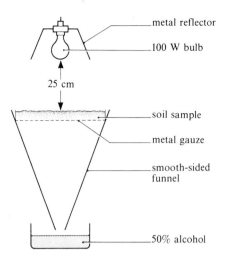

metal reflector
100 W bulb
25 cm
soil sample
metal gauze
smooth-sided funnel
50% alcohol

Fig 13.7 *Tullgren funnel*

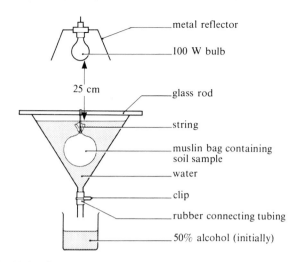

metal reflector
100 W bulb
25 cm
glass rod
string
muslin bag containing soil sample
water
clip
rubber connecting tubing
50% alcohol (initially)

Fig 13.8 *Baermann funnel*

Fig 13.9 *The position of a line transect across a rocky shore*

bias. Therefore in discussion of results it is necessary to state clearly that bias exists.

13.2.2 Methods of sampling an area

In order to standardise the sites where abiotic and biotic aspects of ecosystems are investigated, transects and/or quadrats are commonly used and collecting and sampling is confined to the area of the transect or quadrat.

Line transect. This may be used to sample a uniform area but is particularly useful where it is suspected that there is a transition in habitats and populations through an area (fig. 13.9). For example, a tape or string running along the ground in a straight line between two poles indicates the position of the transect and sampling is rigorously confined to species actually touching the line.

Belt transect. A belt transect is simply a strip of chosen width through the habitat, made by setting up two line transects, say 0.5 m or 1 m apart, between which species are recorded. An easier method of obtaining both qualitative and quantitative data from a belt transect is to use a quadrat frame in conjuction with a line transect.

Height variations recorded along line or belt transects produce a profile of the transect, sometimes known as a **profile transect**, and this is used when presenting data (fig 13.19).

A decision over which type of transect to use depends on the qualitative and quantitative nature of the investigation, the degree of accuracy required, the nature of the organisms present, the size of the area to be investigated and the time available. Over a short distance a line transect might be used and a continuous record kept of each plant species lying immediately beneath it. Alternatively, over a longer distance the species present every metre, or other suitable distance along the transect, may be recorded.

Quadrat. A quadrat frame is a metal or wooden frame, preferably collapsible to facilitate carrying, which forms a square of known area, such as 0.25 m^2 or 1 m^2 (fig 13.11). It is placed to one side of a line transect and sampling carried out. It is then moved along the line transect to different positions. Both the species present within the frame and the numbers or abundance (section 13.2.3) of these may be recorded depending upon the nature of the investigation. In all cases the method of recording the species must be consistent, for example all species partially or completely visible within the quadrat are listed. The structure of the quadrat frame can be modified according to the demands of the investigation. For example, it can be divided by string or wire into convenient sections to assist in counting or estimating numbers or abundance of the species (fig 13.11). This is particularly useful when studying a habitat supporting several species of plants.

A quadrat may be used without a transect when studying an apparently uniform habitat. In this case the quadrat is used randomly. One fairly random sampling technique is to

Fig 13.10 *Pooter in use collecting small non-vertebrates*

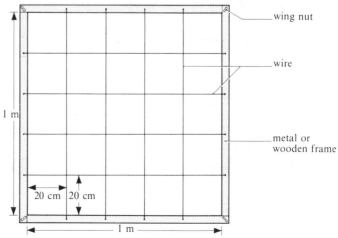

Fig 13.11 *Quadrat frame (1 m²) with wire sub-quadrats (each 400 cm²) forming a graduated quadrat*

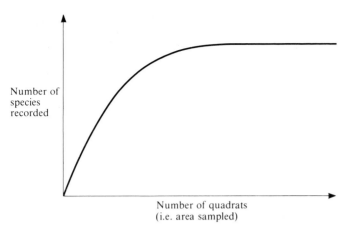

Fig 13.12 *Graph showing the relationship between the number of species recorded in an area and the number of quadrats studied. (In quantitative studies there is no point in sampling more quadrats beyond a certain point as it is unrewarding and uneconomical on time.)*

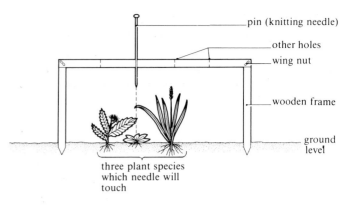

Fig 13.13 *Pin frame or point quadrat*

fling a robust quadrat over the shoulder and record the species within it wherever it falls. This is repeated several times so that a representative sample of the area is covered. Alternatively a sampling point may be chosen by using a table of random numbers, also generated by some calculators, to select a random coordinate on an imaginary grid laid over the area. The sides of the grid may be marked by measuring tapes. Investigations have shown that in a uniform habitat there comes a point beyond which analysing the species within a quadrat becomes unnecessary as it does not increase the number of different species recorded. This relationship is shown in fig 13.12. As a rule of thumb, once five quadrats have failed to show any new species it may be assumed that no further species will be found. However, when an assumption such as this is made it must be stated in the ecological report as it may affect the reliability of the results.

Pin frame (point quadrat). This is a frame bearing a number of holes through which a 'pin', such as a knitting needle can be passed (fig 13.13). It is particularly useful with transect studies of overgrown habitats where several plant species may overlap. All species touched by the pin as it descends to the ground are recorded for each of the holes.

Permanent quadrat. In long-term ecological investigations involving the study of community change (succession) or seasonal changes, a permanent quadrat or transect is used. Metal pegs and nylon rope are used to mark out an area of ground. Periodic samples of abiotic and biotic factors can be taken and the results presented in such a way as to reveal trends and changes and possible factors accounting for, or associated with, these changes.

Computer program. SAMPLE: Two programs on ecological sampling: quadrat and capture–recapture.

13.2.3 Methods of estimating population size

In all studies in quantitative ecology it is essential to be able to estimate, with a degree of accuracy, the number of organisms within a given area of ground or volume of water or air. In most cases this is equivalent to estimating the population size, and methods employed are determined by the size and mode of life of organisms involved and the size of the area under investigation. The numbers of plants and sessile or slow-moving animals in a small area may be counted directly, or their percentage cover or abundance estimated, whereas indirect methods may be required for fast-moving organisms in large open areas. In habitats where organisms are difficult to observe, because of their behaviour and mode of life, it is necessary to estimate numbers of organisms using either the **removal** method or the **capture–recapture** method. Methods of estimating populations may be either objective or subjective.

Objective methods

The use of quadrats, direct observation and photography are known as direct counting methods, whereas the removal and capture–recapture techniques are indirect counting methods.

Quadrat. If the number of organisms within a number of quadrats, representing a known fraction of the total area, are determined, an estimate of the total numbers in the whole area can be obtained by simple multiplication. This method provides a means of calculating three aspects of species distribution.

(1) **Species density.** This is the number of individuals of a given species in a given area, such as 10 m^{-2}. It is obtained by counting the number of organisms in randomly thrown quadrats. The method has the advantages of being accurate, enabling different areas and different species to be compared and providing an absolute measure of abundance. The disadvantages are that it is time-consuming and requires individuals to be defined, for example is a grass tussock counted as one plant or does each plant of the tussock need to be counted?

450

(2) **Species frequency.** This is a measure of the probability (chance) of finding a given species with any one throw of a quadrat in a given area. For example, if the species occurs once in every ten quadrats it has a frequency of 10%. This measure is obtained by recording the presence or absence of the species in a randomly thrown quadrat. (The number present is irrelevant.) In this method the size of the quadrat must be stated since it will influence the results, and also whether the frequency refers to 'shoot' or 'rooted' frequency. (For 'shoot' frequency the species is only recorded as present if foliage overlaps into the quadrat from outside. For 'rooted' frequency the species is only recorded as present if it is actually rooted in the quadrat.) This method has the advantage of being quick and easy and useful in certain large-scale ecosystems such as woodland. The disadvantages are that quadrat size, plant size and spatial distribution (that is random, uniform or clumped) (section A2.8) all affect the species frequency.

13.2 What is the species frequency of the species recorded in 86 quadrats out of 200 thrown?

(3) **Species cover.** This is a measure of the proportion of ground occupied by the species and gives an estimate of the area covered by the species as a percentage of the total area. It is obtained either by observing the species covering the ground at a number of random points, by the subjective estimate of percentage of quadrat coverage or by the use of a pin frame (fig 13.13). This is a useful method for estimating plant species, especially grasses, where individuals are hard to count and are not as important as cover. However it has the disadvantages of being slow and tedious.

13.3 If a pin frame containing ten pins was used ten times and 36 units were recorded for plant X, what is the percentage cover of X?

Direct observation. Direct counting is not only applicable to sessile or slow-moving animals but also to many larger mobile organisms such as deer, wild ponies and lions, and wood pigeons and bats as they leave their roost.

Photography. It is possible to obtain population sizes of larger mammals and sea birds which congregate in open spaces by direct counting from aerial photographs.

Removal method. The removal method is very suitable for estimating numbers of small organisms, particularly insects, within a known area of grassland or volume of water. Using a net in some form of standard sweep, the number of animals captured is recorded and the animals kept. This procedure is repeated a further three

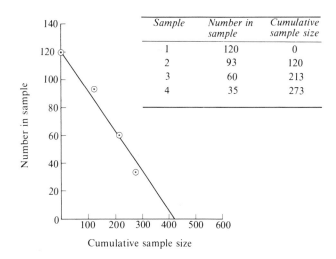

Sample	Number in sample	Cumulative sample size
1	120	0
2	93	120
3	60	213
4	35	273

Fig 13.14 *Graph of number in sample against cumulative sample size. Extrapolation of the line to the point when sample size equals zero gives an estimate of the number in the population*

times and the gradually reducing numbers recorded. A graph is plotted of number of animals captured per sample against the previous cumulative number of animals captured. By extrapolating the line of the graph to the point at which no further animals would be captured (that is number in sample = 0) the total population may be estimated, as shown in fig 13.14.

Capture–recapture method. (See reference to computer program, p. 450.) This method involves capturing the organism, marking it in some way, without causing it any damage, and replacing it so that it can resume a normal role in the population. For example, fish are netted and their operculum tagged with aluminium discs, birds are netted and rings attached to their legs, small mammals may be tagged by dyes, or by clipping the ear or removing a toe, and arthropods are marked with paint. In all cases some form of coding may be adopted so that individual organisms are identified. Having trapped, counted and marked a representative sample of the population the individuals are released in the same area. At a later stage the population is retrapped and counted and the population size estimated using the expression below:

$$\text{Estimated total population} = \frac{\substack{\text{number of organisms} \\ \text{in initial sample}} \times \substack{\text{number of organisms} \\ \text{in second sample}}}{\text{number of marked organisms recaptured}}$$

This estimate of population size is called the **Lincoln index**. It relies on a number of assumptions which are summarised below.

(1) Organisms mix randomly within the population. (This does not always apply since some organisms live in colonies, troops or shoals.)

(2) Sufficient time must elapse between capture and

451

recapture to allow random mixing. The less mobile the species the longer the time lapse must be.

(3) It is only applicable to populations whose movement is restricted geographically.
(4) Organisms disperse evenly within the geographical area of the population.
(5) Changes in population size as a result of immigration, emigration, births and deaths are negligible.
(6) Marking does not hinder the movement of the organisms or make them conspicuous to predators.

Where plants and small animals, such as barnacles, are concerned, direct counting becomes very tedious and, depending upon the degree of accuracy required from the study, may be replaced by estimating percentage cover or abundance within a quadrat frame. In the early stages of estimation it is advisable to use a graduated quadrat frame (fig 13.11) to increase the accuracy of estimation. Various schemes may be adopted for representing percentage cover or abundance, some being totally subjective, others partially, or completely, objective.

> **13.4** In an attempt to estimate the number of trout in a small lake, 625 trout were netted, marked and released. One week later 873 trout were netted and of these 129 had been marked. What was the estimated size of the population?

Subjective methods

These involve some form of frequency assessment, frequency scale or estimate of abundance in terms of cover. For example, an arbitrary scale devised by Crisp and Southward for limpets on a rocky shore uses the following letters, frequencies and percentages.

A abundant > 50%
C common 10–50%
F frequent 1–10%
O occasional < 1%
R rare   present – only a few found in 30 min searching

These assessments and scales are arbitrary and the frequencies can be adjusted to varying percentage values, for example in a particular study, abundant may represent > 90%. The value of using the five categories above is that they can be applied to methods of presenting data, such as in constructing kite diagrams, as described in section A2.7.3. The major disadvantage of this method is that it is subjective and tends to rate small species with poor cover lower than conspicuous species, flowering species and species occurring in clumps.

13.3 Ecological research projects and investigations

Ecological projects are broadly concerned with studying either the organisms in an area (**synecology**)

or a single species (**autecology**). In both cases it is necessary to spend time reading about and discussing the project so as to clarify the aims, nature and extent of the project. All investigations should include problems which have to be solved or hypotheses to be tested.

The aims of the project should be stated clearly and should include both general and specific aims. For example:

(1) to develop and encourage an attitude of curiosity and enquiry;
(2) to develop the ability to plan an investigation, construct hypotheses and design experiments;
(3) to develop the ability to formulate questions and collect relevant qualitative and quantitative data to answer these;
(4) to develop practical and observational skills including the use of apparatus and biological keys;
(5) to develop the ability to record data accurately;
(6) to develop the ability to apply existing knowledge to the interpretation of data;
(7) to develop a critical attitude to data, assessment of their validity and conclusions based on them;
(8) to develop the ability to communicate biological information by means of tables, graphs and the spoken and written word;
(9) to develop an appreciation of organisms and the importance of conservation;
(10) to develop an understanding of the interrelationships between organisms, between organisms and their environment, and the dynamic aspects of ecology. This can be extended to further aims which are specific to the study as described in section 13.5.

13.3.1 Writing up the project or investigations

Irrespective of the quality of both the investigation and the data obtained, the project or investigation is of little use to other scientists until it is presented as a report and this should take the following form.

(1) **Introduction**: including the idea, the problems, hypotheses and aims (that is what you set out to do and why).
(2) **Method**: the strategy of the project (that is what you did (was done), where and how it was done including all practical details of apparatus and techniques employed both in the field and in the laboratory).
(3) **Results and observations**: tabulated data, graphs, histograms, profiles, presence–absence graphs, kite diagrams and any other relevant and realistic way of representing data and relationships clearly and concisely.
(4) **Discussion of results**: this involves an analysis of the results, preferably quantitative if possible, tentative conclusions based on data presented and references to already published material.

(5) Discussion of significance of conclusions: criticisms of the techniques employed, sources of error and suggestions for further study.

(6) List of references consulted.

13.4 A synecological investigation

A synecological investigation involves studying the abiotic and biotic elements associated with a natural community (biotic element of ecosystem) found in a particular defined geographical area (or ecosystem) such as an oak woodland or a rocky shore, which may contain several plant and animal species and possibly several habitats. In such an investigation it is necessary to carry out the following exercises:

(1) map the area and habitat(s) in plan view and, if necessary, in profile;

(2) identify the species and estimate the number of each species present;

(3) measure (possibly collect and analyse) the abiotic factors within the habitat(s).

The overall aim of such an investigation is to determine the qualitative and quantitative relationships between the plant and animal populations within the area being studied and the possible interactions between these and edaphic, topographic and climatic factors. Given this information, it is possible to explain the nature and extent of the factors governing the number and distribution of organisms in terms of a food web and, depending upon the sophistication of the investigation, pyramids of numbers, standing-crop biomass and energy.

13.4.1 Mapping an area

Plan view

The following simple method is designed primarily for mapping a small area, such as a grassland 10 m × 10 m or a small pond, but can be used on a larger scale, for example to map the whole rocky shore of a bay.

(1) Select the approximate area to study and stretch a measuring tape along one side of the area. This marks the base-line AY (fig 13.15).

(2) From the base-line measure the perpendicular distance to certain natural landmarks within the area or marker poles showing the limit of the study area. Record these measurements.

(3) Transfer the measurement of AY and the various perpendicular distances to a sheet of squared paper using a suitable scale.

(4) Using the base-line and measured distances to perpendicular landmarks drawn in (3) above as a guide, complete the map freehand.

(5) If the area is relatively small divide the actual base-line AY into an equal number of sections and from these lay out perpendicular string line transects. Repeat the

procedure using the extreme left transect AF as a new base-line to produce a string-grid, as shown in fig 13.16. Draw these grid lines on the map and label them using A, B, C etc. along one edge and 1, 2, 3 etc. along the other edge.

(6) Mark the positions of obvious structural and vegetational zones.

(7) Using a quadrat frame, pin frame or sweep net, depending on the area, systematically sample the area from, say, left to right and record the species present and their numbers or abundance.

(8) If the area is extremely large and a qualitative and quantitative study is required, belt transects spaced out at set intervals across the area, and set at right-angles to any suspected zonation, can be used in conjunction with quadrats to sample the area at particular points called **stations**. Direct measurements of the abiotic features of the environment should be made as frequently as possible or samples removed for subsequent analysis.

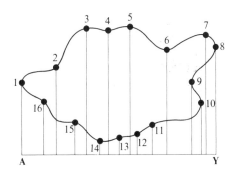

Fig 13.15 *A suggested method of mapping the significant aspects of an area, such as a small, irregularly shaped pond*

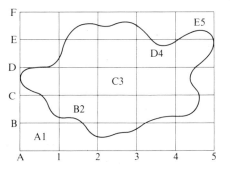

Fig 13.16 *Map of the area under investigation showing the various subsections, for example A1-E5, obtained by the use of a string grid. These provide reference areas for subsequent study*

Plotting a height profile

In some areas, distribution of organisms may be influenced by a factor related to height, such as on a rocky shore. Here the length of time each part of the shore is exposed due to the vertical motion of tides is height dependent. In such cases it is necessary to produce a height profile showing how the height along the transect varies, as from high to low water marks in the rocky shore example used below. At each point (station) along the transect where the community is sampled the height should be obtained accurately by the use of a surveying theodolite and measuring points. Over short distances a simple home-made levelling device attached to a reference pole, and a graduated pole can achieve relatively accurate results as described below (fig 13.17).

(1) Attach the levelling device at a convenient height (h_1), such as 1.5 m, on the measuring pole.
(2) Set out a line transect from high water mark to the water's edge.
(3) Set up the reference pole at a specific point, such as high water mark, on the transect and the marker pole at a known distance (x) further down the shore. Mark these positions on the transect and label them A and B. Keep to one side of the transect line whilst taking readings to avoid trampling on the specimens to be studied.
(4) When the wooden sighting bar is horizontal, look along the sighting tube with a point on the marker pole. The exact position of this point is then located by the person holding the marker pole and the height (h_2) recorded. The height difference between the stations is equal to $h_2 - h_1$.

Table 13.3 Horizontal and vertical distances recorded at stations A–K on a rocky shore. Northumberland 1968.

Station	Horizontal distance/m (x, x_1 etc.)	Height between stations/m (($h_2 - h_1$) etc.)	Height above low water/m (($h_2 - h_1$) etc.)
A	0		9.6
B	20	1.5	8.1
C	40	1.7	6.4
D	60	1.8	4.6
E	80	0.8	3.8
F	100	0.6	3.2
G	120	0.7	2.5
H	140	0.9	1.6
I	160	0.8	0.8
J	180	0.4	0.4
K	200	0.4	0

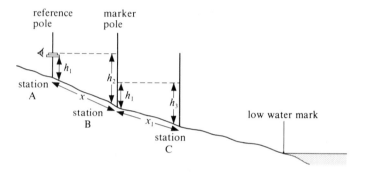

Fig 13.18 *Methods of obtaining heights and horizontal distances of stations above low water mark*

(5) Move the reference pole to station B and the marker pole a known distance (x_1) to station C. Repeat stages (3) and (4) and record the new height (h_3) (fig 13.18).
(6) Continue to obtain readings h_4, h_5 and so on, distances x_3, x_4 and so on and stations D, E and so on to the water's edge at low water. Record all distances as shown in table 13.3, and calculate the heights and horizontal distances of the stations above the low water mark.
(7) Transfer these data to a scale representation of the shore profile and mark on the positions of the stations (fig 13.19).

13.4.2 Identifying and estimating the number of each species present

Line and belt transects, frame quadrats and pin frames are used to sample systematically the area as described in section 13.2.2. Specimens are identified using a key and the number of organisms are either counted directly or estimated as described in section 13.2.3.

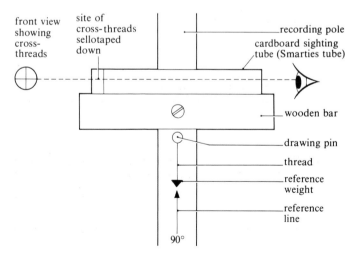

Fig 13.17 *A simple home-made levelling device attached to a reference pole. The position of the pole is adjusted until the sighting bar is shown to be horizontal by the thread indicating 90° on the reference point on the pole. Holding the pole steady, the observer looks along the sighting tube and indicates to the person holding the graduated pole the corresponding level position on his pole, as shown by the cross-wire sites. This height is recorded*

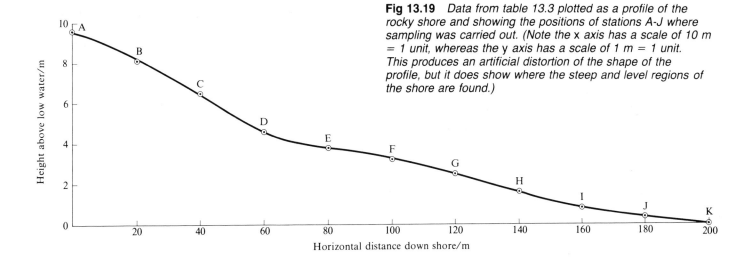

Fig 13.19 *Data from table 13.3 plotted as a profile of the rocky shore and showing the positions of stations A-J where sampling was carried out. (Note the x axis has a scale of 10 m = 1 unit, whereas the y axis has a scale of 1 m = 1 unit. This produces an artificial distortion of the shape of the profile, but it does show where the steep and level regions of the shore are found.)*

13.4.3   Recording and representing data

Data should be recorded directly they are obtained using some form of field booking sheet. In the case of synecological investigations of marine habitats the information shown on the booking sheets illustrated in tables 13.4 and 13.5 has proved successful. These sheets are best attached to a clip-board, completed in pencil and kept in a large polythene bag to protect them from rain. Once all the data have been collected they must be represented in some suitably efficient diagrammatic form that will highlight relationships between organisms and/or the nature of the environment. Methods of representing data are given in section A2.7 and include presence–absence graphs, kite diagrams, and histograms. Trophic pyramids are described in section 12.3.6. Some examples of the use of all four methods of representation are included in figs 13.20–13.22.

Table 13.4 Suggested format of field booking sheet.

Field Booking Sheet – Marine Ecology
(1) Name of site and grid reference ....
(2) Nature of profile (rocky, sandy, muddy, dune) ....
(3) Sketch map of area showing area(s) of study/position(s) of transects.

(4) Special features (exposure, aspect, etc.) ....
(5) Date ....
(6) Weather, conditions ....
(7) Tide data:
    predicted high water ....
    ,,   low water ....
    observed high water ....
    ,,   low water ....
    predicted tidal range ....
(8) Notes and key to recorded data (abundance scales'– % cover, reference height of level, e.g. h_1 etc.)

Table 13.5 Data required on a booking sheet for investigating the synecology of marine habitats.

Station name	Horizontal distance from origin/m	Level reading (h_2)/m etc.	Change in height ($h_2 - h_1$ etc.)/m	Height above low water/m	Time exposed	Time covered	ANIMALS (species and abundance)	PLANTS (species and abundance)	NOTES

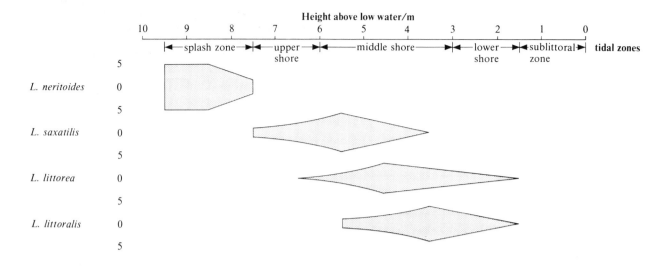

Fig 13.20 *Kite-diagrams showing the frequency and distribution of four common species of periwinkle,* Littorina, *on a rocky shore on the Dale peninsula, Pembrokeshire, April 1976. See table 13.6 for data*

13.4.4 Collecting and analysing abiotic factors

The amount of time spent on this stage of a synecological investigation will depend upon the nature of the area being studied. It is more applicable to areas where edaphic factors predominate, such as woodland, grassland and salt marsh, than to a rocky shore. The abiotic factors to be studied and the methods of study are described in section 13.1.

13.5 An autecological investigation

An autecological investigation involves studying all the ecological factors related to a single plant or animal species throughout its life cycle. The aim of the investigation is to describe as precisely as possible the ecological niche of the species. The species selected for study should be one which is both common and locally available. Initially the investigation should concern itself with undertaking extensive background reading on the species selected. During the reading, notes should be made on all aspects of the biology of the species and also on opportunities for practical work. This may involve either repeating investigations carried out by others or developing new investigations to be undertaken as part of the current study.

A straightforward approach to an autecological study is to prepare a comprehensive list of the questions which must be answered in order to reveal all there is to know about the species under investigation. The study should be undertaken as rigorously as possible and treated as a research project. Therefore it must involve some measure of original investigation including observation, measurement and experimentation. It must not simply be a report based on knowledge gleaned from reading books, journals and magazines. The species under investigation should be studied over a period of at least one full year.

A guide to the sorts of questions to be asked in the investigations of an **animal** is given below.

(1) **Classification.** What is the name of the species? What other groups of organisms does it resemble most closely? What are the similarities and differences between related species? What is its full taxonomic description?

(2) **Habitat.** Where is it found? What are the characteristic abiotic features of the area? How do these factors change over the course of a year?

(3) **Structure.** What is its adult structure? What are its characteristic external features? What are its dimensions and mass?

(4) **Movement.** How does it move from place to place? Which parts of the organism are involved in the movement and what are the functions of these parts? How are these parts adapted to the environment?

(5) **Nutrition.** What are the food sources of the organism? When does the organism feed? How much food is eaten? How is the food captured and ingested? What special features assist ingestion? Are there any unusual features of digestion and absorption?

(6) **Respiration.** Where is the gaseous exchange surface? How does gaseous exchange occur? How much oxygen is required by the organism?

(7) **Excretion.** What are the waste products of metabolism? How are these removed from the organism? What special organs of excretion are present?

(8) **Reproduction.** Are the sexes separate? What visible external differences are there between the sexes? Does any form of courtship occur? Does the organism defend a territory? How does mating occur? When does mating occur? How often does mating occur? How many gametes are produced? Where does fertilisation occur?

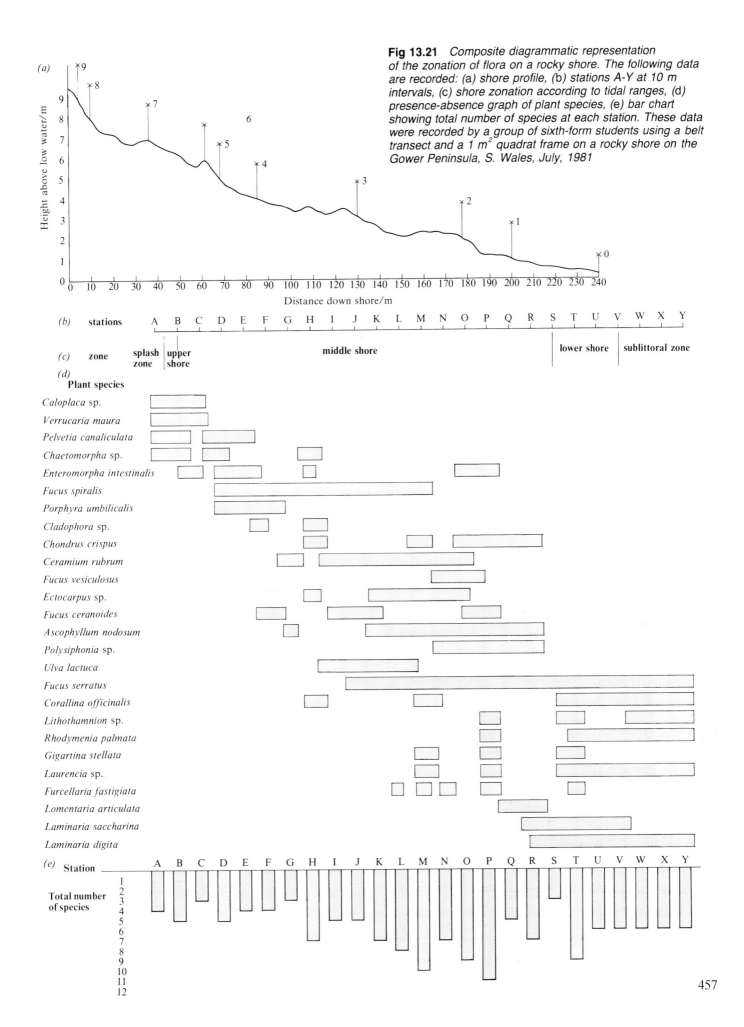

Fig 13.21 *Composite diagrammatic representation of the zonation of flora on a rocky shore. The following data are recorded: (a) shore profile, (b) stations A-Y at 10 m intervals, (c) shore zonation according to tidal ranges, (d) presence-absence graph of plant species, (e) bar chart showing total number of species at each station. These data were recorded by a group of sixth-form students using a belt transect and a 1 m² quadrat frame on a rocky shore on the Gower Peninsula, S. Wales, July, 1981*

457

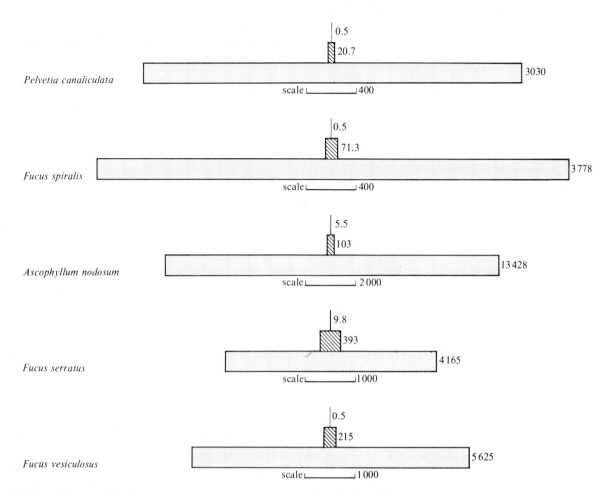

Fig 13.22 *Trophic pyramids of standing crop biomass for five rocky shore communities based on particular algal primary producers. The numerical values represent estimates of standing crop biomass in g m⁻². The stippled areas represent primary producers, the cross-hatched areas represent primary consumers (herbivores and detritus feeders) and the solid areas represent secondary consumers (carnivores). It must not be assumed however, that each trophic level is supported entirely by the level beneath. Northumberland coast, March 1969. (After D. A. S. Smith (1970) School Science Review, ASE).*

Table 13.6 The distribution of four common species of the periwinkle, *Littorina* on a rocky shore on the Dale peninsula, Pembrokeshire. April 1976. These data are represented graphically by the kite-diagrams shown in fig 13.20.

Height above low water/m	L. neritoides		L. saxatalis		L. littorea		L. littoralis	
	number	scale	number	scale	number	scale	number	scale
9–10	63	10	–	–	–	–	–	–
8–9	54	10	–	–	–	–	–	–
7–8	7	4	3	2	–	–	–	–
6–7	–	–	8	4	1	1	–	–
5–6	–	–	17	8	3	2	2	2
4–5	–	–	6	4	13	6	9	4
3–4	–	–	1	1	6	4	16	8
2–3	–	–	–	–	2	2	5	4
1–2	–	–	–	–	1	1	1	1
0–1	–	–	–	–	–	–	–	–

Abundance scale: ≥20 = 10; 19−15 = 8; 14−10 = 6; 9−5 = 4; 4−2 = 2; 1 = 1.

(9) **Life cycle.** How long does development take? What degree of parental care is shown? Are there larval stages? When do adults become sexually mature? What is the typical life span of an individual of the species?

(10) **Behaviour.** How does the organism receive stimuli? To which stimuli does the organism mainly respond? How are the major sense organs adapted to the mode of life of the organism? To what extent does learning occur? How does the organism react to other members of the same species? How does the species react to unfavourable weather conditions? How does the organism communicate?

(11) **Ecology.** How many organisms occur in the population? What other organisms live in the same habitat? How are the various species distributed within the habitat? How is the species related to other species in the same habitat in terms of position in food chains and food webs? Is the organism a host, parasite or symbiont? What is the ecological niche of the species?

Similarly the sorts of questions to be asked in the investigation of a **flowering plant** are given below.

(1) **Classification.** What is the name of the species? What subspecies, varieties and ecotypes of the species exist? What are the similarities and differences between closely related species? What is its full taxonomic description?

(2) **Habitat**

(a) *Edaphic factors* – What is the parent rock type? What type of soil profile is shown? How thick are the various horizons? What is the percentage water content (field capacity) of the soil? What is the percentage organic content of the soil? What is the mineral composition of the soil? What is the pH of the soil? What is the height and seasonal variation of the water table in relation to the life history and distribution of the species?

(b) *Climatic factors* – What are the extremes and mean temperatures in the habitats? What is the annual rainfall in the habitats? What is the mean relative humidity of the air in the habitats? What is the direction of the prevailing wind? How much light is received by the plant?

(c) *Topographical factors* – To which direction is the species normally exposed? Does the species appear to prefer exposed or sheltered sites? Does the species appear to prefer sloping or flat habitats? Does altitude appear to affect the distribution of the species?

(3) **Structure.** How extensive is the root system? What form does the root system take? How does the stem branch? How many leaves are carried on each branch? What shapes are the leaves? What variations in length and breadth exist between the leaves? How tall does the plant grow?

(4) **Physiology.** What pigments are present in the leaves and petals? Which surface of the leaf has the highest transpiration rate? What effect has darkness on transpiration rate? Do diurnal changes in water content of leaves occur?

(5) **Reproduction.**

(a) *Flower* – How many flowers are produced per plant? How many and of what shape and size are the sepals, petals, anthers, carpels or pistil? What variation in petal colour exists? What pigments are present in the leaves? When does flowering begin? How long is the flowering period? How does pollination occur? What adaptations to insect or wind pollination are shown?

(b) *Fruit and seeds* – How are the fruits formed? What is the structure of the fruit? How many seeds are produced per flower? How are fruits and seeds dispersed? How far are fruits and seeds dispersed?

(c) *Perennation* – How does vegetative propagation occur? What are the organs of perennation? At what rate does the species colonise an area?

(6) **Life cycle.** What type of seed is produced? What conditions are required for germination? When do the seeds germinate? What percentage of seeds germinate? Which form of germination occurs? At what rate does the shoot system develop? What is the extent of growth in terms of space and time? (Why do some of the seedlings not become mature?)

(7) **Ecology.** Does the species grow as solitary plants or in patches? What size are the patches? Which species grow in the same habitat? What degree of competition exists between the species being studied and other species? Is the species a parasite, host or symbiont? How is the species related to animals in terms of position in the food web? Does the species offer protection or shelter to animals? If so, which animals and how is this provided? What is the ecological niche of the species?

Fungi, algae, mosses, liverworts or conifers may be used in autecological studies and the questions above may be modified as appropriate to the species under investigation.

Answers and discussion

Chapter 2

2.1

Time (in units of 20 min)	0	1	2	3	4	5	6	7	8	9	10
A Number of bacteria	1	2	4	8	16	32	64	128	256	512	1024
B Log$_{10}$ number of bacteria	0.0	0.3	0.6	0.9	1.2	1.5	1.8	2.1	2.4	2.7	3.0
C Number of bacteria expressed as power of 2	2^0	2^1	2^2	2^3	2^4	2^5	2^6	2^7	2^8	2^9	2^{10}

Graph A (an arithmetic plot) increases in steepness as time progresses. Graph B (a logarithmic plot) is a straight line (increases linearly with time). See fig 2.1(ans).

2.2 The graph would be as fig 2.7.

2.3 See fig 2.3(ans).

Factors responsible for the changes are discussed in section 2.2.4. The difference in the growth curve of living bacteria compared with living and dead bacteria is due to the following:

(a) a few cells die during lag and log phases;

(b) during the stationary phase the combined total of living plus dead cells continues to increase slowly for some time since some cells are still reproducing.

(c) during the phase of decline the combined total of living plus dead cells remains constant, though many are dying.

2.4 Generation time is the time taken for numbers to double during the log phase. This is about 2.5 h.

2.5 (i)(*a*) louse (*b*) human

(ii)(*a*) mosquito or tick (*b*) human (also monkeys)

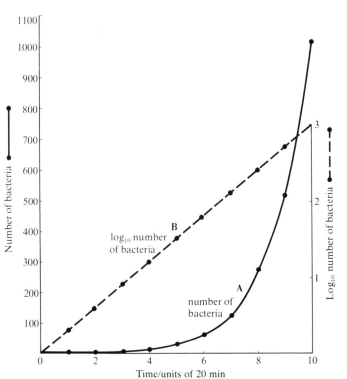

Fig 2.1(ans) *Growth of a model population of bacteria as plotted on arithmetic and logarithmic scales*

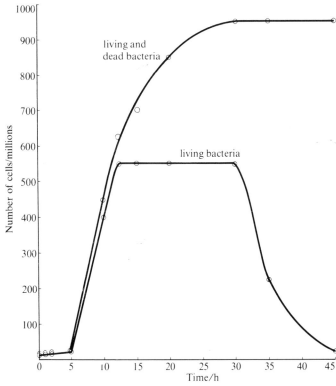

Fig 2.3(ans) *Growth of a bacterial population*

461

2.6 Consult tables 2.6 and 2.7, chapter 2.

2.7 The sporangiophores bear the sporangia above the main mycelium so that the spores are more likely to catch air currents and be dispersed.

Chapter 3

3.1 There is greater nutrient availability in lowland reservoirs because the rivers flowing into them have had longer to accumulate nutrients, particularly from cultivated areas where fertilisers may be used.

3.2 Algae have autotrophic nutrition. Pathogens are parasites or saprotrophs, obtaining food from their hosts.

3.3 Amphibians, like liverworts and mosses, are only partially adapted to life on land, having bodies which easily lose water, and they still rely on water for sexual reproduction. Both groups of organisms are also believed to represent intermediate stages in the evolution towards more advanced forms which are better adapted to life on land.

3.4 The sporophyte has become adapted for life on land although the gametophyte is still dependent on water for swimming gametes. The sporophyte generation has true vascular tissue and true roots, stem and leaves with which to exploit the land environment more successfully.

The sporophyte is the dominant generation, the life of the gametophyte being short.

The mature sporophyte is no longer dependent on the gametophyte.

3.5 (*a*), (*c*) and (*d*)

3.6 Sexual reproduction is dependent on water since it involves free-swimming sperm.

The gametophyte thallus is susceptible to desiccation.

The plants are often relatively intolerant of high light intensities.

3.7 By asexual reproduction, either vegetative (see text) or by dissemination of spores. (Sexual reproduction does not result in spread because the zygote grows from the previous gametophyte generation – contrast seed-bearing plants.)

3.8 (*a*) The zygote is located in the venter of an archegonium of the previous gametophyte generation. Thus it is protected by the archegonium and the surrounding tissues of the gametophyte.

(*b*) The gametophyte generation from which the zygote develops is photosynthetic and provides the zygote with food.

3.9 The *Dryopteris* spore can develop wherever it falls, providing conditions are moist and fertile. Pollen grains must reach the female parts of the sporophyte.

3.10 The megaspore is large because it must contain sufficient food reserves to support the female gametophyte and subsequent development of the embryo sporophyte until the latter becomes self-supporting. Microspores, by being small, can be produced economically in large numbers and are light enough to be carried by air currents, thus increasing the chances of the male gametes that they contain reaching the female parts of the plants.

Chapter 5

5.1 The molecular formula shows the number of each type of atom. The structural formula shows the arrangement of the atoms relative to each other. Note that angles of bonds can also be shown; see figs 5.3 and 5.4, for example.

5.2 (*a*) C_8H_{18}, octane

(*b*) C_6H_6, benzene

5.3

5.4 triose $C_3H_6O_3$ hexose $C_6H_{12}O_6$
tetrose $C_4H_8O_4$ heptose $C_7H_{14}O_7$
pentose $C_5H_{10}O_5$

5.5 (*a*) Valency of C = 4, O = 2, H = 1.

(*b*) Molecular formula is $C_3H_6O_3$ in both cases; the compounds are therefore trioses.

(*c*) Each contains two hydroxyl groups. This could have been predicted, since it has already been explained that in monosaccharides all the carbon atoms except one have a hydroxyl group attached.

(*d*) Glyceraldehyde contains a secondary alcohol group,

$$>CHOH$$

Dihydroxyacetone contains a primary alcohol group —CH_2OH at both ends of the molecule.

(Hydroxyl and carbonyl groups are parts of these larger groups.)

5.6 Pentoses: ribose, ribulose
Hexoses: glucose, mannose, galactose, fructose

5.7 (*a*) See fig 5.7(ans).

(*b*) This is optical isomerism.

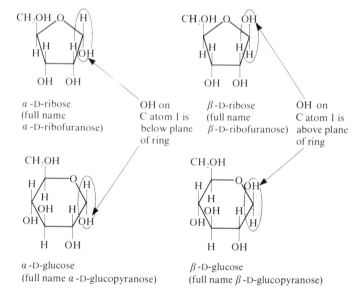

α-D-ribose (full name α-D-ribofuranose)   OH on C atom 1 is below plane of ring

β-D-ribose (full name β-D-ribofuranose)   OH on C atom 1 is above plane of ring

α-D-glucose (full name α-D-glucopyranose)

β-D-glucose (full name β-D-glucopyranose)

Fig 5.7(ans) *The α and β isomers of D-glucose and D-ribose*

5.8 α-glucose

5.9 Cellulose is made of β-glucose residues. The —OH group on C atom 1 is at 180° to the —OH group on C atom 4. Since these groups must come alongside each other in adjacent residues to form a glycosidic link, adjacent residues have to be at 180° to each other.

5.10 Relevant information is provided in sections 5.2.3 and 5.2.4.

5.11 The principal sources of variation are as follows.

(*a*) Use of both pentoses and hexoses. Although normally

only one monosaccharide is used, sometimes two are used in alternating sequence (such as in murein) and in a few complex polysaccharides more may be used.

(b) Two types of linkage, 1,4 and 1,6, are common between residues. Thus branching can occur. (Other linkages, such as 1,2 and 1,3, are possible because every hydroxyl group in a monosaccharide can participate in a condensation.)

(c) Lengths of chains and branches, and extent of branching can vary enormously.

(d) Different optical isomers exist. Most naturally occurring isomers are D- rather than L-isomers so little extra variation comes from this. However α- and β-forms are important. (Compare starch and cellulose.)

(e) Sugars may be ketoses or aldoses. Ketohexoses form five-membered, and aldohexoses six-membered rings.

(f) The high chemical reactivity of sugars (aldehyde, ketone and hydroxyl groups) means that they combine readily with other substances to form related compounds, such as amino sugars and acid sugars. These can participate in building polysaccharides.

5.12 One that occurs when two compounds are joined by the elimination of a water molecule.

5.13 Body temperatures of poikilothermic animals become lower in cold environments. Lipids rich in unsaturated fatty acids (which have low melting points) generally remain liquid at temperatures lower (usually 5 °C or lower) than those rich in saturated fatty acids. This may be necessary if the lipid is to maintain its function, such as a constituent of membranes.

5.14 Triolein because it contains three *unsaturated* oleic acid molecules. Tristearin is a fat, triolein an oil.

5.15 (a) Cell respiration (internal or tissue respiration). Fat undergoes oxidation.

(b) Only the hydrogen part of carbohydrate and fat molecules yields water on oxidation ($2H_2 + O_2 \rightarrow 2H_2O$) and fats contain relatively more hydrogen than carbohydrates on a weight basis (nearly twice as much).

5.16

*peptide bond.

5.17 (a) AAA   AAB   ABA   ABB
        BAA   BAB   BBA   BBB

(b) $2^3 = 8$

(c) $2^{100} = 1.27 \times 10^{30}$

(d) $20^{100} = 1.27 \times 10^{130}$ This is much larger than the number of atoms in the Universe (estimated at about 10^{100})! Thus, there is an effectively infinite potential for variation among protein structures.

(e) 20^n where n is the number of amino acids in the molecule.

5.18 The outstanding feature is that the ratio of adenine to thymine is always about 1.0, and so is the ratio of guanine to cytosine. In other words, the number of adenine molecules equals the number of thymine molecules and guanine = cytosine. Note also that the number of purine residues (adenine + guanine) therefore equals the number of pyrimidine residues (thymine + cytosine). Also revealed is the fact that the DNAs of different organisms have different base compositions, in other words the ratio of A:G or T:C is variable.

5.19 Adenine must pair with thymine and guanine with cytosine to account for the observed base ratios.

5.20 Compare the volume of the unknown sample needed to reduce the dye with the volume of 0.1% ascorbic acid solution needed in the standard described.

Percentage ascorbic acid in unknown sample =

$$\frac{\text{volume 0.1\% ascorbic acid used in standard}}{\text{volume of unknown sample used}} \times \frac{0.1}{100}$$

5.21 (a) Carry out Benedict's test on all three solutions. The sucrose solution would not give a brick-red precipitate on boiling. The glucose and glucose/sucrose solutions could be distinguished by pre-treating both as for hydrolysis (see non-reducing sugar test) and repeating Benedict's test. The glucose/sucrose mix will now show a greater amount of reducing sugar. (In practice, different dilutions of the solutions may have to be tried for convincing results. 0.05% glucose solution, 0.5% sucrose solution and a mixture of equal volumes of 0.1% glucose solution and 1.0% sucrose solution are suitable.)

(i) Paper chromatography or thin-layer chromatography.

(ii) Effect on plane-polarised light using a polarimeter (both sucrose and glucose are dextro-rotatory, but sucrose produces a greater degree of rotation than glucose).

(iii) Sucrose is converted to reducing sugars (glucose + fructose) by the enzyme sucrase (invertase). The reaction may either be followed using a polarimeter or by Benedict's test.

5.22 Dissolve 10 g glucose in distilled water and make up to 100 cm³. (Do not add 10 g glucose to 100 cm³ distilled water because the final volume would be greater than 100 cm³.)

5.23 Add 10 cm³ of 10% glucose to 50 cm³ of 2% sucrose solution and make up to 100 cm³ with distilled water.

Chapter 6

6.1 Excess heat irreversibly changes the secondary and tertiary structure of the enzyme. This means that its specific shape will be changed and that the amino acids normally located close to each other in the functional active site will no longer be in such a position. Therefore the active site will be non-functional. Normally enzyme conformation is easily disrupted by high temperatures.

6.2 (a) Initially the reactions A and B are fast and a lot of product is formed. Later, product formation levels off and there is no further increase. This may be because (i) all substrate has been converted to product, (ii) the enzyme has become inactivated, or (iii) the equilibrium point of a reversible reaction has been reached, and substrate and product are present in balanced concentrations.

(b) When the temperature is raised, (i) initial reaction rate is increased, and (ii) the enzyme becomes less stable and is inactivated more rapidly.

(c) Sensitivity to heat is an indication of the protein nature of the enzymes.

(d) At lower temperatures (as in curve C) rate of formation of product remains constant over 1 h.

6.3 (a) 5.50

(b) (i) pepsin, (ii) salivary amylase

(c) The active site of the enzyme is being destroyed. The ionisable groups of the enzyme, especially those of the active site, are being modified. Hence the substrate no longer fits easily into the active site and catalytic activity is diminished.

(d) A change in pH results in a change in the activity of most enzymes. Each enzyme would have its rate of reaction modified to a different extent as each possesses its own particular pH activity curve. All cells rely on a delicate

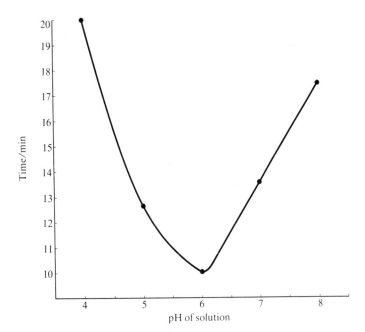

Fig 6.3(ans) *Activity of catalase on hydrogen peroxide at varying pH*

balance between their enzyme systems, and so any changes in enzyme activity could cause the death of the cell or multicellular organism.

(*e*) See fig 6.3(ans).

Optimum pH for enzyme activity is 6.00.

From pH 4–6, ionisable groups of the active site are modified such that the active site becomes more efficient at receiving and complexing with its substrate. The reverse is true when the pH changes from 6–8.

6.4 Increasing the substrate concentration increases the probability of substrate molecules fitting into the active sites rather than inhibitor molecules.

6.5 Increased substrate concentration has no effect on the overall rate as there is no competition for the active site. Inhibition is therefore irreversible.

6.6 (*a*) The two sites are located on different parts of the enzyme, an active site for complexing with A, and another site specific for binding with X.

(*b*) (i) X could inhibit e_1 and therefore only permit production of S along the A–S pathway. This situation could remain until the surplus of X had been used up.

(ii) X could accelerate the catalytic activity of e_5, again enhancing the production of S at the expense of X.

(*c*) Feedback inhibition.

(*d*) Because of enzyme specificity, each step requires specific enzymes and this allows for fine control of the pathway.

6.7 (1) All are proteins and synthesised within living organisms.

(2) They catalyse chemical reactions by lowering the activation energy required to start the reaction.

(3) Only small amounts of enzyme are needed to catalyse reactions.

(4) At the end of a reaction the enzymes are unchanged.

(5) Each enzyme is specific and possesses an active site where enzyme and substrate combine temporarily to form an enzyme/substrate complex before products are released.

(6) Enzymes work best at an optimum pH and optimum temperature.

(7) Being proteins enzymes are denatured by extremes of pH and temperature.

(8) The Q_{10} of enzymes over a temperature range of 0–40 °C is approximately 2.

(9) Some enzymes work in conjunction with cofactors.

(10) Certain chemical reagents inhibit enzyme activity, as can end-products of metabolic pathways.

Chapter 7

7.1 Endoplasmic reticulum, ribosomes, microtubules, microfilaments, microvilli (visible as a 'brush border' in the light microscope).

In addition, small structures that are difficult to identify with certainty using a light microscope can be easily identified with the electron microscope, such as lysosomes and mitochondria.

7.2 (*a*) cell wall with middle lamella and plasmodesmata, chloroplasts (plastids in general), large central vacuole (animal cells do possess small vacuoles, such as food vacuoles, contractile vacuoles)

(*b*) centrioles, microvilli, pinocytotic vesicles are more commonly seen in animal cells.

7.3 (*a*) A: polar head of phospholipid (hydrophilic)

  B: non-polar hydrocarbon tails of phospholipid (hydrophobic)

  C: phospholipid

  D: lipid layer

(*b*) cholesterol (the most common sterol)

7.4 (*a*) A Na^+/K^+ pump operates whereby efflux of Na^+ is linked to influx of K^+. Without K^+, no Na^+ efflux can occur, so Na^+ accumulates within the cells by diffusion and K^+ leaves the cells by diffusion. (*b*) ATP is a source of energy for active transport of Na^+ ions.

7.5 Non-polar amino acids are hydrophobic (repelled by water). Therefore the signal sequence tends to be repelled by the aqueous cytosol around the ribosome until it finds the receptor protein in the ER.

Chapter 9

9.1 Photoautotrophism is the process by which light energy from the Sun is used as an energy source for synthesising organic compounds from inorganic materials, with carbon dioxide as a source of carbon. Chemoheterotrophism is the process by which organic compounds are synthesised from pre-existing organic sources of carbon, using energy from chemical reactions.

9.2 (*a*) autotrophic organisms (*b*) heterotrophic organisms

9.3 75×10^{12} kg carbon per year [$(40 \times 10^{12}) + (35 \times 10^{12})$]

9.4 Solar energy is free; the raw material, water, is abundant; the product of combustion is also water, which is non-toxic and would not pollute the environment (nuclear power involves safety and pollution hazards).

9.5 **Overall form and position**

Large surface area to volume ratio for maximum interception of light and efficient gaseous exchange.

Blade often held at right-angles to incident light, particularly in dicotyledons.

Stomata

Pores in the leaf allow gaseous exchange. Carbon dioxide needed for photosynthesis, with oxygen a waste product.

In dicotyledons, stomata are located mainly in the shady lower epidermis, thus minimising loss of water vapour in transpiration.

Guard cells

Regulate opening of stomata (ensure stomata open only in light when photosynthesis occurs).

Mesophyll

Contains special organelles for photosynthesis, the chloroplasts, containing chlorophyll.

In dicotyledons, palisade mesophylls cells, with more chloroplasts, are located near the upper surface of the leaf for maximum interception of light. Length of the cells increases the chance for light absorption.

Chloroplasts are located near the periphery of the cell for easier gas exchange with intercellular spaces.

Chloroplasts may be phototactic (that is move within the cell towards light).

In dicotyledons, spongy mesophyll has large intercellular spaces for efficient gaseous exchange (monocotyledons also have extensive intercellular spaces).

Vascular system

Supplies water, a reagent in photosynthesis; also mineral salts. Removes the products of photosynthesis.

Supporting skeleton provided together with collenchyma and sclerenchyma.

9.6 Chlorophyll *a* absorption in red light is about twice that of chlorophyll *b* and the absorption peak is at a slightly longer wavelength (lower energy). Absorption in the blue is lower and shifted to a slightly shorter wavelength (higher energy). Note that only very slight differences in chemical structure cause these differences.

9.7 Chlorophyll *a* has the lower energy of excitation. It is possible therefore for chlorophyll *b* to transfer energy to chlorophyll *a* whilst still losing some energy as heat during the transfer.

9.8 The dark blue colour of the dye should disappear as it is reduced, leaving the green of the chloroplasts.

9.9 The DCPIP should have remained blue in tubes (2) and (3), which were controls. Tube (2) shows that light alone cannot induce the colour change, and that chloroplasts must be present for the Hill reaction to occur. Tube (3) shows that light must be present as well as chloroplasts for the Hill reaction to occur.

9.10 The two organelles closest in size to the chloroplasts are nuclei (slightly larger) and mitochondria (slightly smaller). More rigorous differential centrifugation or density gradient centrifugation would be necessary to isolate pure chloroplasts.

9.11 Indirect evidence suggests that nuclei and mitochondria were not involved in reducing DCPIP because light was needed, and these organelles lack chlorophyll or any other conspicuous pigment.

9.12 To reduce enzyme activity. During homogenisation destructive enzymes may be released from other parts of the cell, such as from lysosomes or vacuoles.

9.13 Cell reactions operate efficiently only at certain pHs; any significant change in pH, caused for example by release of acids from other parts of the cell, might have affected chloroplast activity.

9.14 (*a*) water (*b*) DCPIP

9.15 Non-cyclic photophosphorylation only: (i) oxygen was evolved (ii) electrons were accepted by DCPIP, therefore they could not recycle into PSI.

9.16 (*a*) The chloroplasts lack chloroplast envelopes (bounding membranes) and stroma. Only the internal membrane system remains.

(*b*) The medium lacking sucrose was hypotonic to the chloroplasts. Without the protection of the cell walls, broken during homogenisation, chloroplasts absorb water by osmosis, swell and burst. The stroma dissolves, leaving only membranes.

(*c*) The change was desirable because bursting the chloroplasts allows more efficient access of DCPIP to the membranes where the Hill reaction is located.

9.17 The discovery of the Hill reaction was a landmark for several reasons:

(1) it showed that oxygen evolution could occur without reduction of carbon dioxide, providing evidence for separate light and dark reactions and the splitting of water;

(2) it showed that chloroplasts could carry out a light-driven reduction of an electron acceptor;

(3) it gave biochemical evidence that the light reaction of photosynthesis was entirely located in the chloroplast.

9.18 If an isotope has a shorter half-life (for example ^{11}C, 20.5 min) it rapidly decays to the point at which it is undetectable, thus severely restricting its usefulness in biological experiments, which often take hours or days to complete.

9.19 Photosynthesis in *Chlorella* and higher plants is biochemically similar so that *Chlorella* was used for the following reasons:

(1) *Chlorella* culture is virtually a chloroplast culture since a large volume of every cell is occupied by a single chloroplast;

(2) greater uniformity of growth can be achieved;

(3) the cells are very rapidly exposed to radioactive carbon dioxide and also quickly killed, so handling techniques are easier.

9.20 For maximum illumination of algae.

9.21

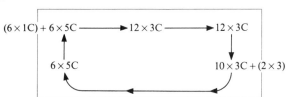

This emphasises the cyclic flow of carbon; the complexity of Calvin's cycle is due mainly to the difficulty of converting $10 \times 3C$ into $6 \times 5C$.

9.22 Availability of carbon dioxide, water, light and chlorophyll.

9.23 (*a*) In region A light intensity is the limiting factor.

(*b*) B: some factor other than light intensity is becoming the limiting factor. In region B, both light intensity and the other factor(s) are limiting. C: light intensity is no longer a limiting factor.

(*c*) D: the 'saturation point' for light intensity under these conditions, that is the point beyond which an increase in light intensity will cause no further increase in the rate of photosynthesis.

(*d*) E: the maximum rate of photosynthesis attainable under the conditions of the experiment.

9.24 X, Y and Z are the points at which light ceases to be the major limiting factor in the four experiments. Up to these points there is a linear relationship between light intensity and rate of photosynthesis.

9.25 Enzymes would start to become denatured.

9.26 Some likely situations would be (*a*) in a shaded community such as a wood; dawn and twilight in a warm climate; (*b*) carbon dioxide is normally limiting, but it would be particularly so in a crowded stand of plants, such as a crop under sunny, warm conditions; (*c*) a bright winter's day.

9.27 The plant continues to use sugars in the dark, for example for respiration. Photosynthesis ceases in the dark, so as sugars are depleted starch reserves are converted to sugars, including sucrose which travels from the leaves to other parts of the plant.

9.28 It could be argued that paper or foil prevents photosynthesis by restricting diffusion of carbon dioxide to the covered parts. This can be disproved by arranging an air gap between paper and leaf as shown below in fig 9.28(ans).

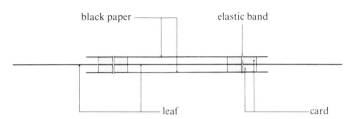

Fig 9.28(ans) *Section through leaf wrapped in black paper*

9.29 It should be placed in an identical flask but with water replacing potassium hydroxide solution. Unsoaked cotton wool should secure the leaf stalk. (The stalk itself could be surface-treated with lime water to check whether possible injury here could affect photosynthesis.)

9.30 Rates of carbon dioxide uptake, oxygen production and carbohydrate production could be used. Rate of increase in the dry mass of leaves may also be measured. This is particularly suitable for crop plants over a growing season when relatively large samples may be taken. An experiment for measuring carbon dioxide uptake is described in section 9.7.

9.31 (*a*) The rate of gas production is directly proportional to LI up to a LI of *x* units. At this point light saturation began to occur and this was complete at *y* units (*x* and *y* values depend on experimental conditions). Thereafter some factor other than light was limiting the rate of gas production.
(*b*) The laboratory was darkened to avoid extra light which could have stimulated extra photosynthesis. Temperature was kept constant because this also affects the rate of photosynthesis.

9.32 (*a*) Temperature may vary as the lamp heats the water (this should be avoided by the water bath).
(*b*) The carbon dioxide concentration of the water may vary during the experiment, especially if potassium hydrogen-carbonate was added earlier.
(*c*) Any stray light which is admitted to the laboratory will affect photosynthesis.

9.33 As the bubble of oxygen rises through the water, some of the dissolved nitrogen will come out of solution and enter the bubble, and some of the oxygen will dissolve. This exchange is due to the different partial pressures (concentrations) of oxygen and nitrogen in the bubble and the water, there being a tendency for them to come to equilibrium with time. Traces of water vapour and carbon dioxide will also be present in the collected gas. Once the gas has been collected, it will tend to come into equilibrium with atmospheric air by diffusion of gases through the water.

9.34 The amount of oxygen produced by photosynthesis in the experiment must all be collected. If the water is not saturated with air, some of the oxygen released in photosynthesis will dissolve in the water and reduce the amount recorded.

9.35 Specimen results are given in the following table.

Time/h	Colour of indicator			
	tube A	*tube B*	*tube C*	*tube D*
0	red	red	red	red
18	yellow	purple	red	red

The control tubes, C and D, were necessary to prove that any changes that took place in tubes A and B were due to the presence of leaves. In tube A conditions became more acidic as a result of carbon dioxide being produced during respiration. Photosynthesis did not take place in the absence of light. In tube B conditions became less acidic, indicating a net uptake of carbon dioxide. The carbon dioxide produced by respiration was used in photosynthesis, together with that already in the air inside the leaf and dissolved in the indicator solution. The rate of photosynthesis was greater than the rate of respiration.

9.36 The carbon dioxide compensation point. At this point rate of photosynthesis equals rate of respiration.

9.37 The higher the concentration of carbon dioxide, up to 0.1%, the greater the rate of photosynthesis. As carbon dioxide concentration increases, it competes more effectively with oxygen for the active site in RuBP carboxylase, thereby increasing the rate of carbon dioxide fixation, that is photosynthesis. Increase in oxygen concentration inhibits photosynthesis for the opposite reason, oxygen tending to exclude carbon dioxide and to stimulate photorespiration, which releases carbon dioxide.

9.38 High oxygen concentration and low carbon dioxide concentration (maximum rate achieved at 100% oxygen). High light intensities are also needed for high rates since the process is light-dependent.

9.39 Mesophyll chloroplasts for light reactions, bundle sheath chloroplasts for dark reactions.

9.40 Oxygen production is associated with grana (major location of PSII) and oxygen would compete with carbon dioxide for RuBP carboxylase and stimulate photorespiration. Also grana occupy a large volume of the chloroplast and in their absence there is more stroma, and hence more RuBP carboxylase and storage space for starch.

9.41 **Carbon dioxide pump.** By acting as a carbon dioxide pump, the malate shunt increases carbon dioxide concentration in the bundle sheath cells, thus increasing the efficiency with which RuBP carboxylase works.
Hydrogen pump. Malate carries hydrogen from $NADPH_2$ in the mesophyll to NADP in the bundle sheath cells, where $NADPH_2$ is regenerated. The advantage is that $NADPH_2$ is generated by the efficient light reaction in the mesophyll chloroplasts (PSII present) and can be used as reducing power in the Calvin cycle of bundle sheath chloroplasts, whose own synthesis of $NADPH_2$ is limited.

9.42 (*a*) Lowering oxygen concentrations stimulates C_3 photosynthesis because it reduces photorespiration.
(*b*) Lowering oxygen concentration does not affect C_4 photosynthesis because photorespiration is already inhibited.

9.43 Mutualistic bacteria in the root nodules of legumes fix nitrogen which leads to increased growth and thus to increased demand for other minerals, notably potassium and phosphorus. (However, ploughing-in of legumes is sometimes done, thus keeping the minerals in the soil.)

9.44 Chemoheterotrophic. They can be classified further as saprotrophic.

9.45 Anywhere there is insufficient oxygen for decomposition of all accumulating organic matter, such as bogs, aquatic sediments like mud deposits, arctic tundra, deeper zones of soil and waterlogged soils.

9.46 Both increase aeration and hence oxygen content of soil. This stimulates decomposition and nitrification. It also inhibits denitrification, oxygen being used instead of nitrate.

Chapter 10

10.1 (1) Decomposes organic matter and therefore helps recycling of elements from dead to living organisms.
(2) Removes organic refuse.
(3) Renders food unfit for human consumption (such as makes bread mouldy).
(4) In the Far East *Mucor* has been used to produce alcohol. A mixture of *Mucor* and yeast was added to rice. *Mucor* converted the rice to sugars which the yeast then converted to alcohol.

10.2 See section 4.5.3.

10.3 Because of the continual heat loss from the relatively larger body surface of the mouse.

10.4 4.18 J raise the temperature of 1 g water through 1 °C,
7.5×4.18 J raise the temperature of 1 g water through 7.5 °C,
$7.5 \times 4.18 \times 500$ J raise the temperature of 500 g water through 7.5 °C,
therefore 15.675 kJ are produced when 1 g of sugar is burned in oxygen.

10.5 Fats are much richer in hydrogen than carbohydrates. As most of the energy that is released in the body arises by the oxidation of hydrogen to water, so fats liberate more heat than carbohydrates.

10.6 Less energy is released in the body because both of these substances are not completely oxidised; for example the nitrogen in protein is excreted as urea in urine rather than being oxidised to nitrogen dioxide, therefore excretion of urea from the body actually expels some energy.

10.7 (*a*) Certain 'factors' (now known as vitamins) are needed in small amounts in the diet, which are essential for healthy growth and development.
(*b*) The growth 'factors' must be contained in the 3 cm³ rations of milk provided for the rats, which confirms that only minute amounts are required. When the milk was stopped, growth was quickly curtailed. Rats without milk did grow initially, therefore they must have had a small store of vitamins in their body initially.
(*c*) It is deficient in iron, vitamin B and roughage.

10.8 Active pepsin would digest cells that produce it, there being no mucus barrier within the zymogen glands.

10.9 (*a*) The small intestine wall is thrown into folds, the presence of villi and the presence of microvilli.
(*b*) It increases tremendously the secretory and absorptive surface of the small intestine and makes it very efficient at these processes.

10.10 Enzyme activity would be impaired as the enzymes would be denatured by the low pH.

10.11 It ensures that even if the soluble food molecules are in concentrations lower than those already in the blood, they will still pass into the blood.

Chapter 11

11.1 Those substances which yield more energy than ATP when hydrolysed are able to transfer their phosphate groups to ADP to form ATP. Those that liberate less energy than ATP when hydrolysed will receive phosphate groups from ATP.

11.2 ATP can be compared with a battery in the sense that its manufacture requires energy and that it is a convenient short-term carrier of energy. It is manufactured during respiration and can move to any part of the cell requiring energy, be 'run down' (converted to ADP), then 'recharged' (converted back to ATP) by respiration.

11.3 dehydrogenation

11.4 dehydrogenases (see table 6.3)

11.5 (*a*) phosphorylation
(*b*) isomerases
(*c*) (i) dehydrogenation/oxidation
(ii) phosphorylation
(*d*) a vitamin of the B complex

11.6 For rapid diffusion of intermediates between cytoplasm and mitochondrion.

11.7

Entering	*Leaving*
pyruvate	
oxygen	carbon dioxide
reduced hydrogen carrier	oxidised hydrogen carrier
ADP	ATP
phosphate	water

11.8 Provides an increased surface area for absorption of digested food materials.

11.9 Large concentrations of dissolved molecules would increase the solute potential of the plasma, which in turn could affect many other physiological processes.

11.10 (1) Within the cell, the pigment is separated from the more variable chemical environment of the plasma.
(2) Enclosing the pigment will decrease the viscosity of the blood and reduce the work the heart has to do to pump the blood around the body.

11.11 Initially when blood and water first meet, the concentration gradient of oxygen between them would be great. However, as blood and water flow along together the gradient will decrease until blood shows a percentage saturation for oxygen equal to that of the water. This would be well below the blood's maximum saturation point and therefore inefficient (fig 11.11(ans)).

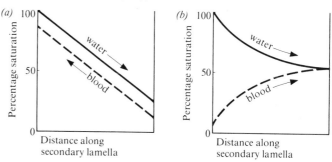

Fig 11.11(ans) *(a) Counterflow of water and blood.*
(b) Parallel flow of water and blood

11.12 Because the dead space air forms part of the pulmonary ventilation volume.

11.13 Smaller mammals have a large surface area to volume ratio from which heat can be lost and therefore must use up more oxygen in order to maintain a constant body temperature.

11.14 By relating oxygen consumption to body mass and

calculating the volume of oxygen consumed per gram of body weight in unit time.

11.15 $RQ = \dfrac{CO_2}{O_2} = \dfrac{102}{145} = 0.70$

11.16 $RQ = \dfrac{CO_2}{O_2} = \dfrac{2}{0} = \infty$

11.17 Because Man generally respires carbohydrate and fat substrates.

11.18 (a) Respiratory rate is about 17 breaths min^{-1}.
(b) Tidal volume is 450 cm^3 (average).
(c) Pulmonary ventilation is $17 \times 450 \text{ cm}^3 = 7.65 \text{ dm}^3 \text{ min}^{-1}$.
(d) Oxygen uptake is given by the slope of the line AB. Therefore oxygen consumption is 1500 cm^3 in 4 min = 375 $\text{cm}^3 \text{ min}^{-1}$.

11.19 (a)

Aerobic respiration	Photosynthesis
This is a catabolic process and results in the breakdown of carbohydrate molecules into simple inorganic compounds	An anabolic process which results in the synthesis of carbohydrate molecules from simple inorganic compounds.
Energy is incorporated into ATP for immediate use.	Energy is accumulated and stored in carbohydrate. Some ATP is formed.
Oxygen is used up.	Oxygen is released.
Carbon dioxide and water are released.	Carbon dioxide and water are used up.
The process results in a decrease in dry mass.	Results in an increase in dry mass.
In eukaryotes the process occurs in mitochondria.	In eukaryotes the process occurs in chloroplasts.
Takes place continuously throughout the lifetime of all cells, and is independent of chlorophyll and light.	Occurs only in cells possessing chlorophyll and only in the presence of light.

(b) *List of similarities between photosynthesis and aerobic respiration*
Both are energy-converting processes.
Both require mechanisms for exchange of carbon dioxide and oxygen.
Both require special organelles in eukaryotes, that is mitochondria for respiration and chloroplasts for photosynthesis; mitochondria and chloroplasts resemble prokaryotic organisms in possessing circular DNA and a prokaryote-type protein-synthesising system.
The light reactions of photosynthesis resemble cell respiration in the following ways:
(i) phosphorylation occurs (that is synthesis of ATP from ADP and P_i);
(ii) this is coupled to flow of electrons along a chain of electron carriers;
(iii) the electron carriers must be organised on membranes for coupling to take place; these are cristae in mitochondria and thylakoids in chloroplasts.
Both also involve cyclic pathways that take place in solution in the matrix around the membranes (Krebs cycle in respiration, Calvin cycle in photosynthesis).
Part of the glycolytic sequence of enzymes is common to both processes.

Chapter 12

12.1 Dry mass is used because the water content of food samples or organisms may vary and water contributes no energy.

12.2 Small birds or mammals have a much higher surface area to volume ratio than humans and therefore lose body heat relatively more rapidly. Since small mammals and birds are endothermic ('warm-blooded') like humans they must consume relatively more energy to maintain body heat. (Birds also have a higher metabolic rate and body temperature than mammals.)

12.3 In pyramid (c) the primary producers are large, such as trees, and therefore fewer in numbers than the herbivores they support. A situation like this would be obtained, for example, with aphids feeding on a rose bush, or caterpillars on a tree.

12.4

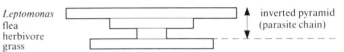

Leptomonas
flea
herbivore
grass
inverted pyramid (parasite chain)

12.5 Pyramid (d) is based on a parasite food chain, where a parasite is itself parasitised. The pyramid of numbers that results is unusual in being inverted since the organisms become progressively smaller and more numerous along the food chain. The first trophic level shown could be a tree or shrub, although parasite food chains are sometimes represented as starting with the host, whether it be plant or animal.

12.6 As conditions become suitable in the spring there is rapid growth and reproduction of the algae that form the phytoplankton (an algal bloom) and their mass exceeds that of their grazers. This is presumably followed by an increase in numbers and biomass of the primary consumers, and then by an increase in secondary consumers as materials and energy flow from one trophic level to the next. The biomass of phytoplankton decreases as the grazers increase and as the unfavourable conditions of winter return. At some point the biomass of the longer-lived consumers exceeds that of the producers. Such seasonal fluctuations in pyramids of biomass are typical of lake and ocean ecosystems based on phytoplankton.

12.7 (a) May, June and July
(b) (i) Increase in light intensity and duration, and increase in temperature coupled with the availability of nutrients. Photosynthesis and growth are therefore favoured.
(ii) Grazing by primary consumers, such as zooplankton, and decrease in production due to depletion of nutrients. (The latter is due to the dead remains of producers sinking through the lake to colder, non-circulating water.)
(iii) Decline in numbers of zooplankton. Increase in nutrients (circulation of nutrients improves in the autumn as the surface layers of water cool and mix more freely with the colder, deeper layers). Temperature and light are still favourable.
(iv) Light and temperature unfavourable for photosynthesis and growth.

12.8 Transfer of energy between trophic levels cannot be 100% efficient, so successive levels have less energy and can support, on average, fewer animals.

12.9 seed – blackbird – hawk; trophic level 3, T3
leaf litter – earthworm – blackbird – hawk, T4
leaf – caterpillar – beetle – insectivorous bird – hawk, T5
rose bush (sap) – aphid – ladybird – spider – insectivorous bird – hawk, T6

12.10 Blue-green bacteria and some other bacteria are also photosynthetic (they are prokaryotes, not plants). Chemosynthetic bacteria are also autotrophic (table 9.1) and therefore make a contribution to primary productivity. The total contribution of all these organisms is small compared with autotrophic eukaryotes (photosynthetic protoctista and plants).

12.11 (a) 24×10^6 kJ m^{-2} yr^{-1}

(b) 1.3%

(c) 800, 200 and 69 700 kJ m^{-2} yr^{-1} respectively

(d) 114×10^3 kJ m^{-2} yr^{-1}

(e) Grasses and herbs (producers), seed-eating birds, common green grasshoppers and field mice (primary consumers), spiders (secondary consumers)

(f) All except grasses and herbs

(g) Other primary or secondary consumers, decomposers, detritivores (also possibly emigration)

12.12 (a) (i) ×2 (ii) ×500 (iii) ×2 500 (iv) ×3 750

(b) DDT is subject to progressive concentration as it passes along the food chain. This suggests that it is a persistent chemical, not easily broken down, and that it is stored rather than metabolised in living organisms. (In fact, it remains active for 10–15 years in soil.)

(c) (i) and (ii) 4th trophic level (top carnivore) (iii) 2nd trophic level (herbivore)

(d) DDT has spread all over the world as a result of two factors. First, it is carried at very low concentrations in water. If it is washed off agricultural land and into rivers some of it reaches the sea and becomes concentrated in marine food chains. Penguins feed on fish and are part of these food chains. Secondly, DDT can be carried in the atmosphere, both because it is volatile and because it is sprayed as a dust which can be carried by wind systems over large distances.

(e) (i) A small proportion of the original midges were resistant to DDD and these were not killed by the spraying procedure. Between sprays their numbers increased and after successive sprays they continued to breed and eventually constituted the greater part of the population. In other words the population had undergone intensive selection pressure (see chapter 25).

(ii) The data given suggest that DDD (and therefore DDT) is stored predominantly in fatty tissues. (This is because DDD and DDT are soluble in fat rather than water.) During times of food shortage, fat is mobilised and used so that the DDD or DDT accumulated over a long period is released into the bloodstream in relatively high concentrations.

(f) It has been suggested that the high death toll of birds in the winter of 1962–3 compared with 1946–7 was due to the additional effects of DDT mobilisation from fatty tissues. In 1946–7 the use of DDT was limited; in the late 1950s and early 1960s its use was widespread.

12.13 (1) Maximum possible specificity so that it has minimal effects on species other than pests. The greater the specificity, however, the smaller the potential market and the more expensive it will be!

(2) Inexpensive to manufacture.

(3) Relatively non-persistent, if toxic to non-pest species.

12.14 (a) Deforestation reduces the total world volume of photosynthetic material and thus reduces consumption of atmospheric carbon dioxide in photosynthesis.

(b) Removal of the tree canopy exposes the forest floor to sunlight and warmer temperatures. In forests or woodlands with significant litter and soil humus contents this exposure will favour accelerated rates of decomposition and carbon dioxide release.

12.15 BOD of discharge
BOD of receiving water

Nature of organic material
Total organic load of the river
Temperature
Extent of aeration from atmosphere (varies with wind etc.)
Dissolved oxygen in stream
Numbers and type of bacteria in effluent and stream
Ammonia content of effluent

12.16 Geographical barriers, such as oceans; ecological barriers, such as unfavourable habitats separating areas of favourable habitats; distance over which dispersal must operate; air and water currents; size and nature of invasion areas

12.17 Various environmental factors may be altered.

(a) *Light* – Light intensity at the forest floor may be only 1–6% of that striking the canopy.

– Light quality may also change: light passing through leaves is enriched in far-red light (shorter wavelengths of red being filtered out). This has physiological implications for the phytochrome system (see chapter 15).

(b) *Temperature* – Daily and seasonal fluctuations of temperature are less within a forest than outside it. Lower maximum and higher minimum temperatures are usual. The mean temperature is relatively lower in summer and higher in winter compared with the air temperature outside the canopy.

(c) *Wind* – Plants below the canopy are protected from wind, and wind speeds on average are only 40–80% of those outside.

(d) *Moisture* – Interception of rainfall and subsequent evaporation from the canopy will reduce the amount of water reaching the vegetation below. Relative humidity is usually greater in the woodland than outside, partly as a result of lower temperatures during the day.

12.18 Birth rate $= \dfrac{10\,000}{500\,000} \times 1\,000 = 20$ per thousand head per year.

12.19 (a) Two eggs from each female must, on average, survive.

(b)

	Number of fertilised eggs that must die for stable population	Pre-reproductive mortality
oyster	$(100 \times 10^6) - 2$	>99.9%
codfish	$(9 \times 10^6) - 2$	>99.9%
plaice	$(35 \times 10^4) - 2$	>99.9%
salmon	$(10 \times 10^4) - 2$	>99.9%
stickleback	498	498/500 = 99.6%
winter moth	198	99.0%
mouse	48	96.0%
dogfish	18	90.0%
penguin	6	75.0%
elephant	3	60.0%
Victorian Englishwoman	8	80.0%

(c) The stickleback and dogfish give birth to live young, that is they are viviparous. Therefore fewer eggs need to be produced owing to the greater degree of parental involvement in the development of offspring. Also, the female parent could not physically support any greater numbers of offspring.

12.20 Population (b), since a high percentage of individuals would die before reproductive age is reached. Population (a) would have to combine its high survival rate with low reproductive rate to maintain a stable population size.

12.21 (*a*) Out of 3 200 eggs, 640 survive, so 2 560 die – a mortality of 80%.

(*b*) Out of 640 fry, 64 survive, so 576 die – a mortality of 90%.

(*c*) Out of 64 smolts, 2 survive, so 62 die – a mortality of about 97%.

The total pre-reproductive mortality for salmon is 3 198 out of 3 200 = 99.97% (see fig 12.21(ans)).

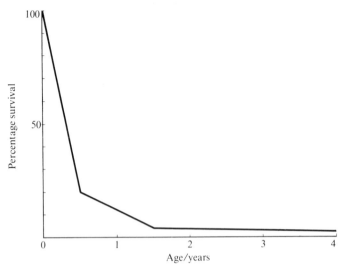

Fig 12.21 (ans)   *Graph showing pre-reproductive mortality for salmon*

12.22 (*a*) *S. nemoralis* (*b*) *S. nemoralis* (*c*) *S. rugosa* (*d*) The results for *S. speciosa* show that it has flexibility of response to its environment. When colonising a woodland site, which is a stable K-selecting habitat, it behaves more like a K-strategist in devoting less biomass to reproduction than when it grows in the disturbed habitat of an earlier successional stage.

A general conclusion that might be drawn is that some species may have variable levels of '*r*' depending on environmental conditions. This in fact appears to be true for some plant and animal populations. If either strategy is selected for over a period of time, speciation might occur as has happened with *S. rugosa* and *S. nemoralis*.

12.23 (*a*) A sigmoid (S-shaped) growth curve.

(*b*) Food and space. Food is more likely in this case.

(*c*) Faster reproductive rate. More efficient feeding. Greater resistance to toxic waste products, either of *Paramecium* or of bacteria growing in the same culture (*P. aurelia* has been shown to be more resistant than *P. caudatum*). Production of a poison or growth inhibitor (allelopathy). Predation.

12.24 (*a*) Natural variation in the target population leaves some resistant survivors.

(*b*) LD$_{50}$ toxicity definition means that not all of the target population will be eliminated, just sufficient of the damaging surplus.

(*a*) and (*b*) together ensure a viable pool of resistant organisms from which the population can recover to be a pest once more.

Thought questions

1 (*a*) E.g. acid rain

Costs	*Benefits*
(i) Local and/or long distance air pollution hazard may be caused.	(i) Cheaper electricity generation gives more profits to generating board.
(ii) Use of high-level dis-	(ii) Cheaper supplies for

persal in the atmosphere distributes pollution over large distances. Countries far removed from the catchment of the generator may experience serious pollution requiring expensive remedial action.

(iii) Crop damage

(iv) Human and animal health hazard

domestic industrial consumers in the catchment of that generator.

(*b*) The answer to (*b*) should promote a lively discussion. The main benefit is a cleaner environment and reduction of the hidden costs of industrial activity. Environmental costs are often not fully accounted in conventional cost:benefit analyses.

The chief snag is that the polluter may pass on the extra costs of pollution control to the consumer. The system will only work if the consumer supports the clean manufacturer by buying the more expensive product, even if cheaper alternatives produced by the manufacturers not using pollution control are available.

An obvious related example currently seen in many British supermarkets is the relative price of organically produced food compared with fruit and vegetables grown under conventional farming systems using chemical fertilisers and pest control. Other difficulties lie with enforcing pollution legislation. A possible benefit is revenue from the manufacture and export of pollution control technology, for example, W. Germany exports desulphurisation technology.

2 A population with a common gene pool (deme) will slowly evolve over time. If the population is very small its members may become inbred and lose vigour. In the recent past the black rhino has been hunted, for its horn, to near extinction. Each local population is now a tiny fraction of the original population and physically isolated. Such animals will be increasingly inbred. Outbreeding strengthens genetic diversity and where populations are very small this may be important for the health of the animals. Semen may be collected from anaesthetised wild or captive males and used to inseminate anaesthetised captive females at oestrus (the time of ovulation). Such artificial insemination technology makes it less necessary to move the animals themselves yet allows the spread of genes. It also makes possible the storage of genetic material (by cryopreservation – deep-frozen semen) in the event of a lack of males, in a local population, where females are still found.

Student action/role-playing game (p. 437)

The main points 'for' and 'against' are summarised below.

For

(*a*) A larger proportion of national timber needs could be supplied. Home-grown timber currently supplies 10% of Britain's timber needs.

(*b*) Job creation in areas of rural unemployment.

(*c*) Increased landscape and habitat (creates woods) diversity.

(*d*) Source of local revenue.

(*e*) Long-term investment for land-owner and nation.

Against

(*a*) Exotic conifer monocultures are vulnerable to disease and support little wildlife at the closed canopy stage. (In the early scrub stages they can be good habitats, but unless managed on a well-planned rotation basis this is short-lived.)

(*b*) Deep ploughing destroys peatlands leading to permanent species loss and alien landscape change (an eyesore to some).

(c) Serious detrimental changes in land drainage and water quality may result.

(d) The promised local jobs may not materialise. Forestry is not very labour-intensive and requires prior expertise.

(e) Conifer forests are poor recreation areas which may deter tourism.

In a role-playing exercise students could argue these, and other points, as representatives of major interested parties. These could include

(a) Nature Conservancy Council
(b) Countryside Commission for Scotland
(c) Council for the Protection of Rural Scotland
(d) Friends of the Earth
(e) Royal Society for the Protection of Birds
(f) Forestry Commission
(g) Estate owner(s)
(h) Scottish Tourist Board
(i) Water Authority
(j) Local and strategic regional planning authorities
(k) Local residents
(l) The local Member of Parliament

An eminent chairperson and supporting experts/dignitaries (two or three) will also be needed. Their role is to report to national government advising an appropriate course of action in the light of evidence presented.

Chapter 13

13.1

Fresh mass of soil	$= 60$ g
dry mass of soil	$= 45$ g
therefore mass of water	$= 60$ g $- 45$ g $= 15$ g

therefore percentage water content of fresh soil $= \dfrac{15}{60} \times 100 = 25\%$

Dry mass of soil	$= 45$ g
dry mass of soil after combustion	$= 30$ g
therefore mass of organic material	$= 15$ g

therefore percentage organic content of fresh soil $= \dfrac{15}{60} \times 100 = 25\%$

13.2 43%

13.3 36%

13.4 4 230

Index

Numbers in italic denote figures.

motor neurone, 244, 245
moulds, 33, 34
mouthparts, of insects, 108, 293, 297
mucilages, 139
mucin, 206, 311
Mucor, 18, 19, 20, 21, 290
mucus, 206
mull, 394
multiple-enzyme complex, 179
mumps, virus, 15, 31
murein, 6, 139
Musca domestica (housefly), *116*, 298, *299*
muscle tissue, 243–4, *244*
    active transport in, 198
muscularis externa, 306–7
mushrooms, *see Agaricus*, 25
mutualism, 13, 290, 291
    fungal, 19
    micro-organisms involving, 292
mycobacteria, *5*
mycology, 4, 17
mycoplasma (pleuro-pneumonia-like
    organisms), *5*
mycelium, 17, 20
mycoprotein, 27
mycorrhizas, 286–7
myelin sheath, 245
myelocytes, 243
myeloid tissue, 243
myoglobin, 147, *152*

NAD (nicotinamide adenine dinucleotide),
    157, 170, *171*, 329–30
NAD dehydrogenase, 330
NADP (nicotinamide adenine dinucleotide
    phosphate), 157, *171*, 329–30
natality (birth rate), 405
negative feedback inhibition (end-product
    inhibition), 179
nematoblast, 88, 90
Nematoda, 100
neoteny, 119
Nereis diversicolor, 102, 102–4
    trochophore larva, 104
nerves, 245–7
nerve cells, active transport in, 198
net assimilation rate (unit leaf rate), 374
neurilemma, 245
neuroglia, 247
neurones, 244, 245
niacin (nicotinic acid; vitamin B$_3$), 305
nicotinic acid (niacin), 305
nitrate(s), 385
nitrate fertilisers, 281–2
nitrification, 282–3
nitrogen, 285
    cycle, 24, 281–2
    fixation, 281–2
nodes of Ranvier, 245
non-sulphur bacteria, 280, 289
non-swimming male gametes, 75
notochord, 119
nuclear envelope, 4, 198–200
nuclease, 314, 315
nucleic acids, 154–60
nucleocapsid, 15
nucleolar organiser, 200
nucleolus, 200
nucleoprotein, 148
nucleoside, 157, 315
nucleotidase, 314
nucleotides, 154, 156, 155–7
nucleus, 4, 186–9, 198–200
nutrients
    cycles (biogeochemical cycles), 24, 364,
        377–80

recommended daily intake, 301
nutrition, 1, 13, 249–304

Obelia, 88–92
obligate symbiont, 291
oceanic thermocline, 381, *383*
octane, *126*
oesophagus, 311–12, 317
oestradiol, *143*
oestrone, *143*
Oligochaeta, 101, *102*, 104–6
oligodendrocytes, 247
omnivores, 289, 367
oogamy, 42
oogonium, 42
oomycetes, 49–51
oosphere, 42
Operophthera brumata (winter moth), 411,
    412–14
Ophiuroidea, 118
optical isomerism, 132–5
organ, 85
organisms
    interactions between, within community,
        404
    interactions with abiotic environment,
        401–4
    methods of collection, 445–9
organ system, 85
organophosphates, 422
osmoregulation, 197
osmosis, 195
Osteichthyes, 121, *122*
osteoblasts, 240, 241
osteoclasts, 208, 241
osteocytes, 241
ostia, 111
ovary, 94, 96
overcrowding, 412–14
overfishing, 418–19
overnutrition, 300
ovule, 74
oxidation ponds, 48
oxidative deamination, 340
oxidative decarboxylation, 328
oxidative phosphorylation, 260, 326, 328–9
oxidoreductase, 180
oxygen
    algae source, 48
    cycle, 284
oxyntic (parietal) cells, 313
ozone layer, 381
    depletion, 429

palisade cells, 219, 252–3
pancreas, 315, *316*
    acinar cells, *204*
pancreatic juice, 314, 315
pancreatic lipase, optimum pH, 174
pancreozymin, 319
pantothenic acid (vitamin B$_5$), 305
paper industry, 181
parallel flow system, in gills, 345, 346
Paramecium, 54–8, 293–4
parasites, 13, 292–3, 366
    facultative, 13, 292
    obligate, 13, 292
    specialisations, 293
parasitic fungi, 18–19
parasitism, 291–2
parenchyma, 215–20, 228, 230
pathogens, 13, 30
pectin(s) (pectic substances), 139, 211
pectinases, 181
pedology, 387
pellagra, 305

Pellia, 60, 61
Pelycopoda (Bivalvia), 107
penicillin, 26
penicillinase, 12
Penicillium, 18, 19, *23*, 23–4, 26
pentosans, 137
pentose(s), 131, 462, 462–3
pentose-phosphate shunt (hexose-
    monophosphate shunt), 339
pepsin(ogen), 313, 314
    optimum pH, 174
peptide bond, 145
perennials, 80
perichondrium, 239
pericycle, 216, 220
periodontal disease, 311
periosteum, 240
peristalsis, 315–16
peritoneum, 99
peroxisomes (microbodies), 208
pesticides, 376–7, 420–5
    ecosystem effects, *424*
    lethal dose, 421
    residues, 422
pH
    and amino acids, 145, 154
    in enzyme reactions, 174–6
    of soil, 393, 441
Phaeophyta (brown algae), 41, 45–7
phage, *see* bacteriophage
phagocytes, 198
pharynx, 349
phenylalanine, 144
pheromones, 425
phloem, 216, 228–30
phosphate levels in lakes, Norfolk Broads,
    384
    removal at sewage works, 384
phosphate compounds, 324
phosphofructokinase, 178
phospholipids, 141, *142*, 193
phosphophenylpyruvate, 324
phosphoprotein, 148
phosphorescence, 257
phosphoric acid, 155
phosphorus, 285
    cycle, *283*, 284
photophosphorylation, 260
    cyclic, 261–2
    non-cyclic, 261
photoautographs, *250*, 251
photoautotrophic (holophytic) organisms, 249
photoheterotrophs, 251, 289
photorespiration, 275–6, 279–80
photosynthesis, 250–75
    aerobic respiration compared, 468
    dark reactions, 264, 267
    factors affecting, 268–70
    light reactions, 260–2, 267
    prokaryotes/eukaryotes compared, 280
photosynthetic pigments, 255–9
photosystems, 259
phototrophs, 249
Phytophthora infestans, 49–51
phytoplankton, 382, 383
Pieris (white butterflies), *117*, 298–9, 422
pigments, 257–9
pili (fimbriae), 6
pin moulds, 21
pitfall trap, 446, *447*
pituitary dwarfism, 30
plagioclimax, 400
plague, 31
Planaria lugubris, 92–5, *93*, *94*
plankton, 48
plankton net, 446